KENDALL'S

ADVANCED
Theory of
STATISTICS

KENDALL'S

ADVANCED
Theory of
STATISTICS

Volume 2

CLASSICAL INFERENCE
AND RELATIONSHIP

FIFTH EDITION

Alan Stuart & J. Keith Ord

Edward Arnold
A division of Hodder & Stoughton
LONDON   MELBOURNE   AUCKLAND

© 1991 Trustees of the late Sir Maurice Kendall, Professor Alan
Stuart and Professor J. Keith Ord

First published in Great Britain 1946
First edition 1946
Second edition 1947
Third edition 1951
Fourth edition 1979
Fifth edition 1991

British Library Cataloguing in Publication Data
Stuart, Alan *1922–*
    Kendall's advanced theory of statistics.–5th. ed.
    Vol. 2
    1. Statistical methods
    I. Title   II. Kendall, Maurice *1907–1983*   III. Ord, J. K.
    (John Keith) *1942*
    519.5

    ISBN 0-340-52923-7

Typeset in Great Britain by J. W. Arrowsmith Ltd, Bristol BS3 2NT.
Printed in Great Britain for Edward Arnold, a division of Hodder
and Stoughton Limited, Mill Road, Dunton Green, Sevenoaks,
Kent TN13 2YA by St Edmondsbury Press Limited, Bury St
Edmonds, Suffolk, and bound by Hartnolls Limited, Bodmin,
Cornwall.

PREFACE TO THE FIFTH EDITION

This volume, unlike its predecessor, has been considerably reshaped. The former Chapters 20 and 21 have been merged and rewritten to form the new Chapter 20, and the former Chapter 34 has been moved forward to become Chapter 24. More radically, the former Chapter 29 has been replaced by a chapter on Analysis of Variance that is essentially the former Chapter 35 from Volume 3, and the former Chapters 31 to 33 are no longer in this volume, which ends with a new Chapter 31 on Comparative Statistical Inference.

With these changes, the two volumes contain the essential theoretical topics for any statistician, and they will be supplemented in due course by a volume on Bayesian Inference. Further volumes will be added to cover more specialized topics, such as those of the chapters omitted here and those in the former Volume 3.

The whole of the text has been revised in detail, and a great deal of new material incorporated in it, particularly in Chapter 28, which has a new title reflecting this. As a result there are now four chapters heavily concentrated on the Linear Model. We shall be grateful to readers who point out misprints and other errors.

For the first time in the history of this work, we have in this volume not had the benefit of the experience of the late E. V. Burke; we mourn his loss, but hope that we have sustained his high standards.

Alan Stuart
Keith Ord

London, March 1991

CONTENTS

GLOSSARY OF ABBREVIATIONS

The following abbreviations are sometimes used:

ARE	asymptotic relative efficiency
ASN	average sample number
AV	analysis of variance
BAN	best asymptotically normal
BCR	best critical region
c.f.	characteristic function
c.g.f.	cumulant-generating function
d.f.	distribution function
d.fr.	degrees of freedom
EM	estimation-maximisation (algorithm)
f.f.	frequency function
f.g.f.	frequency-generating function
f.m.g.f.	factorial moment-generating function
LF	likelihood function
LR	likelihood ratio
LS	least-squares
MCS	minimum chi-square
m.g.f.	moment-generating function
ML	maximum likelihood
MS	mean square
m.s.e.	mean-square-error
MV	minimum variance
MVB	minimum variance bound
$N(a, b)$	(multi-)normal with mean (-vector) a and variance (covariance matrix) b
OC	operating characteristic
s.d.	standard deviation
s.e.	standard error
SPR	sequential probability ratio
SS	sum of squares
UMP	uniformly most powerful
UMPU	uniformly most powerful unbiased
UMVU	uniformly minimum variance unbiased

CHAPTER 17

ESTIMATION

The estimation problem

17.1 On several occasions in previous chapters we have encountered the problem of estimating from a sample the values of the parameters of the population. We have hitherto dealt on somewhat intuitive lines with such questions as arose—for example, in the theory of large samples we have taken the means and moments of the sample to be satisfactory estimates of the corresponding means and moments in the parent.

We now proceed to study this branch of the subject in more detail. In the present chapter, we shall examine the sort of criteria that we require a 'good' estimate to satisfy, and discuss the question whether there exist 'best' esimates in an acceptable sense of the term. In the following chapters, we shall consider methods of obtaining estimates with the required properties.

17.2 It will be evident that if a sample is not random and nothing precise is known about the nature of the bias operating when it was chosen, very little can be inferred from it about the population. Certain conclusions of a trivial kind are sometimes possible—for instance, if we take ten turnips from a pile of 100 and find that they weigh ten pounds altogether, the mean weight of turnips in the pile must be greater than one-tenth of a pound; but such information is rarely of value, and estimation based on biased samples remains very much a matter of individual opinion and cannot be reduced to exact and objective terms. We shall therefore confine our attention to random samples only. Our general problem, in its simplest terms, is then to estimate the value of a parameter of the population from the information given by the sample. Initially, we consider the case when only one parameter is to be estimated. The case of several parameters will be discussed later.

17.3 Let us first consider what we mean by 'estimation'. We know, or assume as a working hypothesis, that the population is distributed in a form that is completely known but for the value of some parameter* θ. We are given a sample of observations $x_1, \ldots, x_n$. We require to determine, from the observations, a number that can be taken to be the value of θ, or a range of numbers that can be taken to include that value.

Now the observations are random variables, and any function of the observations will also be a random variable. A function of the observations alone is called a

---

* The term 'parameter' was discussed in **2.2**, Vol. 1. See also **21.3** below.

statistic. If we use a statistic to estimate θ, it may on occasion differ considerably from the true value of θ. It appears, therefore, that we cannot expect to find any method of estimation that can be guaranteed to give us a close estimate of θ on every occasion and for every sample. We must content ourselves with formulating a rule that will give good results 'in the long run' or 'on the average', or that has 'a high probability of success'—phrases that express the fundamental fact that we have to regard our method of estimation as generating a distribution of estimates and to assess its merits according to the properties of this distribution.

17.4 It will clarify our ideas if we draw a distinction between the method or rule of estimation, which we shall call an estimator, and the value to which it gives rise in a particular case, the estimate. The distinction is the same as that between a function $f(x)$, regarded as defined for a range of the variable x, and the particular value that the function assumes, say $f(a)$, for a specified value of x equal to a. Our problem is not to find estimates, but to find estimators. We do not reject an estimator because it may give a bad result in a particular case (in the sense that the estimate differs materially from the true value). We should only reject it if it gave bad results in the long run, that is to say, if the distribution of possible values of the estimator were seriously discrepant with the true value of θ. The merit of the estimator is judged by the distribution of estimates to which it gives rise, i.e. by the properties of its sampling distribution.

17.5 In the theory of large samples, we have often taken as an estimator of a parameter θ a statistic t calculated from the sample in exactly the same way as θ is calculated from the population: e.g. the sample mean is taken as an estimator of the population mean. Let us examine how this procedure can be justified. Consider the case when the population is

$$\mathrm{d}F(x) = (2\pi)^{-1/2} \exp\left\{-\tfrac{1}{2}(x-\theta)^2\right\} \mathrm{d}x, \qquad -\infty < x < \infty. \tag{17.1}$$

Requiring an estimator for the parameter θ, which is the population mean, we take the sample mean

$$t = \sum_{j=1}^{n} x_j / n. \tag{17.2}$$

The distribution of t is (Example 11.12)

$$\mathrm{d}F(t) = \{n/(2\pi)\}^{1/2} \exp\left\{-\tfrac{1}{2}n(t-\theta)^2\right\} \mathrm{d}t, \tag{17.3}$$

that is to say, t is distributed normally about θ with variance $1/n$. We notice two things about this distribution: (a) it has a mean (and median and mode) at the true value θ, and (b) as n increases, the scatter of possible values of t about θ becomes smaller, so that the probability that a given t differs by more than a fixed amount from θ decreases. We may say that the precision of the estimator increases with n.

Generally, it will be clear that the phrase 'precision increasing with n' has a definite meaning whenever the sampling distribution of t has a variance that decreases with $1/n$ and a central value that is either identical with θ or differs from it by a quantity that also decreases with $1/n$. Many of the estimators with which we are commonly concerned are of this type, but there are exceptions. Consider, for example, the Cauchy population

$$dF(x) = \frac{1}{\pi} \frac{dx}{\{1 + (x - \theta)^2\}}, \qquad -\infty < x < \infty. \tag{17.4}$$

If we estimate θ by the mean statistic t we have, for the distribution of t,

$$dF(t) = \frac{1}{\pi} \frac{dt}{\{1 + (t - \theta)^2\}} \tag{17.5}$$

(cf. Example 11.1). In this case the distribution of t is the same as that of any single value of the sample, and does not increase in precision as n increases.

Identifiability of parameters

17.6 It is clear that as sample size increases indefinitely, we always obtain more information about the underlying distribution $f(x \mid \theta)$, but this will only be useful in estimating θ if θ is a single-valued function of f; if not, i.e. if more than one value of θ corresponds to a given value of f, not even an infinite-sized sample can distinguish between them. Thus, for a normal distribution f with mean θ^2, (17.2) will estimate θ^2 with increasing precision as before, but there is no way to decide whether θ is positive or negative from even an exact knowledge of θ^2, for f is the same whatever the sign of θ. In such a case, we call θ *unidentifiable*. As this instance makes clear, even if θ is unidentifiable, some functions of θ may be identifiable—in this case, any single-valued function of θ^2 is.

Unless otherwise stated, we shall assume that parameters are identifiable, θ being uniquely determined by $f(x \mid \theta)$.

Consistency

17.7 The possession of the property of increasing precision is evidently a very desirable one; and indeed, if the variance of the sampling distribution of an estimator decreases with increasing n, it is necessary that its central value should tend to θ, for otherwise the estimator would have values differing systematically from the true value. We therefore formulate our first criterion for a suitable estimator as follows:

An estimator t_n, computed from a sample of n values, will be said to be a *consistent* estimator of θ if, for any positive ε and η, however small, there is some N such that the probability that

$$|t_n - \theta| < \varepsilon \tag{17.6}$$

is greater than $1 - \eta$ for all $n > N$. In the notation of the theory of probability,

$$P\{|t_n - \theta| < \varepsilon\} > 1 - \eta, \qquad n > N. \tag{17.7}$$

The definition bears an obvious analogy to the definition of convergence in the mathematical sense. Given any fixed small quantity ε, we can find a large enough sample size such that, for all samples over that size, the probability that t differs from the true value by more than ε is as near zero as we please. t_n is said to *converge in probability*, or to *converge stochastically*, to θ. Thus t is a consistent estimator of θ if it converges to θ in probability.

The original definition of consistency by Fisher (1921a) was that when calculated from the whole population the estimators should be equal to the parameter, and it was no doubt this definition that inspired the choice of the term 'consistency'.

Example 17.1

The sample mean is a consistent estimator of the parameter θ in the population (17.1). This we have already established in general argument, but more formally the proof would proceed as follows.

Suppose we are given ε. From (17.3) we see that $(t - \theta)n^{1/2}$ is distributed normally about zero with unit variance. Thus the probability that $|(t - \theta)n^{1/2}| \leq \varepsilon n^{1/2}$ is the value of the normal integral between limits $\pm \varepsilon n^{1/2}$. Given any positive η, we can always take n large enough for this quantity to be greater than $1 - \eta$ and it will continue to be so for any larger n. The number N may therefore be determined and the inequality (17.7) is satisfied.

Example 17.2

Suppose we have a statistic t_n whose mean value differs from θ by terms of order n^{-1}, whose variance v_n is of order n^{-1} and that tends to normality as n increases. Clearly, as in Example 17.1, $(t_n - \theta)/v_n^{1/2}$ will then tend to zero in probability and t_n will be consistent. This covers a great many statistics encountered in practice.

Even if the limiting distribution of t_n is unspecified, the result will still hold, as can be seen from a direct application of the Chebyshev inequality (3.94). In fact, if $E(t_n) = \theta + k_n$, and var $t_n = v_n$, where $\lim_{n \to \infty} k_n = \lim_{n \to \infty} v_n = 0$, we have at once

$$P\{|t_n - (\theta + k_n)| < \varepsilon\} \geq 1 - \frac{v_n}{\varepsilon^2} \underset{n \to \infty}{\to} 1,$$

so that (17.7) will be satisfied.

17.8 The property of consistency is a limiting property, that is to say, it concerns the behaviour of an estimator as the sample size tends to infinity. It requires nothing of the estimator's behaviour for finite n, and if there exists one consistent estimator t_n we may construct infinitely many others: e.g. for fixed a and b, $[(n - a)/(n - b)]t_n$ is also consistent. We have seen that in some circumstances a consistent estimator of the population mean is the sample mean $\bar{x} = (\sum x_j)/n$. But so is $\bar{x}' = (\sum x_j)/(n - 1)$. Why do we prefer one to the other? Intuitively it seems absurd to divide the sum of n quantities by anything other than their number n. We shall see in a moment, however, that intuition is not a very reliable guide in such matters. There is reason for preferring

$$\frac{1}{(n-1)} \sum_{j=1}^{n} (x_j - \bar{x})^2$$

to

$$\frac{1}{n} \sum_{j=1}^{n} (x_j - \bar{x})^2$$

as an estimator of the population variance σ^2, notwithstanding the fact that the latter is the sample variance and is a consistent estimator of σ^2.

Unbiased estimators
17.9　Consider the sampling distribution of an estimator t. If the estimator is consistent, its distribution must, for large samples, have a central value in the neighbourhood of θ. We may choose among the class of consistent estimators by requiring that θ shall be equated to this central value not merely for large, but for all samples.

If we require that for all n and θ the *mean* value of t shall be θ, i.e. that

$$E(t) = \theta, \tag{17.8}$$

we call t an *unbiased* estimator of θ. This is an unfortunate word, like so many in statistics. There is nothing except convenience to exalt the arithmetic mean above other measures of location as a criterion of bias. We might equally well have chosen the median of the distribution of t, or its mode, as determining the 'unbiased' estimator. The mean value is used, as always, for its mathematical convenience. This is perfectly legitimate, and it is only necessary to remark that the term 'unbiased' should not be allowed to convey overtones of a non-technical nature.

Example 17.3
Since

$$E\left\{\frac{1}{n} \sum x\right\} = \frac{1}{n} \sum E(x) = \mu_1',$$

the sample mean is an unbiased estimator of the population mean whenever the latter exists. But the sample variance is not an unbiased estimator of the population variance, for

$$E\{\sum(x_j - \bar{x})^2\} = E\left\{\sum\left[x_j - \sum_j x_j/n\right]^2\right\} = E\left\{\frac{n-1}{n} \sum_j x_j^2 - \frac{1}{n} \sum_{j \neq k} \sum x_j x_k\right\}$$

$$= (n-1)\mu_2' - (n-1)\mu_1'^2 = (n-1)\mu_2.$$

Thus $(1/n) \sum (x - \bar{x})^2$ has a mean value $[(n-1)/n]\mu_2$. It follows that an unbiased estimator is given by

$$s^2 = \frac{1}{n-1} \sum (x - \bar{x})^2,$$

and for this reason it is usually preferred to the sample variance.

Our discussion in **17.8** shows that consistent estimators are not necessarily unbiased. We have already (Example 14.5) encountered an unbiased estimator that is not consistent. Thus neither property implies the other. But a consistent estimator whose asymptotic distribution has finite mean value must be asymptotically unbiased.

In certain circumstances, there may be no unbiased estimator (cf. Exercise 17.12). Even if there is one, it may be forced to give absurd estimates at times, or even always. For example in estimating a parameter θ where $a \leq \theta \leq b$, no estimator t distributed on the same interval can be unbiased, for if $\theta = a$, we must have $E(t) > a$, and if $\theta = b$, $E(t) < b$. Thus 'impossible' estimates must sometimes arise if we insist on unbiasedness in this case. We shall meet an important example of this in **27.34** below; another example is the estimation of the characteristic exponent α of a stable distribution (cf. **4.36**)—when $\alpha = 2$, no estimator in $(0, 2)$ can be unbiased (cf. Du Monchel (1983)). See also Exercises 10.21–2, Vol. 1.

Similarly, a bounded estimator cannot be unbiased for any unbounded function of a parameter, since its expectation is also bounded. This argument is enough to establish the non-existence of an unbiased estimator in Exercise 17.12 for negative a or b.

Further, in the sequential binomial sampling scheme discussed in Example 9.15, Vol. 1, when $k = 1$ there is an unbiased estimator of a probability that only takes the values 0 or 1. An even more extreme instance, where the only unbiased estimate *always* gives impossible estimates, is contained in Exercise 17.26.

Corrections for bias: the jackknife and the bootstrap

17.10 If we have a biased estimator t, and wish to remove its bias, this may be possible by direct evaluation of its expected value and the application of a simple adjustment, as in Example 17.3. But sometimes the expected value is a rather complicated function of the parameter, θ, being estimated, and it is not obvious what the correction should be. Quenouille (1956) proposed an ingenious method of overcoming this difficulty in a fairly general class of situations.

We denote our biased estimator by t_n, its suffix being the number of observations from which t_n is calculated. Now suppose that

$$E(t_n) - \theta = \sum_{r=1}^{\infty} a_r / n^r, \tag{17.9}$$

where the a_r may be functions of θ but not of n. t_{n-1} may be calculated from each of the n possible subsets of $(n-1)$ observations—when the ith observation is omitted, we write it $t_{n-1,i}$. Let $\bar{t}_{n-1}$ denote the average of these n values, and consider the new statistic

$$t'_n = nt_n - (n-1)\bar{t}_{n-1} = t_n + (n-1)(t_n - \bar{t}_{n-1}). \tag{17.10}$$

It may happen that $t'_n \equiv t_n$ (as when t_n is the sample mean, or the sample median when n is even), but if not it follows at once from (17.9) that

$$E(t'_n) - \theta = a_2 \left(\frac{1}{n} - \frac{1}{n-1} \right) + a_3 \left(\frac{1}{n^2} - \frac{1}{(n-1)^2} \right) + \cdots$$

$$= -\frac{a_2}{n^2} - O(n^{-3}). \tag{17.11}$$

Thus t'_n is only biased to order $1/n^2$. Similarly,

$$t''_n = \{n^2 t'_n - (n-1)^2 \bar{t}'_{n-1}\}/\{n^2 - (n-1)^2\}$$

will be biased only to order $1/n^3$, and so on. This method, called the *jackknife*, removes bias to any required degree.

Because t'_n differs from t_n by a quantity of order n^{-1}, we see that if t_n has variance of order n^{-1} the variance of t'_n is asymptotically the same as that of t_n, so that reduction of bias carries no penalty in the variance—cf. Exercise 17.18; but this is not generally true for further-order corrections—cf. Robson and Whitlock (1964). Thorburn (1976) gives conditions for the asymptotic distributions of t'_n and t_n to be the same.

t'_n is the mean of n identically distributed, equally correlated variates $d_i = nt_n - (n-1)t_{n-1,i}$. If this correlation is negligible, the variance of t'_n is estimated by

$$\sum_{i=1}^{n} \{d_i - t'_n\}^2/\{n(n-1)\} = \frac{n-1}{n} \sum (t_{n-1,i} - \bar{t}_{n-1})^2,$$

using the final result in Exercise 17.17—the corresponding estimator of the variance of t_n is shown by Efron and Stein (1981) to be biased upwards; but if not, this estimator may not be useful even as $n \to \infty$; see R. G. Miller (1964), who also discusses asymptotic normality and (1974a) reviews the subject of jackknifing, generally treated in the book by Gray and Schucany (1972). Cressie (1981) discusses transformation of the observations before jackknifing estimators. See also Sharot (1976), Frangos (1980) and P. K. Sen (1977). Parr and Schucany (1980) provide a bibliography.

Arvesen and Layard (1975) discuss the situation where the original observations have different distributions.

Instead of the jackknife's use of each subset of $(n-1)$ distinct observations, Efron (1979) proposed a *bootstrap* resampling method that uses each possible sample of size m drawn at random from the original sample, to estimate the sampling distribution of any random variable, based on $m \leq n$ observations. Thus the bootstrap is not restricted to estimating the mean and variance of such a variable as is the jackknife; indeed, the jackknife is shown to be a linear approximation to the bootstrap in those cases.

Since the number of possible samples, n^n, is prohibitively large for even moderate n, the method proceeds by using B replicates, or random samples with replacement, from the original sample. In general, B should be at least of order $n \log n$; then, as $n \to \infty$, the bootstrap sampling distribution converges to the true sampling distribution under fairly general conditions. Efron (1981) compares the efficiency of these and other methods of estimating sampling variances in the case of the binormal correlation coefficient, and finds that the bootstrap performs best.

K. Singh (1981), Bickel and Freedman (1981), Freedman (1981) and Efron (1985) verify the validity of the bootstrap approximation for many common statistical procedures. See also Parr (1983), Schenker (1985), P. Hall (1986a,b; 1987), Efron (1987). DiCiccio and Tibshirani (1987), P. Hall (1988a), Davidson and Hinkley (1988), and Tibshirani (1988) discuss improved confidence intervals based on the bootstrap. Hinkley (1988), DiCiccio and Romano (1988) and P. Hall (1988b) review the field—see also the discussions following these papers. P. Hall and Martin (1988) provide a unified treatment.

Example 17.4
To find an unbiased estimator of θ^2 in the binomial distribution

$$P\{x = r\} = \binom{n}{r} \theta^r (1 - \theta)^{n-r}, \qquad r = 0, 1, 2, \ldots, n; \ n \geqslant 2.$$

The intuitive estimator is

$$t_n = \left(\frac{r}{n}\right)^2,$$

since r/n is an unbiased estimator of θ, but

$$E(t_n) = \mathrm{var}\,(r/n) + \{E(r/n)\}^2 = \theta^2 + \theta(1 - \theta)/n.$$

Now t_{n-1} can only take the values

$$\left(\frac{r-1}{n-1}\right)^2 \qquad \text{or} \qquad \left(\frac{r}{n-1}\right)^2$$

according to whether a 'success' or a 'failure' is omitted from the sample. Thus

$$\bar{t}_{n-1} = \frac{1}{n}\left\{ r\left(\frac{r-1}{n-1}\right)^2 + (n-r)\left(\frac{r}{n-1}\right)^2 \right\} = \frac{r^2(n-2)+r}{n(n-1)^2}.$$

Hence, from (17.10),

$$\begin{aligned} t'_n &= n t_n - (n-1)\bar{t}_{n-1} \\ &= \frac{r^2}{n} - \frac{r^2(n-2)+r}{n(n-1)} \\ &= \frac{r(r-1)}{n(n-1)}, \end{aligned} \qquad (17.12)$$

which, it may be directly verified, is exactly unbiased for θ^2. Exercise 17.12 gives a general result of which this is a special case.

Exercises 17.17 and 17.13 give further applications of the method.

17.11 Often there will exist more than one consistent estimator of a parameter, even if we confine ourselves to unbiased estimators. Consider once again the estimation of the mean of a normal population with variance σ^2. The sample mean is consistent and unbiased. We will now prove that the same is true of the median.

Consideration of symmetry is enough to show (cf. Exercise 14.7, Vol. 1, 5th edn) that the median is an unbiased estimator of the population mean, which is, of course, the same as the population median. For large n the distribution of the median tends to the normal form (cf. **14.12**)

$$\mathrm{d}F(x) \propto \exp\{-2nf_1^2(x - \theta)^2\}\,\mathrm{d}x, \qquad (17.13)$$

where f_1 is the population median ordinate, in our case equal to $(2\pi\sigma^2)^{-\frac{1}{2}}$. The variance of the sample median is therefore, from (17.13), equal to $\pi\sigma^2/(2n)$ and tends to zero for large n. Hence the estimator is consistent.

17.12 We must therefore seek further criteria to choose between estimators with the common property of consistency. Such a criterion arises naturally if we consider the sampling variances of the estimators. We expect that the estimator with the smaller variance will be distributed more closely round the value θ; this will certainly be so for distributions of the normal type. An unbiased consistent estimator with a smaller variance will therefore deviate less, on the average, from the true value than one with a larger variance. Hence we may reasonably regard it as better.

In the case of the mean and median of normal samples we have, for any n, from (17.3),

$$\text{var (mean)} = \sigma^2/n, \tag{17.14}$$

whereas (14.18) showed that for $n = 2r + 1$, the variance of the median is, putting $\sigma^2 = 1$,

$$\text{var}(x_{(r+1)}) = \frac{\pi}{2(n+2)} + \frac{\pi^2}{4(n+2)(n+4)} + o(n^{-2}),$$

where

$$\text{var (mean)/var (median)} = \frac{2}{\pi} + \frac{1}{n}\left(\frac{4}{\pi} - 1\right) + o(n^{-1}). \tag{17.15}$$

Hodges (1967) gives the following exact values:

n:	1	3	5	7	9	11	13	15	17	19	∞
$\dfrac{\text{var (mean)}}{\text{var (median)}}$:	1	0.743	0.697	0.679	0.669	0.663	0.659	0.656	0.653	0.651	0.637

The approximation (17.15) is accurate to two decimal places for $n \geq 7$. The values for even n are always higher than the contiguous values for odd n—the definition of the median is different in this case—but tend monotonically to the same limit from the value 1 at $n = 2$. (The approximation (17.15) must now have the figure '4' replaced by '6'.) A few values of $\{\text{var (median)/var (mean)}\}^{1/2}$ for even n were given in **14.6**. Thus the mean always has smaller variance than the median in the normal case.

Example 17.5
For the Cauchy distribution

$$\mathrm{d}F(x) = \frac{1}{\pi} \frac{\mathrm{d}x}{\{1 + (x - \theta)^2\}}, \qquad -\infty < x < \infty,$$

we have already seen (**17.6**) that the sample mean is not a consistent estimator of θ, the population median. However, for the sample median, t, we have, since the median ordinate is $1/\pi$, the large-sample variance

$$\text{var } t = \pi^2/4n$$

from (17.13). It is seen that the median is consistent, and although direct comparison with the mean is not possible because the latter does not possess a sampling variance, the median is evidently a better estimator of θ than the mean. This provides an interesting contrast with the case of the normal distribution, particularly in view of the similarity of the shapes of the distributions.

Minimum variance bounds

17.13   It seems natural, then, to use the sampling variance of an estimator as a criterion of its acceptability, and it has, in fact, been so used since the days of Laplace and Gauss. But only in relatively recent times has it been established that, under fairly general conditions, there exists a bound below which the variance of an unbiased estimator cannot fall. In order to establish this bound, we first derive some preliminary results, which will also be useful in other connections later.

17.14   If the frequency function of the continuous or discrete population is $f(x\,|\,\theta)$, we define the *likelihood function** of a sample of n independent observations by

$$L(x_1, x_2, \ldots, x_n\,|\,\theta) = f(x_1\,|\,\theta)f(x_2\,|\,\theta)\cdots f(x_n\,|\,\theta). \tag{17.16}$$

We shall often write this simply as L. Evidently, since L is the joint frequency function of the observations,

$$\int \cdots \int L\,dx_1 \cdots dx_n = 1. \tag{17.17}$$

Now suppose that the first two derivatives of L with respect to θ exist for all θ. If we differentiate both sides of (17.17) with respect to θ, and if we may interchange the operations of differentiation and integration on its left-hand side (see **17.16** below), we obtain

$$\int \cdots \int \frac{\partial L}{\partial \theta}\,dx_1 \cdots dx_n = 0,$$

which we may rewrite

$$E\left(\frac{\partial \log L}{\partial \theta}\right) = \int \cdots \int \left(\frac{1}{L}\frac{\partial L}{\partial \theta}\right) L\,dx_1 \cdots dx_n = 0. \tag{17.18}$$

If we differentiate (17.18) again, we obtain, if we may again interchange operations,

$$\int \cdots \int \left\{\left(\frac{1}{L}\frac{\partial L}{\partial \theta}\right)\frac{\partial L}{\partial \theta} + L\frac{\partial}{\partial \theta}\left(\frac{1}{L}\frac{\partial L}{\partial \theta}\right)\right\} dx_1 \cdots dx_n = 0,$$

---

* R. A. Fisher calls L the likelihood when regarded as a function of θ and the probability of the sample when it is regarded as a function of x for fixed θ. While appreciating this distinction, we use the term likelihood and the symbol L in both cases to preserve a single notation.

which becomes

$$\int \cdots \int \left\{ \left(\frac{1}{L} \frac{\partial L}{\partial \theta} \right)^2 + \frac{\partial^2 \log L}{\partial \theta^2} \right\} L \, \mathrm{d}x_1 \cdots \mathrm{d}x_n = 0$$

or

$$E \left[\left(\frac{\partial \log L}{\partial \theta} \right)^2 \right] = -E \left(\frac{\partial^2 \log L}{\partial \theta^2} \right). \tag{17.19}$$

17.15 Now consider an unbiased estimator, t, of some function of θ, say $\tau(\theta)$. This formulation allows us to consider unbiased and biased estimators of θ itself, and also permits us to consider, for example, the estimation of the standard deviation when the parameter is equal to the variance. We thus have

$$E(t) = \int \cdots \int t L \, \mathrm{d}x_1 \cdots \mathrm{d}x_n = \tau(\theta). \tag{17.20}$$

We now differentiate (17.20), the result being

$$\int \cdots \int t \frac{\partial \log L}{\partial \theta} L \, \mathrm{d}x_1 \cdots \mathrm{d}x_n = \tau'(\theta),$$

which we may rewrite, using (17.18), as

$$\tau'(\theta) = \int \cdots \int \{t - \tau(\theta)\} \frac{\partial \log L}{\partial \theta} L \, \mathrm{d}x_1 \cdots \mathrm{d}x_n. \tag{17.21}$$

By the Cauchy-Schwarz inequality, we have from (17.21)*

$$\{\tau'(\theta)\}^2 \le \int \cdots \int \{t - \tau(\theta)\}^2 L \, \mathrm{d}x_1 \cdots \mathrm{d}x_n \cdot \int \cdots \int \left(\frac{\partial \log L}{\partial \theta} \right)^2 L \, \mathrm{d}x_1 \cdots \mathrm{d}x_n,$$

which, on rearrangement, becomes

$$\mathrm{var} \, t = E\{t - \tau(\theta)\}^2 \ge \{\tau'(\theta)\}^2 / E \left[\left(\frac{\partial \log L}{\partial \theta} \right)^2 \right]. \tag{17.22}$$

This is the fundamental inequality for the variance of an estimator, often known as the Cramér-Rao inequality, after two of its several discoverers (C. R. Rao (1945); Cramér (1946)); it was first given implicitly by Aitken and Silverstone (1942). Using (17.19), it may be written in what is often, in practice, the more convenient form

$$\mathrm{var} \, t \ge -\{\tau'(\theta)\}^2 / E \left(\frac{\partial^2 \log L}{\partial \theta^2} \right). \tag{17.23}$$

---

* (17.21) is the covariance between t and $\partial \log L / \partial \theta$, whose square cannot exceed the product of their variances, a result familiar in the theory of correlation (Chapter 26).

In the case where $\tau(\theta) \equiv \theta$, we have $\tau'(\theta) = 1$ in (17.22), so for an unbiased estimator of θ

$$\operatorname{var} t \geq 1 / E \left[\left(\frac{\partial \log L}{\partial \theta} \right)^2 \right] = -1 / E \left(\frac{\partial^2 \log L}{\partial \theta^2} \right). \tag{17.24}$$

The quantity

$$I = E \left[\left(\frac{\partial \log L}{\partial \theta} \right)^2 \right] \tag{17.25}$$

is sometimes called the *amount of information* or the *Fisher information* in the sample. We may similarly use (17.22) as a bound for I in terms of the moments of t,

$$I \geq \{\tau'(\theta)\}^2 / \operatorname{var} t. \tag{17.26}$$

We shall call (17.22) and (17.23) the minimum variance bound (abbreviated to MVB) for the estimation of $\tau(\theta)$. An estimator which attains this bound for all θ will be called an MVB estimator.

It is only necessary that (17.18) hold for the MVB (17.22) to follow from (17.20). If (17.19) also holds, we may write the MVB in the form (17.23).

17.16 The interchange of the operations of differentiation and integration leading to (17.18) or (17.19) is permissible if, e.g., the limits of integration (i.e. the limits of variation of x) are finite and independent of θ, and also if these limits are infinite, provided that the integral resulting from the interchange is uniformly convergent for all θ and its integrand is a continuous function of x and θ. These are sufficient sets of conditions—Exercises 17.21-22 show that they are not necessary.

17.17 It is very easy to establish the condition under which the MVB is attained. The inequality in (17.22) arose purely from the use of the Cauchy–Schwarz inequality, and the necessary and sufficient condition that the Cauchy–Schwarz inequality becomes an equality is (cf. **2.29**) that $\{t - \tau(\theta)\}$ is proportional to $\partial \log L / \partial \theta$ for all sets of observations. We write this condition

$$\frac{\partial \log L}{\partial \theta} = A(\theta)\{t - \tau(\theta)\} \tag{17.27}$$

where A is independent of the observations. Thus t is an MVB estimator if and only if it is a linear function of $\partial \log L / \partial \theta$.

Multiplying both sides of (17.27) by $\{t - \tau(\theta)\}$ and taking expectations, we find, using (17.21),

$$\tau'(\theta) = A(\theta) \operatorname{var} t. \tag{17.28}$$

Thus $A(\theta)$ has the same sign as $\tau'(\theta)$ and

$$\operatorname{var} t = \tau'(\theta) / A(\theta). \tag{17.29}$$

We thus conclude that if (17.27) is satisfied, t is an MVB estimator of $\tau(\theta)$, with variance (17.29), which is then equal to the right-hand side of (17.23). If $\tau(\theta) \equiv \theta$, var t is just $1 / A(\theta)$, which is then equal to the right-hand side of (17.24).

Example 17.6
To estimate θ in the normal distribution

$$dF(x) = \frac{1}{\sigma(2\pi)^{1/2}} \exp\left\{-\frac{1}{2}\left(\frac{x-\theta}{\sigma}\right)^2\right\} dx, \qquad -\infty < x < \infty,$$

where σ is known.

We have

$$\frac{\partial \log L}{\partial \theta} = \frac{n}{\sigma^2}(\bar{x} - \theta).$$

This is of the form (17.27) with

$$t = \bar{x}, \qquad A(\theta) = n/\sigma^2 \qquad \text{and} \qquad \tau(\theta) = \theta.$$

Thus $\bar{x}$ is the MVB estimator of θ, with variance σ^2/n.

Example 17.7
To estimate θ in the Cauchy distribution

$$dF(x) = \frac{1}{\pi}\frac{dx}{\{1+(x-\theta)^2\}}, \qquad -\infty < x < \infty.$$

We have

$$\frac{\partial \log L}{\partial \theta} = 2\Sigma\frac{x-\theta}{\{1+(x-\theta)^2\}}.$$

This cannot be put in the form (17.27). Thus there is no MVB estimator in this case.

Example 17.8
To estimate θ in the Poisson distribution

$$f(x\,|\,\theta) = e^{-\theta}\theta^x/x!, \qquad x = 0, 1, 2, \ldots,$$

we have

$$\frac{\partial \log L}{\partial \theta} = \frac{n}{\theta}(\bar{x} - \theta).$$

Thus $\bar{x}$ is the MVB estimator of θ, with variance θ/n.

Example 17.9
To estimate θ in the binomial distribution, for which

$$L(r\,|\,\theta) = \binom{n}{r}\theta^r(1-\theta)^{n-r}, \qquad r = 0, 1, 2, \ldots, n.$$

We find

$$\frac{\partial \log L}{\partial \theta} = \frac{n}{\theta(1-\theta)}\left(\frac{r}{n} - \theta\right).$$

Hence r/n is the MVB estimator of θ, with variance $\theta(1-\theta)/n$.

17.18 It follows from our discussion that, where an MVB estimator exists, it will exist for one specific function $\tau(\theta)$ of the parameter θ, and for no other function of θ. The following example makes the point clear.

Example 17.10
To estimate θ in the normal distribution with mean zero

$$dF(x) = \frac{1}{\theta(2\pi)^{1/2}} \exp\left(-\frac{x^2}{2\theta^2}\right) dx, \qquad -\infty < x < \infty.$$

We find

$$\frac{\partial \log L}{\partial \theta} = -\frac{n}{\theta} + \frac{\sum x^2}{\theta^3} = \frac{n}{\theta^3}\left(\frac{1}{n}\sum x^2 - \theta^2\right).$$

We see at once that $(1/n)\sum x^2$ is an MVB estimator of θ^2 (the variance of the population) with sampling variance $(\theta^3/n)d(\theta^2)/d\theta = 2\theta^4/n$, by (17.29). But there is no MVB estimator of θ itself.

Equation (17.27) determines a condition on the frequency function under which an MVB estimator of some function of θ, $\tau(\theta)$, exists. If the frequency is not of this form, there may still be an estimator of $\tau(\theta)$ that has, uniformly in θ, smaller variance than any other estimator; we then call it a minimum variance (MV) estimator. In other words, the least *attainable* variance may be greater than the MVB. Further, if the regularity conditions leading to the MVB do not hold, the least attainable variance may be less than the (in this case inapplicable) MVB. In any case (17.27) demonstrates that there can only be one function of θ for which the MVB is attainable, namely, that function (if any) which is the expectation of a statistic t in terms of which $\partial \log L/\partial \theta$ may be linearly expressed.

17.19 From (17.27) we have on integration the necessary form for the likelihood function (continuing to write $A(\theta)$ for the integral of the arbitrary function $A(\theta)$ in (17.27))

$$\log L = tA(\theta) + P(\theta) + R(x_1, x_2, \ldots, x_n),$$

which we may rewrite in the frequency-function form

$$f(x|\theta) = \exp\{A(\theta)B(x) + C(x) + D(\theta)\}, \tag{17.30}$$

where $t = \sum_{i=1}^{n} B(x_i)$, $R(x_1, \ldots, x_n) = \sum_{i=1}^{n} C(x_i)$ and $P(\theta) = nD(\theta)$. (17.30) is the exponential family of distributions introduced in **5.47**, Vol. 1. We shall return to it in **17.36**.

The regularity conditions required to derive (17.27) limit the generality of this result. In particular, V. M. Joshi (1976) showed that if $\log L$ is not absolutely continuous in θ the family (17.30) is no longer necessary—cf. Exercise 17.29.

Bhattacharyya bounds

17.20   We can find better (i.e. greater) lower bounds than the MVB (17.22) for the variance of an estimator in cases where the MVB is not attainable. The condition (17.27) for (17.22) to be an attainable bound is that there be an estimator t for which $t - \tau(\theta)$ is a linear function of

$$\frac{\partial \log L}{\partial \theta} = \frac{1}{L} \frac{\partial L}{\partial \theta}.$$

But even if no such estimator exists, there may still be one for which $t - \tau(\theta)$ is a linear function of

$$\frac{1}{L} \frac{\partial L}{\partial \theta} \quad \text{and} \quad \frac{1}{L} \frac{\partial^2 L}{\partial \theta^2}$$

or, in general, of $(1/L) \partial^r L / \partial \theta^r$. Following A. Bhattacharyya (1946), we therefore seek such a linear function that most closely approximates $t - \tau(\theta)$.

Estimating $\tau(\theta)$ by a statistic t as before, we write

$$L^{(r)} = \frac{\partial^r L}{\partial \theta^r}$$

and

$$\tau^{(r)} = \frac{\partial^r \tau(\theta)}{\partial \theta^r}.$$

We now construct the function

$$D_s = t - \tau(\theta) - \sum_{r=1}^{s} a_r L^{(r)} / L, \tag{17.31}$$

where the a_r are constants to be determined. Since, on differentiating (17.17) r times, we obtain

$$E(L^{(r)} / L) = 0 \tag{17.32}$$

under the same conditions as before, we have from (17.31) and (17.32)

$$E(D_s) = 0. \tag{17.33}$$

The variance of D_s is therefore

$$E(D_s^2) = \int \cdots \int \left\{ t - \tau(\theta) - \sum_r a_r L^{(r)} / L \right\}^2 L \, dx_1 \cdots dx_n. \tag{17.34}$$

We minimise (17.34) for variation of the a_r by putting

$$\int \cdots \int \left\{ t - \tau(\theta) - \sum_r a_r L^{(r)} / L \right\} \frac{L^{(p)}}{L} \cdot L \, dx_1 \cdots dx_n = 0 \tag{17.35}$$

for $p = 1, 2, \ldots, s$. This gives

$$\int \cdots \int (t - \tau(\theta)) L^{(p)} \, dx_1 \cdots dx_n = \sum_r a_r \int \cdots \int \frac{L^{(r)}}{L} \cdot \frac{L^{(p)}}{L} L \, dx_1 \cdots dx_n. \tag{17.36}$$

The left-hand side of (17.36) is, from (17.32), equal to

$$\int \cdots \int tL^{(p)} \, dx_1 \cdots dx_n = \tau^{(p)}$$

on comparison with (17.20). The right-hand side of (17.36) is simply

$$\sum a_r E \left(\frac{L^{(r)}}{L} \cdot \frac{L^{(p)}}{L} \right).$$

On insertion of these values in (17.36) it becomes

$$\tau^{(p)} = \sum_{r=1}^{s} a_r E \left(\frac{L^{(r)}}{L} \cdot \frac{L^{(p)}}{L} \right), \qquad p = 1, 2, \ldots, s. \tag{17.37}$$

We may invert this set of linear equations, provided that the matrix of coefficients $J_{rp} = E(L^{(r)}/L \cdot L^{(p)}/L)$ is non-singular, to obtain

$$a_r = \sum_{p=1}^{s} \tau^{(p)} J_{rp}^{-1}, \qquad r = 1, 2, \ldots, s. \tag{17.38}$$

Thus, at the minimum of (17.34), (17.31) takes the value

$$D_s = t - \tau(\theta) - \sum_{r=1}^{s} \sum_{p=1}^{s} \tau^{(p)} J_{rp}^{-1} L^{(r)} / L, \tag{17.39}$$

and (17.34) itself has the value, from (17.39),

$$E(D_s^2) = \int \cdots \int \left\{ (t - \tau(\theta)) - \sum_r \sum_p \tau^{(p)} J_{rp}^{-1} L^{(r)} / L \right\}^2 L \, dx_1 \cdots dx_n,$$

which, on using (17.35), becomes

$$= \int \cdots \int \{t - \tau(\theta)\}^2 L \, dx_1 \cdots dx_n$$

$$- \sum_r \sum_p \tau^{(p)} J_{rp}^{-1} \int \cdots \int tL^{(r)} \, dx_1 \cdots dx_n,$$

and finally we have

$$E(D_s^2) = \operatorname{var} t - \sum_r \sum_p \tau^{(p)} J_{rp}^{-1} \tau^{(r)}. \tag{17.40}$$

Since its left-hand side is non-negative, (17.40) gives the required inequality

$$\operatorname{var} t \geq \sum_{r=1}^{s} \sum_{p=1}^{s} \tau^{(p)} J_{rp}^{-1} \tau^{(r)}. \tag{17.41}$$

In the particular case $s = 1$, (17.41) reduces to the MVB (17.22).

17.21 The condition for the bound in (17.41) to be attained is simply that $E(D_s^2) = 0$, which with (17.33) leads to $D_s = 0$ or, from (17.39),

$$t - \tau(\theta) = \sum_{r=1}^{s} \sum_{p=1}^{s} \tau^{(p)} J_{rp}^{-1} L^{(r)} / L, \tag{17.42}$$

which is the generalisation of (17.27). (17.42) requires $t - \tau(\theta)$ to be a linear function of the quantities $L^{(r)}/L$. If it is a linear function of the first s such quantities, there is clearly nothing to be gained by adding further terms. On the other hand, the right-hand side of (17.41) is a non-decreasing function of s, as is easily seen by the consideration that the variance of D_s cannot be increased by allowing the optimum choice of a further coefficient a_r in (17.34). Thus we may expect, in some cases where the MVB (17.22) is not attained for the estimation of a particular function $\tau(\theta)$, to find the greater bound in (17.41) attained for some value of $s > 1$.

> Fend (1959) showed that if (17.27) holds, any polynomial of degree s in t attains the sth but not the $(s-1)$th variance bound (17.41), and is therefore the MV unbiased estimator of its expectation.
>
> If we seek similarly to improve the bound (17.26) for I, the natural analogue of the procedure of **17.20** is to approximate $\partial \log L/\partial\theta$ by a polynomial in $t - \tau(\theta)$, minimising var $\{\partial \log L/\partial\theta - \sum_{r=1}^{s} C_r(t - \tau(\theta))^r\}$.
>
> This yields an increasing series of bounds in terms of the central moments μ_j and the skewness and kurtosis coefficients of t—Jarrett (1984) gives an explicit general result. $s = 1$ yields (17.26) of course; for $s = 2$, the result, previously given by A. Bhattacharyya (1946) is
>
> $$I \geqslant \mu_2^{-1}\{(\tau')^2 + (\mu_2^{-1/2}\,\partial\mu_2/\partial\theta - \beta_1^{1/2}\tau')^2/(\beta_2 - \beta_1 - 1)\},$$
>
> the second term being positive by Exercise 3.19.

17.22 We now investigate the improvement in the bound arising from taking $s = 2$ instead of $s = 1$ in (17.41). Remembering that we have defined

$$J_{rp} = E\left(\frac{L^{(r)}}{L} \cdot \frac{L^{(p)}}{L}\right), \tag{17.43}$$

we find that we may rewrite (17.41) in this case as

$$\begin{vmatrix} \text{var } t & \tau' & \tau'' \\ \tau' & J_{11} & J_{12} \\ \tau'' & J_{12} & J_{22} \end{vmatrix} \geqslant 0, \tag{17.44}$$

which becomes, on expansion,

$$\text{var } t \geqslant \frac{(\tau')^2}{J_{11}} + \frac{(\tau'J_{12} - \tau''J_{11})^2}{J_{11}(J_{11}J_{22} - J_{12}^2)}. \tag{17.45}$$

The second term on the right of (17.45) is the improvement in the bound to the variance of t. It may easily be confirmed that, writing $J_{rp}(n)$ as a function of sample size,

$$\left.\begin{aligned} J_{11}(n) &= nJ_{11}(1), \\ J_{12}(n) &= nJ_{12}(1), \\ J_{22}(n) &= nJ_{22}(1) + 2n(n-1)\{J_{11}(1)\}^2, \end{aligned}\right\} \tag{17.46}$$

and using (17.46), (17.45) becomes

$$\text{var } t \geqslant \frac{(\tau')^2}{nJ_{11}(1)} + \frac{\{\tau'' - \tau'J_{12}(1)/J_{11}(1)\}^2}{2n^2\{J_{11}(1)\}^2} + o\left(\frac{1}{n^2}\right), \tag{17.47}$$

which makes it clear that the improvement in the bound is of order $1/n^2$ compared with the leading term of order $1/n$, and is only of importance when $\tau' = 0$ and the first term disappears. In the case where t is estimating θ itself, (17.45-7) give

$$\text{var } t \geq \frac{1}{J_{11}} + \frac{J_{12}^2}{J_{11}(J_{11}J_{22} - J_{12}^2)}$$

$$= \frac{1}{J_{11}} + \frac{J_{12}^2}{2J_{11}^4} + o\left(\frac{1}{n^2}\right). \tag{17.48}$$

If the bound with $s = 2$ equals the MVB ($s = 1$), this does not generally imply that the latter is attainable. For the exponential family (17.30), however, this equality does imply attainability—cf. Patil and Shorrock (1965).

Example 17.11

To estimate $\theta(1 - \theta)$ in the binomial distribution of Example 17.9, it is natural to take as our estimator an unbiased function of $r/n = p$, which is the MVB estimator of θ. We have seen in Example 17.4 that $r(r-1)/\{n(n-1)\}$ is an unbiased estimator of θ^2. Hence, or from Exercise 17.12,

$$t = \frac{r}{n} - \frac{r(r-1)}{n(n-1)} = \frac{r(n-r)}{n(n-1)} = \left(\frac{n}{n-1}\right) p(1-p)$$

is an unbiased estimator of $\theta(1 - \theta)$. The exact variance of t is obtained by rewriting it as

$$\frac{n-1}{n} t \equiv \tfrac{1}{4} - (p - \tfrac{1}{2})^2$$

so that

$$\left(\frac{n-1}{n}\right)^2 \text{var } t = \text{var }\{(p - \tfrac{1}{2})^2\} = E\{(p - \tfrac{1}{2})^4\} - [E\{(p - \tfrac{1}{2})^2\}]^2$$

$$= \lambda_4' - (\lambda_2')^2$$

where λ_r' is the rth moment of p about the value $\tfrac{1}{2}$. Using (3.8), Vol. 1, we express the λ_r' in terms of the central moments λ_r about $E(p) = \theta$:

$$\lambda_4' - (\lambda_2')^2 = \lambda_4 + 4(\theta - \tfrac{1}{2})\lambda_3 + 4(\theta - \tfrac{1}{2})^2\lambda_2 - \lambda_2^2.$$

Now λ_r is $n^{-r}\mu_r$, where μ_r is the moment of the binomial variable np, so

$$\left(\frac{n-1}{n}\right)^2 \text{var } t = \frac{\mu_4 - \mu_2^2}{n^4} - \frac{2(1-2\theta)\mu_3}{n^3} + (1-2\theta)^2 \frac{\mu_2}{n^2}$$

and we may substitute for the μ_r from (5.4) (where p is written for our θ), obtaining finally

$$\text{var } t = \frac{\theta(1-\theta)}{n} \left\{(1-2\theta)^2 + \frac{2\theta(1-\theta)}{n-1}\right\},$$

which is an order of magnitude in n smaller when $\theta = \frac{1}{2}$, when it equals $1/\{8n(n-1)\}$, than when $\theta \neq \frac{1}{2}$. We cannot attain the MVB for $\tau(\theta) = \theta(1-\theta)$, since it is attained for θ itself (cf. Example 17.9) and (cf. **17.18**) only one function can attain its MVB. However, when $\theta \neq \frac{1}{2}$ the term in n^{-1} in var t is $(\tau')^2/J_{11}$, the MVB, which is therefore *asymptotically* attained—cf. **17.23** below. For any θ, var t exactly attains the bound (17.45) for $s = 2$, as may be seen from substituting the results $J_{12} = 0$, $J_{22} = 2n(n-1)/\{\theta^2(1-\theta)^2\}$ (which are left to the reader in Exercise 17.3) in (17.45). Alternatively, this result follows from the fact that t is a linear function of L'/L and L''/L—cf. Exercise 17.3.

> Exercise 17.28 gives a general result for the bounds (17.41) in the binomial case.

17.23 Example 17.11 brings out one point of some importance. If we have an MVB estimator t for $\tau(\theta)$, i.e.

$$\text{var } t = \{\tau'(\theta)\}^2/J_{11}, \qquad (17.49)$$

and we require to estimate some other function of θ, say $\psi\{\tau(\theta)\}$, we know from (10.14), Vol. 1, that in large samples

$$\text{var }\{\psi(t)\} \sim \left(\frac{\partial\psi}{\partial\tau}\right)^2 \text{var } t$$

$$\sim \left(\frac{\partial\psi}{\partial\tau}\right)^2 \left(\frac{\partial\tau}{\partial\theta}\right)^2 \Big/ J_{11}$$

from (17.49), provided that ψ' is non-zero. But this may be rewritten

$$\text{var }\{\psi(t)\} \sim \left(\frac{\partial\psi}{\partial\theta}\right)^2 \Big/ J_{11},$$

so that *any* such function of θ has an MVB estimator in large samples if some function of θ has such an estimator for all n. Further, the estimator is always the corresponding function of t. We shall, in **17.35**, be able to supplement this result by a more exact one concerning functions of the MVB estimator.

17.24 There is no difficulty in extending the results of this chapter on MVB estimation to the case when the distribution has more than one parameter, and we are interested in estimating a single function of them all, say $\tau(\theta_1, \theta_2, \ldots, \theta_k)$. In this case, the analogue of the simplest result, (17.22), is

$$\text{var } t \geq \sum_{i=1}^{k} \sum_{j=1}^{k} \frac{\partial\tau}{\partial\theta_i} \frac{\partial\tau}{\partial\theta_j} I_{ij}^{-1} \qquad (17.50)$$

where the matrix which has to be inverted to obtain the terms in (17.50) is

$$\{I_{ij}\} = \left\{ E\left(\frac{\partial \log L}{\partial\theta_i} \cdot \frac{\partial \log L}{\partial\theta_j}\right)\right\}. \qquad (17.51)$$

As before, (17.50) only takes account of terms of order $1/n$, and a more complicated inequality is required if we are to take account of lower-order terms.

17.25 Even better lower bounds for the sampling variance of estimators can be obtained without imposing regularity conditions. See Kiefer (1952), Barankin (1949) and Blischke *et al.* (1969). Chapman and Robbins (1951) established that if $E(t) = \theta$,

$$\text{var } t \geqslant \frac{1}{\inf_h \frac{1}{h^2} \left\{ \int \left[\left(L(x \mid \theta + h) \right)^2 \Big/ L(x \mid \theta) \right] dx - 1 \right\}}, \qquad (17.52)$$

the infimum being for all $h \neq 0$ for which $L(x \mid \theta) = 0$ implies $L(x \mid \theta + h) = 0$. (17.52) is in general at least as good a bound as (17.24), though more generally valid. For the denominator of the right-hand side of (17.24) is

$$E\left[\left(\frac{\partial \log L}{\partial \theta} \right)^2 \right] = \int \lim_{h \to 0} \left[\frac{L(x \mid \theta + h) - L(x \mid \theta)}{hL(x \mid \theta)} \right]^2 L(x \mid \theta) \, dx,$$

and provided that we may interchange the integral and the limiting operation, this becomes

$$= \lim_{h \to 0} \frac{1}{h^2} \left[\int \frac{\{L(x \mid \theta + h)\}^2}{L(x \mid \theta)} \, dx - 1 \right]. \qquad (17.53)$$

The denominator on the right of (17.52) is the infimum of this quantity over all permissible values of h, and is consequently no greater than the limit as h tends to zero. Thus (17.52) is at least as good a bound as (17.24)—cf. Exercise 17.6.

Sen and Ghosh (1976) give a condition for the bound (17.52) to be attained, and investigate its relation to the bounds (17.42) with $s > 1$.

Minimum variance estimation

17.26 So far, we have been largely concerned with the uniform attainment of the MVB by a single estimator. In many cases, however, it will not be attained, even when the conditions under which it is derived are satisfied. When this is so, there may still be an MV estimator for all values of θ. The question of the uniqueness of an MV estimator now arises. We can easily show that if an MV estimator exists, it is essentially unique, irrespective of whether any bound is attained.

Let t_1 and t_2 be MV unbiased estimators of $\tau(\theta)$, each with variance V. Consider the new estimator

$$t_3 = \tfrac{1}{2}(t_1 + t_2)$$

which evidently also estimates $\tau(\theta)$ with variance

$$\text{var } t_3 = \tfrac{1}{4}\{\text{var } t_1 + \text{var } t_2 + 2 \text{ cov } (t_1, t_2)\}. \qquad (17.54)$$

Now by the Cauchy–Schwarz inequality

$$\text{cov } (t_1, t_2) = \int \cdots \int (t_1 - \tau)(t_2 - \tau) L \, dx_1 \cdots dx_n$$

$$\leqslant \left\{ \int \cdots \int (t_1 - \tau)^2 L \, dx_1 \cdots dx_n \cdot \int \cdots \int (t_2 - \tau)^2 L \, dx_1 \cdots dx_n \right\}^{1/2}$$

$$\leqslant (\text{var } t_1 \text{ var } t_2)^{1/2} = V. \qquad (17.55)$$

Thus (17.54) and (17.55) give

$$\text{var } t_3 \leqslant V,$$

which contradicts the assumption that t_1 and t_2 have MV unless the equality holds. This implies that the equality sign holds in (17.55). This can only be so if

$$(t_1 - \tau) = k(\theta)(t_2 - \tau), \tag{17.56}$$

i.e. the variables are proportional. But (17.56) implies that

$$\text{cov}(t_1, t_2) = k(\theta)\, \text{var}\, t_2 = k(\theta)\, V$$

and this equals V since the equality sign holds in (17.55). Thus

$$k(\theta) = 1$$

and hence, from (17.56), $t_1 = t_2$ identically. Thus an MV estimator is unique.

17.27 The argument of the preceding section can be generalised to give an interesting relation for the correlation ρ between the MV estimator and any other estimator, which, it will be remembered from **16.23**, is defined as the ratio of their covariance to the square root of the product of their variances.

Suppose that t_1 is the MV unbiased estimator of $\tau(\theta)$, with variance V, and that t_2 is any other unbiased estimator of $\tau(\theta)$ with finite variance V_2 where $V/V_2 = E$. Consider a new estimator

$$t_3 = at_1 + (1 - a)t_2. \tag{17.57}$$

This will also estimate $\tau(\theta)$, with variance

$$\text{var}\, t_3 = a^2\, \text{var}\, t_1 + (1 - a)^2\, \text{var}\, t_2 + 2a(1 - a)\, \text{cov}(t_1, t_2). \tag{17.58}$$

Writing

$$\rho = \frac{\text{cov}(t_1, t_2)}{(\text{var}\, t_1\, \text{var}\, t_2)^{1/2}} = \frac{\text{cov}(t_1, t_2)E^{1/2}}{V},$$

we obtain from (17.58), since $\text{var}\, t_3 \geqslant V$,

$$a^2 + (1 - a)^2 E^{-1} + 2a(1 - a)\rho E^{-1/2} \geqslant 1. \tag{17.59}$$

(17.59) may be rearranged as a quadratic inequality in a,

$$a^2\{1 + E^{-1} - 2\rho E^{-1/2}\} - 2a(E^{-1} - \rho E^{-1/2}) + (E^{-1} - 1) \geqslant 0,$$

and the discriminant of the left-hand side cannot be positive, since the roots of the equation are complex or equal. Thus

$$(E^{-1} - \rho E^{-1/2})^2 \leqslant (1 + E^{-1} - 2\rho E^{-1/2})(E^{-1} - 1)$$

which yields

$$(\rho E^{-1/2} - 1)^2 \leqslant 0. \tag{17.60}$$

Hence

$$\rho = E^{1/2}, \tag{17.61}$$

where E is the reciprocal of the relative variance of the other estimator, and in large samples is usually called its efficiency, for a reason to be discussed in **17.28-9**.

From the definition of ρ, it follows at once using (17.61) that the covariance of any unbiased finite-variance estimator u with an MV unbiased estimator t is exactly var t. Since $u \equiv t + (u - t)$ this implies cov $(t, u - t) = 0$—cf. Exercise 17.11. Hence t has zero covariance with *every* finite-variance function having zero expectation. Conversely, if cov $(t, u - t) = 0$ for all u with finite variance and $E(u) = E(t)$,

$$\text{cov}(t, u) = \text{var } t$$

and their correlation, which must lie between 0 and 1, is $(\text{var } t / \text{var } u)^{1/2}$, so t must be the MV estimator of its expectation.

> Blom (1978) shows that n var t cannot increase with n—see Exercise 17.32.
> Bondesson (1975) shows that for a location parameter, when $f = f(x - \theta)$, no MV estimator of θ exists if f tends to zero too rapidly, while no polynomial in x can be an MV estimator unless f is normal.

Example 17.12
In Example 17.8, we saw that $\bar{x}$ is the MVB estimator of θ in the Poisson distribution. Since by Example 3.10 all the cumulants of the Poisson are equal to θ, $\bar{x}$ is the MV estimator of each of them. If we were instead to estimate the variance unbiasedly by the general formula

$$\frac{1}{n-1} \sum_{j=1}^{n} (x_j - \bar{x})^2,$$

the efficiency of the latter, by (17.61), would be ρ^2, and the correlation ρ was found in Exercise 12.21 to be $(1 + 2\theta)^{-1/2}$. Thus $E = (1 + 2\theta)^{-1}$, tending to 1 as $\theta \to 0$, but tending to 0 as $\theta \to \infty$. At $\theta = 1$, the efficiency is only $\frac{1}{3}$.

Efficiency
17.28 So far, our discussion of MV estimation has been exact, in the sense that it has not restricted sample size in any way. We now turn to consideration of large-sample properties. Even if there is no MV estimator for each value of n, there will often be one as n tends to infinity. Since most of the estimators we deal with are asymptotically normally distributed in virtue of the central limit theorem, the distribution of such an estimator will depend for large samples on only two parameters—its mean and its variance. If it is a consistent estimator it will commonly be asymptotically unbiased—cf. **17.9**. This leaves the variance as the means of discriminating between consistent, asymptotically normal estimators of the same parametric function.

Among such estimators, one with MV in large samples* is called an efficient estimator, or simply efficient, the term being due to Fisher (1921a), although the term is sometimes used for small samples also.

---

* But see the discussion of 'superefficiency' in **18.16**.

Portnoy (1977) shows that for the exponential family (17.30), any MV unbiased estimator is also efficient.

17.29 If we compare asymptotically normal estimators in large samples, we may reasonably set up a *measure* of efficiency, something we have not attempted to do for small n and arbitrarily distributed estimators. We shall define the efficiency of any other estimator, relative to an efficient estimator, as the reciprocal of the ratio of sample sizes required to give the estimators equal sampling variances, i.e. to make them equally precise.

If t_1 is an efficient estimator, and t_2 is another estimator, and (as is commonly the case) the variances are, in large samples, simple inverse functions of sample size, we may translate our definition of efficiency into a simple form. Let us suppose that

$$\left.\begin{array}{l} V_1 = \text{var}\,(t_1 \mid n_1) \sim a_1/n_1^r \qquad (r>0), \\ V_2 = \text{var}\,(t_2 \mid n_2) \sim a_2/n_2^s \qquad (s>0), \end{array}\right\} \tag{17.62}$$

where a_1, a_2 are constants independent of n, and we have shown sample size as an argument in the variances. If we are to have $V_1 = V_2$, we must have

$$1 = \frac{V_1}{V_2} = \lim_{n_1, n_2 \to \infty} \frac{a_1}{a_2} \frac{n_2^s}{n_1^r}.$$

Thus

$$\frac{a_2}{a_1} = \lim \frac{n_2^s}{n_1^r} = \lim \left(\frac{n_2}{n_1}\right)^s \cdot \frac{1}{n_1^{r-s}}. \tag{17.63}$$

Since t_1 is efficient, we must have $r \geqslant s$. If $r > s$, the last factor on the right of (17.63) will tend to zero, and hence, if the product is to remain equal to a_2/a_1, we must have

$$\frac{n_2}{n_1} \to \infty, \qquad r > s, \tag{17.64}$$

and we would thus say that t_2 has zero efficiency. If, in (17.63), $r = s$, we have at once

$$\lim \frac{n_2}{n_1} = \left(\frac{a_2}{a_1}\right)^{1/r},$$

which from (17.62) may be written

$$\lim \frac{n_2}{n_1} = \lim \left(\frac{V_2}{V_1}\right)^{1/r},$$

and the efficiency of t_2 is the reciprocal of this, namely

$$E = \lim \left(\frac{V_1}{V_2}\right)^{1/r}. \tag{17.65}$$

Note that if $r > s$, (17.65) gives the same result as (17.64). If $r = 1$, which is the most common case, (17.65) reduces the inverse variance ratio encountered at the end of **17.27**. Thus, when we are comparing estimators with variances of order $1/n$, we measure efficiency relative to an efficient estimator by the inverse of the variance ratio.

If the variance of an efficient estimator is not of the simple form (17.62)—see, for example, Exercise 18.21 below—the measurement of relative efficiency is not so simple.

Although it follows from the result of **17.26** that efficient estimators tend asymptotically to equivalence, there will in general be a multiplicity of efficient estimators, for if t_1 is efficient so is any $t_2 = t_1 + cn^{-p}$, where p is chosen large enough. If (17.62) holds for two efficient estimators t_1, t_2, we must have $r = s$ and $a_1 = a_2 = a$, say, so that $E = 1$ at (17.65). Now take (17.62) to a further term, so that

$$V_j = \frac{a}{n_j^r} + \frac{b_j}{n_j^{r+q}} + o(n_j^{-(r+q)}), \qquad j = 1, 2,$$

where $q > 0$ and we assume $b_1 \leqslant b_2$. If we now equate V_1 to V_2 we obtain, instead of (17.63),

$$\left(\frac{n_2}{n_1}\right)^r = \left(1 + \frac{b_2}{an_2^q}\right) \Big/ \left(1 + \frac{b_1}{an_1^q}\right)$$

and writing $d_n = n_2 - n_1$, this gives, in expanding the right-hand side,

$$1 + \frac{d_n}{n_1} = 1 + \frac{b_2}{ran_2^q} - \frac{b_1}{ran_1^q}.$$

Since $E = 1$, $n_1/n_2 \to 1$ and we obtain

$$d_n \to \frac{b_2 - b_1}{ra} \qquad \text{for } q = 1,$$

$$\to \infty \qquad \text{for } q < 1,$$

$$\to 0 \qquad \text{for } q > 1.$$

The limiting value of d is called the deficiency of t_2 with respect to t_1 by Hodges and Lehmann (1970). It is the number of additional observations that t_2 requires asymptotically to attain the second-order performance of t_1. The commonest case is $r = q = 1$, when $d = (b_2 - b_1)/a$. Exercise 17.8 applies the deficiency concept to the estimation of a variance.

Example 17.13
We saw in Example 17.6 that the sample mean is an MVB estimator of the mean μ of a normal population, with variance σ^2/n. We saw in Example 11.12 that it is exactly normally distributed. *A fortiori*, it is an efficient estimator. In **17.11–12**, we saw that the sample median is asymptotically normal with mean μ and variance $\pi\sigma^2/(2n)$. Thus, from (17.65) with $r = 1$, the efficiency of the sample median is $2/\pi = 0.637$.

Consider now the estimation of the standard deviation. Two possible estimators are the standard deviation of the sample, say t_1, and the mean deviation of the sample multiplied by $(\pi/2)^{1/2}$ (cf. **5.42**), say t_2. The latter is easier to calculate, as a rule, and if we have plenty of observations (as, for example, if we are using existing records and increasing sample size is merely a matter of turning up more records) it may be worth while using t_2 instead of t_1. Both estimators are asymptotically normally distributed.

In large samples the variance of the mean deviation is (cf. (10.39)) $\sigma^2(1 - 2/\pi)/n$. The variance of t_2 is then asymptotically $V_2 = \sigma^2(\pi - 2)/2n$. The asymptotic variance of the standard deviation (cf. **10.8**(d)) is $V_1 = \sigma^2/(2n)$, and we shall see later that it is an efficient estimator. Thus, using (17.65) with $r = 1$, the efficiency of t_2 is

$$E = \lim V_1/V_2 = 1/(\pi - 2) = 0.876.$$

The precision of the estimator from the mean deviation of a sample of 1000 is then about the same as that from the standard deviation of a sample of 876. If it is easier to calculate the m.d. of 1000 observations than the s.d. of 876 and there is no shortage of observations, it may be more convenient to use the former.

It has to be remembered, nevertheless, that in adopting such a procedure we are deliberately wasting information. By taking greater pains we could improve the efficiency of our estimator from 0.876 to unity.

Minimum mean-square-error estimation

17.30   Our discussions of unbiasedness and the minimisation of sampling variance have been conducted more or less independently. Sometimes, however, it is relevant to investigate both questions simultaneously. It is reasonable to argue that the presence of bias should not necessarily outweigh small sampling variance in an estimator. What we are really demanding of an estimator t is that it should be 'close' to the true value θ. Let us, therefore, consider its mean-square error (m.s.e.) about that true value, instead of about its own expected value. We have at once

$$E(t - \theta)^2 = E\{(t - E(t)) + (E(t) - \theta)\}^2 = \mathrm{var}\, t + \{E(t) - \theta\}^2,$$

the cross-product term on the right being equal to zero. The last term on the right is simply the square of the bias of t in estimating θ. If t is unbiased, this last term is zero, and m.s.e. becomes variance. In general, however, the minimisation of m.s.e. gives different results.

> Exercise 17.15 shows that truncation of an estimator t to the known range of the parameter always improves m.s.e., whether or not t is unbiased—if it is, variance is also always improved.

Example 17.14
What multiple a of an estimator t estimates $E(t) = \theta$ with smallest m.s.e.? We have, from **17.30**,

$$M(a) = E(at - \theta)^2 = a^2 \,\mathrm{var}\, t + \theta^2(a - 1)^2,$$

and this can only be less than var t if $a < 1$: we must 'shrink' the unbiased estimator.

If V is the coefficient of variation of t, defined at (2.28), $M(a) < \mathrm{var}\, t$ if and only if $a > (1 - V^2)/(1 + V^2)$, and is minimised for variation in a when

$$a = \theta^2/(\theta^2 + \mathrm{var}\, t) = 1/(1 + V^2).$$

The reduction in m.s.e. is then

$$\mathrm{var}\, t - E(at - \theta)^2 = \mathrm{var}\, t - \theta^2 \,\mathrm{var}\, t/(\theta^2 + \mathrm{var}\, t) = \mathrm{var}\, t/(1 + 1/V^2),$$

which is a monotone function of V, increasing from 0 to var t as V increases from 0 to ∞ and a decreases from 1 to 0. Large reductions occur when V is large. V is usually an inverse function of sample size, and $\to 0$ as $n \to \infty$, while $a \to 1$.

In general, V (and therefore a) is a function of θ, so at is not a statistic usable for estimation. Thus, e.g., if θ is the population mean and t is the sample mean, var $t = \sigma^2/n$, where σ^2 is the population variance, and $V^2 = \sigma^2/(n\theta^2)$. For the Poisson distribution, $\sigma^2 = \theta$ and we can make no progress, but for the exponential distribution $\theta^{-1} \exp(-x/\theta)$, $x \geq 0$, $\theta > 0$, we have $\sigma^2 = \theta^2$ and $a = n/(n+1)$, independent of θ; the reader may show that if instead we are estimating $\lambda = \theta^{-1}$, $(n-1)/(nt)$ is unbiased with $V^2 = (n-2)^{-1}$, so that $a = (n-2)/(n-1)$ and $(n-2)/(nt)$ is the m.s.e.-minimising multiple.

J. R. Thompson (1968) estimates a, replacing θ by t and σ^2 by its unbiased estimator s^2. Thus modified, at can have larger m.s.e. than t itself for moderate V, as he shows for the normal, binomial, Poisson and Gamma populations.

Exercise 17.16 deals with the estimation of powers of a normal standard deviation.

Minimum mean-square-error estimators are not much used, but it is as well to recognise that the objection to them is a practical, rather than a theoretical one (cf. the comparative study by Johnson (1950)). MV unbiased estimators are more tractable because they assume away the difficulty by insisting on unbiasedness.

Pitman (1938) showed that among estimators $t(\mathbf{x})$ of the location parameter θ in $f(x - \theta)$ that satisfy

$$t(\mathbf{x} + a) = t(\mathbf{x}) + a, \tag{A}$$

minimum m.s.e. is attained by

$$t_L(\mathbf{x}) = \int_{-\infty}^{\infty} \theta L(x \mid \theta)\, d\theta \Big/ \int_{-\infty}^{\infty} L(x \mid \theta)\, d\theta,$$

which is always unbiased and is therefore MV unbiased subject to (A). For the scale parameter θ in $\theta^{-1} f(x \mid \theta)$, $\theta > 0$, $t(\mathbf{x})$ satisfying

$$t(c\mathbf{x}) = ct(\mathbf{x}), \qquad c > 0 \tag{B}$$

minimises m.s.e when

$$t_S(\mathbf{x}) = \int_0^{\infty} \theta^{-2} L(x \mid \theta)\, d\theta \Big/ \int_0^{\infty} \theta^{-3} L(x \mid \theta)\, d\theta,$$

and need not be unbiased—any multiple of t will also satisfy (B). Example 17.14 shows that we can sometimes improve on the unbiased estimator, the exponential distribution there being a case in point. For the location parameter, no multiple of t satisfies (A).

Hoaglin (1975) studies the small-sample variance of t_L in Cauchy and in logistic samples.

Pitman (1937c) defined t to be 'closer' to θ than u is if $P\{|t - \theta| < |u - \theta|\} > \frac{1}{2}$, but this attractive concept is intransitive, as he pointed out. Exercise 17.34 shows that closeness is equivalent to comparing the variances of estimators that are binormally distributed. See also Geary (1944) and Johnson (1950).

Sufficient statistics

17.31 The criteria of estimation that we have so far discussed, namely consistency, unbiasedness, and minimum variance or m.s.e., are reasonable guides in assessing the properties of an estimator. To permit a more fundamental discussion, we now introduce the concept of sufficiency, which is due to Fisher (1921a, 1925).

Consider first the estimation of a single parameter θ by a single statistic t—we generalise in **17.38** below. There is an unlimited number of possible estimators of θ, from among which we must choose. With a sample of $n \geq 2$ observations as before, consider the joint distribution of a set of r functionally independent statistics, $f_r(t, t_1, t_2, \ldots, t_{r-1} | \theta)$, $r = 2, 3, \ldots, n$, where we have selected the statistic t for special consideration. We may write this as the product of the marginal distribution of t and the conditional distribution of the other statistics given t, i.e.

$$f_r(t, t_1, \ldots, t_{r-1} | \theta) = g(t | \theta) h_{r-1}(t_1, \ldots, t_{r-1} | t, \theta). \tag{17.66}$$

Now if the last factor on the right of (17.66) does not contain θ, we clearly have a situation in which, given t, the set $t_1, \ldots, t_{r-1}$ contribute nothing further to our knowledge of θ. If, further, this is true for every r and *any* set of $(r-1)$ statistics t_i, we may fairly say that t contains all the information in the sample about θ, and we therefore call it a (single) *sufficient* statistic for θ. We thus formally define t as sufficient for θ if and only if

$$f_r(t, t_1, \ldots, t_{r-1} | \theta) = g(t | \theta) h_{r-1}(t_1, \ldots, t_{r-1} | t), \tag{17.67}$$

where h_{r-1} is free of θ, for $r = 2, 3, \ldots, n$ and any choice of $t_1, \ldots, t_{r-1}$.[*]

17.32 As it stands, the definition of (17.67) does not enable us to see whether, in any given situation, a sufficient statistic exists. However, we may reduce it to a condition on the likelihood function. For if the latter may be written

$$L(x_1, \ldots, x_n | \theta) = g(t | \theta) k(x_1, \ldots, x_n), \tag{17.68}$$

where $g(t | \theta)$ is a function of t and θ alone[†] and k is free of θ, it is easy to see that (17.67) is deducible from (17.68). For any fixed r, and any set of $t_1, \ldots, t_{r-1}$, insert the differential elements $dx_1, \ldots, dx_n$ on both sides of (17.68) and make the transformation

$$\begin{cases} t = t(x_1, \ldots, x_n), \\ t_i = t_i(x_1, \ldots, x_n), & i = 1, 2, \ldots, r-1, \\ t_i = x_i, & i = r, \ldots, n-1. \end{cases}$$

The Jacobian of the transformation will not involve θ, and (17.68) will be transformed to

$$g(t | \theta) \, dt \, l(t, t_1, \ldots, t_{r-1}, t_r, \ldots, t_{n-1}) \prod_{i=1}^{n-1} dt_i, \tag{17.69}$$

and if we now integrate out the redundant variables $t_r, \ldots, t_{n-1}$, we obtain, for the joint distribution of $t, t_1, \ldots, t_{r-1}$, precisely the form (17.67).

---

[*] This definition is usually given only for $r = 2$, but the definition for all r seems to us more natural. It adds no further restriction to the concept of suffiiency.

[†] We retain the notation $g(t | \theta)$, since the function of t and θ may always be expressed as the marginal distribution of t.

It should be noted that in performing the integration with respect to $t_r, \ldots, t_{n-1}$ for fixed t, no factor in θ is introduced. This is clearly so when the range of the distribution of the underlying variable is independent of θ; if the range depends on θ, this merely introduces a factor into L that forms part of $g(t|\theta)$ on the right of (17.68), while k remains free of θ—see Example 17.16 and **17.40-1** below.

The converse result is also easily established. In (17.67) with $r = n$, put $t_i = x_i$ $(i = 1, 2, \ldots, n-1)$. We then have

$$f_n(t, x_1, x_2, \ldots, x_{n-1}|\theta) = g(t|\theta) h_{n-1}(x_1, \ldots, x_{n-1}|t). \qquad (17.70)$$

On inserting the differential elements $dt, dx_1, \ldots, dx_{n-1}$ on either side of (17.70), the transformation

$$\begin{cases} x_n = x_n(t, x_1, \ldots, x_{n-1}), \\ x_i = x_i, \qquad i = 1, 2, \ldots, n-1 \end{cases}$$

applied to (17.70) yields (17.68) at once. Thus (17.67) is necessary and sufficient for (17.68). This proof deals only with the case when the variates are continuous. In the discrete case the argument simplifies, as the reader will find on retracing the steps of the proof. A very general proof of the equivalence of (17.67) and (17.68) is given by Halmos and Savage (1949).

We have discussed only the case $n \geqslant 2$. For $n = 1$ we take (17.68) as the definition of sufficiency.

Sufficiency and minimum variance
17.33 The necessary and sufficient condition for sufficiency at (17.68) has one immediate consequence of interest. On taking logarithms of both sides and differentiating, we have

$$\frac{\partial \log L}{\partial \theta} = \frac{\partial \log g(t|\theta)}{\partial \theta}. \qquad (17.71)$$

On comparing (17.71) with (17.27), the condition that an MVB estimator of $\tau(\theta)$ exist, we see that such an estimator can only exist if there is a sufficient statistic. In fact, (17.27) is simply the special case of (17.71) when

$$\frac{\partial \log g(t|\theta)}{\partial \theta} = A(\theta)\{t - \tau(\theta)\}. \qquad (17.72)$$

Thus sufficiency, which perhaps at first sight seems a more restrictive criterion than the attainment of the MVB, is in reality a less restrictive one. For whenever (17.27) holds, (17.71) holds also, while even if (17.27) does not hold we may still have a sufficient statistic.

Example 17.15
The argument of **17.33** implies that in all the cases (Examples 17.6, 17.8, 17.9, 17.10) where we have found MVB estimators to exist, they are also sufficient statistics. The reader should verify this in each case by direct factorisation of L as at (17.68).

Example 17.16
Consider the estimation of θ in

$$dF(x) = dx/\theta, \qquad 0 \leqslant x \leqslant \theta.$$

The likelihood function (LF) is

$$\left. \begin{array}{ll} L(x|\theta) = \theta^{-n} & \text{if } x_{(n)} \leqslant \theta, \\ \quad\quad\quad = 0 & \text{otherwise,} \end{array} \right\}$$

since we know that all the observations, including the largest of them, $x_{(n)}$, cannot exceed θ.

We may write this in the form

$$L(x|\theta) = \theta^{-n} u(\theta - x_{(n)}), \tag{17.73}$$

where

$$\left. \begin{array}{ll} u(z) = 1, & z \geqslant 0 \\ \quad\quad = 0, & z < 0 \end{array} \right\}.$$

(17.73) makes it clear at once that the LF can be factorised into a function of $x_{(n)}$ and θ alone,

$$g(x_{(n)}|\theta) = \theta^{-n} u(\theta - x_{(n)}),$$

and a second factor

$$k(x_1, \ldots, x_n) = 1.$$

Thus, from (17.68), $x_{(n)}$ is a sufficient statistic for θ.

17.34 If t is a sufficient statistic for θ, any one-to-one function of t, say u, will also be sufficient. For if $t(u)$ is one-to-one, we may write (17.68) as

$$\begin{aligned} L(x|\theta) &= g(t(u)|\theta)k(x) \\ &= g_1(u|\theta)k_1(x), \end{aligned} \tag{17.74}$$

where k_1 is independent of θ, so that u is also sufficient for θ. To resolve the estimation problem, we choose a function of t that is a consistent, and usually also an unbiased, estimator of θ.

Even if $t(u)$ is not one-to-one, the factorisation (17.74) making u sufficient may still be possible in particular cases—Exercise 23.31 below treats one of these.

Apart from such functional relationships, the sufficient statistic is unique. If there were two distinct sufficient statistics, t_1 and t_2, (17.67) with $r = 2$ would give

$$f_2(t_1, t_2|\theta) = g_1(t_1|\theta)h_1(t_2|t_1) = g_2(t_2|\theta)h_2(t_1|t_2)$$

so that, writing $h_3(t_1, t_2) = h_2(t_1|t_2)/h_1(t_2|t_1)$, we have

$$g_1(t_1|\theta) \equiv h_3(t_1, t_2)g_2(t_2|\theta) \tag{17.75}$$

identically in θ. This cannot hold unless t_1 and t_2 are functionally related.

17.35 We have seen in **17.33** that a sufficient statistic provides the MVB estimator, where there is one. We now prove a more general result, due to C. R. Rao (1945) and Blackwell (1947), that irrespective of the attainability of any variance bound, the MV unbiased estimator of $\tau(\theta)$, if one exists, is always a function of the sufficient statistic.

We first prove a quite general result. Consider a statistic t and another statistic u with finite variance. To find $E(u)$, we carry out the expectation operation in two stages, first holding t constant and then allowing t to vary over its distribution. Symbolically

$$E(u) = E_t\{E(u|t)\}. \tag{17.76}$$

In general, $E(u|t)$ is not a statistic, since it depends on the parameter θ. For the variance of u, we have from (17.76)

$$\mathrm{var}\, u = E\{u - E(u)\}^2$$

$$= EE_t(\{[u - E(u|t)] + [E(u|t) - EE(u|t)]\}^2|t).$$

Now the contents of the second set of square brackets are a constant with respect to the conditional expectation given t. Thus, when the terms inside the braces are squared, the cross-product term is a constant times $u - E(u|t)$, whose conditional expectation is zero. We are left with

$$\mathrm{var}\, u = EE_t\{[u - E(u|t)]^2|t\} + E_t\{[E(u|t) - EE(u|t)]^2\},$$

which by the definition of variance may be written

$$\mathrm{var}\, u = E_t\{\mathrm{var}\,(u|t)\} + \mathrm{var}_t\{E(u|t)\}. \tag{17.77}$$

Thus, quite generally, the unconditional expectation is the expectation of conditional expectation, by (17.76), and the unconditional variance is the expectation of conditional variance plus the variance of conditional expectation. The variable $E(u|t)$ has the same expectation as u, and smaller variance than u by (17.77) unless $\mathrm{var}\,(u|t) \equiv 0$, i.e. u is a function of t.

Now let t be sufficient for θ. (17.67) ensures that $E(u|t)$ cannot depend on θ, so that it is a function of t alone, say $p(t)$. (17.76) gives $E\{p(t)\} = E(u) = \tau(\theta)$, say, while (17.77) shows that

$$\mathrm{var}\,\{p(t)\} \leq \mathrm{var}\, u, \tag{17.78}$$

with equality if and only if $u = p(t)$.

Thus conditioning upon the value of a sufficient statistic t gives an estimator $p(t)$ with smaller variance. It follows that any MV estimator must be a function of t.

We shall see in **22.9–13** that often there is only one function of a sufficient statistic with any given expectation. Then, whatever u we start from, $E(u|t)$ must be the same, and it is the *unique* MV unbiased estimator of $\tau(\theta)$. It is not always easy to use this result constructively, since the conditional expectation $E(u|t)$ may be difficult to evaluate—Exercise 17.24 gives an important class of cases where explicit results can be obtained—but it does assure us that an unbiased estimator (with finite variance) that is a function of a sufficient statistic is the unique MV estimator.

In the case of the exponential family (17.30), we can find the MV estimator based on the sufficient statistic directly in particular cases, as in Exercise 17.14. Since $p(t)$ has the same expectation as, and smaller variance than, u, it follows from **17.30** that its m.s.e. is also smaller than that of u no matter which function of θ is being estimated. We may be able to improve m.s.e. further by using a function of t with a different expectation. Example 17.14 implies that in estimating $\tau(\theta)$, even a constant multiple $ap(t)$, where $a < 1$, can have better m.s.e. than $p(t)$ itself—in the case given there,

$$dF = \theta^{-1} \exp(-x/\theta), \qquad \theta > 0, \qquad x \geq 0,$$

it is easily seen that $t = \bar{x}$ is sufficient for θ and unbiased, while $nt/(n+1)$ has smaller m.s.e.

Distributions possessing sufficient statistics

17.36 We now seek to define the class of distributions in which a (single) sufficient statistic exists for a parameter. We first consider the case where the range of the variate does not depend on the parameter θ. From (17.71) we have, if t is sufficient for θ in a sample of n independent observations,

$$\frac{\partial \log L}{\partial \theta} = \sum_{j=1}^{n} \frac{\partial \log f(x_j|\theta)}{\partial \theta} = K(t, \theta), \tag{17.79}$$

where K is some function of t and θ. Regarding this as an equation in t, we see that since it remains true for any fixed value of θ and any u, t must be expressible in the form

$$t = M\left\{ \sum_{j=1}^{n} k(x_j) \right\} = M(w), \tag{17.80}$$

where $w = \sum k(x_j)$ and M and k are arbitrary functions. Thus $K(M(w), \theta)$ is a function of w and θ only, say $N(w, \theta)$. We have then, from (17.79), if the derivatives exist,

$$\frac{\partial^2 \log L}{\partial \theta \, \partial x_j} = \frac{\partial N}{\partial w} \frac{\partial w}{\partial x_j}. \tag{17.81}$$

Now the left-hand side of (17.81) is a function of θ and x_j only and $\partial w/\partial x_j$ is a function of x_j only. Hence $\partial N/\partial w$ is a function of θ and x_j only. But it must be symmetrical in the xs, since w is, and hence is a function of θ only. Hence, integrating it with respect to w, we have

$$N(\theta, w) = wp(\theta) + q(\theta),$$

where p and q are arbitrary functions of θ. Thus (17.79) becomes

$$\frac{\partial}{\partial \theta} \log L = \frac{\partial}{\partial \theta} \sum_j \log f(x_j | \theta) = p(\theta) \sum k(x_j) + q(\theta), \qquad (17.82)$$

whence

$$\frac{\partial}{\partial \theta} \log f(x | \theta) = p(\theta) k(x) + q(\theta)/n,$$

giving the necessary condition for a sufficient statistic to exist,

$$f(x | \theta) = \exp \{ A(\theta) B(x) + C(x) + D(\theta) \}. \qquad (17.83)$$

This result, which is due to Darmois (1935), Pitman (1936) and Koopman (1936), is precisely the form of the exponential family of distributions, seen at (17.30) to be a condition for the existence of an MVB estimator for some function of θ.

> L. Brown (1964) gives a rigorous treatment of the regularity conditions sufficient for this result to hold, with references to related work. See also Denny (1967, 1972); E. B. Andersen (1970) treats the discrete case.

If (17.83) holds, it is easily verified that if the range of $f(x | \theta)$ is independent of θ, the likelihood function yields a sufficient statistic for θ. Thus, under this condition, (17.83) is sufficient as well as necessary for the distribution to possess a sufficient statistic.

All the distributions of Example 17.15 are of the form (17.83).

17.37   Under regularity conditions, there is therefore a one-to-one correspondence between the existence of a sufficient statistic for θ and the existence of an MVB estimator of some function of θ. If (17.83) holds, a sufficient statistic exists for θ, and there will be just one function, t, of that statistic (itself sufficient) that will satisfy (17.27) and so estimate some function $\tau(\theta)$ with variance equal to the MVB. In large samples, moreover (cf. **17.23**), *any* function of the sufficient statistic will estimate its expected value with MVB precision. Finally, for any n (cf. **17.35**), any function of the sufficient statistic will have the minimum *attainable* variance in estimating its expected value.

Sufficient statistics for several parameters
17.38   All the ideas of the previous sections generalise immediately to the case where the distribution is dependent upon several parameters $\theta_1, \ldots, \theta_k$. It also makes no difference to the essentials if we have a multivariate distribution, rather than a univariate one. Thus if we define each x_i as a vector variate with p (≥ 1) components, $\mathbf{t}$ and $\mathbf{t}_i$ as vectors of statistics, and $\boldsymbol{\theta}$ as a vector of parameters with k components, **17.31–2** remain substantially unchanged. If we can write

$$L(\mathbf{x} | \boldsymbol{\theta}) = g(\mathbf{t} | \boldsymbol{\theta}) h(\mathbf{x}) \qquad (17.84)$$

we call the components of **t** a set of (jointly) sufficient statistics for **θ**. The property (17.67) follows as before.

If **t** has s components, we may have s greater than, equal to, or less than k. If $s = 1$, we call **t** a single sufficient statistic, as we have seen. If we put $\mathbf{t} = \mathbf{x}$ we see that the observations themselves always constitute a set of sufficient statistics for **θ** with $s = n$. In order to reduce the problem of analysing the data as far as possible, we naturally desire s to be as small as possible. Even this is not quite restrictive enough—see Exercises 18.13 and 22.31 for cases in which we have alternative sufficient statistics with the same value of s. In Chapter 22 we shall define the concept of a *minimal* set of sufficient statistics for **θ**, which is a function of all other sets of sufficient statistics.

It evidently does not follow from the joint sufficiency of **t** for **θ** that any particular component of **t**, say $t^{(1)}$, is individually sufficient for θ_1. This will be so only if $g(\mathbf{t}|\mathbf{θ})$ factorises with $g_1(t^{(1)}|\theta_1)$ as one factor. Nor is the converse true always: individual sufficiency of all the $t^{(i)}$ when the others are known does not imply joint sufficiency.

If $k = 1$, the result of **17.35** holds unchanged if t is a vector with $s < n$ components (if $s = n$, the result is an empty one); the result is most easily applied with s as small as possible.

Example 17.17
Consider the estimation of the parameters μ and σ^2 in

$$dF(x) = \frac{1}{\sigma(2\pi)^{1/2}} \exp\left\{-\frac{1}{2}\left(\frac{x-\mu}{\sigma}\right)^2\right\} dx, \qquad -\infty < x < \infty.$$

We have

$$L(x|\mu, \sigma^2) = \frac{1}{\sigma^n(2\pi)^{n/2}} \exp\left\{-\frac{1}{2}\sum_{i=1}^n \left(\frac{x_i-\mu}{\sigma}\right)^2\right\} \tag{17.85}$$

and we have seen (Example 11.7) that the joint distribution of $\bar{x}$ and s^2 in normal samples is

$$g(\bar{x}, s^2|\mu, \sigma^2) \propto \frac{1}{\sigma} \exp\left\{-\frac{n}{2\sigma^2}(\bar{x}-\mu)^2\right\} \cdot \frac{1}{\sigma^{n-1}} s^{n-3} \exp\left\{-\frac{ns^2}{2\sigma^2}\right\},$$

so that, remembering that $\sum(x-\mu)^2 = n\{s^2 + (\bar{x}-\mu)^2\}$, we have

$$L(x|\mu, \sigma^2) = g(\bar{x}, s^2|\mu, \sigma^2)k(x)$$

and therefore $\bar{x}$ and s^2 are jointly sufficient for μ and σ^2. We have already seen (Examples 17.6, 17.15) that $\bar{x}$ is sufficient for μ when σ^2 is known and (Examples 17.10, 17.15) that $(1/n)\sum(x-\mu)^2$ is sufficient for σ^2 when μ is known. It is easily seen directly from (17.85) that s^2 alone is not sufficient for σ^2 when μ is unknown.

17.39   The principal results for sufficient statistics generalise to the k-parameter case in a natural way. The condition (generalising the exponential family (17.83)) for a distribution to possess a set of k jointly sufficient statistics for its k parameters becomes, under similar conditions of continuity and the existence of derivatives,

$$f(x|\boldsymbol{\theta}) = \exp\left\{\sum_{j=1}^{k} A_j(\boldsymbol{\theta})B_j(x) + C(x) + D(\boldsymbol{\theta})\right\}, \qquad (17.86)$$

a result due to Darmois (1935), Koopman (1936) and Pitman (1936). More general results of this kind are given by Barankin and Maitra (1963). The result of **17.35** on the unique MV properties of functions of a sufficient statistic finds its generalisation in a theorem due to C. R. Rao (1947): for the simultaneous estimation of r $(\leq k)$ functions τ_j of the k parameters θ_s, the unbiased functions of a minimal set of k sufficient statistics, say t_i, have the minimum attainable variances, and (if the range is independent of the θ_s) the (not necessarily attainable) lower bounds to their variances are given by

$$\operatorname{var} t_i \geq \sum_{j=1}^{k} \sum_{l=1}^{k} \frac{\partial \tau_i}{\partial \theta_j} \frac{\partial \tau_i}{\partial \theta_l} I_{jl}^{-1}, \qquad i = 1, 2, \ldots, r, \qquad (17.87)$$

where the *information matrix*, the generalisation of (17.25),

$$(I_{jl}) = \left\{ E\left(\frac{\partial \log L}{\partial \theta_j} \cdot \frac{\partial \log L}{\partial \theta_l}\right)\right\} \qquad (17.88)$$

is to be inverted. (17.87), in fact, is a further generalisation of (17.22) and (17.50), its simplest cases. Like them, it takes account only of terms of order $1/n$ in the variance.

> Exercise 17.20 shows that these lower bounds cannot be smaller than those obtained from (17.24) for a single estimator.

In random sampling from the experimental family (17.86), the joint distribution of the sufficient statistics is itself a member of the experimental family as Exercise 17.14 shows. A good approximation to its density in the more general case where the observations need not be independent is outlined in Exercise 17.33.

Sufficiency when the range depends on the parameter

17.40 Now consider the situation when the range of the variable does depend on θ. We omit the trivial case $n = 1$. First, we take the case when only one terminal of the range, say the lower terminal, depends on θ. We then have the frequency function $f(x|\theta)$, $a(\theta) \leq x \leq b$, where $a(\theta)$ is a monotone function of θ, and θ is in some non-degenerate interval.

Just as in Example 17.16, we have

$$L(x|\theta) = \prod_{i=1}^{n} f(x_i|\theta) u(x_{(1)} - a(\theta))$$

where

$$u(z) = 1 \quad \text{if } z \geq 0,$$
$$= 0 \quad \text{otherwise.}$$

It is at once obvious from the likelihood function that the smallest observation $x_{(1)}$ cannot be factored away from θ in the u-function; thus, if there is a single sufficient statistic, it must be $x_{(1)}$. But $x_{(1)}$ can only be sufficient if $f(x_i \mid \theta)$ can be factored into a function of x_i alone and a function of θ alone, i.e. if

$$f(x \mid \theta) = g(x)/h(\theta). \tag{17.89}$$

Then and only then is

$$L(x \mid \theta) = \frac{u(x_{(1)} - a(\theta))}{\{h(\theta)\}^n} \cdot \prod_{i=1}^{n} g(x_i)$$

of the form required at (17.68) for $x_{(1)}$ to be sufficient.

Evidently the same result will hold, with $x_{(n)}$ instead of $x_{(1)}$, if the range of the variate is $a \leqslant x \leqslant b(\theta)$, where $b(\theta)$ is monotone in θ; if and only if (17.89) holds, $x_{(n)}$ is singly sufficient for θ.

In these situations, Exercise 17.24 uses the result of **17.35** to establish an explicit form for the unique MV unbiased estimator of any function $\tau(\theta)$. The reader should notice that there the problem is reparametrised so that the affected terminal is at θ itself.

17.41   If both terminals of the range depend on θ, we have $a(\theta) \leqslant x \leqslant b(\theta)$ and

$$L(x \mid \theta) = \prod_{i=1}^{n} f(x_i \mid \theta) u(x_{(1)} - a(\theta)) u(b(\theta) - x_{(n)}).$$

We see by exactly the argument in **17.40** that the extreme observations $x_{(1)}$ and $x_{(n)}$ are a pair of sufficient statistics for θ if and only if (17.89) holds. We now consider whether there can be a *single* sufficient statistic in this case; if there is it must clearly be a function of $x_{(1)}$ and $x_{(n)}$.

Essentially, we are asking whether we can find a single statistic that will tell us whether the product $u(x_{(1)} - a(\theta)) u(b(\theta) - x_{(n)})$ in $L(x \mid \theta)$ is equal to 1 or 0. It is only equal to 1 if both

$$x_{(1)} \geqslant a(\theta), \qquad b(\theta) \geqslant x_{(n)}. \tag{17.90}$$

There are four possibilities:

(i) $a(\theta), b(\theta)$ are both increasing functions of θ. (17.90) becomes

$$a^{-1}(x_{(1)}) \geqslant \theta \geqslant b^{-1}(x_{(n)}).$$

(ii) They are both decreasing functions of θ. (17.90) becomes

$$b^{-1}(x_{(n)}) \geqslant \theta \geqslant a^{-1}(x_{(1)}).$$

(iii) $a(\theta)$ is increasing, $b(\theta)$ decreasing. (17.90) becomes

$$a^{-1}(x_{(1)}) \geqslant \theta, \qquad b^{-1}(x_{(n)}) \geqslant \theta. \tag{17.91}$$

(iv) $a(\theta)$ is decreasing, $b(\theta)$ increasing. (17.90) becomes

$$a^{-1}(x_{(1)}) \leqslant \theta, \qquad b^{-1}(x_{(n)}) \leqslant \theta. \tag{17.92}$$

In cases (i) and (ii), we need both $x_{(1)}$ and $x_{(n)}$, and no single sufficient statistic exists. But (17.91) shows that in case (iii), $u(t_1 - \theta)$ is equivalent to the product of the original u-functions, where

$$t_1 = \min\{a^{-1}(x_{(1)}), b^{-1}(x_{(n)})\}, \tag{17.93}$$

so t_1 is singly sufficient for θ. Similarly, in case (iv),

$$t_2 = \max\{a^{-1}(x_{(1)}), b^{-1}(x_{(n)})\} \tag{17.94}$$

is singly sufficient since $u(\theta - t_2)$ is equivalent to the product. We may summarise by saying that if the upper terminal is a monotone decreasing function of the lower terminal and (17.89) holds, there is a single sufficient statistic, given by t_1 or by t_2 according as the lower terminal is an increasing or a decreasing function of θ.

Exercise 20.16 gives the distribution of t_1, from which that of t_2 is immediately obtainable (by writing

$$a^*(\theta) = -a(\theta), \qquad b^*(\theta) = -b(\theta)$$

in case (iv), so that we return to case (iii), (17.94) becoming (17.93) in terms of a^* and b^*) and the unique MV unbiased estimator of $\tau(\theta)$ in either case. Further, no generality is lost by making the lower terminal equal to θ itself, so we need only consider the range $(\theta, b(\theta))$ where $b(\theta)$ is a decreasing function and (17.93) is the single sufficient statistic.

These results were originally due to Pitman (1936) and R. C. Davis (1951). The condition that θ lies in a non-degenerate interval is important—cf. Example 17.23.

Example 17.18
For the uniform distribution

$$dF = dx/(2\theta), \qquad -\theta \le x \le \theta$$

we are in case (iv) of **17.41**. The single sufficient statistic (17.94) is

$$t_2 = \max\{-x_{(1)}, x_{(n)}\}$$

and since

$$x_{(1)} \le x_{(n)},$$

this is the same as

$$t_2 = \max\{|x_{(1)}|, |x_{(n)}|\}$$

which is intuitively acceptable.

Example 17.19
The distribution $dF(x) = \exp\{-(x - \alpha)\}\,dx$, $\alpha \le x < \infty$, is of the form (17.89), since it may be written

$$f(x) = \exp(-x)/\exp(-\alpha).$$

Here the smallest observation, $x_{(1)}$, is sufficient for the lower terminal α.

Example 17.20

The distribution

$$dF(x) = \frac{1}{\Gamma(p)} (x-\alpha)^{p-1} \exp\{-(x-\alpha)\}, \qquad \alpha \leqslant x < \infty; \, p > 1,$$

evidently cannot be put in the form (17.89). Thus there is no single sufficient statistic for α when $n \geqslant 2$. Example 17.19 is the case $p = 1$.

Example 17.21

In the two-parameter distribution

$$dF(x) = g(x) \, dx/h(\alpha, \beta), \qquad \alpha \leqslant x \leqslant \beta,$$

it is clear that, given β, $x_{(1)}$ is sufficient for α; and given α, $x_{(n)}$ is sufficient for β. $x_{(1)}$ and $x_{(n)}$ are a set of jointly sufficient statistics for α and β, as is confirmed by observing that the joint distribution of $x_{(1)}$ and $x_{(n)}$ is, by (14.2) with $r = 1$, $s = n$,

$$g(x_{(1)}, x_{(n)}) = p(x_{(1)}, x_{(n)})/\{h(\alpha, \beta)\}^n,$$

so that when g and L are non-zero

$$L(x \,|\, \alpha, \beta) = g(x_{(1)}, x_{(n)}) k(x_1, x_2, \ldots, x_n).$$

Example 17.22

The uniform distribution $dF(x) = dx/\theta$, $k\theta \leqslant x \leqslant (k+1)\theta$; $k > 0$ comes under case (i) of **17.41**. No single sufficient statistic exists, but $(x_{(1)}, x_{(n)})$ are a sufficient pair since (17.89) holds.

Example 17.23

The uniform distribution $dF = dx$, $\theta \leqslant x < \theta + 1$; $\theta = 0, 1, 2, \ldots$, in which θ is confined to integer values, does not satisfy the condition that the upper terminal be a monotone decreasing function of the lower. But, evidently, *any* single observation, x_i, in a sample of n is a single sufficient statistic for θ. In fact, $[x_i]$ estimates θ with zero variance. If the integer restriction on θ is removed, no single sufficient statistic exists, in accordance with **17.41**.

17.42 We have now concluded our discussion of the basic ideas of the theory of estimation. In Chapter 22 we shall be developing the theory of sufficient statistics further. Meanwhile, we continue our study of estimation from another point of view, by studying the properties of the estimators given by the method of maximum likelihood.

EXERCISES

17.1   Show that in samples from

$$dF = \frac{1}{\Gamma(p)\theta^p} x^{p-1} e^{-x/\theta} dx, \qquad p > 0; \ 0 \leqslant x < \infty,$$

the MVB estimator of θ for known p is $\bar{x}/p$, with variance $\theta^2/(np)$, while if $\theta = 1$ that of $\partial \log \Gamma(p)/\partial p$ is $(1/n) \sum_{i=1}^{n} \log x_i$ with variance $\{\partial^2 \log \Gamma(p)/\partial p^2\}/n$.

17.2   A random variable x has f.f.

$$f(x \mid \theta) = \theta f_1(x) + (1 - \theta) f_2(x)$$

where $0 < \theta < 1$ and f_1, f_2 are completely specified f.f.s whose ranges of variation do not involve θ. Show that the MVB in estimating θ from a sample of n observations is not generally attainable, but is equal to $\theta(1 - \theta)/\{n(1 - I)\}$, where

$$I = \int_{-\infty}^{\infty} \frac{f_1(x) f_2(x)}{f(x \mid \theta)} dx.$$

$I = 0$ if the ranges of f_1 and f_2 do not overlap and the MVB then is that for the binomial distribution (Example 17.9), which is the special case $f_1(x) = x$, $f_2(x) = 1 - x$, $x = 0, 1$.

(Hill, 1963b)

17.3   In Example 17.11, show that

$$L'/L = (r - n\theta)/\{\theta(1 - \theta)\}$$

and

$$L''/L = \frac{\partial^2 \log L}{\partial \theta^2} + \left(\frac{\partial \log L}{\partial \theta}\right)^2$$

$$= \frac{1}{\theta^2(1 - \theta)^2} [\{(r - n\theta) - \tfrac{1}{2}(1 - 2\theta)\}^2 - \{\tfrac{1}{4} + (n - 1)\theta(1 - \theta)\}].$$

Hence show that $J_{12} = 0$. Using (17.19), show that $E(L''/L) = 0$ and that

$$J_{22} = \text{var}(L''/L) = 2n(n - 1)/\{\theta^2(1 - \theta)^2\}.$$

Use these results to verify that var t coincides with (17.45). Show that t is a linear function of the uncorrelated variables L'/L and L''/L.

17.4   Writing (17.27) as

$$(t - \tau) = \frac{\tau'(\theta)}{J_{11}} \cdot \frac{L'}{L},$$

and the c.f. of t about its mean as $\phi(z)$ $(z = iu)$, show that for an MVB estimator

$$\frac{\partial \phi(z)}{\partial z} = \frac{\tau'(\theta)}{J_{11}} \left\{ \frac{\partial \phi(z)}{\partial \theta} + z\tau'(\theta)\phi(z) \right\}$$

and that its cumulants are given by

$$\varkappa_{r+1} = \frac{\partial \tau}{\partial \theta} \cdot \frac{\partial \varkappa_r}{\partial \theta} \bigg/ J_{11}, \qquad r = 2, 3, \ldots \tag{A}$$

Hence show that the covariance between t and an unbiased estimator of its rth cumulant is equal to its $(r + 1)$th cumulant.

Establish the inequality (17.50) for an estimated function of several parameters, and show that (A) holds in this case also when the bound is attained.

(A. Bhattacharyya, 1946-47)

17.5 Show that for the estimation of θ in the logistic distribution

$$f(x) = e^{-(x-\theta)}\{1 + e^{-(x-\theta)}\}^{-2},$$

the MVB is exactly $3/n$, whereas the sample mean has exact variance $\pi^2/(3n)$ and the sample median asymptotic variance $4/n$.

17.6 Show that in estimating σ in the distribution

$$dF = \frac{1}{\sigma(2\pi)^{1/2}} \exp\left(-\frac{1}{2}\frac{x^2}{\sigma^2}\right) dx, \qquad -\infty < x < \infty,$$

$$s_1 = \left(\frac{1}{2}\sum_{i=1}^{n} x_i^2\right)^{1/2} \Gamma\left(\frac{1}{2}n\right) \bigg/ \Gamma\left\{\frac{1}{2}(n+1)\right\}$$

and

$$s_2 = \left\{\frac{1}{2}\sum_{i=1}^{n}(x_i - \bar{x})^2\right\}^{1/2} \Gamma\left\{\frac{1}{2}(n-1)\right\} \bigg/ \Gamma\left(\frac{1}{2}n\right)$$

are both unbiased. Hence show that s in Example 17.3 has expectation $\sigma(1 - (1/4n))$ approximately, agreeing with the result of Exercise 10.20, Vol. 1.

Show that (17.52) generally gives a greater bound than (17.24), which gives $\sigma^2/(2n)$, but, by considering the case $n = 2$, that even this greater bound is not attained for small n by s_1.

(Cf. Chapman and Robbins, 1951)

17.7 In estimating μ^2 in

$$dF = \frac{1}{\sqrt{(2\pi)}} \exp\left\{-\tfrac{1}{2}(x-\mu)^2\right\} dx,$$

show that $(\bar{x}^2 - 1/n - \mu^2)$ is a linear function of

$$\frac{1}{L}\frac{\partial L}{\partial \mu} \quad \text{and} \quad \frac{1}{L}\frac{\partial^2 L}{\partial \mu^2},$$

and hence, from (17.42), that $\bar{x}^2 - 1/n$ is an unbiased estimator of μ^2 with minimum attainable variance.

17.8 In estimating the variance σ^2 of a distribution with known mean μ from a sample of n observations x_i, consider the two unbiased estimators

$$t_1 = \frac{1}{n}\sum_{i=1}^{n}(x_i - \mu)^2, \qquad t_2 = \frac{1}{n-1}\sum_{i=1}^{n}(x_i - \bar{x})^2.$$

Using (10.5) and (12.35), show that in the terminology of **17.29** they have the same efficiency, and that the deficiency of t_2 with respect to t_1 is $d = 2/(2 + \gamma_2)$, where $\gamma_2 = \kappa_4/\kappa_2^2$ is the kurtosis coefficient of the distribution. From Exercise 3.19, $\gamma_2 \geq -2$, so that $0 \leq d < \infty$, with $d = 1$ in the normal case, reflecting the single d.fr. lost in passing from t_1 to t_2 (cf. Example 11.7).

(Hodges and Lehmann, 1970)

17.9 For the three-parameter distribution

$$dF(x) = \frac{1}{\Gamma(p)} \exp\left\{-\left(\frac{x-\alpha}{\sigma}\right)\right\}\left(\frac{x-\alpha}{\sigma}\right)^{p-1}\frac{dx}{\sigma}, \qquad p, \sigma > 0; \quad \alpha \leq x < \infty,$$

show that there are single sufficient statistics for p and σ individually when the other two parameters are known; that there is a pair of sufficient statistics for p and σ jointly if α is known; and that if σ is known and $p = 1$, there is a single sufficient statistic for α.

17.10 In samples of size n from a normal distribution with cumulants κ_1, κ_2, show that the sample mean k_1 has moments

$$E(k_1^r) = \sum_{j=0}^{r} \binom{r}{j} \kappa_1^{r-j} E(k_1 - \kappa_1)^j = \sum_{i=0}^{[\frac{1}{2}r]} \frac{r!}{i!(r-2i)!} \kappa_1^{r-2i} \left(\frac{\kappa_2}{2n}\right)^i, \qquad r = 1, 2, \ldots$$

Hence show that

$$\kappa_1^r = \sum_{i=0}^{[\frac{1}{2}r]} (-1)^i \frac{r!}{i!(r-2i)!} E(k_1^{r-2i}) \left(\frac{\kappa_2}{2n}\right)^i,$$

and finally, using (16.7), that for $n \geq 2$ and $m > -\frac{1}{2}(n-1)$ the MV unbiased estimator of $\kappa_1^r \kappa_2^m$ is

$$\sum_{i=0}^{[\frac{1}{2}r]} (-1)^i \frac{r!}{i!(r-2i)!} \frac{\Gamma\{\frac{1}{2}(n-1)\}}{\Gamma\{\frac{1}{2}(n-1)+m+i\}} \frac{k_1^{r-2i}}{(2n)^i} \left\{\frac{(n-1)k_2}{2}\right\}^{m+i}$$

where k_2 is the second k-statistic. (Cf. Hoyle, 1968)

17.11 Show that if t is the MV unbiased estimator, and u another unbiased estimator, of θ, the covariance of t and $(u - t)$ is zero. Hence show that we may regard the variation of $(u - \theta)$ as composed of two orthogonal parts, one being the variation of $(t - \theta)$ and the other a component due to inefficiency of estimation. (Cf. Fisher, 1925)

17.12 In the binomial distribution of Example 17.4, let a and b be integers or zero and define $\tau_{ab} = \theta^a(1-\theta)^b$. Show that if $a, b \geq 0$ and $a + b \leq n$, τ_{ab} has the unbiased estimator $r^{(a)}(n-r)^{(b)}/n^{(a+b)}$, but that otherwise no unbiased estimator of τ_{ab} exists.

17.13 For a sample of n observations from a distribution of form (17.89) with $a \leq x \leq \theta$, show that the sufficient statistic $x_{(n)}$ is not an unbiased estimator of θ. Using the jackknife of **17.10**, show that

$$t' = x_{(n)} + (x_{(n)} - x_{(n-1)})(n-1)/n \sim 2x_{(n)} - x_{(n-1)}$$

is unbiased to order n^{-1} and has the same m.s.e. as $x_{(n)}$ to order n^{-2}, namely

$$\frac{2}{n^2} \left\{\frac{h(\theta)}{h'(\theta)}\right\}^2.$$
(Cf. Robson and Whitlock, 1964)

17.14 For a sample of n independent observations from a distribution with frequency function (17.86), and the range of the variates independent of the parameters, show that the statistics

$$t_j = \sum_{i=1}^{n} B_j(x_i), \qquad j = 1, 2, \ldots, k,$$

are jointly sufficient for the k parameters $\theta_1, \ldots, \theta_k$, and that their joint f.f. is

$$g(t_1, t_2, \ldots, t_k | \theta) = \exp\{nD(\theta)\} h(t_1, t_2, \ldots, t_k) \exp\left\{\sum_{j=1}^{k} A_j(\theta) t_j\right\}$$

which is itself of the form (17.86). When $k = 1$, write this

$$g_n(t | \theta) = \{d(\theta)\}^n \cdot h_n(t) \exp\{tA(\theta)\}$$

and show that if m is an integer, $0 \leq m < n$, and c is a constant,

$$E\{h_{n-m}(t-c)/h_n(t)\} = \{d(\theta)\}^m \exp\{cA(\theta)\}.$$

Show that if

$$f(x|\theta) = \theta\, e^{-\theta x}, \qquad 0 \le x < \infty,$$

the MV unbiased estimator of θ^m is $(n-1)^{(m)}/\sum_{j=1}^{n} x_j$; Example 17.14 gave the result for $m = 1$.

17.15 t is an estimator of a parameter θ known to lie in the interval (a, b). u is defined equal to t if $a \le t \le b$, but equal to a if $t < a$ and to b if $t > b$. Show that u has better m.s.e. than t in estimating θ. If $E(t) = \theta$, show that u also has better variance than t.

Show that if $\nu > 0$ is known in the distribution

$$f(x) = \frac{\exp\{-\frac{1}{2}(x+\theta)\} x^{(\nu-2)/2}}{2^{\nu/2}} \sum_{r=0}^{\infty} \frac{(\theta x)^r}{r!\, 2^{2r}\Gamma(\frac{1}{2}\nu + r)}, \qquad x \ge 0; \quad \theta \ge 0,$$

$t = x - \nu$ is an unbiased estimator of θ, and that

$$u = t, \qquad x \ge \nu$$
$$= 0, \qquad x < \nu$$

has smaller m.s.e. than t. ($f(x)$ is the non-central χ^2 distribution to be treated in Chapter 23).

17.16 In a sample of size $n = \nu + 1$ from a normal population with variance σ^2, show that σ^p is unbiasedly estimated if $\nu + p > 0$ by

$$t_p = S^p 2^{-p/2} \Gamma(\tfrac{1}{2}\nu)/\Gamma\{\tfrac{1}{2}(\nu+p)\},$$

where $S^2 = \sum (x_i - \bar{x})^2$. Verify that $p = 1$ gives the second estimator of σ in Exercise 17.6 and $p = 2$ gives the estimator of σ^2 in Example 17.3.

Using Example 17.14, show that the multiple of t_p with smallest m.s.e. in estimating σ^p is exactly t_p with ν replaced by $(\nu + p)$. Verify that when $p = 2$, this gives $(n+1)$ as the divisor of S^2.

17.17 $\bar{x}$ is the mean and s^2 the variance of a random sample of n observations $x_1, \ldots, x_n$ from a population with mean μ and variance σ^2. Show that use of the jackknife of **17.10** to correct the bias of (a) $t_{na} = s^2$ as an estimator of σ^2 yields the exactly unbiased $t'_{na} = ns^2/(n-1)$ and of (b) $t_{nb} = \bar{x}^2$ as an estimator of μ^2 yields the exactly unbiased

$$t'_{nb} = \sum_{i \ne j=1}^{n} x_j x_j / \{n(n-1)\}.$$

Show that

$$\hat{V}(\bar{x}) = \bar{x}^2 - t'_{nb} = \sum_{i=1}^{n} (x_i - \bar{x})^2 / \{n(n-1)\}$$

is an unbiased estimator of the variance of $\bar{x}$ even if the x_j have different distributions, provided that they all have the same mean and are all uncorrelated in pairs.

17.18 Show that if the jackknife of **17.10** is used to correct for bias, and var $t_n \sim c/n$, var $t'_n \sim$ var t_n. Show that the m.s.e. of t'_n is consequently no greater than that of t_n.

(Cf. Quenouille, 1956)

17.19 In a sample of n observations x_i from a Poisson distribution with parameter λ, we write $X = \sum_{i=1}^{n} x_i$, $\bar{x} = X/n$ and k_p for the pth k-statistic of the sample, with expectation $\varkappa_p$, the pth cumulant. Show using **17.35** that $E(k_p | X) = \bar{x}$ and that var $(k_p - \bar{x}) = \sum_{r=1}^{p} c_{rp} \lambda^r$ is unbiasedly estimated by

$$\hat{V}_p = \text{var}\,(k_p | X) = \sum_{r=1}^{p} c_{rp} X^{(r)} n^{-r}.$$

(cf. Example 11.17 and Exercise 5.26, Vol. 1). Hence, using the results in Chapter 12, show that

$$\operatorname{var}(k_2 \mid X) = \frac{2}{n} \frac{X^{(2)}}{n^{(2)}}, \qquad \operatorname{var}(k_3 \mid X) = \frac{6}{n} \left\{ \frac{3X^{(2)}}{n^{(2)}} + \frac{X^{(3)}}{n^{(3)}} \right\}$$

and similarly that

$$\mu_3(k_2 \mid X) = \frac{4}{n^3(n-1)^2} \{(n-2)X^{(2)} + 2X^{(3)}\}.$$

(Gart and Pettigrew, 1970)

17.20 For a (positive definite) covariance matrix, the product of any diagonal element with the corresponding diagonal element of the inverse matrix cannot be less than unity. Hence show that if, in (17.87), we have $r = k$ and $\tau_i = \theta_i$ (all i), the resulting bound for an estimator of θ_i is not less than the bound given by (17.24). Give a reason for this result.

(C. R. Rao, 1952)

17.21 The MVB (17.22) holds good for a distribution $f(x \mid \theta)$ whose range (a, b) depends on θ, provided that (17.18) remains true. Show that this is so if

$$f(a \mid \theta) = f(b \mid \theta) = 0,$$

and that if in addition

$$\left[\frac{\partial f(x \mid \theta)}{\partial \theta} \right]_{x=a} = \left[\frac{\partial f(x \mid \theta)}{\partial \theta} \right]_{x=b} = 0,$$

(17.19) also remains true and we may write the MVB in the form (17.23).

17.22 Apply the result of Exercise 17.21 to show that the MVB holds for the estimation of θ in

$$dF(x) = \frac{1}{\Gamma(p)} (x - \theta)^{p-1} \exp\{-(x - \theta)\} \, dx, \qquad \theta \leqslant x < \infty; \quad p > 2,$$

and equals $(p-2)/n$, but is not attainable since there is no single sufficient statistic for θ.

17.23 x is a random variable in the range (a, b) (which may depend on θ) whose distribution is $f(x \mid \theta)$. If $E\{t(x, \theta)\} = \tau(\theta)$ and f and $\partial f / \partial x$ vanish at a and at b, show that

$$\operatorname{var} t \geqslant \left[E\left(\frac{\partial t}{\partial x} \right) \right]^2 \Big/ E\left\{ \left(\frac{\partial \log f}{\partial x} \right)^2 \right\} = \left[E\left(\frac{\partial t}{\partial x} \right) \right]^2 \Big/ -E\left(\frac{\partial^2 \log f}{\partial x^2} \right).$$

(Cf. Exercise 17.21 to establish (17.22-3).)

      (B. R. Rao (1985b). The analogue of (17.45) also holds, $\tau^{(r)}$ being replaced by $E(\partial^r t / \partial x^r)$ and $L^{(r)}/L$ by $(1/f) \partial^r f / \partial x^r$ in J_{rp}—see Sankaran (1964).)

17.24 For a distribution of form

$$f(x) = g(x)/h(\theta), \qquad \theta \leqslant x \leqslant b,$$

show that if a function $t(x)$ of a single observation is an unbiased estimator of $\tau(\theta)$, then

$$-t(x) = \{\tau(x)h'(x) + \tau'(x)h(x)\}/g(x).$$

Hence show, using **17.35**, that the unique MV unbiased estimator of $\tau(\theta)$ in samples of size n is

$$p(x_{(1)}) = \tau(x_{(1)}) - \frac{\tau'(x_{(1)})}{n} \frac{h(x_{(1)})}{g(x_{(1)})},$$

and similarly if $a \leq x \leq \theta$ that it is

$$p(x_{(n)}) = \tau(x_{(n)}) + \frac{\tau'(x_{(n)})}{n} \frac{h(x_{(n)})}{g(x_{(n)})}.$$

Show that for $f(x) = 1/\theta$, $0 \leq x \leq \theta$, $([n+s]/n)x_{(n)}^s$ is the estimator of θ^s, while for

$$f(x) = \exp -\{(x-\theta)\}, \qquad \theta \leq x < \infty, \qquad x_{(1)}^s - (s/n)x_{(1)}^{s-1} \text{ estimates } \theta^s.$$

(Tate, 1959; Karakostas (1985) generalises to the two-parameter case of Example 17.21; Lwin (1975) generalises to the case where a scale parameter is also unknown in the truncated exponential family (17.30).)

17.25   If the pair of statistics (t_1, t_2) is jointly sufficient for two parameters (τ_1, τ_2), and t_1 is sufficient for τ_1 when τ_2 is known, show that the conditional distribution of t_2, given t_1, is independent of τ_1. As an illustration, consider c independent binomial distributions with sample sizes n_i $(i = 1, 2, \ldots, c)$ and parameters θ_i connected by the relation

$$\lambda_i = \log\left(\frac{\theta_i}{1-\theta_i}\right) = \alpha + \beta x_i,$$

where the x_i are known constants, and show that if y_i is the number of 'successes' in the ith sample, the conditional distribution of $\sum_i x_i y_i$, given $\sum_i y_i$, is independent of α.

(D. R. Cox, 1958a)

17.26   Show that from a single observation x ($x = 1, 2, \ldots$), on a zero-truncated Poisson distribution with parameter θ (cf. Exercise 3.29), the only unbiased estimator of $1 - e^{-\theta}$ takes the values 0 when x is odd, 2 when x is even, whereas $1 - e^{-\theta}$ always lies between these values. (Cf. Lehmann (1983a) for a discussion and further examples.)

17.27   As in Example 17.13, show that the efficiency of the estimator of σ based on the mean difference discussed in **10.14** is 0.978 in normal samples.

17.28   In Exercise 17.2, show that (17.43) is given by

$$J_{rp} = 0, \qquad r \neq p,$$

$$= (r!)^2 \binom{n}{r}\left\{\frac{1-I}{\theta(1-\theta)}\right\}^r, \qquad r = p,$$

reducing to the binomial case if $I = 0$, when (17.41) becomes

$$\text{var } t \geq \sum_{r=1}^{s} (\tau^{(r)})^2 \{\theta(1-\theta)\}^r \bigg/ \left\{(r!)^2\binom{n}{r}\right\}.$$

Hence verify the result of Exercise 17.3 that t of Example 17.11 attains this bound with $s = 2$.

(Whittaker, 1973)

17.29   From an $N(\theta, 1)$ variate x, form a new variate y by doubling the frequency for $a \leq |x - \theta| \leq b$, where $0 < a < 1 < b$ and $\int_a^b (u^2 - 1) \exp(-\frac{1}{2}u^2) du = 0$, and adjusting the constant to keep total frequency equal to unity. Show that y attains the MVB (which is 1) for θ.

(V. M. Joshi, 1976)

17.30   Using **17.27**, show that if t is an MV unbiased estimator, and u is any other unbiased estimator, of θ, the statistic v defined by $au + (1-a)v = t$, where a is a constant, $0 < a < 1$, has variance

$$\text{var } v \lesseqgtr \text{var } u \quad \text{according as } a \lesseqgtr \tfrac{1}{2}.$$

Hence show that in estimating the variance θ of a normal distribution with zero mean, the statistic

$$v = \left\{ (n-2) \sum_{j=1}^{n} x_j^2 + (n\bar{x})^2 \right\} \Big/ \{n(n-1)\}$$

has the same mean and variance as

$$u = \sum_{j=1}^{n} (x_j - \bar{x})^2 / (n-1).$$

17.31   For the natural exponential family given by (17.30) with $A(\theta) = \theta$, $B(x) = x$, show that the c.f. of x is $\phi(t) = \exp\{D(\theta) - D(\theta + it)\}$ and that $E(x) = -D'(\theta)$. Show further that $-C'(x)$ is the MV unbiased estimator of θ.

17.32 $x_1, x_2, x_3, \ldots$ are independent variates with f.f. $f(x|\theta)$. If t_n is an MV unbiased estimator of θ based on the first n x_j and $\bar{t}_{n-1}$ is defined, as in **17.10**, as the average of n $t_{n-1,i}$, show that

$$\operatorname{cov}(t_{n-1,i}, t_{n-1,j}) / \operatorname{var} t_{n-1} \leq (n-2)/(n-1), \qquad i \neq j,$$

and that $\operatorname{var} \bar{t}_{n-1} \leq (n-1) \operatorname{var} t_{n-1}/n$.
   Hence show that $n \operatorname{var} t_n \leq (n-1) \operatorname{var} t_{n-1}$ with equality only if $t_n = \sum_{j=1}^{n} g(x_j)$.
(Cf. Blom, 1978.)

17.33   Given that $\mathbf{t}$ is sufficient for $\boldsymbol{\theta}$, show that its density when $\boldsymbol{\theta} = \boldsymbol{\theta}_0$

$$g(\mathbf{t}|\boldsymbol{\theta}_0) = g(\mathbf{t}|\boldsymbol{\theta}) L(\mathbf{x}|\boldsymbol{\theta}_0)/L(\mathbf{x}|\boldsymbol{\theta}),$$

so that if $\mathbf{t}$ is close to $\boldsymbol{\theta}$, and we approximate $g(\mathbf{t}|\mathbf{t})$ by its asymptotic multinormal density, we obtain

$$g(\mathbf{t}|\boldsymbol{\theta}_0) \doteq (2\pi)^{-k/2} |\mathbf{V}(\mathbf{t})|^{-1/2} L(\mathbf{x}|\boldsymbol{\theta}_0)/L(\mathbf{x}|\mathbf{t})$$

where $\mathbf{V}(\mathbf{t})$ is the inverse of (17.88).
   (Durbin (1980) gives details of analytical conditions that make this approximation correct to $O(n^{-1})$ or to $O(n^{-3/2})$ if its constant term is properly adjusted, just as in (11.97), which is the special case of a sum of independent univariate observations from an exponential family; independence is not required here.)

17.34 t_1 and t_2 are unbiased estimators of θ with variances σ_1^2 σ_2^2 and correlation coefficient ρ, and $u_j = t_j - \theta$. Show that $|u_1| < |u_2|$ if and only if $V = u_2 + u_1$ and $W = u_2 - u_1$ have the same sign. Show that the correlation between V and W is

$$R = (\sigma_2^2 - \sigma_1^2)/\{(\sigma_2^2 - \sigma_1^2)^2 + 4\sigma_1^2 \sigma_2^2 (1-\rho^2)\}^{1/2}.$$

Hence, using (15.28), show that if t_1 and t_2 are binormally distributed,

$$P = \operatorname{Prob}\{|u_1| < |u_2|\} = \tfrac{1}{2} + \frac{1}{\pi} \arcsin R,$$

and that if $\sigma_1^2 < \sigma_2^2$, $P > \tfrac{1}{2}$, so that the more efficient estimator is closer to θ in Pitman's sense of **17.30**.
   If t_1 is an MV estimator, use (17.61) to show that

$$R = \left(\frac{1 - \rho^2}{1 + 3\rho^2} \right)^{1/2},$$

where $\rho^2 = \sigma_1^2 / \sigma_2^2$.
(Cf. Geary, 1944.)

CHAPTER 18

ESTIMATION: MAXIMUM LIKELIHOOD

18.1 We have already (**8.7–12**) encountered the maximum likelihood (abbreviated ML) principle in its general form. In this chapter we shall be concerned with its properties when used as a method of estimation. We shall confine our discussion for the most part to the case of samples of n independent observations from the same distribution. The joint probability of the observations, regarded as a function of a single unknown parameter θ, is called (cf. **17.14**) the likelihood function (abbreviated LF) of the sample, and is written

$$L(x \mid \theta) = f(x_1 \mid \theta) f(x_2 \mid \theta) \cdots f(x_n \mid \theta), \tag{18.1}$$

where we write $f(x \mid \theta)$ indifferently for a univariate or multivariate, continuous or discrete distribution.

The ML principle, whose extensive use in statistical theory dates from the work of Fisher (1921a), directs us to take as our estimator of θ that value (say, $\hat{\theta}$) within the admissible range of θ which makes the LF as large as possible. That is, we choose $\hat{\theta}$ so that for any admissible value θ

$$L(x \mid \hat{\theta}) \geqslant L(x \mid \theta). \tag{18.2}$$

Unless otherwise specified, we assume that θ may take any real value in an interval (which may be infinite in either or both directions). In determining $\hat{\theta}$, we may clearly ignore factors in L not involving θ.

18.2 The determination of the form of the ML estimator becomes relatively simple in one general situation. If the LF is a twice-differentiable function of θ throughout its range, stationary values of the LF within the admissible range of θ will, if they exist, be given by roots of

$$L'(x \mid \theta) = \frac{\partial L(x \mid \theta)}{\partial \theta} = 0. \tag{18.3}$$

A sufficient (though not a necessary) condition that any of these stationary values (say, $\tilde{\theta}$) be a local maximum is that

$$L''(x \mid \tilde{\theta}) < 0. \tag{18.4}$$

If we find all the local maxima of the LF in this way (and, if there are more than one, choose the largest of them) we shall have found the solution(s) of (18.2), provided that there is no terminal maximum of the LF at the extreme permissible values of θ.

In practice, it is often simpler to work with the logarithm of the LF than with the function itself. Under the conditions above, they will have maxima together, since

$$\frac{\partial}{\partial\theta}\log L = L'/L$$

and $L > 0$. We therefore seek solutions of

$$(\log L)' = 0 \tag{18.5}$$

for which

$$(\log L)'' < 0, \tag{18.6}$$

if these are simpler to solve than (18.3) and (18.4). (18.5) is often called the likelihood equation.

18.3 If $L \to 0$ as θ tends to its permissible extremes, and *all* roots of (18.5) satisfy (18.6), there can be only one such root, since there must be a minimum (contradicting (18.6)) between any two maxima; and there must be one such root because $L > 0$. Thus a unique maximum of the LF must exist.

Maximum likelihood and sufficiency
18.4 If a single sufficient statistic exists for θ, we see at once that the ML estimator of θ, if unique, must be a function of it. For sufficiency of t for θ implies the factorisation of the LF (17.84). That is,

$$L(x|\theta) = g(t|\theta)h(x), \tag{18.7}$$

the second factor on the right of (18.7) being independent of θ. Thus choice of $\hat{\theta}$ to maximise $L(x|\theta)$ is equivalent to choosing $\hat{\theta}$ to maximise $g(t|\theta)$, and hence $\hat{\theta}$ will be a function of t alone.

18.5 If an MVB estimator t exists for $\tau(\theta)$, and the likelihood equation (18.5) has a solution $\hat{\theta}$, then $t = \tau(\hat{\theta})$ and the solution $\hat{\theta}$ is unique, occurring at a maximum of the LF. For we have seen **(17.33)** that, when there is a single sufficient statistic, the LF is of the form in which MVB estimation of some function $\tau(\theta)$ is possible. Thus, as at (17.27), the LF is of the form

$$(\log L)' = A(\theta)\{t - \tau(\theta)\}, \tag{18.8}$$

so that the solutions of (18.5) are of form

$$t = \tau(\hat{\theta}). \tag{18.9}$$

Differentiating (18.8) again, we have

$$(\log L)'' = A'(\theta)\{t - \tau(\theta)\} - A(\theta)\tau'(\theta). \tag{18.10}$$

But since, from (17.29), $\tau'(\theta)/A(\theta) = \text{var } t$, the last term in (18.10) may be written

$$-A(\theta)\tau'(\theta) = -\{A(\theta)\}^2 \text{ var } t. \tag{18.11}$$

Moreover, at $\hat{\theta}$ the first term on the right of (18.10) is zero in virtue of (18.9). Hence (18.10) becomes, on using (18.11),

$$(\log L)''_\theta = -A(\hat{\theta})\tau'(\hat{\theta}) = -\{A(\hat{\theta})\}^2 \operatorname{var} t < 0. \tag{18.12}$$

By (18.12), every solution of (18.5) is a maximum of the LF. But under regularity conditions there must be a minimum between successive maxima. Since there is no minimum, it follows that there cannot be more than one maximum. This is otherwise obvious from the uniqueness of the MVB estimator t.

(18.9) shows that where an MVB (unbiased) estimator exists, it is given by the ML method. This result does not extend to the more general variance bound (17.41)—cf. Exercise 18.37.

18.6   The uniqueness of the ML estimator where a single sufficient statistic exists extends to the case where the range of $f(x|\theta)$ depends upon θ, but the argument is somewhat different in this case. We have seen (**17.40–1**) that a single sufficient statistic can exist only if

$$f(x|\theta) = g(x)/h(\theta). \tag{18.13}$$

The LF is thus also of form

$$L(x|\theta) = \prod_{i=1}^{n} g(x_i)/\{h(\theta)\}^n \tag{18.14}$$

when it is non-zero, and (18.14) is as large as possible if $h(\theta)$ is as small as possible. Now from (18.13)

$$1 = \int f(x|\theta)\, \mathrm{d}x = \int g(x)\, \mathrm{d}x/h(\theta),$$

where integration is over the whole range of x. Hence

$$h(\theta) = \int g(x)\, \mathrm{d}x. \tag{18.15}$$

From (18.15) it follows that to make $h(\theta)$ as small as possible, we must choose $\hat{\theta}$ so that the value of the integral on the right (one or both of whose limits of integration depend on θ) is minimised.

Now a single sufficient statistic t for θ exists (**17.40–1**) only if one terminal of the range is independent of θ or if the upper terminal is a monotone decreasing function of the lower terminal, and is then given by $x_{(1)}$, $x_{(n)}$ or a function of them. In either of these situations, the value of (18.15) is a monotone function of the range of integration on its right-hand side, reaching a unique terminal minimum if that range is as small as is possible, consistent with the observations, when it is a function of t. The ML estimator $\hat{\theta}$ obtained by minimising this range is thus unique and is a function of t, and the LF (18.14) has a terminal maximum at $L(x|\hat{\theta})$.

The results of this and the previous section were originally obtained by Huzurbazar (1948), who used a different method in the 'regular' case of **18.5**.

18.7   Thus we have seen that where a single sufficient statistic t exists for a parameter θ, the ML estimator $\hat{\theta}$ of θ is a function of t alone. Further, $\hat{\theta}$ is unique, the LF having a single maximum in this case. The maximum is a stationary value (under regularity conditions) or a terminal maximum according to whether the range is independent of, or dependent upon, θ.

18.8   It follows from our results that all the optimum properties of single sufficient statistics are conferred upon ML estimators that are one-to-one functions of them. For example, we may obtain the solution of the likelihood equation, and find a function of it which is unbiased for the parameter. If then follows from the results of **17.35** that this will be the unique MV estimator of the parameter, attaining the MVB (17.22) if this is possible.

The sufficient statistics derived in Examples 17.8–10, 17.16, 17.18 and 17.19 are all easily obtained by the ML method.

Example 18.1
To estimate θ in

$$dF(x) = dx/\theta, \qquad 0 \leqslant x \leqslant \theta,$$

we see at once from the LF in Example 17.16 that $\hat{\theta} = x_{(n)}$, the sufficient statistic, the LF having a sharp (non-differentiable) maximum there.

Obviously, $\hat{\theta}$ is not an unbiased estimator of θ, since it can never exceed it. A modified unbiased estimator is easily seen to be

$$t = (n+1)x_{(n)}/n.$$

Example 18.2
To estimate the mean θ of a normal distribution with known variance σ^2, we see that the LF consists of a factor free of θ multiplied by $\exp\{-n(\bar{x} - \theta)^2/(2\sigma^2)\}$, which $\to 0$ as $\theta \to \pm\infty$ and has a single maximum when $\theta = \bar{x}$. We may alternatively see (Example 17.6) that

$$(\log L)' = \frac{n}{\sigma^2}(\bar{x} - \theta)$$

and obtain the ML estimator by equating this to zero, and find

$$\hat{\theta} = \bar{x}.$$

In this case, $\hat{\theta}$ is unbiased for θ.

Suppose now that θ is known to lie in the interval (a, b). If $\bar{x}$ is in this interval, $\hat{\theta} = \bar{x}$ as before; but if $\bar{x} < a$, the LF has a sharp maximum at a, and if $b < \bar{x}$ it has one at b, since L declines monotonically on either side of $\bar{x}$.

Exercise 18.2 gives a general form of distribution f for which $\hat{\theta} = \bar{x}$—if θ is a location parameter, f must be normal, and if θ is a scale parameter, f must be exponential.

The general one-parameter case

18.9   If no single sufficient statistic for θ exists, the LF no longer necessarily has a unique maximum value if the conditions of **18.3** do not hold (cf. Examples 18.7 and 18.9 and Exercises 18.17, 18.33), and we choose the ML estimator to satisfy (18.2).

We now have to consider the properties of the estimators obtained by this method. We shall see that, under very broad conditions, the ML estimator is consistent; and that under regularity conditions, the most important of which is that the range of $f(x|\theta)$ does not depend on θ, the ML estimator is consistent, is asymptotically normally distributed and is an efficient estimator. These, however, are large-sample properties and, important as they are, it should be borne in mind that they are not such powerful recommendations of the ML method as the properties, inherited from sufficient statistics, that we have discussed in Sections **18.4** onwards. Perhaps it would be unreasonable to expect any method of estimation to produce 'best' results under all circumstances and for all sample sizes. However that may be, the fact remains that, outside the field of sufficient statistics, the optimum properties of ML estimators are asymptotic ones.

Shenton and Bowman (1977) deal with small-sample properties.

Example 18.3

As an example of the general situation, consider the estimation of the correlation parameter ρ in samples of n from the standardised binormal distribution

$$dF = \frac{1}{2\pi(1-\rho^2)^{1/2}} \exp\left\{-\frac{1}{2(1-\rho^2)}(x^2 - 2\rho xy + y^2)\right\} dx\, dy, \quad -\infty < x, y < \infty; \quad |\rho| < 1.$$

We find

$$\log L = -n\log(2\pi) - \tfrac{1}{2}n\log(1-\rho^2) - \frac{1}{2(1-\rho^2)}(\textstyle\sum x^2 - 2\rho\sum xy + \sum y^2),$$

and thus the three statistics $(\sum x^2, \sum xy, \sum y^2)$ are sufficient for ρ—indeed, the pair of statistics $(\sum x^2 + \sum y^2, \sum xy)$ is sufficient. There is no single sufficient statistic for ρ. See Exercise 18.13. At the turning-points of L, $\partial \log L/\partial\rho = 0$ and we have the cubic equation

$$g(\rho) = \frac{(1-\rho^2)^2}{n}\frac{\partial \log L}{\partial\rho} = \rho(1-\rho^2) + (1+\rho^2)\frac{1}{n}\sum xy - \rho\left(\frac{1}{n}\sum x^2 + \frac{1}{n}\sum y^2\right) = 0.$$

This has three roots, two of which may be complex. No more than two real roots can be at maxima of $\log L$, since two maxima must be separated by a minimum. There is always a maximum of the LF in the admissible interval $-1 < \rho < 1$, for $L > 0$ in the interval and $\to 0$ as $\rho \to \pm 1$; in fact

$$g(-1) = \frac{1}{n}\sum(x+y)^2 > 0 \qquad \text{and} \qquad g(+1) = -\frac{1}{n}\sum(x-y)^2 < 0,$$

so $g(\rho)$ crosses the axis in the interval and is declining from a maximum at $+1$. Since $g(0) = (1/n) \sum xy$, this maximum's root will have the sign of $\sum xy$.

Moreover,

$$g'(\rho) = \frac{(1-\rho^2)^2}{n} \frac{\partial^2 \log L}{\partial \rho^2} - 4\rho \frac{(1-\rho^2)}{n} \frac{\partial \log L}{\partial \rho}$$

so when

$$\frac{\partial \log L}{\partial \rho} = 0$$

we have

$$g'(\rho) = \frac{(1-\rho^2)^2}{n} \frac{\partial^2 \log L}{\partial \rho^2} = 1 - 3\rho^2 - \frac{1}{n} \left(\sum x^2 - 2\rho \sum xy + \sum y^2 \right)$$

and substitution from $g(\rho) = 0$ makes this

$$= -\left\{ 2\rho^2 + \frac{(1-\rho^2)}{\rho} \sum \frac{xy}{n} \right\}.$$

If a root ρ_1 has the same sign as $\sum xy$, $g'(\rho_1)$ is negative and the root is at a maximum; we have shown that there must be one such root in $[-1, +1]$. We now see that if there is a second maximum there, it must be at a root ρ_3 of opposite sign to $\sum xy$, for the intervening minimum at ρ_2 must be so to make $g'(\rho_2) > 0$, and ρ_3 must be further from ρ_1, satisfying

$$2|\rho_3|^3/(1-\rho_3^2) > \frac{1}{n} |\sum xy|$$

while for ρ_2 this inequality is reversed. If ρ_1 and ρ_3 are both in $[-1, +1]$, that with the larger value of the LF is the ML estimator $\hat{\rho}$ in accordance with (18.2). At any root of the likelihood equation, we find

$$L(x \mid g(\rho) = 0) = \{ 2\pi(1-\rho^2)^{1/2} \}^{-n} \cdot \exp \left\{ -\tfrac{1}{2} \left(n + \frac{\sum xy}{\rho} \right) \right\}.$$

The first factor on the right increases with $|\rho|$, and the second factor increases with $-\sum xy/\rho$, i.e. with $\rho \operatorname{sgn} (\sum xy)$. Since

$$\rho_1 \operatorname{sgn} (\sum xy) = \rho_1 \operatorname{sgn} (\rho_1) = |\rho_1|,$$

and $\rho_3 \operatorname{sgn} (\sum xy) = -|\rho_3|$, we see that if $|\rho_1| > |\rho_3|$, both factors in the LF will be no less at ρ_1 than at ρ_3, so that $\hat{\rho} = \rho_1$; if the real root with the sign of $\sum xy$ is the largest root in absolute value, it is the ML estimator.

If there are three real roots of $g(\rho) = 0$, $g(\rho)$ has two real turning-points satisfying $g'(\rho) = 0$, a quadratic equation that has distinct real roots if and only if its discriminant

$$\left(\frac{1}{n} \sum xy \right)^2 - 3 \left(\frac{1}{n} \sum x^2 + \frac{1}{n} \sum y^2 - 1 \right) > 0, \tag{18.16}$$

one root of $g'(\rho) = 0$ then lying on either side of $\rho_0 = \frac{1}{3}(1/n) \sum xy$. It follows that at least one root of $g(\rho) = 0$ lies on each side of ρ_0, and since we may have $|\rho_0| > 1$, that there may be real roots of $g(\rho) = 0$ outside the interval $(-1, 1)$. Such real roots, and complex roots, are inadmissible.

Since, by the results of **10.3** and **10.9**, the sample moments on the left side of (18.16) are consistent estimators of the corresponding population moments, that left side will converge in probability to $\rho^2 - 3(1 + 1 - 1) < 0$. Thus, as $n \to \infty$, there will tend to be only one real root of the likelihood equation, and by our argument above, it must be at a maximum.

Consistency of ML estimators

18.10 We now show that, under very general conditions, ML estimators are consistent. We begin with a heuristic version of Wald's (1949) argument.

As at (18.2), we consider the case of n independent observations from a distribution $f(x \mid \theta)$, and for each n we choose the ML estimator $\hat{\theta}$ so that, if θ is any admissible value of the parameter, we have*

$$\log L(x \mid \hat{\theta}) \geq \log L(x \mid \theta). \tag{18.17}$$

We denote the true value of θ by θ_0, and let E_0 represent the operation of taking expectations when the true value θ_0 holds. Consider the random variable $L(x \mid \theta)/L(x \mid \theta_0)$. In virtue of the fact that the geometric mean of a non-degenerate distribution cannot exceed its arithmetic mean, we have, for all $\theta^* \neq \theta_0$,

$$E_0 \left\{ \log \frac{L(x \mid \theta^*)}{L(x \mid \theta_0)} \right\} < \log E_0 \left\{ \frac{L(x \mid \theta^*)}{L(x \mid \theta_0)} \right\}. \tag{18.18}$$

Now the expectation on the right of (18.18) is

$$\int \cdots \int \frac{L(x \mid \theta^*)}{L(x \mid \theta_0)} L(x \mid \theta_0) \, dx_1 \cdots dx_n = 1.$$

Thus (18.18) becomes

$$E_0 \left\{ \log \frac{L(x \mid \theta^*)}{L(x \mid \theta_0)} \right\} < 0$$

or, inserting a factor $1/n$,

$$E_0 \left\{ \frac{1}{n} \log L(x \mid \theta^*) \right\} < E_0 \left\{ \frac{1}{n} \log L(x \mid \theta_0) \right\} \tag{18.19}$$

provided that the expectation on the right exists.

> If the right side of (18.19) does not exist, examples can be given in which the ML estimator is not consistent, converging to a fixed value whatever the true value of θ may be—cf. Ferguson (1982).

---

* Because of the equality sign in (18.17), the sequence of values of $\hat{\theta}$ may be determinable in more than one way. See **18.12–13** below.

Now for any value of θ

$$\frac{1}{n} \log L(x \mid \theta) = \frac{1}{n} \sum_{i=1}^{n} \log f(x_i \mid \theta)$$

is the mean of a set of n independent identical variates with expectation

$$E_0\{\log f(x \mid \theta)\} = E_0\left\{\frac{1}{n} \log L(x \mid \theta)\right\}.$$

By the strong law of large numbers of **7.34**, therefore, $(1/n) \log L(x \mid \theta)$ converges with probability one to its expectation, as n increases. Thus as $n \to \infty$ we have, from (18.19), with probability one,

$$\frac{1}{n} \log L(x \mid \theta^*) < \frac{1}{n} \log L(x \mid \theta_0)$$

or

$$\lim_{n \to \infty} \text{prob} \{\log L(x \mid \theta^*) < \log L(x \mid \theta_0)\} = 1, \qquad \theta^* \neq \theta_0. \tag{18.20}$$

On the other hand, (18.17) with $\theta = \theta_0$ gives

$$\log L(x \mid \hat{\theta}) \geqslant \log L(x \mid \theta_0). \tag{18.21}$$

(18.20) and (18.21) imply that, as $n \to \infty$, $L(x \mid \hat{\theta})$ cannot take any other value than $L(x \mid \theta_0)$. If $L(x \mid \theta)$ is identifiable (cf. **17.6**), this implies that

$$\text{prob} \{\lim_{n \to \infty} \hat{\theta} = \theta_0\} = 1. \tag{18.22}$$

Wald's (1949) rigorous proof of the consistency of ML estimators requires further conditions, which are often difficult to verify—see also extensions by Huber (1967) and Perlman (1972).

18.11 Other proofs of consistency, e.g. by Cramér (1946), concentrate on solutions to the likelihood equation, rather than on direct maximisation of the LF—cf. **18.16** below.

18.12 There may be multiple roots of the likelihood equation, but Huzurbazar (1948) has shown under regularity conditions that if there is a consistent root, it is unique as $n \to \infty$.

Suppose that the LF possesses two derivatives. It follows from the convergence in probability of $\hat{\theta}$ to θ_0 that

$$\frac{1}{n}\left[\frac{\partial^2}{\partial \theta^2} \log L(x \mid \theta)\right]_{\theta = \hat{\theta}} \xrightarrow[n \to \infty]{} \frac{1}{n}\left[\frac{\partial^2}{\partial \theta^2} \log L(x \mid \theta)\right]_{\theta = \theta_0} \tag{18.23}$$

Now by the strong law of large numbers, once more,

$$\frac{1}{n} \frac{\partial^2}{\partial \theta^2} \log L(x \mid \theta) = \frac{1}{n} \sum_{i=1}^{n} \frac{\partial^2}{\partial \theta^2} \log f(x_i \mid \theta)$$

is the mean of n independent identical variates and converges with probability one to its expectation. Thus we may write (18.23) as

$$\lim_{n\to\infty} \text{prob} \left\{ \left[\frac{\partial^2}{\partial \theta^2} \log L(x|\theta) \right]_{\theta=\hat\theta} = E_0 \left[\frac{\partial^2}{\partial \theta^2} \log L(x|\theta) \right]_{\theta=\theta_0} \right\} = 1. \qquad (18.24)$$

But we have seen at (17.19) that under regularity conditions

$$E \left[\frac{\partial^2}{\partial \theta^2} \log L(x|\theta) \right] = -E \left\{ \left(\frac{\partial \log L(x|\theta)}{\partial \theta} \right)^2 \right\} < 0. \qquad (18.25)$$

Thus (18.24) becomes

$$\lim_{n\to\infty} \text{prob} \left\{ \left[\frac{\partial^2}{\partial \theta^2} \log L(x|\theta) \right]_{\theta=\hat\theta} < 0 \right\} = 1. \qquad (18.26)$$

Now suppose that the conditions of **18.2** hold, and that two local maxima of the LF, at $\hat\theta_1$ and $\hat\theta_2$, are roots of (18.5) satisfying (18.6). If $\log L(x|\theta)$ has a second derivative everywhere, as we have assumed, there must be a minimum between the maxima at $\hat\theta_1$ and $\hat\theta_2$. If this is at $\hat\theta_3$, we must have

$$\left[\frac{\partial^2 \log L(x|\theta)}{\partial \theta^2} \right]_{\theta=\hat\theta_3} \geqslant 0. \qquad (18.27)$$

If both $\hat\theta_1$ and $\hat\theta_2$ are consistent estimators, $\hat\theta_3$, which lies between them in value, must also be consistent and must satisfy (18.26). Since (18.26) and (18.27) directly contradict each other, it follows that we can only have one consistent estimator $\hat\theta$ obtained as a root of the likelihood equation (18.5).

Non-uniqueness of ML estimators
18.13 A point that should be discussed in connection with the consistency of ML estimators is that, for particular samples, there is the possibility that the LF has two (or more) equal suprema, i.e. that the equality sign holds in (18.2). How can we choose between the values $\hat\theta_1$, $\hat\theta_2$, etc., at which they occur? There is an essential indeterminacy here. The difficulty usually arises only when particular configurations of sample values are realised which have small probability of occurrence, but it may arise in *all* samples—see Examples 18.7, 18.9 and Exercises 18.17, 18.33 below. If the parameter itself is unidentifiable, the difficulty *must* arise in *all* samples, as the following example makes clear.

Example 18.4
In Example 18.3 put

$$\cos \theta = \rho.$$

To each real solution of the cubic likelihood equation, say $\hat\rho$, there will now correspond an infinity of estimators of θ, of form

$$\hat\theta_r = \arccos \hat\rho + 2r\pi$$

where r is any integer. The parameter θ is essentially incapable of estimation, even when $n \to \infty$. Considered as a function of θ, the LF is periodic, with an infinite number of equal maxima at $\hat{\theta}_r$, and the $\hat{\theta}_r$ differ by multiples of 2π. There can be only one consistent estimator of θ_0, the true value of θ, but we have no means of deciding which $\hat{\theta}_r$ is consistent. In such a case, we must recognise that only $\cos\theta$ is directly estimable. Since θ is *unidentifiable*, neither ML nor any other method can be effective.

Consistency and bias of ML estimators

18.14   Although, under the conditions of **18.10**, the ML estimator is consistent, it is not unbiased generally. We have already seen in Example 18.1 that there may be bias even when the ML estimator is a function of a single sufficient statistic. In general, we must expect bias, for if the ML estimator is $\hat{\theta}$ and we seek to estimate a function $\tau(\theta)$, we have seen in **8.10** that the ML estimator of $\tau(\theta)$ is $\tau(\hat{\theta})$. But in general

$$E\{\tau(\hat{\theta})\} \neq \tau\{E(\hat{\theta})\}, \tag{18.28}$$

so that if $\hat{\theta}$ is unbiased for θ, $\tau(\hat{\theta})$ cannot be unbiased for $\tau(\theta)$. If the ML estimator is consistent, the remark concerning asymptotic unbiasedness below Example 17.3 may apply.

Consistency, asymptotic normality and efficiency of ML estimators

18.15   When we turn to the discussion of the normality and efficiency of ML estimators, the following example is enough to show that we must make restrictions before we can obtain optimum results.

Example 18.5
We saw in Example 17.22 that in the distribution

$$dF(x) = dx/\theta, \qquad k\theta \leq x \leq (k+1)\theta; \qquad k > 0,$$

there is no single sufficient statistic for θ, but that the extreme observations $x_{(1)}$ and $x_{(n)}$ are a pair of jointly sufficient statistics for θ. Let us now find the ML estimator of θ. We maximise the LF as in Example 18.1. Here

$$L(x \mid \theta) = \theta^{-n} u(x_{(1)} - k\theta) u((k+1)\theta - x_{(n)})$$

is non-zero only if $x_{(n)}/(k+1) \leq \theta \leq x_{(1)}/k$, and then is monotone decreasing in θ. Thus

$$\hat{\theta} = x_{(n)}/(k+1)$$

is the ML estimator. We see at once that $\hat{\theta}$ is a function of $x_{(n)}$ only, although $x_{(1)}$ and $x_{(n)}$ are both required for sufficiency.

Since $x_{(n)}$ will have an extreme-value distribution as in **14.17**, $\hat{\theta}$ will not be asymptotically normal here. Nor does it have the smallest possible variance asymptotically, for by symmetry, $x_{(1)}$ and $x_{(n)}$ have the same variance, say V. The ML estimator has variance

$$\text{var } \hat{\theta} = V/(k+1)^2,$$

and the estimator

$$\theta^* = x_{(1)}/k$$

has variance

$$\text{var } \theta^* = V/k^2.$$

Since $x_{(1)}$ and $x_{(n)}$ are asymptotically independently distributed (**14.23**), the function

$$\bar{\theta} = a\hat{\theta} + (1-a)\theta^*$$

will, like $\hat{\theta}$ and θ^*, be a consistent estimator of θ, and its variance is

$$\text{var } \bar{\theta} = V\left\{\frac{a^2}{(k+1)^2} + \frac{(1-a)^2}{k^2}\right\}$$

which is minimised for variation in a when

$$a = \frac{(k+1)^2}{k^2 + (k+1)^2}.$$

Then

$$\text{var } \bar{\theta} = V/\{k^2 + (k+1)^2\}.$$

Thus, for all $k > 0$,

$$\frac{\text{var } \bar{\theta}}{\text{var } \hat{\theta}} = \frac{(k+1)^2}{k^2 + (k+1)^2} < 1$$

and the ML estimator has the larger variance. If k is large, the variance of $\hat{\theta}$ is nearly twice that of the other estimator.

18.16 To prove the consistency, asymptotic normality and efficiency of $\hat{\theta}$, we shall assume that the first two derivatives of $\log L(x|\theta)$ exist, and that (17.18–19) hold, i.e. that

$$E\left(\frac{\partial \log L}{\partial \theta}\right) = 0 \qquad\qquad (18.29)$$

and

$$R^2(\theta) = -E\left(\frac{\partial^2 \log L}{\partial \theta^2}\right) = E\left\{\left(\frac{\partial \log L}{\partial \theta}\right)^2\right\}, \qquad\qquad (18.30)$$

where $R^2(\theta) > 0$. As we pointed out in **18.2**, our differentiability assumptions imply that $\hat{\theta}$ is a root of the likelihood equation $\partial \log L/\partial \theta = 0$, and in this section we use the symbol $\hat{\theta}$ to denote such a root.

Using Taylor's theorem, we have

$$\left(\frac{\partial \log L}{\partial \theta}\right)_{\hat{\theta}} = \left(\frac{\partial \log L}{\partial \theta}\right)_{\theta_0} + (\hat{\theta} - \theta_0)\left(\frac{\partial^2 \log L}{\partial \theta^2}\right)_{\theta^*}, \qquad\qquad (18.31)$$

where θ^* is some value between $\hat{\theta}$ and θ_0. Thus the left-hand side of (18.31) is zero. On its right-hand side, both $\partial \log L/\partial \theta$ and $\partial^2 \log L/\partial \theta^2$ are sums of independent identical variates, and as $n \to \infty$ each therefore converges to its expectation by the strong law of large numbers, as in the argument of **18.10**. The first of these expectations is zero by (18.29) and the second non-zero by (18.30). Since the right-hand side of (18.31) as a whole must converge to zero, to remain equal to the left, we see that we must have $(\hat{\theta} - \theta_0)$ converging to zero as $n \to \infty$, so that $\hat{\theta}$ is a consistent estimator under our assumptions.

We now rewrite (18.31) in the form

$$(\hat{\theta} - \theta_0) R(\theta_0) = \frac{\left(\dfrac{\partial \log L}{\partial \theta}\right)_{\theta_0} \Big/ R(\theta_0)}{\left(\dfrac{\partial^2 \log L}{\partial \theta^2}\right)_{\theta^*} \Big/ \{-R^2(\theta_0)\}}. \tag{18.32}$$

In the denominator on the right of (18.32) we have, since $\hat{\theta}$ is consistent for θ_0 and θ^* lies between them, from (18.23–4) and (18.30),

$$\lim_{n \to \infty} \text{prob} \left\{ \left[\frac{\partial^2 \log L}{\partial \theta^2} \right]_{\theta^*} = -R^2(\theta_0) \right\} = 1, \tag{18.33}$$

so that the denominator converges to unity. The numerator on the right of (18.32) is the ratio to $R(\theta_0)$ of the sum of the n independent identical variates $\partial \log f(x_i | \theta_0)/\partial \theta$. This sum has zero mean by (18.29) and variance defined at (18.30) to be $R^2(\theta_0)$. The central limit theorem (**7.35**) therefore applies, and the numerator is asymptotically a standardised normal variate; the same is therefore true of the right side as a whole. Thus the left side of (18.32) is asymptotically standard normal or, in other words, the ML estimator $\hat{\theta}$ is asymptotically normally distributed with mean θ_0 and variance $1/R^2(\theta_0)$.

This result, which gives the ML estimator an asymptotic variance equal to the MVB (17.24), implies that under these regularity conditions the ML estimator is efficient. Since the MVB can only be attained if there is a single sufficient statistic (cf. **17.33**) we are also justified in saying that the ML estimator is 'asymptotically sufficient'.

A more rigorous proof on these lines is given by Cramér (1946). Daniels (1961)—see the note by Williamson (1984)—relaxes the conditions for the asymptotic normality and efficiency of ML estimators. See also Huber (1967) and Lecam (1970). Weiss and Wolfowitz (1973) prove efficiency in a class of non-regular estimators of a location parameter—cf. Exercise 18.5 for an instance.

Although a root of the likelihood equation will be consistent, the ML estimator may be a different root—Kraft and Lecam (1956) give examples in which the ML estimator exists, is unique and is not consistent while a consistent root exists, the conditions above being satisfied. In such a case, we could proceed iteratively starting with another consistent estimator, as in **18.21** below.

Lecam (1953) has objected to the use of the term 'efficient' because it implies absolute minimisation of variance in large samples, and in the strict sense this is not achieved by the

ML (or any other) estimator. For example, consider a consistent estimator t of θ, asymptotically normally distributed with variance of order n^{-1}. Define a new statistic

$$t' = \begin{cases} t & \text{if } |t| \geqslant n^{-1/4}, \\ kt & \text{if } |t| < n^{-1/4}. \end{cases} \tag{18.34}$$

We have

$$\lim_{n \to \infty} \text{var } t'/\text{var } t = \begin{cases} 1 & \text{if } \theta \neq 0, \\ k^2 & \text{if } \theta = 0, \end{cases}$$

and k may be taken very small, so that at one point t' is more efficient than t, and nowhere is it worse. Lecam has shown (cf. also Bahadur (1964)) that such 'superefficiency' can arise only for a set of θ-values of measure zero. In view of this, we shall retain the term 'efficiency' in its ordinary use. However, C. R. Rao (1962b) shows that even this limited paradox can be avoided by redefining the efficiency of an estimator in terms of its correlation with $\partial \log L/\partial \theta$—cf. **17.15–17** and (17.61). Walker (1963) gives sufficient regularity conditions for the asymptotic variances of all asymptotically normal estimators to be bounded by the MVB.

Brillinger (1964) shows that if $\hat{\theta}$ is jackknifed by dividing the sample into a fixed number of groups and omitting one group at a time, then as $n \to \infty$, t'_n at (17.10) leads to normality under regularity conditions; he gives expansions for its reduced bias and its m.s.e. Reeds (1978) shows that under regularity conditions, the ordinary jackknifed $\hat{\theta}$ has the same asymptotic distribution as $\hat{\theta}$ itself, and that the jackknifed estimator of var $\hat{\theta}$ is consistent.

Example 18.6

In Example 18.3 we found that the ML estimator $\hat{\rho}$ of the correlation parameter in a standardised bivariate normal distribution is a root of the cubic equation

$$g(\rho) = \frac{(1-\rho^2)^2}{n} \frac{\partial \log L}{\partial \rho} = 0,$$

and that

$$g'(\rho) = \frac{(1-\rho^2)^2}{n} \frac{\partial^2 \log L}{\partial \rho^2} - \frac{4\rho(1-\rho^2)}{n} \frac{\partial \log L}{\partial \rho}$$

with

$$\frac{(1-\rho^2)^2}{n} \frac{\partial^2 \log L}{\partial \rho^2} = 1 - 3\rho^2 - \frac{1}{n} (\textstyle\sum x^2 - 2\rho \sum xy + \sum y^2).$$

Taking expectations in $g'(\rho)$, (17.18) removes the second term on the right, and we find, since $E(x^2) = E(y^2) = 1$, $E(xy) = \rho$,

$$E\{g'(\rho)\} = \frac{(1-\rho^2)^2}{n} E\left\{\frac{\partial^2 \log L}{\partial \rho^2}\right\} = 1 - 3\rho^2 - 2(1-\rho^2)$$

$$= -(1+\rho^2)$$

so that

$$E\left(\frac{\partial^2 \log L}{\partial \rho^2}\right) = -\frac{n(1+\rho^2)}{(1-\rho^2)^2}.$$

Hence, from **18.16**, we have asymptotically

$$\operatorname{var} \hat{\rho} = \frac{(1-\rho^2)^2}{n(1+\rho^2)}.$$

Exercise 18.12 uses this result to establish the efficiency of the sample correlation coefficient.

Example 18.7
The distribution

$$dF(x) = \tfrac{1}{2}\exp\{-|x-\theta|\}\,dx, \qquad -\infty < x < \infty,$$

yields the log likelihood

$$\log L(x|\theta) = -n\log 2 - \sum_{i=1}^{n} |x_i - \theta|.$$

This is maximised when $\sum_i |x_i - \theta|$ is minimised, and by the result of Exercise 2.1 this occurs when θ is the median of the n values of x. (If n is odd, the value of the middle observation is the median; if n is even, any value in the interval between the two middle observations is a median.) Thus the ML estimator is $\hat{\theta} = \tilde{x}$, the sample median, and is not unique (because the median is not) when n is even. It is easily seen from Example 10.7 that its asymptotic variance in this case is

$$\operatorname{var} \hat{\theta} = 1/n.$$

We cannot use the result of **18.16** to check the efficiency of $\hat{\theta}$, since the differentiability conditions there imposed do not hold for this distribution. But since

$$\frac{\partial \log f(x|\theta)}{\partial \theta} = \begin{cases} +1 & \text{if } x > \theta, \\ -1 & \text{if } x < \theta, \end{cases}$$

fails to exist only at $x = \theta$, we have

$$\left(\frac{\partial \log f(x|\theta)}{\partial \theta}\right)^2 = 1, \qquad x \neq \theta.$$

For $\varepsilon > 0$, we now interpret $E[(\partial \log f(x|\theta)/\partial\theta)^2]$ as

$$\lim_{\varepsilon \to 0} \left\{ \int_{-\infty}^{\theta-\varepsilon} + \int_{\theta+\varepsilon}^{\infty} \right\} \left(\frac{\partial \log f(x|\theta)}{\partial \theta}\right)^2 dF(x) = 1.$$

Thus we have

$$E\left\{\left(\frac{\partial \log L}{\partial \theta}\right)^2\right\} = nE\left\{\left(\frac{\partial \log f(x|\theta)}{\partial \theta}\right)^2\right\} = n,$$

so that the MVB for an estimator of θ is

$$\operatorname{var} t \geq 1/n,$$

which is attained asymptotically by $\hat{\theta}$.

Log L here consists of a continuous set of straight-line segments with joins at the observed values x_i. Exercise 18.38 gives a log L with cusps at these points.

18.17 The result of **18.16** simplifies for a distribution admitting a single sufficient statistic for the parameter. For in that case, from (18.10) and (18.12),

$$E\left(\frac{\partial^2 \log L}{\partial \theta^2}\right) = -A(\theta)\tau'(\theta) = \left(\frac{\partial^2 \log L}{\partial \theta^2}\right)_{\hat{\theta}=\theta}, \qquad (18.35)$$

so that there is no need to evaluate the expectation in this case: the MVB becomes simply $-1/(\partial^2 \log L/\partial \theta^2)_{\hat{\theta}=\theta}$; it is attained exactly when $\hat{\theta}$ is unbiased for θ, and asymptotically in any case under the conditions of **18.16**.

Estimation of the asymptotic variance

18.18 To estimate the variance of the asymptotic distribution of $\hat{\theta}$, there are two obvious estimators

$$\widehat{\mathrm{var}}_1\,(\hat{\theta}) = 1\bigg/\left\{-E\left(\frac{\partial^2 \log L}{\partial \theta^2}\right)\right\}_{\theta=\hat{\theta}} \qquad (18.36)$$

and

$$\widehat{\mathrm{var}}_2\,(\hat{\theta}) = 1\bigg/\left\{-\frac{\partial^2 \log L}{\partial \theta^2}\right\}_{\theta=\hat{\theta}}, \qquad (18.37)$$

each of which is consistent since its denominator converges to the true $R^2(\theta)$ at (18.30). (18.36) and (18.37) will coincide when (18.35) holds, i.e. when there is a single sufficient statistic for θ, but in general they will differ.

> Efron and Hinkley (1978) show that (18.37) is a first approximation to var $(\hat{\theta}|u)$ where u is an ancillary statistic as in **31.00** below (i.e. a component of the minimal sufficient statistics, with marginal distribution free of θ), so that the conditionality principle of **31.00** is satisfied asymptotically; further, a first approximation to u is (18.37) itself. See also Hinkley (1980) and Skovgaard (1985).

Other estimators of var $\hat{\theta}$ may also be used. For example, if $\hat{\theta}$ may be expressed as a function of the sample moments, then whether or not $L(x\,|\,\theta)$ is correctly specified, $\hat{\theta}$ will be a consistent estimator of the corresponding function of the population moments as in Chapter 10, and var $(\hat{\theta})$, another such function, will be consistently estimated by the corresponding function of the sample moments. We thus obtain a more robust estimator of var $(\hat{\theta})$, although of course it is not in general as efficient as (18.36) and (18.37) if $L(x\,|\,\theta)$ is the true LF.

Example 18.8
To estimate the standard deviation σ of a normal distribution with known mean, taken as the origin,

$$\mathrm{d}F(x) = \frac{1}{\sigma(2\pi)^{1/2}}\exp\left(-\frac{x^2}{2\sigma^2}\right)\mathrm{d}x, \qquad -\infty < x < \infty.$$

We have, from Example 17.10,

$$(\log L)' = -n/\sigma + \sum x^2/\sigma^3,$$

so that the ML estimator, which is sufficient, is $\hat{\sigma} = \surd(\sum x^2/n)$.

Exercise 17.6 demonstrated the bias in $\hat{\sigma}$ and other estimators. (However, the ML estimator $\hat{\sigma}^2$ of the variance is unbiased, since $\sum x^2/\sigma^2$ has a χ_n^2 distribution with expectation n—cf. **18.14**.) We have

$$(\log L)'' = n/\sigma^2 - 3\sum x^2/\sigma^4 = \frac{n}{\sigma^2}\left(1 - \frac{3\hat{\sigma}^2}{\sigma^2}\right).$$

Thus, using (18.35), we have as n increases

$$\operatorname{var}\hat{\sigma} \to -1/(\log L)''_{\hat{\sigma}=\sigma} = \sigma^2/(2n),$$

which is estimated from either (18.36) or (18.37) as

$$\operatorname{v\hat{a}r}(\hat{\sigma}) = \hat{\sigma}^2/(2n)$$

In terms of the sample moments, $\hat{\sigma} = (m_2')^{1/2}$ is estimating $\sigma = (\mu_2')^{1/2}$, and from (10.5) and (10.14) we have

$$\operatorname{var}(\hat{\sigma}) = \left(\frac{1}{2\mu_2'^{1/2}}\right)^2 \frac{\mu_4' - \mu_2'^2}{n} = \frac{\mu_4' - \mu_2'^2}{4n\mu_2'}$$

reducing to $\sigma^2/(2n)$ as above if the distribution is normal with $\mu_1' = 0$, when $\mu_4' = 3\mu_2'^2$. The more robust estimator of var $(\hat{\sigma})$ is

$$\frac{m_4' - m_2'^2}{4nm_2'} = \frac{m_2'}{4n}(\hat{\gamma}_2 + 2), \quad \text{where } \hat{\gamma}_2 = m_4'/(m_2')^2 - 3,$$

and is less efficient than $\hat{\sigma}^2/(2n) = m_2'/(2n)$ in the normal case because the latter is based on the sufficient statistic and is not affected by sampling variation in $\hat{\gamma}_2$.

18.19   Even when the regularity conditions for its efficiency do not hold, the ML estimator may have minimum variance properties conferred upon it by the sufficient statistic(s) of which it is always a function. Thus, e.g., in Example 18.1 $\hat{\theta} = x_{(n)}$ is a multiple $n/(n+1)$ of the unbiased estimator t with minimum variance (cf. Exercise 17.24), and consequently $\hat{\theta}$ will be asymptotically unbiased with the same asymptotic variance as t. However, we saw in Example 11.4 (cf. also (19.85) below) that here

$$\operatorname{var} x_{(n)} = \frac{n\theta^2}{(n+1)^2(n+2)}.$$

Thus when we consider the m.s.e. of $x_{(n)}$ as in **17.30**, the square of its bias, $\theta^2/(n+1)^2$, is of the same order of magnitude as its variance. If we apply the result of Example 17.14, we find that

$$\frac{n(n+2)}{(n+1)^2}t = \left(\frac{n+2}{n+1}\right)x_{(n)}$$

has m.s.e. $\theta^2/(n+1)^2$, against $\theta^2/\{n(n+2)\}$ for the unbiased t and $2\theta^2/\{(n+1)(n+2)\}$ for $\hat{\theta}$. Thus $\hat{\theta}$ in this case has asymptotically, and indeed almost exactly, twice as large an m.s.e. as its best multiple has.

The cumulants of an ML estimator
18.20   Haldane and Smith (1956) obtained expressions for the first four cumulants of an ML estimator. Suppose that the distribution sampled, $f(x\mid\theta)$, is grouped into

k classes, and that the probability of an observation falling into the rth class is π_r ($r = 1, 2, \ldots, k$). We thus reduce any distribution to a multinomial distribution (5.49), and if the range of the original distribution is independent of the unknown parameter θ, we seek solutions of the likelihood equation (18.5). Since the probabilities π_r are functions of θ, we write, from (5.123),

$$L(x \mid \theta) \propto \prod_r \pi_r^{n_r},$$

where n_r is the number of observations in the rth class and $\sum_r n_r = n$, the sample size. The likelihood equation is

$$\frac{\partial \log L}{\partial \theta} = \sum_r n_r \frac{\pi_r'}{\pi_r} = 0,$$

where a prime denotes differentiation with respect to θ. Of course, the ML estimator $\hat{\theta}$ will now differ from what it would be for the ungrouped observations, when more information is available. Exercises 18.24–5 discuss the difference, which is important in some contexts—cf. 30.15–19 below. However, as $k \to \infty$ the difference disappears.

Now, using Taylor's theorem, we expand π_r and π_r' about the true value θ_0, and obtain

$$
\begin{aligned}
\pi_r(\hat{\theta}) &= \pi_r(\theta_0) + (\hat{\theta} - \theta_0)\pi_r'(\theta_0) + \tfrac{1}{2}(\hat{\theta} - \theta_0)^2 \pi_r''(\theta_0) + \cdots \\
\pi_r'(\hat{\theta}) &= \pi_r'(\theta_0) + (\hat{\theta} - \theta_0)\pi_r''(\theta_0) + \tfrac{1}{2}(\hat{\theta} - \theta_0)^2 \pi_r'''(\theta_0) + \cdots
\end{aligned}
\qquad (18.38)
$$

If we insert (18.38) into the likelihood equation, expand binomially, and sum the series, we have, writing

$$A_i = \sum_r \{\pi_r'(\theta_0)\}^{i+1}/\{\pi_r(\theta_0)\}^i,$$

$$B_i = \sum_r \{\pi_r'(\theta_0)\}^i \pi_r''(\theta_0)/\{\pi_r(\theta_0)\}^i,$$

$$C_i = \sum_r \{\pi_r'(\theta_0)\}^{i-1}\{\pi_r''(\theta_0)\}^2/\{\pi_r(\theta_0)\}^i,$$

$$D_i = \sum_r \{\pi_r'(\theta_0)\}^i \pi_r'''(\theta_0)/\{\pi_r(\theta_0)\}^i,$$

$$\alpha_i = \sum_r \{\pi_r'(\theta_0)\}^i \left\{\frac{n_r}{n} - \pi_r(\theta_0)\right\} \Big/ \{\pi_r(\theta_0)\}^i,$$

$$\beta_i = \sum_r \{\pi_r'(\theta_0)\}^{i-1} \pi_r''(\theta_0)\left\{\frac{n_r}{n} - \pi_r(\theta_0)\right\} \Big/ \{\pi_r(\theta_0)\}^i,$$

$$\delta_i = \sum_r \{\pi_r'(\theta_0)\}^{i-1} \pi_r'''(\theta_0)\left\{\frac{n_r}{n} - \pi_r(\theta_0)\right\} \Big/ \{\pi_r(\theta_0)\}^i,$$

the expansion

$$
\alpha_1 - (A_1 + \alpha_2 - \beta_1)(\hat{\theta} - \theta_0) + \tfrac{1}{2}(2A_2 - 3B_1 + 2\alpha_3 - 3\beta_2 + \delta_1)(\hat{\theta} - \theta_0)^2 \\
- \tfrac{1}{6}(6A_3 - 12B_2 + 3C_1 + 4D_1)(\hat{\theta} - \theta_0)^3 + \cdots = 0. \qquad (18.39)
$$

For large n, (18.39) may be inverted by Lagrange's theorem to give

$$
\begin{aligned}
(\hat{\theta} - \theta_0) = {} & A_1^{-1}\alpha_1 + A_1^{-3}\alpha_1[(A_2 - \tfrac{3}{2}B_1)\alpha_1 - A_1(\alpha_2 - \beta_1)] \\
& + A_1^{-5}\alpha_1[\{2(A_2 - \tfrac{3}{2}B_1)^2 - A_1(A_3 - 2B_2 + \tfrac{1}{2}C_1 + \tfrac{2}{3}D_1)\}\alpha_1^2 \\
& - 3A_1(A_2 - \tfrac{3}{2}B_1)\alpha_1(\alpha_2 - \beta_1) + \tfrac{1}{2}A_1^2\alpha_1(2\alpha_3 - 3\beta_2 + \delta_1) \\
& + A_1^2(\alpha_2 - \beta_1)^2] + O(n^{-3}).
\end{aligned}
\tag{18.40}
$$

(18.40) enables us to obtain the moments of $\hat{\theta}$ as series in powers of n^{-1}.

Consider the sampling distribution of the sum

$$
W = \sum_r h_r \left\{ \frac{n_r}{n} - \pi_r(\theta_0) \right\},
$$

where the h_r are any constant weights. From the moments of the multinomial distribution (cf. **5.49**), we obtain for the moments of W, writing $S_i = \sum_r h_r^i \pi_r(\theta_0)$,

$$
\left.
\begin{aligned}
\mu_1'(W) &= 0, \\
\mu_2(W) &= n^{-1}(S_2 - S_1^2), \\
\mu_3(W) &= n^{-2}(S_3 - S_1 S_2 + 2S_1^3), \\
\mu_4(W) &= 3n^{-2}(S_2 - S_1^2)^2 + n^{-3}(S_4 - 4S_1 S_3 - 3S_2^2 + 12S_1^2 S_2 - 6S_1^4), \\
\mu_5(W) &= 10n^{-3}(S_2 - S_1^2)(S_3 - 3S_1 S_2 + 2S_1^3) + O(n^{-4}), \\
\mu_6(W) &= 15n^{-3}(S_2 - S_1^2)^3 + O(n^{-4}).
\end{aligned}
\right\}
\tag{18.41}
$$

From (18.41) we can derive the moments and product-moments of the random variables α_i, β_i and δ_i appearing in (18.40), for all of these are functions of form W. Finally, we substitute these moments into the powers of (18.40) to obtain the moments of $\hat{\theta}$. Expressed as cumulants, these are

$$
\left.
\begin{aligned}
\kappa_1 &= \theta_0 - \tfrac{1}{2}n^{-1}A_1^{-2}B_1 + O(n^{-2}), \\
\kappa_2 &= n^{-1}A_1^{-1} + n^{-2}A_1^{-4}[-A_2^2 + \tfrac{7}{2}B_1^2 + A_1(A_3 - B_2 - D_1) - A_1^3] + O(n^{-3}), \\
\kappa_3 &= n^{-2}A_1^{-3}(A_2 - 3B_1) + O(n^{-3}), \\
\kappa_4 &= n^{-3}A_1^{-5}[-12B_1(A_2 - 2B_1) + A_1(A_3 - 4D_1) - 3A_1^3] + O(n^{-4}),
\end{aligned}
\right\}
\tag{18.42}
$$

whence

$$
\left.
\begin{aligned}
\gamma_1 &= \kappa_3 / \kappa_2^{3/2} = n^{-1/2} A_1^{-3/2}(A_2 - 3B_1) + o(n^{-1/2}), \\
\gamma_2 &= \kappa_4 / \kappa_2^2 = n^{-1} A_1^{-3}[-12B_1(A_2 - 2B_1) + A_1(A_3 - 4D_1) - 3A_1^3] + o(n^{-1}).
\end{aligned}
\right\}
\tag{18.43}
$$

The first cumulant in (18.42) shows that the bias in $\hat{\theta}$ is of the order of magnitude n^{-1} unless $B_1 = 0$, when it is of order n^{-2}, as may be confirmed by calculating a further term in the first cumulant. If $k \to \infty$, $A_i \to E\{(\partial \log f / \partial \theta)^{i+1}\}$ and so on, and thus the leading term in the second cumulant corresponds to the asymptotic variance previously established in **18.16**. (18.43) illustrates the tendency to normality, established in **18.16** for ungrouped observations.

If the terms in (18.42) were all evaluated, and unbiased estimates made of each of the first four moments of $\hat{\theta}$, a Pearson distribution or Edgeworth expansion could be fitted and an estimate of the small-sample distribution of $\hat{\theta}$ obtained which would provide a better approximation than the ultimate normal approximation of **18.16**. Pfanzagl (1973) gives an Edgeworth expansion, to improve the normal approximation in the continuous case, which is equivalent to using the cumulants (18.42).

> Winterbottom (1979) shows how to use (18.42) in a Cornish–Fisher expansion to find approximate confidence intervals for θ.
>
> The next higher-order terms in (18.42–3) are derived by Shenton and Bowman (1963).

Peers (1978) obtained the asymptotic joint cumulants of

$$n^{1/2}\hat{\theta} \qquad \text{and} \qquad t_r = n^{-r/2}(\partial^r \log L/\partial\theta^r)_{\theta=\hat{\theta}}, \quad r = 2, 3, 4,$$

in terms of $n^{-s/2}\partial^s \log L/\partial\theta^s$, $s = 1, 2, 3, 4$. The set of statistics $(\hat{\theta}, t_2, t_3, t_4)$ is 'second-order sufficient' for θ, since the Taylor expansion of $\log L$ about $\hat{\theta}$ contains only these statistics to the second order of approximation in n.

Successive approximation to ML estimators

18.21   In most of the examples we have considered, the ML estimator has been obtained in explicit form. An exception was in Example 18.3, where we were left with a cubic equation to solve for the estimator, and this can be done without much trouble when the sample is given. Sometimes, however, the likelihood equation is so complicated that iterative methods must be used to find a root, starting from some trial value t.

As at (18.31), we expand $\partial \log L/\partial\theta$ in a Taylor series, but this time about its value at t, obtaining

$$0 = \left(\frac{\partial \log L}{\partial\theta}\right)_{\hat{\theta}} = \left(\frac{\partial \log L}{\partial\theta}\right)_{t} + (\hat{\theta} - t)\left(\frac{\partial^2 \log L}{\partial\theta^2}\right)_{\theta*},$$

where $\theta*$ lies between $\hat{\theta}$ and t. Thus

$$\hat{\theta} = t - \left(\frac{\partial \log L}{\partial\theta}\right)_{t} \Big/ \left(\frac{\partial^2 \log L}{\partial\theta^2}\right)_{\theta*}. \tag{18.44}$$

If we can choose t so that it is likely to be in the neighbourhood of $\hat{\theta}$, we can replace $\theta*$ in (18.44) by t and obtain

$$\hat{\theta} = t - \left(\frac{\partial \log L}{\partial\theta}\right)_{t} \Big/ \left(\frac{\partial^2 \log L}{\partial\theta^2}\right)_{t}, \tag{18.45}$$

which will give a closer approximation to $\hat{\theta}$. The process can be repeated until no further correction is achieved to the desired degree of accuracy.

The most common method for choice of t is to take it as the value of some (preferably simply calculated) consistent estimator of θ, ideally one with high efficiency, so that, by (17.61), it will be highly correlated with the efficient $\hat{\theta}$. Then, as $n \to \infty$, we shall have the two consistent estimators t and $\hat{\theta}$ converging to θ_0, and $\theta*$ consequently also doing so. The three random variables $(\partial^2 \log L/\partial\theta^2)_{\theta*}$, $(\partial^2 \log L/\partial\theta^2)_{t}$ and $[E(\partial^2 \log L/\partial\theta^2)]_{t}$ will all converge to $[E(\partial^2 \log L/\partial\theta^2)]_{\theta_0}$. Use of the second of these variables, instead of the first, in (18.44) gives (18.45) above: use of the third instead of the first gives the alternative iterative procedure

$$\hat{\theta} = t - \left(\frac{\partial \log L}{\partial \theta}\right)_t \bigg/ \left[E\left(\frac{\partial^2 \log L}{\partial \theta^2}\right)\right]_t = t + \left(\frac{\partial \log L}{\partial \theta}\right)_t (\text{var } \hat{\theta})_t, \qquad (18.46)$$

var $\hat{\theta}$ being the asymptotic variance obtained in **18.16**. (18.45) is the Newton–Raphson iterative process: (18.46) is sometimes known as 'the method of scoring for parameters', due to Fisher (1925), because $(\partial \log L/\partial \theta)_{\hat{\theta}}$ is called the score function. Kale (1961) shows that although (18.45) ultimately converges faster, (18.46) will often give better results for the first few iterations when n is large. It is usually less laborious.

Both (18.45) and (18.46) may fail to converge in particular cases. Even when they do converge, if the likelihood equation has multiple roots there is no guarantee that they will converge to the root corresponding to the absolute maximum of the LF; this should be verified by examining the changes in sign of $\partial \log L/\partial \theta$ from positive to negative and searching the intervals in which these changes occur to locate, evaluate and compare the maxima. V. D. Barnett (1966a) discusses a systematic method of doing this.

It should be observed that the first iterate obtained from either (18.46) or (18.45) is itself an estimator of θ, with the same asymptotic properties as $\hat{\theta}$ itself, for the correction term is of order $n^{-1/2}$, as may be seen from (17.18–19).

Example 18.9
To estimate the location parameter θ in the Cauchy distribution

$$dF(x) = \frac{dx}{\pi\{1 + (x - \theta)^2\}}, \qquad -\infty < x < \infty.$$

The likelihood equation is

$$\frac{\partial \log L}{\partial \theta} = 2 \sum_{i=1}^{n} \frac{(x_i - \theta)}{\{1 + (x_i - \theta)^2\}} = 0,$$

an equation of degree $(2n - 1)$ in θ.

Since $L \to 0$ as $\theta \to \pm\infty$, r of the real roots of the likelihood equation must be at maxima of the LF, alternating with $(r - 1)$ at minima, where $1 \le r \le n$.

For $n = 1$, the equation is linear, and the unique maximum of the LF is at $\hat{\theta} = x_1$. But even for $n = 2$, a problem arises. We assume $x_1 < x_2$ and write $M = \frac{1}{2}(x_1 + x_2)$, $R = x_2 - x_1$. The solutions of the likelihood equation are the real roots of the cubic

$$P = (x_1 - \theta)\{1 + (x_2 - \theta)^2\} + (x_2 - \theta)\{1 + (x_1 - \theta)^2\}$$
$$= 2(M - \theta)\{1 + (x_1 - \theta)(x_2 - \theta)\} \equiv 2LQ,$$

say. The linear factor L changes from positive to negative at its unique root $\theta = M$, while the quadratic Q has real roots only if $R \ge 2$, when they are $\theta = M \pm \{(R/2)^2 - 1\}^{1/2}$, coinciding at M when $R = 2$. If $R < 2$, $Q > 0$, so P has the sign of $(M - \theta)$, and thus there is a unique real root at a maximum

$$\hat{\theta} = M, \qquad R \le 2.$$

If $R > 2$, P has three real roots,

$$\theta_1 = M - \left\{ \left(\frac{R}{2} \right)^2 - 1 \right\}^{1/2}, \qquad \theta_2 = M, \qquad \theta_3 = M + \left\{ \left(\frac{R}{2} \right)^2 - 1 \right\}^{1/2},$$

and Q changes sign from positive to negative at θ_1, and back to positive at θ_3. Thus

$$P \begin{cases} >0, & \theta < \theta_1, \\ <0, & \theta_1 < \theta < \theta_2, \\ >0, & \theta_2 < \theta < \theta_3, \\ <0, & \theta_3 < \theta, \end{cases} \Bigg\} \, R > 2$$

and $\theta_2 = M$ is at a local minimum since it is the middle root. Further, it is easily verified that $L(x \mid \theta_1) \equiv L(x \mid \theta_3)$; the two maxima of the LF are *always* equal, and we have no reason for choosing between them. Unlike the cases in Example 18.7 and Exercises 18.17, 18.33, the equal maxima of the LF are separated by an interval that may be arbitrarily large.

For $n \geqslant 3$, we must iterate. From **18.16** the asymptotic variance of $\hat{\theta}$ is given by

$$-\frac{1}{\text{var } \hat{\theta}} \sim E \left(\frac{\partial^2 \log L}{\partial \theta^2} \right) = nE \left(\frac{\partial^2 \log f}{\partial \theta^2} \right)$$

$$= \frac{n}{\pi} \int_{-\infty}^{\infty} \frac{2(x - \theta)^2 - 2}{\{1 + (x - \theta)^2\}^3} \, dx$$

$$= \frac{4n}{\pi} \int_0^{\infty} \frac{(x^2 - 1)}{(1 + x^2)^3} \, dx$$

$$= -n/2.$$

Hence

$$\text{var } \hat{\theta} = 2/n.$$

For small n, (18.45) or (18.46) may not converge—cf. V. D. Barnett (1966a). For $n \geqslant 9$, however, $\hat{\theta}$ is almost always—cf. Haas *et al.* (1970)—the nearest maximum to the sample median t, which has large-sample variance (Example 17.5) $\text{var } t = \pi^2/(4n)$ and thus has efficiency $8/\pi^2 = 0.81$ approximately. For $n \geqslant 15$, we may confidently use the median as our starting point in seeking the value of $\hat{\theta}$, and solve (18.46), which here becomes

$$\hat{\theta} = t + \frac{4}{n} \sum_i \frac{(x_i - t)}{\{1 + (x_i - t)^2\}}.$$

This is our first approximation to $\hat{\theta}$, which we may improve by further iterations of the process.

Reeds (1985) shows that $r - 1$ is asymptotically a Poisson variate with parameter π^{-1}, so that there is a unique maximum ($r = 1$) in about 70 per cent of samples, as Barnett (1966a) had verified empirically—even for $n = 3$, it was 65 per cent.

More efficient simple starting points for ML iteration are available. Bloch (1966) gives an estimator with efficiency >0.95, namely the linear combination of 5 order-statistics $x_{(r)}$ ($r = 0.13n$, $0.4n$, $0.5n$, $0.6n$, $0.87n$) with weights -0.052, 0.3485, 0.407, 0.3485, -0.052. Rothenberg et al. (1964) show that the mean of the central 24 per cent of a Cauchy sample has asymptotic variance $2.28/n$, its efficiency therefore being 0.88. See also V. D. Barnett (1966b), Balmer et al. (1974), G. J. Cane (1974), Haas et al. (1970) and Chan (1970).

If a scale parameter σ alone is, or both θ and σ are, to be estimated, the LF is generally unimodal, as Exercise 18.39 shows. Exercise 18.40 gives explicit estimators for $n = 3$ and 4.

Example 18.10

We now examine the iterative method of solution in more detail, and for this purpose we use some data due to Fisher (1925, Chapter 9).

Consider a multinomial distribution (cf. **5.49**) with four classes, their probabilities being

$$p_1 = (2 + \theta)/4,$$
$$p_2 = p_3 = (1 - \theta)/4,$$
$$p_4 = \theta/4.$$

The parameter θ, which lies in the range $(0, 1)$, is to be estimated from the observed frequencies (a, b, c, d) falling into the classes, the sample size n being equal to $a + b + c + d$. We have

$$L(a, b, c, d) \mid \theta) \propto (2 + \theta)^a (1 - \theta)^{b+c} \theta^d,$$

so that

$$\frac{\partial \log L}{\partial \theta} = \frac{a}{2 + \theta} - \frac{(b + c)}{1 - \theta} + \frac{d}{\theta},$$

and if this is equated to zero, we obtain the quadratic equation in θ

$$n\theta^2 + \{2(b + c) + d - a\}\theta - 2d = 0.$$

Since the product of the coefficient of θ^2 and the constant term is negative, the product of the roots of the quadratic must also be negative, and only one root can be positive. Only this positive root falls into the permissible range for θ. Its value $\hat{\theta}$ is given by

$$2n\hat{\theta} = \{a - d - 2(b + c)\} + [\{a + 2(b + c) + 3d\}^2 - 8a(b + c)]^{1/2}.$$

The ML estimator $\hat{\theta}$ can very simply be evaluated from this formula. For Fisher's (genetical) example, where the observed frequencies are

$$a = 1997, \quad b = 906, \quad c = 904, \quad d = 32, \quad n = 3839$$

the value of $\hat{\theta}$ is 0.0357.

It is easily verified from a further differentiation that

$$\text{var } \hat{\theta} \sim -\frac{1}{E\left(\dfrac{\partial^2 \log L}{\partial \theta^2}\right)} = \frac{2\theta(1 - \theta)(2 + \theta)}{n(1 + 2\theta)},$$

the value being 0.0000336 in this case, when $\hat{\theta}$ is substituted for θ in var $\hat{\theta}$.

For illustrative purposes, we now suppose that we wish to find $\hat{\theta}$ iteratively in this case, starting from the value of an inefficient estimator. A simple inefficient estimator that was proposed by Fisher is

$$t = \{a + d - (b + c)\}/n,$$

which is easily seen to be consistent and has variance

$$\text{var } t = (1 - \theta^2)/n.$$

The value of t for the genetical data is

$$t = \{1997 + 32 - (906 + 904)\}/3839 = 0.0570.$$

This is a long way from the value of $\hat{\theta}$, 0.0357, which we seek. Using (18.46) we have, for our first approximation to $\hat{\theta}$,

$$\hat{\theta}_1 = 0.0570 + \left(\frac{\partial \log L}{\partial \theta}\right)_{\theta = t} (\text{var } \hat{\theta})_{\theta = t}.$$

Now

$$\left(\frac{\partial \log L}{\partial \theta}\right)_{\theta = 0.0570} = \frac{1997}{2.0570} - \frac{1810}{0.9430} + \frac{32}{0.0570} = -387.1713,$$

$$(\text{var } \hat{\theta})_{\theta = 0.0570} = \frac{2 \times 0.057 \times 0.943 \times 2.057}{3839 \times 1.114} = 0.00005170678,$$

so that our improved estimator is

$$\hat{\theta}_1 = 0.0570 - 387.1713 \times 0.00005170678 = 0.0570 - 0.0200$$
$$= 0.0370,$$

which is in fairly close agreement with the value sought, 0.0357. A second iteration gives

$$\left(\frac{\partial \log L}{\partial \theta}\right)_{\theta = 0.0370} = \frac{1997}{2.037} - \frac{1810}{0.963} + \frac{32}{0.037} = -34.31495,$$

$$(\text{var } \hat{\theta})_{\theta = 0.0370} = \frac{2 \times 0.037 \times 0.963 \times 2.037}{3839 \times 1.074} = 0.00003520681,$$

and hence

$$\hat{\theta}_2 = 0.0370 - 34.31495 \times 0.00003520681 = 0.0370 - 0.0012$$
$$= 0.0358.$$

This is very close to the value sought. At least one further iteration would be required to bring the value to 0.0357 correct to 4 d.p., and a further iteration to confirm that the value of $\hat{\theta}$ arrived at was stable to a sufficient number of decimal places to make further iterations unnecessary. The reader should carry through these further iterations to become familiar with the method, and also verify that use of (18.45) instead of (18.46) gives a much worse value of $\hat{\theta}$, and that it converges much more slowly in this case.

This example makes it clear that care must be taken to carry the iteration process far enough for practical purposes. It is a somewhat unfavourable example, in that t has an efficiency of $2\theta(2+\theta)/[(1+\theta)(1+2\theta)]$, which takes the value of 0.13, or 13 per cent, when $\hat{\theta} = 0.0357$ is substituted for θ. One would usually seek to start from the value of an estimator with greater efficiency than this.

> Exercise 18.35 shows that a single iteration is always enough using (18.46), but not using (18.45), if there is an MVB estimator of θ. See also Exercise 18.36.

ML estimators for several parameters
18.22 We now turn to discussion of the general case, in which more than one parameter are to be estimated simultaneously, whether in a univariate or multivariate distribution. If we interpret θ, and possibly also x, as a vector, the formulation of the ML principle at (18.2) holds good: we have to choose the set of admissible values of the parameters $\theta_1, \ldots, \theta_k$ that makes the LF an absolute maximum. We assume here that the admissible range for any θ_j is independent of that of the other parameters—see **18.38** below for the contrary case. Under the regularity conditions of **18.2–3**, the necessary condition for a local turning point in the LF is that

$$\frac{\partial}{\partial \theta_r} \log L(x \mid \theta_1, \ldots, \theta_k) = 0, \qquad r = 1, 2, \ldots, k, \tag{18.47}$$

and a sufficient condition that this be a maximum is that when (18.47) holds the matrix

$$\left(\frac{\partial^2 \log L}{\partial \theta_r \, \partial \theta_s} \right) \tag{18.48}$$

be negative definite. The k equations (18.47) are to be solved for the k ML estimators $\hat{\theta}_1, \ldots, \hat{\theta}_k$.

Just as in **18.3**, if $L \to 0$ as θ tends to its boundary and each set of roots of (18.47) has negative definite (18.48), only one such set can exist, giving a unique maximum— cf. Mäkeläinen *et al.* (1981).

The case of joint sufficiency
18.23 Just as in **18.4**, we see that if there exists a set of s statistics $t_1, \ldots, t_s$ that are jointly sufficient for the parameters $\theta_1, \ldots, \theta_k$, the ML estimators $\hat{\theta}_1, \ldots, \hat{\theta}_k$, if unique, must be functions of the sufficient statistics. As before, this follows immediately from the factorisation at (17.84),

$$L(x \mid \theta_1, \ldots, \theta_k) = g(t_1, \ldots, t_s \mid \theta_1, \ldots, \theta_k) h(x), \tag{18.49}$$

in virtue of the fact that $h(x)$ does not contain $\theta_1, \ldots, \theta_k$.

However, the ML estimators need not be one-to-one functions of the sufficient set of statistics, and are therefore not necessarily themselves a sufficient set. In Example 18.5, we have already met a case where the ML estimator of a single parameter is a function of only one of the jointly sufficient pair of statistics, there being no single sufficient statistic.

18.24   The uniqueness of the solution of the likelihood equation in the presence of sufficiency (**18.5**) extends to the multiparameter case, as Huzurbazar (1949) pointed out. Under regularity conditions, the most general form of distribution admitting a set of k jointly sufficient statistics (17.86) yields an LF whose logarithm is

$$\log L = \sum_{j=1}^{k} A_j(\boldsymbol{\theta}) \sum_{i=1}^{n} B_j(x_i) + \sum_{i=1}^{n} C(x_i) + nD(\boldsymbol{\theta}), \tag{18.50}$$

where $\boldsymbol{\theta}$ is written for $\theta_1, \ldots, \theta_k$ and x is possibly multivariate. The likelihood equations are therefore

$$\frac{\partial \log L}{\partial \theta_r} = \sum_j \frac{\partial A_j}{\partial \theta_r} \sum_i B_j(x_i) + n \frac{\partial D}{\partial \theta_r} = 0, \qquad r = 1, 2, \ldots, k, \tag{18.51}$$

and a solution $\hat{\boldsymbol{\theta}} = (\hat{\theta}_1, \hat{\theta}_2, \ldots, \hat{\theta}_k)$ of (18.51) is a maximum if

$$\left(\frac{\partial^2 \log L}{\partial \theta_r \partial \theta_s} \right)_{\hat{\theta}} = \sum_j \left(\frac{\partial^2 A_j}{\partial \theta_r \partial \theta_s} \right)_{\hat{\theta}} \sum_i B_j(x_i) + n \left(\frac{\partial^2 D}{\partial \theta_r \partial \theta_s} \right)_{\hat{\theta}} \tag{18.52}$$

forms a negative definite matrix (18.48).

From (17.18) we have

$$E \left(\frac{\partial \log L}{\partial \theta_r} \right) = \sum_j \frac{\partial A_j}{\partial \theta_r} E \left(\sum_i B_j(x_i) \right) + n \frac{\partial D}{\partial \theta_r} = 0, \tag{18.53}$$

and further

$$E \left(\frac{\partial^2 \log L}{\partial \theta_r \partial \theta_s} \right) = \sum_j \frac{\partial^2 A_j}{\partial \theta_r \partial \theta_s} E \left(\sum_i B_j(x_i) \right) + n \frac{\partial^2 D}{\partial \theta_r \partial \theta_s}. \tag{18.54}$$

On their right sides, (18.53) and (18.54) have exactly the same form as (18.51) and (18.52), the difference being only that $T = \sum_i B_j(x_i)$ is replaced by its expectation and $\hat{\theta}$ by the true value θ. If we eliminate T from (18.52), using (18.51), and replace $\hat{\theta}$ by θ, we shall get exactly the same result as if we eliminate $E(T)$ from (18.54), using (18.53). We thus have

$$\left(\frac{\partial^2 \log L}{\partial \theta_r \partial \theta_s} \right)_{\hat{\theta}=\theta} = E \left(\frac{\partial^2 \log L}{\partial \theta_r \partial \theta_s} \right), \tag{18.55}$$

which is the generalisation of (18.35). Moreover, from (17.19),

$$E \left(\frac{\partial^2 \log L}{\partial \theta_r^2} \right) = -E \left\{ \left(\frac{\partial \log L}{\partial \theta_r} \right)^2 \right\}, \tag{18.56}$$

and from (17.18) we find analogously

$$E \left(\frac{\partial^2 \log L}{\partial \theta_r \partial \theta_s} \right) = -E \left\{ \frac{\partial \log L}{\partial \theta_r} \cdot \frac{\partial \log L}{\partial \theta_s} \right\}, \tag{18.57}$$

so that the matrix

$$-\left\{E\left(\frac{\partial^2 \log L}{\partial \theta_r\, \partial \theta_s}\right)\right\} = \left\{E\left(\frac{\partial \log L}{\partial \theta_r} \cdot \frac{\partial \log L}{\partial \theta_s}\right)\right\}$$

$$= \left\{\operatorname{cov}\left(\frac{\partial \log L}{\partial \theta_r}, \frac{\partial \log L}{\partial \theta_s}\right)\right\} \qquad (18.58)$$

is the covariance matrix **D** of the variates $\partial \log L/\partial \theta_r$, and this is non-negative definite by **15.3**. If we rule out linear dependencies among the variates, **D** is positive definite and the matrix on the left of (18.58) is negative definite. Thus, from (18.55), the matrix

$$\left\{\left(\frac{\partial^2 \log L}{\partial \theta_r\, \partial \theta_s}\right)_{\hat\theta = \theta}\right\}$$

is also negative definite, and hence any solution of (18.51) is a maximum. But under regularity conditions, there must be a minimum between any two maxima. Since there is no minimum, there can be only one maximum. Thus, under regularity conditions, joint sufficiency ensures that the likelihood equations have a unique solution, and that this is at a maximum of the LF.

Foutz (1977) gives a more general proof of uniqueness, together with consistency.

Example 18.11
We have seen in Example 17.17 that in samples from a univariate normal distribution the sample mean and variance, $\bar x$ and s^2, are jointly sufficient for the population mean and variance, μ and σ^2. It follows from **18.23** that the ML estimators must be functions of $\bar x$ and s^2. We may confirm directly that $\bar x$ and s^2 are themselves the ML estimators. The LF is given by

$$\log L = -\tfrac{1}{2}n \log (2\pi) - \tfrac{1}{2}n \log (\sigma^2) - \sum_i (x_i - \mu)^2/(2\sigma^2),$$

whence the likelihood equations are

$$\frac{\partial \log L}{\partial \mu} = \frac{\sum (x - \mu)}{\sigma^2} = \frac{n(\bar x - \mu)}{\sigma^2} = 0,$$

$$\frac{\partial \log L}{\partial(\sigma^2)} = -\frac{n}{2\sigma^2} + \frac{\sum (x - \mu)^2}{2\sigma^4} = 0.$$

The solution of these is

$$\hat\mu = \bar x$$

$$\hat\sigma^2 = \frac{1}{n} \sum (x - \bar x)^2 = s^2.$$

$\hat\mu$ is unchanged from Example 18.2, where σ^2 was known, but $\hat\sigma^2$ differs from Example 18.8 with known mean μ (there taken as origin), where $\hat\sigma^2 = (1/n) \sum (x - \mu)^2$. While $\hat\mu$ is unbiased, $\hat\sigma^2$ is now biased, having expected value $(n-1)\sigma^2/n$. As in the one-parameter case (**18.14**), ML estimators need not be unbiased.

18.25   In the case where the terminals of the range of a distribution depend on more than one parameter, there has not, so far as we know, been any general investigation of the uniqueness of the ML estimator in the presence of sufficient statistics, corresponding to that for the one-parameter case in **18.6**. But if the statistics are individually, as well as jointly, sufficient for the parameters on which the terminals of the range depend, the result of **18.6** obviously holds good, as in the following example.

Example 18.12
From Example 17.21, we see that for the distribution

$$dF(x) = \frac{dx}{\beta - \alpha}, \qquad \alpha \leqslant x \leqslant \beta, \tag{18.59}$$

the extreme observations $x_{(1)}$ and $x_{(n)}$ are a pair of jointly sufficient statistics for α and β. In this case, it is clear that the ML estimators

$$\hat{\alpha} = x_{(1)}, \qquad \hat{\beta} = x_{(n)},$$

maximise the LF uniquely, and the same will be true whenever each terminal of the range of a distribution depends on a different parameter.

The general multiparameter case
18.26   In the general case, where there is not necessarily a set of k sufficient statistics for the k parameters, the joint ML estimators may not exist (cf. Exercises 18.23, 18.34), and we saw in **18.9** that even in the one-parameter case, $\hat{\theta}$ may not be unique when finite. However, $\hat{\theta}$ has similar optimum properties, in large samples, to those in the single-parameter case.

In the first place, we note that the consistency results given in **18.10** and **18.16** hold good for the multiparameter case if we there interpret θ as a vector of parameters $\theta_1, \ldots, \theta_k$ and $\hat{\theta}$, θ^* as vectors of estimators of θ. We therefore have the result that under regularity conditions the joint ML estimators converge in probability, as a set, to the true set of parameter values θ_0.

Further, by an immediate generalisation of the method of **18.16**, we may show (see, e.g., Wald (1943a)) that the joint ML estimators tend, under regularity conditions, to a multivariate normal distribution, with covariance matrix whose inverse is given by

$$(V_{rs}^{-1}) = -E\left(\frac{\partial^2 \log L}{\partial \theta_r \, \partial \theta_s}\right) = E\left(\frac{\partial \log L}{\partial \theta_r} \cdot \frac{\partial \log L}{\partial \theta_s}\right) \tag{18.60}$$

using (18.57). We shall only sketch the essentials of the proof. The analogue of the Taylor expansion of (18.31) becomes, on putting the left-hand side equal to zero,

$$\left(\frac{\partial \log L}{\partial \theta_r}\right)_{\theta_0} = \sum_{s=1}^{k} (\hat{\theta}_s - \theta_{s0})\left(-\frac{\partial^2 \log L}{\partial \theta_r \, \partial \theta_s}\right)_{\theta^*}, \qquad r = 1, 2, \ldots, k. \tag{18.61}$$

Since θ^* is a value converging in probability to θ_0, and the second derivatives on the right-hand side of (18.61) converge in probability to their expectations, we may regard (18.61) as a set of linear equations in the quantities $(\hat{\theta}_r - \theta_{r0})$, which we may rewrite

$$\mathbf{y} = \mathbf{V}^{-1}\mathbf{z} \tag{18.62}$$

where $\mathbf{y} = \partial \log \mathbf{L}/\partial \boldsymbol{\theta}$, $\mathbf{z} = \hat{\boldsymbol{\theta}} - \boldsymbol{\theta}_0$ and $\mathbf{V}^{-1}$ is defined at (18.60).

By the multivariate central limit theorem, the vector $\mathbf{y}$ will tend to be multinormally distributed, with zero mean if (18.29) holds for each θ_r, and hence so will the vector $\mathbf{z}$ be. The covariance matrix of $\mathbf{y}$ is $\mathbf{V}^{-1}$ of (18.60), by definition, so that the exponent of its multinormal distribution will be the quadratic form (cf. **15.3**)

$$-\tfrac{1}{2}\mathbf{y}'\mathbf{V}\mathbf{y}. \tag{18.63}$$

The transformation (18.62) gives the quadratic form for $\mathbf{z}$

$$-\tfrac{1}{2}\mathbf{z}'\mathbf{V}^{-1}\mathbf{z},$$

so that the covariance matrix of $\mathbf{z}$ is $(\mathbf{V}^{-1})^{-1} = \mathbf{V}$, as stated at (18.60). From (17.87–8), we see that each $\hat{\theta}_r$ asymptotically attains the minimum attainable variance, which by Exercise 17.20 is no less than if a single parameter were being estimated.

> Cheng and Amin (1983) show that if, instead of the LF, we maximise the product of the spacings of the order-statistics, transformed as in Exercise 14.14, the asymptotic properties of the estimators are no worse than those of the ML estimators.

18.27 If there is a set of k jointly sufficient statistics for the k parameters, we may use (18.55) in (18.60) to obtain, for the inverse of the covariance matrix of the ML estimators in large samples,

$$(V_{rs}^{-1}) = -\left(\frac{\partial^2 \log L}{\partial \theta_r \, \partial \theta_s}\right)_{\hat{\theta}=\theta}. \tag{18.64}$$

(18.64), which is the generalisation of the result of **18.17**, removes the need to find expectations.

If there is no set of k sufficient statistics, the elements of the covariance matrix may be estimated from the sample by standard methods.

> Kale (1962) considers alternative iterative methods for solving the likelihood equations for several parameters. (18.62) is used with $\boldsymbol{\theta}_0$ replaced by a trial vector $\mathbf{t}$ (cf. **18.21**) so that $\hat{\boldsymbol{\theta}} = \mathbf{t} + \mathbf{V}\mathbf{y}$ may be iterated as often as necessary.
>
> Exercise 18.34, where no ML estimator exists, satisfies the two-parameter generalisation of the conditions of **18.16**, so that there are consistent roots of the likelihood equations nonetheless, and we may use iterative methods as in **18.16**.

18.28 We have seen in **18.16** that under regularity conditions the ML estimator of a single parameter is efficient. Now, in the multiparameter case, consider any linear function

$$l = \mathbf{c}'\hat{\boldsymbol{\theta}} \tag{18.65}$$

of the joint ML estimators $\hat{\boldsymbol{\theta}}$, which have covariance matrix $\mathbf{V}$ as before. We may reformulate the problem in terms of a new set of parameters $\boldsymbol{\phi}$ chosen so that $\phi_1 = \mathbf{c}'\boldsymbol{\theta}$. From the invariance of ML estimators under transformations (cf. **8.10**) we see that $\hat{\phi}_1 = l$. It follows from **18.26** that var $(\hat{\phi}_1)$ is asymptotically minimised, and in particular that

$$\text{var}(\mathbf{c}'\hat{\boldsymbol{\theta}}) \leqslant \text{var}(\mathbf{c}'\mathbf{u}), \qquad (18.66)$$

where $\mathbf{u}$ is any other set of consistent estimators of $\boldsymbol{\theta}$, with covariance matrix $\mathbf{D}$. (18.66) holds for any choice of $\mathbf{c}$.

We may rewrite (18.66) as

$$\mathbf{c}'\mathbf{V}\mathbf{c} \leqslant \mathbf{c}'\mathbf{D}\mathbf{c}.$$

We may simultaneously diagonalise $\mathbf{V}$ and $\mathbf{D}$ by a real non-singular transformation $\mathbf{c} = \mathbf{A}\mathbf{b}$ to give

$$\mathbf{b}'(\mathbf{A}'\mathbf{V}\mathbf{A})\mathbf{b} \leqslant \mathbf{b}'(\mathbf{A}'\mathbf{D}\mathbf{A})\mathbf{b},$$

the bracketed matrices being diagonal. Since $\mathbf{b}$ is arbitrary, suitable choice of its elements ensures that any element of $\mathbf{A}'\mathbf{V}\mathbf{A}$ cannot exceed the corresponding element of $\mathbf{A}'\mathbf{D}\mathbf{A}$. Thus their determinants satisfy

$$|A'VA| \leqslant |A'DA|, \qquad \text{i.e.} \quad |A'||V||A| \leqslant |A'||D||A|,$$

or

$$|V| \leqslant |D|. \qquad (18.67)$$

The determinant of a covariance matrix is called the *generalised variance* of the estimators. Thus the ML estimators minimise the generalised variance in large samples, a result due originally to Geary (1942). Our proof was given by Daniels (1951–52).

Example 18.13
Consider again the ML estimators $\bar{x}$ and s^2 in Example 18.11. We have

$$\frac{\partial^2 \log L}{\partial \mu^2} = -\frac{n}{\sigma^2},$$

$$\frac{\partial^2 \log L}{\partial(\sigma^2)^2} = \frac{n}{2\sigma^4} - \frac{\sum(x-\mu)^2}{\sigma^6}, \qquad \frac{\partial^2 \log L}{\partial\mu\,\partial(\sigma^2)} = -\frac{n(\bar{x}-\mu)}{\sigma^4}.$$

Remembering that the ML estimators $\bar{x}$ and s^2 are sufficient for μ and σ^2, we use (18.64) and obtain the inverse of their covariance matrix in large samples by putting $\bar{x} = \mu$ and $\sum(x-\mu)^2 = n\sigma^2$ in these second derivatives. We find

$$\mathbf{V}^{-1} = \begin{pmatrix} n/\sigma^2 & 0 \\ 0 & n/2\sigma^4 \end{pmatrix}, \qquad \text{so that} \qquad \mathbf{V} = \begin{pmatrix} \sigma^2/n & 0 \\ 0 & 2\sigma^4/n \end{pmatrix}.$$

We see from this that $\bar{x}$ and s^2 are asymptotically normally and *independently* distributed with the variances given. However, we know that the independence property and the normality and variance of $\bar{x}$ are exact for any n (Examples 11.3,

11.12); but the normality property and the variance of s^2 are strictly limiting ones, for we have seen (Example 11.7) that ns^2/σ^2 is distributed exactly like χ^2 with $(n-1)$ degrees of freedom, the variance of s^2 therefore, from (16.5), being exactly $(\sigma^2/n)^2 \cdot 2(n-1) = 2\sigma^4(n-1)/n^2$.

Although μ as well as σ^2 is being estimated here, this variance is smaller than the variance of the ML estimator of σ^2 with μ known, which Example 18.8 (with origin at μ) gives as $\hat{\sigma}^2 = (1/n) \sum (x-\mu)^2$, distributed as a multiple σ^2/n of a χ^2_n variate, with exact variance $2\sigma^4/n$. The difference disappears asymptotically, in accordance with the statement at the end of **18.26**.

18.29   Where a distribution depends on k parameters, we may be interested in estimating any number of them from 1 to k, the others being known. Under regularity conditions, the ML estimators of the parameters concerned will be obtained by selecting the appropriate subset of the k likelihood equations (18.47) and solving them. By the nature of this process, it is not to be expected that the ML estimator of a particular parameter will be unaffected by knowledge of the other parameters of the distribution. The form of the ML estimator depends on the company it keeps, as is made clear by the following example.

Example 18.14
For the binormal distribution

$$dF(x, y) = \frac{dx\, dy}{2\pi\sigma_1\sigma_2(1-\rho^2)^{1/2}} \exp\left[-\frac{1}{2(1-\rho^2)}\left\{\left(\frac{x-\mu_1}{\sigma_1}\right)^2 - 2\rho\left(\frac{x-\mu_1}{\sigma_1}\right)\left(\frac{y-\mu_2}{\sigma_2}\right)\right.\right.$$
$$\left.\left. + \left(\frac{y-\mu_2}{\sigma_2}\right)^2\right\}\right], \qquad -\infty < x, y < \infty; \quad \sigma_1, \sigma_2 > 0; \quad |\rho| < 1$$

we obtain the logarithm of the LF

$$\log L(x, y \mid \mu_1, \mu_2, \sigma_1^2, \sigma_2^2, \rho) = -n \log (2\pi) - \tfrac{1}{2}n\{\log \sigma_1^2 + \log \sigma_2^2 + \log (1-\rho^2)\}$$
$$-\frac{1}{2(1-\rho^2)} \sum \left\{\left(\frac{x-\mu_1}{\sigma_1}\right)^2\right.$$
$$\left. -2\rho\left(\frac{x-\mu_1}{\sigma_1}\right)\left(\frac{y-\mu_2}{\sigma_2}\right) + \left(\frac{y-\mu_2}{\sigma_2}\right)^2\right\}$$

from which the five likelihood equations are

$$\frac{\partial \log L}{\partial \mu_1} = \frac{n}{\sigma_1(1-\rho^2)}\left\{\frac{(\bar{x}-\mu_1)}{\sigma_1} - \rho\frac{(\bar{y}-\mu_2)}{\sigma_2}\right\} = 0,$$

$$\frac{\partial \log L}{\partial \mu_2} = \frac{n}{\sigma_2(1-\rho^2)}\left\{\frac{(\bar{y}-\mu_2)}{\sigma_2} - \rho\frac{(\bar{x}-\mu_1)}{\sigma_1}\right\} = 0,$$

(18.68)

$$\frac{\partial \log L}{\partial(\sigma_1^2)} = -\frac{1}{2\sigma_1^2(1-\rho^2)}\left\{n(1-\rho^2) - \frac{\sum (x-\mu_1)^2}{\sigma_1^2} + \rho\frac{\sum (x-\mu_1)(y-\mu_2)}{\sigma_1\sigma_2}\right\} = 0,$$

$$\frac{\partial \log L}{\partial(\sigma_2^2)} = -\frac{1}{2\sigma_2^2(1-\rho^2)}\left\{n(1-\rho^2) - \frac{\sum (y-\mu_2)^2}{\sigma_2^2} + \rho\frac{\sum (x-\mu_1)(y-\mu_2)}{\sigma_1\sigma_2}\right\} = 0,$$

(18.69)

$$\frac{\partial \log L}{\partial \rho} = \frac{1}{(1-\rho^2)} \left\{ n\rho - \frac{1}{(1-\rho^2)} \left[\rho \left(\frac{\sum (x-\mu_1)^2}{\sigma_1^2} + \frac{\sum (y-\mu_2)^2}{\sigma_2^2} \right) \right. \right.$$
$$\left. \left. - (1+\rho^2) \frac{\sum (x-\mu_1)(y-\mu_2)}{\sigma_1 \sigma_2} \right] \right\} = 0. \tag{18.70}$$

(a) Suppose first that we wish to estimate ρ alone, the other four parameters being known. We then solve (18.70) alone. We have already dealt with this case, in standardised form, in Example 18.3. (18.70) yields a cubic equation for the ML estimator $\hat{\rho}$.

(b) Suppose now that we wish to estimate σ_1^2, σ_2^2 and ρ, μ_1 and μ_2 being known. We notice that the same three statistics that were jointly sufficient for ρ alone in Example 18.3 remain sufficient for these three parameters, but that we cannot, as there, find a pair of sufficient statistics, let alone a single sufficient statistic, for them. We have to solve the three likelihood equations (18.69) and (18.70). Dropping the non-zero factors outside the braces, these equations become, after a slight rearrangement,

$$\left. \begin{aligned} n(1-\rho^2) &= \frac{\sum (x-\mu_1)^2}{\sigma_1^2} - \rho \frac{\sum (x-\mu_1)(y-\mu_2)}{\sigma_1 \sigma_2}, \\ n(1-\rho^2) &= \frac{\sum (y-\mu_2)^2}{\sigma_2^2} - \rho \frac{\sum (x-\mu_1)(y-\mu_2)}{\sigma_1 \sigma_2}, \end{aligned} \right\} \tag{18.71}$$

and

$$n(1-\rho^2) = \frac{\sum (x-\mu_1)^2}{\sigma_1^2} + \frac{\sum (y-\mu_2)^2}{\sigma_2^2} - \frac{1+\rho^2}{\rho} \frac{\sum (x-\mu_1)(y-\mu_2)}{\sigma_1 \sigma_2}. \tag{18.72}$$

If we add the equations in (18.71), and subtract (18.72) from this sum, we have

$$n(1-\rho^2) = \frac{1-\rho^2}{\rho} \frac{\sum (x-\mu_1)(y-\mu_2)}{\sigma_1 \sigma_2}$$

or

$$\rho = \frac{\frac{1}{n} \sum (x-\mu_1)(y-\mu_2)}{\sigma_1 \sigma_2}. \tag{18.73}$$

Substituting (18.73) into (18.71) we obtain

$$\hat{\sigma}_1^2 = \frac{1}{n} \sum (x-\mu_1)^2,$$
$$\hat{\sigma}_2^2 = \frac{1}{n} \sum (y-\mu_2)^2, \tag{18.74}$$

and hence, from (18.73),

$$\hat{\rho} = \frac{\frac{1}{n} \sum (x-\mu_1)(y-\mu_2)}{\hat{\sigma}_1 \hat{\sigma}_2}. \tag{18.75}$$

In this case, therefore, all three ML estimators use the known population means, and $\hat{\rho}$ is the sample correlation coefficient calculated about them.

(c) Finally, suppose that we wish to estimate all five parameters of the distribution. We solve (18.68), (18.69) and (18.70) together. (18.68) reduces to

$$\left. \begin{array}{l} \dfrac{(\bar{x} - \mu_1)}{\sigma_1} = \rho \dfrac{(\bar{y} - \mu_2)}{\sigma_2}, \\[3mm] \dfrac{(\bar{y} - \mu_2)}{\sigma_2} = \rho \dfrac{(\bar{x} - \mu_1)}{\sigma_1}, \end{array} \right\} \tag{18.76}$$

a pair of equations whose only solution is

$$\bar{x} - \mu_1 = \bar{y} - \mu_2 = 0. \tag{18.77}$$

Taken with (18.74) and (18.75), which are the solutions of (18.69) and (18.70), (18.77) gives for the set of five ML estimators

$$\left. \begin{array}{ll} \hat{\mu}_1 = \bar{x}, & \hat{\sigma}_1^2 = \dfrac{1}{n} \sum (x - \bar{x})^2, \\[3mm] \hat{\mu}_2 = \bar{y}, & \hat{\sigma}_2^2 = \dfrac{1}{n} \sum (y - \bar{y})^2, \end{array} \quad \hat{\rho} = \dfrac{(1/n) \sum (x - \bar{x})(y - \bar{y})}{\hat{\sigma}_1 \hat{\sigma}_2}. \right\} \tag{18.78}$$

Thus the ML estimators of all five parameters are the corresponding sample moments and its correlation coefficient. They are jointly sufficient statistics, as is easily verified from the LF.

> Exercise 18.14 treats other combinations of parameters of the binormal distribution.

18.30 Since the ML estimator of a parameter is a different function of the observations, according to which of the other parameters of the distribution is known, its large-sample variance will also differ. To facilitate the evaluation of the covariance matrices of ML estimators, we recall that if a distribution admits a set of k sufficient statistics for its k parameters, we may avail ourselves of the form (18.64) for the inverse of the covariance matrix of the ML estimators in large samples.

Example 18.15
We now evaluate the large-sample covariance matrices of the ML estimators in each of the three cases considered in Example 18.14.

(a) When we are estimating ρ alone, $\hat{\rho}$ is not sufficient. But we have already evaluated its large-sample variance in Example 18.6, finding

$$\operatorname{var} \hat{\rho} = \frac{(1 - \rho^2)^2}{n(1 + \rho^2)}. \tag{18.79}$$

The fact that we were there dealing with a standardised distribution is irrelevant, since ρ is invariant under changes of location and scale.

(b) In estimating the three parameters σ_1^2, σ_2^2 and ρ, the three ML estimators given by (18.74) and (18.75) are jointly sufficient, and we therefore make use of (18.64). Writing the parameters in the order above, we find for the 3×3 inverse covariance matrix

$$\mathbf{V}_3^{-1} = (V_{rs}^{-1}) = -\left(\frac{\partial^2 \log L}{\partial \theta_r \, \partial \theta_s}\right)_{\hat{\theta}=\theta}$$

$$= \frac{n}{(1-\rho^2)}\begin{pmatrix} \dfrac{2-\rho^2}{4\sigma_1^4} & \dfrac{-\rho^2}{4\sigma_1^2\sigma_2^2} & \dfrac{-\rho}{2\sigma_1^2} \\[2mm] \dfrac{-\rho^2}{4\sigma_1^2\sigma_2^2} & \dfrac{2-\rho^2}{4\sigma_2^4} & \dfrac{-\rho}{2\sigma_2^2} \\[2mm] \dfrac{-\rho}{2\sigma_1^2} & \dfrac{-\rho}{2\sigma_2^2} & \dfrac{1+\rho^2}{1-\rho^2} \end{pmatrix} \tag{18.80}$$

Inversion of (18.80) gives for the large-sample covariance matrix

$$\mathbf{V}_3 = \frac{1}{n}\begin{pmatrix} 2\sigma_1^4 & 2\rho^2\sigma_1^2\sigma_2^2 & \rho(1-\rho^2)\sigma_1^2 \\ 2\rho^2\sigma_1^2\sigma_2^2 & 2\sigma_2^4 & \rho(1-\rho^2)\sigma_2^2 \\ \rho(1-\rho^2)\sigma_1^2 & \rho(1-\rho^2)\sigma_2^2 & (1-\rho^2)^2 \end{pmatrix}. \tag{18.81}$$

Only if $\rho = 0$ does (18.81) become diagonal, when the three ML estimators are uncorrelated and asymptotically independent because of their approach to normality.

(c) In estimating all five parameters, the ML estimators (18.78) are a sufficient set. Moreover, the 3×3 matrix $\mathbf{V}_3^{-1}$ at (18.80) will form part of the 5×5 inverse variance matrix $\mathbf{V}_5^{-1}$ which we now seek. Writing the parameters in the order μ_1, μ_2, σ_1^2, σ_2^2, ρ, (18.80) will be the lower 3×3 principal minor of $\mathbf{V}_5^{-1}$. For the elements involving derivatives with respect to μ_1 and μ_2, we find from (18.68)

$$\frac{\partial^2 \log L}{\partial \mu_1^2} = \frac{-n}{\sigma_1^2(1-\rho^2)}, \qquad \frac{\partial^2 \log L}{\partial \mu_1 \, \partial \mu_2} = \frac{n\rho}{\sigma_1\sigma_2(1-\rho^2)}, \qquad \frac{\partial^2 \log L}{\partial \mu_2^2} = \frac{-n}{\sigma_2^2(1-\rho^2)}, \tag{18.82}$$

while

$$\frac{\partial^2 \log L}{\partial \mu_i \, \partial \sigma_1^2} = \frac{\partial^2 \log L}{\partial \mu_i \, \partial \sigma_2^2} = \frac{\partial^2 \log L}{\partial \mu_i \, \partial \rho} = 0, \qquad i = 1, 2 \tag{18.83}$$

at $\bar{x} = \mu_1$, $\bar{y} = \mu_2$. Thus if we write, for the inverse of the covariance matrix of the ML estimators of μ_1 and μ_2,

$$\mathbf{V}_2^{-1} = \frac{n}{(1-\rho^2)}\begin{pmatrix} \dfrac{1}{\sigma_1^2} & \dfrac{-\rho}{\sigma_1\sigma_2} \\[2mm] \dfrac{-\rho}{\sigma_1\sigma_2} & \dfrac{1}{\sigma_2^2} \end{pmatrix}, \tag{18.84}$$

we have

$$\mathbf{V}_5^{-1} = \begin{pmatrix} \mathbf{V}_2^{-1} & \mathbf{0} \\ \mathbf{0} & \mathbf{V}_3^{-1} \end{pmatrix}, \tag{18.85}$$

and we may invert $\mathbf{V}_2^{-1}$ and $\mathbf{V}_3^{-1}$ separately to obtain the non-zero elements of the inverse of $\mathbf{V}_5^{-1}$. We have already inverted $\mathbf{V}_3^{-1}$ at (18.81). The inverse of $\mathbf{V}_2^{-1}$ is, from (18.84),

$$\mathbf{V}_2 = \frac{1}{n} \begin{pmatrix} \sigma_1^2 & \rho\sigma_1\sigma_2 \\ \rho\sigma_1\sigma_2 & \sigma_2^2 \end{pmatrix}, \tag{18.86}$$

so

$$\mathbf{V}_5 = \begin{pmatrix} \mathbf{V}_2 & \mathbf{0} \\ \mathbf{0} & \mathbf{V}_3 \end{pmatrix}, \tag{18.87}$$

with $\mathbf{V}_2$ and $\mathbf{V}_3$ defined at (18.86) and (18.81).

We see from this result, what we have already observed (cf. **16.25**) to be true for any sample size, that the sample means are distributed independently of the variances and covariance in bivariate normal samples, and that the correlation between the sample means is ρ; and that the correlation between the sample variances is ρ^2 (Example 13.1).

It follows from the changes in covariance matrices with the set of parameters being estimated, that if an inappropriate ML estimator is used, it may be inefficient— Exercise 18.12 treats the binormal correlation coefficient.

> Shenton and Bowman (1977) give methods of evaluating the performance of ML estimators in small samples, using series expansions.

Non-identical distributions

18.31 We have now largely completed our general survey of the ML method. Throughout our discussions so far, we have been considering problems of estimation when all the observations come from the same underlying distribution. We now briefly examine the behaviour of ML estimators when this condition no longer holds. In fact, we replace (18.1) by the more general LF

$$L(x \mid \theta_1, \ldots, \theta_k) = f_1(x_1 \mid \theta) f_2(x_2 \mid \theta) \cdots f_n(x_n \mid \theta), \tag{18.88}$$

where the different factors f_i on the right of (18.88) depend on possibly different functions of the set of parameters $\theta_1, \ldots, \theta_k$.

It is not now necessarily true even that ML estimators are consistent—A. P. Basu and Ghosh (1980) give sufficient conditions for asymptotic normality and consistency. In one particular class of cases, in which the number of parameters increases with the number of observations (k being a function of n), the ML method may become ineffective as in Example 18.16 below.

Example 18.16
(a) Suppose that x_i is a normal variate with mean θ_i and variance $\sigma^2 > 0$ ($i = 1, 2, \ldots, n$). The LF (18.88) gives

$$\log L = -\tfrac{1}{2} n \log (2\pi) - \tfrac{1}{2} n \log (\sigma^2) - \frac{1}{2\sigma^2} \sum_i (x_i - \theta_i)^2. \tag{18.89}$$

The final term in (18.89) has its numerator reduced to zero if we put $\theta_i = x_i$, all i, so that as $\sigma^2 \to 0$ the preceding term makes $\log L \to \infty$. Thus no ML estimators of (θ_i, σ^2) exist. Nonetheless, we can estimate the θ_i since $E(x) = \theta_i$, while since $E(x_i^2) = \sigma^2 + \theta_i^2$, $(1/n) \sum_i x_i^2$ is an upward-biased estimator of σ^2, so that the sample does yield some information concerning σ^2, even if $n = 1$.

(b) The ML estimator exists, but is still seriously deficient if we have two observations from each of the n normal distributions. We then have

$$\log L = -n \log (2\pi) - n \log (\sigma^2) - \frac{1}{2\sigma^2} \sum_{i=1}^{n} \sum_{j=1}^{2} (x_{ij} - \theta_i)^2$$

and

$$\left. \begin{aligned} \hat{\theta}_i &= \tfrac{1}{2}(x_{i1} + x_{i2}) = \bar{x}_i, \\ \hat{\sigma}^2 &= \frac{1}{2n} \sum_{i=1}^{n} \sum_{j=1}^{2} (x_{ij} - \bar{x}_i)^2. \end{aligned} \right\} \tag{18.91}$$

But since

$$E \left\{ \frac{1}{2} \sum_{j=1}^{2} (x_{ij} - \bar{x}_i)^2 \right\} = \tfrac{1}{2}\sigma^2,$$

(cf. Example 17.3) we have from (18.91), for all n,

$$E(\hat{\sigma}^2) = \tfrac{1}{2}\sigma^2,$$

so that $\hat{\sigma}^2$ is not consistent, as Neyman and Scott (1948) pointed out. What has happened is that the small-sample bias of ML estimators (**18.14**) persists in this example as n increases, for the number of parameters also increases with n.

(c) Finally, suppose that we interchange the roles of mean and variance in (a), and consider $n = 2$ independent normal observations x_i with the same mean θ and variances σ_i^2, $i = 1, 2$. As in (a), we see that the ML estimator does not exist by putting $\theta = x_i$ and letting $\sigma_i^2 \to 0$ for either value of i. Here, however, we can estimate not only θ (say, by $t = \tfrac{1}{2}(x_1 + x_2)$), but also the σ_i^2, for

$$E\{x_1(x_1 - x_2)\} = \sigma_1^2, \qquad E\{x_2(x_2 - x_1)\} = \sigma_2^2.$$

The information about the common mean contained in the other observation thus permits us to estimate each variance from only a single observation. These estimators are not free from problems, for one of them must be negative if x_1 and x_2 have the same sign, but they illustrate the fact that all the parameters may be estimated when the ML method will estimate none of them. It is worth adding that we may unbiasedly estimate $\sigma_1^2 + \sigma_2^2$, which is 4 var t, by $(x_1 - x_2)^2$, and there is clearly no difficulty here.

The situation is not essentially changed if $n > 2$, as the reader may verify.

Other situations of the type of Example 18.16 have been discussed in the literature in connection with applications to which they are relevant. We shall discuss these when they arise. Here we need only emphasise that careful investigation of the properties of ML estimators is necessary in non-standard situations—it cannot be assumed that the large-sample optimum properties will persist.

Dependent observations

18.32 Hitherto, we have assumed the observations to be independent. However, in both stochastic processes and in time-series analysis, dependence between successive observations is assumed. That is, the log-likelihood becomes

$$\log L = \sum_{j=1}^{n} \log f_j(x_j / \theta, X_{(j-1)}) \tag{18.92}$$
$$= \sum l_j, \quad \text{say},$$

where $X_{(j-1)} = (x_1, \ldots, x_{j-1})$ denotes the previous history. $X_{(j-1)}$ might also include starting values $(x_0, x_{-1}, \ldots)$ but we define $\log L$ conditional upon these. Provided the number of initial values is finite, the asymptotic arguments are not affected.

The ML estimator for θ remains consistent under the conditions given in **18.16** plus the requirement that

$$\lim_{|t| \to \infty} \text{cov} \, (l_j, l_{j+t}) = 0; \tag{18.93}$$

that is, we require the dependence between observations to diminish sufficiently quickly. Furthermore, asymptotic normality can be demonstrated under the additional conditions that

$$\lim_{|t| \to \infty} \text{cov} \left[\frac{\partial l_j}{\partial \theta}, \frac{\partial l_{j+t}}{\partial \theta} \right] = 0 \tag{18.94}$$

and

$$\lim_{n \to \infty} \frac{1}{n} \sum_{j=1}^{n} \frac{\partial^2 l_j}{\partial \theta^2} > 0. \tag{18.95}$$

When θ is a vector, the matrix in (18.95) must be non-singular. These conditions can be stated more formally in terms of *strong mixing* conditions; see Lehmann (1983b).

Example 18.17

Consider the two-state Markov chain with transition probabilities

$$P(X_j = r \mid X_{j-1} = s) = q_{sr}; \qquad r = 0, 1; \qquad s = 0, 1$$

usually written as the matrix of transition probabilities

$$Q = \begin{bmatrix} q_{00} & q_{01} \\ q_{10} & q_{11} \end{bmatrix}. \tag{18.96}$$

Let

$$Z_j(s, r) = 1, \quad (\text{if } X_j = r, X_{j-1} = s)$$
$$= 0, \quad \text{otherwise},$$

so that the log-likelihood (18.92) is

$$\log L = \sum_{j=1}^{n} (\sum \sum Z_j(s, r) \log q_{sr}) \tag{18.97}$$

which reduces to

$$Z(0, 0) \log q_{00} + Z(0, 1) \log q_{01} + Z(1, 0) \log q_{10} + Z(1, 1) \log q_{11}, \qquad (18.98)$$

where

$$Z(s, r) = \sum_{j=1}^{n} Z_j(s, r).$$

It is clear from (18.98) that the $Z(s, r)$ are jointly sufficient for the q_{sr}. Further, it follows directly that the ML estimators are

$$\hat{q}_{s1} = \frac{Z(s, 1)}{Z(s, 0) + Z(s, 1)}, \qquad s = 0, 1. \qquad (18.99)$$

The covariances in (18.93) and (18.94) both depend upon the covariances between $Z_j(s, r)$ and $Z_{j+t}(s', r')$. By way of example we consider $s = s' = 1$, $r = r' = 1$. It suffices to consider $j = 1$ and to condition on the initial event $X_0 = 1$. Then

$$P[Z_1(1, 1) = 1] = q_{11}$$
$$P[Z_{t+1}(1, 1) = 1] = P(X_{t+1} = 1 | X_t = 1) P(X_t = 1 | X_0 = 1)$$
$$= q_{11} q_{11}^{(t)}$$

where $q_{11}^{(t)}$ is the $(1, 1)$th element of Q^t. Finally,

$$P[Z(1, 1) = 1, Z_{t+1}(1, 1) = 1]$$
$$= P(X_{t+1} = 1 | X_t = 1) P(X_t = 1 | X_1 = 1) P(X_1 = 1 | X_0 = 1)$$
$$= q_{11}^2 q_{11}^{(t-1)},$$

so that the covariance is

$$q_{11}^2 [q_{11}^{(t-1)} - q_{11}^{(t)}].$$

It is readily established that this tends to zero as $t \to \infty$ provided that $0 < q_{sr} < 1$, all s, r; that is, the Markov chain must be *ergodic*. Finally, we have from (18.98) that

$$\frac{\partial^2 \log L}{\partial q_{s1}^2} = \frac{-Z(s, 0)}{(1 - q_{s1})^2} - \frac{Z(s, 1)}{q_{s1}^2}, \qquad s = 0, 1,$$

so that, conditionally upon $Z(s, 0) + Z(s, 1) = n_s$,

$$-E\left(\frac{\partial^2 \log L}{\partial q_{s1}^2}\right) = \frac{n_s}{q_{s1}(1 - q_{s1})}.$$

Since $\partial^2 \log L / \partial q_{01} \partial q_{11} = 0$, conditions (18.95) are satisfied and $\hat{q}_{s1}$ is asymptotically $N(q_{s1}, q_{s1}(1 - q_{s1})/n_s)$. (For a general discussion of estimation in stochastic processes, see Billingsley (1961) and for time-series, Fuller (1976)).

ML estimation of location and scale parameters
18.33 We may use ML methods to solve, following Fisher (1921a), the problem of finding efficient estimators of location and scale parameters for any given form of distribution.

Consider a frequency function

$$dF(x) = f\left(\frac{x-\alpha}{\beta}\right) d\left(\frac{x-\alpha}{\beta}\right), \qquad \beta > 0. \tag{18.100}$$

The parameter α locates the distribution and β is a scale parameter. We rewrite (18.100) as

$$dF = \exp\{g(y)\} \, dy = \exp\{g(y)\} \, dx/\beta, \tag{18.101}$$

where

$$y = (x-\alpha)/\beta, \qquad g(y) = \log f(y).$$

In samples of size n, the LF is

$$\log L(x \mid \alpha, \beta) = \sum_{i=1}^{n} g(y_i) - n \log \beta. \tag{18.102}$$

(18.102) yields the likelihood equations

$$\left.\begin{aligned} \frac{\partial \log L}{\partial \alpha} &= -\frac{1}{\beta} \sum_i g'(y_i) = 0, \\ \frac{\partial \log L}{\partial \beta} &= -\frac{1}{\beta} \left\{ \sum_i y_i g'(y_i) + n \right\} = 0, \end{aligned}\right\} \tag{18.103}$$

where $g'(y) = \partial g(y)/\partial y$. Under regularity conditions, solution of (18.103) gives the ML estimators $\hat{\alpha}$ and $\hat{\beta}$.

If, in $\hat{y} = (x - \hat{\alpha})/\hat{\beta}$, we replace x by $(x - \alpha)/\beta$, $\hat{\alpha}$ by $(\hat{\alpha} - \alpha)/\beta$ and $\hat{\beta}$ by $\hat{\beta}/\beta$, $\hat{y}$ is unaffected. Hence, whatever the true values (α, β), the solutions of (18.103) will be the same. Thus the joint distribution of the variables $((\hat{\alpha} - \alpha)/\beta, \hat{\beta}/\beta)$, and that of their ratio $(\hat{\alpha} - \alpha)/\hat{\beta}$, cannot depend upon the parameters (α, β)—cf. Haas *et al.* (1970)—but can be functions of n alone. This gives a direct method of removing the biases in $\hat{\alpha}$ and $\hat{\beta}$, for if

$$E\left(\frac{\hat{\alpha} - \alpha}{\beta}\right) = a(n), \qquad E\left(\frac{\hat{\beta}}{\beta}\right) = b(n),$$

we find at once

$$E\left(\frac{\hat{\beta}}{b(n)}\right) = \beta, \qquad E\left(\hat{\alpha} - \frac{a(n)\hat{\beta}}{b(n)}\right) = \alpha.$$

We now assume that for all permissible values of α and β, (17.18) holds, so that

$$\left.\begin{aligned} E\left(\frac{\partial \log L}{\partial \alpha}\right) &= -\frac{n}{\beta} E\{g'(y)\} = 0, \\ E\left(\frac{\partial \log L}{\partial \beta}\right) &= -\frac{n}{\beta} [E\{yg'(y)\} + 1] = 0, \end{aligned}\right\} \tag{18.104}$$

and that (17.19) and its generalisation (18.57) also hold. We may rewrite (18.104) as

$$E\{g'(y)\} = 0, \tag{18.105}$$
$$E\{yg'(y)\} = -1. \tag{18.106}$$

We now evaluate the elements of the inverse covariance matrix (18.60). From (18.103), dropping the argument of $g(y)$, we have

$$E\left(\frac{\partial^2 \log L}{\partial \alpha^2}\right) = \frac{n}{\beta^2} E(g''), \tag{18.107}$$

$$E\left(\frac{\partial^2 \log L}{\partial \beta^2}\right) = \frac{n}{\beta^2} E(g''y^2 + 2g'y + 1),$$

which on using (18.106) becomes

$$E\left(\frac{\partial^2 \log L}{\partial \beta^2}\right) = \frac{n}{\beta^2} E(g''y^2 - 1). \tag{18.108}$$

Also

$$E\left(\frac{\partial^2 \log L}{\partial \alpha \, \partial \beta}\right) = \frac{n}{\beta^2} E(g' + g''y),$$

which from (18.105) becomes

$$E\left(\frac{\partial^2 \log L}{\partial \alpha \, \partial \beta}\right) = \frac{n}{\beta^2} E(g''y). \tag{18.109}$$

(18.107–9) give, for the matrix (18.60),

$$\mathbf{V}^{-1} = -\frac{n}{\beta^2} E\begin{pmatrix} g'' & g''y \\ g''y & g''y^2 - 1 \end{pmatrix}, \tag{18.110}$$

from which the variances and covariance may be determined by inversion. Of course, if α or β alone is being estimated, the variance of the ML estimator will be the reciprocal of the appropriate term in the leading diagonal of $\mathbf{V}^{-1}$ and by the first sentence of Exercise 17.20 cannot be larger than when both parameters are being estimated.

18.34   If $g(y)$ is an even function of its argument, i.e. the distribution is symmetric about α, (18.110) simplifies. For then

$$g(y) = g(-y),$$
$$g'(y) = -g'(-y),$$
$$g''(y) = g''(-y). \tag{18.111}$$

Using (18.111), we see that the off-diagonal term in (18.110)

$$E\{g''(y)y\} = 0, \tag{18.112}$$

so that (18.110) is a diagonal matrix for symmetric distributions. Hence the ML estimators of the location and scale parameters of any symmetric distribution obeying our regularity conditions will be asymptotically uncorrelated, and (since they are asymptotically bivariate normally distributed by **18.26**) asymptotically *independent*. In particular, this applies to the normal distribution, for which we have already derived this result directly in Example 18.13.

18.35 Even for asymmetrical distributions, we can make the off-diagonal term in (18.110) zero by a simple change of origin. Put

$$z = y - \frac{E(g''y)}{E(g'')}. \tag{18.113}$$

Then

$$B2E(g''y) = E\left\{ g''\left[z + \frac{E(g''y)}{E(g'')} \right] \right\}$$
$$= E(g''z) + E(g''y),$$

so that

$$E(g''z) = 0. \tag{18.114}$$

Thus if instead of y we use z as in (18.113), we reduce (18.110) to a diagonal matrix, and we obtain the variances of the estimators easily by taking the reciprocals of the terms in the diagonal.

The variance of $\hat{\beta}$ is unaffected by the change of origin since β is a scale parameter. The origin that makes the estimators uncorrelated is called the *centre of location* of the distribution. A consequence of choosing the centre of location as origin is that, where iterative procedures are necessary, the estimation process may be simpler. Parameters whose estimators are uncorrelated (orthogonal) are sometimes themselves called orthogonal parameters.

Example 18.18
The distribution

$$dF(x) = \frac{1}{\Gamma(p)} \left(\frac{x-\alpha}{\beta} \right)^{p-1} \exp\left\{ -\frac{(x-\alpha)}{\beta} \right\} d\left(\frac{x-\alpha}{\beta} \right), \qquad \alpha \leq x < \infty; \quad \beta > 0; \quad p > 2,$$

has its range dependent upon α, but is zero and has a zero first derivative with respect to α at its lower terminal for $p > 2$ (cf. Exercise 17.22), so our regularity conditions hold. Here

$$g(y) = -\log \Gamma(p) + (p-1) \log y - y,$$

and

$$E(g'') = E\left\{ -\frac{(p-1)}{y^2} \right\} = -\frac{1}{(p-2)},$$

$$E(g''y) = E\left\{ -\frac{(p-1)}{y} \right\} = -1,$$

$$E(g''y^2) = E\{-(p-1)\} = -(p-1).$$

Thus the centre of location is, from (18.113),

$$\frac{E(g''y)}{E(g'')} = p - 2.$$

The inverse covariance matrix (18.110) is

$$\mathbf{V}^{-1} = \frac{n}{\beta^2} \begin{pmatrix} \dfrac{1}{p-2} & 1 \\ 1 & p \end{pmatrix}, \tag{18.115}$$

and its inverse, the covariance matrix of $\hat{\alpha}$ and $\hat{\beta}$, is easily obtained directly as

$$\mathbf{V} = \frac{(p-2)\beta^2}{2n} \begin{pmatrix} p & -1 \\ -1 & \dfrac{1}{p-2} \end{pmatrix}. \tag{18.116}$$

If we measure from the centre of location, we have for the uncorrelated estimators $\hat{\alpha}_u$, $\hat{\beta}_u$, as the reader should verify,

$$\left. \begin{array}{l} \text{var } \hat{\alpha}_u = (p-2)\beta^2/n, \\ \text{var } \hat{\beta}_u = \beta^2/(2n). \end{array} \right\} \tag{18.117}$$

Comparing (18.117) with (18.116), we see that var $\hat{\beta}$ is unaffected by the change of origin, while var $\hat{\alpha}_u$ is reduced from its value in (18.116) to what is seen from (18.115) to be its value when α alone is being estimated. If β alone is estimated, (18.115) shows that var $\hat{\beta}$ is smaller than in (18.116–17).

> D. R. Cox and Reid (1987) consider the general (i.e. not only the location–scale) problem of choosing a parametrisation so that the ML estimators of the parameters of interest, $\boldsymbol{\theta}$, shall be uncorrelated with those of unwanted ('nuisance') parameters $\boldsymbol{\psi}$. In general, this cannot be achieved for all values of the parameters, but it can be if θ is a scalar, the case they discuss in detail.

Efficiency of the method of moments

18.36   In Chapter 6 we discussed distributions of the Pearson family. We were there mainly concerned with the properties of populations only and no question of the reliability of estimates arose. If, however, the observations are a *sample* from a population, the question arises whether fitting by moments provides the most efficient estimators of the unknown parameters. As we shall see presently, in general it does not.

Consider a distribution dependent on four parameters. If the ML estimators of these parameters are to be obtained in terms of linear functions of the moments (as in the fitting of Pearson curves), we must have

$$\frac{\partial \log L}{\partial \theta_r} = a_0 + a_1 \sum x + a_2 \sum x^2 + a_3 \sum x^3 + a_4 \sum x^4, \qquad r = 1, \ldots, 4, \tag{18.118}$$

and consequently

$$f(x \mid \theta_1, \ldots, \theta_4) = \exp(b_0 + b_1 x + b_2 x^2 + b_3 x^3 + b_4 x^4), \tag{18.119}$$

where the bs depend on the θs. This is the most general form for which the method of moments gives ML estimators. The bs are, of course, conditioned by the requirements that the total frequency shall be unity and the distribution function converge.

Without loss of generality we may take $b_1 = 0$. If, then, b_3 and b_4 are zero, the distribution is normal and the method of moments is efficient. In other cases, (18.119) does not yield a Pearson distribution except as an approximation. For example, consider

$$\frac{\partial \log f}{\partial x} = 2b_2 x + 3b_3 x^2 + 4b_4 x^3.$$

If b_3 and b_4 are small, this is approximately

$$\frac{\partial \log f}{\partial x} = \frac{2b_2 x}{1 - \frac{3b_3}{2b_2} x - \frac{2b_4}{b_2} x^2}, \tag{18.120}$$

which is one form of the equation defining Pearson distributions (cf. (6.1)). Only when b_3 and b_4 are small compared with b_2 can we expect the method of moments to give estimators of high efficiency.

18.37 A detailed discussion of the efficiency of moments in determining the parameters of a Pearson distribution has been given by Fisher (1921a). Exercise 18.16 deals with the Type IV distribution and Exercises 18.26–8 with some discrete distributions. As an illustration, we discuss the three-parameter Gamma (Type III) distribution.

Example 18.19
Consider the Gamma distribution with three parameters, α, σ, p,

$$dF = \frac{1}{\sigma \Gamma(p)} \left(\frac{x - \alpha}{\sigma}\right)^{p-1} \exp\left\{-\left(\frac{x - \alpha}{\sigma}\right)\right\} dx, \qquad \alpha \leqslant x < \infty; \quad \sigma > 0; \quad p > 2.$$

For the LF, we have

$$\log L = -np \log \sigma - n \log \Gamma(p) + (p - 1) \sum \log (x - \alpha) - \sum (x - \alpha)/\sigma.$$

The three likelihood equations are

$$\frac{\partial \log L}{\partial \alpha} = -(p - 1) \sum \frac{1}{(x - \alpha)} + n/\sigma = 0,$$

$$\frac{\partial \log L}{\partial \sigma} = -np/\sigma + \sum (x - \alpha)/\sigma^2 = 0,$$

$$\frac{\partial \log L}{\partial p} = -n \log \sigma - n \frac{d}{dp} \log \Gamma(p) + \sum \log (x - \alpha) = 0.$$

Taking the parameters in the above order, the inverse covariance matrix (18.60) is

$$\mathbf{V}^{-1} = n \begin{pmatrix} \dfrac{1}{\sigma^2(p-2)} & \dfrac{1}{\sigma^2} & \dfrac{1}{\sigma(p-1)} \\[2mm] \dfrac{1}{\sigma^2} & \dfrac{p}{\sigma^2} & \dfrac{1}{\sigma} \\[2mm] \dfrac{1}{\sigma(p-1)} & \dfrac{1}{\sigma} & \dfrac{d^2 \log \Gamma(p)}{dp^2} \end{pmatrix}.$$

The (2×2) leading diagonal of $\mathbf{V}^{-1}$ is, of course, (18.115) with σ written for β. The determinant

$$|\mathbf{V}^{-1}|/n^3 = \Delta = \frac{1}{(p-2)\sigma^4} \left\{ 2\frac{d^2 \log \Gamma(p)}{dp^2} - \frac{2}{p-1} + \frac{1}{(p-1)^2} \right\}.$$

From this the sampling variances are found to be

$$\text{var } \hat{\alpha} = \frac{1}{n\Delta\sigma^2} \left\{ p\frac{d^2 \log \Gamma(p)}{dp^2} - 1 \right\},$$

$$\text{var } \hat{\sigma} = \frac{1}{n\Delta\sigma^2} \left\{ \frac{1}{p-2}\frac{d^2 \log \Gamma(p)}{dp^2} - \frac{1}{(p-1)^2} \right\},$$

$$\text{var } \hat{p} = \frac{2}{n\Delta(p-2)\sigma^4} = \frac{2}{n} \bigg/ \left\{ 2\frac{d^2 \log \Gamma(p)}{dp^2} - \frac{2}{p-1} + \frac{1}{(p-1)^2} \right\}. \qquad (18.121)$$

Now for large p, using Stirling's series (3.63), we find when p is large

$$2\frac{d^2}{dp^2} \log \Gamma(1+p) - \frac{2}{p} + \frac{1}{p^2} = \frac{1}{3} \left\{ \frac{1}{p^3} - \frac{1}{5p^5} + \frac{1}{7p^7} - \cdots \right\}$$

and hence approximately, from (18.121),

$$\text{var } \hat{p} \doteq \frac{6}{n} \left\{ (p-1)^3 + \frac{1}{5}(p-1) \right\}. \qquad (18.122)$$

If instead we estimate the parameters by equating sample moments to the population moments in terms of parameters, we find

$$\alpha + \sigma p = m_1,$$
$$\sigma^2 p = m_2,$$
$$2\sigma^3 p = m_3,$$

so that we shall have in the sample

$$b_1 = m_3^2 / m_2^3 = 4/p, \qquad (18.123)$$

which equates the sample skewness coefficient to its population value $\beta_1 = 4/p$. Now, from Exercise 10.26 and (10.14), we have

$$\operatorname{var} b_1 = \frac{4\beta_1}{n}\left\{\frac{\mu_6}{\mu_2^3} - 6\beta_2 + 9 + \frac{\beta_1}{4}(9\beta_2 + 35) - \frac{3\mu_5\mu_3}{\mu_2^4}\right\},$$

which using (16.5) with $\nu = 2p$ reduces to

$$\operatorname{var} b_1 = \frac{96(p+1)(p+5)}{np^3}. \tag{18.124}$$

Hence, from (18.123) and (10.14) we have for $\tilde{p}$, the estimator by the method of moments,

$$\operatorname{var} \tilde{p} \sim \left(\frac{4}{\beta_1^2}\right)^2 \operatorname{var} b_1 = \frac{6}{n} p(p+1)(p+5).$$

For large p the efficiency of this estimator is then, from (18.122),

$$\frac{\operatorname{var} \hat{p}}{\operatorname{var} \tilde{p}} = \frac{\{(p-1)^3 + \frac{1}{5}(p-1)\}}{p(p+1)(p+5)},$$

which is evidently less than 1. When p exceeds 39.1 ($\beta_1 = 0.102$), the efficiency is over 80 per cent. For $p = 20$ ($\beta_1 = 0.20$), it is 65 per cent. For $p = 5$, a more exact calculation based on the tables of the trigamma function $\mathrm{d}^2 \log \Gamma(1+p)/\mathrm{d}p^2$ shows that the efficiency is only 22 per cent.

> Bowman and Shenton (1968, 1969, 1970, 1988) study the ML estimators in detail. Example 18.17 treated the case where p is known, and Exercise 18.15 treats the case $\alpha = 0$, while Exercise 17.1 in effect considered two single-parameter cases.

Order restrictions on parameters: isotonic estimation
18.38   We have assumed since **18.22** that the admissible values of any parameter do not depend on those of another. However, problems do arise in which we know that the parameters occur in a certain order, e.g. in the simplest two-parameter case that $\theta_1 \leqslant \theta_2$, and we must require their estimators to satisfy the same order restrictions.

Example 18.20
Independent random samples of equal size n are taken from two normal distributions with the same known variance (say, unity) and means θ_1, θ_2 where $\theta_1 \leqslant \theta_2$. The LF to be maximised is given by

$$\log L = \text{constant} - \frac{1}{2}\left\{\sum_{j=1}^{n}(x_{1j} - \theta_1)^2 + \sum_{j=1}^{n}(x_{2j} - \theta_2)^2\right\}$$

and on factorising the sums of squares this is

$$\log L = \text{constant} - \tfrac{1}{2}n\{(\bar{x}_1 - \theta_1)^2 + (\bar{x}_2 - \theta_2)^2\}.$$

We thus require to minimise the function in braces for choice of θ_1, θ_2, subject to the order restriction $\theta_1 \leq \theta_2$. Clearly the unconditional minimum ignoring this order restriction is

$$\theta_1 = \bar{x}_1, \qquad \theta_2 = \bar{x}_2,$$

so that if $\bar{x}_1 \leq \bar{x}_2$ we have the ML estimators

$$\hat{\theta}_1 = \bar{x}_1, \qquad \hat{\theta}_2 = \bar{x}_2, \qquad \bar{x}_1 \leq \bar{x}_2. \qquad (18.125)$$

However, if $\bar{x}_1 > \bar{x}_2$, this solution breaches the order restriction. Since $\log L$ is constant on any circle with centre $(\bar{x}_1, \bar{x}_2)$ and decreases as the circle's radius increases, the circle that touches the line $\theta_1 = \theta_2$ will have a single point in the region $\theta_1 \leq \theta_2$ and the maximum value of $\log L$ in that region. It is easy to see that it touches at the point $(\frac{1}{2}(\bar{x}_1 + \bar{x}_2), \frac{1}{2}(\bar{x}_1 + \bar{x}_2))$. Thus the ML estimators are

$$\hat{\theta}_1 = \hat{\theta}_2 = \tfrac{1}{2}(\bar{x}_1 + \bar{x}_2), \qquad \bar{x}_1 > \bar{x}_2, \qquad (18.126)$$

and (18.125-6) are the complete ML solution.

18.39 In Example 18.20, the ML solution is an intuitively acceptable one: if $\bar{x}_1$ and $\bar{x}_2$ are in the wrong order, we pool and average them. If we now generalise to three normal distributions, with the order restriction

$$\theta_1 \leq \theta_2 \leq \theta_3$$

the analogous geometrical representation is of $\log L$ constant on spheres centred at $(\bar{x}_1, \bar{x}_2, \bar{x}_3)$ and decreasing as the radius increases. When $\bar{x}_1 \leq \bar{x}_2 \leq \bar{x}_3$, the ML estimator is $\theta_j = \bar{x}_j$, at the unconditional maximum of the LF as before. If $\bar{x}_1 > \bar{x}_2$ or $\bar{x}_2 > \bar{x}_3$ holds, the result (18.126) applies to the pair of θs involved, while the other θ is unaffected by the single inversion of order and may be estimated unconditionally as in (18.125). But if $\bar{x}_1 > \bar{x}_3$ we must also have $\bar{x}_1 > \bar{x}_2$ or $\bar{x}_2 > \bar{x}_3$ or both, and there is a new problem. We have to determine the sphere of smallest radius that has a point in the region $\theta_1 \leq \theta_2 \leq \theta_3$.

General methods of solving this type of problem are given by Barlow *et al.* (1972), who show that the ML solution can be found for the class of distributions

$$f(x) = \exp\{A(\theta) + A'(\theta)(x - \theta) + B(x)\}, \qquad (18.127)$$

of which the normal distribution is a special case, by finding the *isotonic regression**
of the sample means (i.e. the unconditional ML estimators—cf. Exercise 18.2), which is the upper envelope of all convex functions lying below the cumulative graph of the sample means. Barlow *et al.* (1972) give algorithms for carrying out the computations—the pooling of adjacent inverted-order means in Example 18.20 and its generalisation above correspond to the simplest examples of such a convex envelope.

---

* 'Isotonic' is used here to mean 'order-preserving', since 'monotonic' has an ambiguity of direction; its earlier and etymological sense 'having the same tension' has particular reference to osmotic pressure.

ML estimation for incomplete data: the EM algorithm

18.40 Thus far, we have assumed the data set to be complete, that is, none of the observations is missing. Yet there are many reasons why particular values may not be available: responses to a questionnaire may be omitted, experiments may fail or the records get mixed up, values may be reported incorrectly or not at all or the data may be abridged in some way (censored, truncated or grouped). In addition it may sometimes be desirable to consider observations that, were they to be available, would simplify the analysis, either by providing a more regular structure (in an unbalanced experiment design) or by simplifying algebraic or numerical procedures. Whether such observations could be made or whether the investigator planned to make them is of no consequence. Example 18.21 below and Exercise 18.42 illustrate these points. All we need to know is that the complete data set x is an augmented form of the observed, or incomplete, set y such that the frequency function for y may be determined from that for x, as a marginal f.f.; that is,

$$g(\mathbf{y}|\boldsymbol{\theta}) = \int f(\mathbf{x}|\boldsymbol{\theta})\,d\mathbf{x} \qquad (18.128)$$

where the integral is taken over $\{\mathbf{x}: \mathbf{y} = y(\mathbf{x})\}$ and the integral in (18.128) is replaced by a sum when the variates are discrete. In general, there are many possible sets $\mathbf{x}$ in which we may choose to imbed $\mathbf{y}$, but there may be a natural choice in some situations.

18.41 These problems may be treated in a unified way through the EM algorithm, first presented in full generality by Dempster *et al.* (1977). The algorithm, an iterative procedure, consists of two conceptually distinct steps at each stage of the iteration, those of estimation (E) and maximisation (M), which may be kept distinct or merged depending on the particular application. We assume that

$$Q(\boldsymbol{\theta}^*|\boldsymbol{\theta}) = E\{\log L(\mathbf{x}|\boldsymbol{\theta}^*)|\mathbf{y}, \boldsymbol{\theta}\} \qquad (18.129)$$

exists for all $\mathbf{x}$ and $\boldsymbol{\theta}$, where L is the likelihood to be maximised. Let $\boldsymbol{\theta}_{(k)}$ denote the estimate of $\boldsymbol{\theta}$ obtained at the kth iteration. Then we update it as follows:

E-step: Evaluate $Q(\boldsymbol{\theta}|\boldsymbol{\theta}_{(k)})$ by estimating the unknown elements of $\mathbf{x}$. This may be performed algebraically, when feasible, or numerically.

M-step: Select $\boldsymbol{\theta}_{(k+1)}$ as a value of $\boldsymbol{\theta}$ that maximises $Q(\boldsymbol{\theta}|\boldsymbol{\theta}_{(k)})$.

The rationale is that we wish to maximise $\log L$ to find the ML estimator of $\boldsymbol{\theta}$. Since we cannot do this because of the incompleteness of the data we maximise instead the conditional expectation of $\log L$, given the observed $\mathbf{y}$ and the current estimate of $\boldsymbol{\theta}$. The process may be illustrated by Example 18.21.

Example 18.21

In Example 18.10, suppose that we now have the additional information that the first class, with observed frequency $a = 1997$, is actually the merger of two classes with unknown frequencies a_1 and a_2 and probabilities $p_{11} = \theta/4$, $p_{12} = \frac{1}{2}$ adding to the previous p_1. We now use the EM algorithm to evaluate the ML estimator $\hat{\theta}$, which we know to be equal to 0.0357. We start from $\theta_{(1)} = 0.0570$ as before.

(18.129) gives, at the E-step,

$$E\{\log L(x \mid \theta) \mid \mathbf{y}, \theta_{(1)}\}$$

$$= E\left\{ \text{constant} + a_1 \log \frac{\theta_{(1)}}{4} + (1997 - a_1) \log \tfrac{1}{2} \mid a = 1997 \right\}$$

$$+ (b+c) \log \frac{1 - \theta_{(1)}}{4} + d \log \frac{\theta_{(1)}}{4} \qquad (18.130)$$

since b, c and d are components of y. Now clearly

$$E\{a_1 \mid a = 1997\} = 1997 \times \frac{\theta_{(1)}/4}{\theta_{(1)}/4 + \frac{1}{2}} = 1997 \times \frac{0.0570}{2.0570} = 55.3373.$$

For the M-step, we have to maximise (18.130) for this value of a_1. Dropping terms free of θ, (18.130) becomes

$$(a_1 + d) \log \theta + (b+c) \log (1 - \theta).$$

Differentiating and equating to zero, we find, with $b = 906$, $c = 904$, $d = 32$

$$\theta_{(2)} = \frac{a_1 + d}{a_1 + d + b + c} = \frac{55.3373 + 32}{87.3373 + 1810} = 0.04603.$$

Successive further iterations give

$$\theta_{(3)} = 0.04077$$
$$\theta_{(4)} = 0.03820$$
$$\theta_{(5)} = 0.03694$$
$$\theta_{(6)} = 0.03631$$
$$\theta_{(7)} = 0.03601$$
$$\theta_{(8)} = 0.03586$$
$$\theta_{(9)} = 0.03579$$
$$\theta_{(10)} = 0.03571$$

It has taken 10 iterations to reach the correct value, illustrating the fact that convergence is slow. The scoring method in Example 18.10 was much faster.

18.42   In addition to its generality, an attraction of the method is its simplicity, since it is a gradient method that requires only first derivatives. Unfortunately, as we have seen in Example 18.21, this property implies that the rate of convergence is linear, rather than the quadratic rate achieved by the Newton–Raphson procedure in **18.21**. Modifications to improve the rate of convergence in particular cases were noted in the discussion in Dempster *et al.* (1977). A general procedure to speed up the algorithm is given in Louis (1982); this approach has been simplified and extended by Laird *et al.* (1987).

Wu (1983) and Boyles (1983) discuss the convergence properties of the algorithm; when the function $Q(\phi, \theta)$ is continuous in both arguments, the procedure converges to a stationary point of the LF. However, a global maximum is guaranteed only when the LF is unimodal.

Titterington (1984) discusses alternatives to the EM algorithm for incomplete data.

The use of the likelihood function

18.43　It is natural, as Fisher (1956) recommends, to examine the course of the LF throughout the permissible range of variation of θ, and to draw a graph of the LF. This may be generally informative, but is not of any immediate value in the estimation problem. The LF contains all the information in the sample, and is thus a comprehensive summary of the data, precisely in the sense that, as we remarked in **17.38**, the observations themselves constitute a set of jointly sufficient statistics for the parameters of any problem. This way of putting it has the merit of drawing attention to the fact that the functional form of the distribution(s) generating the observations must be specified before the LF can be used at all, whether for ML estimation or otherwise. In other words, some information (in a general sense) must be supplied by the statistician: if he or she is unable or unwilling to supply it, resort must be had to the non-parametric hypotheses to be discussed elsewhere, and the quite different methods to which they lead.

Efron (1982) discusses the virtues of the *ML summary* $L(x | \hat{\theta})$ of the data, and distinguishes them from those of ML estimation that we have discussed.

In **31.00** below, we discuss some recent developments linking the use of the LF with conditional inferential procedures.

EXERCISES

18.1 In Example 18.7 show by considering the case $n = 1$ that the ML estimator does not attain the MVB for small samples; and deduce that for this distribution the efficiency of the sample mean compared with the ML estimator is $\frac{1}{2}$.

18.2 If the ML estimator $\hat{\theta}$ is a root of $\partial \log L/\partial\theta = 0$, show that the most general form of distribution differentiable in θ, for which $\hat{\theta} = \bar{x}$, the sample arithmetic mean, is

$$f(x \mid \theta) = \exp\{A(\theta) + A'(\theta)(x - \theta) + B(x)\},$$

a member of the natural exponential family (5.117), and hence that $\bar{x}$ is sufficient for θ, with MVB variance of $\{nA''(\theta)\}^{-1}$. Show that if θ is a location parameter, f is a normal distribution with mean θ (a result going back to Gauss), while if θ is a scale parameter, $f = \theta^{-1} \exp(-x/\theta)$.

(Cf. Keynes (1911) and Teicher (1961))

18.3 Show that the most general continuous distribution for which the ML estimator of a parameter θ is the geometric mean of the sample is

$$f(x \mid \theta) = \left(\frac{x}{\theta}\right)^{\theta A'(\theta)} \exp\{A(\theta) + B(x)\}.$$

Show further that the corresponding distribution having the harmonic mean as ML estimator of θ is

$$f(x \mid \theta) = \exp\left[\frac{1}{x}\{\theta A'(\theta) - A(\theta)\} - A'(\theta) + B(x)\right].$$

(Keynes, 1911)

18.4 In Exercise 18.3, show in each case that the ML estimator is sufficient for θ, but that it is not an MVB estimator of θ, in contrast to the case of the arithmetic mean in Exercise 18.2. Find in each case the function of θ that is estimable with variance equal to its MVB, and evaluate the MVB.

18.5 In samples of n observations from

$$f(x \mid \theta) = \frac{1}{\Gamma(p)}(x - \theta)^{p-1} \exp\{\theta - x\}, \qquad \theta \leqslant x < \infty; p \geqslant 2,$$

show that the smallest observation $x_{(1)}$ is neither a sufficient statistic for θ nor the ML estimator of $\hat{\theta}$ of θ, but that $\hat{\theta}$ is unique and satisfies

$$x_{(1)} - (p - 1) \leqslant \hat{\theta} < x_{(1)},$$

the equality holding when $n = 1$.

(For general f of asymptotic form $\alpha c(x - \theta)^{\alpha-1}$ as $x \to \theta$ (where α and c may be unknown parameters) R. L. Smith (1985) shows that for $\alpha > 2$, $\hat{\theta}$ has the standard ML properties, that for $\alpha = 2$, $\hat{\theta}$ is asymptotically normal and efficient with variance of lower order, but that for $\alpha < 2$, the normality is lost, while for $\alpha \leqslant 1$, ML estimators may not exist.)

18.6 Show that for samples of n from the extreme-value distribution (cf. (14.66))

$$dF(x) = \alpha \exp\{-\alpha(x - \mu) - \exp[-\alpha(x - \mu)]\} \, dx, \qquad -\infty < x < \infty,$$

the ML estimators $\hat{\alpha}$ and $\hat{\mu}$ are given by

$$\frac{1}{\hat{\alpha}} = \bar{x} - \frac{\sum x \, e^{-\hat{\alpha}x}}{\sum e^{-\hat{\alpha}x}},$$

$$e^{-\hat{\alpha}\hat{\mu}} = \frac{1}{n}\sum e^{-\hat{\alpha}x},$$

and that in large samples

$$n \text{ var } \hat{\alpha} = \alpha^2/(\pi^2/6),$$

$$n \text{ var } \hat{\mu} = \frac{1}{\alpha^2}\left\{1 + \frac{(1-\gamma)^2}{\pi^2/6}\right\},$$

$$n \text{ cov } (\hat{\alpha}, \hat{\mu}) = -(1-\gamma)/(\pi^2/6),$$

where γ is Euler's constant $0.5772\ldots$ (B. F. Kimball, 1946)

18.7   If x is distributed in the normal form

$$dF(x) = \frac{1}{\sigma\sqrt{(2\pi)}} \exp\left\{-\frac{1}{2}\left(\frac{x-\mu}{\sigma}\right)^2\right\} dx, \qquad -\infty < x < \infty,$$

the lognormal distribution of $y = e^x$ has mean $\theta_1 = \exp(\mu + \frac{1}{2}\sigma^2)$, and its variance is $\theta_2 = \exp(2\mu + \sigma^2)\{\exp(\sigma^2) - 1\}$. (Cf. (6.63).)
    Show that the ML estimator of θ_1 is

$$\hat{\theta}_1 = \exp(\bar{x} + \tfrac{1}{2}s^2),$$

where $\bar{x}$ and s^2 are the sample mean and variance of x, and that

$$E(\hat{\theta}_1) = E\{\exp(\bar{x})\}E\{\exp(\tfrac{1}{2}s^2)\} = \theta_1 \exp\left\{-\frac{(n-1)}{n}\frac{\sigma^2}{2}\right\}\left(1 - \frac{\sigma^2}{n}\right)^{-(n-1)/2} > \theta_1,$$

so that $\hat{\theta}_1$ is biased upwards. Show that $E(\hat{\theta}_1) \to_{n\to\infty} \theta_1$, so $\hat{\theta}_1$ is asymptotically unbiased.

18.8   In Exercise 18.7, define the series

$$f(t) = 1 + t + \frac{n-1}{n+1}\frac{t^2}{2!} + \frac{(n-1)^2}{(n+1)(n+3)}\frac{t^3}{3!} + \cdots.$$

Show that the adjusted ML estimator

$$\bar{\theta}_1 = \exp(\bar{x})f(\tfrac{1}{2}s^2)$$

is strictly unbiased and that it is an MV unbiased estimator. Show further that $\hat{\theta}_1 > \bar{\theta}_1$ for all samples, so that the bias of $\hat{\theta}_1$ over $\bar{\theta}_1$ is uniform.

> (Rukhin (1986) gives estimates with smaller m.s.e. than both $\bar{\theta}_1$ and $\hat{\theta}_1$, and similarly improved estimates for moments about zero, the median and the mode, which are all of form $\exp(a\mu + b\alpha^2)$—cf. 6.29–30.)

18.9   In Exercise 18.7, show that

$$\text{var } \hat{\theta}_1 = E\{\exp(2\bar{x})\}E\{\exp(s^2)\} - \{E(\hat{\theta}_1)\}^2$$

$$= \exp(2\mu + \sigma^2/n)\left[\exp\{\sigma^2/n\}\left(1 - \frac{2\sigma^2}{n}\right)^{-(n-1)/2} - \left(1 - \frac{\sigma^2}{n}\right)^{-(n-1)}\right]$$

exactly, with asymptotic variance

$$\text{var } \hat{\theta}_1 \sim \exp(2\mu + \sigma^2)\cdot\frac{1}{n}(\sigma^2 + \tfrac{1}{2}\sigma^4),$$

and that this is also the asymptotic variance of $\bar{\theta}_1$ in Exercise 18.8. Hence show that the unbiased moment-estimator of θ_1,

$$\bar{y} = \frac{1}{n}\sum y,$$

has efficiency

$$(\sigma^2 + \tfrac{1}{2}\sigma^4)/\{\exp(\sigma^2) - 1\}.$$

(Exercises 18.7–9 are due to Finney (1941) and H. S. Sichel (1951–2))

18.10   A multinomial distribution has n classes, each of which has equal probability $1/n$ of occurring. In a sample of N observations, k classes occur. Show that the LF for the estimation of n is

$$L(k \mid n) = \frac{N!}{\prod_{i=1}^{n}(r_i!)} \left(\frac{1}{n}\right)^{N} \cdot \binom{n}{k} \cdot \frac{k!}{\prod_{j=1}^{N}(m_j!)},$$

where $r_i (\geq 0)$ is the number of observations in the ith class and m_j is the number of classes with $j (\geq 1)$ observations in the sample. Show that the ML estimator of n is $\hat{n}$ where

$$\frac{N}{\hat{n}} = \sum_{j=\hat{n}-k+1}^{\hat{n}} \frac{1}{j},$$

and hence that approximately

$$\frac{N}{\hat{n}} = \log\left(\frac{\hat{n}}{\hat{n} - k + 1}\right),$$

and that k is sufficient for n. Show that for large N,

$$\operatorname{var} \hat{n} \sim \frac{n}{\exp\left(\dfrac{N}{n}\right) - \left(1 + \dfrac{N}{n}\right)}.$$

(Lewontin and Prout, 1956)

18.11   In Example 18.14, verify that the ML estimators (18.78) are jointly sufficient for the five parameters of the distribution, and that the ML estimators (18.74–75) are jointly sufficient for σ_1^2, σ_2^2 and ρ and also sufficient for ρ alone.

18.12   In estimating the correlation parameter ρ of the bivariate normal population, the other four parameters being known (Examples 18.14 and 18.15, case (a)), show that the sample correlation coefficient (which is the ML estimator of ρ if all five parameters are being estimated—cf. (18.78)) has estimating efficiency $1/(1+\rho^2)$. Show further that if the estimator

$$r' = \frac{\dfrac{1}{n}\sum(x - \mu_1)(y - \mu_2)}{\sigma_1 \sigma_2}$$

based on the true means and variances is used, the efficiency drops even further, to

$$\left(\frac{1 - \rho^2}{1 + \rho^2}\right)^2.$$

(Stuart, 1955a)

18.13   In Example 18.3, show that $(\sum x^2 + \sum y^2, \sum xy)$ is a pair of sufficient statistics for the single parameter ρ, and that the ML estimator $\hat{\rho}$ is a function of this pair. Show that when $n = 1$, this sufficient statistic is a single-valued function of (x, y), itself sufficient, but not conversely.

18.14   In Examples 18.14 and 18.15, find the ML estimators of μ_1 when all other parameters are known, and of σ_1^2 similarly, and show that their large-sample variances are respectively $\sigma_1^2(1 - \rho^2)/n$ and $4\sigma_1^4(1 - \rho^2)/[n(2 - \rho^2)]$. Find the joint ML estimators of μ_1 and σ_1^2 when the other three parameters are known, and evaluate their large-sample covariance matrix.

18.15 In Example 18.19, show that if $\alpha = 0$, $\bar{x}$ and g are jointly sufficient statistics for σ and p, and that the ML estimators of σ and p are roots of the equations

$$\bar{x} = \sigma p, \qquad \log(\bar{x}/g) = \log p - \frac{d}{dp} \log \Gamma(p),$$

where $\bar{x}$, g are respectively the arithmetic and geometric means of the observations. Find the asymptotic covariance matrix of $\hat{\sigma}$ and $\hat{p}$.

> (Choi and Wette (1969) and M. J. Box (1971) discuss the biases of $\hat{p}$ and $1/\hat{\sigma}$. The *Biometrika Tables*, Vol. II, give tables for the computation of $\hat{p}$, its bias and its variance. See also Grice and Bain (1980) for inferences concerning the mean σp. For the c.f. of $\log(g/\bar{x})$, see Exercise 11.3.)

18.16 Show that the centre of location of the Pearson Type IV distribution

$$dF \propto \exp\left\{-\nu \arctan\left(\frac{x-\alpha}{\beta}\right)\right\}\left\{1 + \left(\frac{x-\alpha}{\beta}\right)^2\right\}^{-(\rho+2)/2} dx, \qquad -\infty < x < \infty, \rho > 1,$$

where ν and ρ are assumed known, is distant $\nu\beta/(\rho+4)$ below the mode of the distribution; that the variance of the ML estimator $\hat{\alpha}$ in large samples is

$$\frac{\beta^2}{n} \cdot \frac{(\rho+4)^2 + \nu^2}{(\rho+1)(\rho+2)(\rho+4)};$$

and that the efficiency of the method of moments in locating the curve is therefore

$$\frac{\rho^2(\rho-1)\{(\rho+4)^2 + \nu^2\}}{(\rho+1)(\rho+2)(\rho+4)(\rho^2+\nu^2)}. \qquad \text{(Fisher, 1921a)}$$

18.17 For n independent observations from the uniform distribution

$$dF = dx, \qquad \theta - \tfrac{1}{2} \le x \le \theta + \tfrac{1}{2},$$

show that $(x_{(1)}, x_{(n)})$ is sufficient for θ, that no single sufficient statistic exists, and that the LF is maximum at any value in the interval $(x_{(n)} - \tfrac{1}{2}, x_{(1)} + \tfrac{1}{2})$. Show that the mid-point of this interval is an unbiased estimator of θ.

18.18 Members are drawn from an infinite population in which the proportion bearing a given attribute is π, the drawing proceeding until a members bearing that attribute have appeared. The sample number then attained is n. Show that the distribution of n is given by

$$\binom{n-1}{a-1} \pi^a (1-\pi)^{n-a}, \qquad n = a, a+1, a+2, \ldots,$$

and that the ML estimator of π is a/n. Show also that this is biased and that its asymptotic variance is $\pi^2(1-\pi)/a$.

18.19 In the lognormal distribution of Exercises 18.7–8, consider the estimation of the variance θ_2. Show that the ML estimator

$$\hat{\theta}_2 = \exp(2\bar{x} + s^2)\{\exp(s^2) - 1\}$$

is biased and that the adjusted ML estimator

$$\bar{\theta}_2 = \exp(2\bar{x})\left\{f(2s^2) - f\left(\frac{n-2}{n-1} s^2\right)\right\}$$

is unbiased with minimum variance.

18.20 In Exercise 18.19, show that asymptotically

$$\text{var } \hat{\theta}_2 \sim \frac{2\sigma^2}{n} \exp(4\mu + 2\sigma^2)[2\{\exp(\sigma^2) - 1\}^2 + \sigma^2\{2\exp(\sigma^2) - 1\}^2],$$

and hence that the efficiency of the unbiased moment-estimator $s_y^2 = \sum (y - \bar{y})^2/(n-1)$ is

$$\frac{2\sigma^2[2\{\exp(\sigma^2) - 1\}^2 + \sigma^2\{2\exp(\sigma^2) - 1\}^2]}{\{\exp(\sigma^2) - 1\}^2\{\exp(4\sigma^2) - 2\exp(3\sigma^2) + 3\exp(2\sigma^2) - 4\}}$$

(Finney, 1941)

18.21 Show that if θ is an integer-valued parameter and the LF is unimodal, the ML estimator $\hat{\theta}$ is the integer part of θ^*, where $L(x|\theta^*) = L(x|\theta^* - 1)$. Given that the LF has m modes and tends to zero at extreme values of θ, show that there are $(2m-1)$ solutions of θ_j^*, the ML estimates being the integer part of one of the θ_{2i-1}^*, $i = 1, 2, \ldots, m$. For a normal distribution with integer mean μ, show that $\mu = [\bar{x} + \frac{1}{2}]$, the integer nearest to $\bar{x}$.

(Cf. Dahiya (1981); Hammersley (1950) showed that if the variance is known $\hat{\mu}$ is unbiased and consistent, with asymptotic variance.

$$\text{var } \hat{\mu} \sim \left(\frac{8\sigma^2}{n\pi}\right)^{1/2} \exp\left(-\frac{n}{8\sigma^2}\right),$$

decreasing exponentially as n increases. Cf. also Khan (1973) and M. Ghosh (1975). Lindsay and Roeder (1987) treat a general class of integer-parameter distributions.)

18.22 In the previous Exercise, show that for an integer Poisson parameter

$$\hat{\lambda} = [\{1 - \exp(-\bar{x}^{-1})\}^{-1}]$$

(Cf. Hammersley (1950) and Stark (1975).)

18.23 Log $(t - \gamma)$ is normally distributed with mean μ, variance σ^2 where $\gamma < t < \infty$, and the three parameters (γ, μ, σ^2) are to be estimated. Defining

$$\hat{\mu}(\gamma) = \frac{1}{n} \sum_{i=1}^n \log(t_i - \gamma), \qquad \hat{\sigma}^2(\gamma) = \frac{1}{n} \sum_{i=1}^n \{\log(t_i - \gamma) - \hat{\mu}(\gamma)\}^2,$$

show that

$$L^{**}(\gamma) = \sup_{\mu, \sigma^2} L(x|\gamma, \mu, \sigma^2) \propto \{\hat{\sigma}(\gamma)\}^{-n} \prod_{i=1}^n (t_i - \gamma)^{-1}$$

and that if $t_{(1)}$ is the smallest observed value of t,

$$\lim_{\gamma \to t_{(1)}} L^{**}(\gamma) = +\infty,$$

so that the ML estimator of (γ, μ, σ^2) for this three-parameter lognormal distribution is always $(t_{(1)}, -\infty, +\infty)$.

(Hill, 1963a; Voorn (1981) explains the source of the unbounded LF.)

18.24 For a sample of n observations from $f(x|\theta)$ grouped into intervals of width h, write

$$f(x|\theta, h) = \int_{x-h/2}^{x+h/2} f(y|\theta) \, dy.$$

Show by a Taylor expansion that

$$f(x|\theta, h) = hf(x|\theta) \left\{ 1 + \frac{h^2}{24} \frac{\partial^2 f(x|\theta)/\partial x^2}{f(x|\theta)}, + \cdots \right\}$$

and hence, to the first approximation, that the correction to be made to the ML estimator $\hat{\theta}$ to allow for grouping is

$$\Delta = -\frac{1}{24} h^2 \frac{\sum_{i=1}^{n} \partial((\partial^2 f/\partial x^2)/f)/\partial\theta}{\sum_{i=1}^{n} \partial^2(\log f)/\partial\theta^2},$$

the value of the right-hand side being taken at $\hat{\theta}$.

(Lindley, 1950)

18.25 Using the previous Exercise, show that for estimating the mean of a normal population with known variance, $\Delta = 0$, while in estimating the variance with known mean, $\Delta = -h^2/12$.

Each of these corrections is exactly the Sheppard grouping correction (cf. (3.54–5)) to the corresponding population moment. To show that the ML grouping correction does not generally coincide with the Sheppard correction, consider the distribution

$$dF = e^{-x/\theta} \, dx/\theta, \qquad \theta > 0; \quad 0 \leqslant x < \infty,$$

where $\hat{\theta} = \bar{x}$, the sample mean, and the correction to it is

$$\Delta = -\frac{1}{12} \frac{h^2}{\bar{x}},$$

whereas the Sheppard correction to the population mean is zero.

(Lindley, 1950; the normal case is studied in detail by Gjeddebaek (1949–61) and Kulldorff (1958) and the general theory by Kulldorf (1961) and Stadje (1985).)

18.26 Rewriting the negative binomial distribution (5.43) with m for the mean $k(1-p)/p$

$$f_r = \left(1 + \frac{m}{k}\right)^{-k} \binom{k+r-1}{r} \left(\frac{m}{m+k}\right)^r, \qquad r = 0, 1, 2, \ldots,$$

show that for a sample of n independent observations, with n_r observations at the value r and $n_0 < n$, the ML estimator of m is the sample mean

$$\hat{m} = \bar{r},$$

while that of k is a root of

$$n \log\left(1 + \frac{\bar{r}}{k}\right) = \sum_{r=1}^{\infty} n_r \sum_{i=0}^{r-1} \frac{1}{k+i}.$$

Show that as k decreases towards zero, the right-hand side of this equation exceeds the left, and that if the sample variance s_r^2 exceeds $\bar{r}$ the left-hand side exceeds the right as $k \to \infty$, and hence that the equation has at least one finite positive root. On the other hand, given that $s_r^2 < \bar{r}$, show that the two sides of the equation tend to equality as $k \to \infty$, so that $\hat{k} = \infty$, and f_r reduces to a Poisson distribution with parameter m.

(Anscombe, 1950)

18.27 In the previous Exercise, show that

$$\text{var } \hat{m} = (m + m^2/k)/n,$$

$$\text{var } \hat{k} \sim \left\{ \frac{2k(k+1)}{n\left(\dfrac{m}{m+k}\right)^2} \right\} \Bigg/ \left\{ 1 + 2 \sum_{j=2}^{\infty} \frac{\left(\dfrac{j}{j+1}\right)\left(\dfrac{m}{m+k}\right)^{j-1}}{\left(\dfrac{k+j}{j-1}\right)} \right\},$$

$$\text{cov } (\hat{m}, \hat{k}) \sim 0.$$

(Anscombe (1950); Fisher (1941) investigated the efficiency of the method of moments in this case.)

18.28 For the Neyman Type A contagious distribution of Exercise 5.7, with frequency function f_r, show that the ML estimators of λ_1, λ_2 are given by the roots of the equations

$$\hat{\lambda}_1 \hat{\lambda}_2 = \bar{r} = \sum_r n_r(r+1)f_{r+1}/(nf_r),$$

where $n_r, \bar{r}$ have the same meanings as in Exercise 18.26.

(Cf. Shenton (1949), who investigated the efficiency of the method of moments in this case and found it to be relatively low ($<70\%$) for small λ_1 (<3) and large λ_2 ($\geqslant 1$), and examined (1950, 1951) the efficiency of using moments in the general case and in the particular case of the Gram-Charlier Type A series. See also Katti and Gurland (1962).)

18.29 For n observations from the logistic distribution

$$F(x) = \frac{1}{1 + \exp\{-(x-\alpha)/\beta\}}, \qquad -\infty < x < \infty, \quad \sigma > 0,$$

show that the ML estimators of α and β are roots of the equations

$$\sum_{i=1}^{n} \frac{1}{1 + e^{-y_i}} = \tfrac{1}{2}n, \tag{A}$$

$$\sum_{i=1}^{n} \frac{y_i}{1 + e^{-y_i}} - \tfrac{1}{2} \sum_{i=1}^{n} y_i = \tfrac{1}{2}n, \tag{B}$$

where $y_i = (x_i - \alpha)/\beta$, and that the elements of their asymptotic inverse covariance matrix (18.110) are $E(g'') = -\tfrac{1}{3}$, $E(g'y) = 0$, $E(g''y^2 - 1) = -\tfrac{1}{3}(\pi^2/3 + 1)$.

Show that when β is known, there is a unique solution of (A) for α. When $n = 2$, show that the unique solutions of (A) and (B) are $\hat{\alpha} = \tfrac{1}{2}(x_1 + x_2)$, $\hat{\beta} = |x_1 - x_2|/(4L)$, where $L \doteqdot 0.7717$ is the unique positive root of $\tanh L = 1/(2L)$.

(Cf. Antle et al. (1970), who show that the solutions of (A) and (B) are unique for any n if the x_i are not all equal.)

18.30 Independent samples of sizes n_1, n_2 are taken from two normal populations with equal means μ and variances respectively equal to $\lambda\sigma^2, \sigma^2$. Find the ML estimator of μ, and show that its large-sample variance is

$$\text{var } (\hat{\mu}) = \sigma^2 \Bigg/ \left(\frac{n_1}{\lambda} + n_2 \right).$$

Hence show that the unbiased estimator

$$t = (n_1 \bar{x}_1 + n_2 \bar{x}_2)/(n_1 + n_2)$$

has efficiency

$$\frac{\lambda(n_1 + n_2)^2}{(n_1 \lambda + n_2)(n_1 + n_2 \lambda)},$$

which attains the value 1 if and only if $\lambda = 1$.

18.31 For a sample of n observations from a normal distribution with mean θ and variance $V(\theta)$, show that the ML estimator $\hat{\theta}$ is a root of

$$V' = 2(\bar{x} - \theta) + \frac{V'}{V} \frac{1}{n} \sum (x - \theta)^2,$$

and hence that if $V(\theta) = \sigma^2 \theta^k$, where σ^2 is known, $\hat{\theta}$ is a function of both $\bar{x}$ and $\sum x^2$ unless $k = 0$ (when $\hat{\theta} = \bar{x}$, the single sufficient statistic) or $k = 1$ (when $\hat{\theta}$ is a function of $\sum x^2$ only). When $k = 2$, show that the distribution of $\sum x^2 / (\sum x)^2$ does not depend on θ.

18.32 A parameter θ is estimated by θ^*, a root of the equation $g(x, \theta) = 0$. Given that regularity conditions for the MVB in **17.14–15** hold, and $E\{g(x, \theta)\} = 0$, show that

$$\frac{E(g^2)}{\left\{ E\left(\dfrac{\partial g}{\partial \theta}\right) \right\}^2} \geq \frac{1}{E\left\{ \left(\dfrac{\partial \log L}{\partial \theta}\right)^2 \right\}},$$

the equality holding only when $g(x, \theta)$ is a constant multiple of $\partial \log L / \partial \theta$, when $\theta^* = \hat{\theta}$, the ML estimator. This generalization reduces to (17.22) when $g(x, \theta) = t - \tau(\theta)$.

(Godambe, 1960; Durbin, 1960)

18.33 For a random sample from the distribution

$$\begin{aligned} f(x \mid \theta) &= \tfrac{1}{3} \exp\{-|x - \theta|\}, & x &< \theta, \\ &= \tfrac{1}{3}, & \theta &\leq x \leq \theta + 1, \\ &= \tfrac{1}{3} \exp\{-|x - (\theta + 1)|\}, & \theta + 1 &< x, \end{aligned}$$

show that the ML estimator of θ is never unique. (Consider the cases $n = 1$, $n = 2$ in particular.)

(Daniels, 1961)

18.34 A sample of n observations x_i is drawn from a normal population with mean μ and a positive variance that has equal probabilities of being 1 or σ^2. By considering $L(x \mid \mu, \sigma^2)$ when $\mu = x_j$ and $\sigma^2 \to 0$, show that as $n \to \infty$ no ML estimator of (μ, σ^2) exists.

(Kiefer and Wolfowitz, 1956)

18.35 Show that if an MVB estimator exists for a parameter θ, the method of scoring for parameters given at (18.46) will reach the ML estimator $\hat{\theta}$ in a single iteration, no matter what trial value is used, but that this is not true of (18.45).

18.36 In Example 18.8, show that if (18.46) were used to evaluate $\hat{\sigma}$ from a trial value $t > 0$, the first iteration would give the value $\theta_1 = \tfrac{1}{2}(t + \hat{\theta}^2/t)$, that $\theta_1 > \hat{\theta}$ always, and that θ_1 is closer to $\hat{\theta}$ than t is if $t > \tfrac{1}{3}\hat{\theta}$.

18.37 Given that a single observation x is distributed in the form

$$f(x \mid \theta) = \theta^{-1/m} \exp(-x\theta^{-1/m}), \qquad x > 0, \theta > 0, m = 1, 2, 3, \ldots,$$

show that the ML estimator of θ is $\hat{\theta} = x^m$, that it is biased unless $m = 1$, and that $u = \hat{\theta}/m!$ is unbiased. Show from **17.21** that u is the MV unbiased estimator of θ, attaining the bound (17.41) with $s = m$.

(Fend, 1959)

18.38 A sample of n observations from the triangular distribution

$$f(x) = 2x/\theta, \quad 0 \le x \le \theta,$$
$$= 2(1-x)/(1-\theta), \quad \theta \le x \le 1,$$

is ordered so that $0 \equiv x_{(0)} < x_{(1)} < \cdots < x_{(n)} < x_{(n+1)} \equiv 1$. Show that $\log L(x \mid \theta)$ is continuous in θ, and differentiable except where θ is equal to some $x_{(r)}$. Show that in the open interval $(x_{(r)}, x_{(r+1)})$, $r = 0, 1, \ldots, n$,

$$\frac{\partial \log L}{\partial \theta} = \frac{n\theta - r}{\theta(1-\theta)}, \qquad \frac{\partial^2 \log L}{\partial \theta^2} = \frac{r}{\theta^2} + \frac{n-r}{(1-\theta)^2},$$

and hence that $\log L$ has no regular maximum, but has a cusp at each $x_{(r)}$, one of which is the ML estimator $\hat{\theta}$.

Show that as $n \to \infty$, $\hat{\theta} \sim x_{(n\theta)}$. (Cf. Oliver 1972))

18.39 For the two-parameter Cauchy distribution

$$dF = \frac{dx}{\pi\sigma\left\{1 + \left(\dfrac{x-\theta}{\sigma}\right)^2\right\}}, \qquad -\infty < x < \infty; \quad \sigma > 0,$$

show that the ML estimators of θ and σ from a sample of n observations are given by the roots of the equations

$$\frac{1}{2}\frac{\partial \log L}{\partial \theta} = \sum_{i=1}^{n} \frac{x_i - \theta}{\sigma^2 + (x_i - \theta)^2} = 0, \tag{A}$$

$$\frac{\sigma}{n}\frac{\partial \log L}{\partial \sigma} = 1 - \frac{2\sigma^2}{n}\sum_{i=1}^{n} \frac{1}{\sigma^2 + (x_i - \theta)^2} = 0, \tag{B}$$

provided that (B) has a solution (as it always has if fewer than $\frac{1}{2}n$ observations coincide in value, and thus almost certainly when $n \ge 3$) and that otherwise $\hat{\sigma} = 0$ is to be used in (A). Show that $\partial^2 \log L/\partial\sigma^2 < 0$ when (B) holds, and hence that if θ is known, the solution of (B) for $\hat{\sigma}$ is at a unique maximum of the LF.

Writing $y_i = x_i - \theta$; $D_i = y_i^2 - \sigma^2$; $S_i = y_i^2 + \sigma^2$, show by considering the positive and the negative values of D_i separately that

$$\frac{\partial^2 \log L}{\partial \theta^2} \le \frac{1}{\sigma^2}\sum_{i=1}^{n} \frac{D_i}{S_i},$$

this upper bound being zero when (B) holds. Further, show that when (A) and (B) both hold,

$$-2\sigma^2 \sum_i y_i S_i^{-2} = \sum_i y_i D_i S_i^{-2}$$

and

$$-2\sigma^2 \sum_i D_i S_i^{-2} = \sum_i D_i^2 S_i^{-2},$$

so that

$$\left(\frac{\partial^2 \log L}{\partial\theta\,\partial\sigma}\right)^2 - \frac{\partial^2 \log L}{\partial\theta^2}\cdot\frac{\partial^2 \log L}{\partial\sigma^2} = \frac{4}{\sigma^2}\left\{\left(\sum_i y_i D_i S_i^{-2}\right)^2 - \left(\sum_i D_i^2 S_i^{-2}\right)\left(\sum y_i^2 S_i^{-2}\right)\right\} < 0,$$

and hence that the solution of (A) and (B) for $(\hat{\theta}, \hat{\sigma})$ is at a unique maximum of the LF.

(Copas, 1975)

18.40   In Exercise 18.39, when $n = 3$ and $x_{(1)} < x_{(2)} < x_{(3)}$, show (writing (i) for $x_{(i)}$ and $(i-j)$ for $(x_{(i)} - x_{(j)})$) that the ML estimators are

$$\hat{\theta} = \{(1)(3-2)^2 + (2)(3-1)^2 + (3)(2-1)^2\}/\{(3-2)^2 + (3-1)^2 + (2-1)^2\},$$
$$3^{-1/2}\hat{\sigma}(3-2)(3-1)(2-1)/\{(3-2)^2 + (3-1)^2 + (2-1)^2\}.$$

When $n = 4$, show similarly that

$$\hat{\theta} = \{(2)(4) - (1)(3)\}/\{(4-3) + (2-1)\}$$
$$\hat{\sigma}^2(4-3)(3-2)(2-1)(4-1)/\{(4-3) + (2-1)\}^2.$$

Show that as $x_{(n)} \to \infty$ with the other observations held fixed,

$$\hat{\theta} \to \tfrac{1}{2}\{(1) + (2)\} \text{ for } n = 3 \text{ and } \hat{\theta} \to (2) \text{ for } n = 4.$$

(Cf. Ferguson, 1978)

18.41   Show that

$$\sum_{j=1}^{n} (x_j - \mu)^2/(x_j \mu^2) \equiv n(\bar{x} - \mu)^2/(\bar{x}\mu^2) + \sum_{j=1}^{n} (x_j^{-1} - \bar{x}^{-1})$$
$$\equiv u \qquad\qquad\qquad + v, \text{ say,}$$

so that the LF of a sample from the inverse Gaussian distribution of Exercise 11.28 may be written

$$L(x \,|\, \mu, \lambda) = \left(\frac{\lambda}{2\pi}\right)^{-n/2} \left(\prod_{j=1}^{n} x_j\right)^{-3/2} \exp\left\{\frac{-\lambda}{2}(u + v)\right\}.$$

Hence show that if λ is known, $\bar{x}$ is a complete sufficient statistic for μ, and also its ML estimator, λu having an exact χ_1^2 distribution (cf. Exercise 11.29). Show that λv is distributed independently of $\bar{x}$, having an exact χ_{n-1}^2 distribution, and that $\bar{x}$ and v are jointly complete sufficient statistics for μ and λ, the ML estimators then being

$$\hat{\mu} = \bar{x}, \qquad \hat{\lambda} = n v^{-1}.$$

Finally, show that $\hat{\mu}$ is unbiased, while $(n-3)\hat{\lambda}/n$ is unbiased for λ.

(Cf. Folks and Chhikara, 1978. Iwase and Seto (1983) give MV unbiased estimates of the cumulants—cf. Exercise 11.28—and of other functions of the parameters and find their estimator of κ_2 has smaller m.s.e. than the ML estimator of κ_2 to which it is asymptotically equivalent.)

18.42   A sample of n observations is drawn from a Poisson population with f.f.

$$\pi_j = e^{-\lambda}\lambda^j/j!, \qquad j = 0, 1, \dots.$$

However, the counts are recorded only for m groups $\{y_1, \dots, y_m\}$; let x_j correspond to the number of times j occurred, which is now merged into the rth group ($j \in r$). Applying the EM algorithm of 18.40–1, show that at the kth iteration

$$x_{j(k)} = y_r \pi_{j(k)} \Big/ \sum_{i \in r} \pi_{i(k)}$$

and

$$\lambda_{(k+1)} = \sum_j x_{j(k)} / n.$$

(Note that the mth group must be closed off at some appropriate upper value.)

CHAPTER 19

ESTIMATION: LEAST SQUARES IN THE LINEAR MODEL AND OTHER METHODS

19.1 In this chapter we shall consider principles of estimation other than that of maximum likelihood (ML), to which Chapter 18 was devoted. The chief of these, the principle (or method) of least squares (LS), while conceptually quite distinct from the ML method and possessing its own optimum properties, coincides with the ML method in the important case of normally distributed observations. The other methods, to be discussed later in the chapter, are essentially competitors to the ML method, and are equivalent to it, if at all, only in an asymptotic sense.

The method of least squares

19.2 We have seen (Examples 18.2, 17.6) that the ML estimator of the mean μ in a sample of n from a normal distribution

$$\mathrm{d}F(y) = \frac{1}{\sigma\sqrt{(2\pi)}} \exp\left\{-\frac{1}{2}\left(\frac{y-\mu}{\sigma}\right)^2\right\} \mathrm{d}y \tag{19.1}$$

is obtained by maximising the likelihood function

$$\log L(y|\mu) = -\tfrac{1}{2}n \log (2\pi\sigma^2) - \frac{1}{2\sigma^2} \sum_{j=1}^{n} (y_j - \mu)^2 \tag{19.2}$$

with respect to μ. From inspection of (19.2) it is maximised when

$$\sum_{j=1}^{n} (y_j - \mu)^2 \tag{19.3}$$

is minimised. The ML principle therefore tells us to choose $\hat{\mu}$ so that (19.3) is at its minimum.

Now suppose that the population mean, μ, is itself a linear function of parameters θ_i $(i = 1, 2, \ldots, k)$. We write

$$\mu = \sum_{i=1}^{k} x_i\theta_i, \tag{19.4}$$

where the x_i in (19.4) are not random variables but known constant coefficients combining the unknown parameters θ_i. If we now wish to estimate the θ_i individually, we have, from (19.3) and (19.4), to minimise

$$\sum_{j=1}^{n} \left(y_j - \sum_{i=1}^{k} x_i\theta_i\right)^2 \tag{19.5}$$

with respect to the θ_i. We may now generalise a stage further: suppose that, instead of the n observations coming from identical normal distributions, the means of these distributions differ. In fact, let

$$\mu_j = \sum_{i=1}^{k} x_{ij}\theta_i, \qquad j = 1, 2, \ldots, n. \tag{19.6}$$

We now have to minimise

$$\sum_j \left(y_j - \sum_i x_{ij}\theta_i \right)^2 \tag{19.7}$$

with respect to the θ_i.

19.3   The LS method gets its name from the minimisation of a sum of squares as in (19.7). As a general principle, it states that if we wish to estimate the parameter vector $\boldsymbol{\theta}$ from a set of n observations y_j, $j = 1, 2, \ldots, n$, which are related to $\boldsymbol{\theta}$ through their expectations

$$E(y_j) = p_j(\boldsymbol{\theta}),$$

then we should choose our estimator $\hat{\boldsymbol{\theta}}$ to minimise the sum of squared deviations between the y_j and their estimated expectations, i.e. to minimise

$$S = \sum_{j=1}^{n} \{y_j - p_j(\hat{\boldsymbol{\theta}})\}^2.$$

If the distributions of the y_j about their expectations are independently normal with the same variance σ^2, then exactly as in **19.2**, the LS estimator will be identical with the ML estimator of $\boldsymbol{\theta}$. Exercise 19.19 shows that the identity persists if the y_j are distributed in a sub-class of the exponential family (17.30), which includes the normal case. In general, however, the estimators will differ.

As with any other systematic principle of estimation, the acceptability of the LS method depends on the properties of the estimators to which it leads. Like the ML method, it has some asymptotic optimum properties (cf. Hartley and Booker (1965) and Villegas (1969)) but does not generally yield unbiased estimators—M. J. Box (1971) gives a general expression for the LS bias—see also Bates and Watts (1980, 1981), Clarke (1980, 1987a, b), H. White (1981) and R. D. Cook *et al.* (1986). However, in an extremely important class of situation, it has the property, even in small samples, that it provides unbiased estimators, linear in the observations, that have minimum variance (MV). This situation is usually described as the *linear model*, in which observations are distributed with constant variance about (possibly differing) mean values which are linear functions of the unknown parameters, and in which the observations are all uncorrelated in pairs. This is just the situation we had at (19.6) above, but we shall now abandon the normal distribution assumption, since it is quite unnecessary for this MV property of LS estimators. We now proceed to formalise the problem, and we shall find it convenient, as in Chapter 15, to use the notation and terminology of matrix and vector theory.

The LS estimator in the linear model

19.4   We write the linear model in the form

$$\mathbf{y} = \mathbf{X}\boldsymbol{\theta} + \boldsymbol{\varepsilon}, \tag{19.8}$$

where $\mathbf{y}$ is an $(n \times 1)$ vector of observations, $\mathbf{X}$ is an $(n \times k)$ matrix of known coefficients (with $n > k$), $\boldsymbol{\theta}$ is a $(k \times 1)$ vector of parameters, and $\boldsymbol{\varepsilon}$ is an $(n \times 1)$ vector of 'error' random variables ε_j with

$$E(\boldsymbol{\varepsilon}) = \mathbf{0} \tag{19.9}$$

and covariance matrix

$$\mathbf{V}(\boldsymbol{\varepsilon}) = E(\boldsymbol{\varepsilon}\boldsymbol{\varepsilon}') = \sigma^2 \mathbf{I} \tag{19.10}$$

where $\mathbf{I}$ is the $(n \times n)$ identity matrix. (19.9) and (19.10) thus embody the assumptions that the ε_j are uncorrelated, and all have zero means and the same variance σ^2. (19.9) is less restrictive than it appears, for if $E(\varepsilon_i) = \mu$ for all i, we can adjoin an $(n \times 1)$ column vector of units to $\mathbf{X}$, making it $(n \times (k+1))$, and a further element $\theta_{k+1} \equiv \mu$ to $\boldsymbol{\theta}$, to absorb the value of μ into $\mathbf{X}\boldsymbol{\theta}$; the essential assumption is that the ε_i have *equal* means. (19.10) is restrictive; however, it can be generalised (cf. **19.19** and Exercises 19.2 and 19.5), with the results correspondingly changed.

The reader should notice particularly that no restriction is placed upon the elements of $\mathbf{X}$. By defining these suitably, we shall see in Chapter 29 and elsewhere that the present linear model may serve for the analysis of categorised and classified observations.

The LS method requires that we minimise the scalar sum of squares

$$S = (\mathbf{y} - \mathbf{X}\boldsymbol{\theta})'(\mathbf{y} - \mathbf{X}\boldsymbol{\theta}) \tag{19.11}$$

for variation in the components of $\boldsymbol{\theta}$. A necessary condition that (19.11) be minimised is that $\partial S / \partial \boldsymbol{\theta} = \mathbf{0}$. Differentiating, we have $2\mathbf{X}'(\mathbf{y} - \mathbf{X}\boldsymbol{\theta}) = \mathbf{0}$, which gives for our LS estimator the vector

$$\hat{\boldsymbol{\theta}} = (\mathbf{X}'\mathbf{X})^{-1}\mathbf{X}'\mathbf{y}, \tag{19.12}$$

if we assume that $\mathbf{X}'\mathbf{X}$, the matrix of sums of squares and products of the elements of the column vectors composing $\mathbf{X}$, is non-singular and can therefore be inverted. This assumption will be relaxed in **19.13** below.

Example 19.1
Consider the simplest case, where $\boldsymbol{\theta}$ is a (1×1) vector, i.e. a single parameter. We may then write (19.8) as

$$\mathbf{y} = \mathbf{x}\theta + \boldsymbol{\varepsilon}$$

where $\mathbf{x}$ is now an $(n \times 1)$ vector. The LS estimator is, from (19.12),

$$\hat{\theta} = (\mathbf{x}'\mathbf{x})^{-1}\mathbf{x}'\mathbf{y}$$

$$= \sum_{j=1}^{n} x_j y_j \bigg/ \sum_{j=1}^{n} x_j^2.$$

Example 19.2

Suppose now that $\boldsymbol{\theta}$ has two components. The matrix $\mathbf{X}$ now consists of two vectors $\mathbf{x}_1, \mathbf{x}_2$. The model (19.8) becomes

$$\mathbf{y} = (\mathbf{x}_1\mathbf{x}_2) \begin{pmatrix} \theta_1 \\ \theta_2 \end{pmatrix} + \boldsymbol{\varepsilon},$$

and the LS estimator is now, from (19.12),

$$\hat{\boldsymbol{\theta}} = \begin{pmatrix} \mathbf{x}_1'\mathbf{x}_1 & \mathbf{x}_1'\mathbf{x}_2 \\ \mathbf{x}_2'\mathbf{x}_1 & \mathbf{x}_2'\mathbf{x}_2 \end{pmatrix}^{-1} \begin{pmatrix} \mathbf{x}_1'\mathbf{y} \\ \mathbf{x}_2'\mathbf{y} \end{pmatrix},$$

$$= \begin{pmatrix} \sum x_1^2 & \sum x_1 x_2 \\ \sum x_1 x_2 & \sum x_2^2 \end{pmatrix}^{-1} \begin{pmatrix} \sum x_1 y \\ \sum x_2 y \end{pmatrix},$$

where all summations are over the suffix $j = 1, 2, \ldots, n$. Since $\mathbf{X}'\mathbf{X}$ is the matrix of sums of squares and cross-products of the elements of the column vectors of $\mathbf{X}$, and $\mathbf{X}'\mathbf{y}$ the vector of cross-products of the elements of $\mathbf{y}$ with each of the $\mathbf{x}$-vectors in turn, the generalisation of this example to a $\boldsymbol{\theta}$ with more than two components is obvious.

Example 19.3

In Example 19.2 we specialise so that $\mathbf{x}_1 = \mathbf{1}$, a vector of units. Hence,

$$\hat{\boldsymbol{\theta}} = \begin{pmatrix} n & \sum x_2 \\ \sum x_2 & \sum x_2^2 \end{pmatrix}^{-1} \begin{pmatrix} \sum y \\ \sum x_2 y \end{pmatrix}$$

and we now invert the first matrix directly, obtaining

$$\begin{pmatrix} n & \sum x_2 \\ \sum x_2 & \sum x_2^2 \end{pmatrix}^{-1} = \frac{1}{\{n \sum x_2^2 - (\sum x_2)^2\}} \begin{pmatrix} \sum x_2^2 & -\sum x_2 \\ -\sum x_2 & n \end{pmatrix}.$$

Multiplying this by $\begin{pmatrix} \sum y \\ \sum x_2 y \end{pmatrix}$, we have

$$\hat{\boldsymbol{\theta}} = \frac{1}{\{n \sum x_2^2 - (\sum x_2)^2\}} \cdot \begin{pmatrix} \sum x_2^2 \sum y - \sum x_2 \sum x_2 y \\ -\sum x_2 \sum y + n \sum x_2 y \end{pmatrix},$$

so that

$$\hat{\theta}_2 = \frac{n \sum x_2 y - \sum x_2 \sum y}{n \sum x_2^2 - (\sum x_2)^2} = \frac{\sum (x_2 - \bar{x}_2)(y - \bar{y})}{\sum (x_2 - \bar{x}_2)^2},$$

and

$$\hat{\theta}_1 = \frac{\sum x_2^2 \sum y - \sum x_2 \sum x_2 y}{n \sum x_2^2 - (\sum x_2)^2} = \bar{y} - \hat{\theta}_2 \bar{x}_2.$$

It will be seen that $\hat{\theta}_2$ is exactly of the form of $\hat{\theta}$ in Example 19.1, with deviations from means replacing those from origins. This is a general effect of the introduction of a new parameter whose $\mathbf{x}$-values are the unit vector (see Exercise 19.1).

19.5   We may now establish the unbiasedness of the LS estimator (19.12). Using (19.8), it may be written

$$\hat{\boldsymbol{\theta}} = (\mathbf{X}'\mathbf{X})^{-1}\mathbf{X}'(\mathbf{X}\boldsymbol{\theta} + \boldsymbol{\varepsilon}) = \boldsymbol{\theta} + (\mathbf{X}'\mathbf{X})^{-1}\mathbf{X}'\boldsymbol{\varepsilon}. \tag{19.13}$$

Since **X** is constant, we have, on using assumption (19.9),

$$E(\hat{\boldsymbol{\theta}}) = \boldsymbol{\theta} \tag{19.14}$$

as stated. The reader should observe that assumption (19.10) has not been necessary so far. It is needed, however, to evaluate the covariance matrix of $\hat{\boldsymbol{\theta}}$, which is quite generally

$$\mathbf{V}(\hat{\boldsymbol{\theta}}) = E\{(\hat{\boldsymbol{\theta}} - \boldsymbol{\theta})(\hat{\boldsymbol{\theta}} - \boldsymbol{\theta})'\},$$

and this, on substitution from (19.13), becomes

$$\mathbf{V}(\hat{\boldsymbol{\theta}}) = E\{[(\mathbf{X}'\mathbf{X})^{-1}\mathbf{X}'\boldsymbol{\varepsilon}][(\mathbf{X}'\mathbf{X})^{-1}\mathbf{X}'\boldsymbol{\varepsilon}]'\}$$
$$= (\mathbf{X}'\mathbf{X})^{-1}\mathbf{X}'E(\boldsymbol{\varepsilon}\boldsymbol{\varepsilon}')\mathbf{X}(\mathbf{X}'\mathbf{X})^{-1}. \tag{19.15}$$

Using (19.10), (19.15) becomes

$$\mathbf{V}(\hat{\boldsymbol{\theta}}) = \sigma^2(\mathbf{X}'\mathbf{X})^{-1}. \tag{19.16}$$

(19.12) and (19.16) show that the computation of the vector of LS estimators and of their covariance matrix depends on the inversion of $\mathbf{X}'\mathbf{X}$. In the simpler cases (cf. Example 19.3) this can be done algebraically, but usually it must be done numerically, and often there is the risk of large computational error because of rounding-off of the elements of $\mathbf{X}'\mathbf{X}$, which should therefore be avoided as far as possible.

The *Householder decomposition* of **X** defines an orthogonal matrix **P** such that

$$\mathbf{P}'\mathbf{X} = \begin{pmatrix} \mathbf{U} \\ \mathbf{0} \end{pmatrix},$$

where **U** is a non-singular $(k \times k)$ upper triangular matrix, so that (19.12) becomes

$$\hat{\boldsymbol{\theta}} = (\mathbf{U}'\mathbf{U})^{-1}\begin{pmatrix} \mathbf{U} \\ \mathbf{0} \end{pmatrix}' \mathbf{P}'\mathbf{y} = \mathbf{U}^{-1}\begin{pmatrix} \mathbf{I} \\ \mathbf{0} \end{pmatrix} \mathbf{P}'\mathbf{y},$$

and **U** may be inverted more accurately than $\mathbf{U}'\mathbf{U} = \mathbf{X}'\mathbf{X}$. See Fletcher (1975) and Björck (1978).

Example 19.4
The variance of $\hat{\theta}$ in Example 19.1 is, from (19.16),

$$\text{var}(\hat{\theta}) = \sigma^2 / \sum x_j^2.$$

Example 19.5
In Example 19.2 we have

$$(\mathbf{X}'\mathbf{X})^{-1} = \begin{pmatrix} \sum x_1^2 & \sum x_1 x_2 \\ \sum x_1 x_2 & \sum x_2^2 \end{pmatrix}^{-1}$$

$$= \frac{1}{\{\sum x_1^2 \sum x_2^2 - (\sum x_1 x_2)^2\}} \begin{pmatrix} \sum x_2^2 & -\sum x_1 x_2 \\ -\sum x_1 x_2 & \sum x_1^2 \end{pmatrix},$$

so that, from (19.16),

$$\text{var}(\hat{\theta}_1) = \frac{\sigma^2 \sum x_2^2}{\{\sum x_1^2 \sum x_2^2 - (\sum x_1 x_2)^2\}} = \frac{\sigma^2}{\sum x_1^2 \left\{1 - \frac{(\sum x_1 x_2)^2}{\sum x_1^2 \sum x_2^2}\right\}};$$

and var $(\hat{\theta}_2)$ is the same expression with suffixes 1 and 2 interchanged. The covariance term is

$$\text{cov}(\hat{\theta}_1, \hat{\theta}_2) = \frac{-\sigma^2 \sum x_1 x_2}{\{\sum x_1^2 \sum x_2^2 - (\sum x_1 x_2)^2\}},$$

and this is zero if and only if $\sum x_1 x_2 = 0$.

Example 19.6

We found in Example 19.3 that

$$(\mathbf{X'X})^{-1} = \frac{1}{\sum (x_2 - \bar{x}_2)^2} \begin{pmatrix} (1/n) \sum x_2^2 & -\bar{x}_2 \\ -\bar{x}_2 & 1 \end{pmatrix}.$$

From (19.16), we therefore have

$$\text{var}(\hat{\theta}_1) = \sigma^2 \sum x_2^2 / \{n \sum (x_2 - \bar{x}_2)^2\}, \qquad \text{var}(\hat{\theta}_2) = \sigma^2 / \sum (x_2 - \bar{x}_2)^2,$$
$$\text{cov}(\hat{\theta}_1, \hat{\theta}_2) = -\sigma^2 \bar{x}_2 / \sum (x_2 - \bar{x}_2)^2.$$

Var $(\hat{\theta}_2)$ is, as is to be expected, var $(\hat{\theta})$ in Example 19.4, with deviations from the mean in the denominator. $\hat{\theta}_1$ and $\hat{\theta}_2$ are uncorrelated if and only if $\bar{x}_2 = 0$.

Optimum properties of LS

19.6 We now show that the MV unbiased linear estimators of any set of linear functions of the parameters θ_i are given by the LS method.* Plackett (1949), in a discussion of the origins of LS theory, makes it clear that the fundamental results are due to Gauss, as does a more detailed historical review by Seal (1967)—see also Plackett (1972).

Let **t** be any vector of estimators, linear in the observations **y**, i.e. of form

$$\mathbf{t} = \mathbf{Ty}. \tag{19.17}$$

If **t** is to be unbiased for a set of linear functions of the parameters, say **Cθ**, where **C** is a known matrix of coefficients, we must have $E(\mathbf{t}) = E(\mathbf{Ty}) = \mathbf{C\theta}$, which on using (19.8) gives

$$E\{\mathbf{T}(\mathbf{X\theta} + \boldsymbol{\varepsilon})\} = \mathbf{C\theta} \tag{19.18}$$

or, using (19.9), since (19.18) must hold identically in **θ**,

$$\mathbf{TX} = \mathbf{C}. \tag{19.19}$$

(19.19) is a necessary and sufficient condition for **Cθ** to be unbiasedly estimable by **Ty**.

---

* Barnard (1963) gives alternative optimum properties of the LS method.

The covariance matrix of **t** is

$$\mathbf{V(t)} = E\{(\mathbf{t} - \mathbf{C\theta})(\mathbf{t} - \mathbf{C\theta})'\} \tag{19.20}$$

and since, from (19.17), (19.8) and (19.19), $\mathbf{t} - \mathbf{C\theta} = \mathbf{T\epsilon}$, (19.20) becomes

$$\mathbf{V(t)} = E(\mathbf{T\epsilon\epsilon'T'}) = \sigma^2\mathbf{TT'} \tag{19.21}$$

from (19.10). We wish to minimise the diagonal elements of (19.21), which are the variances of our set of estimators.

We now write $\mathbf{t} = (\mathbf{t} - \hat{\mathbf{t}}) + \hat{\mathbf{t}}$, where

$$\hat{\mathbf{t}} = \mathbf{C\hat{\theta}} = \mathbf{C(X'X)}^{-1}\mathbf{X'y}. \tag{19.22}$$

Comparing (19.22) with (19.17), we see that $\hat{\mathbf{t}}$ has $\mathbf{T} = \mathbf{C(X'X)}^{-1}\mathbf{X'}$, so that (19.19) is always satisfied and *any* linear function $\mathbf{C\theta}$ is unbiasedly estimable by $\hat{\mathbf{t}}$. We have

$$\mathbf{V(t)} = \mathbf{V(t} - \hat{\mathbf{t}}) + \mathbf{V}(\hat{\mathbf{t}}) + 2\mathbf{D}(\mathbf{t} - \hat{\mathbf{t}}, \hat{\mathbf{t}}), \tag{19.23}$$

where $\mathbf{D(u, w)}$ is the matrix of covariances of the elements of **u** and the elements of **w**. Just as at (19.21),

$$\begin{aligned}
\mathbf{D(t} - \hat{\mathbf{t}}, \hat{\mathbf{t}}) &= E\{(\mathbf{t} - \hat{\mathbf{t}})(\hat{\mathbf{t}} - \mathbf{C\theta})'\} \\
&= E\{[\mathbf{T} - \mathbf{C(X'X)}^{-1}\mathbf{X'}]\mathbf{\epsilon\epsilon'}[\mathbf{C(X'X)}^{-1}\mathbf{X'}]'\} \\
&= \sigma^2[\mathbf{T} - \mathbf{C(X'X)}^{-1}\mathbf{X'}]\mathbf{X(X'X)}^{-1}\mathbf{C'} = \mathbf{0},
\end{aligned} \tag{19.24}$$

using (19.19). From (19.23) and (19.24),

$$\mathbf{V(t)} = \mathbf{V(t} - \hat{\mathbf{t}}) + \mathbf{V}(\hat{\mathbf{t}}), \tag{19.25}$$

so that each diagonal element of $\mathbf{V(t)}$ cannot be less than the corresponding element of $\mathbf{V}(\hat{\mathbf{t}})$. Thus (19.22) has MV among linear estimators of $\mathbf{C\theta}$. Its covariance matrix is, from (19.21),

$$\mathbf{V}(\hat{\mathbf{t}}) = \sigma^2\mathbf{C(X'X)}^{-1}\mathbf{C'}. \tag{19.26}$$

The reader should observe the resemblance of (19.24–5) to the result at the end of **17.27** and in Exercise 17.11, which is not confined to linear estimators.

If one or more of the latent roots of $\mathbf{X'X}$ are near zero, their product $|\mathbf{X'X}|$ will also be, so that the elements of $(\mathbf{X'X})^{-1}$ will tend to be large, and estimation therefore imprecise. This situation, known as *near-multicollinearity*, is discussed expositorily by Silvey (1969) and Stewart (1987).

Although the LS estimators are MV unbiased in the linear model among linear functions of **y**, they are not so in general if non-linear functions of **y** are admitted as estimators—this depends on the distribution of the ϵ_i. If the latter are normal (and hence independent), the stronger property follows from the fact that the LS estimators (which are then ML, by **19.3**) are functions of the $(k+1)$ minimal sufficient statistics $(\hat{\mathbf{\theta}}, s^2)$ for the $(k+1)$ parameters $(\mathbf{\theta}, \sigma^2)$—see **17.39**. T. W. Anderson (1962b) shows that if all possible distributions of the ϵ_i are considered, LS estimators are very rarely MV unbiased among all estimators.

Cox and Hinkley (1968) give the efficiency of the LS estimators for some special cases—see also Exercise 19.16. Bradu and Mundlak (1970) discuss the lognormal case—cf. also Evans and Shaban (1974).

If biased estimators are permitted, it does not follow (cf. **17.30**) that LS estimators will have minimum m.s.e. properties—cf. **19.12** and Exercise 28.14 below.

19.7 It is instructive to display the MV property geometrically, following Durbin and Kendall (1951). We shall here discuss only their simplest case, where we are estimating a single parameter θ from a sample of n observations, all with mean θ and variance σ^2. Thus $y_j = \theta + \varepsilon_j$, $j = 1, 2, \ldots, n$, which is (19.8) with $k = 1$ and $\mathbf{X}$ an $(n \times 1)$ vector of units. We consider linear estimators

$$t = \sum_j c_j y_j, \tag{19.27}$$

the simplest case of (19.17). The unbiasedness condition (19.19) here becomes

$$\sum_j c_j = 1. \tag{19.28}$$

Consider an n-dimensional Euclidean space with one co-ordinate for each c_j. We call this the estimator space. (19.28) is a hyperplane in this space, and any point P in the hyperplane corresponds uniquely to one unbiased estimator. Now since the y_j are uncorrelated, we have from (19.27)

$$\mathrm{var}\, t = \sigma^2 \sum_j c_j^2 \tag{19.29}$$

so that the variance of t is $\sigma^2 OP^2$, where O is the origin in the estimator space. It follows at once that t has MV when P is the foot of the perpendicular from O to the hyperplane. By symmetry, we must then have every $c_j = 1/n$ and $t = \bar{y}$, the sample mean.

Now consider the usual n-dimensional sample space, with one co-ordinate for each y_j. The bilinear form (19.27) establishes a duality between this and the estimator space. For any fixed t, a point in one space corresponds to a hyperplane in the other, while for varying t a point in one space corresponds to a family of parallel hyperplanes in the other. To the hyperplane (19.28) in the estimator space there corresponds the point $(t, t, \ldots, t)$ lying on the equiangular vector in the sample space. If a vector through the origin is orthogonal to a hyperplane in one space, the corresponding hyperplane and vector are orthogonal in the other space.

It now follows that the MV unbiased estimator will be given in the sample space by the hyperplane orthogonal to the equiangular vector at the point $L = (\bar{y}, \bar{y}, \ldots, \bar{y})$. If Q is the sample point, we drop a perpendicular from Q on to the equiangular vector to find L, i.e. we minimise $QL^2 = \sum_j (y_j - t)^2$. Thus we minimise a sum of squares in the sample space and consequently minimise the variance (another sum of squares) in the estimator space, as a result of the duality established between them.

W. H. Kruskal (1961) gives a completely geometrical approach to LS theory.

19.8 A direct consequence of the result of **19.6** is that the LS estimator $\hat{\theta}$ minimises the value of the generalised variance for linear estimators of $\boldsymbol{\theta}$. This result, which is due to Aitken (1948), is exact, unlike the equivalent asymptotic result proved for ML estimators in **18.28**.

The result of **19.6**, specialised to the estimation of a single linear function $\mathbf{c}'\boldsymbol{\theta}$, where $\mathbf{c}'$ is a $(1 \times k)$ vector, is that

$$\mathrm{var}\,(\mathbf{c}'\hat{\boldsymbol{\theta}}) \leq \mathrm{var}\,(\mathbf{c}'\mathbf{u}), \tag{19.30}$$

where $\hat{\boldsymbol{\theta}}$ is the LS estimator, and $\mathbf{u}$ any other linear estimator, of $\boldsymbol{\theta}$. Thus, by the argument leading from (18.66) to (18.67), we have

$$|V| \le |D|, \tag{19.31}$$

the required result.

Estimation of variance

19.9   The result of **19.6** is the first part of the Gauss theorem on LS, often called the Gauss–Markov theorem. Its second part deals with the estimation of the variance, σ^2, from the observations.

Consider the set of *residuals* in LS estimation,

$$\mathbf{y} - \mathbf{X}\hat{\boldsymbol{\theta}} = \{\mathbf{I}_n - \mathbf{X}(\mathbf{X}'\mathbf{X})^{-1}\mathbf{X}'\}\mathbf{y}, \tag{19.32}$$

where we have substituted from (19.12) and written $\mathbf{I}_n$ for the identity matrix of order n. Using (19.8), (19.32) reduces to

$$\mathbf{y} - \mathbf{X}\hat{\boldsymbol{\theta}} = \{\mathbf{I}_n - \mathbf{X}(\mathbf{X}'\mathbf{X})^{-1}\mathbf{X}'\}\boldsymbol{\varepsilon}, \tag{19.33}$$

so that the residuals are the same linear functions of the errors as of the observations.

Now the matrix in curly brackets on the right of (19.33) is symmetric and idempotent, as may easily be verified by transposing and by squaring it. Thus the sum of squared residuals is

$$(\mathbf{y} - \mathbf{X}\hat{\boldsymbol{\theta}})'(\mathbf{y} - \mathbf{X}\hat{\boldsymbol{\theta}}) = \boldsymbol{\varepsilon}'\{\mathbf{I}_n - \mathbf{X}(\mathbf{X}'\mathbf{X})^{-1}\mathbf{X}'\}\boldsymbol{\varepsilon}. \tag{19.34}$$

Now any quadratic form $\boldsymbol{\varepsilon}'\mathbf{B}\boldsymbol{\varepsilon}$ is a scalar and therefore identical with its trace, $\mathrm{tr}\,(\boldsymbol{\varepsilon}'\mathbf{B}\boldsymbol{\varepsilon})$. Moreover, matrices may be commuted under the trace operator. Thus

$$E(\boldsymbol{\varepsilon}'\mathbf{B}\boldsymbol{\varepsilon}) = E(\mathrm{tr}\,\boldsymbol{\varepsilon}'\mathbf{B}\boldsymbol{\varepsilon}) = E(\mathrm{tr}\,\mathbf{B}\boldsymbol{\varepsilon}\boldsymbol{\varepsilon}') = \mathrm{tr}\,\{\mathbf{B}E(\boldsymbol{\varepsilon}\boldsymbol{\varepsilon}')\} = \sigma^2\,\mathrm{tr}\,\mathbf{B}, \tag{19.35}$$

using (19.10). Applying (19.35), we have from (19.34)

$$\begin{aligned}
E\{(\mathbf{y} - \mathbf{X}\hat{\boldsymbol{\theta}})'(\mathbf{y} - \mathbf{X}\hat{\boldsymbol{\theta}})\} &= \sigma^2\,\mathrm{tr}\,\{\mathbf{I}_n - \mathbf{X}(\mathbf{X}'\mathbf{X})^{-1}\mathbf{X}'\} \tag{19.36}\\
&= \sigma^2[\mathrm{tr}\,\mathbf{I}_n - \mathrm{tr}\,\{\mathbf{X}(\mathbf{X}'\mathbf{X})^{-1}\mathbf{X}'\}], \\
&= \sigma^2\{\mathrm{tr}\,\mathbf{I}_n - \mathrm{tr}\,\mathbf{X}' \cdot \mathbf{X}(\mathbf{X}'\mathbf{X})^{-1}\} \\
&= \sigma^2(\mathrm{tr}\,\mathbf{I}_n - \mathrm{tr}\,\mathbf{I}_k), \\
&= \sigma^2(n - k). \tag{19.37}
\end{aligned}$$

Thus an unbiased estimator of σ^2 is

$$\frac{1}{n-k}(\mathbf{y} - \mathbf{X}\hat{\boldsymbol{\theta}})'(\mathbf{y} - \mathbf{X}\hat{\boldsymbol{\theta}}) = s^2, \tag{19.38}$$

the sum of squared residuals divided by the number of observations minus the number of parameters estimated.

This result permits unbiased estimators to be formed of the covariance matrices at (19.16) and (19.26), simply by putting s^2 of (19.38) for σ^2 in those expressions.

> The sufficiency argument given near the end of **19.6** ensures that s^2 is the MV unbiased estimator of σ^2 when the ε_i are normally distributed; the multiple of s^2 with smallest m.s.e. is $(n-k)s^2/(n-k+2)$ by exactly the argument of Exercise 17.16.

R. G. Miller (1974b) applies the jackknife method of **17.10** to estimation and variance estimation in the linear model (19.8).

Example 19.7
The unbiased estimator of σ^2 in Examples 19.1 and 19.4 is, from (19.38),

$$s^2 = \frac{1}{n-1} \sum_{j=1}^{n} (y_j - x_j \hat{\theta})^2 \qquad (19.39)$$

so that var $(\hat{\theta})$ is estimated unbiasedly by $s^2 / \sum x_j^2$.

Example 19.8
In Examples 19.2 and 19.5, the unbiased estimator of σ^2 is, from (19.38),

$$s^2 = \frac{1}{n-2} \sum_{j=1}^{n} (y_j - x_{1j} \hat{\theta}_1 - x_{2j} \hat{\theta}_2)^2, \qquad (19.40)$$

where

$$\begin{pmatrix} \hat{\theta}_1 \\ \hat{\theta}_2 \end{pmatrix} = \frac{1}{\{\sum x_1^2 \sum x_2^2 - (\sum x_1 x_2)^2\}} \begin{pmatrix} \sum x_2^2, & -\sum x_1 x_2 \\ -\sum x_1 x_2, & \sum x_1^2 \end{pmatrix} \begin{pmatrix} \sum x_1 y \\ \sum x_2 y \end{pmatrix}$$

$$= \frac{1}{\{\sum x_1^2 \sum x_2^2 - (\sum x_1 x_2)^2\}} \begin{pmatrix} \sum x_2^2 \sum x_1 y - \sum x_1 x_2 \sum x_2 y \\ \sum x_1^2 \sum x_2 y - \sum x_1 x_2 \sum x_1 y \end{pmatrix}, \qquad (19.41)$$

and we reduce this to the situation of Examples 19.3 and 19.6 by putting all $x_{1j} = 1$.

The normality assumption
19.10   All our results so far have assumed nothing concerning the distribution of the errors, ε_i, except the conditions (19.9) and (19.10) concerning their first- and second-order moments: we obtain unbiased estimators of the parameters and, further, unbiased estimators of the sampling variances and covariances of these estimators, without assumptions concerning the forms of the error distributions. However, if we wish to test hypotheses concerning the parameters, distributional assumptions are necessary. We shall be discussing the problems of testing hypotheses in the linear model in Chapter 23; here we shall only point out some fundamental features of the situation.
   19.11   If we postulate that the ε_i are normally distributed, the fact that they are uncorrelated implies their independence (cf. **15.3**). The identity

$$\mathbf{y}'\mathbf{y} \equiv (\mathbf{y} - \mathbf{X}\hat{\boldsymbol{\theta}})'(\mathbf{y} - \mathbf{X}\hat{\boldsymbol{\theta}}) + (\mathbf{X}\hat{\boldsymbol{\theta}})'(\mathbf{X}\hat{\boldsymbol{\theta}}), \qquad (19.42)$$

is easily verified using (19.12). The second term on the right of (19.42) is

$$\hat{\boldsymbol{\theta}}'\mathbf{X}'\mathbf{X}\hat{\boldsymbol{\theta}} = \mathbf{y}'\mathbf{X}(\mathbf{X}'\mathbf{X})^{-1}\mathbf{X}'\mathbf{y} = (\boldsymbol{\varepsilon}' + \boldsymbol{\theta}'\mathbf{X}')\mathbf{X}(\mathbf{X}'\mathbf{X})^{-1}\mathbf{X}'(\mathbf{X}\boldsymbol{\theta} + \boldsymbol{\varepsilon}). \qquad (19.43)$$

From (19.43) it follows that if $\boldsymbol{\theta} = \mathbf{0}$,

$$\hat{\boldsymbol{\theta}}'\mathbf{X}'\mathbf{X}\hat{\boldsymbol{\theta}} = \boldsymbol{\varepsilon}'\mathbf{X}(\mathbf{X}'\mathbf{X})^{-1}\mathbf{X}'\boldsymbol{\varepsilon}, \qquad (19.44)$$

and (19.42) may then be rewritten, using (19.34) and (19.44),

$$\boldsymbol{\varepsilon}'\boldsymbol{\varepsilon} = \boldsymbol{\varepsilon}'\{\mathbf{I}_n - \mathbf{X}(\mathbf{X}'\mathbf{X})^{-1}\mathbf{X}'\}\boldsymbol{\varepsilon} + \boldsymbol{\varepsilon}'\{\mathbf{X}(\mathbf{X}'\mathbf{X})^{-1}\mathbf{X}'\}\boldsymbol{\varepsilon}. \qquad (19.45)$$

The matrices in braces on the right of (19.45) are symmetric and idempotent, so that their ranks are equal to their traces, found in **19.9**, and Cochran's theorem (**15.16**) applies. Thus the two quadratic forms on the right of (19.45) are independently distributed (after division by σ^2 in each case to adjust the scale) as χ^2 with $(n-k)$ and k degrees of freedom if $\boldsymbol{\theta} = \mathbf{0}$.

Whatever the value of $\boldsymbol{\theta}$, we find from (19.43), using (19.9),

$$E\{(\mathbf{X}\hat{\boldsymbol{\theta}})'(\mathbf{X}\hat{\boldsymbol{\theta}})\} = E\{\boldsymbol{\varepsilon}'\mathbf{X}(\mathbf{X}'\mathbf{X})^{-1}\mathbf{X}'\boldsymbol{\varepsilon}\} + \boldsymbol{\theta}'\mathbf{X}'\mathbf{X}\boldsymbol{\theta}. \tag{19.46}$$

We saw in **19.9** that the first term on the right has the value $k\sigma^2$. Thus

$$E\{(\mathbf{X}\hat{\boldsymbol{\theta}})'(\mathbf{X}\hat{\boldsymbol{\theta}})\} = k\sigma^2 + (\mathbf{X}\boldsymbol{\theta})'(\mathbf{X}\boldsymbol{\theta}), \tag{19.47}$$

which exceeds $k\sigma^2$ unless $\mathbf{X}\boldsymbol{\theta} = \mathbf{0}$, which requires $\boldsymbol{\theta} = \mathbf{0}$ unless $\mathbf{X}$ takes special values. Thus it is intuitively reasonable to use the ratio $(\mathbf{X}\hat{\boldsymbol{\theta}})'(\mathbf{X}\hat{\boldsymbol{\theta}})/(ks^2)$ (where s^2, defined at (19.38), *always* has expected value σ^2) to test the hypothesis that $\boldsymbol{\theta} = \mathbf{0}$. We shall be returning to the justification of this and similar procedures from a less intuitive point of view in Chapter 23.

Ridge regression

19.12   The LS estimators are MV unbiased among linear estimators, but they are not in general minimum m.s.e. in that class, and hence the squared Euclidean distance between $\hat{\boldsymbol{\theta}}$ and $\boldsymbol{\theta}$, say $D^2(\hat{\boldsymbol{\theta}})$, which is the sum of the m.s.e.s of the components of $\hat{\boldsymbol{\theta}}$, is not minimised. Hoerl and Kennard (1970) generalise $\hat{\boldsymbol{\theta}}$ to the class of *ridge regression* estimators

$$\boldsymbol{\theta}_c^* = (\mathbf{X}'\mathbf{X} + c\mathbf{I})^{-1}\mathbf{X}'\mathbf{y},$$

(where $\boldsymbol{\theta}_0^* = \hat{\boldsymbol{\theta}}$) and choose the positive scalar c to minimise $D^2(\boldsymbol{\theta}_c^*)$. Since

$$E(\boldsymbol{\theta}_c^*) = \{\mathbf{I} - c(\mathbf{X}'\mathbf{X} + c\mathbf{I})^{-1}\}\boldsymbol{\theta}$$

and

$$V(\boldsymbol{\theta}_c^*) = \sigma^2(\mathbf{X}'\mathbf{X} + c\mathbf{I})^{-1}\mathbf{X}'\mathbf{X}(\mathbf{X}'\mathbf{X} + c\mathbf{I})^{-1},$$

the m.s.e.s of the elements of $\boldsymbol{\theta}_c^*$ are in the main diagonal of

$$\begin{aligned}
\mathbf{M}(\boldsymbol{\theta}_c^*) &= V(\boldsymbol{\theta}_c^*) + (E(\boldsymbol{\theta}_c^*) - \boldsymbol{\theta})(E(\boldsymbol{\theta}_c^*) - \boldsymbol{\theta})' \\
&= (\mathbf{X}'\mathbf{X} + c\mathbf{I})^{-1}(\sigma^2\mathbf{X}'\mathbf{X} + c^2\boldsymbol{\theta}\boldsymbol{\theta}')(\mathbf{X}'\mathbf{X} + c\mathbf{I})^{-1},
\end{aligned}$$

and

$$\begin{aligned}
D^2(\boldsymbol{\theta}_c^*) &= \operatorname{tr} \mathbf{M}(\boldsymbol{\theta}_c^*) = \operatorname{tr}\{(\mathbf{X}'\mathbf{X} + c\mathbf{I})^{-2}(\sigma^2\mathbf{X}'\mathbf{X} + c^2\boldsymbol{\theta}\boldsymbol{\theta}')\} \\
&= \sum_{j=1}^{k} \frac{\sigma^2\lambda_j + c^2 P_j}{(\lambda_j + c)^2},
\end{aligned}$$

where the λ_j are the latent roots of $\mathbf{X}'\mathbf{X}$ and $P_j \geqslant 0$. This could be made less than $D^2(\boldsymbol{\theta}_0^*) = \sigma^2 \sum \lambda_j^{-1}$ by choosing $c < \sigma^2/\max P_j$ if the latter were known. If instead $\boldsymbol{\theta}_c^*$ is computed for a range of values from 0 upwards, we can *estimate* the minimising

c, subject of course to sampling error, which may lead in some cases to a value worse than the zero of LS theory. $D^2(\theta_0^*)$ is large when there are some small roots λ_j, e.g., when the columns of X are highly correlated, the near-multicollinear situation of **19.6**.

The stability of the components of θ_c^* can be examined by plotting them against c in the so-called *ridge trace*. See also Theobald (1974), Farebrother (1976) and McDonald and Galarneau (1975), and modifications by Guilkey and Murphy (1975), Vinod (1976), Obenchain (1978) and Oman (1981).

Alldredge and Gilb (1976) provide a useful annotated bibliography of the subject. Dempster *et al.* (1977) and Gunst and Mason (1977) give the results of extensive sampling experiments comparing this and other methods, including *shrunken* estimators (cf. Example 17.14), to LS—see also Causey (1980), D. G. Gibbons (1981), Lawless (1981) and Copas (1983).

G. Smith and Campbell (1980) criticise ridge regression methods on several grounds, including the use of a simple sum of m.s.e.s as a criterion. See also the discussion following their paper, and Exercise 19.23 below, which demonstrates the importance of the parametrisation adopted.

The singular case

19.13   In **19.4** we assumed $X'X$ to be non-singular, so that (19.12) was valid, and $n > k$, so that (19.38) could be valid. If $n = k$, (19.12) still holds if $(X'X)^{-1}$ exists, but (19.38) is useless since the sum of squared residuals is identically zero, as (19.34) shows. If $n < k$, the rank of X (and that of $X'X$, which is the same) is less than k, so $X'X$ has no inverse.

We now let X (and $X'X$) have rank $r < k$ and suppose that $n \geq r$. The LS estimation problem must be discussed afresh, since $X'X$ has no inverse and (19.12) is invalid. The treatment follows Plackett (1950).

The condition (19.19) is still necessary and sufficient for $C\theta$ to be unbiasedly estimable by Ty. In this singular case, it cannot be satisfied if we wish to estimate θ itself, when it becomes

$$TX = I. \tag{19.48}$$

For, remembering that X is of rank r, we partition it into

$$X = \left(\begin{array}{c:c} X_{r,r} & X_{r,k-r} \\ \hdashline X_{n-r,r} & X_{n-r,k-r} \end{array} \right), \tag{19.49}$$

the suffixes of the matrix elements of (19.49) indicating the numbers of rows and columns. We assume, without loss of generality, that $X_{r,r}$ is non-singular, and therefore has inverse $X_{r,r}^{-1}$. The last $n - r$ rows of X are linearly dependent upon the first r rows, so that $X_{n-r,r} = CX_{r,r}$ and $X_{n-r,k-r} = CX_{r,k-r}$ for some $(n - r) \times r$ matrix C. Define a new matrix, of order $k \times (k - r)$,

$$D = \left(\begin{array}{c} X_{r,r}^{-1} \cdot X_{r,k-r} \\ \hdashline -I_{k-r} \end{array} \right), \tag{19.50}$$

where $\mathbf{I}_{k-r}$ is the identity matrix of that order. Evidently, $\mathbf{D}$ is of rank $k-r$. If we form the product $\mathbf{XD}$, we see at once that

$$\mathbf{XD} = \mathbf{0}. \tag{19.51}$$

If we postmultiply (19.48) by $\mathbf{D}$, we obtain, using (19.51),

$$\mathbf{D} = \mathbf{TXD} = \mathbf{0}. \tag{19.52}$$

This contradicts the fact that $\mathbf{D}$ has rank $k-r$. Hence (19.48) cannot hold.

Of course, as (19.19) implies, some linear functions $\mathbf{C\theta}$ may still be estimated unbiasedly, for (19.52) is then generalised to $\mathbf{CD} = \mathbf{0}$, which does not contradict the rank $(k-r)$ of $\mathbf{D}$ if $\mathbf{C} \neq \mathbf{I}$.

Alalouf and Styan (1979) give various characterisations of estimability. Kourouklis and Paige (1981) give numerically stable methods of deriving estimators, their variances and their covariances.

19.14 We may proceed as in the non-singular case if we first introduce a set of $(k-r)$ linear constraints upon the parameters

$$\mathbf{a} = \mathbf{B\theta}, \tag{19.53}$$

where $\mathbf{a}$ is a $(k-r) \times 1$ vector of constants and $\mathbf{B}$ is a known $(k-r) \times k$ matrix, of rank $(k-r)$. We now seek an estimator of the form $\mathbf{t} = \mathbf{Ly} + \mathbf{Na}$. The condition (19.48) now becomes

$$\mathbf{I} = \mathbf{LX} + \mathbf{NB}. \tag{19.54}$$

Provided that

$$|BD| \neq 0, \tag{19.55}$$

the matrix $\mathbf{B}$, of rank $(k-r)$, makes up the deficiency in rank of $\mathbf{X}$. In fact, we treat $\mathbf{a}$ as a vector of dummy observations and solve (19.8) and (19.53) together, in the augmented model

$$\begin{pmatrix} \mathbf{y} \\ \mathbf{a} \end{pmatrix} = \begin{pmatrix} \mathbf{X} \\ \mathbf{B} \end{pmatrix} \mathbf{\theta} + \begin{pmatrix} \mathbf{\epsilon} \\ \mathbf{0} \end{pmatrix}. \tag{19.56}$$

The matrix

$$\begin{pmatrix} \mathbf{X} \\ \mathbf{B} \end{pmatrix}' \begin{pmatrix} \mathbf{X} \\ \mathbf{B} \end{pmatrix} = \mathbf{X'X} + \mathbf{B'B}$$

is positive definite, for a non-null vector $\mathbf{d}$ makes $\mathbf{d'X'Xd} = (\mathbf{Xd})'\mathbf{Xd}$ equal to zero only if $\mathbf{Xd} = \mathbf{0}$, whence $\mathbf{d}$ must be a column of $\mathbf{D}$. But (19.55) ensures that $\mathbf{Bd} \neq \mathbf{0}$, so that $\mathbf{d'B'Bd} > 0$. Thus $\mathbf{X'X} + \mathbf{B'B}$ is strictly positive definite and may be inverted. (19.56) therefore yields, as at (19.22), the solution

$$\mathbf{C\hat{\theta}} = \mathbf{C}(\mathbf{X'X} + \mathbf{B'B})^{-1}(\mathbf{X'y} + \mathbf{B'a}), \tag{19.57}$$

which as before is the MV linear unbiased estimator of θ. Using (19.21), its covariance matrix, is, since $\mathbf{a}$ is constant,

$$\mathbf{V}(\mathbf{C}\hat{\boldsymbol{\theta}}) = \sigma^2 \mathbf{C}(\mathbf{X'X} + \mathbf{B'B})^{-1}\mathbf{X'X}(\mathbf{X'X} + \mathbf{B'B})^{-1}\mathbf{C'}. \tag{19.58}$$

The matrix $\mathbf{B}$ in (19.53) is arbitrary, subject to (19.55). In fact, if for $\mathbf{B}$ we substitute $\mathbf{UB}$, where $\mathbf{U}$ is any non-singular $(k-r) \times (k-r)$ matrix, (19.57) and (19.58) are unaltered in value. Thus we may choose $\mathbf{B}$ for convenience in computation in any particular case.

19.15   Exercise 19.8 shows that σ^2 is estimated unbiasedly by the sum of squared residuals divided by $(n-r)$ if $n > r$.

Chipman (1964) gives a detailed discussion of LS theory in the singular case. See also T. O. Lewis and Odell (1966) and C. R. Rao (1974).

Example 19.9

As a simple example of a singular situation suppose that we have

$$\boldsymbol{\theta} = \begin{pmatrix} \theta_1 \\ \theta_2 \\ \theta_3 \end{pmatrix} \qquad \mathbf{X} = \begin{pmatrix} 1 & 1 & 0 \\ 1 & 0 & 1 \\ 1 & 1 & 0 \\ 1 & 0 & 1 \end{pmatrix}.$$

Here $n = 4$, $k = 3$ and $\mathbf{X}$ has rank $2 < k$ because of the linear relation between its column vectors

$$\mathbf{x}_1 - \mathbf{x}_2 - \mathbf{x}_3 = \mathbf{0}.$$

We first verify that $\boldsymbol{\theta}$ cannot be unbiasedly estimated, as we saw in **19.13**.

The matrix $\mathbf{D}$ at (19.50) is of order 3×1 being

$$\mathbf{D} = \begin{pmatrix} \begin{pmatrix} 1 & 1 \\ 1 & 0 \end{pmatrix}^{-1} \cdot \begin{pmatrix} 0 \\ 1 \end{pmatrix} \\ -1 \end{pmatrix} = \begin{pmatrix} 1 \\ -1 \\ -1 \end{pmatrix},$$

expressing the linear relation. We now introduce the matrix of order 1×3

$$\mathbf{B} = (1 \quad 0 \quad 0),$$

which satisfies (19.55) since $\mathbf{BD} = 1$, a scalar in this case of a single linear relation.

Hence (19.53) is

$$c = (1 \quad 0 \quad 0) \begin{pmatrix} \theta_1 \\ \theta_2 \\ \theta_3 \end{pmatrix} = \theta_1,$$

again a scalar in this simple case. From (19.57), the LS estimator is

$$
\begin{pmatrix} \hat\theta_1 \\ \hat\theta_2 \\ \hat\theta_3 \end{pmatrix} = \left[\begin{pmatrix} 1 & 1 & 1 & 1 \\ 1 & 0 & 1 & 0 \\ 0 & 1 & 0 & 1 \end{pmatrix} \begin{pmatrix} 1 & 1 & 0 \\ 1 & 0 & 1 \\ 1 & 1 & 0 \\ 1 & 0 & 1 \end{pmatrix} + \begin{pmatrix} 1 \\ 0 \\ 0 \end{pmatrix}(1 \quad 0 \quad 0) \right]^{-1}
$$

$$
\times \left[\begin{pmatrix} 1 & 1 & 1 & 1 \\ 1 & 0 & 1 & 0 \\ 0 & 1 & 0 & 1 \end{pmatrix} \begin{pmatrix} y_1 \\ y_2 \\ y_3 \\ y_4 \end{pmatrix} + \begin{pmatrix} 1 \\ 0 \\ 0 \end{pmatrix} c \right]
$$

$$
= \begin{pmatrix} 5 & 2 & 2 \\ 2 & 2 & 0 \\ 2 & 0 & 2 \end{pmatrix}^{-1} \begin{pmatrix} y_1+y_2+y_3+y_4+c \\ y_1+y_3 \\ y_2+y_4 \end{pmatrix}
$$

$$
= \begin{pmatrix} 1 & -1 & -1 \\ -1 & \frac{3}{2} & 1 \\ -1 & 1 & \frac{3}{2} \end{pmatrix} \begin{pmatrix} \sum y + c \\ y_1+y_3 \\ y_2+y_4 \end{pmatrix} = \begin{pmatrix} c \\ \frac{1}{2}(y_1+y_3)-c \\ \frac{1}{2}(y_2+y_4)-c \end{pmatrix}.
$$

Since we chose $\mathbf{B}$ so that $c = \theta_1$, we can obviously make no progress in estimating $\boldsymbol{\theta}$ itself. However, by (19.19), any set of linear functions $\mathbf{C}\boldsymbol{\theta}$ are unbiasedly estimated by $\mathbf{T}\mathbf{y}$ if $\mathbf{T}\mathbf{X} = \mathbf{C}$. Thus $(\theta_1 + \theta_2)$ and $(\theta_1 + \theta_3)$ are estimable, since $\mathbf{C} = \begin{pmatrix} 1 & 1 & 0 \\ 1 & 0 & 1 \end{pmatrix}$ satisfies (19.19) with $\mathbf{T} = \begin{pmatrix} \frac{1}{2} & 0 & \frac{1}{2} & 0 \\ 0 & \frac{1}{2} & 0 & \frac{1}{2} \end{pmatrix}$. The estimator of $(\theta_1 + \theta_2)$ is therefore $\frac{1}{2}(y_1 + y_3)$ and that of $(\theta_1 + \theta_3)$ is $\frac{1}{2}(y_2 + y_4)$. It may be verified that $\mathbf{CD} = \mathbf{0}$ as stated in **19.13**.

From (19.58),

$$
\mathbf{V}(\hat{\boldsymbol{\theta}}) = \sigma^2 \begin{pmatrix} 1 & -1 & -1 \\ -1 & \frac{3}{2} & 1 \\ -1 & 1 & \frac{3}{2} \end{pmatrix} \begin{pmatrix} 4 & 2 & 2 \\ 2 & 2 & 0 \\ 2 & 0 & 2 \end{pmatrix} \begin{pmatrix} 1 & -1 & -1 \\ -1 & \frac{3}{2} & 1 \\ -1 & 1 & \frac{3}{2} \end{pmatrix}
$$

$$
= \sigma^2 \begin{pmatrix} 0 & 0 & 0 \\ 0 & \frac{1}{2} & 0 \\ 0 & 0 & \frac{1}{2} \end{pmatrix}
$$

so that

$$
\text{var}\,(\hat\theta_1 + \hat\theta_2) = \text{var}\,\hat\theta_2 = \text{var}\,(\hat\theta_1 + \hat\theta_3) = \text{var}\,\hat\theta_3 = \sigma^2/2,
$$

as is evident from the fact that each estimator is a mean of two observations with variance σ^2. Also

$$
\text{cov}\,(\hat\theta_1 + \hat\theta_2, \hat\theta_1 + \hat\theta_3) = 0,
$$

a useful property which is due to the orthogonality of the second and third columns of $\mathbf{X}$. When we come to discuss the application of LS theory to the analysis of variance in Chapter 29, we shall be returning to this subject.

LS with known linear constraints

19.16   Suppose now that, when $\mathbf{X'X}$ is singular, (19.53) represents a set of p (instead of $(k-r)$) linear relations among the k parameters $\boldsymbol{\theta}$ that are known *a priori* to hold. Provided that the augmented model (19.56) is of full rank k, (19.57–8) follow as before, while $\boldsymbol{\theta}$ itself may now be unbiasedly estimable. The analysis of variance, to be treated in Chapter 29, often involves situations of this kind, which also arise naturally in the Bayesian framework—see Exercise 31. In particular, we see that the ridge regression estimator is given by putting $\mathbf{a} = \boldsymbol{\theta}$ and $\mathbf{B'B} = c\mathbf{I}$ in **19.14**, although its motivation in **19.12** was different.

In the non-singular case, Exercise 19.17 expresses $\boldsymbol{\theta}$ as an adjustment to the unconstrained LS estimator (19.12).

Extension of a linear model to include further parameters

19.17   Suppose that a linear model (singular or non-singular)

$$\mathbf{y} = \mathbf{X}\boldsymbol{\theta} + \boldsymbol{\varepsilon} \tag{19.59}$$

has been fitted to the observations, and that the LS estimator of $\boldsymbol{\theta}$ is

$$\hat{\boldsymbol{\theta}} = \mathbf{T}\mathbf{y},$$

where $\mathbf{T} = (\mathbf{X'X})^{-1}\mathbf{X'}$ in the non-singular case. We find that the sum of squared residuals is

$$
\begin{aligned}
(\mathbf{y} - \mathbf{X}\hat{\boldsymbol{\theta}})'(\mathbf{y} - \mathbf{X}\hat{\boldsymbol{\theta}}) &= \{(\mathbf{I} - \mathbf{XT})\mathbf{y}\}'\{(\mathbf{I} - \mathbf{XT})\mathbf{y}\} \\
&= \mathbf{y}'(\mathbf{I} - \mathbf{XT})\mathbf{y},
\end{aligned}
\tag{19.60}
$$

since $\mathbf{X'y} = \mathbf{X'X}\hat{\boldsymbol{\theta}}$ from below (19.11). The matrix $(\mathbf{I} - \mathbf{XT})$ is idempotent.

Now consider the extended model

$$\mathbf{y} = \mathbf{X}\boldsymbol{\theta} + \mathbf{Z}\boldsymbol{\beta} + \boldsymbol{\varepsilon}. \tag{19.61}$$

The LS estimators, $\hat{\boldsymbol{\beta}}$ and $\hat{\boldsymbol{\theta}}$, are given by the solutions of the two estimating equations

$$
\left.
\begin{aligned}
\mathbf{X'X}\boldsymbol{\theta} + \mathbf{X'Z}\boldsymbol{\beta} - \mathbf{X'y} = \mathbf{0} \\
\mathbf{Z'X}\boldsymbol{\theta} + \mathbf{Z'Z}\boldsymbol{\beta} - \mathbf{Z'y} = \mathbf{0}
\end{aligned}
\right\}
\tag{19.62}
$$

whence

$$\hat{\boldsymbol{\theta}} = \mathbf{T}(\mathbf{y} - \mathbf{Z}\hat{\boldsymbol{\beta}})$$

so that, provided the model is non-singular,

$$\hat{\boldsymbol{\beta}} = \{\mathbf{Z'}(\mathbf{I} - \mathbf{XT})\mathbf{Z}\}^{-1}\mathbf{Z'}(\mathbf{I} - \mathbf{XT})\mathbf{y}, \tag{19.63}$$

$$\mathbf{V}(\hat{\boldsymbol{\beta}}) = \sigma^2\{\mathbf{Z'}(\mathbf{I} - \mathbf{XT})\mathbf{Z}\}^{-1}. \tag{19.64}$$

19.18   The residuals from (19.61) are

$$\mathbf{e} = \mathbf{y} - \mathbf{X}\hat{\boldsymbol{\theta}} - \mathbf{Z}\hat{\boldsymbol{\beta}} = (\mathbf{I} - \mathbf{XT})\mathbf{y} - \mathbf{A}(\mathbf{A'A})^{-1}\mathbf{A}^{-1}\mathbf{y}, \tag{19.65}$$

where $\mathbf{A} = (\mathbf{I} - \mathbf{XT})\mathbf{Z}$. It follows that the residual sum of squares is

$$\mathbf{e}'\mathbf{e} = \mathbf{y}'(\mathbf{I} - \mathbf{XT})\mathbf{y} - \mathbf{y}'\mathbf{A}(\mathbf{A}'\mathbf{A})^{-1}\mathbf{A}'\mathbf{y}. \tag{19.66}$$

Compared with (19.59), the reduction in the sum of squares is

$$\mathbf{y}'\mathbf{A}(\mathbf{A}'\mathbf{A})^{-1}\mathbf{A}'\mathbf{y} = \hat{\boldsymbol{\beta}}'\mathbf{Z}'(\mathbf{I} - \mathbf{XT})\mathbf{y}. \tag{19.67}$$

On comparing this with (19.60), we see that they embody the same matrix $(\mathbf{I} - \mathbf{XT})$; (19.67) differs only in that $\hat{\boldsymbol{\beta}}'\mathbf{Z}'$ replaces $\mathbf{y}'$ in premultiplying this matrix. This simplifies the computation of (19.67), since we have to replace a quadratic form in $\mathbf{y}$ by a set of corresponding bilinear forms, obtained by each column of $\mathbf{Z}$ in turn replacing $\mathbf{y}'$ in (19.60). These bilinear forms, assembled into a column vector, are premultiplied by $\hat{\boldsymbol{\beta}}'$ to obtain (19.67).

The difference between (19.60) and (19.67),

$$(\mathbf{y} - \mathbf{Z}\hat{\boldsymbol{\beta}})'(\mathbf{I} - \mathbf{XT})\mathbf{y}, \tag{19.68}$$

is the sum of squared residuals when the extended model (19.61) has been fitted.

It is easy to see that a reduction of exactly the same form as (19.67) applies to the minimised SS under any constraints upon the elements of $\boldsymbol{\theta}$: the only change is that $(\mathbf{I} - \mathbf{XT})$ is replaced by the matrix $\mathbf{Q}$ of the quadratic form of the minimised SS. The analogues of (19.67–8) are then $\hat{\boldsymbol{\beta}}'\mathbf{Z}'\mathbf{Qy}$ and $(\mathbf{y} - \mathbf{Z}\hat{\boldsymbol{\beta}})'\mathbf{Qy}$ respectively.

A more general linear model
19.19   The LS theory which we have been developing assumes throughout that (19.10) holds, i.e. that the errors are uncorrelated and have constant variance. There is no difficulty in generalising the linear model to the situation where the covariance matrix of errors is $\sigma^2\mathbf{V}$, $\mathbf{V}$ being positive definite (it is always non-negative definite by **15.3**), and we find (the details we left to the reader in Exercises 19.2 and 19.5–6) that (19.22) generalises to

$$\hat{\mathbf{t}} = \mathbf{C}(\mathbf{X}'\mathbf{V}^{-1}\mathbf{X})^{-1}\mathbf{X}'\mathbf{V}^{-1}\mathbf{y}, \tag{19.69}$$

and that this is the MV unbiased linear estimator of $\mathbf{C\theta}$. (19.25) remains true and (19.26) becomes

$$\mathbf{V}(\hat{\mathbf{t}}) = \sigma^2\mathbf{C}(\mathbf{X}'\mathbf{V}^{-1}\mathbf{X})^{-1}\mathbf{C}'. \tag{19.70}$$

In particular, if $\mathbf{V}$ is diagonal but not equal to $\mathbf{I}$, so that the ε_i are uncorrelated but with unequal variances, (19.69) provides the required set of estimators—cf. Exercise 19.6.

(19.69) coincides with the simple form (19.22) whenever $\mathbf{y} = \mathbf{Xz}$, i.e. $\mathbf{y}$ is in the space spanned by the columns of $\mathbf{X}$. Thus a necessary and sufficient condition that (19.69) and (19.22) be identical is that they coincide for $\mathbf{y}$ orthogonal to $\mathbf{X}$, i.e. that $\mathbf{X}'\mathbf{y} = \mathbf{0}$ implies $\mathbf{X}'\mathbf{V}^{-1}\mathbf{y} = \mathbf{0}$. McElroy (1967) shows that if $\mathbf{X}$ contains a unit vector (so that one parameter is a constant term, as in Exercise 19.1), (19.69) coincides with (19.22) if the errors have equal variances and are all equally non-negatively correlated. See also Mitra and Rao (1969), Zyskind (1969), Lowerre (1974) and Mathew (1983).

It should be observed that (19.22) will remain unbiased in the present more general model, since (19.9) alone is sufficient to establish (19.14). However, the variance estimator (19.38) is generally biased—Swindel (1968) gives attainable bounds for the bias. Kariya (1980) shows that the variance estimator, as well as $\hat{\mathbf{t}}$, remains unchanged if and only if

$$\mathbf{V} = \mathbf{X}\mathbf{P}\mathbf{X}' + \mathbf{I} - \mathbf{X}(\mathbf{X}'\mathbf{X})^{-1}\mathbf{X}'$$

for some positive definite matrix $\mathbf{P}$. Bloomfield and Watson (1975) and Knott (1975) obtain the minimum efficiency of (19.22) compared with (19.69)—see also Haberman (1975).

Baksalary and Kala (1983) discuss optimum combination of two vectors of estimators.

To use (19.69–70), of course, we need to know $\mathbf{V}$. In practical cases this is usually unknown; if an estimate of $\mathbf{V}$ is available from past experience, or from replicated observations on the model, it can be substituted for $\mathbf{V}$, but the optimum properties of (19.69) no longer necessarily hold—cf. C. R. Rao (1967). Bement and Williams (1969) discuss the effect of such estimation on the variances in (19.70) in the case when $\mathbf{V}$ is diagonal—see also Fuller and Rao (1978) and J. N. K. Rao (1980). C. R. Rao (1970) and Chew (1970) discuss the estimation of the elements of $\mathbf{V}$ in the diagonal and other structured cases. Strand (1974) gives a bound for the m.s.e. incurred by using a wrong $\mathbf{V}$—see also Horn *et al.* (1975).

S. F. Arnold (1979) shows that if the errors are multinormal with equal variances and equal correlations ρ, neither σ^2 nor ρ is separately estimable, although $\sigma^2(1-\rho)$ is, and that the general mean in the linear model is also not estimable. Procedures concerning $\boldsymbol{\theta}$ are otherwise unaffected, and retain optimum properties in this more general model. Jobson and Fuller (1980) discuss the case when the error variances are functions of $\boldsymbol{\theta}$. Pfefferman (1984) reviews results in the case when $\boldsymbol{\theta}$ is a random vector.

Ordered LS estimation of location and scale parameters

19.20 A particular situation in which (19.69) and (19.70) are of value is in the estimation of location and scale parameters from the order-statistics, i.e. the sample observations ordered according to magnitude. The results are due to Lloyd (1952) and Downton (1953).

We denote the order-statistics, as previously (see **14.1**), by $y_{(1)}, y_{(2)}, \ldots, y_{(n)}$. As usual, we write μ and σ for the location and scale parameters to be estimated and

$$z_{(r)} = (y_{(r)} - \mu)/\sigma, \qquad r = 1, 2, \ldots, n. \tag{19.71}$$

Let

$$\left.\begin{array}{l} E(\mathbf{z}) = \boldsymbol{\alpha}, \\ V(\mathbf{z}) = \mathbf{V}, \end{array}\right\} \tag{19.72}$$

where $\mathbf{z}$ is the $(n \times 1)$ vector of the $z_{(r)}$. Since $\mathbf{z}$ has been defined by (19.71) $\boldsymbol{\alpha}$ and $\mathbf{V}$ are independent of location and scale.

Now, from (19.71) and (19.72),

$$E(\mathbf{y}) = \mu\mathbf{1} + \sigma\boldsymbol{\alpha}, \tag{19.73}$$

where **y** is the vector of $y_{(r)}$ and **1** is the unit vector, while

$$V(\mathbf{y}) = \sigma^2 \mathbf{V}. \tag{19.74}$$

We may now apply (19.69) and (19.70) to find the LS estimators of μ and σ. We have

$$\begin{pmatrix} \hat{\mu} \\ \hat{\sigma} \end{pmatrix} = \{(1\,|\,\alpha)'\mathbf{V}^{-1}(1\,|\,\alpha)\}^{-1}(1\,|\,\alpha)'\mathbf{V}^{-1}\mathbf{y} \tag{19.75}$$

and

$$\mathbf{V}\begin{pmatrix} \hat{\mu} \\ \hat{\sigma} \end{pmatrix} = \sigma^2 \{(1\,|\,\alpha)'\mathbf{V}^{-1}(1\,|\,\alpha)\}^{-1}. \tag{19.76}$$

Now

$$\{(1\,|\,\alpha)'\mathbf{V}^{-1}(1\,|\,\alpha)\}^{-1} = \begin{pmatrix} \mathbf{1}'\mathbf{V}^{-1}\mathbf{1} & \mathbf{1}'\mathbf{V}^{-1}\alpha \\ \mathbf{1}'\mathbf{V}^{-1}\alpha & \alpha'\mathbf{V}^{-1}\alpha \end{pmatrix}^{-1}$$

$$= \frac{1}{\Delta}\begin{pmatrix} \alpha'\mathbf{V}^{-1}\alpha & -\mathbf{1}'\mathbf{V}^{-1}\alpha \\ -\mathbf{1}'\mathbf{V}^{-1}\alpha & \mathbf{1}'\mathbf{V}^{-1}\mathbf{1} \end{pmatrix} \tag{19.77}$$

where

$$\Delta = \{(\mathbf{1}'\mathbf{V}^{-1}\mathbf{1})(\alpha'\mathbf{V}^{-1}\alpha) - (\mathbf{1}'\mathbf{V}^{-1}\alpha)^2\}. \tag{19.78}$$

From (19.75) and (19.77),

$$\left.\begin{array}{l} \hat{\mu} = -\alpha'\mathbf{V}^{-1}(\mathbf{1}\alpha' - \alpha\mathbf{1}')\mathbf{V}^{-1}\mathbf{y}/\Delta, \\ \hat{\sigma} = \mathbf{1}'\mathbf{V}^{-1}(\mathbf{1}\alpha' - \alpha\mathbf{1}')\mathbf{V}^{-1}\mathbf{y}/\Delta. \end{array}\right\} \tag{19.79}$$

From (19.76) and (19.77)

$$\left.\begin{array}{l} \operatorname{var} \hat{\mu} = \sigma^2\alpha'\mathbf{V}^{-1}\alpha/\Delta, \\ \operatorname{var} \hat{\sigma} = \sigma^2\mathbf{1}'\mathbf{V}^{-1}\mathbf{1}/\Delta, \\ \operatorname{cov}(\hat{\mu}, \hat{\sigma}) = -\sigma^2\mathbf{1}'\mathbf{V}^{-1}\alpha/\Delta. \end{array}\right\} \tag{19.80}$$

19.21   Now since **V** and $\mathbf{V}^{-1}$ are positive definite, we may write

$$\left.\begin{array}{l} \mathbf{V} = \mathbf{TT}', \\ \mathbf{V}^{-1} = (\mathbf{T}^{-1})'\mathbf{T}^{-1}, \end{array}\right\} \tag{19.81}$$

so that for an arbitrary vector **b**

$$\mathbf{b}'\mathbf{V}\mathbf{b} = \mathbf{b}'\mathbf{TT}'\mathbf{b} = (\mathbf{T}'\mathbf{b})'(\mathbf{T}'\mathbf{b}) = \sum_{i=1}^{n} h_i^2,$$

where h_i is the ith row element of the vector $\mathbf{T}'\mathbf{b}$.

Similarly, for a vector $\mathbf{c}$,

$$\mathbf{c}'\mathbf{V}^{-1}\mathbf{c} = (\mathbf{T}^{-1}\mathbf{c})'(\mathbf{T}^{-1}\mathbf{c}) = \sum_{i=1}^{n} k_i^2,$$

k_i being the element of $\mathbf{T}^{-1}\mathbf{c}$. Now by the Cauchy inequality,

$$\sum h_i^2 \sum k_i^2 = \mathbf{b}'\mathbf{V}\mathbf{b} \cdot \mathbf{c}'\mathbf{V}^{-1}\mathbf{c} \geq (\sum h_i k_i)^2 = \{(\mathbf{T}'\mathbf{b})'(\mathbf{T}^{-1}\mathbf{c})\}^2 = (\mathbf{b}'\mathbf{c})^2. \tag{19.82}$$

In (19.82), put

$$\left.\begin{array}{l} \mathbf{b} = (\mathbf{V}^{-1} - \mathbf{I})\mathbf{1}, \\ \mathbf{c} = \boldsymbol{\alpha}. \end{array}\right\} \tag{19.83}$$

We obtain

$$\mathbf{1}'(\mathbf{V}^{-1} - \mathbf{I})\mathbf{V}(\mathbf{V}^{-1} - \mathbf{I})\mathbf{1} \cdot \boldsymbol{\alpha}'\mathbf{V}^{-1}\boldsymbol{\alpha} \geq \{\mathbf{1}'(\mathbf{V}^{-1} - \mathbf{I})\boldsymbol{\alpha}\}^2. \tag{19.84}$$

If μ and σ^2 are the mean and variance, we find that

$$\left.\begin{array}{l} E(\mathbf{1}'\mathbf{z}) = \mathbf{1}'\boldsymbol{\alpha} = 0, \\ V(\mathbf{1}'\mathbf{z}) = \mathbf{1}'\mathbf{V}\mathbf{1} = n = \mathbf{1}'\mathbf{1}. \end{array}\right\} \tag{19.85}$$

Using (19.85) in (19.84), it becomes

$$(\mathbf{1}'\mathbf{V}^{-1}\mathbf{1} - n)\boldsymbol{\alpha}'\mathbf{V}^{-1}\boldsymbol{\alpha} \geq (\mathbf{1}'\mathbf{V}^{-1}\boldsymbol{\alpha})^2,$$

which we may rewrite, using (19.80) and (19.78),

$$\operatorname{var} \hat{\mu} \leq \sigma^2/n = \operatorname{var} \bar{y}. \tag{19.86}$$

(19.86) is obvious enough, since $\bar{y}$, the sample mean, is a linear estimator and therefore cannot have variance less than the MV estimator $\hat{\mu}$. But the point of the above argument is that it enables us to determine when (19.86) becomes a strict equality. This happens when the Cauchy inequality (19.82) becomes an equality, i.e. when $h_i = \lambda k_i$ for some constant λ, or

$$\mathbf{T}'\mathbf{b} = \lambda \mathbf{T}^{-1}\mathbf{c}.$$

From (19.83) this is

$$\mathbf{T}'(\mathbf{V}^{-1} - \mathbf{I})\mathbf{1} = \lambda \mathbf{T}^{-1}\boldsymbol{\alpha},$$

or

$$\mathbf{T}\mathbf{T}'(\mathbf{V}^{-1} - \mathbf{I})\mathbf{1} = \lambda \boldsymbol{\alpha}. \tag{19.87}$$

Using (19.81), (19.87) finally becomes

$$(\mathbf{I} - \mathbf{V})\mathbf{1} = \lambda \boldsymbol{\alpha}, \tag{19.88}$$

the condition that var $\hat{\mu} = \text{var } \bar{y} = \sigma^2/n$. If (19.88) holds, we must also have, by the uniqueness of the LS solution,

$$\hat{\mu} = \bar{y}, \tag{19.89}$$

and this may be verified by using (19.88) on $\hat{\mu}$ in (19.79).

   If the distribution is symmetrical, the situation simplifies. For then the vector of expectations

$$\boldsymbol{\alpha} = \begin{pmatrix} E(z_{(1)}) \\ \vdots \\ E(z_{(n)}) \end{pmatrix} = \begin{pmatrix} \alpha_1 \\ \vdots \\ \alpha_n \end{pmatrix}$$

has

$$\alpha_i = -\alpha_{n+1-i} \quad \text{all } i \tag{19.90}$$

as follows immediately from (14.2). Hence

$$\boldsymbol{\alpha}'\mathbf{V}^{-1}\mathbf{1} = \mathbf{1}'\mathbf{V}^{-1}\boldsymbol{\alpha} = 0 \tag{19.91}$$

and thus (19.79) becomes

$$\left. \begin{array}{l} \hat{\mu} = \mathbf{1}'\mathbf{V}^{-1}\mathbf{y}/\mathbf{1}'\mathbf{V}^{-1}\mathbf{1}, \\ \hat{\sigma} = \boldsymbol{\alpha}'\mathbf{V}^{-1}\mathbf{y}/\boldsymbol{\alpha}'\mathbf{V}^{-1}\boldsymbol{\alpha}, \end{array} \right\} \tag{19.92}$$

while (19.80) simplifies to

$$\begin{array}{l} \text{var } \hat{\mu} = \sigma^2/\mathbf{1}'\mathbf{V}^{-1}\mathbf{1}, \\ \text{var } \hat{\sigma} = \sigma^2/\boldsymbol{\alpha}'\mathbf{V}^{-1}\boldsymbol{\alpha}, \\ \text{cov } (\hat{\mu}, \hat{\sigma}) = 0. \end{array} \tag{19.93}$$

Thus the ordered LS estimators $\hat{\mu}$ and $\hat{\sigma}$ are uncorrelated if the distribution is symmetrical, an analogous result to that for ML estimators obtained in **18.34**.

Example 19.10

To estimate the mid-range (or mean) μ and range σ of the uniform distribution

$$dF(y) = dy/\sigma, \qquad \mu - \tfrac{1}{2}\sigma \leqslant y \leqslant \mu + \tfrac{1}{2}\sigma.$$

Using (19.41), it is easy to show that, standardising as in (19.71),

$$\alpha_r = E(z_{(r)}) = \{r/(n+1)\} - \tfrac{1}{2}, \tag{19.94}$$

and from (14.2) that the elements of the covariance matrix $\mathbf{V}$ of the $z_{(r)}$ are (cf. Example 11.4 for the variances)

$$V_{rs} = r(n-s+1)/\{(n+1)^2(n+2)\}, \qquad r \leqslant s. \tag{19.95}$$

The inverse of $\mathbf{V}$ is

$$\mathbf{V}^{-1} = (n+1)(n+2) \begin{pmatrix} 2 & -1 & & & & \\ -1 & 2 & & & \mathbf{0} & \\ & & & & & \\ & \mathbf{0} & & & & \\ & & & & & -1 \\ & & & & -1 & 2 \end{pmatrix}. \tag{19.96}$$

From (19.96),

$$\mathbf{1}'\mathbf{V}^{-1} = (n+1)(n+2) \begin{pmatrix} 1 \\ 0 \\ 0 \\ \vdots \\ 0 \\ 0 \\ 1 \end{pmatrix}. \tag{19.97}$$

and, from (19.94) and (19.96),

$$\boldsymbol{\alpha}'\mathbf{V}^{-1} = \tfrac{1}{2}(n+1)(n+2) \begin{pmatrix} -1 \\ 0 \\ 0 \\ \vdots \\ 0 \\ 0 \\ 1 \end{pmatrix}. \tag{19.98}$$

Using (19.97) and (19.98), (19.92) and (19.93) give

$$\left.\begin{aligned} \hat{\mu} &= \tfrac{1}{2}(y_{(1)} + y_{(n)}), \\ \hat{\sigma} &= (n+1)(y_{(n)} - y_{(1)})/(n-1), \\ \operatorname{var} \hat{\mu} &= \sigma^2/\{2(n+1)(n+2)\}, \\ \operatorname{var} \hat{\sigma} &= 2\sigma^2/\{(n-1)(n+2)\}, \\ \operatorname{cov}(\hat{\mu}, \hat{\sigma}) &= 0. \end{aligned}\right\} \tag{19.99}$$

Apart from the bias correction to $\hat{\sigma}$, these are essentially the results we obtain by the ML method. The agreement is to be expected, since $y_{(1)}$ and $y_{(n)}$ are a pair of jointly sufficient statistics for μ and σ, as we saw in effect in Example 17.21.

As will have been made clear by Example 19.10, in order to use the theory in **19.20–21**, we must determine the covariance matrix $\mathbf{V}$ of the standardised order-statistics, and this is a function of the form of the underlying distribution. This is in direct contrast with the general LS theory using unordered observations, discussed earlier in this chapter, which does not presuppose knowledge of the distributional form.

The general LS theory developed in this chapter is fundamental in many branches of statistical theory, and we shall use it repeatedly in later chapters.

Generalised linear models

19.22   For the linear models in **19.4** and **19.17**, we assumed that the expectation of **y** is linear in **θ** with known coefficients, and for testing purposes (cf. **19.10**) that the distribution of **y** is normal, with a specified form of covariance matrix. We now consider generalising these assumptions.

Clearly, the linearity in **θ** is an essential element in the relatively simple analysis of the model, but there is no obvious reason to choose the expectation of **y** as the linear function—more generally, we may allow a function of it to be linear in **θ**. Further, as we saw in Example 5.8, the normal distribution is a member of the natural exponential family, which may therefore be used as a generalisation of normality. Thus we arrive at the *generalised linear model* (so called by Nelder and Wedderburn (1972)) in which

(a)   writing $E(y) = \mu$, there is a function **g** such that $g(\mu) = \eta$ and $\eta = X\theta$;
(b)   each observation $y/a(\phi)$ is independently distributed in the same natural exponential form, where $a(\phi)$ is a scaling constant, possibly depending on a further parameter ϕ.

The function **g** is called the *link function*, because it connects the expectation of **y** with the function **η**, which is linear in **θ**. It is assumed to be monotonic and differentiable.

19.23   We write the distribution of each y in the form

$$\exp\left\{\frac{y\tau - b(\tau)}{a(\phi)} + c(y, \phi)\right\}, \tag{19.100}$$

where we use τ here instead of θ as in (5.117) to avoid confusion with the parameters in **θ** above. $y/a(\phi)$ in (19.100) plays the role of x in (5.117), and $-b(\tau)/a(\phi)$ that of $D(\theta)$ there. Thus from (5.121) we have

$$\kappa_r\{y/a(\phi)\} = \frac{\partial^r}{\partial \tau^r} b(\tau)/a(\phi)$$

so that for y we have

$$\kappa_r = b^{(r)}(\tau)\{a(\phi)\}^{r-1}, \tag{19.101}$$

whence

$$\left. \begin{array}{l} \mu = E(y) = b'(\tau) \\ \mathrm{var}\, y = b''(\tau)a(\phi). \end{array} \right\} \tag{19.102}$$

Thus the mean of y does not depend on the scaling constant, and the variance does so only as a multiplicative factor of the variance function

$$V = b''(\tau).$$

The substitutions

$$b(\tau) = \tfrac{1}{2}\tau^2, \qquad a(\phi) = \sigma^2$$

reduce (19.100) to the $N(\tau, \sigma^2)$ distribution, as the cumulants (19.102) make clear.

19.24 Ignoring constants,

$$\log L(y \mid \tau) = \{\textstyle\sum y_i \tau_i - \sum b(\tau_i)\} / a(\phi) \qquad (19.103)$$

with

$$\boldsymbol{\eta} = g(\boldsymbol{\mu}) = g\{b'(\tau_1), b'(\tau_2), \ldots, b'(\tau_n)\} = \mathbf{X}\boldsymbol{\theta}$$

using (19.102). In general, we cannot simplify $\log L$ further. However, if the link function g has the property that $\boldsymbol{\eta} = (\tau_1, \tau_2, \ldots, \tau_n)'$ (in which case the link is called *canonical*), the first term in (19.103) is equal to $\mathbf{y}'\mathbf{X}\boldsymbol{\theta}$, whence it is clear that the k statistics $\mathbf{y}'\mathbf{x}_j$ (where $\mathbf{x}_j$ is the jth column of $\mathbf{X}$) are (minimal) sufficient for the k parameters in $\boldsymbol{\theta}$. If $b(\tau) = \tfrac{1}{2}\tau^2$ (the normal case), (19.103) may be written

$$\log L = \{\mathbf{y}'\mathbf{X}\boldsymbol{\theta} - \tfrac{1}{2}(\mathbf{X}\boldsymbol{\theta})'\mathbf{X}\boldsymbol{\theta}\} / a(\phi)$$

so that putting $\partial \log L / \partial \boldsymbol{\theta} = \mathbf{0}$ leads back to the usual LS solution (19.12), as it must since we are now satisfying the conditions that we used there.

19.25 Because of its greater generality, we cannot expect the generalised linear model to give closed-form solutions as the ordinary linear model does. However, iterative ML estimation may be carried out using the methods given in **18.21** and **18.27**. The iterative procedure here becomes a weighted form of LS analysis—general details are given by McCullagh and Nelder (1983)—and it will be discussed in later volumes.

19.26 The generalised linear model, by making explicit the function linking a linear function of the parameters to the mean of an observed variable, is applicable to a large class of problems. In many cases, the single link function $\eta = \mu$ is used, but, for example, a set of variables may determine the value of a response variable. If the response variable is the probability of a certain event, we require estimates of it to be in $[0, 1]$. There is nothing in the ordinary linear model that would ensure this. In the generalised linear model, however, we could choose the link function $\eta = \log p$. η would then be on the negative real line, and we should investigate its dependence on $\mathbf{X}\boldsymbol{\theta}$. An alternative link function here is $\eta \equiv \log\{p/(1-p)\}$, using the whole real line. These generalised linear models, respectively known as log-linear and linear logistic models, will be considered in later volumes, which will also treat the theory generally.

19.27 McCullagh and Nelder (1983) summarise the history and development of the subject, which mainly took place in the 1970s, and give a general exposition of the theory, together with details of the computer programs available (notably GLIM) and a bibliography.

Other methods of estimation

19.28 We saw in the preceding chapter that, apart from the fact that they are functions of sufficient statistics for the parameters being estimated, the desirable properties of the ML estimators are all asymptotic ones, namely:

1. consistency;
2. asymptotic normality; and
3. efficiency.

As we saw in **17.29**, the ML estimator, $\hat{\theta}$, cannot be unique in the possession of these properties. For example, the addition to $\hat{\theta}$ of an arbitrary constant C/n^r will make no difference to its first-order properties if r is large enough. It is thus natural to inquire, as Neyman (1949) did, concerning the class of estimators which share the asymptotic properties of $\hat{\theta}$. Added interest is lent to the inquiry by the numerical tedium sometimes involved (cf. Examples 18.3, 18.9, 18.10) in evaluating the ML estimator.

19.29 Suppose that we have s (≥ 1) samples, with n_i observations in the ith sample. As at **18.19**, we simplify the problem by supposing that each observation in the ith sample is classified into one of k_i mutually exclusive and exhaustive classes. If π_{ij} is the probability of an observation in the ith sample falling into the jth class, we therefore have

$$\sum_{j=1}^{k_i} \pi_{ij} = 1, \qquad (19.104)$$

and we have reduced the problem to one concerning a set of s multinomial distributions. Let n_{ij} be the number of ith sample observations actually falling into the jth class, and $p_{ij} = n_{ij}/n_i$ the corresponding relative frequency. The probabilities π_{ij} are functions of a set of unknown parameters $(\theta_1, \ldots, \theta_r)$.

A continuous function T of the random variables p_{ij} is called a best asymptotically normal estimator (abbreviated BAN estimator) of θ_1, one of the unknown parameters, if

1. $T(\{p_{ij}\})$ is consistent for θ_1;
2. T is asymptotically normal as $N = \sum_{i=1}^{s} n_i \to \infty$;
3. T is efficient; and
4. $\partial T / \partial p_{ij}$ exists and is continuous in p_{ij} for all i, j.

The first three of these conditions are precisely those we have already proved for the ML estimator in Chapter 18. It is easily verified that the ML estimator also possesses the fourth property in this multinomial situation. Thus the class of BAN estimators contains the ML estimator as a special case.

19.30 Neyman showed that a set of necessary and sufficient conditions for an estimator to be BAN is that

1. $T(\{\pi_{ij}\}) \equiv \theta_1$;

2.   condition (4) of **19.29** holds; and

3. $\sum_{i=1}^{s} (1/n_i) \sum_{j=1}^{k_i} [(\partial T/\partial p_{ij})_{p_{ij}=\pi_{ij}}]^2 \pi_{ij}$ be minimised for variation in $\partial T/\partial p_{ij}$.

Condition (1) is enough to ensure consistency: it is, in general, a stronger condition than consistency.* For since the statistic T is a continuous function of the p_{ij}, and the p_{ij} converge in probability to the π_{ij}, T converges in probability to $T(\{\pi_{ij}\})$, i.e. to θ_1.

Condition (3) is simply the efficiency condition, for the function there to be minimised is simply the variance of T subject to the necessary condition for a minimum $\sum_j (\partial T/\partial p_{ij})_{p_{ij}=\pi_{ij}} \pi_{ij} = 0$.

As they stand, these three conditions are not of much practical value. However, Neyman also showed that a sufficient set of conditions is obtainable by replacing (3) by a direct condition on $\partial T/\partial p_{ij}$, which we shall not give here. From this he deduced that

(a)   the ML estimator is a BAN estimator, as we have already seen;
(b)   the class of estimators known as minimum chi-square estimators are also BAN estimators.

We now proceed to examine this second class of estimators.

For later work on BAN estimators, see Bemis and Bhapkar (1983).

Minimum chi-square estimators

19.31   Referring to the situation described in **19.29**, a statistic T is called a minimum chi-square (abbreviated MCS) estimator of θ_1, if it is obtained by minimising, with respect to θ_1, the expression

$$\chi^2 = \sum_{i=1}^{s} n_i \sum_{j=1}^{k_i} \frac{(p_{ij} - \pi_{ij})^2}{\pi_{ij}} = \sum_{i=1}^{s} n_i \left(\sum_{j=1}^{k_i} \frac{p_{ij}^2}{\pi_{ij}} - 1 \right), \tag{19.105}$$

where the π_{ij} are functions of $\theta_1, \ldots, \theta_r$. To minimise (19.105), we put

$$\frac{\partial \chi^2}{\partial \theta_1} = -\sum_i n_i \sum_j \left(\frac{p_{ij}}{\pi_{ij}} \right)^2 \frac{\partial \pi_{ij}}{\partial \theta_1} = 0, \tag{19.106}$$

and a root of (19.106), regarded as an equation in θ_1, is the MCS estimator of θ_1. Evidently, we may generalise (19.106) to a set of r equations to be solved together to find the MCS estimators of $\theta_1, \ldots, \theta_r$.

The procedure for finding MCS estimators is quite analogous to that for finding ML estimators, discussed in Chapter 18. Moreover, the (asymptotic) properties of MCS estimators are similar to those of ML estimators. In fact, there is, with probability 1, a unique consistent root of the MCS equations, and this corresponds to the absolute minimum value (infimum) of (19.105). The proofs are given, for the commonest case $s = 1$, by C. R. Rao (1957).

---

* In fact (1) is the form in which consistency was originally defined by Fisher (1921a), as we mentioned in **17.7**.

19.32   A modified form of MCS estimator is obtained by minimising

$$(\chi')^2 = \sum_{i=1}^{s} n_i \sum_{j=1}^{k_i} \frac{(p_{ij} - \pi_{ij})^2}{p_{ij}} = \sum_i n_i \left(\sum_j \frac{\pi_{ij}^2}{p_{ij}} - 1 \right) \qquad (19.107)$$

instead of (19.105). In (19.107), we assume that no $p_{ij} = 0$. To minimise it for variation in θ_1, we put

$$\frac{\partial(\chi')^2}{\partial \theta_1} = 2 \sum_i n_i \sum_j \left(\frac{\pi_{ij}}{p_{ij}} \right) \frac{\partial \pi_{ij}}{\partial \theta_1} = 0 \qquad (19.108)$$

and solve for the estimator of θ_1. These modified MCS estimators have also been shown to be BAN estimators by Neyman (1949).

Choice between methods
19.33   Since the ML, the MCS and the modified MCS methods all have the same asymptotic properties, the choice between them must rest, in any particular case, either on the grounds of computational convenience, or on those of superior sampling properties in small samples, or on both. As to the first ground, there is little that can be said in general. Sometimes the ML, and sometimes the MCS, equation is the more difficult to solve. But when dealing with a continuous distribution, the observations *must* be grouped in order to make use of the MCS method, and it seems rather wasteful to impose an otherwise unnecessary grouping for estimation purposes. Furthermore, there is, especially for continuous distributions, preliminary inconvenience in having to determine the π_{ij} in terms of the parameters to be estimated. Our own view is therefore that the now traditional leaning towards ML estimation is fairly generally justifiable on computational grounds. The following example illustrates the MCS computational procedure in a relatively simple case.

Example 19.11
Consider the estimation, from a single sample of n observations, of the parameter θ of a Poisson distribution. We have seen (Examples 17.8, 17.15) that the sample mean $\bar{x}$ is an MVB sufficient estimator of θ, and it follows from **18.5** that $\bar{x}$ is also the ML estimator.

The MCS estimator of θ, however, is not equal to $\bar{x}$, illustrating the point that MCS methods do not necessarily yield a single sufficient statistic if one exists.

The theoretical probabilities here are

$$\pi_j = e^{-\theta} \theta^j / j!, \qquad j = 0, 1, 2, \ldots,$$

so that

$$\frac{\partial \pi_j}{\partial \theta} = \pi_j \left(\frac{j}{\theta} - 1 \right).$$

The minimising equation (19.106) is therefore, dropping the factor n,

$$\sum_j \frac{p_j^2}{\pi_j} \left(1 - \frac{j}{\theta} \right) = 0. \qquad (19.109)$$

This is the equation to be solved for θ, and we use an iterative method of solution similar to that used for the ML estimator at **18.21**. We expand the left-hand side of (19.109) in a Taylor series as a function of θ about the sample mean $\bar{x}$, regarded as a trial value. We obtain to the first order of approximation

$$\sum_j \frac{p_j^2}{\pi_j}\left(1 - \frac{j}{\theta}\right) = \sum_j \frac{p_j^2}{m_j}\left(1 - \frac{j}{\bar{x}}\right) + (\theta - \bar{x})\sum_j \frac{p_j^2}{m_j}\left\{\frac{j}{\bar{x}^2} + \left(1 - \frac{j}{\bar{x}}\right)^2\right\}, \qquad (19.110)$$

where we have written $m_j = e^{-\bar{x}}\bar{x}^j/j!$. From (19.109), we find

$$(\theta - \bar{x}) = \bar{x} \cdot \frac{\displaystyle\sum_j \frac{p_j^2}{m_j}(j - \bar{x})}{\displaystyle\sum_j \frac{p_j^2}{m_j}\{j + (j - \bar{x})^2\}}. \qquad (19.111)$$

We use (19.111) to find an improved estimate of θ from $\bar{x}$, and repeat the process as necessary—cf. (18.45) for the ML estimator.

As a numerical example, we use Whitaker's (1914) data on the number of deaths of women over 85 years old reported in *The Times* newspaper for each day of 1910–12, 1096 days in all. The distribution is given in the first two columns of Table 19.1.

Table 19.1

No. of deaths (j)	Frequency reported (np_j)	nm_j	$\dfrac{np_j^2}{m_j}$	$(j - \bar{x})$	$\dfrac{np_j^2}{m_j}(j - \bar{x})$	$\{j + (j - \bar{x})^2\}$	$\dfrac{np_j^2}{m_j}\{j + (j - \bar{x})^2\}$
0	364	336.25	394.1	−1.1816	−465.7	1.396	551.1
1	376	397.30	355.8	−0.1816	−64.6	1.033	365.9
2	218	234.72	202.5	0.8184	165.8	2.670	540.6
3	89	92.45	85.69	1.8184	155.8	6.307	540.4
4	33	27.31	39.87	2.8184	112.4	11.943	476.1
5	13	6.45	26.20	3.8184	100.0	19.580	512.9
6	2	1.27	3.15	4.8184	15.2	29.217	92.0
7	1	0.25	4.00	5.8184	23.3	40.854	163.4
Total	$n = 1096$	1096.00			+42.2		3242.4

The mean number of deaths reported is found to be $\bar{x} = 1295/1096 = 1.181569$. This is therefore the ML estimator, and we use it as our first trial value for the MCS estimator. The third column of the table gives the expected frequencies in a Poisson distribution with parameter equal to $\bar{x}$, and the necessary calculations for (19.111) are set out in the remaining five columns.

Thus, from (19.111), we have

$$\theta = 1.1816\left\{1 + \frac{42.2}{3242.4}\right\}$$

$$= 1.198$$

as our improved value. K. Smith (1916) reported a value of 1.196903 when working to greater accuracy, with more than one iteration of this procedure.

Smith also gives details of the computational procedure when we are estimating the parameters of a continuous distribution. This is considerably more laborious.

19.34 Small-sample properties, the second ground for choice between the ML and MCS methods, are more amenable to general inquiry. C. R. Rao (1961, 1962a) defines a concept of second-order efficiency—cf. the concept of deficiency in **17.29**— and shows that in the multinomial model of **18.20**, the ML is the only BAN estimator with optimum second-order efficiency, under regularity conditions; essentially, the $O(n^{-2})$ term in the variance in (18.42) is minimised.

We have already seen an instance of an ML estimator with poor m.s.e. properties in **18.19**. Berkson (1955, 1956) has given another, in which the ML estimator is sometimes infinite, while another BAN estimator has smaller m.s.e. (These papers should be read with Silverstone (1957), which points out some errors in them.) J. K. Ghosh and Subramanyam (1974) extend Rao's results to the exponential family, and show that if the ML estimator in Berkson's problem is adjusted to have the same first-order bias as the other BAN estimator, it has the smaller variance. Amemiya (1980) confirmed both Berkson's and Ghosh and Subramanyam's results in detail. J. K. Ghosh and Sinha (1981) give a necessary and sufficient condition for it to be possible to improve the m.s.e. of the ML estimator, and L. Davis (1984) gives examples in which the ML estimator has the smaller m.s.e.

Statistical curvature
19.35 Efron (1975)—see also Efron (1978) and Amari (1982a, b)—obtains closely related results using a concept of *statistical curvature* that measures departure from the exponential family by

$$\gamma^2(\theta) = \frac{\operatorname{var} f''}{\{\operatorname{var}(f')\}^2} \left[1 - \frac{\{\operatorname{cov}(f', f'')\}^2}{\operatorname{var}(f') \operatorname{var}(f'')} \right] \tag{19.112}$$

where $f^{(r)} = \partial^r \log f(x|\theta)/\partial \theta^r$.

By the Cauchy–Schwarz inequality, the term in square brackets lies between 0 and 1, attaining 0 if and only if f' is a linear function of f''. Exercise 19.24 asks the reader to show that this is true for any member of the exponential family (17.30). Interpreting (19.112) as in Exercise 26.23 or (27.27) below, we see that $\gamma^2\{\operatorname{var}(f')\}^2$ is the variance of f'' after regression upon f'.

If we reparametrise from θ to a monotone function $\tau(\theta)$, γ^2 is invariant, while if we replace $f(x|\theta)$ by the LF $L(x|\theta)$ based on n observations, γ^2 becomes γ^2/n—cf. Exercise 19.24.

The Bhattacharyya bound to order n^{-2} for the estimation of θ at (17.47) may be written

$$\operatorname{var} t \geqslant \mathrm{MVB} + \frac{B(\theta)}{n^2} + o\left(\frac{1}{n^2}\right). \tag{19.113}$$

As we saw in **17.18** and Example 17.11, attainability of the MVB and of the Bhattacharyya bound depends on the parametrisation used—for the binomial discussed in Example 17.11, the MVB is attained for θ itself, while the Bhattacharyya bound is attained for $\theta(1-\theta)$, and neither is attained for an arbitrary function of θ. Efron (1975) obtains an essentially different variance bound for efficient estimators t of θ biased to at most $O(n^{-2})$,

$$\text{var } t \geqslant \text{MVB} + \frac{1}{n^2}\{B(\theta) + \gamma^2(\theta) + \Delta(t,\theta)\} + o\left(\frac{1}{n^2}\right) \tag{19.114}$$

where γ^2 is as at (19.112) and Δ is also non-negative. Since γ^2 is invariant under reparametrisation, its effect cannot be removed from (19.114), but Δ may be reduced to zero if we put $t = \hat{\theta}$, the ML estimator. The $O(n^{-2})$ term in (19.114) is then minimised, giving the ML estimator optimum second-order efficiency.

Although the bound for efficient estimators (19.114) is at least as great as the MVB and the Bhattacharyya bounds, it does not prevent attainment of these, any more than the second of these prevents the attainment of the first. In the exponential family, (19.114) reduces to the Bhattacharyya bound if $\hat{\theta}$ is used.

EXERCISES

19.1 In the linear model (19.8), suppose that a further parameter θ_0 is introduced, so that we have the new model

$$\mathbf{y} = \mathbf{X\theta} + \mathbf{1}\theta_0 + \mathbf{\varepsilon},$$

where $\mathbf{1}$ is an $(n \times 1)$ vector of units. Show that the LS estimator in the new model of $\mathbf{\theta}$, the original vector of k parameters, remains of exactly the same form (19.12) as in the original model, with y_i replaced by $(y_i - \bar{y})$ and x_{ij} by $(x_{ij} - \bar{x}_j)$ for $i = 1, 2, \ldots, n$ and $j = 1, 2, \ldots, k$.

19.2 If, in the linear model (19.8), we replace the simple covariance matrix (19.10) by a non-singular covariance matrix $\sigma^2\mathbf{V}$ that allows correlations and unequal variances among the ε_i, show by putting $\mathbf{w} = \mathbf{T'y}$, where $\mathbf{TT'} = \mathbf{V}^{-1}$, that the LS and MV unbiased estimator of $\mathbf{C\theta}$ is

$$\mathbf{C\hat{\theta}} = \mathbf{C}(\mathbf{X'V}^{-1}\mathbf{X})^{-1}\mathbf{X'V}^{-1}\mathbf{y}.$$

(Cf. Aitken, 1935; Plackett, 1949)

19.3 Generalising (19.35), show that if $E(\mathbf{\varepsilon\varepsilon'}) = \sigma^2\mathbf{V}$, $E(\mathbf{\varepsilon'B\varepsilon}) = \sigma^2 \operatorname{tr}(\mathbf{BV})$. Show further that $\operatorname{var}(\mathbf{\varepsilon'B\varepsilon}) = 2\sigma^4 \operatorname{tr}(\mathbf{BVBV})$ if $\mathbf{\varepsilon}$ is multinormal.

19.4 Show that in **19.11** the ratio $(\mathbf{X\hat{\theta}})'(\mathbf{X\hat{\theta}})/(ks^2)$ is distributed in Fisher's F distribution with k and $(n - k)$ degrees of freedom if $\mathbf{\theta} = \mathbf{0}$.

19.5 In Exercise 19.2, show that, generalising (19.26),

$$\mathbf{V}(\mathbf{C\hat{\theta}}) = \sigma^2 \cdot \mathbf{C}(\mathbf{X'V}^{-1}\mathbf{X})^{-1}\mathbf{C'},$$

and that the generalisation of (19.37) is

$$E\{(\mathbf{y} - \mathbf{X\hat{\theta}})'\mathbf{V}^{-1}(\mathbf{y} - \mathbf{X\hat{\theta}})\} = E\{\mathbf{\varepsilon}'[\mathbf{V}^{-1} - \mathbf{V}^{-1}\mathbf{X}(\mathbf{X'V}^{-1}\mathbf{X})^{-1}\mathbf{X'V}^{-1}]\mathbf{\varepsilon}\} = (n - k)\sigma^2.$$

19.6 In Exercises 19.2 and 19.5, show that (19.25) remains true.

When $k = 1$, $\mathbf{X} = \mathbf{1}$, $\mathbf{C} = 1$ and the covariance matrix $\mathbf{V}$ is diagonal with elements V_j, show that the LS estimator of θ, the common mean of the uncorrelated observations y_j, is

$$\hat{t} = \sum V_j^{-1} y_j / \sum V_j^{-1} = \sum w_j y_j, \quad \text{say},$$

with variance $(\sum V_j^{-1})^{-1} = w_j V_j$.

Let $t = \mathbf{l'y}$ be any other linear estimator with $\mathbf{l}$ a random variable distributed completely independently of $\mathbf{y}$ and satisfying (19.19), $\mathbf{l'1} = 1$. Using (19.25), show that

$$\operatorname{var} t = \operatorname{var} \hat{t}\{1 + \sum_j w_j^{-1} E(l_j - w_j)^2\},$$

depending on the distribution of $\mathbf{l}$ only through the individual m.s.e. of the l_j, even though the latter may be dependent. Show that if the l_j are consistent estimators of the w_j, $\operatorname{var} t \sim \operatorname{var} \hat{t}$.

Given that all V_j are equal, show that

$$\operatorname{var} t = \operatorname{var} \hat{t}(1 + U^2),$$

where U is the coefficient of variation of the l_j.

(Cf. Rubin and Weisberg, 1975)

19.7 In **19.14**, show using (19.51) that $(\mathbf{X'X} + \mathbf{B'B})^{-1}\mathbf{B'B} = \mathbf{D}(\mathbf{BD})^{-1}\mathbf{B}$ and hence that (19.58) gives

$$\mathbf{V}(\mathbf{\hat{\theta}})(\mathbf{X'X}) = \sigma^2\{\mathbf{I}_k - \mathbf{D}(\mathbf{BD})^{-1}\mathbf{B}\}.$$

(Plackett, 1950)

19.8 Using the first result of Exercise 19.7, modify the argument of **19.9** to show that the unbiased estimator of σ^2 in the singular case is $[1/(n - r)](\mathbf{y} - \mathbf{X\hat{\theta}})'(\mathbf{y} - \mathbf{X\hat{\theta}})$.

(Plackett, 1950)

19.9   For the linear model (19.8-10), show using **19.9** that the quadratic form $\theta'A\theta$ is unbiasedly estimated by $\hat{\theta}'A\hat{\theta} - s^2 \operatorname{tr}\{(X'X)^{-1}A\}$.

19.10   Show that in the case of a symmetrical distribution, the condition that the ordered LS estimator $\hat{\mu}$ in (19.92) is equal to the sample mean $\bar{y} = 1'y/1'1$ is that

$$V1 = 1,$$

i.e. that the sum of each row of the covariance matrix be unity. Show (cf. Exercise 14.15) that this property holds for the univariate normal distribution.

(Lloyd, 1952; Stephens (1975) shows that asymptotically
$V\alpha = \frac{1}{2}\alpha$ in the normal case.)

19.11   For the exponential distribution

$$dF(y) = \exp\left\{-\left(\frac{y-\mu}{\sigma}\right)\right\} dy/\sigma, \qquad \sigma > 0; \quad \mu \leqslant y < \infty,$$

show that in (19.72) the elements of α are

$$\alpha_r = \sum_{i=1}^{r} (n-i+1)^{-1}$$

and that those of V are

$$V_{rs} = \sum_{i=1}^{m} (n-i+1)^{-2} \quad \text{where } m = \min(r, s).$$

Hence verify that V^{-1} has elements all zero except

$$V_{r,r+1}^{-1} = V_{r+1,r}^{-1} = -(n-r)^2,$$
$$V_{r,r}^{-1} = (n-r)^2 + (n-r+1)^2.$$

19.12   In Exercise 19.11, show from (19.79) that the MV unbiased estimators are

$$\hat{\mu} = y_{(1)} - (\bar{y} - y_{(1)})/(n-1) \quad \text{and} \quad \hat{\sigma} = n(\bar{y} - y_{(1)})/(n-1).$$

Compare these with the ML estimators of the same parameters.

(Sarhan, 1954)

19.13   Show that when all the n_i are large the minimisation of the chi-squared expression (19.105) or (19.107) gives the same estimator as the ML method.

19.14   In the case $s = 1$, show that the first two moments of the statistic (19.105) are given by

$$n_1^2 E(\chi^2) = k_1 - 1,$$

$$n_1^4 \operatorname{var}(\chi^2) = 2(k_1 - 1)\left(1 - \frac{1}{n_1}\right) + \frac{1}{n_1}\sum_{j=1}^{k_1}\frac{1}{\pi_{1j}} - \frac{k_1^2}{n_1},$$

and that for any $c > 0$, the generalisation of (19.107) has expectation

$$E\left\{\sum_{j=1}^{k_1}\frac{(n_{1j} - n_1)^2 + b}{n_{1j} + c}\right\} = k_1 - 1 + \frac{1}{n_1}\left[(b - c + 2)\sum_{j=1}^{k_1}\frac{1}{\pi_{1j}} - (3 - c)k_1 + 1\right] + O\left(\frac{1}{n_1^2}\right).$$

Thus, to the second order at least, the π_{1j} disappear from the expectation if $b = c - 2$. If $b = 0$, $c = 2$, it is $(k_1 - 1)(1 - (1/n_1))$ and if $b = 1$, $c = 3$, it is $(k_1 - 1) + (1/n_1)$, which for $k_1 > 2$ is even closer to the expectation of (19.105).

(F. N. David, 1950; Haldane, 1955)

19.15   For a binomial distribution with probability of success equal to π, show that the MCS estimator of π obtained from (19.105) is identical with the ML estimator for any n, and that if the number of successes is not 0 or n, the modified MCS estimator obtained from (19.107) is also identical with the ML estimator.

19.16 y_i is a Poisson variable with parameter θx_i, where x_i is a constant observed with y_i, ($i = 1, 2, \ldots, n$). Show that the ML estimator of θ is $\sum y_i / \sum x_i$, with asymptotic variance $\theta / \sum x_i$, and that the LS estimator $\sum y_i x_i / \sum x_i^2$ has exact variance $\theta \sum x_i^3 / (\sum x_i^2)^2$. Hence show that the LS estimator is inefficient unless the x_i are all equal. Explain the result.

19.17 Show by using a vector of Lagrangian undetermined multipliers that the minimisation of (19.11) subject to p linear constraints (19.53) yields the LS estimator

$$\hat{\boldsymbol{\theta}} = (\mathbf{X}'\mathbf{X})^{-1}[\mathbf{X}'\mathbf{y} - \mathbf{B}'\{\mathbf{B}(\mathbf{X}'\mathbf{X})^{-1}\mathbf{B}'\}^{-1}\{\mathbf{B}(\mathbf{X}'\mathbf{X})^{-1}\mathbf{X}'\mathbf{y} - \mathbf{a}\}].$$

(Kreuger and Neudecker (1977) treat the singular case.)

19.18 In **19.9**, show that the LS residuals $\mathbf{y} - \mathbf{X}\hat{\boldsymbol{\theta}}$ have covariance matrix $\{\mathbf{I}_n - \mathbf{X}(\mathbf{X}'\mathbf{X})^{-1}\mathbf{X}'\}\sigma^2$, so that they are correlated. Show that if the linear model (19.8) is fitted to the first m of N observations only, the m residuals thus obtained are uncorrelated with the $(N - m)$ residuals obtained from the other observations when (19.8) is fitted to all N observations, and that this result holds also if the model is singular. Hence, if we successively put $m = 1, 2, \ldots, n-1$ and $N = m + 1$, we generate a set of n uncorrelated residuals (independent if the ε_i are normal).

(Cf. Hedayat and Robson, 1970)

19.19 Show that if $\mathbf{y}$ is distributed in the form $f(y \mid \theta) = \exp(\theta y) B(y) C(\theta)$, its c.f. is $C(\theta)/C(\theta + it)$ with cumulants $\kappa_1 = -C'(\theta)/C(\theta) = p(\theta)$, $\kappa_2 = \partial \kappa_1 / \partial \theta$. Hence show that if for n such independent y_j with differing means $p_j(\theta)$ the distributions are re-scaled to have the same variance, the LS estimator of θ is identical with the ML estimator.

(Charnes *et al.* (1976) generalise this.)

19.20 Let $\mathbf{A} = \mathbf{X}'\mathbf{X}$ in (19.12), and the row vector $\mathbf{x}$ contain one further observation on each variable. Show that the inverse of $\mathbf{B} = \mathbf{A} + \mathbf{x}'\mathbf{x}$ is

$$\mathbf{B}^{-1} = \mathbf{A}^{-1} - (\mathbf{A}^{-1}\mathbf{x}'\mathbf{x}\mathbf{A}^{-1})/(1 + \mathbf{x}\mathbf{A}^{-1}\mathbf{x}').$$

Hence show that a quadratic form $\mathbf{x}\mathbf{A}^{-1}\mathbf{x}'$ may be evaluated by
$$1 + \mathbf{x}\mathbf{A}^{-1}\mathbf{x}' = |\mathbf{A} + \mathbf{x}'\mathbf{x}|/|\mathbf{A}|$$
and that the estimator $\hat{\boldsymbol{\theta}}$ at (19.12) is adjusted by

$$\hat{\boldsymbol{\theta}}_{n+1} = \hat{\boldsymbol{\theta}}_n + \mathbf{A}^{-1}\mathbf{x}'(y_{n+1} - \mathbf{x}\hat{\boldsymbol{\theta}}_n)/(1 + \mathbf{x}\mathbf{A}^{-1}\mathbf{x}').$$

(Cf. Bartlett, 1951)

19.21 Show that the linear function $\mathbf{c}_1\hat{\boldsymbol{\theta}}$ ($\mathbf{c}_1\mathbf{c}_1' = 1$) has smallest minimised variance (19.26) when $\mathbf{c}_1$ is the latent vector corresponding to the largest latent root of $\mathbf{X}'\mathbf{X}$, and similarly that each of the set of linear functions $\{\mathbf{c}_j\hat{\boldsymbol{\theta}}\}$, $j = 1, 2, \ldots, p$, has smallest minimised variance subject to orthogonality to the others when the $\mathbf{c}_j$ are the latent vectors corresponding to the p largest latent roots of $\mathbf{X}'\mathbf{X}$.

Show that $\mathbf{c}_j\hat{\boldsymbol{\theta}}$ is the LS estimator of λ_j in the model

$$\mathbf{y} = \mathbf{X}\mathbf{C}'\boldsymbol{\lambda} + \boldsymbol{\eta},$$

where $\boldsymbol{\lambda}$ is ($p \times 1$) and $\mathbf{C}$ is the ($p \times k$) matrix whose rows are the $\mathbf{c}_j$.

(E. Greenberg, 1975; the $\mathbf{X}\mathbf{c}_j'$ are the first p *principal components* of $\mathbf{X}$, to be discussed in a later volume.)

19.22 Generalising Exercise 19.1, show that if we extend (19.8) to

$$\mathbf{y} = \mathbf{X}\boldsymbol{\theta} + \mathbf{Z}\boldsymbol{\theta}_0 + \boldsymbol{\varepsilon},$$

where $\boldsymbol{\theta}_0$ is a vector of additional parameters, the new LS estimator of the original parameter $\boldsymbol{\theta}$ is the solution of

$$(\mathbf{I} - \mathbf{M})\mathbf{y} = (\mathbf{I} - \mathbf{M})\mathbf{X}\boldsymbol{\theta},$$

where $\mathbf{M} = \mathbf{Z}(\mathbf{Z'Z})^{-1}\mathbf{Z'}$. Show that if θ_0 is a scalar and $\mathbf{Z}$ is a vector whose first n_1 elements are unity and the remaining $n_2 = n - n_1$ elements zero, $\hat{\boldsymbol{\theta}}$ is as in the original model, except that for $i = 1, 2, \ldots, n_1$, y_i is replaced by

$$\left(y_i - \frac{1}{n_1}\sum_{i=1}^{n_1} y_i\right) \quad \text{and } x_{ij} \text{ by } \left(x_{ij} - \frac{1}{n_1}\sum_{i=1}^{n_1} x_{ij}\right).$$

In Example 19.3, show that $\hat{\theta}_2$ is thus changed from

$$\frac{\text{cov}(x_2, y)}{\text{var } x_2}$$

to

$$\hat{\theta}_2 = \frac{n_1 \text{ cov}_1(x_2, y) + n_2 \text{ cov}_2(x_2, y)}{n_1 \text{ var}_1 x_2 + n_2 \text{ var}_2 x_2},$$

where cov_1, var_1 refer to the first n_1 observations and cov_2, var_2 to the other n_2.

19.23 In **19.12** show that the ridge repression estimate of $\mathbf{A\theta} = \boldsymbol{\phi}$ obtained from $\mathbf{y} = \mathbf{XA}^{-1}\boldsymbol{\phi} + \boldsymbol{\varepsilon}$ is

$$(\mathbf{A'}^{-1}\mathbf{X'XA}^{-1} + c\mathbf{I})^{-1}\mathbf{A'}^{-1}\mathbf{X'y}$$

and hence that $\boldsymbol{\theta}$ is estimated by

$$\boldsymbol{\theta}_c^* = (\mathbf{X'X} + c\mathbf{A'A})^{-1}\mathbf{X'y}$$

agreeing with $\boldsymbol{\theta}_c^*$ in **19.12** if and only if $\mathbf{A'A} = \mathbf{I}$, i.e. $\mathbf{A}$ is orthogonal. Show that $\boldsymbol{\theta}_c^*$ for any choice of $\mathbf{A}$ satisfies

$$(\boldsymbol{\theta}_c^* - \tfrac{1}{2}\boldsymbol{\theta}_0^*)'\mathbf{X'X}(\boldsymbol{\theta}_c^* - \tfrac{1}{2}\boldsymbol{\theta}_0^*) \leq \tfrac{1}{4}(\mathbf{X\theta}_0^*)'(\mathbf{X\theta}_0^*),$$

where $\boldsymbol{\theta}_0^*$ is the LS estimator.

(Leamer (1981) shows that within this bound, choice of parametrisation can seriously affect inference.)

19.24 Show that the statistical curvature defined at (19.112) is zero for the exponential family (17.30), that quite generally it is invariant under monotone reparametrisation, and that for an LF based on n independent observations from $f(x|\theta)$, it becomes equal to γ^2/n.

19.25 Show that we may obtain correct results for the linear model (19.8) by using the augmented model in Exercise 19.1 with each observation counted twice, the signs of $\mathbf{y}$ and $\mathbf{X}$ being changed on the second occasion. Show that if we analyse this model with $2n$ observations, then $\hat{\theta}_0 = 0$, $\mathbf{X'y}$ and $\mathbf{X'X}$ are doubled and therefore $\hat{\boldsymbol{\theta}}$ is given by (19.12) as before, while the estimator of error variance, (19.38), must be multiplied by $[2n - (k+1)]/[2(n-k)]$ and the estimator of the covariance matrix (19.16) by $[2n - (k+1)]/(n-k)$.

CHAPTER 20

INTERVAL ESTIMATION

20.1 In the previous three chapters we have been concerned with methods that provide an estimate of the value of one or more unknown parameters; and the methods gave functions of the sample values—the estimators—which, for any given sample, provided a unique estimate. It was recognised that the estimate might differ from the parameter in any particular case, and hence that there was a margin of uncertainty, which was expressed in terms of the sampling variance of the estimator. With the somewhat intuitive approach that has served our purpose up to this point, we might have said that it is more probable than not that θ lies in the range $t \pm \sqrt{(\text{var } t)}$, very probable that it lies in the range $t \pm 2\sqrt{(\text{var } t)}$, and so on. We thus locate θ in a range and not at a particular point, although one point in the range, t itself, is the essential estimate of θ.

20.2 In the present chapter we shall examine this procedure more closely and look at the problem of estimation from a different point of view. We now abandon attempts to estimate θ by a function which, for a specified sample, gives a unique estimate. Instead, we shall consider the specification of a range in which θ lies. The method we shall first discuss, which uses *confidence intervals*, relies only on the theory of probability without importing any new principle of inference. Other methods, to be discussed later, explicitly require something beyond this.

Confidence intervals
20.3 Consider first a distribution dependent on a single unknown parameter θ and suppose that we are given a random sample of n values $x_1, \ldots, x_n$ from it. Let z be a random variable dependent on the xs and on θ, with d.f. $F(z)$ which we shall initially assume does not depend on θ. Then (at least in the case where z is continuously distributed) we can find a fixed value z_1 such that

$$F(z_1) = P\{z(x_1, \ldots, x_n, \theta) \leq z_1\} = 1 - \alpha. \tag{20.1}$$

The inequality $z(x_1, \ldots, x_n, \theta) \leq z_1$ implies a restriction on the value of θ. Now it may happen that this restriction takes the form $t_0 \leq \theta \leq t_1$ where t_0 and t_1 are values (possibly infinite) of a statistic $t(x_1, \ldots, x_n)$ not dependent on θ. Then (20.1) is equivalent to

$$P(t_0 \leq \theta \leq t_1) = 1 - \alpha. \tag{20.2}$$

741

For example, if $z = \bar{x} - \theta$, the inequality $\bar{x} - \theta \leqslant z_1$ is equivalent to $\bar{x} - z_1 \leqslant \theta$, so (20.1) implies

$$P(\bar{x} - z_1 \leqslant \theta < \infty) = 1 - \alpha. \tag{20.3}$$

Here t_1 is infinite, just as t_0 would be negatively infinite if we were to reverse the inequality in (20.1).

More generally, even if $F(z)$ depends on θ, we may still construct probabilistic statements like (20.2), as we shall see in Example 20.3.

20.4 In (20.2), we have the probability that two randomly varying statistics, t_0 and t_1, will have θ between them, i.e. that the interval (t_0, t_1) covers θ. Instead of seeking a single statistic to estimate θ, we seek a pair that will cover it.

Thus, if we state that the inequality $t_0 \leqslant \theta \leqslant t_1$ in (20.2) is true for any sample, we shall be correct in a proportion $(1 - \alpha)$ of samples in the long run. We stress that this remains true however θ may vary—not merely for repeated samples from a population with fixed θ, but for repeated samples from populations with varying θ.

20.5 The interval (t_0, t_1) is called a *confidence interval* for θ, while t_0 and t_1 are called the lower and upper *confidence limits*. Intervals like (20.2) are called two-sided, and those like (20.3) are called one-sided. For fixed *confidence coefficient* (or *confidence level*) $1 - \alpha$, the totality of the confidence intervals for θ over all possible samples determines a zone called a *confidence belt*—it will be graphically illustrated in Example 20.3.

The ideas and methods of confidence interval estimation are due to J. Neyman—see especially Neyman (1937).

Example 20.1
Suppose we have a sample of size n from the normal distribution with known variance (taken without loss of generality to be unity)

$$dF = \frac{1}{\sqrt{(2\pi)}} \exp\left\{-\tfrac{1}{2}(x - \mu)^2\right\} dx, \qquad -\infty < x < \infty.$$

The distribution of the sample mean $\bar{x}$ is exactly normal,

$$dF = \left(\frac{n}{2\pi}\right)^{1/2} \exp\left\{-\frac{n}{2}(\bar{x} - \mu)^2\right\} d\bar{x}, \qquad -\infty < \bar{x} < \infty.$$

Here $\bar{x} - \mu$ has a distribution that does not depend on μ.

From the tables of the normal integral we know that the probability of a positive deviation from the mean not greater than twice the standard deviation is 0.97725. We have then

$$P(\bar{x} - \mu \leqslant 2/\sqrt{n}) = 0.97725,$$

which is equivalent to

$$P(\bar{x} - 2/\sqrt{n} \leqslant \mu) = 0.97725.$$

Thus, if we assert that μ is greater than or equal to $\bar{x} - 2/\sqrt{n}$ we shall be right in about 97.725 per cent of cases in the long run.

Similarly we have

$$P(\bar{x} - \mu \geqslant -2/\sqrt{n}) = P(\mu \leqslant \bar{x} + 2/\sqrt{n}) = 0.97725.$$

Thus, combining the two results,

$$P(\bar{x} - 2/\sqrt{n} \leqslant \mu \leqslant \bar{x} + 2/\sqrt{n}) = 1 - 2(1 - 0.97725)$$
$$= 0.9545. \tag{20.4}$$

Hence, if we assert that μ lies in the range $\bar{x} \pm 2/\sqrt{n}$, we shall be right in about 95.45 per cent of cases in the long run, the confidence coefficient being 0.9545.

Conversely, given the confidence coefficient, we can easily find from the tables of the normal integral the deviation d such that

$$P(\bar{x} - d/\sqrt{n} \leqslant \mu \leqslant \bar{x} + d/\sqrt{n}) = 1 - \alpha.$$

For instance, if $1 - \alpha = 0.8$, $d = 1.28$, so that if we assert that μ lies in the range $\bar{x} \pm 1.28/\sqrt{n}$ the odds are 4 to 1 that we shall be right.

The reader to whom this approach is new will probably ask: but is this not a roundabout method of using the standard error to set limits to an estimate of the mean? In a way, it is. Effectively, what we have done in this example is to show how the use of the standard error of the mean in normal samples may be justified. We must remember, however, that we are dealing here with a very special case where $\bar{x}$ is exactly normally distributed, something that is certainly not required by the theory of confidence intervals.

Another point of interest in this example is that the upper and lower confidence limits derived above are equidistant from the mean $\bar{x}$. This is not by any means necessary, and it is easy to see that we can derive any number of alternative limits for the same confidence coefficient $1 - \alpha$. Suppose, for instance, we take $1 - \alpha = 0.9545$, and select two numbers α_0 and α_1 that satisfy the condition

$$\alpha_0 + \alpha_1 = \alpha = 0.0455,$$

say $\alpha_0 = 0.01$ and $\alpha_1 = 0.0355$. From the tables of the normal integral we have

$$P(\bar{x} - \mu \geqslant 2.326/\sqrt{n}) = 0.01 = \alpha_0,$$
$$P(\bar{x} - \mu \leqslant -1.806/\sqrt{n}) = 0.0355 = \alpha_1,$$

and hence

$$P\left(\bar{x} - \frac{2.326}{\sqrt{n}} \leqslant \mu \leqslant \bar{x} + \frac{1.806}{\sqrt{n}}\right) = 1 - \alpha_0 - \alpha_1 = 0.9545. \tag{20.5}$$

Thus, with the same confidence coefficient we may assert that μ lies in the interval $\bar{x} - 2/\sqrt{n}$ to $\bar{x} + 2/\sqrt{n}$, or in the interval $\bar{x} - 2.326/\sqrt{n}$ to $\bar{x} + 1.806/\sqrt{n}$. In either case we shall be right in about 95.45 per cent of cases in the long run.

We note that in the first case the interval has length $4/\sqrt{n}$, while in the second case its length is $4.132/\sqrt{n}$. Intuitively, we should choose the first set of limits since they locate the parameter in a narrower range with the same confidence coefficient. We shall consider this point in more detail later in this chapter.

Graphical representation

20.6 In a number of simple cases, including that of Example 20.1, the confidence limits can be represented in a useful graphical form. We take two orthogonal axes, OX relating to the observed $\bar{x}$ and OY to μ (see Fig. 20.1), and for simplicity consider $n = 1$ initially.

The two straight lines shown have as their equations

$$\mu = \bar{x} + 2, \qquad \mu = \bar{x} - 2.$$

Consequently, for any point between the lines,

$$\bar{x} - 2 \leqslant \mu \leqslant \bar{x} + 2.$$

Hence, if for any observed $\bar{x}$ we read off the two ordinates on the lines corresponding to that value, we obtain the two confidence limits. The vertical interval between the limits is the confidence interval (shown in the diagram for $\bar{x} = 1$), and the total zone between the lines is the confidence belt. We may refer to the two lines as upper and lower confidence lines respectively.

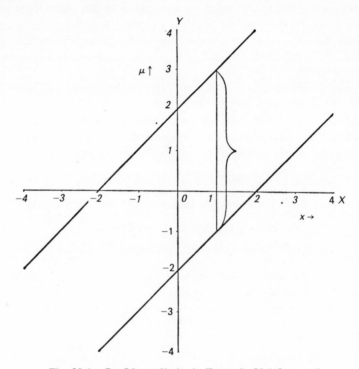

Fig. 20.1 Confidence limits in Example 20.1 for $n = 1$

For different values of n and a fixed confidence coefficient $1-\alpha$, there will be different confidence lines, all parallel to $\mu = \bar{x}$, and getting closer to each other as n increases. They may be shown on a single diagram for selected values of n, and a chart so constructed provides a useful method of reading off confidence limits in practical work.

Alternatively, we may wish to vary the confidence coefficient, and we may show a series of pairs of confidence lines, each pair corresponding to a selected value of $1-\alpha$, on a single diagram relating to some fixed value of n. In this case, of course, the lines become further apart with increasing $1-\alpha$. In fact, in many practical situations, we are interested in the variation of the confidence interval with $1-\alpha$, and we may validly make assertions of the form (20.2) simultaneously for a number of values of α: each will be true in the corresponding proportion of cases in the long run. Indeed, this procedure may be taken to its extreme form, when we consider *all* values of $1-\alpha$ in $(0, 1)$ simultaneously, and thus generate a *confidence distribution* of the parameter—the term is due to D. R. Cox (e.g. 1958b): we then have an infinite sequence of simultaneous confidence statements, each containing the preceding one, with increasing values of $1-\alpha$.

Example 20.2

Surprisingly little changes in Example 20.1 if we now assume that the variance σ^2 is unknown. The distribution of

$$t = (n-1)^{1/2}(\bar{x}-\mu)/s$$

where s^2 is the sample variance, is known to be of Student's form (11.42) in Example 11.8. Given α, we can now find t_0 and t_1, where $\frac{1}{2}\alpha$ of the distribution lies below $-t_1$ and $\frac{1}{2}\alpha$ above t_0, so that

$$P(-t_1 \leqslant t \leqslant t_0) = 1-\alpha,$$

which is equivalent to

$$P\{\bar{x} - st_0(n-1)^{-1/2} \leqslant \mu \leqslant \bar{x} + st_1(n-1)^{-1/2}\} = 1-\alpha.$$

Thus we may assert that μ lies in this interval with confidence coefficient $1-\alpha$, and the interval is independent of μ and σ^2. Because of the symmetry of Student's distribution, the confidence limits are equidistant from $\bar{x}$, but this is an accidental circumstance not required by the argument.

As in Example 20.1, the confidence limits are linear in the statistic $\bar{x}$, and the confidence lines are parallel straight lines as in Fig. 20.1. The difference is that whereas, with known variance, the vertical distance between the lines is the same for all samples, here that distance is a random variable, being a multiple of s. Thus we cannot here fix the length of the confidence interval in advance of the observations.

Central and non-central intervals

20.7 In Examples 20.1 and 20.2, the sampling distribution on which the confidence intervals were based was symmetrical, and by taking equal deviations from the mean we obtained equal values of

$$1 - \alpha_0 = P(t_0 \leqslant \theta)$$

and

$$1 - \alpha_1 = P(\theta \leqslant t_1).$$

In general, we cannot achieve this result with equal deviations, but subject to the condition $\alpha_0 + \alpha_1 = \alpha$, α_0 and α_1 may still be chosen arbitrarily.

If α_0 and α_1 are taken to be equal, we shall say that the intervals are *central*. In such a case we have

$$P(t_0 > \theta) = P(\theta > t_1) = \alpha/2. \tag{20.6}$$

In the contrary case the intervals with be called *non-central*. Centrality in this sense does not mean that the confidence limits are equidistant from the sample statistic, unless its sampling distribution is symmetrical.

20.8 In the absence of other considerations it is usually convenient to employ central intervals, but circumstances sometimes arise in which non-central intervals are more serviceable. Suppose, for instance, we are estimating the proportion of some drug in a medicinal preparation and the drug is toxic in large doses. We must then clearly err on the safe side, an excess of the true value over our estimate being more serious than a deficiency. In such a case we might like to take α_1 equal to zero, so that

$$P(\theta \leqslant t_1) = 1$$
$$P(t_0 \leqslant \theta) = 1 - \alpha,$$

in order to be certain that θ is not greater than t_1. But if our statistic has a sampling distribution with infinite range, this is only possible with t_1 infinite, so we must content ourselves with making α_1 very close to zero.

Again, if we are estimating the proportion of viable seed in a sample of material that is to be placed on the market, we are more concerned with the accuracy of the lower limit than that of the upper limit, for a deficiency of germination is more serious than an excess from the grower's point of view. In such circumstances we should probably take α_0 as small as conveniently possible so as to be near to certainty about the minimum value of viability. This kind of situation often arises in the specification of the quality of a manufactured product, the seller wishing to guarantee a minimum standard but being much less concerned with whether his product exceeds expectation.

Conservative confidence intervals and discontinuities
20.9 On a somewhat similar point, it may be remarked that in certain circumstances it is enough to know that $P(t_0 \leqslant \theta \leqslant t_1) \geqslant 1 - \alpha$. Such an interval is often called *conservative*. We then know that in asserting θ to lie in the range t_0 to t_1 we shall be right in *at least* a proportion $1 - \alpha$ of the cases. Mathematical difficulties in ascertaining confidence limits exactly for given $1 - \alpha$, or theoretical difficulties when the distribution

is discrete may, for example, lead us to be content with this inequality rather than the equality of (20.2).

Example 20.3

To find confidence intervals for the probability π of 'success' in a binomial sample.

We shall determine the limits for the case $n = 20$ and confidence coefficient 0.95. The obvious statistic to use is the observed proportion of successes, p, but we see that here we cannot find a function of p whose distribution does not depend on π. Nonetheless, confidence intervals can be constructed.

We require first the distribution function of the binomial. Table 20.1 shows the d.f. for $\pi = 0.1(0.1)0.5$; the values for $1 - \pi$ follow by symmetry. For the accurate construction of the confidence belt we require more detailed information such as is obtainable from the comprehensive tables of the binomial function referred to in **5.7**. Our table, however, will serve for purposes of illustration. The final figures may be a unit or two in error owing to rounding up, but that need not bother us to the degree of approximation here considered.

Although the variate p is discrete, π can take any value in $(0, 1)$. For given π we cannot in general find limits to p for which $1 - \alpha$ is exactly 0.95; but we will take p to be the sample proportion that gives a confidence coefficient at least equal to 0.95, so the intervals will be conservative. We will consider only central intervals, so that

Table 20.1

Proportion of successes p	$\pi = 0.1$	$\pi = 0.2$	$\pi = 0.3$	$\pi = 0.4$	$\pi = 0.5$
0.00	0.1216	0.0115	0.0008	—	—
0.05	0.3918	0.0691	0.0076	0.0005	—
0.10	0.6770	0.2060	0.0354	0.0036	0.0002
0.15	0.8671	0.4114	0.1070	0.0159	0.0013
0.20	0.9569	0.6296	0.2374	0.0509	0.0059
0.25	0.9888	0.8042	0.4163	0.1255	0.0207
0.30	0.9977	0.9133	0.6079	0.2499	0.0577
0.35	0.9997	0.9678	0.7722	0.4158	0.1316
0.40	1.0000	0.9900	0.8866	0.5955	0.2517
0.45		0.9974	0.9520	0.7552	0.4119
0.50		0.9994	0.9828	0.8723	0.5881
0.55		0.9999	0.9948	0.9433	0.7483
0.60		1.0000	0.9987	0.9788	0.8684
0.65			0.9997	0.9934	0.9423
0.70			1.0000	0.9983	0.9793
0.75				0.9996	0.9941
0.80				1.0000	0.9987
0.85					0.9998
0.90					1.0000
0.95					

for given π we have to find π_0 and π_1 such that

$$P(p \geq \pi_0) \geq 0.975$$
$$P(p \leq \pi_1) \geq 0.975,$$

the inequalities for P being as near to equality as we can make them.

Consider the diagram in Fig. 20.2.

Note that the distribution function gives the probability of obtaining a proportion p or less of successes, so that its complement gives the probability of a proportion strictly greater than p. Here, for example, on the horizontal through $\pi = 0.1$ we find $\pi_0 = 0$ and $\pi_1 = 0.25$ from our table; and for $\pi = 0.4$ we have $\pi_0 = 0.15$ and $\pi_1 = 0.60$. The points so obtained lie on stepped lines which have been drawn in. For example, when $\pi = 0.3$ the greatest value of π_0 such that $P(p \geq \pi_0) \geq 0.975$ is 0.10. By the time π has increased to 0.4 the value of π_0 has increased to 0.20. Somewhere between is the value of π such that $P(p \geq 0.1)$ is exactly 0.975. If we tabulate the probabilities for finer intervals of π these stepped lines become closer and the steps shallower; and in the limit, if we calculate values of π such that $P(p \geq \pi_0) = 0.975$ exactly we obtain points lying inside our present stepped lines. These points have been joined by dotted lines in Fig. 20.2.

The zone between the stepped lines is the confidence belt. It will be observed that it was constructed horizontally. In applying it, however, we read it *vertically*, that is to say, with observed abscissa p we read off two values p_0 and p_1 where the vertical

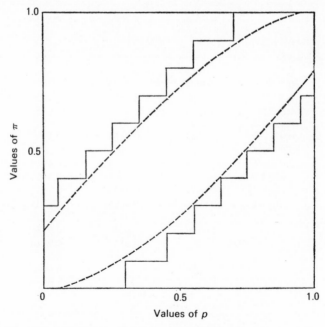

Fig. 20.2   Confidence limits for a binomial parameter ($n = 20$)

through p intersects the boundaries of the confidence belt, and assert that $p_0 \leqslant \pi \leqslant p_1$. We may see at once that this gives the required confidence interval. Considering the diagram horizontally we see that, for any given π, an observation falls in the confidence belt with probability $\geqslant 1 - \alpha$. If and only if the observation is in the belt, the pair of values (p_0, p_1) will contain between them the true value of π. Thus the latter event has probability $\geqslant 1 - \alpha$, whatever the true value of π.

20.10 In Example 20.2, we remarked that, as the number of successes (say, c) is necessarily integral, and the proportion of successes p $(=c/n)$ therefore discrete, the confidence belt yields conservative confidence statements. By a randomisation device, we can always make exact statements of form $P = 1 - \alpha$ even in the presence of discontinuity. The method was given by Stevens (1950).

In fact, after we have drawn our sample and observed c successes, let us independently draw a random number x from the uniform distribution $dF = dx$, $0 \leqslant x \leqslant 1$, e.g. by selecting a random number of four digits from the usual tables and putting a decimal point in front. Then the variate

$$y = c + x \tag{20.7}$$

can take all values in the range 0 to $n+1$. If y_0 is some given value $c_0 + x_0$, we have, writing π for the probability to be estimated,

$$P(y \geqslant y_0) = P(c > c_0) + P(c = c_0)P(x \geqslant x_0)$$

$$= \sum_{j=c_0+1}^{n} \binom{n}{j} \pi^j (1-\pi)^{n-j} + \binom{n}{c_0} \pi^{c_0}(1-\pi)^{n-c_0}(1-x_0)$$

$$= x_0 \sum_{j=c_0+1}^{n} \binom{n}{j} \pi^j (1-\pi)^{n-j} + (1-x_0) \sum_{j=c_0}^{n} \binom{n}{j} \pi^j (1-\pi)^{n-j}. \tag{20.8}$$

This defines a continuous probability distribution for y. It is clearly continuous as x_0 moves from $0+$ to $1-$, for c_0 is then constant. And at the points where $x_0 = 0$ the probability approaches the same value from the left and from the right. We can therefore use this distribution to set confidence limits for π and our confidence statements based upon them will be exact.

The confidence intervals are of the type exhibited in Fig. 20.3. The upper limit is now shifted to the right by amounts which, in effect, join up the discontinuities by a series of arcs. The lower limit also has a series of arcs, but there is no displacement to the right, and we have therefore shown on the diagram only the (dotted) approximate upper limit of Fig. 20.2. On our scale the lower approximate limit would almost coincide with the lower series of arcs. The general effect is to shorten the confidence interval.

20.11 It is at first sight surprising that the intervals set up in this way lie inside the conservative step-intervals of Fig. 20.2, and are therefore no less accurate; for by taking an additional random number x we have imported additional uncertainty into the situation. A little reflection will show, however, that we have not got something

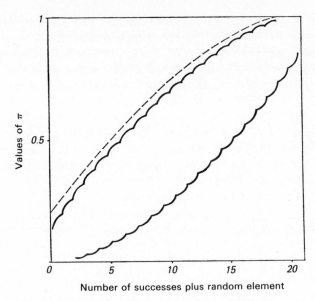

Fig. 20.3 Randomised confidence intervals for a binomial parameter

for nothing. We have removed one uncertainty, associated with the inequality in $P \geqslant 1 - \alpha$, by bringing in another so as to make statements of the kind $P = 1 - \alpha$; and what we lose on the second we more than offset by removing the first.

20.12 In **20.3** we saw that a central element in the construction of a confidence interval is that the probability statement

$$P\{z(x_1, \ldots, x_n, \theta) \leqslant z_1\} = 1 - \alpha \tag{20.9}$$

be equivalent to

$$P\{t_0(x_1, \ldots, x_n) \leqslant \theta \leqslant t_1(x_1, \ldots, x_n)\} = 1 - \alpha. \tag{20.10}$$

In the cases of Examples 20.1–3, this caused us no difficulty, the function z being essentially linear in θ. In general, however, it may be far from straightforward if z is a more complicated function. Even in the relatively simple quadratic case, the results are somewhat unexpected. Suppose

$$z(x_1, \ldots, x_n, \theta) - z_1$$

is quadratic in θ, and call it $Q(\theta, x)$. To pass from (20.9) to (20.10) we need to solve for θ the relation $Q(\theta, x) < 0$. The nature of the solution depends upon whether the roots of Q are real and upon the sign S of θ^2 in Q. If S is positive, the values of θ satisfying (20.9) lie between the real roots of $Q = 0$ (which are therefore t_0 and t_1 in (20.10)), but if S is negative, (20.9) requires θ to lie *outside* the interval between the real roots; while if roots of $Q = 0$ are complex, θ may take any value on the real line if S is negative, and if S is positive there is no real θ that satisfies (20.9).

In Example 20.4, this last event alone cannot occur.

Example 20.4 Estimation of the ratio of means of two normal variables

Let x, y be bivariate normally distributed as in Example 18.14, all five parameters of the distribution being unknown, as in case (c) of that Example. We make a convenient re-parametrisation and write $\rho\sigma_1\sigma_2 = \sigma_{12}$ in order to use the covariance σ_{12} as the fifth parameter instead of the correlation coefficient ρ.

From a sample of n observations on (x, y), we now wish to estimate the ratio of means $\theta = \mu_2/\mu_1$, where we assume $\mu_1 \neq 0$. From Example 18.14(c) it follows from the fact that ML estimators are unaffected by re-parametrisation (cf. **8.9** and **18.14**) that the ML estimator of σ_{12} is

$$\hat{\sigma}_{12} = \hat{\sigma}_1\hat{\sigma}_2\hat{\rho} = \frac{1}{n}\Sigma\,(x - \bar{x})(y - \bar{y}) = s_{12},$$

say. Similarly, we now write s_1^2, s_2^2 for the ML estimators of σ_1^2, σ_2^2 found at (18.78). The ML estimator of θ, which is our primary interest here, is $\hat{\theta} = \hat{\mu}_2/\hat{\mu}_1 = \bar{y}/\bar{x}$. It is intuitively clear that if $|\bar{x}|$ is small, we must expect poor precision in estimating θ. We now turn to the problem of finding confidence intervals for θ, where we shall see that if $\bar{x}^2/s_1^2$ is small, difficulties of the kind discussed above arise.

Consider the new variable $z_i = y_i - \theta x_i$ $(i = 1, 2, \ldots, n)$, which is normally distributed (since it is a linear function of bivariate normal variables—cf. **15.4**) with zero expectation. The sample mean of the z_i is

$$\bar{z} = \bar{y} - \theta\bar{x},$$

and also has zero mean. For fixed θ, the variance of $\bar{z}$ is

$$V_\theta(\bar{z}) = \sigma_z^2/n = \frac{1}{n}(\sigma_2^2 - 2\theta\sigma_{12} + \theta^2\sigma_1^2)$$

with unbiased estimator

$$\hat{V}_\theta = \frac{1}{n(n-1)}\sum_{i=1}^{n}(z_i - \bar{z})^2 = \frac{1}{n-1}(s_2^2 - 2\theta s_{12} + \theta^2 s_1^2)$$

and $\hat{V}_\theta/V_\theta(\bar{z})$ is distributed, independently of $\bar{z}$, as a $\chi_{n-1}^2/(n-1)$ variable. Thus

$$\bar{z}/\hat{V}_\theta^{1/2} = (\bar{y} - \theta\bar{x})\Big/\left\{\frac{1}{n-1}(s_2^2 - 2\theta s_{12} + \theta^2 s_1^2)\right\}^{1/2} \tag{20.11}$$

has a Student's t-distribution with $\nu = n - 1$ degrees of freedom, a result due to Fieller (1940). Equivalently, $\bar{z}^2/\hat{V}_\theta$ has a t^2 distribution, i.e. an F-distribution with $(1, n-1)$ degrees of freedom (cf. **16.15**). From the tables of Student's t, we may now find critical values t_α such that, for any θ,

$$P\left\{\frac{\bar{z}^2}{\hat{V}_\theta} \leq t_\alpha^2\right\} = 1 - \alpha. \tag{20.12}$$

We now examine the nature of the confidence intervals for θ which emerge from (20.12).

The event $\bar{z}^2/\hat{V}_\theta \leqslant t_\alpha^2$ in (20.12) can be rewritten in the form of a quadratic inequality in θ:

$$\left(\bar{x}^2 - \frac{t_\alpha^2}{n-1} s_1^2\right)\theta^2 - 2\left(\bar{x}\bar{y} - \frac{t_\alpha^2}{n-1} s_{12}\right)\theta + \left(\bar{y}^2 - \frac{t_\alpha^2}{n-1} s_2^2\right) \leqslant 0 \qquad (20.13)$$

and the problem is to determine the values of θ that satisfy this inequality—these will constitute the $100(1-\alpha)$ per cent confidence interval for θ.

The left side of (20.13) is a parabola in θ, of standard type $a\theta^2 - 2b\theta + c$. We require the values of θ for which this parabola lies on or below the θ-axis, and we shall see that not only the confidence interval, but also its *form*, will change with the relationships between the randomly varying coefficients a, b, c. We ignore the degenerate case $a = 0$, which occurs with probability zero.

The parabola has its unique turning point at $\theta = b/a$, where the value of the left side of (20.13) is seen to be $(ac - b^2)/a$. If $a > 0$, the turning point is a minimum, and if the minimum value $(ac - b^2)/a$ were positive, there would be no value of θ at all which satisfied (20.13), the parabola having no real roots since $b^2 < ac$. This situation cannot arise, however, for we see that the value $\theta = \hat{\theta} = \bar{y}/\bar{x}$ always satisfies (20.13), whose left-hand side upon this substitution becomes equal to $-t_\alpha^2 \hat{V}_{\hat{\theta}}$, which cannot be positive. There are thus always some values of θ satisfying (20.13), and if $a > 0$ we must also have $b^2 \geqslant ac$, with real roots of the parabola at

$$\theta = \{b \pm (b^2 - ac)^{1/2}\}/a. \qquad (20.14)$$

(20.13) will then be satisfied by all values of θ lying between these roots, in the finite interval determined by (20.14)—the reader should draw the parabola roughly to help him see this. If $b^2 = ac$, the interval contains only the point $\theta = b/a$, which then must be equal to $\hat{\theta} = \bar{y}/\bar{x}$.

On the other hand, if $a < 0$, the turning point of the parabola is a maximum. If also $b^2 \geqslant ac$, so that (20.14) gives the real roots as before, we see that (20.13) is satisfied by all values of θ lying *outside* the finite interval defined by (20.14), i.e. by its infinite complement on the real line—again, a drawing may help. Further, if $a < 0$ and $b^2 < ac$, the whole of the parabola lies below the θ-axis, and (20.14) is satisfied by any real value of θ. These results, given by Fieller (1954), are less startling than they first seem. If the confidence interval for θ lies outside a finite interval, this can be looked on loosely as saying that θ^{-1} lies inside one. (Nevertheless, this does not imply that we only need to work in terms of θ^{-1} to avoid the problem—the coefficient c in (20.13) is the same function of the ys as a there is of the xs, while b is symmetric in x and y. We could thus find $a < 0$ and $c < 0$ together, with both θ and θ^{-1} poorly determined by confidence intervals.) Similarly, it is not so difficult to accept a confidence interval consisting of all possible values of θ if we reflect that this simply means that the observations are completely uninformative with respect to θ. Accept it we must, for Koschat (1987) has shown that for this problem there is no procedure yielding confidence intervals that are always bounded.

Finally, we should observe that the condition $a < 0$ which led to our difficulties is, from (20.13), simply $\bar{x}^2/s_1^2 < t_\alpha^2/(n-1)$, so that, as we anticipated at the beginning of this Example, it is the relative smallness of $|\bar{x}|$ that produces the trouble.

Scheffé (1970a) avoids the whole-real-line confidence interval by reformulating the problem slightly. In (20.13), he replaces $t_\alpha^2 = F_\alpha(1, n-1)$ by a positive monotone function of an F-statistic which decreases from $2F$ at $F = F_\alpha(2, n-1)$ to $F_\alpha(1, n-1)$ as $F \to \infty$. As a result, one obtains *either* a finite interval or its complement for θ *or* an elliptical region for (μ_1, μ_2), with preassigned overall probability.

Shortest sets of confidence intervals

20.13   We have seen in Example 20.1 that for some problems there may be many sets of confidence intervals, and we must now consider what criteria to use in choosing among the possibilities. The problem is analogous to that of estimation, where we found that in general there are many different estimators for a parameter, but that we could sometimes find one (such as that with minimum variance) which was superior to the rest.

In Example 20.1 the problem presented itself in rather a limited form. We found that for intervals based on a normal sample mean $\bar{x}$ there were infinitely many sets of intervals with confidence coefficient $(1 - \alpha)$ according to the way in which we selected α_0 and α_1 (subject to the condition that $\alpha_0 + \alpha_1 = \alpha$). Among these the central intervals are obviously the shortest, for since the distribution of $\bar{x}$ is unimodal and symmetric, a given range will include the greatest area of the normal distribution if it is centred at the mean. We might reasonably say that the central intervals are the best among those determined by $\bar{x}$.

But it does not follow that they are the shortest of all possible intervals, or even that a set exists that is uniformly the shortest for all values of θ. In general, for two sets of intervals c_1 and c_2, those of c_1 may be shorter than those of c_2 for some samples and longer in others.

Exercise 20.5 shows that the simplicity of Example 20.1 does not persist even in the case of the normal variance, discussed in Example 20.6; essentially this is because s^2, unlike $\bar{x}$, is asymmetrically distributed. Thus we are led to consider intervals that are shortest on the average, so that the expectation of the interval length $E(t_1 - t_0)$ is minimised.

20.14   Although the idea of calculating the expected length of a confidence interval is intuitively appealing, and often useful for comparative purposes (e.g. in Exercise 20.5 and in **20.30** below), expected length is not the principal criterion used in the theory of confidence intervals, for it is not merely the physical length of the interval that matters. The probability that an interval covers the true value of the parameter θ is fixed by the confidence coefficient, but we must also consider the probability that it covers other values of θ than the true one. If we can minimise this, we shall be taking account of the distribution of the interval over all values of θ. If there is a confidence interval that has smallest probability of covering every value of θ but the true one, we call it the *most selective* interval—Neyman (1937), who introduced this criterion, rather confusingly called it the 'shortest' interval; in the more modern literature, 'accurate' is often used in place of our 'selective', but we believe the latter term conveys the interval's property better. As we shall see, most selective intervals rarely exist.

Further, it seems highly desirable that a good confidence interval should cover a value of θ with higher probability when it is the true value than when it is not, so that the confidence coefficient will exceed the probability of covering any false value. Such a confidence interval is called *unbiased*—this use of the term is unconnected with estimation bias, discussed in **17.9**.

20.15   These ideas of selectivity and bias for intervals are, in effect, translations into confidence interval terms of fundamental notions in the theory of testing hypotheses, which will be treated in Chapters 21 and 22—cf. **21.9**. The results to be obtained in those chapters on various optimum properties of tests will therefore translate into equivalent results for confidence intervals. Here we only add that their mathematical tractability ensured that the selectivity and bias criteria became central to the theory of confidence intervals.

Exercises 20.10–11 deal with a relationship between interval length and selectivity, and Madansky (1962) illuminates their difference with an example. Harter (1964) discusses other criteria for intervals.

Choice of confidence intervals
20.16   The confidence intervals that we discussed in Examples 20.1 and 20.3 were in each case based on an unbiased sufficient statistic for the parameter, while in Examples 20.2 and 20.4 they were based on a set of jointly sufficient statistics. We shall see in **22.3** that (translated into confidence interval terms) there is no loss of selectivity if we restrict ourselves to functions of a sufficient statistic.

However, the exact distribution of the sufficient statistics may not be tractable, and it is therefore useful to consider procedures for large samples that will be generally applicable.

Confidence intervals for large samples
20.17   We have seen (**18.16**) that the first derivative of the logarithm of the likelihood function is, under regularity conditions, asymptotically normally distributed with zero mean and

$$\operatorname{var}\left(\frac{\partial \log L}{\partial \theta}\right) = E\left\{\left(\frac{\partial \log L}{\partial \theta}\right)^2\right\} = -E\left\{\frac{\partial^2 \log L}{\partial \theta^2}\right\}. \tag{20.15}$$

We may use this fact to set confidence intervals for θ in large samples. Writing

$$\psi(x, \theta) = \frac{\partial \log L}{\partial \theta}\bigg/\left[E\left\{\left(\frac{\partial \log L}{\partial \theta}\right)^2\right\}\right]^{1/2}, \tag{20.16}$$

so that ψ is a standardised normal variate in large samples, we may from the normal integral determine confidence limits for θ in large samples if ψ is a monotonic function of θ, so that inequalities in one may be transformed to inequalities in the other. The following example illustrates the procedure.

Example 20.5

Consider the Poisson distribution whose general term is

$$f(x, \lambda) = \frac{e^{-\lambda}\lambda^x}{x!}, \qquad x = 0, 1, \ldots .$$

Here, as in Example 20.3 for the binomial, there is no function of x whose distribution does not depend on λ, but we could as there construct a conservative confidence belt—the analogue of Fig. 20.2 would extend over the whole positive quadrant for x and λ. We shall instead discuss the large-sample case as an illustration.

We have seen in Example 17.8 that

$$\frac{\partial \log L}{\partial \lambda} = \frac{n}{\lambda}(\bar{x} - \lambda). \tag{20.17}$$

Hence

$$-\frac{\partial^2 \log L}{\partial \lambda^2} = \frac{n\bar{x}}{\lambda^2}$$

and

$$E\left(-\frac{\partial^2 \log L}{\partial \lambda^2}\right) = \frac{n}{\lambda}. \tag{20.18}$$

Thus, from (20.15–16),

$$\psi = (\bar{x} - \lambda)\sqrt{(n/\lambda)}. \tag{20.19}$$

For example, with $1 - \alpha = 0.95$, corresponding to a normal deviate ± 1.96, we have, for the central confidence limits,

$$(\bar{x} - \lambda)\sqrt{(n/\lambda)} = \pm 1.96,$$

giving, on squaring,

$$\lambda^2 - \left(2\bar{x} + \frac{3.84}{n}\right)\lambda + \bar{x}^2 = 0, \tag{20.20}$$

a quadratic with roots always real. Thus

$$\lambda = \bar{x} + \frac{1.92}{n} \pm \sqrt{\left(\frac{3.84\bar{x}}{n} + \frac{3.69}{n^2}\right)}, \tag{20.21}$$

the ambiguity in the square root giving upper and lower limits respectively.

To order $n^{-1/2}$ this is equivalent to

$$\lambda = \bar{x} \pm 1.96\sqrt{(\bar{x}/n)}, \tag{20.22}$$

from which the upper and lower limits are seen to be equidistant from the mean $\bar{x}$, as we should expect. Essentially, (20.22) is the ± 1.96 standard errors band obtained by estimating the Poisson mean and variance λ by $\bar{x}$. A similar large-sample result follows in the binomial case of Example 20.3—see Exercise 20.3.

P. Hall (1982) gives improved one-sided intervals based on use of a continuity correction.

20.18 We may obtain a closer approximation, following Bartlett (1953), by finding higher cumulants of ψ and using a Cornish–Fisher expansion as in **6.25-6**. Writing

$$\frac{\partial^r \log L}{\partial \theta^r} = d_r \quad \text{and} \quad E(d_r^s) = (r^s),$$

$$E(d_r^s d_t^u) = (r^s t^u),$$

(17.19) and (17.25) give

$$I = (1^2), \tag{20.23}$$
$$= (-2) \tag{20.24}$$

and from (17.18) and (17.19) the first two cumulants of $d_1 = \partial \log L / \partial \theta$ are

$$\kappa_1 = 0, \tag{20.25}$$
$$\kappa_2 = I. \tag{20.26}$$

We now seek

$$\kappa_3 = \mu_3 = (1^3) \quad \text{and} \quad \kappa_4 = \mu_4 - 3\mu_2^2 = (1^4) - 3I^2.$$

Differentiating (20.23) with respect to θ, we find

$$I' = 2(21) + (1^3) \tag{20.27}$$

while differentiating (20.24) gives

$$I' = -(3) - (21). \tag{20.28}$$

Eliminating (21) from (20.27-28) gives

$$\kappa_3 = (1^3) = 3I' + 2(3) = -(3) - 3(21). \tag{20.29}$$

Differentiating each of (20.27) and (20.28) once more, we have respectively

$$I'' = 2(31) + 2(2^2) + 5(21^2) + (1^4) \tag{20.30}$$

and

$$I'' = -(4) - 2(31) - (2^2) - (21^2). \tag{20.31}$$

Eliminating (21^2) from (20.30) and (20.31), we find

$$(1^4) = 6I'' + 5(4) + 8(31) + 3(2^2) \tag{20.32}$$

while eliminating I'' instead gives the alternative form

$$(1^4) = -(4) - 4(31) - 3(2^2) - 6(21^2) \tag{20.33}$$

and

$$\kappa_4 = (1^4) - 3I^2.$$

Using the first four terms in (6.54) on $\psi = d_1/I^{1/2}$ with $l_1 = l_2 = 0$, we then have the statistic

$$T(\theta) = \frac{1}{\sqrt{I}}\left[\frac{\partial \log L}{\partial \theta} - \frac{1}{6}\frac{\kappa_3(d_1)}{I^2}\left\{\left(\frac{\partial \log L}{\partial \theta}\right)^2 - I\right\} - \frac{1}{24}\frac{\kappa_4(d_1)}{I^3}\left\{\left(\frac{\partial \log L}{\partial \theta}\right)^3 - 3I\frac{\partial \log L}{\partial \theta}\right\}\right.$$

$$\left. + \frac{1}{36}\frac{\kappa_3^2(d_1)}{I^4}\left\{4\left(\frac{\partial \log L}{\partial \theta}\right)^3 - 7I\frac{\partial \log L}{\partial \theta}\right\}\right], \tag{20.34}$$

which is, to the next order of approximation, normal with zero mean and unit variance. The first term is the quantity we have called ψ. The correction terms involve the standardised cumulants of d_1 which are equivalent to the cumulants of ψ.

Example 20.6
Let us consider the problem of setting confidence intervals for the variance of a normal distribution with mean μ and variance θ. The distribution of the sample variance is known to be skew, and we can compare the exact results with those given by the foregoing approximation. We first derive exact results. We know that

$$s^2 = \frac{1}{n}\sum (x - \bar{x})^2 \quad \text{and} \quad u^2 = \frac{1}{n}\sum (x - \mu)^2$$

are distributed as multiples θ/n of a chi-square variable with $n-1$ and n degrees of freedom respectively. If we take $n = 10$, Appendix Table 3 gives the lower and upper 5 per cent points of a χ_9^2 variate as 3.325 and 16.919, so

$$P\left(3.325 \leq \frac{10s^2}{\theta} \leq 16.919\right) = 0.90,$$

which we may invert to give

$$P(0.591s^2 \leq \theta \leq 3.008s^2) = 0.90.$$

Turning now to the large-sample approximation, we recall from Example 17.40 (where our present θ was called θ^2)

$$\frac{\partial \log L}{\partial \theta} = d_1 = \frac{n}{2\theta^2}(u^2 - \theta). \tag{20.35}$$

Differentiating twice further we find

$$d_2 = \frac{-n}{\theta^3}(u^2 - \tfrac{1}{2}\theta), \tag{20.36}$$

$$d_3 = \frac{n}{\theta^4}(3u^2 - \theta), \tag{20.37}$$

so that by (20.24) and (20.36)

$$I = -(2) = \frac{n}{2\theta^2}$$

whence

$$I' = -\frac{n}{\theta^3}.$$

From (20.37) on taking expectations,

$$(3) = \frac{2n}{\theta^3}.$$

Using these results in (20.29), we find

$$\kappa_3 = 3I' + 2(3) = \frac{n}{\theta^3}. \tag{20.38}$$

Taking the expansion in (20.34) as far as κ_3 only, we find

$$T(\theta) = \left(\frac{2\theta^2}{n}\right)^{1/2}\left[\frac{n}{2\theta^2}(u^2 - \theta) - \frac{n}{6\theta^3}\left(\frac{2\theta^2}{n}\right)^2\left\{\left(\frac{n}{2\theta^2}\right)^2(u^2 - \theta)^2 - \frac{n}{2\theta^2}\right\}\right]$$

$$= \left(\frac{2}{n}\right)^{1/2}\left[\frac{n}{2\theta}(u^2 - \theta) - \frac{n}{6\theta^2}\{(u^2 - \theta)^2\} + \frac{1}{3}\right]. \tag{20.39}$$

Replacing u^2 by $ns^2/(n-1)$, which has the same expected value, we obtain

$$T = \left(\frac{2}{n}\right)^{1/2}\left[\frac{n}{2}\left(\frac{ns^2}{(n-1)\theta} - 1\right) - \frac{n}{6}\left(\frac{ns^2}{(n-1)\theta} - 1\right)^2 + \frac{1}{3}\right]. \tag{20.40}$$

The first term gives us the confidence limits for θ based on ψ alone. The other terms will be corrections of lower order in n. We then have approximately from (20.40)

$$T = \left(\frac{2}{n}\right)^{1/2}\left[\frac{n}{2}\left(\frac{ns^2}{(n-1)\theta} - 1\right) - \frac{n}{6}\frac{2}{n}T^2 + \frac{1}{3}\right],$$

giving

$$\frac{ns^2}{(n-1)\theta} = 1 + T\left(\frac{2}{n}\right)^{1/2} + \frac{2}{3n}T^2 - \frac{2}{3n}. \tag{20.41}$$

For example, with $n = 10$, $1 - \alpha = 0.90$, $s^2 = 1$, and $T = \pm 1.6449$ (the 5 per cent points of the standardised normal distribution), we find for the limits of s^2/θ the values 0.3403 to 1.6644 and hence limits for θ of 0.601 and 2.939. The true values, as we saw above, are 0.591 and 3.008. For so low a value as $n = 10$ the approximation seems very fair.

20.19   We now prove a result due to Wilks (1938b), which is analogous to that of **18.16** showing that ML estimators have minimum variance asymptotically. Let x have density $f(x|\theta)$ and let the monotone function of θ

$$\zeta(x, \theta) = \frac{\sum_{j=1}^{n} h(x_j, \theta)}{\{n \operatorname{var}[h(x, \theta)]\}^{1/2}} \tag{20.42}$$

be a standardised sum of independent variables $h(x_j, \theta)$ with $E(h) = 0$, asymptotically normally distributed by the central limit theorem. We shall show that

$$h(x, \theta) = \frac{\partial \log f(x|\theta)}{\partial \theta},$$

which gives $\psi(x, \theta)$ of (20.16), yields confidence limits for θ that are asymptotically the shortest obtainable from any member of the class (20.42).

First, we differentiate (20.42) and find

$$\frac{\partial \zeta}{\partial \theta} = \sum_j \frac{\partial h(x_j, \theta)}{\partial \theta} \bigg/ \{n \operatorname{var}[h(x, \theta)]\}^{1/2} - \frac{\partial \operatorname{var}[h(x, \theta)]/\partial \theta}{2 \operatorname{var}[h(x, \theta)]} \zeta(x, \theta). \quad (20.43)$$

Taking expectations, the second term on the right disappears since $E(\zeta) = 0$ and

$$E\left(\frac{\partial \zeta}{\partial \theta}\right) = \sum_j E\left\{\frac{\partial h(x, \theta)}{\partial \theta}\right\} \bigg/ \{n \operatorname{var}[h(x, \theta)]\}^{1/2}. \quad (20.44)$$

Now, just as in **17.14**, we differentiate

$$0 = E\{h(x, \theta)\} = \int_{-\infty}^{\infty} h(x, \theta) f(x|\theta) \, dx$$

under the integral sign to obtain

$$0 = E\left\{\frac{\partial h(x, \theta)}{\partial \theta}\right\} + E\left\{h(x, \theta) \frac{\partial \log f(x|\theta)}{\partial \theta}\right\},$$

so that (20.44) becomes

$$E\left(\frac{\partial \zeta}{\partial \theta}\right) = -\left(\frac{n}{\operatorname{var} h}\right)^{1/2} E\left\{h \frac{\partial \log f}{\partial \theta}\right\}. \quad (20.45)$$

By the Cauchy-Schwarz inequality (2.42), (20.45) becomes

$$\left|E\left(\frac{\partial \zeta}{\partial \theta}\right)\right| \leq \left(\frac{n}{\operatorname{var} h}\right)^{1/2} \left\{E(h^2) E\left[\left(\frac{\partial \log f}{\partial \theta}\right)^2\right]\right\}^{1/2},$$

which since $E(h) = E(\partial \log f/\partial \theta) = 0$ gives

$$\left|E\left(\frac{\partial \zeta}{\partial \theta}\right)\right| \leq \left\{nE\left[\left(\frac{\partial \log f}{\partial \theta}\right)^2\right]\right\}^{1/2}, \quad (20.46)$$

the equality holding only when $h \propto \partial \log f/\partial \theta$. Apart from this case, we have

$$\left|E\left(\frac{\partial \zeta}{\partial \theta}\right)\right| < \left|E\left(\frac{\partial \psi}{\partial \theta}\right)\right|, \quad (20.47)$$

where ψ is defined at (20.16). Thus the absolute average rate of change of ψ with respect to θ is greater than that of ζ.

20.20 Because ψ is the most sensitive member of the class ζ to change in θ, we expect that the interval for θ based on it will be shorter. We now show that this is true asymptotically in n.

The central range of the distribution of ζ, with probability $1 - \alpha$, is

$$-d_{\alpha/2} \leqslant \zeta(x, \theta) \leqslant +d_{\alpha/2}, \tag{20.48}$$

where G is the standardised normal d.f. and $G(-d_\alpha) = \alpha$. From this, we obtain the confidence interval

$$u_0 \leqslant \theta \leqslant u_1.$$

In the special case where $\zeta \equiv \psi$, we write the confidence interval as

$$t_0 \leqslant \theta \leqslant t_1.$$

At the end-points of (20.48), which correspond to u_0 and u_1, we expand ζ in Taylor series about the true value θ_0:

$$\zeta(x, u_j) = \zeta(x, \theta_0) + (u_j - \theta_0)\left(\frac{\partial \zeta}{\partial \theta}\right)_{\theta'} = \pm d_{\alpha/2} \tag{20.49}$$

where θ' lies between u and θ_0; similarly, for ψ we have

$$\psi(x, t_j) = \psi(x, \theta_0) + (t_j - \theta_0)\left(\frac{\partial \psi}{\partial \theta}\right)_{\theta''} = \pm d_{\alpha/2} \tag{20.50}$$

where θ'' lies between t and θ_0. We now find, subtracting the lower end-point result from the upper in each of (20.49) and (20.50),

$$\zeta(x, u_1) - \zeta(x, u_0)$$

$$= (u_1 - \theta_0)\left(\frac{\partial \zeta}{\partial \theta}\right)_{\theta'_{(1)}} - (u_0 - \theta_0)\left(\frac{\partial \zeta}{\partial \theta}\right)_{\theta'_{(0)}}$$

$$= \psi(x, t_1) - \psi(x, t_0) = (t_1 - \theta_0)\left(\frac{\partial \psi}{\partial \theta}\right)_{\theta''_{(1)}} - (t_0 - \theta_0)\left(\frac{\partial \psi}{\partial \theta}\right)_{\theta''_{(0)}}$$

$$= 2d_{\alpha/2}. \tag{20.51}$$

Here we have used suffixes (1) and (0) to identify θ' and θ'' at upper and lower end-points.

As $n \to \infty$, u_1 and u_0, t_1 and t_0 all converge to the true value θ_0, and carry the θ's and θ''s with them since these are bounded by u and t. Moreover, the derivatives at θ_0, being sums of identical independent variates, tend to their expectations by the weak law of large numbers (**7.32**). Thus we find from (20.51)

$$(u_1 - u_0) E\left(\frac{\partial \zeta}{\partial \theta}\right) = (t_1 - t_0) E\left(\frac{\partial \psi}{\partial \theta}\right). \tag{20.52}$$

Together with (20.47), (20.52) implies

$$u_1 - u_0 > t_1 - t_0, \tag{20.53}$$

so that asymptotically ψ gives shorter intervals.

Simultaneous confidence intervals for several parameters

20.21   We have already, in Examples 20.2 and 20.4, discussed cases involving several parameters when we require a confidence interval for one, or for a single function, of them, and we shall shortly be treating another such problem, that of the difference between two normal means. First, we briefly discuss the construction of intervals for several parameters simultaneously.

What we should like to be able to do, given two parameters θ_1 and θ_2 and two statistics t and u, is to make simultaneous interval assertions of the type

$$P\{t_0 \leqslant \theta_1 \leqslant t_1 \text{ and } u_0 \leqslant \theta_2 \leqslant u_1\} = 1 - \alpha. \tag{20.54}$$

This, however, is rarely possible. Sometimes we can make a statement giving a *confidence region* for the two parameters together, such as

$$P\{w_0 \leqslant \theta_1^2 + \theta_2^2 \leqslant w_1\} = 1 - \alpha. \tag{20.55}$$

Even for large samples the problems are severe. We may then find that we can determine intervals of the type

$$P\{t_0(\theta_2) \leqslant \theta_1 \leqslant t_1(\theta_2)\} = 1 - \alpha$$

and substitute a (large-sample) estimate of θ_2 in the limits $t_0(\theta_2)$ and $t_1(\theta_2)$. This is very like the familiar procedure in the theory of standard errors, where we replace parameters occurring in the error variances by estimates obtained from the samples.

We shall not attempt to develop the theory of simultaneous confidence intervals any further here. The reader who is interested may consult papers by S. N. Roy and Bose (1953) and S. N. Roy (1954) on the theoretical aspect. Bartlett (1953, 1955) discussed the generalisation of the method of **20.18** to the case of two or more unknown parameters. Beale (1960), Halperin and Mantel (1963) and Halperin (1964) consider intervals for non-linear functions of parameters, especially in large samples. Bates and Watts (1980, 1981) gave curvature measures related to Beale's (1960) measures of non-linearity—see also Clarke (1987a, b), Hamilton *et al.* (1982) and Cook and Witmer (1985).

The theorem of **20.19–20** concerning shortest intervals was generalised by Wilks and Daly (1939). Under fairly general conditions the large-sample regions for k parameters that are smallest on the average are given by

$$\sum_{i=1}^{k} \sum_{j=1}^{k} \left\{ I_{ij}^{-1} \frac{\partial \log L}{\partial \theta_i} \frac{\partial \log L}{\partial \theta_j} \right\} \leqslant d_\alpha \tag{20.56}$$

where $\mathbf{I}^{-1}$ is the inverse of the information matrix (17.88) and $P\{\chi_k^2 \leqslant d_\alpha\} = 1 - \alpha$. This is clearly related to the result of **17.39** giving the minimum attainable variances (and, by a simple extension, covariances) of a set of unbiased estimators of several parametric functions.

The problem of two means

20.22   We now turn to the problem of finding a confidence interval for the difference between the means of two normal distributions.

Suppose that we have two normal distributions, the first with mean and variance parameters μ_1, σ_1^2 and the second with parameters μ_2, σ_2^2. Samples of size n_1, n_2 respectively are taken, and the sample means and variances observed are $\bar{x}_1$, s_1^2 and $\bar{x}_2$, s_2^2. Without loss of generality, we assume $n_1 \leqslant n_2$.

Now if $\sigma_1^2 = \sigma_2^2 = \sigma^2$, the problem of finding an interval for $\mu_1 - \mu_2 = \delta$ is simple. For in this case $d = \bar{x}_1 - \bar{x}_2$ is normally distributed with

$$\left. \begin{aligned} E(d) &= \delta, \\ \operatorname{var} d &= \sigma^2 \left(\frac{1}{n_1} + \frac{1}{n_2} \right), \end{aligned} \right\} \tag{20.57}$$

and $n_1 s_1^2 / \sigma_1^2$, $n_2 s_2^2 / \sigma_2^2$ are each distributed like χ^2 with $n_1 - 1$, $n_2 - 1$ d.fr. respectively. Since the two samples are independent, $(n_1 s_1^2 + n_2 s_2^2)/\sigma^2$ will be distributed like χ^2 with $n_1 + n_2 - 2$ d.fr., and hence, writing

$$s^2 = (n_1 s_1^2 + n_2 s_2^2)/(n_1 + n_2 - 2)$$

we have

$$E(s^2) = \sigma^2.$$

Now

$$\begin{aligned} y &= \frac{d - \delta}{\left\{ \sigma^2 \left(\frac{1}{n_1} + \frac{1}{n_2} \right) \right\}^{1/2}} \bigg/ \left(\frac{s^2}{\sigma^2} \right)^{1/2} \\ &= \frac{d - \delta}{\left\{ s^2 \left(\frac{1}{n_1} + \frac{1}{n_2} \right) \right\}^{1/2}} \end{aligned} \tag{20.58}$$

is a ratio of a standardised normal variate to the square root of an unbiased estimator of its sampling variance, which is distributed independently of it (since s_1^2 and s_2^2 are independent of $\bar{x}_1$ and $\bar{x}_2$). Moreover, $(n_1 + n_2 - 2)s^2/\sigma^2$ is a χ^2 variable with $n_1 + n_2 - 2$ d.fr. Thus y is of exactly the same form as the one-sample ratio

$$\frac{\bar{x}_1 - \mu_1}{\{s_1^2/(n_1 - 1)\}^{1/2}} = \frac{\bar{x}_1 - \mu_1}{(\sigma^2/n_1)^{1/2}} \bigg/ \left\{ \frac{n_1 s_1^2/(n_1 - 1)}{\sigma^2} \right\}^{1/2}$$

which we have on several occasions (e.g. Example 11.8) seen to be distributed in Student's distribution with $n_1 - 1$ d.fr. Hence (20.58) is also a Student's variable, but with $n_1 + n_2 - 2$ d.fr., a result which may easily be proved directly.

There is therefore no difficulty in setting confidence intervals for δ in this case—it is a straightforward analogue of the one-sample case treated in Example 20.2.

20.23 When we leave the case $\sigma_1^2 = \sigma_2^2$, complications arise. The variate distributed in Student's form, with $n_1 + n_2 - 2$ d.fr., by analogy with (20.58), is now

$$
t = \frac{d - \delta}{\left\{ \dfrac{\sigma_1^2}{n_1} + \dfrac{\sigma_2^2}{n_2} \right\}^{1/2}} \Bigg/ \left\{ \frac{\dfrac{n_1 s_1^2}{\sigma_1^2} + \dfrac{n_2 s_2^2}{\sigma_2^2}}{n_1 + n_2 - 2} \right\}^{1/2}.
\tag{20.59}
$$

The numerator of (20.59) is a standardised normal variate, and its denominator is the square root of an independently distributed χ^2 variate divided by its degrees of freedom, as for (20.58). The difficulty is that (20.59) involves the unknown ratio of variances $\theta = \sigma_1^2/\sigma_2^2$. If we also define $u = s_1^2/s_2^2$, $N = n_1/n_2$, we may rewrite (20.59) as

$$
t = \frac{(d - \delta)(n_1 + n_2 - 2)^{1/2}}{s_2 \left\{ \left(1 + \dfrac{\theta}{N} \right)\left(1 + \dfrac{Nu}{\theta} \right) \right\}^{1/2}},
\tag{20.60}
$$

which clearly displays its dependence upon the unknown θ. If $\theta = 1$, of course, (20.60) reduces to (20.58).

We now have to consider methods by which the 'nuisance parameter' θ can be eliminated from interval statements concerning δ. We must clearly seek some statistic other than t of (20.60). One possibility suggests itself immediately from inspection of (20.58). The statistic

$$
z = \frac{d - \delta}{\left(\dfrac{s_1^2}{n_1 - 1} + \dfrac{s_2^2}{n_2 - 1} \right)^{1/2}}
\tag{20.61}
$$

is, like (20.58), the ratio of a normal variate with zero mean to the square root of an independently distributed unbiased estimator of its sampling variance. However, as we shall see in **20.24**, that estimator is not a multiple of a χ^2 variate, and hence z is not distributed in Student's form. The statistic z is the basis of an approximate confidence interval approach to this problem, as we shall see below, as well as of another (fiducial) method, to be discussed later.

An alternative possibility is to investigate the distribution of (20.58) itself, i.e. to see how far the statistic appropriate to the case $\theta = 1$ retains its properties when $\theta \neq 1$. This, too, has been investigated from the confidence interval standpoint.

However, before discussing these approaches, we examine an exact confidence interval solution to this problem, based on Student's distribution, and its properties. The results are due to Scheffé (1943, 1944).

Exact confidence intervals based on Student's distribution
20.24 If we desire an exact confidence interval for δ based on Student's distribution, it will be sufficient if we can find a linear function of the observations, L, and a quadratic function of them, Q, such that, for all values of σ_1^2, σ_2^2,

(1) L and Q are independently distributed;

(2) $E(L) = \delta$ and var $L = V$; and

(3) Q/V has a χ^2 distribution with k d.fr.

Then

$$t = \frac{L - \delta}{(Q/k)^{1/2}} \tag{20.62}$$

has Student's distribution with k d.fr.

Scheffé (1944) showed that no statistic of the form (20.62) can be a symmetric function of the observations in each sample; that is to say, t cannot be invariant under permutation of the first sample members x_{1i} ($i = 1, 2, \ldots, n_1$) among themselves and of the second sample members x_{2i} ($i = 1, 2, \ldots, n_2$) among themselves.

For suppose that t is symmetric in the sense indicated. Then we must have

$$\left. \begin{array}{l} L = c_1 \sum_i x_{1i} + c_2 \sum_i x_{2i}, \\[2mm] Q = c_3 \sum x_{1i}^2 + c_4 \sum_{i \neq j} x_{1i} x_{1j} + c_5 \sum x_{2i}^2 + c_6 \sum_{i \neq j} x_{2i} x_{2j} + c_7 \sum_{i,j} x_{1i} x_{2j}, \end{array} \right\} \tag{20.63}$$

where the cs are constants independent of the parameters.

Now from (2) above,

$$E(L) = \delta = \mu_1 - \mu_2, \tag{20.64}$$

while from (20.63)

$$E(L) = c_1 n_1 \mu_1 + c_2 n_2 \mu_2. \tag{20.65}$$

(20.64) and (20.65) are identities in μ_1 and μ_2; hence

$$c_1 n_1 \mu_1 = \mu_1, \qquad c_2 n_2 \mu_2 = -\mu_2$$

so that

$$c_1 = 1/n_1, \qquad c_2 = -1/n_2. \tag{20.66}$$

From (20.66) and (20.63),

$$L = \bar{x}_1 - \bar{x}_2 = d, \tag{20.67}$$

and hence

$$\text{var } L = V = \sigma_1^2/n_1 + \sigma_2^2/n_2. \tag{20.68}$$

Since Q/V has a χ^2 distribution with k d.fr.,

$$E(Q/V) = k,$$

so that, using (20.68),

$$E(Q) = k(\sigma_1^2/n_1 + \sigma_2^2/n_2), \tag{20.69}$$

while, from (20.63),

$$E(Q) = c_3 n_1(\sigma_1^2 + \mu_1^2) + c_4 n_1(n_1 - 1)\mu_1^2 + c_5 n_2(\sigma_2^2 + \mu_2^2)$$
$$+ c_6 n_2(n_2 - 1)\mu_2^2 + c_7 n_1 n_2 \mu_1 \mu_2. \tag{20.70}$$

Equating (20.69) and (20.70), we obtain expressions for the cs, and thence find

$$Q = k\left\{\frac{s_1^2}{n_1 - 1} + \frac{s_2^2}{n_2 - 1}\right\}. \tag{20.71}$$

(20.67) and (20.71) reduce (20.62) to (20.61). Now a linear function of two independent χ^2 variates can only itself have a χ^2 distribution if it is a simple sum of them, and $n_1 s_1^2 / \sigma_1^2$ and $n_2 s_2^2 / \sigma_2^2$ are independent χ^2 variates. Thus, from (20.71), Q will only be a χ^2 variate if

$$\frac{k\sigma_1^2}{n_1(n_1 - 1)} = \frac{k\sigma_2^2}{n_2(n_2 - 1)} = 1$$

or

$$\theta = \frac{\sigma_1^2}{\sigma_2^2} = \frac{n_1(n_1 - 1)}{n_2(n_2 - 1)}. \tag{20.72}$$

Given n_1, n_2, this is only true for special values of σ_1^2, σ_2^2; but we require it to be true for *all* values of σ_1^2, σ_2^2 and thus t cannot be a symmetric function in the sense stated.

20.25 Since we cannot find a symmetric function of the desired type having Student's distribution, we now consider others. We specialise (20.62) to the situation where

$$\left.\begin{aligned}
L &= \sum_{i=1}^{n_1} d_i / n_1, \\
Q &= \frac{1}{n_1} \sum_{i=1}^{n_1} (d_i - L)^2,
\end{aligned}\right\} \tag{20.73}$$

and the d_i are independent identical normal variates with

$$E(d_i) = \delta, \qquad \text{var } d_i = \sigma^2, \quad \text{all } i. \tag{20.74}$$

It will be remembered that we have taken $n_1 \leq n_2$. (20.62) now becomes

$$t = \frac{L - \delta}{\{Q/(n_1 - 1)\}^{1/2}} = (L - \delta)\left\{\frac{n_1(n_1 - 1)}{\sum (d_i - L)^2}\right\}^{1/2}, \tag{20.75}$$

which is a Student variate with $(n_1 - 1)$ d.fr.

Suppose now that in terms of the original observations

$$d_i = x_{1i} - \sum_{j=1}^{n_2} c_{ij} x_{2j}. \tag{20.76}$$

The d_i are multinormally distributed, since they are linear functions of normal variates (cf. **15.4**). Necessary and sufficient conditions that (20.74) holds are

$$\left.\begin{array}{c} \sum\limits_j c_{ij} = 1, \\[2mm] \sum\limits_j c_{ij}^2 = c^2, \\[2mm] \sum\limits_j c_{ij}c_{kj} = 0, \end{array}\right\} \quad i \neq k. \tag{20.77}$$

Thus, from (20.76) and (20.77)

$$\text{var } d_i = \sigma^2 = \sigma_1^2 + c^2\sigma_2^2. \tag{20.78}$$

20.26 The central confidence interval, with confidence coefficient $1-\alpha$, derived from (20.75) is

$$|L - \delta| \leq t_{n_1-1,\alpha}\{Q/(n_1-1)\}^{1/2}, \tag{20.79}$$

where $t_{n_1-1,\alpha}$ is the appropriate deviate for $n_1 - 1$ d.fr. The interval length l has expected value, from (20.79),

$$E(l) = 2t_{n_1-1,\alpha} \frac{\sigma}{\{n_1(n_1-1)\}^{1/2}} E\left\{\left(\frac{n_1Q}{\sigma^2}\right)^{1/2}\right\}, \tag{20.80}$$

the last factor on the right being found, from (16.7), since $n_1 Q/\sigma^2$ has a χ^2 distribution with $n_1 - 1$ d.fr., to be

$$E\left\{\left(\frac{n_1Q}{\sigma^2}\right)^{1/2}\right\} = \frac{\sqrt{2}\,\Gamma(\tfrac{1}{2}n_1)}{\Gamma\{\tfrac{1}{2}(n_1-1)\}}. \tag{20.81}$$

To minimise the expected length (20.80), we must minimise σ, or equivalently, minimise c^2 in (20.78), subject to (20.77). The problem may be visualised geometrically as follows: consider a space of n_2 dimensions, with one axis for each *second* suffix of the c_{ij}. Then $\sum_j c_{ij} = 1$ is a hyperplane, and $\sum_j c_{ij}^2 = c^2$ is an n_2-dimensional hypersphere which is intersected by the plane in an (n_2-1)-dimensional hypersphere. We require to locate $n_1 \leq n_2$ vectors through the origin which touch this latter hypersphere and (to satisfy the last condition of (20.77)) are mutually orthogonal, in such a way that the radius of the n_2-dimensional hypersphere is minimised. This can be done by making our vectors coincide with n_1 axes, and then $c^2 = 1$. But if $n_1 < n_2$, we can improve upon this procedure, for we can, while keeping the vectors orthogonal, space them symmetrically about the equiangular vector, and reduce c^2 from 1 to its minimum value n_1/n_2, as we shall now show.

20.27 Written in vector form, the conditions (20.77) are

$$\left.\begin{array}{rl} \mathbf{c}_i\mathbf{u}' = 1 & \\ \mathbf{c}_i\mathbf{c}_k' = c^2 & i = k, \\ = 0 & i \neq k, \end{array}\right\} \tag{20.82}$$

where c_i is the ith row vector of the matrix $\{c_{ij}\}$ and $\mathbf{u}$ is the unit row vector.

If the n_1 vectors $\mathbf{c}_i$ satisfy (20.82), we can add another $(n_2 - n_1)$ vectors, satisfying the normalising and orthogonalising condition of (20.82), so that the augmented set forms a basis for an n_2-space. We may therefore express $\mathbf{u}$ as a linear function of the n_2 $\mathbf{c}$-vectors,

$$\mathbf{u} = \sum_{k=1}^{n_2} g_k \mathbf{c}_k, \tag{20.83}$$

where the g_k are scalars. Now, using (20.82) and (20.83),

$$1 = \mathbf{c}_i \mathbf{u}' = \mathbf{c}_i \sum_{k=1}^{n_2} g_k \mathbf{c}'_k = \sum g_k \mathbf{c}_i \mathbf{c}'_k$$
$$= g_i c^2.$$

Thus

$$g_i = 1/c^2, \quad i = 1, 2, \ldots, n_1. \tag{20.84}$$

Also, since $\mathbf{u}$ is the unit row vector,

$$n_2 = \mathbf{u}\mathbf{u}' = \left(\sum_k g_k \mathbf{c}_k \right) \left(\sum_k g_k \mathbf{c}'_k \right),$$

which, on using (20.82), becomes

$$n_2 = \sum_{k=1}^{n_2} g_k^2 \mathbf{c}_k \mathbf{c}'_k$$
$$= c^2 \left(\sum_{k=1}^{n_1} + \sum_{n_1+1}^{n_2} \right) g_k^2. \tag{20.85}$$

Use of (20.84) gives, from (20.85),

$$n_2 = c^2 \left\{ n_1/c^4 + \sum_{n_1+1}^{n_2} g_k^2 \right\},$$

so

$$n_2 \geqslant \frac{n_1}{c^2}.$$

Hence

$$c^2 \geqslant n_1/n_2, \tag{20.86}$$

the required result.

20.28 The equality sign holds in (20.86) whenever $g_k = 0$ for $k = n_1 + 1, \ldots, n_2$. Then the equiangular vector $\mathbf{u}$ lies entirely in the space spanned by the original n_1 $\mathbf{c}$-vectors. From (20.84), these will be symmetrically disposed around it. Evidently, there is an infinite number of ways of determining c_{ij}, merely by rotating the set of n_1 vectors. Scheffé (1943) obtained the particularly appealing solution

$$
\begin{aligned}
c_{jj} &= (n_1/n_2)^{1/2} - (n_1 n_2)^{-1/2} + 1/n_2, & j &= 1, 2, \ldots, n_1, \\
c_{ij} &= -(n_1 n_2)^{-1/2} + 1/n_2, & j(\neq i) &= 1, 2, \ldots, n_1, \\
c_{ij} &= 1/n_2, & j &= n_1 + 1, \ldots, n_2.
\end{aligned}
\tag{20.87}
$$

It may easily be confirmed that (20.87) satisfies the conditions (20.77) with $c^2 = n_1/n_2$. Substituted into (20.76), (20.87) gives

$$
d_i = x_{1i} - (n_1/n_2)^{1/2} x_{2i} + (n_1 n_2)^{-1/2} \sum_{j=1}^{n_1} x_{2j} + (1/n_2) \sum_{j=1}^{n_2} x_{2j},
\tag{20.88}
$$

which yields in (20.73)

$$
\left.
\begin{aligned}
L &= \bar{x}_1 - \bar{x}_2, \\
Q &= \frac{1}{n_1} \sum_{i=1}^{n_1} (u_i - \bar{u})^2,
\end{aligned}
\right\}
\tag{20.89}
$$

where

$$
\left.
\begin{aligned}
u_i &= x_{1i} - (n_1/n_2)^{1/2} x_{2i}, \\
\bar{u} &= \sum_{i=1}^{n_1} u_i / n_1.
\end{aligned}
\right\}
\tag{20.90}
$$

Hence, from (20.75) and (20.88)–(20.90),

$$
\{\bar{x}_1 - \bar{x}_2 - \delta\} \left\{ \frac{n_1(n_1 - 1)}{\sum (u_i - \bar{u})^2} \right\}^{1/2}
\tag{20.91}
$$

is a Student's variate with $n_1 - 1$ d.fr., and we may proceed to set confidence limits for $\delta = \mu_1 - \mu_2$.

Bain (1967) derives this procedure by a purely algebraic method, and also applies the method to other problems.

20.29 It is rather remarkable that we have been able to find an exact solution of the confidence interval problem in this case only by abandoning the natural requirement of symmetry. (20.91) holds for *any* randomly selected subset of n_1 of the n_2 variates in the second sample. Just as, in **20.10**, we resorted to randomisation to remove the difficulty in making exact confidence interval statements about a discrete variable, so we find here that randomisation allows us to bypass the nuisance parameter θ. But the extent of the randomisation should not be exaggerated. The numerator of (20.91) uses the sample means of both samples, complete; only the denominator varies with different random selections of the subset in the second sample. It is impossible to assess intuitively how much efficiency is lost by this procedure. We now examine the length of the confidence intervals it provides.

20.30 From (20.78) and (20.86), we have, for the solution (20.88),

$$\operatorname{var} d_i = \sigma^2 = \sigma_1^2 + (n_1/n_2)\sigma_2^2. \tag{20.92}$$

Putting (20.92) into (20.80), and using (20.81), we have for the expected length of the confidence interval

$$E(l) = 2t_{n_1-1,\alpha}\left\{\frac{\sigma_1^2 + (n_1/n_2)\sigma_2^2}{n_1(n_1-1)}\right\}^{1/2}\frac{\sqrt{2}\,\Gamma(\tfrac{1}{2}n_1)}{\Gamma\{\tfrac{1}{2}(n_1-1)\}}. \tag{20.93}$$

We now compare this interval l with the interval L obtained from (20.59) if $\theta = \sigma_1^2/\sigma_2^2$ is known. The latter has expected length

$$E(L) = 2t_{n_1+n_2-2,\alpha}\left\{\frac{\sigma_1^2/n_1 + \sigma_2^2/n_2}{n_1+n_2-2}\right\}^{1/2}E\left\{\frac{n_1 s_1^2}{\sigma_1^2} + \frac{n_2 s_2^2}{\sigma_2^2}\right\}^{1/2}, \tag{20.94}$$

the last factor being evaluated by (16.7) from the χ^2 distribution with (n_1+n_2-2) d.fr. as

$$\frac{\sqrt{2}\,\Gamma\{\tfrac{1}{2}(n_1+n_2-1)\}}{\Gamma\{\tfrac{1}{2}(n_1+n_2-2)\}}. \tag{20.95}$$

(20.93)–(20.95) give for the ratio of expected lengths

$$E(l)/E(L) = \frac{t_{n_1-1,\alpha}}{t_{n_1+n_2-2,\alpha}}\left(\frac{n_1+n_2-2}{n_1-1}\right)^{1/2}\frac{\Gamma(\tfrac{1}{2}n_1)\Gamma\{\tfrac{1}{2}(n_1+n_2-2)\}}{\Gamma\{\tfrac{1}{2}(n_1-1)\}\Gamma\{\tfrac{1}{2}(n_1+n_2-1)\}}. \tag{20.96}$$

As $n_1 \to \infty$, with n_2/n_1 fixed, each of the three factors of (20.96) tends to 1, and therefore the ratio of expected interval length does so, as is intuitively reasonable. For small n_1, the first two factors exceed 1, but the last is less than 1. Table 20.2 gives the exact values of (20.96) for $1-\alpha = 0.95$, 0.99 and a few sample sizes.

Table 20.2 $E(l)/E(L)$ (from Scheffé, 1943)

| | | $1-\alpha = 0.95$ | | | | | $1-\alpha = 0.99$ | | | |
n_2-1	5	10	20	40	∞	5	10	20	40	∞
n_1-1 5	1.15	1.20	1.23	1.25	1.28	1.27	1.36	1.42	1.47	1.52
10		1.05	1.07	1.09	1.11		1.10	1.13	1.16	1.20
20			1.03	1.03	1.05			1.05	1.06	1.09
40				1.01	1.02				1.02	1.04
∞					1					1

Evidently, l is a very efficient interval even for moderate sample sizes, having an expected length no more than 11 per cent greater than that of L for $n_1-1 \geq 10$ at $1-\alpha = 0.95$, and no more than 9 per cent greater than for $n_1-1 \geq 20$ at $1-\alpha = 0.99$. Furthermore, we are comparing it with an interval *based on knowledge of θ*. Taking this into account, we may fairly say that the element of randomisation cannot have resulted in very much loss of efficiency.

In addition to this solution to the two-means problem there are also approximate confidence-interval solutions, which we shall now summarise.

Approximate confidence-interval solutions

20.31   Welch (1938) has investigated the approximate distribution of the statistic (20.58), which is a Student's variate when $\sigma_1^2 = \sigma_2^2$, in the case $\sigma_1^2 \neq \sigma_2^2$. In this case, the sampling variance of its numerator is

$$\mathrm{var}\,(d - \delta) = \sigma_1^2/n_1 + \sigma_2^2/n_2,$$

so that, writing

$$\left.\begin{aligned} u &= (d - \delta)/(\sigma_1^2/n_1 + \sigma_2^2/n_2)^{1/2} \\ w^2 &= s^2\left(\frac{1}{n_1} + \frac{1}{n_2}\right)\Big/\left(\frac{\sigma_1^2}{n_1} + \frac{\sigma_2^2}{n_2}\right), \end{aligned}\right\} \tag{20.97}$$

(20.58) may be written

$$y = u/w. \tag{20.98}$$

The difficulty now is that w^2, although distributed independently of u, is not a multiple of a χ^2 variate when $\theta = \sigma_1^2/\sigma_2^2 \neq 1$. However, by equating its first two moments to those of a χ^2 variate, we can determine a number of degrees of freedom, ν, for which it is *approximately* a χ^2 variate. Its mean and variance are, from (20.97),

$$\left.\begin{aligned} E(w^2) &= b(\nu_1\theta + \nu_2), \\ \mathrm{var}\,(w^2) &= 2b^2(\nu_1\theta^2 + \nu_2), \end{aligned}\right\} \tag{20.99}$$

where we have written

$$\left.\begin{aligned} \nu_1 &= n_1 - 1, \quad \nu_2 = n_2 - 1, \\ b &= (n_1 + n_2)\sigma_2^2/\{(n_1 + n_2 - 2)(n_2\sigma_1^2 + n_1\sigma_2^2)\}. \end{aligned}\right\} \tag{20.100}$$

If we identify (20.99) with the moments of a multiple g of a χ^2 variate with ν d.fr.,

$$\mu_1' = g\nu, \quad \mu_2 = 2g^2\nu, \tag{20.101}$$

we find

$$\left.\begin{aligned} g &= b(\theta^2\nu_1 + \nu_2)/(\theta\nu_1 + \nu_2), \\ \nu &= (\theta\nu_1 + \nu_2)^2/(\theta^2\nu_1 + \nu_2). \end{aligned}\right\} \tag{20.102}$$

With these values of g and ν, w^2/g is approximately a χ^2 variate with ν degrees of freedom, and hence, from (20.97),

$$t = u\Big/\left\{\frac{w^2}{g\nu}\right\}^{1/2} \tag{20.103}$$

is a Student's variate with ν d.fr. If $\theta = 1$, $\nu = \nu_1 + \nu_2 = n_1 + n_2 - 2$, $g = b = 1/\nu$, and (20.103) reduces to (20.58), as it should. But in general, g and ν depend upon θ.

ν is never greater than at $\theta = 1$—cf. Exercise 20.22.

20.32   Welch (1938) investigated the extent to which the assumption that $\theta = 1$ in (20.103), when in reality it takes some other value, leads to erroneous conclusions.

(His discussion was couched in terms of testing hypotheses rather than of interval estimation, which is our present concern.) He found that, so long as $n_1 = n_2$, no great harm was done by ignorance of the true value of θ, but that if $n_1 \neq n_2$, serious errors could arise. To overcome this difficulty, he used the technique of **20.31** to approximate the distribution of the statistic z of (20.61). In this case he found that, whatever the values of n_1 and n_2, z was approximately distributed in Student's form with

$$\nu = \left(\frac{\theta}{n_1} + \frac{1}{n_2}\right)^2 \bigg/ \left(\frac{\theta^2}{n_1^2(n_1 - 1)} + \frac{1}{n_2^2(n_2 - 1)}\right) \tag{20.104}$$

degrees of freedom, and that the influence of a wrongly assumed value of θ was now very much smaller. This is what we should expect, since the denominator of z at (20.61) estimates the variances σ_1^2, σ_2^2 separately, while that of t at (20.103) uses a 'pooled' estimate s^2, that is clearly not appropriate when $\sigma_1^2 \neq \sigma_2^2$.

Mickey and Brown (1966) show that the exact distribution of z is bounded by Student's t-distributions with $n_1 + n_2 - 2$ and min $(n_1 - 1, n_2 - 1)$ d.fr.—cf. Exercise 20.22.

Lawton (1965) extends earlier work by J. Hájek to obtain close bounds upon the confidence coefficient and the coverage probability of central intervals based on z.

20.33 Welch (1947) refined the approximate approach of **20.32**. His argument is a general one, but for the present problem may be summarised as follows. Defining s_1^2, s_2^2 with $n_1 - 1$, $n_2 - 1$ as divisors respectively, so that they are unbiased estimators of variances, we seek a statistic $h(s_1^2, s_2^2, P)$ such that

$$P\{(d - \delta) < h(s_1^2, s_2^2, P)\} = P \tag{20.105}$$

whatever the value of θ. Now since $(d - \delta)$ is normally distributed independently of s_1^2, s_2^2, with zero mean and variance $\sigma_1^2/n_1 + \sigma_2^2/n_2 = D^2$, we have

$$P\{(d - \delta) \leqslant h(s_1^2, s_2^2, P) \mid s_1^2, s_2^2\} = I\left(\frac{h}{D}\right) \tag{20.106}$$

where $I(x) = \int_{-\infty}^{x} (2\pi)^{-1/2} \exp\left(-\tfrac{1}{2}t^2\right) dt$. Thus, from (20.105)–(20.106),

$$P = \int\int I(h/D) f(s_1^2) f(s_2^2) \, ds_1^2 \, ds_2^2. \tag{20.107}$$

Now we may expand $I(h/D)$, which is a function of s_1^2, s_2^2, in a Taylor series about the true values σ_1^2, σ_2^2. We write this symbolically

$$I\left\{\frac{h(s_1^2, s_2^2, P)}{D}\right\} = \exp\left\{\sum_{i=1}^{2} (s_i^2 - \sigma_i^2)\partial_i\right\} I\left\{\frac{h(s_1^2, s_2^2, P)}{s}\right\}, \tag{20.108}$$

where the operator ∂_i represents differentiation with respect to s_i^2, and then putting $s_i^2 = \sigma_i^2$, and $s^2 = s_1^2/n_1 + s_2^2/n_2$. We may put (20.108) into (20.107) to obtain

$$P = \prod_{i=1}^{2} \left[\int \exp\{(s_i^2 - \sigma_i^2)\partial_i\} f(s_i^2) \, ds_i^2\right] \times I\left\{\frac{h(s_1^2, s_2^2, P)}{s}\right\}. \tag{20.109}$$

Now since we have

$$f(s_i^2) \, ds_i^2 = \frac{1}{\Gamma(\frac{1}{2}\nu_i)} \left(\frac{\nu_i s_i^2}{2\sigma_i^2}\right)^{(\nu_i/2)-1} \exp\left(-\frac{\nu_i s_i^2}{2\sigma_i^2}\right) d\left(\frac{\nu_i s_i^2}{\sigma_i^2}\right),$$

on carrying out each integration in the symbolic expression (20.109) we find

$$\int \exp\{(s_i^2 - \sigma_i^2)\partial_i\} f(s_i^2) \, ds_i^2 = \left(1 - \frac{2\sigma_i^2 \partial_i}{\nu_i}\right)^{-\nu_i/2} \exp\left(-\sigma_i^2 \partial_i\right)$$

so that (20.109) becomes

$$P = \prod_{i=1}^{2} \left[\left(1 - \frac{2\sigma_i^2 \partial_i}{\nu_i}\right)^{-\nu_i/2} \exp\left(-\sigma_i^2 \partial_i\right)\right] I\left\{\frac{h(s_1^2, s_2^2, P)}{s}\right\}. \qquad (20.110)$$

We can solve (20.110) to obtain the form of the function h, and hence find $h(s_1^2, s_2^2, P)$, for any known P.

Welch gave a series expansion for h, which in our special case becomes

$$\frac{h(s_1^2, s_2^2, P)}{s} = \xi\left[1 + \frac{(1+\xi)^2}{4} \sum_{i=1}^{2} c_i^2/\nu_i - \frac{(1+\xi^2)}{2} \sum_{i=1}^{2} c_i^2/\nu_i^2 + \cdots\right], \qquad (20.111)$$

where $c_i = (s_i^2/n_i)((s_1^2/n_1) + (s_2^2/n_2))^{-1}$, $\nu_i = n_i - 1$ and ξ is defined by $I(\xi) = P$.

Asymptotic expressions of this type have been justified by Chernoff (1949) and D. L. Wallace (1958)—each succeeding term on the right is an order lower in ν_i.

Since $(d - \delta)/s = z$ of (20.61), (20.111) gives the distribution function of z.

Following further work by Welch, Aspin (1948, 1949) and Trickett *et al.* (1956), (20.111) was tabled as a function of ν_1, ν_2 and c_1, for $P = 0.95, 0.975, 0.99$ and 0.995.

Press (1966) shows that for $1 - \alpha = 0.90$, $n_1 \leq n_2 \leq 30$, Welch's intervals have smaller expected length than (20.93) if θ is small (when (20.91) discards information about the more variable population) but not if θ is large. The two sets of intervals are shown to be asymptotically equivalent, and never differ by more than 10 per cent in expected length when $n_1 > 10$ if $0.01 \leq \theta \leq 100$. See also some comparisons with other methods by Mehta and Srinivasan (1970) and Scheffé (1970b). Lee and Gurland (1975) show that Welch's method has almost exactly the prescribed value of α, irrespective of θ, and has very high selectivity; they propose a simpler method with similar properties. Pfanzagl (1974) proves an asymptotic optimum property of (20.111).

> In Chapter 29, where we discuss the analysis of variance, we shall consider setting confidence limits for an arbitrary number of normal means simultaneously.

Tables and charts of confidence intervals

20.34 (1) *Binomial distribution*—Clopper and Pearson (1934) give two central confidence interval charts for the parameter, for $\alpha = 0.01$ and 0.05; each gives contours for $n = 8$ (2) 12 (4) 24, 30, 40, 60, 100, 200, 400 and 1000. The charts are reproduced in the *Biometrika Tables*. Upper or lower one-sided intervals can be obtained for half these values of α. Incomplete B-function tables may also be used—see **5.7** and the *Biometrika Tables*.

Pachares (1960) gives central limits for $\alpha = 0.01$, 0.02, 0.05, 0.10 and $n = 55$ (5) 100, and references to other tables, including those of Clark (1953) for the same values of α and $n = 10$ (1) 50.

Blyth (1986) compares several normal approximations in detail and gives exact tables of upper one-sided limits for $\alpha = 0.005$, 0.01, 0.025, 0.05 and $n = 3$ (1) 20.

Sterne (1954) has proposed an alternative method of setting confidence limits for a proportion. Instead of being central, the belt contains the values of p with the largest probabilities of occurrence. Since the distribution of p is skew in general, we clearly shorten the interval in this way. Walton (1970) extended Sterne's tables. Crow (1956) has shown that these intervals constitute a confidence belt with minimum total area, and has tabulated a slightly modified set of intervals for $n \leqslant 30$ and $1 - \alpha = 0.90$, 0.95 and 0.99. Blyth and Still (1983) give intervals of Crow's type that are approximately central and approximately unbiased and have other desirable regularity properties, for $\alpha = 0.01$, 0.05.

Central intervals for the binomial parameter may easily be derived by use of (20.8), but they are not the most selective unbiased randomised intervals, which are tabulated by Blyth and Hutchinson (1960) for $\alpha = 0.01$, 0.05 and $n = 2$ (1) 24 (2) 50.

(2) *Poisson distribution*—(a) The *Biometrika Tables*, using the work of Garwood (1936), give central confidence intervals for the parameter, for observed values $x = 0$ (1) 30 (5) 50 and $\alpha = 0.002$, 0.01, 0.02, 0.05, 0.10. As in (1), one-sided intervals are available for $\alpha/2$. (b) Woodcock and Eames (1970) give similar tables for $x = 0$ (1) 100 (2) 200 (5) 500 (10) 1200 and α, 10α, 100α or 1000α equal to 0.1, 0.2 and 0.5. (c) Przyborowski and Wilénski (1935) give upper confidence limits only for $x = 0$ (1) 50, $\alpha = 0.001$, 0.005, 0.01, 0.02, 0.05, 0.10. (d) Walton (1970) tabulates Sterne-type intervals and Crow and Gardner (1959) tabulate modified intervals of the Sterne–Crow binomial type for $x = 0$ (1) 300 and $\alpha = 0.001$, 0.01, 0.05, 0.10, 0.20. (e) Blyth and Hutchinson (1961) give most selective unbiased randomised intervals for observed x up to 250 and $\alpha = 0.01$, 0.05.

(3) *Variance of a normal distribution*—Tate and Klett (1959) give the most selective unbiased confidence intervals, and the physically shortest intervals (cf. Exercise 20.5), based on the sample variance s^2 for $\alpha = 0.001$, 0.005, 0.01, 0.05, 0.10 and $n = 3$ (1) 30. The former are also given by Pachares (1961) for $\alpha = 0.01$, 0.05, 0.10 and $n - 1 = 1$ (1) 20, 24, 30, 40, 60, 120; by Lindley *et al* (1960) for $\alpha = 0.001$, 0.01, 0.05 and $n - 1 = 1$ (1) 100; and by G. R. Murdak and W. O. Williford in Owen and Odeh (1977).

(4) *Ratio of normal variances*—S. John (1975) gives the most selective unbiased intervals to 3 d.p. for $\alpha = 0.001$, 0.01, 0.05, 0.1 and $n_1 - 1$, $n_2 - 1 = 1$ (1) 20 (2) 30, 36, 45, 60, 90, 180, ∞. He tabulates in terms of the ratio of one estimator of variance to the pooled estimator.

(5) *Correlation parameter*—F. N. David (1938) gives four central confidence interval charts for the correlation parameter ρ of a bivariate normal population, for $\alpha = 0.01$, 0.02, 0.05, 0.10; each gives contours for $n = 3$ (1) 8, 10, 12, 15, 20, 25, 50, 100, 200 and 400. The *Biometrika Tables* reproduce the $\alpha = 0.01$ and $\alpha = 0.05$ charts. One-sided intervals may be obtained as in (1).

R. E. Odeh (in Shah and Odeh (1986)) provides comprehensive tables of intervals for $\alpha = 0.005$, 0.01, 0.025, 0.05, 0.10, 0.25; $n = 3$ (1) 60 (2) 80 (5) 100 (10) 200 (25) 300 (50) 600 (100) 1000 and observed $r = -0.98$ (0.02) 0.98.

(6) *Difference of two normal means*—The *Biometrika Tables* give tables for setting central confidence intervals based on the expressions (20.110)–(20.111), for $1 - \alpha = 0.90, 0.95, 0.98, 0.99$.

Tolerance intervals

20.35 Throughout this chapter we have been discussing the setting of confidence intervals for parameters, but such intervals can be set for other quantities than parameters. We shall see in **20.37–8** that intervals can be found for any quantile of a distribution without any further assumption on it than continuity, and we shall see in **30.38** that we can find distribution-free intervals for the entire d.f. of a continuous distribution.

There is another type of problem, met in practical sampling, which may be solved by these methods. Suppose that, on the basis of a sample of n independent observations from a distribution, we wish to find two limits, L_1 and L_2, between which at least a given proportion γ of the distribution may be asserted to lie. Clearly, we can only make such an assertion in probabilistic form, i.e. we assert that, with given probability β, at least a proportion γ of the distribution lies between L_1 and L_2. L_1 and L_2 are called *tolerance limits* for the distribution; we shall call them the (β, γ) tolerance limits. The interval (L_1, L_2) is called a *tolerance interval.* In **20.39–40** we shall see that tolerance limits, also, may be set without assumptions (except continuity) on the form of the underlying distribution. First, however, we shall discuss the derivation of tolerance limits for a normal distribution, due to Wald and Wolfowitz (1946).

Tolerance intervals for a normal distribution

20.36 Since the sample mean and variance are a pair of sufficient statistics for the parameters of a normal distribution (Example 17.17), it is natural to base tolerance limits for the distribution upon them. In a sample of size n, we work with the unbiased statistics

$$\bar{x} = \sum x/n, \qquad s'^2 = \sum (x - \bar{x})^2/(n-1),$$

and define

$$A(\bar{x}, s', \lambda) = \int_{\bar{x}-\lambda s'}^{\bar{x}+\lambda s'} f(t)\, dt, \tag{20.112}$$

where $f(t)$ is the normal density. We now seek to determine the value λ so that

$$P\{A(\bar{x}, s', \lambda) > \gamma\} = \beta. \tag{20.113}$$

$L_1 = \bar{x} - \lambda s'$ and $L_2 = \bar{x} + \lambda s'$ will then be a pair of central (β, γ) tolerance limits for the normal distribution. Since we are concerned only with the proportion of that distribution covered by the interval (L_1, L_2), we may without any loss of generality standardise the population mean at 0 and its variance at 1. Thus

$$f(t) = (2\pi)^{-1/2} \exp\left(-\tfrac{1}{2}t^2\right).$$

Consider first the conditional probability, given $\bar{x}$, that $A(\bar{x}, s', \lambda)$ exceeds γ. We denote this by $P\{A > \gamma | \bar{x}\}$. Now A is a monotone increasing function of s', and the equation in s'

$$A(\bar{x}, s', \lambda) = \gamma \qquad (20.114)$$

has just one root, which we denote by $s'(\bar{x}, \gamma, \lambda)$. Let

$$\lambda s'(\bar{x}, \gamma, \lambda) = r. \qquad (20.115)$$

From a table of the normal distribution, the value r satisfying

$$\int_{\bar{x}-r}^{\bar{x}+r} f(t)\, \mathrm{d}t = \gamma \qquad (20.116)$$

may be obtained for any $\bar{x}$ and γ, and we write it $r(\bar{x}, \gamma)$. Moreover, since A is monotone increasing in s', the inequality $A > \gamma$ is equivalent to

$$s' > s'(\bar{x}, \gamma, \lambda) = r(\bar{x}, \gamma)/\lambda.$$

Thus we may write

$$P\{A > \gamma | \bar{x}\} = P\left\{ s' > \frac{r}{\lambda} \Big| \bar{x} \right\}, \qquad (20.117)$$

and since $\bar{x}$ and s' are independently distributed, the right side of (20.117) may have the conditioning on $\bar{x}$ suppressed and (20.117) becomes

$$P\{A > \gamma | \bar{x}\} = P\{(n-1)s'^2 > (n-1)r^2/\lambda^2\}. \qquad (20.118)$$

Since $(n-1)s'^2 = \sum (x - \bar{x})^2$ is a χ^2_{n-1} variate, we have

$$P\{A > \gamma | \bar{x}\} = P\{\chi^2_{n-1} > (n-1)r^2/\lambda^2\}. \qquad (20.119)$$

To obtain the unconditional probability $P(A > \gamma) = \beta$ from (20.119), we must integrate it over the distribution of $\bar{x}$, which is normal with zero mean and variance $1/n$. This is a tedious numerical operation, but fortunately an excellent approximation is available. We expand $P(A > \gamma | \bar{x})$ in a Taylor series about $\bar{x} = \mu = 0$, and since it is an even function of $\bar{x}$, the odd powers in the expansion vanish,[*] leaving

$$P(A > \gamma | \bar{x}) = P(A > \gamma | 0) + \frac{\bar{x}^2}{2!} P''(A > \gamma | 0) + \cdots. \qquad (20.120)$$

Integrating over the distribution of $\bar{x}$, we have, using the moments of $\bar{x}$,

$$P(A > \gamma) = P(A > \gamma | 0) + \frac{1}{2n} P''(A > \gamma | 0) + O(n^{-2}) \cdots. \qquad (20.121)$$

But from (20.120) with $\bar{x} = 1/\sqrt{n}$, we also have

$$P\left(A > \gamma \Big| \frac{1}{\sqrt{n}}\right) = P(A > \gamma | 0) + \frac{1}{2n} P''(A > \gamma | 0) + O(n^{-2}). \qquad (20.122)$$

---

[*] This is because the interval is symmetric about $\bar{x}$; it would not happen otherwise.

(20.121) and (20.122) give the approximation

$$P(A > \gamma) \doteq P\left(A > \gamma \left| \frac{1}{\sqrt{n}}\right.\right) \tag{20.123}$$

and (20.113), (20.123) and (20.119) yield finally

$$\beta = P\{A > \gamma\} \doteq P\{\chi^2_{n-1} > (n-1)r^2/\lambda^2\}, \tag{20.124}$$

where r is defined from (20.116) by

$$\int_{n^{-(1/2)}-r}^{n^{-(1/2)}+r} f(t)\, dt = \gamma. \tag{20.125}$$

Given γ, β, n and $\bar{x}$, we can determine λ approximately from (20.124) and (20.125), and hence the tolerance limits $\bar{x} \pm \lambda s'$. Wald and Wolfowitz (1946) showed that the approximation is extremely good even for values of n as low as 2 if β and γ are $\geqslant 0.95$, as they usually are in practice. Bowker (1947) gave tables of λ (his k) for β (his γ) = 0.75, 0.90, 0.95, 0.99 and γ (his P) = 0.75, 0.90, 0.99 and 0.999, for sample sizes n = 2 (1) 102 (2) 180 (5) 300 (10) 400 (25) 750 (50) 1000.

On examination of the argument above, it will be seen to hold if $\bar{x}$ is replaced by any estimator $\hat{\mu}$ of the mean, and s'^2 by any independent estimator $\hat{\sigma}^2$ of the variance, of a normal population, as pointed out by Wallis (1951). Ellison (1964) has shown that if the mean is estimated from n observations and the variance estimate has ν degrees of freedom, the approximation corresponding to (20.124) is valid only to order ν/n^2 (reducing to $1/n$ for (20.124) itself, where $\nu = n-1$), and Howe (1969) gives better tolerance limits for the case where ν/n^2 is large. In the contrary case, there are some useful tables. Taguti (1958) gives tables of λ (his k) for β (his $1 - \alpha$) and γ (his P) = 0.90, 0.95 and 0.99; and n = 0.5 (0.5) 2 (1) 10 (2) 20 (5) 30 (10) 60 (20) 100, 200, 500, 1000, ∞; ν = 1 (1) 20 (2) 30 (5) 100 (100) 1000, ∞. The small fractional values of n are useful in some applications discussed by Taguti. Weissberg and Beatty (1960) give tables of λ/r (their u) for ν (their f) = 1 (1) 150 (2) 250 (5) 500 (10) 1000 (1000) 10,000, ∞ and β (their γ) = 0.90, 0.95, 0.99; and of r for n = 1 (1) 100 (5) 200 (10) 300 (20) 500 (100) 1000 (1000) 10,000, ∞ and γ (their P) = 0.5, 0.75, 0.9, 0.95, 0.99, 0.999.

Fraser and Guttman (1956) and Guttman (1957) consider tolerance intervals that cover a given proportion of a normal distribution *on the average*. Sharpe (1970) studies the robustness of both kinds of tolerance intervals.

Zacks (1970) considers tolerance limits, and their relationship with confidence limits, for a large class of discrete distributions including the binomial, the negative binomial, and the Poisson.

Distribution-free confidence intervals for quantiles

20.37 The joint distribution of the order-statistics depends very directly upon the parent d.f. (cf., e.g., (14.1) and (14.2)) and therefore point estimation of the population quantiles by order statistics also does so. Remarkably enough, however, pairs of order-statistics may be used to set population confidence intervals for any population quantile that are *distribution-free*, i.e. that do not depend on the form of the underlying distribution, provided that it is continuous.

Consider the pair of order-statistics $x_{(r)}$ and $x_{(s)}$, $r < s$, in a sample of n observations from the continuous d.f. $F(x)$. (14.2) gives the joint distribution of $F_r = F(x_{(r)})$ and $F_s = F(x_{(s)})$ as

$$dG_{r,s} = \frac{F_r^{r-1}(F_s - F_r)^{s-r-1}(1 - F_s)^{n-s} \, dF_r \, dF_s}{B(r, s-r)B(s, n-s+1)}. \tag{20.126}$$

X_p, the p-quantile of $F(x)$, is defined by $F(X_p) = p$. For the probability that the interval $(x_{(r)}, x_{(s)})$ covers X_p, we have

$$P\{x_{(r)} \leq X_p \leq x_{(s)}\} = P\{x_{(r)} \leq X_p\} - P\{x_{(s)} < X_p \mid x_{(r)} \leq X_p\}. \tag{20.127}$$

Now since $x_{(r)} \leq x_{(s)}$, we can *only* have $x_{(s)} < X_p$ if $x_{(r)} < X_p$. Thus

$$P\{x_{(s)} < X_p \mid x_{(r)} \leq X_p\} = P\{x_{(s)} \leq X_p\}, \tag{20.128}$$

the equality on the right having zero probability because of the continuity of F. (20.127) and (20.128) give

$$P\{x_{(r)} \leq X_p \leq x_{(s)}\} = P\{x_{(r)} \leq X_p\} - P\{x_{(s)} \leq X_p\}. \tag{20.129}$$

The d.f. of $x_{(r)}$ given at (14.4) and (14.5) here yields, since $F(X_p) = p$,

$$P\{x_{(r)} \leq X_p\} = G_r(X_p) = I_p(r, n-r+1), \tag{20.130}$$

and similarly for $x_{(s)}$. (20.129) therefore becomes

$$P\{x_{(r)} \leq X_p \leq x_{(s)}\} = I_p(r, n-r+1) - I_p(s, n-s+1) = 1 - \alpha, \tag{20.131}$$

say, independent of the form of the continuous distribution F.

20.38 We see from (20.131) that the interval $(x_{(r)}, x_{(s)})$ covers the quantile X_p with a confidence coefficient that does not depend on $F(x)$ at all, and we thus have a distribution-free confidence interval for X_p. Since $I_p(a, b) = 1 - I_{1-p}(b, a)$, we may also write the confidence coefficient as

$$1 - \alpha = I_{1-p}(n-s+1, s) - I_{1-p}(n-r+1, r). \tag{20.132}$$

By the Incomplete Beta relation with the binomial expansion given at (5.16), (20.132) may be expressed as

$$1 - \alpha = \left\{ \sum_{i=0}^{s-1} - \sum_{i=0}^{r-1} \right\} \binom{n}{i} p^i q^{n-i} = \sum_{i=r}^{s-1} \binom{n}{i} p^i q^{n-i}, \tag{20.133}$$

where $q = 1 - p$. The confidence coefficient is therefore the sum of the terms in the binomial $(q + p)^n$ from the $(r+1)$th to the sth inclusive.

If we choose a pair of symmetrically placed order-statistics we have $s = n - r + 1$, and find in (20.131)–(20.133)

$$1 - \alpha = I_p(r, n - r + 1) - I_p(n - r + 1, r)$$
$$= 1 - \{I_{1-p}(n - r + 1, r) + I_p(n - r + 1, r)\}, \tag{20.134}$$

$$= \sum_{i=r}^{n-r} \binom{n}{i} p^i q^{n-i}, \tag{20.135}$$

so that the confidence coefficient is the sum of the central $(n - 2r + 1)$ terms of the binomial, r terms at each end being omitted.

For any values of r and n, the confidence coefficient attaching to the interval $(x_{(r)}, x_{(n-r+1)})$ may be calculated from (20.134)–(20.135), if necessary using the *Tables of the Incomplete Beta Function*. The tables of the binomial distribution listed in **5.7** may also be used. Exercise 20.23 gives the reader an opportunity to practise the computation.

Scheffé and Tukey (1945) show that if the parent distribution is discrete, the confidence intervals above cover X_p with probability $\geq 1 - \alpha$.

In the special case of the parent median $X_{0.5}$, (20.134) and (20.135) reduce to

$$1 - \alpha = 1 - 2I_{0.5}(n - r + 1, r) = 2^{-n} \sum_{i=r}^{n-r} \binom{n}{i}, \tag{20.136}$$

a particularly simple form. This confidence interval procedure for the median was first proposed by W. R. Thompson (1936).

MacKinnon (1964) gives tables of r for $n = 1$ (1) 1000 and α as little as possible below 0.50, 0.10, 0.05, 0.02, 0.01 and 0.001. Van der Parren (1970) gives similar tables for $n = 3$ (1) 150 and $\alpha = 0.3, 0.2, 0.1, 0.05, 0.02$ and 0.01, together with the exact value of $\alpha/2$ in each case.

One may also obtain confidence intervals for the quantile interval (X_p, X_q), $p < q$; cf. Reiss and Rüschendorf (1976) and Sathe and Lingras (1981).

Distribution-free tolerance intervals
20.39   In **20.36** we discussed the problem of finding tolerance intervals for a normal d.f. Suppose now that we require such intervals without making assumptions beyond continuity on the underlying distributional form. We require to calculate a randomly varying interval (l, u) such that

$$P\left\{ \int_l^u f(x)\, dx \geq \gamma \right\} = \beta, \tag{20.137}$$

where $f(x)$ is the unknown continuous frequency function. It is not obvious that such a distribution-free procedure is possible, but Wilks (1941, 1942) showed that the order-statistics $x_{(r)}, x_{(s)}$ provide distribution-free tolerance intervals, and Robbins (1944) showed that *only* the order-statistics do so.

If we write $l = x_{(r)}$, $u = x_{(s)}$ in (20.137), we may rewrite it

$$P[\{F(x_{(s)}) - F(x_{(r)})\} \geq \gamma] = \beta. \tag{20.138}$$

We may obtain the exact distribution of the random variable $F(x_{(s)}) - F(x_{(r)})$ from (20.126) by the transformation $y = F(x_{(s)}) - F(x_{(r)})$, $z = F(x_{(r)})$, with Jacobian 1. (20.126) becomes

$$dH_{y,z} = \frac{z^{r-1}y^{s-r-1}(1-y-z)^{n-s}\,dy\,dz}{B(r, s-r)B(s, n-s+1)}, \qquad 0 \leqslant y+z \leqslant 1. \qquad (20.139)$$

In (20.139) we integrate out z over its range $(0, 1-y)$, obtaining for the marginal distribution of y

$$dJ_{r,s} = \frac{y^{s-r-1}\,dy}{B(r, s-r)B(s, n-s+1)} \int_0^{1-y} z^{r-1}(1-y-z)^{n-s}\,dz. \qquad (20.140)$$

We put $z = (1-y)t$, reducing (20.140) to

$$\begin{aligned} dJ_{r,s} &= \frac{y^{s-r-1}(1-y)^{n-s+r}\,dy}{B(r, s-r)B(s, n-s+1)} \int_0^1 t^{r-1}(1-t)^{n-s}\,dt \\ &= y^{s-r-1}(1-y)^{n-s+r}\,dy\frac{B(r, n-s+1)}{B(r, s-r)B(s, n-s+1)} \\ &= \frac{y^{s-r-1}(1-y)^{n-s+r}\,dy}{B(s-r, n-s+r+1)}, \qquad 0 \leqslant y \leqslant 1. \end{aligned} \qquad (20.141)$$

Thus $y = F(x_{(s)}) - F(x_{(r)})$ is distributed as a beta variate of the first kind with parameters depending only on the difference $(s-r)$. If we put $r = 0$ in (20.141) and interpret $F(x_{(0)})$ as zero (so that $x_{(0)} = -\infty$), (20.141) reduces to (14.1), with s written for r.

20.40 From (20.141), we see that (20.138) becomes

$$P\{y \geqslant \gamma\} = \int_\gamma^1 \frac{y^{s-r-1}(1-y)^{n-s+r}\,dy}{B(s-r, n-s+r+1)} = \beta, \qquad (20.142)$$

which we may rewrite in terms of the Incomplete Beta function as

$$P\{F(x_{(s)}) - F(x_{(r)}) \geqslant \gamma\} = 1 - I_\gamma(s-r, n-s+r+1) = \beta.$$

The relationship (20.143) for the distribution-free tolerance interval $(x_{(r)}, x_{(s)})$ contains five quantities: γ (the minimum proportion of $F(x)$ it is desired to cover), β (the probability with which we desire to do this), the sample size n, and the order-statistics' positions in the sample, r and s. Given any four of these, we can solve (20.143) for the fifth. In practice, β and γ are usually fixed at levels required by the problem, and r and s symmetrically chosen, so that $s = n - r + 1$. (20.143) then reduces to

$$I_\gamma(n - 2r + 1, 2r) = 1 - \beta. \qquad (20.144)$$

The left side of (20.144) is a monotone increasing function of n, and for any fixed β, γ, r we can choose n large enough so that (20.144) is satisfied. In practice, we must choose n as the nearest integer above the solution of (20.144). If $r = 1$, so that the extreme values in the sample are being used, (20.144) reduces to

$$I_\gamma(n - 1, 2) = 1 - \beta, \qquad (20.145)$$

which gives the probability β with which the range of the sample of n observations covers at least a proportion γ of the population d.f.

The solution of (20.143) (and of its special cases (20.144)-(20.145)) has to be carried out numerically with the aid of the *Tables of the Incomplete Beta Function*, or equivalently (cf. **5.7**) of the binomial d.f. Murphy (1948) gives graphs of γ as a function of n for $\beta = 0.90$, 0.95 and 0.99 and $r + (n - s + 1) = 1$ (1) 6 (2) 10 (5) 30 (10) 60 (20) 100; these are exact for $n \leqslant 100$, and approximate up to $n = 500$.

Example 20.7
We consider the numerical solution of (20.145) for n. It may be rewritten

$$1 - \beta = \frac{1}{B(n-1, 2)} \int_0^\gamma y^{n-2}(1-y)\, dy = n(n-1)\left\{ \frac{\gamma^{n-1}}{n-1} - \frac{\gamma^n}{n} \right\}$$
$$= n\gamma^{n-1} - (n-1)\gamma^n. \tag{20.146}$$

For the values of β, γ that are required in practice (0.90 or larger, usually), n is so large that we may write (20.146) approximately as

$$1 - \beta = n\gamma^{n-1}(1 - \gamma),$$
$$\gamma = \left\{ \left(\frac{1-\beta}{1-\gamma} \right) \frac{1}{n} \right\}^{1/(n-1)} \tag{20.147}$$

or

$$\log n + (n-1)\log \gamma = \log\{(1-\beta)/(1-\gamma)\}. \tag{20.148}$$

The derivative of the left side of (20.148) with respect to n is $(1/n) + \log \gamma$ and for large n the left side of (20.148) is a monotone decreasing function of n. Thus we may guess a trial value of n, compare the left with the (fixed) right side of (20.148), and increase (decrease) n if the left (right) is greater. The value of n satisfying the approximation (20.148) will be somewhat too large to satisfy the exact relationship (20.146), since a positive term γ^n was dropped from the right of the latter, and we may safely use (20.148) unadjusted. Alternatively, we may put the solution of (20.148) into (20.146) and adjust to obtain the correct value.

Example 20.8
We illustrate Example 20.7 with a particular computation. Let us put $\beta = \gamma = 0.99$. (20.148) is then

$$\log n + (n-1)\log 0.99 = 0,$$

the right-hand side, of course, being zero whenever $\beta = \gamma$. We may use logs to base 10, since the adjustment to natural logs cancels through (20.148). Thus we have to solve

$$\log_{10} n - 0.00436 (n-1) = 0.$$

We first guess $n = 1000$. This makes the left side negative, so we reduce n to 500, which makes it positive. We then progress iteratively as follows:

n	$\log_{10} n$	$0.00436(n-1)$
1000	3	4.36
500	2.6990	2.18
700	2.8451	3.05
650	2.8129	2.83
600	2.7782	2.61
640	2.8062	2.79
645	2.8096	2.81

We now put the value $n = 645$ into the exact (20.146). Its right side is

$$645(0.99)^{644} - 644(0.99)^{645} = 1.004 - 0.992 = 0.012.$$

Its left side is $1 - \beta = 0.01$, so the agreement is good and we may for all practical purposes take $n = 645$ in order to get a 99 per cent tolerance interval for 99 per cent of the parent d.f. Exercise 20.24 gives further results for the reader to verify.

> We have discussed only the simplest case of setting distribution-free tolerance intervals for a univariate continuous distribution. Extensions to multivariate tolerance regions, including the discontinuous case, have been made by Wald (1943b), Scheffé and Tukey (1945), Tukey (1947, 1948), Fraser and Wormleighton (1951), Fraser (1951, 1953), and Kemperman (1956). Walsh (1962) considers symmetrical continuous distributions. Goodman and Madansky (1962) consider parameter-free and distribution-free tolerance limits for the exponential distribution. Guttman (1970) reviews the subject as a whole.
>
> Scheffé and Tukey (1945) and Tukey (1948) show that if the underlying distribution is discrete, the above tolerance intervals and regions have probability $\geqslant 1 - \alpha$.

Fiducial intervals

20.41 Let us reconsider Example 20.1 from a different standpoint. The distribution of the statistic $\bar{x}$ given there,

$$dF = \left(\frac{n}{2\pi}\right)^{1/2} \exp\left\{-\tfrac{1}{2}n(\bar{x} - \mu)^2\right\} d\bar{x} \tag{20.149}$$

is equivalent to use of the likelihood function since $\bar{x}$ is sufficient for μ. If we are prepared intuitively to use (20.149) not only, as in Example 20.1, as the distribution of $\bar{x}$ for a fixed unknown value of μ, but as expressing our credence in a particular value of μ given a fixed observed value $\bar{x}$, we are regarding (20.149) as

$$dF = \left(\frac{n}{2\pi}\right)^{1/2} \exp\left\{-\tfrac{1}{2}n(\bar{x} - \mu)^2\right\} d\mu, \tag{20.150}$$

which, following R. A. Fisher, we shall call the *fiducial distribution* of the parameter μ. We note that the integral of (20.150) with respect to μ over the real line is 1, so that the constant needs no adjustment.

This fiducial distribution is not a probability distribution in the usual sense. It is a new concept, expressing the intensity of our belief in the various possible values of a parameter. It may be regarded as a distribution of probability in the sense of degree of belief; the consequent link with interval estimation based on the use of Bayes' theorem will be discussed below. Or it may be regarded as a new concept, giving formal expression to somewhat intuitive ideas about the extent to which we place credence in various values of μ. It so happens, in this case, that the non-differential element in (20.150) is the same as that in (20.149). This is not essential, though it is not infrequent.

20.42 The fiducial distribution may now be used to determine intervals within which μ is located. We select some arbitrary numbers, say 0.02275 and 0.97725, and decide to regard those values as critical in the sense that any acceptable value of μ must not give to the observed $\bar{x}$ a (cumulative) probability less than 0.02275 or greater than 0.97725. Then, since these values correspond to deviations of $\pm 2\sigma$ from the mean of a normal distribution, and $\sigma = 1/\sqrt{n}$, we have

$$-2 \leqslant (\bar{x} - \mu)\sqrt{n} \leqslant 2,$$

which is equivalent to

$$\bar{x} - 2/\sqrt{n} \leqslant \mu \leqslant \bar{x} + 2/\sqrt{n}. \tag{20.151}$$

This, as it happens, is the same inequality as in the central confidence intervals (20.4) in Example 20.1. But it is essential to note that it is not reached by the same line of thought. The confidence approach says that if we assert (20.151) we shall be right in about 95.45 per cent of the cases *in the long run*. Under the fiducial approach the assertion of (20.151) is that (in some sense not defined) we are 95.45 per cent sure of being right *in this particular case*. The shift of emphasis is evidently the one we encountered in considering the likelihood function itself, where the function $L(x|\theta)$ can be considered as a probability in which θ is fixed and x varies, or as a likelihood in which x is fixed and θ varies. So here, we can make an inference about the range of θ either by regarding it as a constant and setting up intervals that are random variables, or by regarding the observations as fixed and setting up intervals based on some undefined intensity of belief in the values of the parameter generating those observations.

There is one further fundamental distinction between the two methods. We have seen earlier in this chapter that in confidence theory it is possible to have different sets of intervals for the same parameter based on different statistics (although we may discriminate between the different sets, and choose the shortest or most selective set). This is explicitly ruled out in fiducial theory (even in the sense that we may choose central or non-central intervals for the same distribution when using both its tails). We must, in fact, use all the information about the parameter that the likelihood function contains. This implies that if we are to set limits to θ by a single statistic t, the latter must be sufficient for θ. For confidence intervals, on the other hand, sufficiency is desirable for the maximisation of selectivity—cf. **20.16**.

As we pointed out in **17.38**, there is always a *set* of jointly sufficient statistics for an unknown parameter, namely the n observations themselves. But this tautology offers little consolation: even a sufficient set of two statistics would be difficult enough to handle; a larger set is almost certainly practically useless. As to what should be done to construct an interval for a single parameter θ where a single sufficient statistic does not exist, writers on fiducial theory are for the most part silent.

20.43 Let $f(t, \theta)$ be a continuous density and $F(t, \theta)$ the d.f. of a statistic t that is sufficient for θ. Consider the behaviour of f for some fixed t, as θ varies. Suppose also that we know beforehand that θ must lie in a certain range, which may in particular be $(-\infty, \infty)$. Take some critical probability $1 - \alpha$ (analogous to a confidence coefficient) and let θ_α be the value of θ for which $F(t, \theta) = 1 - \alpha$.

Now suppose also that over the permissible range of θ, $f(t, \theta)$ is a monotonic non-increasing function of θ for any t. Then for all $\theta \leqslant \theta_\alpha$ the observed t has at least as large a density as $f(t, \theta_\alpha)$, and for $\theta > \theta_\alpha$ it has a lower density. We then choose $\theta \leqslant \theta_\alpha$ as our fiducial interval. It includes all those values of the parameter that give to the density a value greater than or equal to $f(t, \theta_\alpha)$.

If we require a fiducial interval of type

$$\theta_{\alpha_1} \leqslant \theta \leqslant \theta_{\alpha_2}$$

we look for two values of θ such that $f(t, \theta_{\alpha_1}) = f(t, \theta_{\alpha_2})$ and $F(t, \theta_{\alpha_2}) - F(t, \theta_{\alpha_1}) = 1 - \alpha$. If, between these values, $f(t, \theta)$ is greater than the extreme values $f(t, \theta_{\alpha_1})$ and $f(t, \theta_{\alpha_2})$, and is less than those values outside it, the interval again comprises values for which the density is at least as great as the density at the critical points.

If the distribution of t is symmetrical this involves taking a range that cuts off equal tail areas on it. For a non-symmetrical distribution the tails are to be such that their total probability content is α, but the contents of the two tails are not equal. It is the extreme ordinates of the interval that must be equal. Similar considerations have already been discussed in connection with central confidence intervals in **20.7**.

On this understanding, if our fiducial interval is increased by an element $d\theta$ at each end, the probability ordinate at the end decreases by $(\partial F(t, \theta)/\partial\theta)\, d\theta$. For the fiducial distribution we then have

$$dF = -\frac{\partial F(t, \theta)}{\partial\theta}\, d\theta. \tag{20.152}$$

This formula, however, requires that $f(t, \theta)$ shall be a non-decreasing function of θ at the lower end and a non-increasing function of θ at the upper end of the interval.

Example 20.9
As an example of a non-symmetrical sampling distribution, consider the gamma distribution

$$dF = \frac{x^{p-1}\, e^{-x/\theta}}{\theta^p \Gamma(p)}\, dx, \qquad p > 0; \quad 0 \leqslant x < \infty. \tag{20.153}$$

If p is known, $t \equiv \bar{x}/p$ is sufficient for θ (cf. Exercise 17.1) and its sampling distribution is easily seen to be

$$dF = \left(\frac{\beta}{\theta}\right)^{\beta} \frac{t^{\beta-1} e^{-\beta t/\theta}}{\Gamma(\beta)} \, dt, \qquad (20.154)$$

where $\beta = np$. Now in this case θ may vary only from 0 to ∞. As it does so the ordinate of (20.154) for fixed t rises monotonically from zero to a maximum and then falls again to zero. Thus, if we determine θ_{α_1} and θ_{α_2} so that the ordinates at these two values are equal and the integral of (20.154) between them has the assigned value $1 - \alpha$, the fiducial range is $\theta_{\alpha_1} \leq \theta \leq \theta_{\alpha_2}$.

Integrating (20.154) and putting $u = \beta t/\theta$, we have

$$F(t, \theta) = \int_0^{\beta t/\theta} \frac{u^{\beta-1} e^{-u}}{\Gamma(\beta)} \, du. \qquad (20.155)$$

Thus the fiducial distribution of θ is

$$-\frac{\partial F}{\partial \theta} \, d\theta = -\left[\frac{u^{\beta-1} e^{-u}}{\Gamma(\beta)}\right]_{u=\beta t/\theta} \frac{\partial}{\partial \theta}\left(\frac{\beta t}{\theta}\right) d\theta$$

$$= \left(\frac{\beta t}{\theta}\right)^{\beta-1} \frac{e^{-\beta t/\theta}}{\Gamma(\beta)} \frac{\beta t}{\theta^2} \, d\theta = \left(\frac{\beta t}{\theta}\right)^{\beta} \frac{e^{-\beta t/\theta}}{\Gamma(\beta)} \frac{d\theta}{\theta}. \qquad (20.156)$$

This is a Pearson Type V distribution (Exercise 6.6), so its integral from $\theta = 0$ to $\theta = \infty$ is unity.

In comparing (20.156) with (20.154) it should be noticed that we have replaced dt, not by $d\theta$, but by $t \, d\theta/\theta$; i.e. we have replaced dt/t by $d\theta/\theta$. Essentially, this is because (20.154) has the form $g(t/\theta) \, dt/\theta$ with θ a scale parameter, whence $z = \log t - \log \theta$ has the density $g(e^z) e^z \, dz$, free of θ. The fiducial argument is centred on the variable $\log \theta - \log t$ and we thus pass from $d(\log t) = dt/t$ to $d(\log \theta) = d\theta/\theta$.

The distribution (20.153) was treated from the confidence interval viewpoint in Example 20.6, where $ns^2/2\theta$ was a gamma variate with parameter $\frac{1}{2}(n-1)$, whereas in (20.154) and (20.156), $\beta t/\theta$ is a gamma variate with parameter β.

20.44 When two or more parameters are involved, the fiducial argument begins to meet difficulties. We shall concentrate on two important standard cases, the estimation of the mean in normal samples where the variance is unknown, and the estimation of the difference of two means in samples from two normal populations with unequal variances.

Example 20.10 Fiducial interval for a normal mean with population variance unknown

From Example 17.17, the sample mean $\bar{x}$ and the sample variance $s^2 \ (=\sum (x - \bar{x})^2/n)$ are jointly sufficient for the parent mean μ and variance σ^2 of the normal population, with joint distribution

$$dF \propto \frac{1}{\sigma} \exp\left\{-\frac{n}{2\sigma^2}(\bar{x}-\mu)^2\right\} d\bar{x} \left(\frac{s}{\sigma}\right)^{n-2} \exp\left\{-\frac{ns^2}{2\sigma^2}\right\} \frac{ds}{\sigma}, \qquad (20.157)$$

which we have expressed in terms of $\bar{x}$ and s. If we were considering fiducial limits for μ with known σ, we should use the first factor on the right of (20.157); but if we were considering limits for σ with known μ we should *not* use the second factor, since σ enters into the first factor. In fact (cf. Example 17.17), the sufficient statistic in this case is not s^2 but $\sum (x-\mu)^2/n = s'^2$, whose distribution is obtained by merging the two factors in (20.157).

For known σ, we should, as in **20.41**, replace $d\bar{x}$ by $d\mu$ to obtain the fiducial distribution of μ. For known μ, we should use the fact that s'^2 is distributed like t in (20.154), with $\beta = n$ and $\theta = \sigma^2$, and hence, as in Example 20.9, replace ds'/s' by $d\sigma/\sigma$. In (20.157), s is distributed as s', but with $\beta = n-1$. The question is, can we here replace $d\bar{x}\, ds/s$ in (20.157) by $d\mu\, d\sigma/\sigma$ to obtain the joint fiducial distribution of μ and σ?

Fiducialists assume that this is so. The question appears to us to be very debatable.* However, let us make the assumption and see where it leads us. For the fiducial distribution we shall then have

$$dF \propto \frac{1}{\sigma} \exp\left\{-\frac{n}{2\sigma^2}(\bar{x}-\mu)^2\right\} d\mu \left(\frac{s}{\sigma}\right)^{n-1} \exp\left\{-\frac{ns^2}{2\sigma^2}\right\} \frac{d\sigma}{\sigma}.$$

We now integrate for σ to obtain the fiducial distribution of μ and arrive at

$$dF \propto \frac{d\mu/s}{\left\{1+\dfrac{(\bar{x}-\mu)^2}{s^2}\right\}^{n/2}}.$$

This is Student's distribution, with $t = (\mu - \bar{x})(n-1)^{1/2}/s$ and $n-1$ d.fr. Thus, given α, we can find two values of t, t_0 and t_1, so that

$$P\{-t_0 \leqslant t \leqslant t_1\} = 1-\alpha,$$

where

$$P\{\bar{x} - st_0/(n-1)^{1/2} \leqslant \mu \leqslant \bar{x} + st_1/(n-1)^{1/2}\} = 1-\alpha.$$

In Example 20.2, we interpreted this in the confidence-interval sense, but this is by no means essential to the fiducial argument, as Example 20.11 will show.

Example 20.11 Fiducial intervals for the problem of two means
The fiducial solution of the problem of two means starts from the joint distribution of sample means and variances, which may be written, from Example 20.10,

$$dF \propto \frac{1}{\sigma_1 \sigma_2} \exp\left\{-\frac{n_1}{2\sigma_1^2}(\bar{x}_1-\mu_1)^2 - \frac{n_2}{2\sigma_2^2}(\bar{x}_2-\mu_2)^2\right\} d\bar{x}_1\, d\bar{x}_2$$

$$\times \frac{s_1^{n_1-2}}{\sigma_1^{n_1-2}} \frac{s_2^{n_2-2}}{\sigma_2^{n_2-2}} \exp\left\{-\frac{n_1}{2}\frac{s_1^2}{\sigma_1^2} - \frac{n_2}{2}\frac{s_2^2}{\sigma_2^2}\right\} ds_1\, ds_2.$$

---

* Although $\bar{x}$ and s are statistically independent, μ and σ are not independent in any fiducial sense. Some support for the replacement can be derived *a posteriori* from the reflection that it leads to Student's distribution, but if fiducial theory is to be accepted on its own merits, something more is required.

In accordance with the fiducial argument, we replace $d\bar{x}_1$, $d\bar{x}_2$ by $d\mu_1$, $d\mu_2$ and ds_1/s_1, ds_2/s_2 by $d\sigma_1/\sigma_1$, $d\sigma_2/\sigma_2$, as in Example 20.10. Then for the fiducial distribution (omitting powers of s_1 and s_2, which are now constants) we have, as in Example 20.10,

$$dF \propto \frac{1}{\sigma_1^{n_1+1}\sigma_2^{n_2+1}} \exp\left\{-\frac{n_1}{2\sigma_1^2}(\bar{x}_1-\mu_1)^2 - \frac{n_2}{2\sigma_2^2}(\bar{x}_2-\mu_2)^2\right\} d\mu_1\, d\mu_2$$

$$\times \exp\left\{-\frac{n_1 s_1^2}{2\sigma_1^2} - \frac{n_2 s_2^2}{2\sigma_2^2}\right\} d\sigma_1\, d\sigma_2.$$

Writing

$$t_1 = \frac{(\mu_1-\bar{x}_1)\surd(n_1-1)}{s_1}, \qquad t_2 = \frac{(\mu_2-\bar{x}_2)\surd(n_2-1)}{s_2}, \tag{20.158}$$

we find, as in Example 20.10, the joint distribution of μ_1 and μ_2

$$dF \propto \frac{d_\mu t_1}{\{1+t_1^2/(n_1-1)\}^{n_1/2}} \frac{d_\mu t_2}{\{1+t_2^2/(n_2-1)\}^{n_2/2}}, \tag{20.159}$$

where we write $d_\mu t_1$ to remind ourselves that the differential element is $\surd(n_1-1)\,d\mu_1/s_1$ and similarly for the second sample.

We cannot proceed at once to find an interval for $\delta = \mu_1 - \mu_2$. In fact, from (20.158) we have

$$(\mu_1-\bar{x}_1)-(\mu_2-\bar{x}_2) = \delta - d = t_1 s_1/\surd(n_1-1) - t_2 s_2/\surd(n_2-1), \tag{20.160}$$

and to set limits to δ we require the fiducial distribution of the right side of (20.160) or some convenient function of it. This is a linear function of t_1 and t_2, whose fiducial distribution is given by (20.159). In actual fact Fisher (1935b, 1939), following Behrens, chose the statistic (20.61)

$$z = \frac{d-\delta}{\left(\dfrac{s_1^2}{n_1-1}+\dfrac{s_2^2}{n_2-1}\right)^{1/2}}$$

as the most convenient function. We have

$$z = t_1 \cos\psi - t_2 \sin\psi,$$

where

$$\tan^2\psi = \frac{s_2^2}{n_2-1}\bigg/\frac{s_1^2}{n_1-1}.$$

For given ψ the distribution of z (usually known as the Fisher–Behrens distribution) can be found from (20.160). It has no simple form, but Rahman and Saleh (1974) express it in series forms and provide 4 d.p. tables of its 97.5 and 95 percentage points for $\nu_1 = 6$ (1) 15 and $\nu_2 = 6$ (1) 9. Earlier tables had been given by Fisher and Yates' *Statistical Tables*. In using these tables (and in consulting Fisher's papers generally) the reader should note that our $s^2/(n-1)$ is written by him as s^2.

In this case, the most important yet noticed, the fiducial argument does not give the same result as the approach from confidence intervals. That is to say, if we determine from a probability $1 - \alpha$ the corresponding points of z, say z_0 and z_1, and then assert

$$\bar{x}_1 - \bar{x}_2 - z_0 \left(\frac{s_1^2}{n_1 - 1} + \frac{s_2^2}{n_2 - 1} \right)^{1/2} \leq \mu_1 - \mu_2 \leq \bar{x}_1 - \bar{x}_2 + z_1 \left(\frac{s_1^2}{n_1 - 1} + \frac{s_2^2}{n_2 - 1} \right)^{1/2}, \quad (20.161)$$

we shall not be correct in a proportion $1 - \alpha$ of cases in the long run, whatever the values of σ_1^2, σ_2^2, as is obvious from the fact that z may be expressed from (20.60) and (20.61) as

$$z = t \left\{ \frac{s_2^2 (1 + \theta / N)(1 + Nu/\theta)}{(n_2 - 1)s_1^2 + (n_1 - 1)s_2^2} \right\}^{1/2} \left\{ \frac{(n_1 - 1)(n_2 - 1)}{n_1 + n_2 - 2} \right\}^{1/2}$$

where t, defined by (20.60), has an exact Student's distribution. t is distributed independently of θ, but z is not.

The fiducialist view is that there is no particular reason why such statements should be correct in a known proportion of cases in the long run; and that to impose such a desideratum is to miss the point of the fiducial approach. We return to this point later.

> A. W. Davis and Scott (1971) use the symbolic method of **20.33** to obtain an asymptotic expansion of the confidence coefficient corresponding to (20.161), which always exceeds $1 - \alpha$; see also Robinson (1976).

Bayesian intervals

20.45 We now consider the relation between fiducial theory and interval estimation based on Bayes' theorem, as developed by Jeffreys (1961). The theorem, given at (8.3), may be rewritten in our present context as at (8.8),

$$g(\theta | x, H) \propto f(\theta | H) L(x | \theta, H), \quad (20.162)$$

where g is the posterior distribution, f is the prior distribution and L is the likelihood.

The major problem, as we saw in Chapter 8, is to choose a prior distribution $f(\theta | H)$. Jeffreys extended Bayes' postulate (which stated that if nothing is known about θ and its range is finite, the prior distribution should be proportional to $d\theta$) to take account of various situations. In particular, (1) if the range of θ is infinite in both directions the prior probability is still taken as proportional to $d\theta$; (2) if θ ranges from 0 to ∞ the prior distribution is taken as proportional to $d\theta/\theta$.

> Paradoxes that can arise from the use of such improper prior distributions for more than one parameter are discussed, with many examples, by Dawid et al. (1973) and by Stone (1976). Diaconis and Freedman (1986a, b) show that problems also arise with respect to estimation consistency, especially when there are many parameters.

Example 20.12
In the case of the normal distribution considered in **20.41** we have, with $\bar{x}$ sufficient

for μ,

$$L(\bar{x} \,|\, \mu, H) = \frac{n^{1/2}}{(2\pi)^{1/2}} \exp\left\{-\frac{n}{2}(\bar{x}-\mu)^2\right\},$$

and if μ can lie anywhere in $(-\infty, +\infty)$, the prior distribution is taken as

$$f(\mu \,|\, H) = d\mu.$$

Hence, for the posterior distribution of μ,

$$g(\mu \,|\, \bar{x}, H) \propto \frac{n^{1/2}}{(2\pi)^{1/2}} \exp\left\{-\frac{n}{2}(\bar{x}-\mu)^2\right\} d\mu.$$

Integration shows that the constant of proportionality is unity. Thus we may, for any given level of probability, determine the range of μ. The result is the same as that given by confidence-interval theory or fiducial theory.

On the other hand, for the distribution of Example 20.9 we take the prior distribution of θ, which is in $(0, \infty)$, to be

$$f(\theta \,|\, H) = d\theta / \theta.$$

The essential similarity to the fiducial procedure in Example 20.9 will be evident. Using (20.154) we have the posterior distribution

$$g(\theta \,|\, t, H) \propto \left(\frac{\beta}{\theta}\right)^{\beta} \frac{t^{\beta-1} e^{-\beta t/\theta}}{\Gamma(\beta)} \frac{d\theta}{\theta}.$$

The constant of proportionality is t, so that

$$g(\theta \,|\, t, H) = \left(\frac{\beta}{\theta}\right)^{\beta} \frac{t^{\beta} e^{-\beta t/\theta}}{\theta \Gamma(\beta)} d\theta,$$

which is (20.156) again.

Example 20.13 Bayesian interval for mean with variance unknown
Let us now consider the case of setting limits to the mean in normal samples when the variance is unknown. For Student's distribution we have

$$L(t \,|\, \mu, \sigma, H) = \frac{k \, dt}{(1 + t^2/\nu)^{(\nu+1)/2}},$$

where k is some constant and $\nu = n - 1$. The parameters μ and σ do not appear on the right and hence are irrelevant and may be suppressed. Thus we have

$$L(t \,|\, H) = \frac{k \, dt}{(1 + t^2/\nu)^{(\nu+1)/2}}. \tag{20.163}$$

Suppose now that we *assume* that

$$L(t \,|\, \bar{x}, s, H) = f(t) \, dt. \tag{20.164}$$

Then, as before, $\bar{x}$ and s may be suppressed, and we have

$$L(t|H) = f(t)\, dt,$$

and hence, by comparison with (20.162),

$$L(t|\bar{x}, s, H) = \frac{k\, dt}{(1 + t^2/v)^{(v+1)/2}}.$$

We can then find limits to t, given $\bar{x}$ and s, in the usual way. Jeffreys emphasised, however, that this depends on a new postulate expressed by (20.164) which, though natural, is not trivial. It amounts to an assumption that if we are comparing different distributions, samples from which give different $\bar{x}$s and ss, the scale of the distribution of μ must be taken proportional to s and its mean displaced by the difference of sample means.

Example 20.14 Bayesian interval for the problem of two means
In a similar way to Example 20.13, it will be found that to arrive at the Fisher–Behrens distribution it is necessary to postulate that

$$L(t_1, t_2 | \bar{x}_1, \bar{x}_2, s_1, s_2, H) = f_1(t_1) f_2(t_2)\, dt_1\, dt_2.$$

Jeffreys' derivation of the Fisher–Behrens form from Bayes' theorem is as follows.
    The prior probability of the four parameters is

$$f(\mu_1, \mu_2, \sigma_1, \sigma_2 | H) \propto \frac{d\mu_1\, d\mu_2\, d\sigma_1\, d\sigma_2}{\sigma_1 \sigma_2}.$$

The likelihood (denoting the data by D) is

$$L\{D | \mu_1, \mu_2, \sigma_1, \sigma_2, H\} \propto \frac{1}{\sigma_1^{n_1} \sigma_2^{n_2}} \exp\left[-\frac{n_1}{2\sigma_1^2}\{(\mu_1 - \bar{x}_1)^2 + s_1^2\} - \frac{n_2}{2\sigma_2^2}\{(\mu_2 - \bar{x}_2)^2 + s_2^2\} \right].$$

Hence, by Bayes' theorem, the posterior distribution is

$$g(\mu_1, \mu_2, \sigma_1, \sigma_2 | D, H) = \frac{1}{\sigma_1^{n_1+1} \sigma_2^{n_2+1}}$$
$$\times \exp\left[-\frac{n_1}{2\sigma_1^2}\{(\mu_1 - \bar{x}_1)^2 + s_1^2\} - \frac{n_2}{2\sigma_2^2}\{(\mu_2 - \bar{x}_2)^2 + s_2^2\} \right].$$

Integrating out the values of σ_1 and σ_2, we find for the posterior distribution of μ_1 and μ_2 a form that is easily reducible to (20.159).

20.46   We saw in Examples 20.11–14 that in special circumstances there is a close correspondence between fiducial intervals and those derived by the Bayes–Jeffreys argument, and that the latter is more explicit about the postulates on which it depends. In recent years, fiducial theory has faded from the statistical literature to the point where Pedersen (1978), reviewing it, described it as 'now essentially dead'. By contrast, Bayesian methods have experienced a renaissance, while confidence intervals remain, for most statisticians, the prevailing orthodoxy. We shall return to these methods in Chapter 31.

EXERCISES

20.1   In setting confidence limits for the variance of a normal population by the use of the distribution of the sample variance (Example 20.6), sketch the confidence belt for some value of the confidence coefficient, and show graphically that it always provides a connected range within which σ^2 is located.

20.2   Show how to set confidence limits to the ratio of variances σ_1^2/σ_2^2 in two normal populations, based on independent samples of n_1 observations from the first and n_2 observations from the second. (Use the distribution of the ratio of sample variances at (16.24).)

20.3   Use the method of **20.17** to show that large-sample 95 per cent confidence limits for π in the binomial distribution of Example 20.2 are given by

$$\frac{1}{1+(1.96)^2/n}\left(p+\frac{(1.96)^2}{2n}\pm1.96\sqrt{\left(\frac{p(1-p)}{n}+\frac{(1.96)^2}{4n^2}\right)}\right).$$

> (B. K. Ghosh (1979) shows that this interval has high probability of being shorter than the simpler one $p\pm$ $1.96\{[p(1-p)/n]\}^{1/2}$ obtained by using only the leading terms above, as well as controlling α better. Fujino (1980) confirms this and gives an even better approximation. P. Hall (1982) gives improved one-sided corrections using a continuity correction.)

20.4   Using Geary's theorem (Exercise 11.11), show that large-sample 95 per cent confidence limits for the ratio π_2/π_1 of the parameters of two binomial distributions based on independent samples of size n_2, n_1 respectively, are given by

$$\frac{p_2/p_1}{1+(1.96)^2/n_2}\left\{1+\frac{(1.96)^2}{2n_2p_2}+1.96\sqrt{\left[\frac{1-p_1}{n_1p_1}+\frac{1-p_2}{n_2p_2}+\frac{(1.96)^2}{4}\left(\frac{1}{n_2^2p_2^2}+\frac{4(1-p_1)}{n_1n_2p_1}\right)\right]}\right\}.$$

> (Noether, 1957)

20.5   In Example 20.6, show that a confidence interval of the form

$$P\left\{\frac{ns^2}{b_n}\leqslant\sigma^2\leqslant\frac{ns^2}{a_n}\right\}=1-\alpha,$$

where $t=ns^2/\sigma^2$ has the χ^2 distribution $f_{n-1}(t)$ with $(n-1)$ d.fr., has minimum length if a_n, b_n satisfy $f_{n+3}(a_n)=f_{n+3}(b_n)$, and hence does not coincide with the central interval used in Example 20.6.

> (Tate and Klett (1959); A. Cohen (1972) showed that no other interval based on s^2 is shorter than this, but that if the pair of sufficient statistics $(\bar{x}, s^2)$ is used, small improvements can be made.)

20.6   Given that $f(x|\theta)=g(x)/h(\theta)$, $(a(\theta)\leqslant x\leqslant b(\theta))$, $a(\theta)$ is a monotone increasing function and $b(\theta)$ is a monotone decreasing function of θ, show (cf. **17.40-1**) that the extreme observations $x_{(1)}$ and $x_{(n)}$ are a pair of jointly sufficient statistics for θ. From their joint distribution, show that the single sufficient statistic for θ,

$$t=\min\{a^{-1}(x_{(1)}), b^{-1}(x_{(n)})\},$$

has distribution

$$dF=\frac{n\{h(t)\}^{n-1}}{\{h(\theta)\}^n}|h'(t)|\,dt,\qquad \theta\leqslant t\leqslant c,$$

where the constant c is defined by $a(c) = b(c)$. Show that this is of form $f(t|\theta) = g^*(t)/h^*(\theta)$, $\theta \leq t \leq c$, with

$$g^*(t) = -n\{h(t)\}^{n-1}h'(t), \quad h^*(\theta) = \{h(\theta)\}^n,$$

so that from Exercise 17.14 the unique MV unbiased estimator of $\tau(\theta)$ is

$$\tau(t) - \tau'(t)h^*(t)/g^*(t) = \tau(t) + \tau'(t)h(t)/\{nh'(t)\}.$$

Show that these results are unchanged (except that $c \leq t \leq \theta$) when $a(\theta)$ is decreasing, $b(\theta)$ is increasing and $t = \max\{a^{-1}(x_{(1)}), b^{-1}(x_{(n)})\}$.

20.7   In Exercise 20.6, show that $\psi = h(t)/h(\theta)$ has distribution

$$dF = n\psi^{n-1}\,d\psi, \quad 0 \leq \psi \leq 1.$$

Show that

$$P\{\alpha^{1/n} \leq \psi \leq 1\} = 1 - \alpha,$$

and hence set a confidence interval for θ. Show that this is shorter than any other interval based on the distribution of ψ.   (Huzurbazar, 1955)

20.8   Apply the result of Exercise 20.7 to show that a confidence interval for θ in

$$dF = dx/\theta, \quad 0 \leq x \leq \theta,$$

is obtainable from

$$P\{x_{(n)} \leq \theta \leq x_{(n)}\alpha^{-1/n}\} = 1 - \alpha.$$

20.9   Use the result of Exercise 20.7 to show that a confidence interval for θ in

$$dF = e^{-(x-\theta)}\,dx, \quad \theta \leq x < \infty$$

is obtainable from

$$P\left\{x_{(1)} + \frac{1}{n}\log \alpha \leq \theta \leq x_{(1)}\right\} = 1 - \alpha.$$ (Huzurbazar, 1955)

20.10 Given that $I(x)$ is a confidence interval for θ calculated from the distribution of a sample, $f(x|\theta)$, and that θ_0 is the true value of θ, show that the expected length of $I(x)$ may be written as

$$E(L) = \int\left\{\int_{\theta \in I(x)} d\theta\right\}dF(x|\theta_0)$$

and that

$$E(L) = \int_{\theta \neq \theta_0} \text{prob}\,\{\theta \in I(x)|\theta_0\}\,d\theta,$$

the integral over all false values of θ of the probability of inclusion in the confidence interval.
                                                                    (Pratt, 1961, 1963)

20.11 Given that $x \in A(\theta)$ if and only if $\theta \in I(x)$ in Exercise 20.10, show that $E(L)$ is minimised by choosing $A(\theta)$ for each θ so that prob $\{x \in A(\theta)|\theta_0\}$ is minimised. (This is equivalent to choosing the most powerful test for each θ against the alternative value θ_0—cf. **21.9** below.)
                                                                    (Pratt, 1961, 1963)

20.12 x and y have a bivariate normal distribution with variances σ_1^2, σ_2^2, and correlation parameter ρ. Show that the variables

$$u = \frac{x}{\sigma_1} + \frac{y}{\sigma_2}, \quad v = \frac{x}{\sigma_1} - \frac{y}{\sigma_2},$$

are independently normally distributed. In a sample of n observations with sample variances s_x^2 and s_y^2 and correlation coefficient r_{xy}, show that the sample correlation coefficient of u and v may be written

$$r_{uv}^2 = \frac{(l-\lambda)^2}{(1+\lambda)^2 - 4r_{xy}^2\, l\lambda},$$

where $l = s_1^2/s_2^2$ and $\lambda = \sigma_1^2/\sigma_2^2$. Hence show that, whatever the value of ρ, confidence limits for λ are given by

$$l\{K - (K^2 - 1)^{1/2}\}, \qquad l\{K + (K^2 - 1)^{1/2}\}$$

where

$$K = 1 + \frac{2(1 - r_{xy}^2)}{n-2}\, t_\alpha^2$$

and t_α^2 is the 100α per cent point of Student's t^2 distribution.

(Pitman, 1939a)

20.13   Using the asymptotic multivariate normal distribution of maximum likelihood estimators (**18.26**) and the χ^2 distribution of the exponent of a multivariate normal distribution (**15.10**), show that (20.56) gives a large-sample confidence region for a set of parameters. From it, derive a confidence region for the mean and variance of a univariate normal distribution.

20.14   In **20.36**, show that $r(\bar{x}, \gamma)$ defined at (20.116) is, asymptotically in n,

$$r(\bar{x}, \gamma) \sim r(0, \gamma)\left(1 + \frac{1}{2n}\right). \tag{Bowker, 1946}$$

20.15   Using the method of Example 6.4, show that for a χ^2 distribution with ν degrees of freedom, the value above which 100β per cent of the distribution lies is χ_β^2 where

$$\frac{\chi_\beta^2}{\nu} \sim 1 + \left(\frac{2}{\nu}\right)^{1/2} d_{1-\beta} + \frac{2}{3\nu}(d_{1-\beta}^2 - 1) + o\left(\frac{1}{\nu}\right),$$

where

$$\int_{-\infty}^{d_\alpha} (2\pi)^{-1/2} \exp\left(-\tfrac{1}{2}t^2\right) dt = \alpha.$$

20.16   Combine the results of Exercises 20.14–15 to show that, from (20.119),

$$\lambda \sim r(0, \gamma)\left\{1 + \frac{d_\beta}{(2n)^{1/2}} + \frac{5(d_\beta^2 + 2)}{12n}\right\}. \tag{Bowker, 1946}$$

20.17   Let $l_{11}, l_{12}, \ldots, l_{1,n-1}$ be $(n-1)$ linear functions of the observations in a sample of size n that are orthogonal to one another and to $\bar{x}_1$, and let them have zero mean and variance σ_1^2. Similarly define $l_{21}, l_{22}, \ldots, l_{2,n-1}$.

Then, in two samples of size n from normal populations with equal means and variances σ_1^2 and σ_2^2, the function

$$\frac{(\bar{x}_1 - \bar{x}_2)n^{1/2}}{\{\sum (l_{1j} + l_{2j})^2/(n-1)\}^{1/2}}$$

is distributed as Student's t with $n-1$ degrees of freedom. Show how to set confidence intervals to the difference of two means by this result, and show that the solution (20.91) is a member of this class of statistics when $n_1 = n_2$.

20.18 Given two samples of n_1, n_2 members from normal populations with unequal variances, show that by picking n_1 members at random from the n_2 (where $n_1 \leqslant n_2$) and pairing them at random with the members of the first sample, confidence intervals for the difference of means can be based on Student's distribution independently of the variance ratio in the populations. Show that this is equivalent to putting $c_{ij} = 0 (i \neq j); = 1 (i = j)$ in (20.76), and hence that this is an inefficient solution of the two-means problem.

20.19 Use the method of **20.31** to show that the statistic z of (20.61) is distributed approximately in Student's form with degrees of freedom given by (20.104), and show that if we take the first two terms in the expansion on the right of (20.111), (20.105) gives the same approximation to order n^{-1}.

20.20 From Fisher's F distribution (16.24), find the fiducial distribution of $\theta = \sigma_1^2/\sigma_2^2$, and show that if we regard the Student's distribution of the statistic (20.60) as the joint fiducial distribution of δ and θ, and integrate out θ over its fiducial distribution, we arrive at the result of Example 20.11 for the distribution of z.

(Fisher, 1939)

20.21 Prove the statement in **20.24** to the effect that if $ax + by = z$, where x and y are independent random variables and x, y, z are all χ^2 variates, the constants $a = b = 1$.

(Scheffé, 1944)

20.22 Show that ν, the approximate degrees of freedom of z given at (20.104), increases steadily from the value $n_2 - 1$ at $\theta = 0$ to its unique maximum value $n_1 + n_2 - 2$ at $\theta = n_1(n_1 - 1)/\{n_2(n_2 - 1)\}$ and then decreases steadily to $n_1 - 1$ as $\theta \to \infty$. Show that the approximate d.fr. for t at (20.103) has the same extreme values and maximum, but that the latter is now at $\theta = 1$.

20.23 In setting confidence intervals for the median of a continuous distribution using the symmetrically spaced order-statistics $x_{(r)}$ and $x_{(n-r+1)}$, show from (20.136) that for $n = 30$, the values of r shown below give the confidence coefficients shown:

r	$1 - \alpha$	r	$1 - \alpha$
8	0.995	12	0.80
9	0.98	13	0.64
10	0.96	14	0.42
11	0.90	15	0.14

20.24 Show from Example 20.7 that the range of a sample of size $n = 100$ from a continuous distribution has probability exceeding 0.95 of covering at least 95 per cent of the parent d.f., but that if wish to cover at least 99 per cent with probability 0.95, n must be about 475 or more.

CHAPTER 21

TESTS OF HYPOTHESES: SIMPLE HYPOTHESES

21.1 We now pass from the problems of estimating parameters to those of testing hypotheses concerning parameters. Instead of seeking the best (unique or interval) estimator of an unknown parameter, we shall now be concerned with deciding whether some pre-designated value is acceptable in the light of the observations.

In a sense, the testing problem is logically prior to that of estimation. If, for example, we are examining the difference between the means of two normal populations, our first question is whether the observations indicate that there is *any* true difference between the means. In other words, we have to compare the observed differences between the two samples with what might be expected on the hypothesis that there is no true difference at all, but only random sampling variation. If this hypothesis is not sustained, we proceed to the second step of estimating the *magnitude* of the difference between the population means.

Quite obviously, the problems of testing hypotheses and of estimation are closely related (cf. **21.9** below) but it is nevertheless useful to preserve a distinction between them, if only for expository purposes. The monograph by Lehmann (1986) treats the testing problem in full detail. Many of the ideas are due to Neyman and E. S. Pearson, whose remarkable series of papers (1928, 1933a, b, 1936a, b, 1938) is fundamental.

21.2 The kind of hypotheses that we test in statistics is more restricted than the general scientific hypothesis. It is a scientific hypothesis that every particle of matter in the universe attracts every other particle, or that life exists on Mars; but these are not hypotheses that we can test statistically. Statistical hypotheses concern the behaviour of observable random variables. More precisely, suppose that we have a set of random variables $x_1, \ldots, x_n$. As before, we may represent them as the co-ordinates of a point ($\mathbf{x}$, say) in the n-dimensional sample space, one of whose axes corresponds to each variable. Since $\mathbf{x}$ is a random variable, it has a probability distribution, and if we select any region, say w, in the sample space W, we may (at least in principle) calculate the probability that the sample point $\mathbf{x}$ falls in w, say $P(\mathbf{x} \in w)$. We shall say that any hypothesis concerning $P(\mathbf{x} \in w)$ is a statistical hypothesis. In other words, any hypothesis concerning the behaviour of observable random variables is a statistical hypothesis.

For example, the hypothesis (a) that a normal distribution has a specified mean and variance is statistical; so is the hypothesis (b) that it has a given mean but unspecified variance; so is the hypothesis (c) that a distribution is of normal form, both mean and variance remaining unspecified; and so, finally, is the hypothesis (d) that two unspecified continuous distributions are identical. Each of these four examples implies certain behaviour of random variables in the sample space, which may be tested by comparison with observation.

Parametric and non-parametric hypotheses

21.3   It will have been noticed that in the examples (a) and (b) in **21.2** the distribution underlying the observations was taken to be of a certain form (the normal) and the hypothesis was concerned entirely with the value of one or both of its parameters. Such a hypothesis, for obvious reasons, is called *parametric.*

Hypothesis (c) was of a different nature. It may be expressed in an alternative way, since it is equivalent to the hypothesis that the distribution has all cumulants finite, and all cumulants above the second equal to zero (cf. Example 3.11). The cumulants are not, however, parameters in our sense of the term, which (cf. **2.2**) is restricted to constants appearing in the specified probability distribution of our random variable.

With this understanding, hypothesis (c), and also (d), of **21.2** are *non-parametric* hypotheses. Non-parametric hypotheses will be discussed in a later volume, but most of the theoretical discussion in this and the next chapter is equally applicable to the parametric and the non-parametric case. However, our particularised discussions will mostly be of parametric hypotheses.

Simple and composite hypotheses

21.4   There is a distinction between the hypotheses (a) and (b) in **21.2**. In (a), the values of *all* the parameters of the distribution were specified by the hypothesis; in (b) only a subset of the parameters was specified by the hypothesis. This distinction is important for the theory. To formulate it generally, if we have a distribution depending upon l parameters, and a hypothesis specifies unique values for k of these parameters, we call the hypothesis *simple* if $k = l$ and we call it *composite* if $k < l$. In geometrical terms, we can represent the possible values of the parameters as a region in a space of l dimensions, one for each parameter. If the hypothesis considered selects a unique point in this parameter space, it is a simple hypothesis; if the hypothesis selects a subregion of the parameter space that contains more than one point, it is composite.

k is called the number of *constraints* imposed by the hypothesis, and $l - k$ is known as the number of *degrees of freedom* of the hypothesis, a terminology obviously related to the geometrical picture in the previous paragraph.

Critical regions and alternative hypotheses

21.5   To test any hypothesis, say H_0, on the basis of a random sample of observations, we must divide the sample space (i.e. all possible sets of observations) into two regions. If the observed sample point **x** falls into one of these regions, say w, we shall reject H_0; if **x** falls into the complementary region, $W - w$, we shall accept H_0. w is known as the *critical region* of the test, and $W - w$ is called the *acceptance region.*

The rather peremptory terms 'reject' and 'accept' that we have used of a hypothesis under test are now conventional usage, to which we shall adhere, and are not intended to imply that any hypothesis is ever finally accepted or rejected in science. If the reader cannot overcome his philosophical dislike of these admittedly inapposite expressions, he will perhaps agree to regard them as code words, 'reject' standing for 'decide that the observations are unfavourable to' and 'accept' for the opposite. We are concerned to investigate procedures that make such decisions with calculable probabilities of error, in a sense to be explained.

21.6 Now if we know the probability distribution of the observations under H_0, we can determine w so that, given H_0, the probability of rejecting H_0 (i.e. the probability that x falls in w) is equal to a pre-assigned value α, i.e.

$$\text{Prob } \{x \in w \,|\, H_0\} = \alpha. \tag{21.1}$$

If we are dealing with a discrete distribution, it may not be possible to satisfy (21.1) for every α in the interval $(0, 1)$. The value α is called the *size* of the test.* For the moment, we shall regard α as determined in some way. We shall discuss the choice of α later.

Evidently, we can in general find many, and often even an infinity, of subregions w of the sample space, all obeying (21.1). Which of them should we prefer to the others? This is the problem of the theory of testing hypotheses. To put it in everyday terms, which sets of observations are we to regard as favouring, and which as disfavouring, a given hypothesis?

It is of no use whatever to know merely what properties a critical region will have when H_0 holds. What happens when some other hypothesis holds? In other words, we cannot say whether a given body of observations favours a given hypothesis unless we know to what alternative(s) H_0 is being compared. It is perfectly possible for a sample of observations to be a rather 'unlikely' one if H_0 were true; but it may be much more 'unlikely' on another hypothesis. If the situation is such that we are forced to choose one hypothesis or the other, we shall obviously choose H_0, notwithstanding the 'unlikeliness' of the observations. The problem of testing a hypothesis is essentially one of choice between H_0 and some alternative(s). Whether or not we accept H_0 depends crucially upon the alternatives against which it is being tested.

The power of a test
21.7 The discussion of **21.6** leads us to the recognition that a critical region (or, synonymously, a test) must be judged by its properties both when H_0 is true and when it is false. Thus we may say that the errors made in testing a statistical hypothesis are of two types:

(I)   We may wrongly reject H_0, when it is true;
(II) We may wrongly accept H_0, when it is false.

---

* The hypothesis under test is often called the 'null hypothesis', and the size of the test the 'level of significance'.

These are known as Type I and Type II errors respectively. The probability of a Type I error is equal to the size of the critical region used, α. The probability of a Type II error is, of course, a function of the alternative hypothesis (say, H_1) considered, and is usually denoted by β. Thus

$$\text{Prob}\ \{\mathbf{x} \in W - w \,|\, H_1\} = \beta$$

or

$$\text{Prob}\ \{\mathbf{x} \in w \,|\, H_1\} = 1 - \beta. \tag{21.2}$$

This complementary probability, $1 - \beta$ is called the *power* of the test of the hypothesis H_0 against the alternative hypothesis H_1. The specification of H_1 in the last sentence is essential, since power is a function of H_1.

Example 21.1
Consider the problem of testing a hypothetical value for the mean of a normal distribution with unit variance. Formally, in

$$dF(x) = (2\pi)^{-1/2} \exp\{-\tfrac{1}{2}(x - \mu)^2\}\, dx, \qquad -\infty < x < \infty,$$

we test the hypothesis

$$H_0: \mu = \mu_0.$$

This is a simple hypothesis, since it specifies $F(x)$ completely. The alternative hypothesis will also be taken as the simple

$$H_1: \mu = \mu_1 > \mu_0.$$

Thus, essentially, we are to choose between a smaller given value (μ_0) and a larger (μ_1) for the mean of our distribution.

We may represent the situation diagrammatically for a sample of $n = 2$ observations. In Fig. 21.1 we show the scatters of sample points that would arise, the lower cluster arising when H_0 is true, and the higher when H_1 is true.

In this case, of course, the sampling distributions are continuous, but the dots indicate roughly the sample densities around the true means.

To choose a critical region, we need, in accordance with (21.1), to choose a region in the plane containing a proportion α of the distribution given H_0. One such region is represented by the area above the line PQ, which is perpendicular to the line AB connecting the hypothetical means. (A is the point (μ_0, μ_0), and B the point (μ_1, μ_1).) Another possible critical region of size α is the region CAD.

We see at once from the circular symmetry of the clusters that the first of these critical regions contains a very much larger proportion of the H_1 cluster than does the CAD region. The first region will reject H_0 rightly, when H_1 is true, in a higher proportion of cases than will the second region. Consequently, its value of $1 - \beta$ in (21.2), or in other words its power, will be the greater.

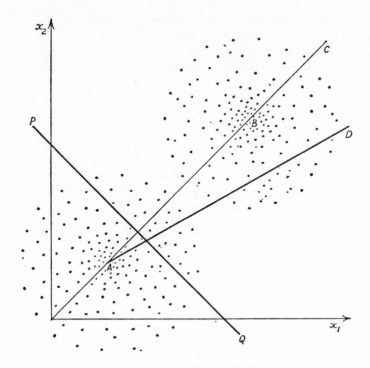

Fig. 21.1   Critical regions for $n = 2$ (see text)

21.8   Example 21.1 directs us to an obvious criterion for choosing among critical regions, all satisfying (21.1). We seek a critical region w such that its power, defined at (21.2), is as large as possible. Then, in addition to having controlled the probability of Type I errors at α, we shall have minimised the probability of a Type II error, β. This is the fundamental idea, first expressed explicitly by J. Neyman and E. S. Pearson, that underlies the theory of this and following chapters.

A critical region, whose power is no smaller than that of any other region of the same size for testing a hypothesis H_0 against the alternative H_1, is called a best critical region (abbreviated BCR), and a test based on a BCR is called a most powerful (abbreviated MP) test.

Tests and confidence intervals
21.9   We may now take the opportunity of translating the ideas of the theory of hypothesis-testing into those of the theory of confidence intervals, as promised in **20.15**.

If a sample is observed, we may ask the question: for which values of θ does the sample point $\mathbf{x}$ form part of the acceptance region A complementary to the size-α critical region for a certain test on the parameter θ? If we aggregate these 'acceptable' values of θ, we obtain the level-$(1 - \alpha)$ confidence interval C for θ corresponding to that test, for θ is in C if and only if $\mathbf{x}$ is in A, i.e. with probability $1 - \alpha$. We used this method of constructing confidence intervals in **20.3**, and indeed throughout Chapter 20.

There is thus no need to derive optimum properties separately for tests and for intervals: there is a one-to-one correspondence between the problems as in the dictionary in Table 21.1.

Table 21.1

Property of test	Property of corresponding confidence interval
Size = α	Confidence coefficient = $1 - \alpha$
Power = probability of rejecting a false value of $\theta = p$	Probability of not covering a false value of $\theta = p$
Most powerful	Most selective
$\longleftarrow \quad \left\{ \begin{matrix} \text{Unbiased} \\ p \geqslant \alpha \end{matrix} \right\} \quad \longrightarrow$	
Equal-tails test $\alpha_1 = \alpha_2 = \frac{1}{2}\alpha$	Central interval

Testing a simple H_0 against a simple H_1

21.10   If we are testing a simple hypothesis against a simple alternative hypothesis, i.e. choosing between two completely specified distributions, the problem of finding a BCR of size α is particularly straightforward. Its solution is given by a lemma due to Neyman and Pearson (1933b), which we now prove.

As in earlier chapters, we write $L(x|H_i)$ for the likelihood function given the hypothesis H_i ($i = 0, 1$), and write a single integral to represent n-fold integration in the sample space. Our problem is to maximise, for choice of w, the integral form of (21.2),

$$1 - \beta = \int_w L(x|H_1)\, \mathrm{d}x, \tag{21.3}$$

subject to the condition (21.1), which we here write

$$\alpha = \int_w L(x|H_0)\, \mathrm{d}x. \tag{21.4}$$

The critical region w should obviously include all points **x** for which $L(x|H_0) = 0$, $L(x|H_1) > 0$; these points contribute nothing to the integral in (21.4). For the other points in w, we may rewrite (21.3) as

$$1 - \beta = \int_w \frac{L(x|H_1)}{L(x|H_0)} L(x|H_0)\, \mathrm{d}x. \tag{21.5}$$

From (21.4)-(21.5), we see that $(1-\beta)/\alpha$ is the average within w, when H_0 holds, of $L(x|H_1)/L(x|H_0)$. Clearly this will be maximised if and only if w consists of that fraction α of the sample space containing the largest values of $L(x|H_1)/L(x|H_0)$. Thus the BCR consists of the points in W satisfying

$$\frac{L(x|H_0)}{L(x|H_1)} \leq k_\alpha \tag{21.6}$$

when H_0 holds. To any constant k_α in (21.6) there corresponds a value α for the size (21.4). If the xs are continuously distributed, we can also find a k_α for any α.

21.11 If the distribution of the xs is not continuous, we may effectively render it so by a randomisation device (cf. **20.10**). In this case,

$$\frac{L(x|H_0)}{L(x|H_1)} = k_\alpha \tag{21.7}$$

with some non-zero probability p, while in general, owing to discreteness, we can only choose k_α in (21.6) to make the size of the test equal to $\alpha - q$ $(0 < q < p)$. To convert the test into one of exact size α, we simply arrange that, whenever (21.7) holds, we use a random device (e.g. a table of random sampling numbers) so that with probability q/p we reject H_0, while with probability $1 - (q/p)$ we accept H_0. The overall probability of rejection will then be $(\alpha - q) + p \cdot q/p = \alpha$, as required, whatever the value of α desired. In this case, the BCR is clearly not unique, being subject to random sampling fluctuation.

Example 21.2
Consider again the normal distribution of Example 21.1,

$$dF(x) = (2\pi)^{-1/2} \exp\left[-\tfrac{1}{2}(x-\mu)^2\right] dx, \qquad -\infty < x < \infty, \tag{21.8}$$

where we are now to test $H_0: \mu = \mu_0$ against the alternative $H_1: \mu = \mu_1(\neq \mu_0)$. We have

$$L(x|H_i) = (2\pi)^{-n/2} \exp\left[-\tfrac{1}{2} \sum_{j=1}^{n} (x_j - \mu_i)^2\right], \qquad i = 0, 1,$$

$$= (2\pi)^{-n/2} \exp\left[-\frac{n}{2}\{s^2 + (\bar{x} - \mu_i)^2\}\right] \tag{21.9}$$

where $\bar{x}$, s^2 are the sample mean and variance respectively. Thus, for the BCR, we have from (21.6)

$$\frac{L(x|H_0)}{L(x|H_1)} = \exp\left[\frac{n}{2}\{(\bar{x} - \mu_1)^2 - (\bar{x} - \mu_0)^2\}\right]$$

$$= \exp\left[\frac{n}{2}\{(\mu_0 - \mu_1)2\bar{x} + (\mu_1^2 - \mu_0^2)\}\right] \leq k_\alpha, \tag{21.10}$$

or

$$(\mu_0 - \mu_1)\bar{x} \leq \tfrac{1}{2}(\mu_0^2 - \mu_1^2) + \frac{1}{n} \log k_\alpha. \tag{21.11}$$

Thus, given μ_0, μ_1 and α, the BCR is determined by the value of the sample mean $\bar{x}$ alone. This is what we might have expected from the fact (cf. Examples 17.6, 17.15) that $\bar{x}$ is an MVB sufficient statistic for μ. Further, from (21.11), we see that if $\mu_0 > \mu_1$ the BCR is

$$\bar{x} \le \tfrac{1}{2}(\mu_0 + \mu_1) + \log k_\alpha / \{n(\mu_0 - \mu_1)\}, \qquad (21.12)$$

while if $\mu_0 < \mu_1$ it is

$$\bar{x} \ge \tfrac{1}{2}(\mu_0 + \mu_1) - \log k_\alpha / \{n(\mu_1 - \mu_0)\}, \qquad (21.13)$$

which is again intuitively reasonable: in testing the hypothetical value μ_0 against a smaller value μ_1, we reject μ_0 if the sample mean falls below a certain value, which depends on α, the size of the test; in testing μ_0 against a larger value μ_1, we reject μ_0 if the sample mean exceeds a certain value.

21.12   A feature of Example 21.2 which is worth remarking, since it occurs in a number of problems, is that the BCR turns out to be determined by a single statistic, rather than by the whole configuration of sample values. This simplification permits us to carry on our discussion entirely in terms of the sampling distribution of that statistic, called a test statistic, and to avoid the complexities of n-dimensional distributions.

Example 21.3
In Example 21.2, we know (cf. Example 11.12) that whatever the value of μ, $\bar{x}$ is itself exactly normally distributed with mean μ and variance $1/n$. Thus, to obtain the BCR (21.13) of size α for testing μ_0 against $\mu_1 > \mu_0$, we determine $\bar{x}_\alpha$ so that

$$\int_{\bar{x}_\alpha}^{\infty} \left(\frac{n}{2\pi}\right)^{1/2} \exp\left\{-\frac{n}{2}(\bar{x} - \mu_0)^2\right\} d\bar{x} = \alpha.$$

Writing

$$G(x) = \int_{-\infty}^{x} (2\pi)^{-1/2} \exp\left(-\tfrac{1}{2}y^2\right) dy \qquad (21.14)$$

for the standardised normal d.f., we see by substituting $y = n^{1/2}(\bar{x} - \mu_0)$ that

$$\alpha = 1 - G\{n^{1/2}(\bar{x}_\alpha - \mu_0)\} = G\{-n^{1/2}(\bar{x}_\alpha - \mu_0)\}$$

since $G(x) = 1 - G(-x)$ by symmetry. If we now write

$$d_\alpha = n^{1/2}(\bar{x}_\alpha - \mu_0) \qquad (21.15)$$

we have

$$G(-d_\alpha) = \alpha. \qquad (21.16)$$

For example, from a table of the normal d.f., we find for $\alpha = 0.05$ that $d_{0.05} = 1.6449$, so that when $\mu_0 = 2$ and $n = 25$, we have from (21.15)

$$\bar{x}_\alpha = 2 + 1.6449/5 = 2.3290.$$

In this normal case, the power of the test may be written down explicitly. It is

$$\int_{\bar{x}_\alpha}^{\infty} \left(\frac{n}{2\pi}\right)^{1/2} \exp\left\{-\frac{n}{2}(\bar{x} - \mu_1)^2\right\} d\bar{x} = 1 - \beta. \tag{21.17}$$

Substituting $y = n^{1/2}(\bar{x} - \mu_1)$, this becomes

$$1 - \beta = 1 - G\{n^{1/2}(\mu_0 - \mu_1) + d_\alpha\} = G\{n^{1/2}(\mu_1 - \mu_0) - d_\alpha\}, \tag{21.18}$$

again using the symmetry. From (21.18) it is clear that the power is a monotone increasing function both of n, the sample size, and of $(\mu_1 - \mu_0)$, the difference between the hypothetical values between which the test has to choose. It should be observed that although (21.13) is only the BCR for $\mu_1 > \mu_0$, the expression for its power at (21.18) remains valid for any μ_1 whatever. As $\mu_1 \to -\infty$, the power $\to 0$, and it remains less than α for all $\mu_1 < \mu_0$.

Example 21.4

As a contrast, consider the Cauchy distribution

$$dF(x) = \frac{dx}{\pi\{1 + (x - \theta)^2\}}, \qquad -\infty < x < \infty,$$

and suppose that we wish to test

$$H_0: \theta = 0$$

against

$$H_1: \theta = \theta_1.$$

For simplicity, we shall confine ourselves to the case $n = 1$. According to (21.6), the BCR is given by

$$\frac{L(x \mid H_0)}{L(x \mid H_1)} = \frac{1 + (x - \theta_1)^2}{1 + x^2} \leqslant k_\alpha.$$

This is equivalent to

$$x^2(k_\alpha - 1) + 2\theta_1 x + (k_\alpha - 1 - \theta_1^2) \geqslant 0. \tag{21.19}$$

The form of the BCR thus defined depends upon the value of α chosen. If $k_\alpha = 1$, (21.19) reduces to $\theta_1 x \geqslant \frac{1}{2}\theta_1^2$, i.e. to $x \geqslant \frac{1}{2}\theta_1$ if $\theta_1 > 0$ and to $x \leqslant \frac{1}{2}\theta_1$ if $\theta_1 < 0$ (cf. Example 21.2). But if $k_\alpha < 1$, the quadratic on the left of (21.19) can be non-negative only within an interval for x, while for $k_\alpha > 1$ it can be so only outside an interval for x. Thus the BCR changes its form with α. Indeed, if $(k_\alpha - 1)^2/k_\alpha > \theta_1^2$, the quadratic has no real root and (21.19) is satisfied either for all or for no values of x.

Since the Cauchy distribution is a Student's distribution with one degree of freedom, and accordingly $F(x) = \frac{1}{2} + (1/\pi) \arctan(x - \theta)$ by Example 16.1, we may calculate the size of the test for any k_α and θ_1 by putting $\theta = 0$ in $F(x)$. Thus, for $\theta_1 = 1$ and $k_\alpha = 1$, the size is

$$P(x \geq \tfrac{1}{2}) = 0.352,$$

while for $\theta_1 = 1$, $k_\alpha = 0.5$, (21.19) holds when $1 \leq x \leq 3$ and

$$P(1 \leq x \leq 3) = 0.148.$$

This method may also be used to determine the powers of these tests. We leave this to the reader as Exercise 21.4.

21.13 The examples we have given so far of the use of the Neyman–Pearson lemma have related to the testing of a parametric hypothesis for some given form of distribution. But, as will be seen on inspection of the proof in **21.10–11**, (21.6) gives the BCR for *any* test of a simple hypothesis against a simple alternative. For instance, we might be concerned to test the *form* of a distribution with known location parameter, as in the following example.

Example 21.5
Suppose that we know that a distribution is standardised, but wish to investigate its form. We wish to choose between the alternative forms

$$\left. \begin{array}{l} H_0 : dF = (2\pi)^{-1/2} \exp\left(-\tfrac{1}{2}x^2\right) dx, \\ H_1 : dF = 2^{-1/2} \exp\left(-2^{1/2}|x|\right) dx. \end{array} \right\} \quad -\infty < x < \infty,$$

For simplicity, we again take sample size $n = 1$.
    Using (21.6), the BCR is given by

$$\frac{L(x \mid H_0)}{L(x \mid H_1)} = \pi^{-1/2} \exp\left(2^{1/2}|x| - \tfrac{1}{2}x^2\right) \leq k_\alpha.$$

Thus we reject H_0 when

$$2^{1/2}|x| - \tfrac{1}{2}x^2 \leq \log\left(k_\alpha \pi^{1/2}\right) = c_\alpha.$$

The BCR therefore consists of extreme positive and negative values of the observation, supplemented, if $k_\alpha > \pi^{-1/2}$ (i.e. $c_\alpha > 0$), by values in the neighbourhood of $x = 0$. Just as in Example 21.4, the form of the BCR depends upon α. The reader should verify this by drawing a diagram.

BCR and sufficient statistics
21.14 If both hypotheses being compared refer to the value of a parameter θ, and there is a single sufficient statistic t for θ, it follows from the factorisation of the likelihood function at (17.68) that (21.6) becomes

$$\frac{L(x \mid \theta_0)}{L(x \mid \theta_1)} = \frac{g(t \mid \theta_0)}{g(t \mid \theta_1)} \leq k_\alpha, \tag{21.20}$$

so that the BCR is a function of the value of the sufficient statistic t, as might be expected. We have already encountered an instance of this in Example 21.2. (The same result evidently holds if θ is a set of parameters for which t is a jointly sufficient set of statistics.) Exercise 21.13 shows that the ratio of likelihoods on the left of (21.20) that determines the BCR is itself a sufficient statistic.

However, it will not always be the case that the BCR will, as in Example 21.2, be of the form $t \geq a_\alpha$ or $t \leq b_\alpha$: Example 21.4, in which the single observation x is a sufficient statistic for θ, is a counter-example. Inspection of (21.20) makes it clear that the BCR will be of this particularly simple form if $g(t|\theta_0)/g(t|\theta_1)$ is a non-decreasing function of t for $\theta_0 > \theta_1$. This will certainly be true if

$$\frac{\partial^2}{\partial\theta\,\partial t}\log g(t|\theta) \geq 0, \tag{21.21}$$

a condition which is satisfied by nearly all the distributions met with in statistics.

Example 21.6
For the exponential distribution

$$dF(x) = \begin{cases} \exp\{-(x-\theta)\}\,dx, & \theta \leq x < \infty, \\ 0 & \text{elsewhere,} \end{cases}$$

the smallest sample observation $x_{(1)}$ is sufficient for θ (cf. Example 17.19). For a sample of n observations, we have, for testing θ_0 against $\theta_1 > \theta_0$,

$$\frac{L(x|\theta_0)}{L(x|\theta_1)} = \begin{cases} \infty & \text{if } x_{(1)} < \theta_1 \\ \exp\{n(\theta_0-\theta_1)\} & \text{otherwise.} \end{cases}$$

Thus we require for a BCR

$$\exp\{n(\theta_0-\theta_1)\} \leq k_\alpha. \tag{21.22}$$

Now the left-hand side of (21.22) does not depend on the observations at all, being a constant, and (21.22) will therefore be satisfied by *every* critical region of size α with $x_{(1)} \geq \theta_1$. Thus every such critical region is of equal power, and is therefore a BCR.

If we allow θ_1 to be greater or less than θ_0, we find

$$\frac{L(x|\theta_0)}{L(x|\theta_1)} = \begin{cases} \infty & \text{if } \theta_0 \leq x_{(1)} < \theta_1, \\ \exp\{n(\theta_0-\theta_1)\} > 1 & \text{if } x_{(1)} \geq \theta_0 > \theta_1, \\ \exp\{n(\theta_0-\theta_1)\} < 1 & \text{if } x_{(1)} \geq \theta_1 > \theta_0, \\ 0 & \text{if } \theta_1 \leq x_{(1)} < \theta_0. \end{cases}$$

Thus the BCR is given by

$$(x_{(1)}-\theta_0) < 0, \quad (x_{(1)}-\theta_0) > c_\alpha.$$

The first of these events has probability zero on H_0. The value of c_α is determined to give probability α that the second event occurs when H_0 is true.

Estimating efficiency and power

21.15 The use of a statistic which is efficient in estimation (cf. **17.28–9**) does not imply that a more powerful test will be obtained than if a less efficient estimator had been used for testing purposes. This result is due to Sundrum (1954).

Let t_1 and t_2 be two asymptotically normally distributed estimators of a parameter θ, and suppose that, at least asymptotically,

$$\left. \begin{aligned} E(t_1) &= E(t_2) = \theta, \\ \mathrm{var}\,(t_i \,|\, \theta = \theta_0) &= \sigma_{i0}^2, \\ \mathrm{var}\,(t_i \,|\, \theta = \theta_1) &= \sigma_{i1}^2. \end{aligned} \right\} \qquad i = 1, 2$$

We now test H_0: $\theta = \theta_0$ against H_1: $\theta = \theta_1 > \theta_0$. Exactly as at (21.15) in Example 21.3, we have the critical regions, one for each test,

$$t_i \geq \theta_0 + d_\alpha \sigma_{i0}, \qquad i = 1, 2, \tag{21.23}$$

where d_α is the normal deviate defined by (21.14) and (21.16). The powers of the tests are (generalising (21.18) which dealt with a case where $\sigma_{i0} = \sigma_{i1} = n^{-1}$)

$$1 - \beta(t_i) = G \left\{ \frac{(\theta_1 - \theta_0) - d_\alpha \sigma_{i0}}{\sigma_{i1}} \right\}. \tag{21.24}$$

Since $G(x)$ is a monotone increasing function of its argument, t_1 will provide a more powerful test than t_2 if and only if, from (21.24),

$$\frac{(\theta_1 - \theta_0) - d_\alpha \sigma_{10}}{\sigma_{11}} > \frac{(\theta_1 - \theta_0) - d_\alpha \sigma_{20}}{\sigma_{21}},$$

i.e. if

$$\theta_1 - \theta_0 > d_\alpha \left(\frac{\sigma_{10}\sigma_{21} - \sigma_{20}\sigma_{11}}{\sigma_{21} - \sigma_{11}} \right). \tag{21.25}$$

If we put $E_j = \sigma_{2j}/\sigma_{1j}\,(j = 0, 1)$, (21.25) becomes

$$\theta_1 - \theta_0 > d_\alpha \left(\frac{E_1 - E_0}{E_1 - 1} \right) \sigma_{10}. \tag{21.26}$$

E_0, E_1 are simply powers (usually square roots) of the estimating efficiency of t_1 relative to t_2 when H_0 and H_1 respectively hold. Now if

$$E_0 = E_1 > 1, \tag{21.27}$$

the right-hand side of (21.25) is zero, and (21.26) always holds. Thus if the estimating efficiency of t_1 exceeds that of t_2 by *the same amount* on both hypotheses, the more efficient statistic t_1 always provides a more powerful test, whatever value α or $\theta_1 - \theta_0$ takes. But if

$$E_1 > E_0 \geq 1 \tag{21.28}$$

we can always find a test size α small enough for (21.26) to be falsified. Hence, the less efficient estimator t_2 will provide a more powerful test if (21.28) holds, i.e. if its relative efficiency is greater on H_0 than on H_1. Alternatively if $E_0 > E_1 > 1$, we can find σ *large* enough to falsify (21.26). If E_1 is continuous in θ, $E_1 \to E_0$ as $\theta_1 \to \theta_0$, so that (21.26) is not falsified in the immediate neighbourhood of θ_0.

This result, though a restrictive one, is enough to show that the relation between estimating efficiency and test power is rather loose. In Chapter 25 we shall again consider this relationship when we discuss the measurement of test efficiency.

Example 21.7

In Examples 18.3 and 18.6 we saw that in estimating the parameter ρ of a standardised bivariate normal distribution, the ML estimator $\hat{\rho}$ is a root of a cubic equation, with large-sample variance equal to $(1-\rho^2)^2/\{n(1+\rho^2)\}$, while the sample correlation coefficient r has large-sample variance $(1-\rho^2)^2/n$. Both estimators are consistent and asymptotically normal, and the ML estimator is efficient. In the notation of **21.15**,

$$E = (1+\rho^2)^{1/2}.$$

If we test $H_0: \rho = 0$ against $H_1: \rho = 0.1$, we have $E_0 = 1$, and (21.26) simplifies to

$$0.1 > d_\alpha \sigma_{10} = d_\alpha n^{-1/2}. \tag{21.29}$$

If we choose n to be, say, 400, so that the normal approximations are adequate, we require

$$d_\alpha > 2$$

to falsify (21.29). This corresponds to $\alpha < 0.023$, so that for tests of size < 0.023, the inefficient estimator r has greater power asymptotically in this case than the efficient $\hat{\rho}$. Since tests of size $0.01, 0.05$ are quite commonly used, this is not merely a theoretical example: it cannot be assumed in practice that 'good' estimators are 'good' test statistics.

Testing a simple H_0 against a class of alternatives

21.16 So far we have been discussing the most elementary problem, where in effect we have only to choose between two completely specified competitive hypotheses. For such a problem, there is a certain symmetry about the situation—it is only a matter of convention or convenience which of the two hypotheses we regard as being under test and which as the alternative. As soon as we proceed to the generalisation of the testing situation, this symmetry disappears.

Consider now the case where H_0 is simple, but H_1 is composite and consists of a class of simple alternatives. The most frequently occurring case is the one in which we have a class Ω of simple parametric hypotheses of which H_0 is one and H_1 comprises the remainder; for example, the hypothesis H_0 may be that the mean of a certain distribution has some value μ_0 and the hypothesis H_1 that it has some other value unspecified.

For each of these other values we may apply the foregoing results and find, for given α, corresponding to any particular member of H_1 (say H_t) a BCR w_t. But this region in general will vary from one H_t to another. It is clearly impossible to use a different region for all the unspecified possibilities and we are therefore led to inquire whether there exists one BCR which is the best for all H_t in H_1. Such a region is called uniformly most powerful (UMP) and the test based on it a UMP test.

21.17   Unfortunately, as we shall find below, a UMP test does not usually exist unless we restrict our alternative class Ω in certain ways. Consider, for instance, the case dealt with in Example 21.2. We found there that for $\mu_1 < \mu_0$ the BCR for a simple alternative was defined by

$$\bar{x} \leq a_\alpha. \tag{21.30}$$

Now so long as $\mu_1 < \mu_0$, the regions determined by (21.30) do not depend on μ_1 and can be found directly from the sampling distribution of $\bar{x}$ when the test size, α, is given. Consequently the test based on (21.30) is UMP for the class of hypotheses $\mu_1 < \mu_0$.

However, from Example 21.2, if $\mu_1 > \mu_0$, the BCR is defined by $\bar{x} \geq b_\alpha$. Here again, if our class Ω is confined to the values of μ_1 greater than μ_0 the test is UMP. But if μ_1 can be either greater or less than μ_0, no UMP test is possible, for one or other of the two BCR that we have just discussed will be better than any other region against this class of alternatives.

21.18   We now prove that for a simple $H_0 : \theta = \theta_0$ concerning a parameter θ defining a class of hypotheses, no UMP test exists in general against an interval including positive and negative values of $\theta - \theta_0$, under regularity conditions, in particular that the derivative of the likelihood with respect to θ is continuous in θ.

We expand the likelihood function in a Taylor series about θ_0, getting

$$L(x \mid \theta_1) = L(x \mid \theta_0) + (\theta_1 - \theta_0) L'(x \mid \theta^*) \tag{21.31}$$

where θ^* is some value between θ_0 and θ_1. For the BCR, if any, we must have, from (21.6) and (21.31),

$$\frac{L(x \mid \theta_1)}{L(x \mid \theta_0)} = 1 + \frac{(\theta_1 - \theta_0) L'(x \mid \theta^*)}{L(x \mid \theta_0)} \geq k_\alpha(\theta_1). \tag{21.32}$$

Thus the BCR is defined by

$$\frac{L'(x \mid \theta^*)}{L(x \mid \theta_0)} \geq a_\alpha, \qquad \theta_1 > \theta_0, \tag{21.33}$$

$$\leq b_\alpha, \qquad \theta_1 < \theta_0. \tag{21.34}$$

Now consider what happens as θ_1 approaches θ_0. θ^* necessarily does the same, and in the immediate neighbourhood of θ_0, (21.33–4) become, in virtue of the continuity of L' in θ,

$$\frac{L'(x|\theta_0)}{L(x|\theta_0)} = \left[\frac{\partial \log L}{\partial \theta}\right]_{\theta=\theta_0} \geq a_\alpha, \qquad \theta > \theta_0, \tag{21.35}$$

$$\leq b_\alpha, \qquad \theta < \theta_0. \tag{21.36}$$

We thus establish, incidentally, that in the immediate neighbourhood of θ_0, one-sided tests based on $[\partial \log L/\partial \theta]_{\theta=\theta_0}$ are UMP. This is a testing analogue of the confidence interval result obtained in **20.19–20**.

Our main result now follows at once. If we are considering an interval of alternatives including positive and negative values of $(\theta_1 - \theta_0)$, (21.35) and (21.36) cannot both hold (and there can therefore be no BCR) unless

$$\left[\frac{\partial \log L}{\partial \theta}\right]_{\theta=\theta_0} = \text{constant}. \tag{21.37}$$

(21.37) is the essential condition for the existence of a two-sided BCR. It cannot be satisfied if (17.18) holds (e.g. for distributions with range independent of θ) unless the constant is zero, for the condition $E(\partial \log L/\partial \theta) = 0$ with (21.37) implies $[\partial \log L/\partial \theta]_{\theta=\theta_0} = 0$.

In Example 21.6, we have already encountered an instance where a two-sided BCR exists. The reader should verify that for that distribution $[\partial \log L/\partial \theta]_{\theta=\theta_0} = n$ exactly, so that (21.37) is satisfied.

UMP tests of more than one parameter
21.19   If the distribution considered has more than one parameter, and we are testing a simple hypothesis, it remains possible that a common BCR exists for a class of alternatives varying with these parameters. The following two examples discuss the case of the two-parameter normal distribution, where we might expect to find such a BCR, but where none exists, and the two-parameter exponential distribution, where a BCR does exist.

Example 21.8
Consider the normal distribution with mean μ and variance σ^2. The hypothesis to be tested is

$$H_0: \mu = \mu_0, \sigma = \sigma_0,$$

and the alternative, H_1, is restricted only in that it must differ from H_0. For any such

$$H_1: \mu = \mu_1, \sigma = \sigma_1,$$

the BCR is, from (21.6), given by

$$\frac{L(x|H_0)}{L(x|H_1)} = \left(\frac{\sigma_1}{\sigma_0}\right)^n \exp\left[-\frac{1}{2}\left\{\Sigma\left(\frac{x-\mu_0}{\sigma_0}\right)^2 - \Sigma\left(\frac{x-\mu_1}{\sigma_1}\right)^2\right\}\right] \leq k_\alpha.$$

This may be written in the form

$$s^2\left(\frac{1}{\sigma_1^2} - \frac{1}{\sigma_0^2}\right) + \frac{(\bar{x}-\mu_1)^2}{\sigma_1^2} - \frac{(\bar{x}-\mu_0)^2}{\sigma_0^2} \leq \frac{2}{n}\log\left\{\left(\frac{\sigma_0}{\sigma_1}\right)^n k_\alpha\right\} \tag{21.38}$$

where $\bar{x}$, s^2 are sample mean and variance respectively. If $\sigma_0 \neq \sigma_1$, we may further simplify this to

$$\left(\frac{\sigma_1^2 - \sigma_0^2}{\sigma_0^2 \sigma_1^2}\right) \sum (x - \rho)^2 \geq c_\alpha, \qquad (21.39)$$

where c_α is independent of the observations, and

$$\rho = \frac{\mu_0 \sigma_1^2 - \mu_1 \sigma_0^2}{\sigma_1^2 - \sigma_0^2}.$$

(21.39), when a strict equality, is the equation of a hypersphere, centred at $x_1 = x_2 = \cdots = x_n = \rho$. Thus the BCR is always bounded by a hypersphere. When $\sigma_1 > \sigma_0$, (21.39) yields

$$\sum (x - \rho)^2 \geq a_\alpha,$$

so that the BCR lies outside the sphere; when $\sigma_1 < \sigma_0$, we find from (21.39)

$$\sum (x - \rho)^2 \leq b_\alpha,$$

and the BCR is inside the sphere.

Since ρ is a function of μ_1 and σ_1, it is clear that there will not generally be a common BCR for different members of H_1, even if we limit ourselves by $\sigma_1 < \sigma_0$ and $\mu_1 < \mu_0$ or similar restrictions. We may illustrate the situation by a diagram of the $(\bar{x}, s)$ plane, for

$$\sum (x - \rho)^2 = \sum (x - \bar{x})^2 + n(\bar{x} - \rho)^2$$
$$= n\{s^2 + (\bar{x} - \rho)^2\},$$

and if this is held constant, we obtain a circle with centre $(\rho, 0)$ and fixed radius a function of α.

Figure 21.2 (adapted from Neyman and Pearson, 1933b) illustrates some of the contours for particular cases. A single curve, corresponding to a fixed value of k_α in (21.38), is shown in each case.

Cases (1) and (2): $\sigma_1 = \sigma_0$ and $\rho = \pm\infty$. The BCR lies on the right of the line (1) if $\mu_1 > \mu_0$ and on the left of (2) if $\mu_1 < \mu_0$. This is the case discussed in Example 21.2 where we put $\sigma_1 = \sigma_0 = 1$.

Case (3): $\sigma_1 < \sigma_0$, say $\sigma_1 = \frac{1}{2}\sigma_0$. Then $\rho = \mu_0 + \frac{4}{3}(\mu_1 - \mu_0)$ and the BCR lies inside the semicircle marked (3).

Case (4): $\sigma_1 < \sigma_0$ and $\mu_1 = \mu_0$. The BCR is inside the semicircle (4).

Case (5): $\sigma_1 > \sigma_0$ and $\mu_1 = \mu_0$. The BCR is outside the semicircle (5).

There is evidently no common BCR for these cases. The regions of acceptance, however, may have a common part, centred round the value (μ_0, σ_0), and we should expect them to do so. Let us find the envelope of the BCR, which is, of course, the same as that of the regions of acceptance. The ratio of likelihoods (21.38) is differentiated with respect to μ_1 and to σ_1, and these derivatives equated to zero. This gives

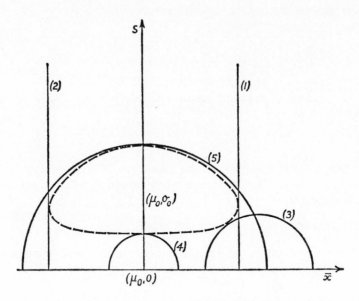

Fig. 21.2   Contours of constant likelihood ratio k (see text)

precisely the ML solutions of Example 18.11

$$\hat{\mu}_1 = \bar{x},$$

$$\hat{\sigma}_1 = s.$$

Substituting in (21.38), we find for the envelope

$$1 - \frac{s^2}{\sigma_0^2} - \left(\frac{\bar{x} - \mu_0}{\sigma_0}\right)^2 = \frac{2}{n}\log\left\{\left(\frac{\sigma_0}{s}\right)^n k_\alpha\right\}$$

or

$$\left(\frac{\bar{x} - \mu_0}{\sigma_0}\right)^2 - \log\left(\frac{s^2}{\sigma_0^2}\right) + \frac{s^2}{\sigma_0^2} = 1 - \frac{2}{n}\log k_\alpha. \tag{21.40}$$

The dotted curve in Fig. 21.2 shows one such envelope. It touches the boundaries of all the BCR that have the same k (and hence are not of the same size α). The space inside may be regarded as a 'good' region of acceptance and the space outside accordingly as a good critical region. There is no BCR for all alternatives, but the regions determined by (21.40) effect a compromise by picking out and amalgamating parts of critical regions that are best for individual alternatives.

Example 21.9
To test the simple hypothesis

$$H_0: \theta = \theta_0, \sigma = \sigma_0$$

against the alternative

$$H_1: \theta = \theta_1 < \theta_0, \sigma = \sigma_1 < \sigma_0,$$

for the exponential distribution

$$dF = \exp\left\{-\left(\frac{x - \theta}{\sigma}\right)\right\} dx / \sigma, \qquad \theta \leq x < \infty; \sigma > 0. \qquad (21.41)$$

From (21.6), the BCR is given by

$$\frac{L_0}{L_1} = \left(\frac{\sigma_1}{\sigma_0}\right)^n \exp\left\{-\frac{n(\bar{x} - \theta_0)}{\sigma_0} + \frac{n(\bar{x} - \theta_1)}{\sigma_1}\right\} \leq k_\alpha,$$

so that whatever the values of θ_1, σ_1 in H_1, the BCR is of form

$$x_{(1)} \leq \theta_0, \quad \bar{x} \leq \frac{\frac{1}{n} \log\left\{k_\alpha \left(\frac{\sigma_0}{\sigma_1}\right)^n\right\} + \left(\frac{\theta_1}{\sigma_1} - \frac{\theta_0}{\sigma_0}\right)}{\left(\frac{1}{\sigma_1} - \frac{1}{\sigma_0}\right)}. \qquad (21.42)$$

The first of these events has probability zero when H_0 holds. There is therefore a common BCR for the whole class of alternatives H_1, on which a UMP test may be based.

We have already dealt with the case $\sigma_1 = \sigma_0 = 1$ in Example 21.6.

UMP tests and sufficient statistics

21.20   In **21.14** we saw that in testing a simple parametric hypothesis against a simple alternative, the BCR is necessarily a function of the value of any (jointly) sufficient statistics for the parameter(s). In testing a simple H_0 against a composite H_1 consisting of a class of simple parametric alternatives, it evidently follows from the argument of **21.14** that if a common BCR exists, providing a UMP test against H_1, and if t is a sufficient statistic for the parameter(s), then the BCR will be a function of t. But, since a UMP test does not always exist, new questions now arise. Does the existence of a UMP test imply the existence of a corresponding sufficient statistic? And, conversely, does the existence of a sufficient statistic guarantee the existence of a corresponding UMP test?

21.21   The first of these questions may be affirmatively answered if an additional condition is imposed. In fact, as Neyman and Pearson (1936a) showed, if (1) there is a common BCR for, and therefore a UMP test of, H_0 against H_1 for every size α in an interval $0 < \alpha \leq \alpha_0$ (where α_0 is not necessarily equal to 1); and (2) if every point in the sample space W (save possibly a set of measure zero) forms part of the boundary of the BCR for at least one value of α, and then corresponds to a value of $L(x|H_0) > 0$; then a single sufficient statistic exists for the parameter(s) whose variation provides the class of admissible alternatives H_1.

To establish this result, we first note that, if a common BCR exists for H_0 against H_1 for two test sizes α_1 and $\alpha_2 < \alpha_1$, a common BCR of size α_2 can always be formed as a sub-region of that of size α_1. This follows from the fact that any common BCR satisfies (21.6). We may therefore, without loss of generality, take it that as α decreases, the BCR is adjusted simply by exclusion of some of its points.*

Now, suppose that conditions (1) and (2) are satisfied. If a point (say, x) of w forms part of the boundary of the BCR for only one value of α, we define the statistic

$$t(x) = \alpha. \tag{21.43}$$

If a point x forms part of the BCR boundary for more than one value of α, we define

$$t(x) = \tfrac{1}{2}(\alpha_1 + \alpha_2), \tag{21.44}$$

where α_1 and α_2 are the smallest and largest values of α for which it does so: it follows from the remark of the previous paragraph that x will also be part of the BCR boundary for all α in the interval (α_1, α_2). The statistic t is thus defined by (21.43) and (21.44) for all points in W (except possibly a zero-measure set). Further, if t has the same value at two points, they must lie on the same boundary. Thus, from (21.6), we have

$$\frac{L(x \mid \theta_0)}{L(x \mid \theta_1)} = k(t, \theta),$$

where k does not contain the observations except in the statistic t. Thus we must have

$$L(x \mid \theta) = g(t \mid \theta) h(x) \tag{21.45}$$

so that the single statistic t is sufficient for θ, the set of parameters concerned.

21.22   We have already considered in Example 21.2 a situation where single sufficiency and a UMP test exist together. Exercises 21.1–3 give further instances. But condition (2) of **21.21** is not always fulfilled, and then the existence of a single sufficient statistic may not follow from that of a UMP test. The following example illustrates the point.

Example 21.10
In Example 21.9, we showed that the distribution (21.41) admits a UMP test of the H_0 against the H_1 there described. The UMP test is based on the BCR (21.42), depending on $x_{(1)}$ and $\bar{x}$.

We have already seen (cf. Example 17.19 and Exercise 17.9) that the smallest observation $x_{(1)}$ is sufficient for θ if σ is known, and that $\bar{x}$ is sufficient for σ if θ is known. The pair of statistics $x_{(1)}$ and $\bar{x}$ are jointly sufficient, but there is no *single* sufficient statistic for θ and σ.

---

* This is not true of critical regions in general—see, e.g., Chernoff (1951).

21.23 On the other hand, the possibility that a single sufficient statistic exists without a one-sided UMP test, even where only a single parameter is involved, is made clear by Example 21.11.

Example 21.11
Consider the multinormal distribution of n variates $x_1, \ldots, x_n$, with

$$E(x_1) = n\theta, \qquad \theta > 0,$$
$$E(x_r) = 0, \qquad r > 1;$$

and covariance matrix

$$\mathbf{V} = \begin{pmatrix} n - 1 + \theta^2 & -1 \cdots -1 \\ -1 & \\ \vdots & \mathbf{I}_{n-1} \\ -1 & \end{pmatrix} \qquad (21.46)$$

where $\mathbf{I}_{n-1}$ is the identity matrix of order $n - 1$. The determinant of (21.46) is easily seen to be

$$|\mathbf{V}| = \theta^2$$

and its inverse matrix is

$$\mathbf{V}^{-1} = \frac{1}{\theta^2} \begin{pmatrix} 1 & & 1 \\ & 1 + \theta^2 & \\ 1 & & \ddots \\ & & 1 + \theta^2 \end{pmatrix} \qquad (21.47)$$

with every off-diagonal element equal to unity. Thus, from **15.3**, the joint distribution is

$$dF = \frac{1}{\theta(2\pi)^{n/2}} \exp\left\{ -\frac{1}{2}\left[\frac{n^2}{\theta^2}(\bar{x} - \theta)^2 + \sum_{i=2}^{n} x_i^2 \right] \right\} dx_1 \cdots dx_n. \qquad (21.48)$$

Consider now the testing of the hypothesis $H_0: \theta = \theta_0 > 0$ against $H_1: \theta = \theta_1 > 0$ on the basis of a single observation. From (21.6), the BCR is given by

$$\frac{L(x \mid \theta_0)}{L(x \mid \theta_1)} = \left(\frac{\theta_1}{\theta_0}\right) \exp\left\{ -\frac{n^2}{2}\left[\frac{(\bar{x} - \theta_0)^2}{\theta_0^2} - \frac{(\bar{x} - \theta_1)^2}{\theta_1^2} \right] \right\} \leq k_\alpha,$$

which reduces to

$$\frac{(\bar{x} - \theta_1)^2}{\theta_1^2} - \frac{(\bar{x} - \theta_0)^2}{\theta_0^2} \leq \frac{2}{n^2} \log (k_\alpha \theta_0 / \theta_1)$$

or

$$\bar{x}^2(\theta_0^2 - \theta_1^2) - 2\bar{x}\theta_0\theta_1(\theta_0 - \theta_1) \leq \frac{2\theta_0^2\theta_1^2}{n^2} \log (k_\alpha \theta_0 / \theta_1).$$

If $\theta_0 > \theta_1$, this is of form

$$\bar{x}^2(\theta_0 + \theta_1) - 2\bar{x}\theta_0\theta_1 \leq a_\alpha, \qquad (21.49)$$

which implies

$$b_\alpha \leq \bar{x} \leq c_\alpha. \tag{21.50}$$

If $\theta_0 < \theta_1$, the BCR is of form

$$\bar{x}^2(\theta_0 + \theta_1) - 2\bar{x}\theta_0\theta_1 \geq d_\alpha, \tag{21.51}$$

implying

$$\bar{x} \leq e_\alpha \quad \text{or} \quad \bar{x} \geq f_\alpha. \tag{21.52}$$

In both (21.50) and (21.52), the limits between which (or outside which) $\bar{x}$ has to lie are functions of the exact value of θ_1. This difficulty, which arises from the fact that θ_1 appears in the coefficient of $\bar{x}^2$ in the quadratics (21.49) and (21.51), means that there is no BCR even for a one-sided set of alternatives, and therefore no UMP test.

It is easily verified from (21.48) that $\bar{x}$ is a single sufficient statistic for θ, and this completes the demonstration that single sufficiency does not imply the existence of a UMP test.

The power function
21.24 Now that we are considering the testing of a simple H_0 against a composite H_1, we generalise the idea of the power of a test defined at (21.2). As we stated there, the power is an explicit function of H_1. If H_1 specifies a set of alternative values for θ, we may consider the power of a test of H_0: $\theta = \theta_0$ against the simple alternative H_1': $\theta = \theta_1 > \theta_0$ for each θ_1. The power for each H_1' depends on the value of θ_1, so that the set of all such values defines the *power function* of the test of H_0 against the class of alternatives H_1: $\mu > \mu_0$. We indicate the compositeness of H_1 by writing it thus, instead of the form used for a simple H_1: $\mu = \mu_1 > \mu_0$. For instance, we saw in Example 21.3 that the power of the most powerful test of the hypothesis that the mean μ of a normal population is μ_0 against the alternative value $\mu_1 > \mu_0$, is given by the power function (21.18), a monotone increasing function of μ_1.

The evaluation of a power function is rarely so easy as in Example 21.3, since even if the sampling distribution of the test statistic is known exactly for both H_0 and the class of alternatives H_1 (and more commonly only approximations are available, especially for H_1), there is still the problem of evaluating (21.2) for each value of θ in H_1, which usually is a matter of numerical integration: only rarely is the power function exactly obtainable from a tabulated integral, as at (21.18). Asymptotically, however, the central limit theorem comes to our aid: the distributions of many test statistics tend to normality, given either H_0 or H_1, as sample size increases, and then the asymptotic power function will be of the form (21.18), as we shall see when we come to the comparison of tests in Chapter 25.

Example 21.12
The general shape of the power function (21.18) in Example 21.3 is simply that of the normal distribution function. It increases from the value

$$G\{-d_\alpha\} = \alpha$$

at $\mu = \mu_0$ (in accordance with the size requirement) to the value

$$G\{0\} = 0.5$$

at $\mu = \mu_0 + d_\alpha/n^{1/2}$, the first derivative G' increasing up to this point; as μ increases beyond it, G' declines to its limiting value of zero as G increases to its asymptote 1.

21.25 Once the power function of a test has been determined, it is of obvious value in determining how large the sample should be in order to test H_0 with given size and power. The procedure is illustrated in the next example.

Example 21.13
How many observations should be taken in Example 21.3 so that we may test H_0: $\mu = 3$ with $\alpha = 0.05$ (i.e. $d_\alpha = 1.6449$) and power of at least 0.75 against the alternatives that $\mu \geqslant 3.5$? Put otherwise, how large should n be to ensure that the probability of a Type I error is 0.05, and that of a Type II error at most 0.25 for $\mu \geqslant 3.5$?
   From (21.18), we require n large enough to make

$$G\{n^{1/2}(3.5 - 3) - 1.6449\} = 0.75, \tag{21.53}$$

it being obvious that the power will be greater than this for $\mu > 3.5$. Now, from a table of the normal distribution

$$G\{0.6745\} = 0.75, \tag{21.54}$$

and hence, from (21.53) and (21.54),

$$0.5 n^{1/2} - 1.6449 = 0.6745,$$

whence

$$n = (4.6388)^2 = 21.5 \text{ approx.},$$

so that $n = 22$ will suffice to give the test the required power.

One- and two-sided tests
21.26 We have seen in **21.18** that in general no UMP test exists when we test a parametric hypothesis H_0: $\theta = \theta_0$ against a two-sided alternative hypothesis, i.e. one in which $\theta - \theta_0$ changes sign. Nevertheless, situations often occur when departures from H_0 in either direction are considered important. In such circumstances, it is tempting to continue to use as our test statistic one which is known to give a UMP test against one-sided alternatives ($\theta > \theta_0$ or $\theta < \theta_0$) but to modify the critical region in the distribution of the statistic by compromising between the BCR for $\theta > \theta_0$ and the BCR for $\theta < \theta_0$.

   21.27 For instance, in Example 21.2 and in **21.17** we saw that the mean $\bar{x}$, used to test H_0: $\mu = \mu_0$ for the mean μ of a normal population, gives a UMP test against $\mu_1 < \mu_0$ with common BCR $\bar{x} \leqslant a_\alpha$, and a UMP test for $\mu_1 > \mu_0$ with common BCR $\bar{x} \geqslant b_\alpha$. Suppose, then, that for the alternative H_1: $\mu \neq \mu_0$, which is two-sided, we construct a compromise *equal-tails* critical region defined by

$$\left.\begin{array}{c} \bar{x} \leqslant a_{\alpha/2}, \\ \bar{x} \geqslant b_{\alpha/2}, \end{array}\right\} \tag{21.55}$$

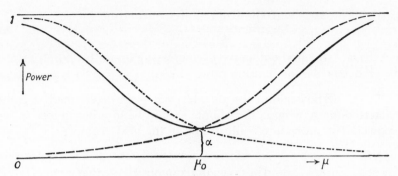

Fig. 21.3 Power functions of three tests based on $\bar{x}$

——— Critical region in both tails equally.
– – – – Critical region in upper tail.
– · – · – Critical region in lower tail.

in other words, combining the one-sided critical regions and making each of them of size $\frac{1}{2}\alpha$, so that the critical region as a whole remains one of size α.

We know that the critical region defined by (21.55) will always be less powerful than one or other of the one-sided BCR, but we also know that it will always be more powerful than the other. For its power function will be, exactly as in Example 21.3,

$$G\{n^{1/2}(\mu - \mu_0) - d_{\alpha/2}\} + G\{n^{1/2}(\mu_0 - \mu) - d_{\alpha/2}\}. \tag{21.56}$$

(21.56) is an even function of $(\mu - \mu_0)$, with a minimum at $\mu = \mu_0$ and thus is $\geq \alpha$; in the terminology of **21.9**, the equal-tails test is unbiased. Hence it is always intermediate in value between $G\{n^{1/2}(\mu - \mu_0) - d_\alpha\}$ and $G\{n^{1/2}(\mu_0 - \mu) - d_\alpha\}$, which are the power functions of the one-sided BCR, except when $\mu = \mu_0$, when all three expressions are equal. The comparison is worth making diagrammatically, in Fig. 21.3, where a single fixed value of n and of α is illustrated.

21.28 We shall see in **22.23** onwards that other, less intuitive, justifications can be given for splitting the critical region in this way between the tails of the distribution of the test statistic. For the moment, the procedure is to be regarded as simply a common-sense way of insuring against the risk of extreme loss of power which, as Fig. 21.3 makes clear, would be the result if we located the critical region in the tail of the statistic's distribution that turned out to be the wrong one for the true value of θ.

Choice of test size
21.29 Throughout our exposition so far we have assumed that the test size α has been fixed in some way. All our results are valid however that choice was made. We now turn to the question of how α is to be determined.

In the first place, it is natural to suggest that α should be made 'small' according to some acceptable criterion, and indeed it is customary to use certain conventional values of α, such as $0.05, 0.01$ or 0.001. But we must take care not to go too far in this direction. We can only fix two of the quantities n, α and β, even in testing a simple H_0 against a simple H_1. If n is fixed, we can only in general decrease the value of α, the probability of Type I error, by increasing the value of β, the probability of Type II error. In other words, reduction in the size of a test decreases its power.

This point is well illustrated in Example 21.3 by the expression (21.18) for the power of the BCR in a one-sided test for a normal population mean. We see there that as $\alpha \to 0$, by (21.16) $d_\alpha \to \infty$, and consequently the power (21.18) $\to 0$.

Thus, for fixed sample size, we have essentially to reconcile the size and power of the test. If the practical risks attaching to a Type I error are great, while those attaching to a Type II error are small, there is a case for reducing α, at the expense of increasing β, if n is fixed. If, however, sample size is at our disposal, we may, as in Example 21.13, ensure that n is large enough to reduce both α and β to any pre-assigned levels. These levels have still to be fixed, but unless we have supplementary information in the form of the *costs* (in money or other common terms) of the two types of error, and the costs of making observations, we cannot obtain an optimum combination of α, β and n for any given problem. It is sufficient for us to note that, however α is determined, we shall obtain a *valid* test.

21.30 The point discussed in **21.29** is reflected in another, which has sometimes been made the basis of criticism of the theory of testing hypotheses.

Suppose that we carry out a test with α fixed, no matter how, and n extremely large. The power of a reasonable test will be very near 1 in detecting departure of any sort from the hypothesis tested. Now, the argument (formulated by Berkson (1938)) runs: nobody really supposes that any hypothesis holds precisely; we are simply setting up an abstract model of real events that is bound to be some way, if only a little, from the truth. Nevertheless, as we have seen, an enormous sample would almost certainly (i.e. with probability approaching 1 as n increases beyond any bound) reject the hypothesis tested at any pre-assigned size α. Why, then, do we bother to test the hypothesis at all with a smaller sample, whose verdict is less reliable than the larger one's?

This paradox is really concerned with two points. In the first place, if n is fixed, and we are not concerned with the exactness of the hypothesis tested, but only with its approximate validity, our alternative hypothesis would embody this fact by being sufficiently distant from the hypothesis tested to make the difference of practical interest. This in itself would tend to increase the power of the test. But if we had no wish to reject the hypothesis tested on the evidence of small deviations from it, we should want the power of the test to be very low against these small deviations, and this would imply a small α and a correspondingly high β and low power.

But the crux of the paradox is the argument from increasing sample size. The hypothesis tested will only be rejected with probability near 1 if we keep α fixed as n increases. There is no reason why we should do this: we can determine α in any way we please, and it is rational, in the light of the discussion of **21.29**, to apply the gain in sensitivity arising from increased sample size to the reduction of α as well as of β. It is only the habit of fixing α at certain conventional levels that leads to the paradox. If we allow α to decline as n increases, it is no longer certain that a very small departure from H_0 will cause H_0 to be rejected: this now depends on the rate at which α declines. A reasonable, though arbitrary, solution is to make α equal to the β at the smallest departure from H_0 that is of practical importance.

21.31   There is a converse to the paradox discussed in **21.30**. Just as, for large n, inflexible use of conventional values of α will lead to very high power, which may possibly be too high for the problem in hand, so for very small fixed n their use will lead to very low power, perhaps too low. Again, the situation can be remedied by allowing α to rise and consequently reducing β. It is always incumbent upon the statistician to satisfy herself that, for the conditions of her problem, she is not sacrificing sensitivity in one direction to sensitivity in another.

Example 21.14
E. S. Pearson (discussion on Lindley (1953)) calculated a few values of the power function (21.56) of the two-sided test for a normal mean, which we reproduce to illustrate our discussion.

Table 21.2   Power function calculated from (21.56)

The entries in the first row of the table give the sizes of the tests.

Value of $|\mu - \mu_0|$	Sample size (n)						
	10	10	10	20	100	100	100
0	0.050	0.072	0.111	0.050	0.050	0.019	0.0056
0.1				0.073	0.170	0.088	0.038
0.2	0.097	0.129	0.181	0.145	0.516	0.362	0.221
0.3				0.269	0.851	0.741	0.592
0.4	0.244	0.298	0.373	0.432	0.979	0.950	0.891
0.5				0.609	0.999	0.996	0.987
0.6	0.475	0.539	0.619	0.765			

It will be seen from the table that when sample size is increased from 20 to 100, the reductions of α from 0.050 to 0.019 and 0.0056 successively reduce the power of the test for each value of $|\mu - \mu_0|$. In fact, for $\alpha = 0.0056$ and $|\mu - \mu_0| = 0.1$, the power actually falls below the value attained at $n = 20$ with $\alpha = 0.05$. Conversely, on reduction of sample size from 20 to 10, the increase in α to 0.072 and 0.111 increases the power correspondingly, though only in the case $\alpha = 0.111$, $|\mu - \mu_0| = 0.2$, does it exceed the power at $n = 20$, $\alpha = 0.5$.

21.32   Bartholomew (1967) discusses the choice of α and β when n is a random variable, distributed free of θ. (21.6) then remains valid, but the distribution of n enters into the determination of k_α, which remains the same whatever the value of n observed. See also the discussion following Bartholomew's paper.

EXERCISES

21.1 Show directly by use of (21.6) that the BCR for testing a simple hypothesis $H_0: \mu = \mu_0$ concerning the mean μ of a Poisson distribution against a simple alternative $H_1: \mu = \mu_1$ is of the form

$$\bar{x} \leqslant a_\alpha \quad \text{if } \mu_0 > \mu_1,$$
$$\bar{x} \geqslant b_\alpha \quad \text{if } \mu_0 < \mu_1,$$

where $\bar{x}$ is the sample mean and a_α, b_α are constants.

21.2 Show similarly that for the parameter π of a binomial distribution, the BCR is of the form

$$p \leqslant a_\alpha \quad \text{if } \pi_0 > \pi_1,$$
$$p \geqslant b_\alpha \quad \text{if } \pi_0 < \pi_1,$$

where p is the observed proportion of 'successes' in the sample.

21.3 Show that for the normal distribution with zero mean and variance σ^2, the BCR for $H_0: \sigma = \sigma_0$ against the alternative $H_1: \sigma = \sigma_1$ is of form

$$\sum_{i=1}^{n} x_i^2 \leqslant a_\alpha \quad \text{if } \sigma_0 > \sigma_1,$$

$$\sum_{i=1}^{n} x_i^2 \geqslant b_\alpha \quad \text{if } \sigma_0 < \sigma_1.$$

Show that the power of the BCR when $\sigma_0 > \sigma_1$ is $F\{(\sigma_0^2/\sigma_1^2)\chi_{\alpha,n}^2\}$, where $\chi_{\alpha,n}^2$ is the lower 100α per cent point and F is the d.f. of the χ^2 distribution with n degrees of freedom.

21.4 In Example 21.4, show that the power of the test when $\theta_1 = 1$ is 0.648 when $k_\alpha = 1$ and 0.352 when $k_\alpha = 0.5$. Draw a diagram of the two Cauchy distributions to illustrate the power and size of each test. Show that when $\theta_1 = 1$, $k_\alpha = 1.5$, the BCR consists of values of x outside the interval $(-2 - \sqrt{5}, -2 + \sqrt{5})$.

21.5 In Exercise 21.3, verify that the power is a monotone increasing function of σ_0^2/σ_1^2, and also verify numerically from a table of the χ^2 distribution that the power is a monotone increasing function of n.

21.6 Confirm that (21.21) holds for the sufficient statistics on which the BCR of Example 21.2, and Exercises 21.1–21.3 are based.

21.7 In **21.15** show that the more efficient estimator always gives the more powerful test if its test power exceeds 0.5.   (Sundrum, 1954)

21.8 Show that for testing $H_0: \mu = \mu_0$ in samples from the distribution

$$dF = dx, \qquad \mu \leqslant x \leqslant \mu + 1,$$

there is a pair of UMP one-sided tests, and hence no UMP test for all alternatives.

21.9 In Example 21.11, show that $\bar{x}$ is normally distributed with mean θ and variance θ^2/n^2, and that it is a sufficient statistic for θ.

21.10 Verify that the distribution of Example 21.9 does not satisfy condition (2) of **21.21**.

21.11 In Example 21.9, let σ be any positive increasing function of θ. Show that there is still a common BCR in testing $H_0: \theta = \theta_0$ against $H_1: \theta = \theta_1 < \theta_0$.
                              (Neyman & Pearson, 1936a)

21.12 Generalising the discussion of **21.27**, write down the power function of any test based on the distribution of $\bar{x}$ with its critical region of form

$$\bar{x} \leqslant a_{\alpha_1},$$
$$\bar{x} \geqslant b_{\alpha_2},$$

where $\alpha_1 + \alpha_2 = \alpha$ (α_1 and α_2 not necessarily being equal). Show that the power function of any such test lies completely between those for the cases $\alpha_1 = 0$, $\alpha_2 = 0$ illustrated in Fig. 21.3.

21.13 Referring to the discussion of **21.14**, show that the likelihood ratio (for testing a simple $H_0: \theta = \theta_0$ against a simple $H_1: \theta = \theta_1$) is a sufficient statistic for θ on either hypothesis by writing the likelihood function as

$$L(x \mid \theta) = L(x \mid \theta_1) \left[\frac{L(x \mid \theta_0)}{L(x \mid \theta_1)} \right]^{(\theta - \theta_1)/(\theta_0 - \theta_1)} \qquad \text{(Pitman, 1957)}$$

21.14 For the most powerful test based on the BCR (21.6), show from (21.5) that $(1 - \beta)/\alpha \geqslant 1/k_\alpha$ and hence by interchanging H_0 and H_1 that $\beta/(1 - \alpha) \leqslant 1/k_\alpha \leqslant (1 - \beta)/\alpha$.

21.15 From Exercise 21.14, show that as $n \to \infty$ with α fixed, the inequality $\beta \leqslant (1 - \alpha)/k_\alpha$ becomes an equality and $\log \beta \sim E\{\log (L_1/L_0)\}$.

(Cf. C. R. Rao (1962a). Efron (1967) gets a fuller result.)

CHAPTER 22

TESTS OF HYPOTHESES: COMPOSITE HYPOTHESES

22.1 We have seen in Chapter 21 that, when the hypothesis tested is simple (specifying the distribution completely), there is *always* a BCR, providing a most powerful test, against a simple alternative hypothesis; that there *may* be a UMP test against a class of simple hypotheses constituting a composite parametric alternative hypothesis; and that there will not, in general, be a UMP test if the parameter whose variation generates the alternative hypothesis is free to vary in both directions from the value tested.

If the hypothesis tested is composite, leaving at least one parameter value unspecified, it is to be expected that UMP tests will be even rarer than for simple hypotheses, but we shall find that progress can be made if we are prepared to restrict the class of tests considered in certain ways.

Composite hypotheses

22.2 First, we formally define the problem. We suppose that the n observations have a distribution dependent upon the values of $l(\leq n)$ parameters which we shall write

$$L(x \mid \theta_1, \ldots, \theta_l)$$

as before. The hypothesis to be tested is

$$H_0: \theta_1 = \theta_{10}; \quad \theta_2 = \theta_{20}; \ldots; \quad \theta_k = \theta_{k0}, \tag{22.1}$$

where $k \leq l$, and the second suffix 0 denotes the value specified by the hypothesis. We lose no generality by thus labelling the k parameters whose values are specified by H_0 as the first k of the set of l parameters. H_0 as defined at (22.1) is said to impose k constraints.

Hypotheses of the form

$$H_0: \theta_1 = \theta_2; \theta_3 = \theta_4; \ldots,$$

which do not specify the values of parameters whose equality we are testing, are transformable into the form (22.1) by re-parametrising the problem in terms of $\theta_1 - \theta_2$, $\theta_3 - \theta_4$, etc., and testing the hypothesis that these new parameters have zero values. Thus (22.1) is a more general composite hypothesis than at first appears.

To keep our notation simple, we shall write $L(x \mid \theta_r, \theta_s)$ and

$$H_0: \theta_r = \theta_{r0}, \tag{22.2}$$

where it is to be understood that θ_r, θ_s may each consist of more than one parameter, the 'nuisance parameter' θ_s being left unspecified by the hypothesis tested.

An optimum property of sufficient statistics

22.3 This is a convenient place to prove an optimum test property of sufficient statistics analogous to the estimation result proved in **17.35**. There we saw that if u is an unbiased estimator of θ and t is a sufficient statistic for θ, then the statistic $E(u|t)$ is unbiased for θ with variance no greater than that of u. We now prove a result due to Lehmann (1950): if w is a critical region for testing H_0, a hypothesis concerning θ in $L(x|\theta)$, against some alternative H_1, and t is a sufficient statistic, both on H_0 and on H_1, for θ, then there is a test of the same size, based on a function of t, which has the same power as w.

We first define a function*

$$c(w)\begin{cases} =1 & \text{if the sample point is in } w, \\ =0 & \text{otherwise.} \end{cases} \tag{22.3}$$

Then the integral

$$\int c(w)L(x|\theta)\,\mathrm{d}x = E\{c(w)\} \tag{22.4}$$

gives the probability that the sample point falls into w, and is therefore equal to the size (α) of the test when H_0 is true and to the power of the test when H_1 is true. Using the factorisation property (17.68) of the likelihood function in the presence of a sufficient statistic, (22.4) becomes

$$E\{c(w)\} = \int c(w)h(x|t)g(t|\theta)\,\mathrm{d}x$$
$$= E\{E(c(w)|t)\}, \tag{22.5}$$

the expectation operation outside the braces being with respect to the distribution of t. Thus the particular function of t, $E(c(w)|t)$, not dependent upon θ since t is sufficient, has the same expectation as $c(w)$. There is therefore a test based on the sufficient statistic t that has the same size and power as the original region w. We may therefore without loss of power confine the discussion of any test problem to functions of a sufficient statistic.

This result is quite general, and therefore also covers the case of a simple H_0 discussed in Chapter 21.

Test size for composite hypotheses: similar regions

22.4 Since a composite hypothesis leaves some parameter values unspecified, a new problem immediately arises, for the size of the test of H_0 will obviously be a function, in general, of these unspecified parameter values, θ_s.

---

* $c(w)$ is known in measure theory as the characteristic function of the set of points w. We shall avoid this terminology, since there is some possibility of confusion with the use of 'characteristic function' for the Fourier transform of a distribution function, with which we have been familiar since Chapter 4.

If we wish to keep Type I errors down to some pre-assigned level, we must seek critical regions whose size can be kept down to that level for all possible values of θ_s. Thus we require

$$\alpha(\theta_s) \leqslant \alpha. \tag{22.6}$$

If a critical region has

$$\alpha(\theta_s) = \alpha \tag{22.7}$$

as a strict equality for all θ_s, it is called a (critical) region *similar to the sample space** with respect to θ_s, or, more briefly, a similar (critical) region. The test based on a similar critical region is called a similar size-α test.

22.5   It is not obvious that similar regions exist at all generally, but in one sense they exist whenever we are dealing with a set of n independent identically distributed observations on a continuous variate x. For no matter what the distribution or its parameters, we have

$$P\{x_1 < x_2 < x_3 < \cdots < x_n\} = 1/n! \tag{22.8}$$

(cf. **11.4**), since any of the $n!$ permutations of the x_i is equally likely. Thus, for α an integral multiple of $1/n!$, there are similar regions based on the $n!$ hypersimplices in the sample space obtained by permuting the n suffixes in (22.8).

22.6   If we confine ourselves to regions defined by symmetric functions of the observations (so that similar regions based on (22.8) are excluded) it is easy to see that, where similar regions do exist, they need not exist for all sample sizes. For example, for a sample of n observations from the normal distribution

$$dF(x) = (2\pi)^{-1/2} \exp\{-\tfrac{1}{2}(x - \theta)^2\}\, dx,$$

there is no similar region with respect to θ for $n = 1$, but for $n \geqslant 2$ the fact that $ns^2 = \sum_{i=1}^{n}(x_i - \bar{x})^2$ has a chi-squared distribution with $(n-1)$ degrees of freedom, whatever the value of θ, ensures that similar regions of any size can be found from the distribution of s^2. This is because $\bar{x}$ is a single sufficient statistic for θ, and to find a similar region we must, by Exercise 22.3, find a statistic uncorrelated with

$$\frac{\partial \log L}{\partial \theta} = n(\bar{x} - \theta).$$

This is impossible when $n = 1$, since $\bar{x} = x$ is then the whole sample, but for $n \geqslant 2$, $\sum (x - \bar{x})^2$ is distributed independently of $\bar{x}$ and thus gives similar regions. The same argument holds in Exercise 22.1, where there is a pair of sufficient statistics for two parameters, and at least three observations are required so that we may have a statistic independent of both.

---

* The term arose because, trivially, the entire sample space is a similar region with $\alpha = 1$.

22.7 Even if n is large, symmetric similar regions will not exist if each observation brings a new parameter with it, as in the following example, due to Feller (1938).

Example 22.1
Consider a sample of n observations, where the ith observation has distribution

$$dF(x_i) = (2\pi)^{-1/2} \exp\{-\tfrac{1}{2}(x_i - \theta_i)^2\} \, dx_i,$$

so that

$$L(x|\theta) = (2\pi)^{-n/2} \exp\{-\tfrac{1}{2}\sum (x_i - \theta_i)^2\}.$$

For a similar region w of size α, we require, identically in θ,

$$\int_w L(x|\theta) \, dx = \alpha.$$

Using (22.3), we may re-write this size condition as

$$\int_W L(x|\theta) \frac{c(w)}{\alpha} \, dx = 1, \tag{22.9}$$

where W is the whole sample space. Differentiating (22.9) with respect to θ_i, we find

$$\int_W L(x|\theta) \frac{c(w)}{\alpha} (x_i - \theta_i) \, dx = 0. \tag{22.10}$$

A second differentiation with respect to θ_i gives

$$\int_W L(x|\theta) \frac{c(w)}{\alpha} \{(x_i - \theta_i)^2 - 1\} \, dx = 0. \tag{22.11}$$

Now from the definition of $c(w)$,

$$g(x|\theta) = L(x|\theta) \frac{c(w)}{\alpha} \tag{22.12}$$

is a (joint) frequency function. (22.10) and (22.11) express the facts that the marginal distribution of x_i in $g(x|\theta)$ has

$$E(x_i) = \theta_i, \qquad \text{var } x_i = 1,$$

just as it has in the initial distribution of x_i.

If we examine the form of $g(x|\theta)$, we see that if we were to proceed with further differentiations, we should find that *all* the moments and product-moments of $g(x|\theta)$ are identical with those of $L(x|\theta)$, which is uniquely determined by its moments. Thus, from (22.12), $c(w)/\alpha = 1$ identically. But since $c(w)$ is either 0 or 1, we see finally that the trivial values $\alpha = 0$ or 1 are the only values for which similar regions can exist. The difficulty here is that all n observations are required to form a sufficient set for the n parameters, and we can find no statistic independent of them all.

22.8 It is nevertheless true that for many problems of testing composite hypotheses, similar regions exist for any size α and any sample size n. We now have to consider how they are to be found.

Let t be a sufficient statistic for the parameter θ_s unspecified by the hypothesis H_0, and suppose that we have a critical region w such that for all values of t, when H_0 is true, w is composed of a fraction α of the probability content of each contour of constant t, i.e.

$$E\{c(w)|t\} = \alpha. \tag{22.13}$$

Then, on taking expectations with respect to t, we have, as at (22.5),

$$E\{c(w)\} = E\{E(c(w)|t)\} = \alpha \tag{22.14}$$

so that the original critical region w is similar of size α, as Neyman (1937) and Bartlett (1937) pointed out.

It should be noticed that here t need be sufficient only for the unspecified parameter θ_s, and only when H_0 is true. This should be contrasted with the more demanding requirements of **22.3**.

Our argument has shown that (22.13) is a sufficient condition that w be similar. We shall show in **22.19** that it is necessary and sufficient, provided that a further condition is fulfilled, and in order to state that condition we must now introduce, following Lehmann and Scheffé (1950), the concept of the *completeness* of a parametric family of distributions, a concept that also permits us to supplement the discussion of sufficient statistics in Chapter 17.

> We shall see in **22.22** that restriction to similar tests may involve a loss of power compared with other tests satisfying (22.6).

Complete parametric families and complete statistics

22.9 Consider a parametric family of (univariate or multivariate) distributions, $f(x|\theta)$, depending on the value of a vector of parameters θ. Let $h(x)$ be any statistic, independent of θ. If

$$E\{h(x)\} = \int h(x)f(x|\theta)\,\mathrm{d}x = 0 \tag{22.15}$$

for all θ implies that

$$h(x) = 0 \tag{22.16}$$

identically (save possibly on a zero-measure set), then the family $f(x|\theta)$ is called *complete*. The term is apt, since no non-zero function can be found that is orthogonal to all members of the family. If (22.15) implies (22.16) only for all bounded $h(x)$, $f(x|\theta)$ is called *boundedly complete*.

In the statistical applications of the concept of completeness, the family of distributions we are interested in is often the sampling distribution of a (possibly vector-) statistic t, say $g(t|\theta)$. We then call t a complete (or boundedly complete) statistic if, for all θ, $E\{h(t)\} = 0$ implies $h(t) = 0$ identically, for all functions (or bounded functions) $h(t)$. In other words, we label the statistic t with the completeness property of its distribution.

An evident immediate consequence of the completeness of a statistic t is that only one function of that statistic can have a given expected value. Thus if one function of t is an unbiased estimator of a certain function of θ, no other function of t will be. Completeness confers a uniqueness property upon an estimator.

> Bounded completeness of f implies that its c.f. $\phi(u)$ can have no zeros, for otherwise $0 = \phi(u) = E(e^{iux})$. If θ is a location parameter, i.e. $f = f(x - \theta)$, the converse also holds—cf. J. K. Ghosh and Singh (1966). Thus, for example, the Cauchy distribution of Example 17.7 is boundedly complete—the c.f. is given in Example 4.2, Vol. 1.

The completeness of sufficient statistics

22.10   The special case of the exponential family (17.83) with $A(\theta) = \theta$, $B(x) = x$ has

$$f(x \mid \theta) = \exp \{\theta x + C(x) + D(\theta)\}, \qquad -\infty < x < \infty. \tag{22.17}$$

If, for all θ,

$$\int h(x) f(x \mid \theta)\, \mathrm{d}x = 0,$$

we must have

$$\int [h(x) \exp \{C(x)\}] \exp (\theta x)\, \mathrm{d}x = 0. \tag{22.18}$$

The integral in (22.18) is the two-sided Laplace transform* of the function in square brackets in the integrand. By the uniqueness property of the transform, the only function having a transform of zero value is zero itself; i.e.

$$h(x) \exp \{C(x)\} = 0$$

identically, whence

$$h(x) = 0$$

identically. Thus $f(x \mid \theta)$ is complete.

This result generalises to the multi-parameter case, as shown by Lehmann and Scheffé (1955): the k-parameter, k-variate exponential family

$$f(\mathbf{x} \mid \boldsymbol{\theta}) = \exp \left\{ \sum_{j=1}^{k} \theta_j x_j + C(\mathbf{x}) + D(\boldsymbol{\theta}) \right\} \tag{22.19}$$

---

* The two-sided Laplace transform of a function $g(x)$ is defined by

$$\lambda(\theta) = \int_{-\infty}^{\infty} \exp (\theta x) g(x)\, \mathrm{d}x.$$

The integral converges in a strip of the complex plane $\alpha < R(\theta) < \beta$, where one or both of α, β may be infinite. (The strip may degenerate to a line.) Except possibly for a zero-measure set, there is a one-to-one correspondence between $g(x)$ and $\lambda(\theta)$. See, e.g., D. V. Widder (1941), *The Laplace Transform*, Princeton University Press, and compare also the inversion theorem for c.f.s in **4.3**.

is a complete family. We have seen (Exercise 17.14) that the joint distribution of the set of k sufficient statistics for the k parameters of the general univariate exponential form (17.86) takes a form of which (22.19) is the special case, with $A_j(\theta) = \theta_j$. (We have replaced nD and h of the Exercise by D and $\exp(C)$ respectively.) By **22.3**, we may confine ourselves, in testing hypotheses about the parameters, to the sufficient statistics.

Example 22.2

Consider the family of normal distributions

$$f(x \mid \theta_1, \theta_2) = (2\pi\theta_2)^{-1/2} \exp\left\{-\frac{1}{2\theta_2}(x - \theta_1)^2\right\}, \qquad -\infty < x < \infty; \; \theta_2 > 0.$$

(a) If θ_2 is known (say $= 1$), the family is complete with respect to θ_1, for we are then considering a special case of (22.17) with

$$\theta = \theta_1, \; \exp\{C(x)\} = (2\pi)^{-1/2} \exp\left(-\tfrac{1}{2}x^2\right)$$

and

$$D(\theta) = -\tfrac{1}{2}\theta_1^2.$$

(b) If, on the other hand, θ_1 is known (say $= 0$), the family is not even boundedly complete with respect to θ_2, for $f(x \mid 0, \theta_2)$ is an even function of x, so that any odd function $h(x)$ will have zero expectation without being identically zero. However, if we transform to $y = x^2$, we see that the distribution of y is complete, since $g(y \mid \theta_2) = (2\pi\theta_2 y)^{-1/2} \exp\{-y/(2\theta_2)\}$ is again a special case of (22.17). Cf. the remark in **22.16** below.

22.11 In **22.10** we discussed the completeness of the characteristic form of the joint distribution of sufficient statistics in samples from a distribution with range independent of the parameters. Hogg and Craig (1956) have established the completeness of the sufficient statistic for distributions whose range is a function of a single parameter θ and that possess a single sufficient statistic for θ. We recall from **17.40–1** that the distribution must then be of form

$$f(x \mid \theta) = g(x)/h(\theta) \tag{22.20}$$

and that

(i) if a single terminal of $f(x \mid \theta)$ (which may be taken to be θ itself without loss of generality) is a function of θ, the corresponding extreme order-statistic is sufficient;

(ii) if both terminals are functions of θ, the upper terminal $(b(\theta))$ must be a monotone decreasing function of the lower terminal (θ) for a single sufficient statistic

$$\min\{x_{(1)}, b^{-1}(x_{(n)})\}. \tag{22.21}$$

to exist. We consider the cases (i) and (ii) in turn.

22.12   In case (i), take the upper terminal equal to θ, the lower equal to a constant a. $x_{(n)}$ is then sufficient for θ. Its distribution is, from (11.34) and (22.20),

$$dG(x_{(n)}) = n\{F(x_{(n)})\}^{n-1}f(x_{(n)})\,dx_{(n)}$$

$$= \frac{n\{\int_a^{x_{(n)}} g(x)\,dx\}^{n-1} g(x_{(n)})}{\{h(\theta)\}^n}\,dx_{(n)}, \qquad a \leqslant x_{(n)} \leqslant \theta. \qquad (22.22)$$

Now suppose that for a statistic $u(x_{(n)})$ we have

$$\int_a^\theta u(x_{(n)})\,dG(x_{(n)}) = 0,$$

or, substituting from (22.22), and dropping the factor in $h(\theta)$,

$$\int_a^\theta u(x_{(n)})\left\{\int_a^{x_{(n)}} g(x)\,dx\right\}^{n-1} g(x_{(n)})\,dx_{(n)} = 0. \qquad (22.23)$$

If we differentiate (22.23) with respect to θ, we find

$$u(\theta)\left\{\int_a^\theta g(x)\,dx\right\}^{n-1} g(\theta) = 0, \qquad (22.24)$$

and since the integral in braces equals $h(\theta)$, while $g(\theta) \neq 0 \neq h(\theta)$, (22.24) implies

$$u(\theta) = 0$$

for any θ. Hence the function $u(x_{(n)})$ is identically zero, and the distribution of $x_{(n)}$, given at (22.22), is complete. Exactly the same argument holds for the lower terminal and $x_{(1)}$.

22.13   In case (ii), the distribution function of the sufficient statistic (22.21) is

$$\begin{aligned} G(t) &= 1 - P\{x_{(1)}, b^{-1}(x_{(n)}) \geqslant t\} \\ &= 1 - P\{x_{(1)} \geqslant t, x_{(n)} \leqslant b(t)\} \\ &= 1 - \left\{\int_t^{b(t)} \frac{g(x)}{h(\theta)}\,dx\right\}^n. \end{aligned} \qquad (22.25)$$

Differentiating (22.25) with respect to t, we obtain the density of the sufficient statistic,

$$g(t) = \frac{-n}{\{h(\theta)\}^n}\left\{\int_t^{b(t)} g(x)\,dx\right\}^{n-1}[g\{b(t)\}b'(t) - g(t)], \qquad \theta \leqslant t \leqslant c \quad (22.26)$$

where c is determined by $c = b(c)$—this is the result of Exercise 20.6. If there is a statistic $u(t)$ for which

$$\int_\theta^c u(t)g(t)\,dt = 0, \qquad (22.27)$$

we find, on differentiating (22.27) with respect to θ and by following through the argument of **22.12**, that $u(\theta) = 0$ for any θ, as before. Thus $u(t) = 0$ identically and $g(t)$ at (22.26) is complete.

22.14 The following example is of a non-complete single sufficient statistic.

Example 22.3
Consider a sample of a single observation x from the uniform distribution

$$dF = dx, \qquad \theta \leqslant x \leqslant \theta + 1.$$

x is evidently a sufficient statistic. (There would be no single sufficient statistic for $n \geqslant 2$, since the condition (ii) of **22.11** is not satisfied.)
Any bounded periodic function $h(x)$ of period 1 which satisfies

$$\int_0^1 h(x)\,dx = 0$$

will give us

$$\int_\theta^{\theta+1} h(x)\,dF = \int_\theta^{\theta+1} h(x)\,dx = \int_0^1 h(x)\,dx = 0,$$

so that the distribution is not even boundedly complete, since $h(x)$ is not identically zero.

Minimal sufficiency
22.15 We recall from **17.38** that, when we consider the problem of sufficient statistics in general (i.e. without restricting ourselves, as we did initially in Chapter 17, to the case of a single sufficient statistic), we have to consider the choice between alternative sets of sufficient statistics. In a sample of n observations we *always* have a set of n sufficient statistics (namely, the observations themselves) for the $k(\geqslant 1)$ parameters of the distribution we are sampling from. Sometimes, though not always, there will be a set of $s(<n)$ statistics sufficient for the parameters. Often, $s = k$; e.g. all the cases of sufficiency discussed in Examples 17.15–16 have $s = k = 1$, while in Example 17.17 we have $s = k = 2$. By contrast, the following is an example in which $s < k$.*

Example 22.4
Consider again the problem of Example 21.11, with the alteration that

$$E(x_1) = n\mu$$

instead of $n\theta$ as previously. Exactly as at (21.48), we find for the joint distribution

$$dF = \frac{1}{\theta(2\pi)^{n/2}} \exp\left\{-\frac{1}{2}\left[\frac{n^2}{\theta^2}(\bar{x} - \mu)^2 + \sum_{i=2}^n x_i^2\right]\right\} dx_1 \cdots dx_n.$$

Here it is clear that the single statistic $\bar{x}$ is sufficient for the parameters μ, θ.
22.16 We thus have to ask ourselves: what is the *smallest* number s of statistics that constitute a sufficient set in any problem? With this in mind, Lehmann and

---

* Fisher (e.g. 1956) called a set of statistics 'sufficient' only if $s = k$ and 'exhaustive' if $s > k$.

Scheffé (1950) define a vector of statistics as *minimal sufficient* if it is a single-valued function of all other vectors of statistics that are sufficient for the parameters of the distribution.* The problems which now raise themselves are: how can we be sure, in any particular situation, that a sufficient vector is the minimal sufficient vector? And can we find a construction that yields the minimal sufficient vector?

A partial answer to the first of these questions is supplied by the following result: if the vector t is a boundedly complete sufficient statistic for θ, and the vector u is a minimal sufficient statistic for θ, then t is equivalent to u, i.e. they are identical, except possibly for a zero-measure set.

The proof is simple. Let w be a region in the sample space for which

$$D = E(c(w)|t) - E(c(w)|u) \neq 0, \tag{22.28}$$

where the function $c(w)$ is defined at (22.3). From (22.28), we find, on taking expectations over the entire sample space,

$$E(D) = 0. \tag{22.29}$$

Now since u is minimal sufficient, it is a function of t, another sufficient statistic, by definition. Hence we may write (22.28)

$$D = h(t) \neq 0. \tag{22.30}$$

Since D is a bounded function, (22.29) and (22.30) contradict the assumed bounded completeness of t, and thus there can be no region w for which (22.28) holds. Hence t and u are equivalent statistics, i.e. t is minimal sufficient.

The converse does not hold: while bounded completeness implies minimal sufficiency, we can have minimal sufficiency without bounded completeness. An important instance is discussed in Example 22.10 below.

A consequence of the result of this section is that there cannot be more than one boundedly complete sufficient statistic for a parameter. Thus in Example 22.2(b) x^2 is minimal sufficient and complete, while x is sufficient and not complete—cf. also Exercises 18.13 and 22.31 for other instances of minimal and non-minimal single sufficient statistics.

An alternative formulation of the problem of minimal sufficiency is given by Dynkin (1951), who uses the term 'necessary' rather than 'minimal'.

22.17   In view of the results of **22.10–13** concerning the completeness of sufficient statistics, a consequence of **22.16** is that all the examples of sufficient statistics we have discussed in earlier chapters are minimal sufficient, as one would expect on intuitive grounds.

Since, by **18.23**, a unique ML estimator $\hat{\theta}$ of θ is a function of any sufficient statistic for θ, it will necessarily be a function of the minimal sufficient statistic. Hence, if $\hat{\theta}$ is itself sufficient, it must be minimal sufficient.

---

* That this is for practical purposes equivalent to a sufficient statistic with minimum number of components is shown by Barankin and Katz (1959). See also Barankin (1960a, 1960b, 1961) and Fraser (1963).

22.18 The result of section **22.16**, though useful, is less direct than the following procedure for finding a minimal sufficient statistic, given by Lehmann and Scheffé (1950).

We have seen in **21.14** and **21.20** that in testing a simple hypothesis, the ratio of likelihoods is a function of the sufficient (set of) statistic(s). We may now, so to speak, put this result into reverse, and use it to find the minimal sufficient set. Writing $L(x|\theta)$ for the LF as before, where x and θ may be vectors, consider a particular set of values x_0 and select all those values of x within the permissible range for which $L(x|\theta)$ is non-zero and

$$\frac{L(x|\theta)}{L(x_0|\theta)} = k(x, x_0) \tag{22.31}$$

is independent of θ. Now any sufficient statistic t (possibly a vector) will satisfy (17.68), whence

$$\frac{L(x|\theta)}{L(x_0|\theta)} = \frac{g(t|\theta)}{g(t_0|\theta)} \cdot \frac{h(x)}{h(x_0)}, \tag{22.32}$$

so that if $t = t_0$, (22.32) reduces to the form (22.31). Conversely, if (22.31) holds for all θ, this implies the constancy of the sufficient statistic t at the value t_0. This may be used to identify sufficient statistics, and to select the minimal set, in the manner of the following examples.

Example 22.5
We saw in Example 17.17 that the set of statistics $(\bar{x}, s^2)$ is jointly sufficient for the parameters (μ, σ^2) of a normal distribution. For this distribution, $L(x|\theta)$ is non-zero for all $\sigma^2 > 0$, and the condition (22.31) is, on taking logarithms, that

$$-\frac{1}{2\sigma^2}\left\{\left(\sum_i x_i^2 - \sum_i x_{0i}^2\right) - 2\mu n(\bar{x} - \bar{x}_0)\right\} \tag{22.33}$$

be independent of (μ, σ^2), i.e. that the term in braces be equal to zero. This will be so, for example, if every x_i is equal to the corresponding x_{0i}, confirming that the set of n observations is a jointly sufficient set, as we have remarked that they always are.

It will also be so if the x_i are any rearrangement (permutation) of the x_{0i}: thus the set of *order-statistics* is sufficient, as it is again obvious that they always are. But in this example, we can go further, for (22.33) will be zero if we divide the observations into any l subsets and within each subset have

$$\sum x_i = \sum x_{0i} \quad \text{and} \quad \sum x_i^2 = \sum x_{0i}^2.$$

Thus the $2l$-dimensional statistic formed of the l subset pairs $(\sum x_i, \sum x_i^2)$ will be jointly sufficient. In particular, with $l = 1$, the condition on (22.33) will be satisfied if

$$\bar{x} = \bar{x}_0, \quad \sum_i x_i^2 = \sum_i x_{0i}^2, \tag{22.34}$$

and clearly, from inspection, nothing less than this will do. Thus the pair $(\bar{x}, \sum x^2)$ is minimal sufficient: equivalently, since $ns^2 = \sum x^2 - n\bar{x}^2$, $(\bar{x}, s^2)$ is minimal sufficient.

Example 22.6

As a contrast, consider the Cauchy distribution of Example 17.7. $L(x | \theta)$ is everywhere non-zero and (22.31) requires that

$$\prod_{i=1}^{n} \{1 + (x_{0i} - \theta)^2\} \Big/ \prod_{i=1}^{n} \{1 + (x_i - \theta)^2\} \tag{22.35}$$

be independent of θ. As in the previous example, the set of order-statistics is sufficient, but nothing less will do here, for (22.35) is the ratio of two polynomials, each of degree $2n$ in θ. If the ratio is to be independent of θ, each polynomial must have the same set of roots, possibly permuted. Thus we are thrown back on the order-statistics as the minimal sufficient set.

Completeness and similar regions

22.19   After our lengthy excursus on completeness, we return to the discussion of similar regions in **22.8**. We may now show that if, given H_0, the sufficient statistic t is boundedly complete, *all* size-α similar regions must satisfy (22.13). For any such region, (22.14) holds and may be re-written

$$E\{E(c(w) | t) - \alpha\} = 0. \tag{22.36}$$

The expression in braces in (22.36) is bounded. Thus if t is boundedly complete, (22.36) implies that $E(c(w) | t) - \alpha = 0$ identically, i.e. that (22.13) holds.

The converse result also holds: if all similar regions satisfy (22.13), then Lehmann and Scheffé (1950) proved that t must be boundedly complete. Thus the bounded completeness of a sufficient statistic is equivalent to the condition that all similar regions w satisfy (22.13).

The choice of most powerful similar regions

22.20   The importance of the result of **22.19** is that it permits us to reduce the problem of finding most powerful similar regions for a composite hypothesis to the familiar problem of finding a BCR for a simple hypothesis.

By **22.19**, the bounded completeness of the statistic t, sufficient for θ_s on H_0, implies that all similar regions w satisfy (22.13), i.e. every similar region is composed of a fraction α of the probability content of each contour of constant t. We therefore may conduct our discussion with t held constant. Constancy of the sufficient statistic, t, for θ_s implies from (17.68) that the conditional distribution of the observations in the sample space will be independent of θ_s. Thus the composite H_0 with θ_s unspecified is reduced to a simple H_0 with t held constant. If t is also sufficient for θ_s when H_1 holds, the composite H_1 is also reduced to a simple H_1 with t constant (and, incidentally, the power of any critical region with t constant, as well as its size, will be independent of θ_s). If, however, t is not sufficient for θ_s when H_1 holds, we consider H_1 as a class of simple alternatives to the simple H_0, in just the manner of the previous chapter.

Thus, by keeping t constant, we reduce the problem to that of testing a simple H_0 concerning θ_r against a simple H_1 (or a class of simple alternatives constituting

H_1). We use the methods of the previous chapter, based on the Neyman–Pearson lemma (21.6), to seek a BCR (or common BCR) for H_0 against H_1. If there is such a BCR for each fixed value of t, it will evidently be an unconditional BCR, and gives the most powerful similar test of H_0 against H_1. Just as previously, if this test remains most powerful against a class of alternative values of θ_r, it is a UMP similar test.

Example 22.7
To test $H_0: \mu = \mu_0$ against $H_1: \mu = \mu_1$ for the normal distribution

$$\mathrm{d}F = \frac{1}{\sigma(2\pi)^{1/2}} \exp\left\{-\frac{1}{2}\left(\frac{x-\mu}{\sigma}\right)^2\right\} \mathrm{d}x, \qquad -\infty < x < \infty.$$

H_0 and H_1 are composite, σ^2 being unspecified.

From Examples 17.10 and 17.15, the statistic (calculated from a sample of n independent observations) $u = \sum_{i=1}^{n} (x_i - \mu_0)^2$ is sufficient for σ^2 when H_0 holds, but not otherwise. From **22.10**, u is a complete statistic. All similar regions for H_0 therefore consist of fractions α of each contour of constant u.

Holding u fixed, we now test

$$H_0: \mu = \mu_0 \quad \text{against} \quad H_1': \mu = \mu_1, \sigma = \sigma_1,$$

both hypotheses being simple. The BCR obtained from (21.6) is that for which

$$\frac{L(x \mid H_0)}{L(x \mid H_1')} \leqslant k_\alpha.$$

This reduces, on simplification, to the condition

$$\bar{x}(\mu_1 - \mu_0) \geqslant C(\mu_0, \mu_1, \sigma^2, \sigma_1^2, k_\alpha, u) \tag{22.37}$$

where C is a constant containing no function of x except u. Thus the BCR consists of large values of $\bar{x}$ if $\mu_1 - \mu_0 > 0$ and of small values of $\bar{x}$ if $\mu_1 - \mu_0 < 0$, and this is true whatever the values of σ^2 and σ_1^2, and whatever the magnitude of $|\mu_1 - \mu_0|$. Thus we have a common BCR for the class of alternatives $H_1: \mu = \mu_1$ for each one-sided situation $\mu_1 > \mu_0$ and $\mu_1 < \mu_0$.

We have been holding u fixed. Now

$$u = \sum (x - \mu_0)^2 = \sum (x - \bar{x})^2 + n(\bar{x} - \mu_0)^2 \tag{22.38}$$

$$= \sum (x - \bar{x})^2 \left\{1 + \frac{n(\bar{x} - \mu_0)^2}{\sum (x - \bar{x})^2}\right\}. \tag{22.39}$$

Since the BCR for fixed u consists of extreme values of $\bar{x}$, (22.38) implies that the BCR consists of small values of $\sum (x - \bar{x})^2$, which by (22.39) implies large values of

$$\frac{t^2}{n-1} = \frac{n(\bar{x} - \mu_0)^2}{\sum (x - \bar{x})^2}. \tag{22.40}$$

t^2 as defined by (22.40) is the square of the Student's t statistic whose distribution was derived in Example 11.8. By Exercise 22.7, t, which is distributed free of σ^2, is distributed independently of the complete sufficient statistic, u, for σ^2. Remembering the necessary sign of $\bar{x}$, we have finally that the unconditional UMP similar test of H_0 against H_1 is to reject the largest or smallest 100α per cent of the distribution of t according to whether $\mu_1 > \mu_0$ or $\mu_1 < \mu_0$.

As we have seen, the distribution of t does not depend on σ^2. The power of the UMP similar test, however, does depend on σ^2, for u is not sufficient for σ^2 when H_0 does not hold. Since every similar region for H_0 consists of fractions α of each contour of constant u, and the distribution on any such contour is a function of σ^2 when H_1 holds, there can be no similar region for H_0 with power independent of σ^2, a result first established by Dantzig (1940). It is for this reason that we found in Example 20.2 that the length of the corresponding confidence interval for μ could not be fixed in advance.

Example 22.8
For two normal distributions with means μ, $\mu + \theta$ and common variance σ^2, to test

$$H_0: \theta = \theta_0 \ (=0, \text{ without loss of generality})$$

against

$$H_1: \theta = \theta_1$$

on the basis of independent samples of size n_1, n_2 with means $\bar{x}_1$, $\bar{x}_2$.
Write

$$\left. \begin{array}{l} n = n_1 + n_2, \\ n\bar{x} = n_1\bar{x}_1 + n_2\bar{x}_2, \\ s^2 = \sum\limits_{i=1}^{2} \sum\limits_{j=1}^{n_i} (x_{ij} - \bar{x})^2 = \sum\sum x_{ij}^2 - n\bar{x}^2. \end{array} \right\} \qquad (22.41)$$

The hypotheses are composite, two parameters being unspecified. When H_0 holds, but not otherwise, the pair of statistics $(\bar{x}, s^2)$ is sufficient for the unspecified parameters (μ, σ^2), and it follows from **22.10** that $(\bar{x}, s^2)$ is complete. Thus all similar regions for H_0 satisfy (22.13), and we hold $(\bar{x}, s^2)$ fixed, and test the simple

$$H_0: \theta = 0$$

against

$$H_1': \theta = \theta_1, \mu = \mu_1, \sigma = \sigma_1.$$

Our original H_1 consists of the class of H_1' for all μ_1, σ_1. The BCR obtained from (21.6) reduces, on simplification, to

$$\bar{x}_2\theta_1 \geqslant g_\alpha$$

where g_α is a constant function of all the parameters, and of $\bar{x}$ and s^2, but not otherwise of the observations. For fixed $\bar{x}$, s^2, the BCR is therefore characterised by extreme values of $\bar{x}_2$ of the same sign as θ_1, and this is true whatever the values of the other parameters. (22.41) then implies that for each fixed $(\bar{x}, s^2)$, the BCR will consist of large values of $(\bar{x}_1 - \bar{x}_2)^2/s^2$, and hence of the equivalent monotone increasing function

$$\frac{(\bar{x}_1 - \bar{x}_2)^2}{\sum(x_1 - \bar{x}_1)^2 + \sum(x_2 - \bar{x}_2)^2} = \frac{t^2}{(n-2)} \cdot \frac{n}{n_1 n_2}. \tag{22.42}$$

(22.42) is the definition of the usual Student's t^2 statistic for this problem, which we have encountered in the corresponding interval estimation problem in **20.22**. By Exercise 22.7, t^2, which is distributed free of μ and σ^2, is distributed independently of the complete sufficient statistic $(\bar{x}, s^2)$ for (μ, σ^2). Thus, unconditionally, the UMP similar test of H_0 against H_1 is given by rejecting the 100α per cent largest or smallest values in the distribution of t, according to whether θ_1 (or, more generally, $\theta_1 - \theta_0$) is positive or negative.

Here, as in the previous example, the power of the BCR depends on (μ, σ^2), since $(\bar{x}, s^2)$ is not sufficient when H_0 does not hold.

Example 22.9

To test the composite $H_0: \sigma = \sigma_0$ against $H_1: \sigma = \sigma_1$ for the distribution

$$dF = \exp\left\{-\left(\frac{x-\theta}{\sigma}\right)\right\} dx/\sigma, \qquad \theta \le x < \infty; \sigma > 0.$$

We have seen (Example 17.19) that $x_{(1)}$, the smallest of a sample of n independent observations, is sufficient for the unspecified parameter θ, whether H_0 or H_1 holds. By **22.12** it is also complete. Thus all similar regions consist of fractions α of each contour of constant $x_{(1)}$.

The comprehensive sufficiency of $x_{(1)}$ renders both H_0 and H_1 simple when $x_{(1)}$ is fixed. The BCR obtained from (21.6) consists of points satisfying

$$\sum_{i=1}^{n} x_i \left(\frac{1}{\sigma_1} - \frac{1}{\sigma_0}\right) \le g_\alpha,$$

where g_α is a constant, a function of σ_0, σ_1. For each fixed $x_{(1)}$, we therefore have the BCR defined by

$$\begin{aligned} \sum_{i=1}^{n} x_i &\le a_\alpha \quad \text{if } \sigma_1 < \sigma_0, \\ &\ge b_\alpha \quad \text{if } \sigma_1 > \sigma_0. \end{aligned} \tag{22.43}$$

The statistic in (22.43), $\sum x_i$, is not distributed independently of $x_{(1)}$. To put (22.43) in a form of more practical value, we observe that the statistic

$$z = \sum_{i=1}^{n} (x_{(i)} - x_{(1)})$$

is distributed independently of $x_{(1)}$. (This is a consequence of the completeness and sufficiency of $x_{(1)}$—see Exercise 22.7 below.) Thus if we rewrite (22.43) for fixed $x_{(1)}$ as

$$\left.\begin{array}{ll} z \leq c_\alpha, & \sigma_1 < \sigma_0, \\ \geq d_\alpha, & \sigma_1 > \sigma_0, \end{array}\right\} \tag{22.44}$$

where $c_\alpha = a_\alpha - n x_{(1)}$, $d_\alpha = b_\alpha - n x_{(1)}$, we have on the left of (22.44) a statistic which for every fixed $x_{(1)}$ determines the BCR by its extreme values and whose distribution does not depend on $x_{(1)}$. Thus (22.44) gives an unconditional BCR for each of the one-sided situations $\sigma_1 < \sigma_0$, $\sigma_1 > \sigma_0$, and we have the usual pair of UMP tests.

Note that in this example, the comprehensive sufficiency of $x_{(1)}$ makes the power of the UMP tests independent of the location parameter θ.

22.21 Examples 22.7 and 22.8 afford a sophisticated justification for two of the standard normal distribution test procedures for means. Exercises 22.13 and 22.14 at the end of this chapter, by following through the same argument, similarly justify two other standard procedures for variances, arriving in each case at a pair of UMP similar one-sided tests. Unfortunately, not all the problems of normal test theory are so tractable: the thorniest of them, the problem of two means that we discussed at length in Chapter 20, does not yield to the present approach, as the next example shows.

Example 22.10
For two normal distributions with means and variances (θ, σ_1^2), $(\theta + \mu, \sigma_2^2)$, to test $H_0 : \mu = 0$ on the basis of independent samples of n_1 and n_2 observations.

Given H_0, the sample means and variances $(\bar{x}_1, \bar{x}_2, s_1^2, s_2^2) = \mathbf{t}$ form a set of four jointly sufficient statistics for the three parameters $\theta, \sigma_1^2, \sigma_2^2$ left unspecified by H_0. They may be seen to be minimal sufficient by use of (22.31)—cf. Lehmann and Scheffé (1950). But $\mathbf{t}$ is not boundedly complete, since $\bar{x}_1, \bar{x}_2$ are normally distributed independently of s_1^2, s_2^2 and of each other, so that any bounded odd function of $(\bar{x}_1 - \bar{x}_2)$ alone will have zero expectation. We therefore cannot rely on (22.13) to find all similar regions, though regions satisfying (22.13) would certainly be similar, by **22.8**. But it is easy to see, from the fact that the likelihood function contains the four components of $\mathbf{t}$ and no other functions of the observations, that any region consisting entirely of a fraction α of each surface of constant $\mathbf{t}$ will have the same probability content in the sample space *whatever the value of μ*, and will therefore be an ineffective critical region with power exactly equal to its size. This disconcerting aspect of a familiar and useful property of normal distributions was pointed out by Watson (1957a).

No useful exact unrandomised similar regions exist for this problem—see Linnik (1964, 1967) and Pfanzagl (1974)—but for all practical purposes α is constant for Welch's method, expounded in **20.33** as an interval estimation technique; if we are prepared to introduce an element of randomisation, Scheffé's method of **20.24–30** is available.

22.22 The discussion of **22.20** and Examples 22.8–10 make it clear that, if there is a complete sufficient statistic for the unspecified parameter, the problem of selecting a most powerful test for a composite hypothesis is considerably reduced if we restrict our choice to similar regions. But something may be lost by this—for specific alternatives there may be a non-similar test, satisfying (22.6), with power greater than the most powerful similar test.

Lehmann and Stein (1948) considered this problem for the composite hypotheses considered in Example 22.7 and Exercise 22.13. In the former, where we are testing the mean of a normal distribution, they found that if $\alpha \geqslant \frac{1}{2}$ there is no non-similar test more powerful than Student's t, whatever the true values μ_1, σ_1, but that for $\alpha < \frac{1}{2}$ (as in practice it always is) there is a more powerful critical region, which is of form

$$\sum_i \{x_i - c_\alpha(\mu_1, \sigma_1)\}^2 \leqslant k_\alpha(\mu_1, \sigma_1). \tag{22.45}$$

Similarly, for the variance of a normal distribution (Exercise 22.13 below), they found that if $\sigma_1 > \sigma_0$ no more powerful non-similar test exists, but if $\sigma_1 < \sigma_0$ the region

$$\sum_i (x_i - \mu_1)^2 \leqslant k_\alpha \tag{22.46}$$

is more powerful than the best similar critical region.

Thus if we restrict the alternative class H_1 sufficiently, we can sometimes improve the power of the test, while reducing the average value of the Type I error below the size α, by abandoning the requirement of similarity. In practice, this is not a very strong argument against using similar regions, precisely because we are not usually in a position to be very restrictive about the alternatives to a composite hypothesis. At the same time we must admit that one of the reasons why the criterion of similarity was so widely adopted was that it simplified the mathematical problems, just as did use of the arithmetic mean and variance rather than other measures of location and dispersion.

Bias in tests

22.23 In **21.26–8**, we briefly discussed the problem of testing a simple H_0 against a two-sided class of alternatives, where no UMP test generally exists. We now return to this subject from another viewpoint, although the two-sided nature of the alternative hypothesis is not essential to our discussion, as we shall see.

Example 22.11
Consider again the problem of Examples 21.2–3 and of **21.27**, that of testing the mean μ of a normal population with known variance, taken as unity for convenience. Suppose that we restrict ourselves to tests based on the distribution of the sample mean $\bar{x}$, as we may do by **22.3** since $\bar{x}$ is sufficient. Generalising (21.55), consider the size-α region defined by

$$\bar{x} \leqslant a_{\alpha_1}, \quad \bar{x} \geqslant b_{\alpha_2}, \tag{22.47}$$

where $\alpha_1 + \alpha_2 = \alpha$, and α_1 is not now necessarily equal to α_2. a and b are defined, as at (21.15), by

$$a_\alpha = \mu_0 - d_\alpha/n^{1/2} \qquad b_\alpha = \mu_0 + d_\alpha/n^{1/2},$$

and

$$G(-d_\alpha) = \int_{-\infty}^{-d_\alpha} (2\pi)^{-1/2} \exp\left(-\tfrac{1}{2}y^2\right) dy = \alpha.$$

We take $d_\alpha > 0$ without loss of generality.

Exactly as at (21.56), the power of the critical region (22.47) is seen to be

$$P = G\{n^{1/2}\Delta - d_{\alpha_2}\} + G\{-n^{1/2}\Delta - d_{\alpha_1}\}, \tag{22.48}$$

where $\Delta = \mu_1 - \mu_0$.

We consider the power (22.48) as a function of Δ. Its first two derivatives are

$$P' = \left(\frac{n}{2\pi}\right)^{1/2} \left[\exp\left\{-\tfrac{1}{2}(n^{1/2}\Delta - d_{\alpha_2})^2\right\} - \exp\left\{-\tfrac{1}{2}(n^{1/2}\Delta + d_{\alpha_1})^2\right\}\right] \tag{22.49}$$

and

$$P'' = \frac{n}{(2\pi)^{1/2}} \left[(d_{\alpha_2} - n^{1/2}\Delta) \exp\left\{-\tfrac{1}{2}(n^{1/2}\Delta - d_{\alpha_2})^2\right\}\right.$$
$$\left. + (n^{1/2}\Delta + d_{\alpha_1}) \exp\left\{-\tfrac{1}{2}(n^{1/2}\Delta + d_{\alpha_1})^2\right\}\right]. \tag{22.50}$$

From (22.49), we can only have $P' = 0$ if

$$\Delta = (d_{\alpha_2} - d_{\alpha_1})/(2n^{1/2}). \tag{22.51}$$

When (22.51) holds, we have from (22.50)

$$P'' = \frac{n}{(2\pi)^{1/2}} (d_{\alpha_1} + d_{\alpha_2}) \exp\left\{-\tfrac{1}{8}(d_{\alpha_2} + d_{\alpha_1})^2\right\}. \tag{22.52}$$

Since we have taken d_α always positive, we therefore have $P'' > 0$ at the stationary value, which is therefore a minimum. From (22.51), it occurs at $\Delta = 0$ only when $\alpha_1 = \alpha_2$, the case discussed in **21.27**. Otherwise, the unique minimum occurs at some value μ_m where

$$\mu_m > \mu_0 \quad \text{if } \alpha_1 > \alpha_2, \qquad \mu_m < \mu_0 \quad \text{if } \alpha_1 < \alpha_2.$$

22.24 The implication of Example 22.11 is that, except when $\alpha_1 = \alpha_2$, there exist values of μ in the alternative class H_1 for which the probability of rejecting H_0 is actually smaller when H_0 is false than when it is true. (Note that if we were considering a one-sided class of alternatives (say, $\mu_1 > \mu_0$), the same situation would arise if we used the critical region located in the wrong tail of the distribution of $\bar{x}$ (say, $\bar{x} \leq a_\alpha$).) It is clearly undesirable to use a test which is more likely to reject the hypothesis when it is true than when it is false. In fact, we can improve on such a test by using a table of random numbers to reject the hypothesis with probability α—the power of this procedure will always be α.

We may now generalise our discussion. If a size-α critical region w for $H_0: \theta = \theta_0$ against the simple $H_1: \theta = \theta_1$ is such that its power

$$P\{\mathbf{x} \in w \mid \theta_1\} \geq \alpha, \tag{22.53}$$

it is said to give an *unbiased** test of H_0 against H_1; in the contrary case, the region w, and the test it provides, are said to be *biased.** If H_1 is composite, and (22.53) holds for every member of H_1, w is said to be an unbiased critical region against H_1. It should be noted that unbiasedness does not require that the power function should actually have a regular minimum at θ_0, as we found to be the case in Example 22.11 when $\alpha_1 = \alpha_2$, although this is often found to be so in practice. Figure 21.3 in **21.27** illustrates the appearance of the power function for an unbiased test (the full line) and two biased tests.

> If no unbiased test exists, there may be a 'locally unbiased Type M' test (Krishnan, 1966) which has average power $\geq \alpha$ in a neighbourhood of H_0.

The criterion of unbiasedness for tests has such strong intuitive appeal that it is natural to restrict oneself to the class of unbiased tests when investigating a problem, and to seek UMP unbiased (UMPU) tests, which may exist even against two-sided alternative hypotheses, for which we cannot generally hope to find UMP tests without some restriction on the class of tests considered. Thus, in Example 22.11, the equal-tails test based on $\bar{x}$ is at once seen to be UMPU in the class of tests there considered. That it is actually UMPU among all tests of H_0 will be seen in **22.33**.

Example 22.12
We have left over to Exercise 22.13 the result that, for a normal distribution with mean μ and variance σ^2, the statistic $z = \sum_{i=1}^{n} (x_i - \bar{x})^2$ gives a pair of one-sided UMP similar tests of the hypothesis $H_0: \sigma^2 = \sigma_0^2$, the BCR being

$$z \geq a_{1-\alpha} \quad \text{if } \sigma_1 > \sigma_0, \qquad z \leq a_\alpha \quad \text{if } \sigma_1 < \sigma_0.$$

Now consider the two-sided alternative hypothesis

$$H_1: \sigma^2 \neq \sigma_0^2.$$

By **21.18** there is no UMP test of H_0 against H_1, but we are intuitively tempted to use the statistic z, splitting the critical region equally between its tails in the hope of achieving unbiasedness, as in Example 22.11. Thus we reject H_0 if

$$z \geq a_{1-\alpha/2} \quad \text{or} \quad z \leq a_{\alpha/2}.$$

This critical region is certainly similar, for the distribution of z is not dependent on μ, the nuisance parameter. Since z/σ^2 has a chi-squared distribution with $(n-1)$ d.fr., whether H_0 or H_1 holds, we have

$$a_{1-\alpha/2} = \sigma_0^2 \chi^2_{1-\alpha/2}, \quad a_{\alpha/2} = \sigma_0^2 \chi^2_{\alpha/2},$$

---

* This use of 'bias' is unconnected with that of the theory of estimation, and is only prevented from being confusing by the fortunate fact that the context rarely makes confusion possible.

where χ_{α}^2 is the 100α per cent point of that chi-squared distribution. When H_1 holds, it is z/σ_1^2 that has this distribution, and H_0 will then be rejected when

$$\frac{z}{\sigma_1^2} \geqslant \frac{\sigma_0^2}{\sigma_1^2}\chi_{1-\alpha/2}^2 \quad \text{or} \quad \frac{z}{\sigma_1^2} \leqslant \frac{\sigma_0^2}{\sigma_1^2}\chi_{\alpha/2}^2.$$

The power of the test against any alternative value σ_1^2 is the sum of the probabilities of these two events. We thus require the probability that a chi-squared variable will fall outside its $100(\tfrac{1}{2}\alpha)$ per cent and $100(1-\tfrac{1}{2}\alpha)$ per cent points each multiplied by a constant σ_0^2/σ_1^2. For each value of α and $(n-1)$, the degrees of freedom, this probability can be calculated from a table of the distribution for each value of σ_0^2/σ_1^2. Figure 22.1 shows the power function resulting from such calculations by Neyman and Pearson (1936b) for the case $n=3$, $\alpha=0.02$. The power is less than α in this case when $0.5 < \sigma_1^2/\sigma_0^2 < 1$, and the test is therefore biased.

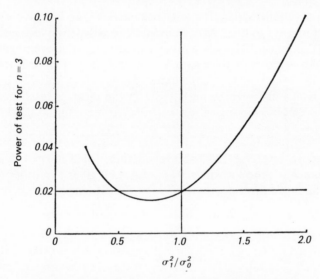

Fig. 22.1   Power function of a test for a normal distribution variance (see text)

We now enquire whether, by modifying the apportionment of the critical region between the tails of the distribution of z, we can remove the bias. Suppose that the critical region is

$$z \geqslant a_{1-\alpha_1} \quad \text{or} \quad z \leqslant a_{\alpha_2},$$

where $\alpha_1 + \alpha_2 = \alpha$. As before, the power of the test is the probability that a chi-squared variable with $(n-1)$ degrees of freedom, say y_{n-1}, falls outside the range of its $100\alpha_2$ per cent and $100(1-\alpha_1)$ per cent points, each multiplied by the constant $\theta = \sigma_0^2/\sigma_1^2$. Writing F for the distribution function of y_{n-1}, we have

$$P = F(\theta\chi_{\alpha_2}^2) + 1 - F(\theta\chi_{1-\alpha_1}^2). \tag{22.54}$$

Regarded as a function of θ, this is the power function. We now choose α_1 and α_2 so that this power function has a regular minimum at $\theta = 1$, where it equals the size of the test. Differentiating (22.54), we have

$$P' = \chi^2_{\alpha_2} f(\theta \chi^2_{\alpha_2}) - \chi^2_{1-\alpha_1} f(\theta \chi^2_{1-\alpha_1}), \tag{22.55}$$

where f is the density of y_{n-1}. If this is to be zero when $\theta = 1$, we require

$$\chi^2_{\alpha_2} f(\chi^2_{\alpha_2}) = \chi^2_{1-\alpha_1} f(\chi^2_{1-\alpha_1}). \tag{22.56}$$

Substituting

$$f(y) \propto e^{-y/2} y^{(n-3)/2}, \tag{22.57}$$

we have finally from (22.56) the condition for unbiasedness

$$\left\{ \frac{\chi^2_{1-\alpha_1}}{\chi^2_{\alpha_2}} \right\}^{(n-1)/2} = \exp\left\{ \tfrac{1}{2}(\chi^2_{1-\alpha_1} - \chi^2_{\alpha_2}) \right\}. \tag{22.58}$$

Values of α_1 and α_2 satisfying (22.58) will give a test whose power function has zero derivative at the origin. To investigate whether it is strictly unbiased, we write (22.55), using (22.57) and (22.58), as

$$P' = c\theta^{(n-3)/2} \chi^{n-1}_{\alpha_2} \exp\left(-\tfrac{1}{2}\chi^2_{\alpha_2}\right)[\exp\{\tfrac{1}{2}\chi^2_{\alpha_2}(1-\theta)\} - \exp\{\tfrac{1}{2}\chi^2_{1-\alpha_1}(1-\theta)\}], \tag{22.59}$$

where c is a positive constant. Since $\chi^2_{1-\alpha_1} > \chi^2_{\alpha_2}$, we have from (22.59)

$$P' \begin{cases} <0, & \theta < 1, \\ =0, & \theta = 1, \\ >0, & \theta > 1. \end{cases} \tag{22.60}$$

(22.60) shows that the test with α_1, α_2 determined by (22.58) is unbiased in the strict sense, for the power function is monotonic decreasing as θ increases from 0 to 1 and monotonic increasing as θ increases from 1 to ∞.

Tables of the values of $\chi^2_{\alpha_2}$ and $\chi^2_{1-\alpha_1}$ satisfying (22.58) are given by Ramachandran (1958) for $\alpha = 0.05$ and $n - 1 = 2$ (1) 8 (2) 24, 30, 40 and 60; other tables are described in **20.34**(3), where the terminology of confidence intervals is used. Table 22.1 compares

Table 22.1   Limits outside which the chi-squared variable $\sum (x - \bar{x})^2/\sigma^2_0$ must fall for $H_0: \sigma^2 = \sigma^2_0$ to be rejected ($\alpha = 0.05$)

Degrees of freedom $(n-1)$	Unbiased test limits	Equal-tails test limits	Differences
2	(0.08, 9.53)	(0.05, 7.38)	(0.03, 2.15)
5	(0.99, 14.37)	(0.83, 12.83)	(0.16, 1.54)
10	(3.52, 21.73)	(3.25, 20.48)	(0.27, 1.25)
20	(9.96, 35.23)	(9.59, 34.17)	(0.37, 1.06)
30	(17.21, 47.96)	(16.79, 46.98)	(0.42, 0.98)
40	(24.86, 60.32)	(24.43, 59.34)	(0.43, 0.98)
60	(40.93, 84.23)	(40.48, 83.30)	(0.45, 0.93)

some of Ramachandran's values with the corresponding limits for the biased equal-tails test that we have considered, obtained from the *Biometrika Tables*.

It will be seen that the differences in both limits are proportionately large for small n, that the lower limit difference increases steadily with n, and the larger limit difference decreases steadily with n. At $n - 1 = 60$, both differences are just over 1 per cent of the values of the limits.

We defer the question whether the unbiased test is UMPU to Example 22.14 below.

Unbiased tests and similar tests

22.25   There is a close connection between unbiasedness and similarity, which often leads to the best unbiased test emerging directly from an analysis of the similar regions for a problem.

We consider a more general form of hypothesis than (22.2), namely

$$H_0': \theta_r \leq \theta_{r0}, \tag{22.61}$$

which is to be tested against

$$H_1: \theta_r > \theta_{r0}. \tag{22.62}$$

If we can find a critical region w satisfying (22.6) for all θ_r in H_0' as well as for all values of the unspecified parameters θ_s, i.e.

$$P(H_0', \theta_s) \leq \alpha \tag{22.63}$$

(where P is the power function whose value is the probability of rejecting H_0), the test based on w will be of size α as before. If it is also unbiased, we have from (22.53)

$$P(H_1, \theta_s) \geq \alpha. \tag{22.64}$$

Now if the power function P is a continuous function of θ_r, (22.63) and (22.64) imply, in view of the form of H_0' and H_1,

$$P(\theta_{r0}, \theta_s) = \alpha, \tag{22.65}$$

i.e. that w is a similar critical region for the 'boundary' hypothesis

$$H_0: \theta_r = \theta_{r0}.$$

All unbiased tests of H_0' are similar tests of H_0. If we confine our discussions to similar tests of H_0, using the methods we have encountered, and find a test with optimum properties—e.g. a UMP similar test—then *provided that this test is unbiased* it will retain the optimum properties in the class of unbiased tests of H_0'—e.g. it will be a UMPU test.

Exactly the same argument holds if H_0' specifies that the parameter point θ_r lies within a certain region R (which may consist of a number of subregions) in the parameter space, and H_1 that the θ_r lies in the remainder of that space: if the power function is continuous in θ_r, then if a critical region w is unbiased for testing H_0', it is a similar region for testing the hypothesis H_0 that θ_r lies on the boundary of R. If

w gives an unbiased test of H'_0, it will carry over into the class of unbiased tests of H'_0 any optimum properties it may have as a similar test of H_0. There will not always be a UMP similar test of H_0 if the alternatives are two-sided: a UMPU test may exist against such alternatives, but it must be found by other methods.

The early work on unbiased tests was largely carried out by Neyman and Pearson in the series of papers mentioned in **21.1**, and by Neyman (1935, 1938b), Scheffé (1942a) and Lehmann (1947). Much of the detail of their work has now been superseded, as pioneering work usually is, but their terminology is still commonly used.

Example 22.13
We return to the hypothesis of Example 22.12. One-sided critical regions based on the statistic $z \geqslant a_{1-\alpha}$, $z \leqslant a_\alpha$, give UMP similar tests against one-sided alternatives. Each of them is easily seen to be unbiased in testing one of

$$H'_0: \sigma^2 \leqslant \sigma_0^2, \qquad H''_0: \sigma^2 \geqslant \sigma_0^2$$

respectively against

$$H'_1: \sigma^2 > \sigma_0^2, \qquad H''_1: \sigma^2 < \sigma_0^2.$$

Thus they are, by the argument of **22.25**, UMPU tests for these one-sided situations.

For the two-sided alternative $H_1: \sigma^2 \neq \sigma_0^2$, the unbiased test based on (22.58) cannot be shown to be UMPU by this method, since we have not shown it to be UMP similar—see Example 22.14 below.

UMPU tests for the exponential family
22.26 We now give an account of some remarkably comprehensive results, due to Lehmann and Scheffé (1955), which establish the existence of, and give a construction for, UMPU tests for a variety of parametric hypotheses in distributions belonging to the exponential family (17.86).

We write the joint distribution of n independent observations from such a distribution as

$$f(\mathbf{x}) = D(\boldsymbol{\tau})h(\mathbf{x}) \exp \left\{ \sum_{j=1}^{r+1} b_j(\boldsymbol{\tau})u_j(\mathbf{x}) \right\}, \qquad (22.66)$$

where $\mathbf{x}$ is the column vector $(x_1, \ldots, x_n)$ and $\boldsymbol{\tau}$ is a vector of $(r+1)$ parameters $(\tau_1, \ldots, \tau_{r+1})$. In matrix notation, the exponent in (22.66) may be concisely written $\mathbf{u}'\mathbf{b}$, where $\mathbf{u}$ and $\mathbf{b}$ are column vectors.

Suppose now that we are interested in the particular linear function of the parameters

$$\theta = \sum_{j=1}^{r+1} a_{j1}b_j(\boldsymbol{\tau}), \qquad (22.67)$$

where $\sum_{j=1}^{r+1} a_{j1}^2 = 1$. Write $\mathbf{A}$ for an orthogonal matrix (a_{uv}) whose first column contains the coefficients in (22.67), and transform to a new vector of $(r+1)$ parameters $(\theta, \boldsymbol{\psi})$, where $\boldsymbol{\psi}$ is the column vector $(\psi_1, \ldots, \psi_r)$, by the equation

$$\begin{pmatrix} \theta \\ \boldsymbol{\psi} \end{pmatrix} = \mathbf{A'b}. \tag{22.68}$$

The first row of (22.68) is (22.67). We now suppose that there is a column vector of statistics $\mathbf{T} = (s, t_1, \ldots, t_r)$ defined by the relation

$$\mathbf{T'} \begin{pmatrix} \theta \\ \boldsymbol{\psi} \end{pmatrix} = \mathbf{u'b}, \tag{22.69}$$

i.e. we suppose that the exponent in (22.66) may be expressed as $\theta s(\mathbf{x}) + \sum_{j=1}^{r} \psi_j t_j(\mathbf{x})$. Using (22.68), (22.69) becomes

$$\mathbf{T'} \begin{pmatrix} \theta \\ \boldsymbol{\psi} \end{pmatrix} = \mathbf{u'A} \begin{pmatrix} \theta \\ \boldsymbol{\psi} \end{pmatrix}. \tag{22.70}$$

(22.70) is an identity in $(\theta, \boldsymbol{\psi})$, so we have $\mathbf{T'} = \mathbf{u'A}$ or

$$\mathbf{T} = \mathbf{A'u}. \tag{22.71}$$

Comparing (22.71) with (22.68), we see that each component of $\mathbf{T}$ is the same function of the $u_j(\mathbf{x})$ as the corresponding component of $(\theta, \boldsymbol{\psi})$ is of the $b_j(\boldsymbol{\tau})$. In particular, the first component is, from (22.67),

$$s(x) = \sum_{j=1}^{r+1} a_{j1} u_j(\mathbf{x}) \tag{22.72}$$

while the $t_j(\mathbf{x})$, $j = 1, 2, \ldots, r$, are orthogonal to $s(\mathbf{x})$.

Note that the orthogonality condition $\sum_{j=1}^{r+1} a_{j1}^2 = 1$ does not hamper us in testing hypotheses about θ defined by (22.67), since only a constant factor need be changed and the hypothesis adjusted accordingly.

22.27   If, therefore, we can reduce a hypothesis-testing problem (usually through its sufficient statistics) to the standard form of one concerning θ in

$$f(\mathbf{x} \mid \theta, \boldsymbol{\psi}) = C(\theta, \boldsymbol{\psi}) h(\mathbf{x}) \exp \left\{ \theta s(\mathbf{x}) + \sum_{i=1}^{r} \psi_i t_i(\mathbf{x}) \right\}, \tag{22.73}$$

by the device of the previous section, we can avail ourselves of the results summarised in **22.10**: given a hypothesis value for θ, the r-component vector $\mathbf{t} = (t_1, \ldots, t_r)$ will be a complete sufficient statistic for the r-component parameter $\boldsymbol{\psi} = (\psi_1, \ldots, \psi_r)$, and we now consider the problem of using s and $\mathbf{t}$ to test various composite hypotheses concerning θ, $\boldsymbol{\psi}$ being an unspecified ('nuisance') parameter. Simple hypotheses are the special case when $r = 0$, with no nuisance parameter.

22.28   For this purpose we shall need an extended form of the Neyman–Pearson lemma of **21.10**. Let $f(\mathbf{x} \mid \boldsymbol{\theta})$ be a frequency function, and $\boldsymbol{\theta}_i$ a subset of admissible values of the vector of parameters $\boldsymbol{\theta}$ $(i = 1, 2, \ldots, k)$. A specific element of $\boldsymbol{\theta}_i$ is written

θ_i^0. θ^* is a particular value of θ. The vector $\mathbf{u}_i(\mathbf{x})$ is sufficient for θ when θ is in θ_i and its distribution is $g_i(\mathbf{u}_i | \theta_i)$. Since the likelihood function factorises in the presence of sufficiency, the conditional value of $f(\mathbf{x}|\theta_i)$, given $\mathbf{u}_i$, will be independent of θ_i, and we write it $f(\mathbf{x}|\mathbf{u}_i)$. Finally, we define $l_i(\mathbf{x})$, $m_i(\mathbf{u}_i)$ to be non-negative functions of the observations and of $\mathbf{u}_i$ respectively.

22.29   Now suppose we have a critical region w for which

$$\int_w \{l_i(\mathbf{x})f(\mathbf{x}|\mathbf{u}_i)\}\, d\mathbf{x} = \alpha_i. \qquad (22.74)$$

Since the product in braces in (22.74) is non-negative, it may be regarded as a frequency function, and we may say that the conditional size of w, given $\mathbf{u}_i$, is α_i with respect to this distribution. We now write

$$\beta_i = \alpha_i \int m_i(\mathbf{u}_i)g_i(\mathbf{u}_i | \theta_i^0)\, d\mathbf{u}_i$$

$$= \int_w l_i(\mathbf{x})m_i(\mathbf{u}_i)\left\{ \int f(\mathbf{x}|\mathbf{u}_i)g_i(\mathbf{u}_i | \theta_i^0)\, d\mathbf{u}_i \right\} d\mathbf{x}$$

$$= \int_w \{l_i(\mathbf{x})m_i(\mathbf{u}_i)f(\mathbf{x}|\theta_i^0)\}\, d\mathbf{x}. \qquad (22.75)$$

The product in braces is again essentially a frequency function, say $p(\mathbf{x}|\theta_i^0)$. To test the simple hypothesis that $p(\mathbf{x}|\theta_i^0)$ holds against the simple alternative that $f(\mathbf{x}|\theta^*)$ holds, we use (21.6) and find that the BCR w of size β_i consists of points satisfying

$$[f(\mathbf{x}|\theta^*)]/[p(\mathbf{x}|\theta_i^0)] \geqslant c_i(\beta_i), \qquad (22.76)$$

where c_i is a non-negative constant. (22.76) will hold for every value of i. Thus for testing the composite hypothesis that *any* of $p(\mathbf{x}|\theta_i^0)$ holds $(i = 1, 2, \dots, k)$, we require all k of the inequalities (22.76) to be satisfied by w. If we now write $km_i(\mathbf{u}_i)/c_i(\beta_i)$ for $m_i(\mathbf{u}_i)$ in $p(\mathbf{x}|\theta_i^0)$, as we may since $m_i(\mathbf{u}_i)$ is arbitrary, we have from (22.76), adding the inequalities for $i = 1, 2, \dots, k$, the necessary and sufficient condition for a BCR

$$f(\mathbf{x}|\theta^*) \geqslant \sum_{i=1}^{k} l_i(\mathbf{x})m_i(\mathbf{u}_i)f(\mathbf{x}|\theta_i^0). \qquad (22.77)$$

This is the required generalisation. (21.6) is its special case with $k = 1$, $l_1(\mathbf{x}) = k_\alpha$ (constant), $m_1(u_1) \equiv 1$. (22.77) will play a role for composite hypotheses similar to that of (21.6) for simple hypotheses.

One-sided alternatives
22.30   Reverting to (22.73), we now investigate the problem of testing

$$H_0^{(1)}: \theta \leqslant \theta_0$$

against

$$H_1^{(1)}: \theta > \theta_0,$$

which we discussed in general terms in **22.25**. Now that we are dealing with the exponential form (22.73), we can show that there is always a UMPU test of $H_0^{(1)}$ against $H_1^{(1)}$. By our discussion in **22.25**, if a size-α critical region is unbiased for $H_0^{(1)}$ against $H_1^{(1)}$, it is a similar region for testing $\theta = \theta_0$.

Consider testing the simple

$$H_0: \theta = \theta_0, \qquad \boldsymbol{\psi} = \boldsymbol{\psi}^0$$

against the simple

$$H_1: \theta = \theta^* > \theta_0, \qquad \boldsymbol{\psi} = \boldsymbol{\psi}^*.$$

We now apply the result of **22.29**. Putting $k = 1$, $l_1(\mathbf{x}) \equiv 1$, $\alpha_1 = \alpha$, $\boldsymbol{\theta} = (\theta, \boldsymbol{\psi})$, $\boldsymbol{\theta}_1 = (\theta_0, \boldsymbol{\psi})$, $\boldsymbol{\theta}^* = (\theta^*, \boldsymbol{\psi}^*)$, $\boldsymbol{\theta}_1^0 = (\theta_0, \boldsymbol{\psi}^0)$, $u_1 = \mathbf{t}$, we have the result that the BCR for testing H_0 against H_1 is defined from (22.77) and (22.73) as

$$\frac{C(\theta^*, \boldsymbol{\psi}^*) \exp\{\theta^* s(\mathbf{x}) + \sum_{i=1}^{r} \psi_i^* t_i(\mathbf{x})\}}{C(\theta_0, \boldsymbol{\psi}^0) \exp\{\theta_0 s(\mathbf{x}) + \sum_{i=1}^{r} \psi_i^0 t_i(\mathbf{x})\}} \geq m_1(\mathbf{t}). \tag{22.78}$$

This may be rewritten

$$s(\mathbf{x})(\theta^* - \theta_0) \geq c_\alpha(\mathbf{t}, \theta^*, \theta_0, \boldsymbol{\psi}^*, \boldsymbol{\psi}^0). \tag{22.79}$$

We now see that c_α is not a function of $\boldsymbol{\psi}$, for since, by **22.27**, $\mathbf{t}$ is a sufficient statistic for $\boldsymbol{\psi}$ when H_0 holds, the value of c_α for given $\mathbf{t}$ will be independent of $\boldsymbol{\psi}^0, \boldsymbol{\psi}^*$. Further, from (22.79) we see that so long as the sign of $(\theta^* - \theta_0)$ does not change, the BCR will consist of the largest 100α per cent of the distribution of $s(\mathbf{x})$ given θ_0. We thus have a BCR for $\theta = \theta_0$ against $\theta > \theta_0$, giving a UMP test. This UMP test cannot have smaller power than a randomised test against $\theta > \theta_0$ which ignores the observations. The latter test has power equal to its size α, so the UMP test is unbiased against $\theta > \theta_0$, i.e. by **22.25** it is UMPU. Its size for $\theta < \theta_0$ will not exceed its size at θ_0, as is evident from the consideration that the critical region (22.79) has *minimum* power against $\theta < \theta_0$ and therefore its power (size) there is less than α. Thus finally we have shown that the largest 100α per cent of the conditional distribution of $s(\mathbf{x})$, given $\mathbf{t}$, gives a UMPU size-α test of $H_0^{(1)}$ against $H_1^{(1)}$.

Two-sided alternatives

22.31   We now consider the problem of testing

$$H_0^{(2)}: \theta = \theta_0$$

against

$$H_1^{(2)}: \theta \neq \theta_0.$$

Our earlier examples stopped short of establishing UMPU tests for two-sided hypotheses of this kind (cf. Examples 22.12 and 22.13). Nevertheless a UMPU test does exist for the exponential form (22.73).

From **22.25** we have that if the power function of a critical region is continuous in θ, and unbiased, it is similar for $H_0^{(2)}$. Now for any region w, the power function is

$$P(w \mid \theta) = \int_w f(\mathbf{x} \mid \theta, \boldsymbol{\psi}) \, d\mathbf{x}, \qquad (22.80)$$

where f is defined by (22.73). (22.80) is continuous and differentiable under the integral sign with respect to θ. For the test based on the critical region w to be unbiased we must therefore have, for each value of $\boldsymbol{\psi}$, the necessary condition

$$P'(w \mid \theta_0) = 0. \qquad (22.81)$$

Differentiating (22.80) under the integral sign and using (22.73) and (22.81), we find the condition for unbiasedness

$$0 = \int_w \left[s(\mathbf{x}) + \frac{C'(\theta_0, \boldsymbol{\psi})}{C(\theta_0, \boldsymbol{\psi})} \right] f(\mathbf{x} \mid \theta_0, \boldsymbol{\psi}) \, d\mathbf{x}$$

or

$$E\{s(\mathbf{x}) c(w)\} = -\alpha C'(\theta_0, \boldsymbol{\psi}) / C(\theta_0, \boldsymbol{\psi}). \qquad (22.82)$$

Since, from (22.73),

$$1/C(\theta, \boldsymbol{\psi}) = \int h(\mathbf{x}) \exp \left\{ \theta s(\mathbf{x}) + \sum_i \psi_i t_i(\mathbf{x}) \right\} d\mathbf{x}$$

we have

$$\frac{C'(\theta, \boldsymbol{\psi})}{C(\theta, \boldsymbol{\psi})} = -E\{s(\mathbf{x})\}, \qquad (22.83)$$

and putting (22.83) into (22.82) gives

$$E\{s(\mathbf{x}) c(w)\} = \alpha E\{s(\mathbf{x})\}. \qquad (22.84)$$

Taking the expectation first conditionally upon the value of $\mathbf{t}$, and then unconditionally, (22.84) gives

$$E_t[E\{s(\mathbf{x}) c(w) - \alpha s(\mathbf{x}) \mid \mathbf{t}\}] = 0. \qquad (22.85)$$

Since $\mathbf{t}$ is complete, (22.85) implies

$$E\{s(\mathbf{x}) c(w) - \alpha s(\mathbf{x}) \mid \mathbf{t}\} = 0 \qquad (22.86)$$

and since all similar regions for H_0 satisfy

$$E\{c(w) \mid \mathbf{t}\} = \alpha, \qquad (22.87)$$

(22.86) and (22.87) combine into

$$E\{s^{i-1}(\mathbf{x}) c(w) \mid \mathbf{t}\} = \alpha E\{s^{i-1}(\mathbf{x}) \mid \mathbf{t}\} = \alpha_i, \qquad i = 1, 2. \qquad (22.88)$$

All our expectations are taken when θ_0 holds.

Now consider a simple

$$H_0: \theta = \theta_0, \qquad \psi = \psi^0$$

against the simple

$$H_1: \theta = \theta^* \neq \theta_0, \qquad \psi = \psi^*,$$

and apply the result of **22.29** with $k = 2$, α_i as in (22.88), $\theta = (\theta, \psi)$, $\theta_1 = \theta_2 = (\theta_0, \psi)$, $\theta^* = (\theta^*, \psi^*)$, $\theta_1^0 = \theta_2^0 = (\theta_0, \psi^0)$, $l_i(x) = s^{i-1}(x)$, $u_1 = u_2 = t$. We find that the BCR w for testing H_0 against H_1 is given by (22.77) and (22.73) as

$$\frac{C(\theta^*, \psi^*) \exp\{\theta^* s(x) + \sum_{i=1}^r \psi_i^* t_i(x)\}}{C(\theta_0, \psi^0) \exp\{\theta_0 s(x) + \sum_{i=1}^r \psi_i^0 t_i(x)\}} \geq m_1(t) + s(x) m_2(t). \qquad (22.89)$$

(22.89) reduces to

$$\exp\{s(x)(\theta^* - \theta_0)\} \geq c_1(t, \theta^*, \theta_0, \psi^*, \psi^0) + s(x) c_2(t, \theta^*, \theta_0, \psi^*, \psi^0)$$

or

$$\exp\{s(x)(\theta^* - \theta_0)\} - s(x) c_2 \geq c_1. \qquad (22.90)$$

(22.90) is equivalent to $s(x)$ lying outside an interval, i.e.

$$s(x) \leq v(t), \qquad s(x) \geq w(t), \qquad (22.91)$$

where $v(t) < w(t)$ are possibly functions also of the parameters. We now show that they are not dependent on the parameters other than θ_0. As before, the sufficiency of t for ψ rules out the dependence of v and w on ψ when t is given. That they do not depend on θ^* follows at once from (22.86), which states that when H_0 holds

$$\int_w \{s(x) | t\} f \, dx = \alpha \int_W \{s(x) | t\} f \, dx. \qquad (22.92)$$

The right-hand side of (22.92), which is integrated over the whole sample space, clearly does not depend on θ^* at all. Hence the left-hand side is also independent of θ^*, so that the BCR w defined by (22.91) depends only on θ_0, as it must. The BCR therefore gives a UMP test of $H_0^{(2)}$ against $H_1^{(2)}$. Its unbiasedness follows by precisely the argument at the end of **22.30**. Thus, finally, we have established that the BCR defined by (22.91) gives a UMPU test of $H_0^{(2)}$ against $H_1^{(2)}$. If we determine from the conditional distribution of $s(x)$, given t, an interval which excludes 100α per cent of the distribution when $H_0^{(2)}$ holds, and take the excluded values as our critical region, then if the region is unbiased it gives the UMPU size-α test.

Finite-interval hypotheses
22.32   We may also consider the hypothesis

$$H_0^{(3)}: \theta_0 \leq \theta \leq \theta_1$$

against

$$H_1^{(3)}: \theta < \theta_0 \quad \text{or} \quad \theta > \theta_1,$$

or the complementary

$$H_0^{(4)}: \theta \leqslant \theta_0 \quad \text{or} \quad \theta \geqslant \theta_1$$

against

$$H_1^{(4)}: \theta_0 < \theta < \theta_1.$$

We now set up two hypotheses

$$H_0': \theta = \theta_0, \qquad \boldsymbol{\psi} = \boldsymbol{\psi}^0, \qquad H_0'': \theta = \theta_1, \quad \boldsymbol{\psi} = \boldsymbol{\psi}^1,$$

to be tested against

$$H_1: \theta = \theta^*, \quad \boldsymbol{\psi} = \boldsymbol{\psi}^*, \quad \text{where } \theta_0 \neq \theta^* \neq \theta_1.$$

We use the result of **22.29** again, this time with $k = 2$, $\alpha_1 = \alpha_2 = \alpha$, $\boldsymbol{\theta} = (\theta, \boldsymbol{\psi})$, $\boldsymbol{\theta}_1 = (\theta_0, \boldsymbol{\psi})$, $\boldsymbol{\theta}_2 = (\theta_1, \boldsymbol{\psi})$, $\boldsymbol{\theta}^* = (\theta^*, \boldsymbol{\psi}^*)$, $\boldsymbol{\theta}_1^0 = (\theta_0, \boldsymbol{\psi}^0)$, $\boldsymbol{\theta}_2^0 = (\theta_1, \boldsymbol{\psi}^1)$, $l_i(\mathbf{x}) \equiv 1$, $u_1 = u_2 = \mathbf{t}$. We find that the BCR w for testing H_0' or H_0'' against H_1 is defined by

$$f(\mathbf{x} \mid \theta^*, \boldsymbol{\psi}^*) \geqslant m_1(\mathbf{t}) f(\mathbf{x} \mid \theta_0, \boldsymbol{\psi}^0) + m_2(\mathbf{t}) f(\mathbf{x} \mid \theta_1, \boldsymbol{\psi}^1). \tag{22.93}$$

On substituting $f(\mathbf{x})$ from (22.73), (22.93) is equivalent to

$$H(s) = c_1 \exp\{(\theta_0 - \theta^*) s(\mathbf{x})\} + c_2 \exp\{(\theta_1 - \theta^*) s(\mathbf{x})\} < 1, \tag{22.94}$$

where c_1, c_2 may be functions of all the parameters and of $\mathbf{t}$. If $\theta_0 < \theta^* < \theta_1$, (22.94) requires that $s(\mathbf{x})$ lie inside an interval, i.e.

$$v(\mathbf{t}) \leqslant s(\mathbf{x}) \leqslant w(\mathbf{t}). \tag{22.95}$$

On the other hand, if $\theta^* < \theta_0$ or $\theta^* > \theta_1$, (22.94) requires that $s(\mathbf{x})$ lie *outside* the interval $(v(\mathbf{t}), w(\mathbf{t}))$. The proof that the end-points of the interval are not dependent on the values of the parameters, other than θ_0 and θ_1, follows the same lines as before, as does the proof of unbiasedness. Thus we have a UMPU test for $H_0^{(3)}$ and another for $H_0^{(4)}$. The test is similar at values θ_0 and θ_1, as follows from **22.25**. To obtain a UMPU test for $H_0^{(3)}$ (or $H_0^{(4)}$), we determine an interval in the distribution of $s(\mathbf{x})$ for given $\mathbf{t}$ which excludes (for $H_0^{(4)}$ includes) 100α per cent of the distribution both when $\theta = \theta_0$ and $\theta = \theta_1$. The excluded (or included) region, if unbiased, will give a UMPU test of $H_0^{(3)}$ (or $H_0^{(4)}$).

22.33   We now turn to some applications of the fundamental results of **22.30-2** concerning UMPU tests for the exponential family of distributions. We first mention briefly that in Example 22.11 and Exercises 21.1–3 above, UMPU tests for all four types of hypothesis are obtained directly from the distribution of the single sufficient statistic, no conditional distribution being involved since there is no nuisance parameter.

Example 22.14

For n independent observations from a normal distribution, the statistics $(\bar{x}, s^2)$ are jointly sufficient for (μ, σ^2), with joint distribution (cf. Example 17.17)

$$g(\bar{x}, s^2 | \mu, \sigma^2) \propto \frac{s^{n-3}}{\sigma^n} \exp\left\{-\frac{\sum(x-\mu)^2}{2\sigma^2}\right\}. \tag{22.96}$$

(22.96) may be written

$$g \propto C(\mu, \sigma^2) \exp\left\{\left(-\tfrac{1}{2}\sum x^2\right)\left(\frac{1}{\sigma^2}\right) + (\sum x)\left(\frac{\mu}{\sigma^2}\right)\right\}, \tag{22.97}$$

which is of form (22.73). Remembering the discussion of **22.26**, we now consider a linear form in the parameters of (22.97). We put

$$\theta = A\left(\frac{1}{\sigma^2}\right) + B\left(\frac{\mu}{\sigma^2}\right), \tag{22.98}$$

where A and B are arbitrary known constants. We specialise A and B to obtain from the results of **22.30-2** UMPU tests for the following hypotheses:

(1) Put $A = 1$, $B = 0$ and test hypotheses concerning $\theta = 1/\sigma^2$, with $\psi = \mu/\sigma^2$ as nuisance parameter. Here $s(\mathbf{x}) = -\tfrac{1}{2}\sum x^2$ and $t(\mathbf{x}) = \sum x$. From (22.97) there is a UMPU test of $H_0^{(1)}$, $H_0^{(2)}$, $H_0^{(3)}$ and $H_0^{(4)}$ concerning $1/\sigma^2$, and hence concerning σ^2, based on the conditional distribution of $\sum x^2$ given $\sum x$, i.e. of $\sum(x-\bar{x})^2$ given $\sum x$. Since these two statistics are independently distributed, we may use the unconditional distribution of $\sum(x-\bar{x})^2$, or of $\sum(x-\bar{x})^2/\sigma^2$, which is a χ^2 distribution with $(n-1)$ degrees of freedom. $H_0^{(2)}$ was discussed in Example 22.12, where the UMP similar test was given for $\theta = \theta_0$ against one-sided alternatives and an unbiased test based on $\sum(x-\bar{x})^2$ given for $H_0^{(2)}$; it now follows that this is a UMPU test for $H_0^{(2)}$, while the one-sided test is UMPU for $H_0^{(1)}$, as we saw in Example 22.13.

Graphs of the critical values of the UMPU tests of $H_0^{(3)}$ and $H_0^{(4)}$ for $\alpha = 0.05, 0.10$ are given by Guenther and Whitcomb (1966).

(2) To test hypotheses concerning μ, invert (22.98) into $\mu = (\theta\sigma^2 - A)/B$.

If we specify a value μ_0 for μ, we cannot choose A and B to make this correspond uniquely to a value θ_0 for θ (without knowledge of σ^2) if $\theta_0 \neq 0$. But if $\theta_0 = 0$ we have $\mu_0 = -A/B$. Thus from our UMPU tests for $H_0^{(1)}: \theta \leq 0$, $H_0^{(2)}: \theta = 0$, we get UMPU tests of $\mu \leq \mu_0$ and of $\mu = \mu_0$. We use (22.71) to see that the test statistic $s(\mathbf{x})|\mathbf{t}$ is here $(-\tfrac{1}{2}\sum x^2)A + (\sum x)B$ given an orthogonal function, say $(-\tfrac{1}{2}\sum x^2)B - (\sum x)A$. This reduces to the conditional distribution of $\sum x$ given $\sum x^2$. Clearly we cannot get tests of $H_0^{(3)}$ or $H_0^{(4)}$ for μ in this case.

The test of $\mu = \mu_0$ against one-sided alternatives has been discussed in Example 22.7, where we saw that the Student's t-test to which it reduces is the UMP similar test of $\mu = \mu_0$ against one-sided alternatives. This test is now seen to be UMPU for $H_0^{(1)}$. It also follows that the two-sided equal-tails Student's t-test, which is unbiased for $H_0^{(2)}$ against $H_1^{(2)}$, is the UMPU test of $H_0^{(2)}$.

For the power of these tests, see **22.35** below.

Example 22.15
Consider k independent samples of n_i $(i=1,2,\ldots,k)$ observations from normal distributions with means μ_i and common variance σ^2. Write $n=\sum_{i=1}^k n_i$. It is easily confirmed that the k sample means $\bar{x}_i$ and the pooled sum of squares $S^2 = \sum_{i=1}^k \sum_{j=1}^{n_i} (x_{ij}-\bar{x}_i)^2$ are jointly sufficient for the $(k+1)$ parameters. The joint distribution of the sufficient statistics is

$$g(\bar{x}_1,\ldots,\bar{x}_k,S^2) \propto \frac{S^{n-k-2}}{\sigma^n} \exp\left\{-\frac{1}{2\sigma^2}\sum_i\sum_j (x_{ij}-\mu_i)^2\right\}. \tag{22.99}$$

(22.99) is a simple generalisation of (22.96), obtained by using the independence of the $\bar{x}_i$ of each other and of S^2, and the fact that S^2/σ^2 has a χ^2 distribution with $(n-k)$ degrees of freedom. (22.99) may be written

$$g \propto C(\mu_i,\sigma^2) \exp\left\{\left(-\frac{1}{2}\sum_i\sum_j x_{ij}^2\right)\left(\frac{1}{\sigma^2}\right) + \sum_i\left(\sum_j x_{ij}\right)\left(\frac{\mu_i}{\sigma^2}\right)\right\}, \tag{22.100}$$

in the form (22.73). We now consider the linear function

$$\theta = A\left(\frac{1}{\sigma^2}\right) + \sum_{i=1}^k B_i\left(\frac{\mu_i}{\sigma^2}\right). \tag{22.101}$$

(1) Put $A=1$, $B_i=0$ (all i). Then $\theta=1/\sigma^2$, and $\psi_i=\mu_i/\sigma^2$ $(i=1,\ldots,k)$ is the set of nuisance parameters. There is a UMPU test of each of the four $H_0^{(r)}$ discussed in **22.30**-2 for $1/\sigma^2$ and therefore for σ^2. The tests are based on the conditional distribution of $\sum\sum x_{ij}^2$ given the vector $(\sum_j x_{ij}, \sum_j x_{2j},\ldots,\sum_j x_{kj})$, i.e. of $S^2 = \sum_i\sum_j (x_{ij}-\bar{x}_i)^2$ given that vector. Just as in Example 22.14, this leads to the use of the unconditional distribution of S^2 to obtain the UMPU tests.

(2) Exactly analogous considerations to those of Example 22.14 (2) show that by putting $\theta_0=0$, we obtain UMPU tests of $\sum_{i=1}^k c_i\mu_i \leqslant c_0$, $\sum c_i\mu_i = c_0$, where c_0 is any constant. (Cf. Exercise 22.19). Just as before, no 'interval' hypothesis can be tested, using this method, concerning the linear form $\sum c_i\mu_i$.

(3) The substitution $k=2$, $c_1=1$, $c_2=-1$, $c_0=0$, reduces (2) to testing $H_0^{(1)}: \mu_1-\mu_2\leqslant 0$, $H_0^{(2)}: \mu_1-\mu_2=0$. The test of $\mu_1-\mu_2=0$ has been discussed in Example 22.8, where it was shown to reduce to a Student's t-test and to be UMP similar. It is now seen to be UMPU for $H_0^{(1)}$. The equal-tails two-sided Student's t-test, which is unbiased, is also seen to be UMPU for $H_0^{(2)}$.

Example 22.16
We generalise the situation in Example 22.15 by allowing the variances of the k normal distributions to differ. We now have a set of $2k$ sufficient statistics for the $2k$ parameters, which are the sample sums and sums of squares $\sum_{j=1}^{n_i} x_{ij}$, $\sum_{j=1}^{n_i} x_{ij}^2$, $i=1,2,\ldots,k$. We write

$$\theta = \sum_{i=1}^k A_i\left(\frac{1}{\sigma_i^2}\right) + \sum_{i=1}^k B_i\left(\frac{\mu_i}{\sigma_i^2}\right). \tag{22.102}$$

(1) Put $B_i = 0$ (all i). We get UMPU tests for all four hypotheses concerning

$$\theta = \sum_i A_i \left(\frac{1}{\sigma_i^2} \right),$$

a weighted sum of the reciprocals of the population variances. The case $k = 2$ reduces this to

$$\theta = \frac{A_1}{\sigma_1^2} + \frac{A_2}{\sigma_2^2}.$$

If we want to test hypotheses concerning the variance ratio σ_2^2 / σ_1^2, then just as in (2) of Examples 22.14–15, we have to put $\theta = 0$ to make any progress. If we do this, the UMPU tests of $\theta = 0$, ≤ 0 reduce to those of

$$\frac{\sigma_2^2}{\sigma_1^2} = -\frac{A_2}{A_1}, \qquad \leq -\frac{A_2}{A_1},$$

and we therefore have UMPU tests of $H_0^{(1)}$ and $H_0^{(2)}$ concerning the variance ratio. The joint distribution of the four sufficient statistics may be written

$$g(\sum x_{1j}, \sum x_{2j}, \sum x_{1j}^2, \sum x_{2j}^2) \propto C(\mu_i, \sigma_i^2)$$

$$\times \exp \left\{ -\frac{1}{2} \left(\frac{1}{\sigma_1^2} \sum x_{1j}^2 + \frac{1}{\sigma_2^2} \sum x_{2j}^2 \right) + \frac{\mu_1}{\sigma_1^2} \sum x_{1j} + \frac{\mu_2}{\sigma_2^2} \sum x_{2j} \right\}.$$

By **22.26**, the coefficients $s(\mathbf{x})$ of θ when this is transformed to make θ one of its parameters, will be the same function of $-\frac{1}{2} \sum x_{1j}^2$, $-\frac{1}{2} \sum x_{2j}^2$ as θ is of $1/\sigma_1^2$, $1/\sigma_2^2$, i.e.

$$-2s(\mathbf{x}) = A_1 \sum x_{1j}^2 + A_2 \sum x_{2j}^2,$$

and the UMPU tests will be based on the conditional distribution of $s(\mathbf{x})$ given any three functions of the sufficient statistics, orthogonal to $s(\mathbf{x})$ and to each other, say

$$\sum x_{1j}, \sum x_{2j} \quad \text{and} \quad A_2 \sum x_{1j}^2 - A_1 \sum x_{2j}^2.$$

This is equivalent to holding $\bar{x}_1$, $\bar{x}_2$ and

$$t = \sum (x_{1j} - \bar{x}_1)^2 - \frac{A_1}{A_2} \sum (x_{2j} - \bar{x}_2)^2$$

fixed, so that $s(\mathbf{x})$ is equivalent to

$$\sum (x_{1j} - \bar{x}_1)^2 + \frac{A_2}{A_1} \sum (x_{2j} - \bar{x}_2)^2$$

for fixed t. In turn, this is equivalent to considering the distribution of the ratio $\sum (x_{1j} - \bar{x}_1)^2 / \sum (x_{2j} - \bar{x}_2)^2$, so that the UMPU tests of $H_0^{(1)}$, $H_0^{(2)}$ are based on the distribution of the sample variance ratio—cf. Exercises 22.14 and 22.17.

(2) We cannot get UMPU tests concerning functions of the μ_i free of the σ_i^2, as is obvious from (22.102). In the case $k = 2$, this precludes us from finding a solution to the problem of two means by this method.

Geometrical interpretation

22.34 The results of **22.26–33** may be better appreciated with the help of a partly geometrical explanation. From **5.48**, the characteristic function of $s(\mathbf{x})$ in (22.73) is

$$\phi(u) = E\{\exp(ius)\} = \frac{C(\theta)}{C(\theta + iu)} \tag{22.103}$$

(where we suppress the second argument of C), its c.g.f. is

$$\psi(u) = \log \phi(u) = \log C(\theta) - \log C(\theta + iu), \tag{22.104}$$

its rth cumulant is

$$\kappa_r = \left[\frac{\partial^r}{\partial(iu)^r} \psi(u)\right]_{u=0} = -\frac{\partial^r}{\partial\theta^r} \log C(\theta), \tag{22.105}$$

and

$$E(s) = \kappa_1 = -\frac{\partial}{\partial\theta} \log C(\theta), \tag{22.106}$$

$$\kappa_r = \frac{\partial^{r-1}}{\partial\theta^{r-1}} E(s), \qquad r \geqslant 2. \tag{22.107}$$

Consider the derivative

$$\mathbf{D}^q f \equiv \frac{\partial^q}{\partial\theta^q} f(x \mid \theta, \boldsymbol{\psi}).$$

From (22.73) and (22.106),

$$\mathbf{D} f = \left\{ s + \frac{C'(\theta)}{C(\theta)} \right\} f = \{s - E(s)\} f. \tag{22.108}$$

By Leibniz's rule, we have from (22.108)

$$\mathbf{D}^q f = \mathbf{D}^{q-1}[\{s - E(s)\} f]$$

$$= \{s - E(s)\}\mathbf{D}^{q-1} f + \sum_{i=1}^{q-1} \binom{q-1}{i} [\mathbf{D}^i \{s - E(s)\}][\mathbf{D}^{q-1-i} f], \tag{22.109}$$

which, using (22.107), may be written

$$\mathbf{D}^q f = \{s - E(s)\}\mathbf{D}^{q-1} f - \sum_{i=1}^{q-1} \binom{q-1}{i} \kappa_{i+1} \mathbf{D}^{q-1-i} f. \tag{22.110}$$

22.35 Now consider any critical region w of size α. Its power function is defined at (22.80), and we may alternatively express this as an integral in the sample space of the sufficient statistics $(s, \mathbf{t})$ by

$$P(w \mid \theta) = \int_w f \, ds \, d\mathbf{t}, \tag{22.111}$$

where f now stands for the joint frequency function of (s, t), which is of the form (22.73) as we have seen. The derivatives of the power function (22.111) are

$$P^{(q)}(w|\theta) = \int_w \mathbf{D}^q f \, ds \, dt, \qquad (22.112)$$

since we may differentiate under the integral sign in (22.111). Using (22.108) and (22.110), (22.111) gives

$$P'(w|\theta) = \int_w \{s - E(s)\} f \, ds \, dt = \text{cov} \{s, c(w)\}, \qquad (22.113)$$

and

$$P^{(q)}(w|\theta) = \int_w \{s - E(s)\} \mathbf{D}^{q-1} f \, ds \, dt - \sum_{i=1}^{q-1} \binom{q-1}{i} \kappa_{i+1} P^{(q-1-i)}(w|\theta), \qquad q \geqslant 2, \qquad (22.114)$$

a recurrence relation which enables us to build up the value of any derivative from lower derivatives. In particular, (22.114) gives

$$P''(w|\theta) = \text{cov} \{[s - E(s)]^2, c(w)\}, \qquad (22.115)$$

$$\left. \begin{array}{l} P'''(w|\theta) = \text{cov} \{[s - E(s)]^3, c(w)\} - 3\kappa_2 P'(w|\theta), \\ P^{(iv)}(w|\theta) = \text{cov} \{[s - E(s)]^4, c(w)\} - 6\kappa_2 P''(w|\theta) - 4\kappa_3 P'(w|\theta). \end{array} \right\} \qquad (22.116)$$

(22.113) and (22.115) show that the first two derivatives are simply the covariances of $c(w)$ with s, and with the squared deviation of s from its mean, respectively. The third and fourth derivatives given by (22.116) are more complicated functions of covariances and of the cumulants of s, as are the higher derivatives.

22.36 We are now in a position to interpret geometrically some of the results of **22.26–33**. To maximise the power we must choose w to maximise, for all admissible alternatives, the covariance of $c(w)$ with s, or some function of $s - E(s)$, in accordance with (22.113) and (22.114). In the $(r+1)$-dimensional space of the sufficient statistics, (s, t), it is obvious that this will be done by confining ourselves to the subspace orthogonal to the r co-ordinates corresponding to the components of t, i.e. by confining ourselves to the conditional distribution of s given t.

If we are testing $\theta = \theta_0$ against $\theta > \theta_0$ we maximise $P(w|\theta)$ for all $\theta > \theta_0$ by maximising $P'(w|\theta)$, i.e. by maximising cov $(s, c(w))$ for all $\theta > \theta_0$. This is easily seen to be done if w consists of the 100α per cent *largest* values of the distribution of s given t. Similarly for testing $\theta = \theta_0$ against $\theta < \theta_0$, we maximise P by *minimising* P', and this is done if w consists of the 100α per cent *smallest* values of the distribution of s given t. Since $P'(w|\theta)$ is always of the same sign, the one-sided tests are unbiased.

For the two-sided $H_0^{(2)}$ of **22.31**, (22.81) and (22.115) require us to maximise $P''(w|\theta)$, i.e. cov $\{[s - E(s)]^2, c(w)\}$. By exactly the same argument as in the one-sided case, we choose w to include the 100α per cent largest values of $\{s - E(s)\}^2$ given $\mathbf{t}$, so that we obtain a two-sided test, which is an equal-tails test only if the distribution of s given $\mathbf{t}$ is symmetrical. It follows that the boundaries of the UMPU critical region are equidistant from $E(s|\mathbf{t})$.

Spjøtvoll (1968) gives results on most powerful tests for some non-exponential families of distributions.

EXERCISES

22.1   Show that for samples of n observations from a normal distribution with mean θ and variance σ^2, no symmetric similar region with respect to θ and σ^2 exists for $n \leqslant 2$, but that such regions do exist for $n \geqslant 3$.   (Feller, 1938)

22.2   Show, as in Example 22.1, that for a sample of n observations, the ith of which has distribution

$$\frac{1}{\Gamma(\theta_i)} e^{-x_i} x_i^{\theta_i - 1} \, dx_i, \qquad 0 \leqslant x_i < \infty; \quad \theta_i > 0,$$

no similar size-α region exists for $0 < \alpha < 1$.   (Feller, 1938)

22.3   Given that $L(x|\theta)$ is a likelihood function and $E(\partial \log L/\partial\theta) = 0$, show that if the distribution of a statistic z does not depend on θ, then $\operatorname{cov}(z, \partial \log L/\partial\theta) = 0$. As a corollary, show that no similar region with respect to θ exists if no statistic exists that is uncorrelated with $\partial \log L/\partial\theta$.   (Neyman, 1938a)

22.4   Show, using the c.f. of z, that the converses of the result and the corollary of Exercise 22.3 are true.

Together, this and the previous exercise state that $\operatorname{cov}(z, \partial \log L/\partial\theta) = 0$ is a necessary and sufficient condition for $\operatorname{cov}(e^{iuz}, \partial \log L/\partial\theta) = 0$, where u is a dummy variable.
                                                                    (Neyman, 1938a)

22.5   Show that the Cauchy family of distributions

$$dF = \frac{dx}{\left\{ \pi\theta^{1/2} \left(1 + \dfrac{x^2}{\theta} \right) \right\}}, \qquad -\infty < x < \infty,$$

is not complete.   (Lehmann and Scheffé, 1950)

22.6   Show that if θ is fixed and a statistic z is distributed independently of t, a sufficient statistic for θ, then the distribution of z does not depend on θ.

In Example 17.23, with $n = 1$, show that $t = [x]$ is sufficient for θ and that, *for each fixed* θ, $z = x$ is distributed independently of t, but that the distribution of z depends on θ.
                              (Here the range of z is different for every θ; the difficulty does not arise if the ranges are identical for all θ—cf. D. Basu (1955, 1958).)

22.7   In Exercise 22.6, write $H_1(z)$ for the d.f. of z, $H_2(z|t)$ for its conditional d.f. given t, and $g(t|\theta)$ for the frequency function of t. Show that

$$\int \{ H_1(z) - H_2(z|t) \} g(t|\theta) \, dt = 0$$

for all θ. Hence show that if t is a *boundedly complete* sufficient statistic for θ, the converse of the result of Exercise 22.6 holds, namely, if the distribution of z does not depend upon θ, z is distributed independently of t.   (D. Basu, 1955)

22.8   Use the result of Exercise 22.7 to show directly that, in univariate normal samples:
(a) any moment about the sample mean $\bar{x}$ is distributed independently of $\bar{x}$;
(b) the quadratic form $\mathbf{x}'\mathbf{A}\mathbf{x}$ is distributed independently of $\bar{x}$ if and only if the elements of each row of the matrix $\mathbf{A}$ add to zero (cf. **15.15**);
(c) the sample range is distributed independently of $\bar{x}$;
(d) $(x_{(n)} - \bar{x})/(x_{(n)} - x_{(1)})$ is distributed independently both of $\bar{x}$ and of s^2, the sample variance.   (Hogg and Craig, 1956)

22.9   Use Exercise 22.7 to show that:
   (a) in samples from a bivariate normal distribution with $\rho = 0$, the sample correlation coefficient is distributed independently of the sample means and variances (cf. **16.28**);
   (b) in independent samples from two univariate normal populations with the same variance σ^2, the statistic

$$F = \frac{\sum_j (x_{1j} - \bar{x}_1)^2 / (n_1 - 1)}{\sum_j (x_{2j} - \bar{x}_2)^2 / (n_2 - 1)}$$

is distributed independently of the set of three jointly sufficient statistics

$$\bar{x}_1, \bar{x}_2, \sum_j (x_{1j} - \bar{x}_1)^2 + \sum_j (x_{2j} - \bar{x}_2)^2$$

and therefore of the statistic

$$t^2 = \frac{(\bar{x}_1 - \bar{x}_2)^2}{\sum (x_{1j} - \bar{x}_1)^2 + \sum (x_{2j} - \bar{x}_2)^2} \left\{ \frac{n_1 n_2 (n_1 + n_2 - 2)}{n_1 + n_2} \right\}$$

which is a function of the sufficient statistics. This holds whether or not the population means are equal.   (Hogg and Craig, 1956)

22.10   Use Exercise 22.7 to show that (a) in samples of size n from the distribution

$$dF = \exp\{-(x - \theta)\}\, dx, \qquad \theta \leqslant x < \infty,$$

$x_{(1)}$ is distributed independently of

$$z = \sum_{i=1}^{r} (x_{(i)} - x_{(1)}) + (n - r)(x_{(r)} - x_{(1)}), \qquad r \leqslant n;$$

   (b) if $f(x_j) = \sigma_j \exp\{-\sigma_j(x_j - \theta)\}$, x_j, $\sigma_j > 0$, $j = 1, 2$, $x_{(2)} - x_{(1)}$ is distributed independently of $x_{(1)}$, and $(\sigma_1 + \sigma_2)(x_{(1)} - \theta)$ is exponentially distributed.
                                                        (cf. Epstein and Sobel, 1954)

22.11   In Exercise 18.15, use Exercise 22.7 to show that $\bar{x}$ and $\bar{x}/g$ are independently distributed.

22.12   Show that for the binomial distribution with parameter π, the sample proportion p is minimal sufficient for π, while for the uniform distribution

$$dF = dx, \qquad \theta - \tfrac{1}{2} \leqslant x \leqslant \theta + \tfrac{1}{2},$$

the pair of statistics $(x_{(1)}, x_{(n)})$ is minimal sufficient for θ.
                                                      (Lehmann and Scheffé, 1950)

22.13   For a normal distribution with variance σ^2 and unspecified mean μ, show by the method of **22.20** that the UMP similar test of $H_0 : \sigma^2 = \sigma_0^2$ against $H_1 : \sigma^2 = \sigma_1^2$ takes the form

$$\sum (x - \bar{x})^2 \geqslant a_{1-\alpha} \quad \text{if } \sigma_1^2 > \sigma_0^2,$$
$$\sum (x - \bar{x})^2 \leqslant a_\alpha \quad \text{if } \sigma_1^2 < \sigma_0^2.$$

22.14   Two normal distributions have unspecified means and variances σ^2, $\theta\sigma^2$. From independent samples of sizes n_1, n_2, show by the method of **22.20** that the UMP similar test of $H_0 : \theta = 1$ against $H_1 : \theta = \theta_1$ takes the form

$$s_1^2 / s_2^2 \geqslant a_{1-\alpha} \quad \text{if } \theta_1 < 1,$$
$$s_1^2 / s_2^2 \leqslant a_\alpha \quad \text{if } \theta_1 > 1,$$

where s_1^2, s_2^2 are the sample variances.
                                (Harter (1963) shows that a test based on the ratio of sample ranges is almost as powerful as the UMP similar test.)

22.15 Independent samples, each of size n, are taken from the distributions

$$dF = \exp\left(-\frac{x}{\theta_1}\right) dx/\theta_1, \quad \theta_1, \theta_2 > 0,$$
$$dG = \exp\left(-y\theta_2\right)\theta_2\, dy, \quad 0 \leqslant x, y < \infty.$$

Show that $t = (\sum x, \sum y) = (X, Y)$ is minimal sufficient for (θ_1, θ_2) and remains so if $H_0: \theta_1 = \theta_2 = \theta$ holds. By considering the function $XY - E(XY)$ show that the distribution of t is not boundedly complete given H_0, so that not all similar regions satisfy (22.13). Finally, show that the statistic XY is then distributed independently of θ, so that H_0 may be tested by similar regions from it. (Watson, 1957a)

22.16 In the problem of the ratio of two normal means (Example 20.4), assume that $\sigma_{xy} = 0$, $\sigma_x^2 = \sigma_y^2 = 1$. To test the composite $H_0: \theta = \theta_0$ against $H_1: \theta = \theta_1$, show that when H_0 holds, the nuisance parameter μ_1 has a complete sufficient statistic $u = \theta_0 \bar{y} + \bar{x}$. Hence show, using **22.20**, that the UMP similar test of H_0 against H_1 is based on large (small) values of $\bar{y} - \theta_0 \bar{x}$ if $(\theta_1 - \theta_0)\mu_1 > 0$ (<0). (Cf. D. R. Cox (1967).)

22.17 In Exercise 22.14, show that the critical region

$$s_1^2/s_2^2 \geqslant a_{1-\alpha/2}, \quad \leqslant a_{\alpha/2},$$

is biased against the two-sided alternative $H_1: \theta \neq 1$ unless $n_1 = n_2$. By exactly the same argument as in Example 22.12, show that an unbiased critical region

$$t = s_1^2/s_2^2 \geqslant a_{1-\alpha_1}, \quad \leqslant a_{\alpha_2}, \quad \alpha_1 + \alpha_2 = \alpha,$$

is determined by the condition (cf. 22.56))

$$V_{\alpha_2} f(V_{\alpha_2}) = V_{1-\alpha_1} f(V_{1-\alpha_1}),$$

where f is the density of the variance-ratio statistic t and V_α its 100α per cent point. Show that the power function of the unbiased test is monotone increasing for $\theta > 1$, monotone decreasing for $\theta < 1$.

(S. John's tables described in **20.34** (4) give critical values for a simple function of s_1^2/s_2^2.)

22.18 In Exercise 22.17, show that the unbiased confidence interval for θ given by $(t/V_{\alpha_2}, t/V_{1-\alpha_1})$ minimises the expectation of $(\log U - \log L)$ for confidence intervals (L, U) based on the tails of the distribution of t. (Scheffé, 1942b)

22.19 In Example 22.15, use **22.26** to show that the UMPU tests for $\sum_i c_i \mu_i$ are based on the distribution of

$$t = \sum c_i(\bar{x}_i - \mu_i) \left/ \left\{ \frac{S^2}{n-k} \sum \frac{c_i^2}{n_i} \right\}^{1/2} \right.,$$

which is a Student's t with $(n-k)$ degrees of freedom.

22.20 In Example 22.16, show that there is a UMPU test of the hypothesis

$$\frac{\mu_i}{\mu_j} = a\frac{\sigma_i^2}{\sigma_j^2}, \quad i \neq j, \quad \mu_j \neq 0.$$

22.21 For independent samples from two Poisson distributions with parameters μ_1, μ_2, show that there are UMPU tests for all four hypotheses considered in **22.30**-**32** concerning μ_1/μ_2, and that the test of $\mu_1/\mu_2 = 1$ consists of testing whether the sum of the observations is binomially distributed between the samples with equal probabilities.

(Lehmann and Scheffé, 1955)

22.22 For independent binomial distributions with parameters θ_1, θ_2, find the UMPU tests for all four hypotheses in **22.30–32** concerning the 'odds ratio' $(\theta_1/(1-\theta_1))/(\theta_2/(1-\theta_2))$, and the UMPU tests for $\theta_1 = \theta_2$, $\theta_1 \leqslant \theta_2$. (Lehmann and Scheffé, 1955)

22.23 For the uniform distribution

$$dF = dx, \qquad \theta - \tfrac{1}{2} \leqslant x \leqslant \theta + \tfrac{1}{2},$$

the conditional distribution of the mid-range M given the range R, and the marginal distribution of M, are given by the results of Exercise 14.12. For testing $H_0 : \theta = \theta_0$ against the two-sided alternative $H_1 : \theta \neq \theta_0$ show that the equal-tails test based on M given R, when integrated over all values of R, gives uniformly less power than the equal-tails test based on the marginal distribution of M; use the value $\alpha = 0.08$ for convenience. (Welch, 1939)

22.24 In Example 22.9, show that the UMPU test of $H_0 : \sigma = \sigma_0$ against $H_1 : \sigma \neq \sigma_0$ is of the form

$$\sum_{i=1}^{n} x_i \geqslant a_{\alpha_1}, \quad \leqslant b_{\alpha_2}. \qquad \text{(Lehmann, 1947)}$$

22.25 For the distribution of Example 22.9, show that the UMP similar test of $H_0 : \theta = \theta_0$ against $H_1 : \theta \neq \theta_0$ is of the form

$$\frac{x_{(1)} - \theta_0}{\bar{x} - x_{(1)}} < 0, \quad \geqslant c_\alpha.$$

(Lehmann, 1947; see also Takeuchi (1969) for $H_1 : \theta < \theta_0$).

22.26 For the uniform distribution

$$dF = dx/\theta, \qquad \mu \leqslant x \leqslant \mu + \theta,$$

show that the UMP similar test of $H_0 : \mu = \mu_0$ against $H_1 : \mu \neq \mu_0$ is of the form

$$\frac{x_{(1)} - \mu_0}{x_{(n)} - x_{(1)}} < 0, \quad \geqslant c_\alpha.$$

Cf. the simple hypothesis with $\theta = 1$, where it was seen in Exercise 21.8 that no UMP test exists. (Lehmann, 1947)

22.27 Given that $x_1, \ldots, x_n$ are independent observations from the distribution

$$dF = \frac{1}{\theta^p \Gamma(p)} \exp(-x/\theta) x^{p-1} \, dx, \qquad p > 0, \quad 0 \leqslant x < \infty,$$

use Exercises 22.6 and 22.7 to show that a necessary and sufficient condition that a statistic $h(x_1, \ldots, x_n)$ be independent of $S = \sum_{i=1}^{n} x_i$ is that $h(x_1, \ldots, x_n)$ be homogeneous of degree zero in x. (Cf. refs. to Exercise 15.22.)

22.28 From (22.113) and (22.114), show that if the first non-zero derivative of the power function is the mth, then

$$P^{(m)}(w \mid \theta) = \text{cov}\{[s - E(s)]^m, c(w)\}$$

and

$$\frac{\{P^{(m)}(w \mid \theta)\}^2}{\mu_{2m}} \leqslant \frac{1}{4},$$

where μ_r is the rth central moment of s. In particular,

$$|P'(w \mid \theta)| \leqslant \tfrac{1}{2} \mu_2^{1/2}.$$

22.29 From **22.35**, show that w is a similar region for a hypothesis for which θ is a nuisance parameter if and only if

$$\text{cov}\,\{s, c(w)\} = 0$$

identically in θ. Cf. Exercises 22.3–4.

22.30 Generalise the argument of the last paragraph of Example 22.7 to show that for any distribution of form

$$dF = f\left(\frac{x-\mu}{\sigma}\right)\frac{dx}{\sigma},$$

admitting a complete sufficient statistic for σ when μ is known, there can be no similar critical region for $H_0: \mu = \mu_0$ against $H_1: \mu = \mu_1$ with power independent of σ.

22.31 For a normal distribution with mean and variance both equal to θ, show that for a single observation, x and x^2 are each singly sufficient for θ, x^2 being minimal. Hence it follows that single sufficiency does not imply minimal sufficiency. (Exercise 18.13 gives a bivariate instance of the same phenomenon.)

22.32 Let (T_r, T_s) be the set of $(r+s)$ minimal sufficient statistics for $(k+l)$ parameters (θ_k, θ_l) where $k \geq 1$, $l \geq 0$. Use Exercise 22.7 to show that the subset T_s can only be distributed free of θ if (T_r, T_s) is not boundedly complete. In Exercise 22.23, where $l = 0$, show that the pair of statistics (M, R) is minimal sufficient for θ, but not boundedly complete, and that R is distributed free of θ.

CHAPTER 23

LIKELIHOOD RATIO TESTS AND THE GENERAL LINEAR HYPOTHESIS

The LR statistic

23.1 The ML method discussed in Chapter 18 is a constructive method of obtaining estimators which, under certain conditions, have desirable properties. A method of test construction closely allied to it is the likelihood ratio (LR) method, proposed by Neyman and Pearson (1928). It has played a role in the theory of tests analogous to that of the ML method in the theory of estimation.

As before, we have an LF

$$L(x \mid \theta) = \prod_{i=1}^{n} f(x_i \mid \theta),$$

where $\theta = (\theta_r, \theta_s)$ is a vector of $r + s = k$ parameters ($r \geq 1$, $s \geq 0$) and x may also be a vector. We wish to test the hypothesis

$$H_0: \theta_r = \theta_{r0}, \tag{23.1}$$

which is composite unless $s = 0$, against

$$H_1: \theta_r \neq \theta_{r0}.$$

We know that there is generally no UMP test in this situation, but that there may be a UMPU test—cf. **22.31**.

The LR method first requires us to find the ML estimators of (θ_r, θ_s), giving the unconditional maximum of the LF

$$L(x \mid \hat{\theta}_r, \hat{\theta}_s), \tag{23.2}$$

and also to find the ML estimators of θ_s, when H_0 holds,* giving the conditional maximum of the LF

$$L(x \mid \theta_{r0}, \hat{\hat{\theta}}_s). \tag{23.3}$$

$\hat{\hat{\theta}}_s$ in (23.3) has been given a double circumflex to emphasise that it does not in general coincide with $\hat{\theta}_s$ in (23.2). Now consider the likelihood ratio†

$$l = \frac{L(x \mid \theta_{r0}, \hat{\hat{\theta}}_s)}{L(x \mid \hat{\theta}_r, \hat{\theta}_s)}. \tag{23.4}$$

---

* When $s = 0$, H_0 being simple, no maximisation process is needed, for L is uniquely determined.

† The ratio is often denoted by λ, and the LR statistic is sometimes called 'the lambda criterion', but we use the Roman letter in accordance with the convention that Greek symbols are reserved for parameters.

Since (23.4) is the ratio of a conditional maximum of the LF to its unconditional maximum, we clearly have

$$0 \leq l \leq 1. \tag{23.5}$$

Intuitively, l is a reasonable test statistic for H_0: it is the maximum likelihood under H_0 as a fraction of its largest possible value, and large values of l signify that H_0 is reasonably acceptable. The critical region for the test statistic is therefore

$$l \leq c_\alpha, \tag{23.6}$$

where c_α is determined from the distribution $g(l)$ of l to give a size-α test, i.e.

$$\int_0^{c_\alpha} g(l) \, dl = \alpha. \tag{23.7}$$

Neither maximum value of the LF is affected by a change of parameter from θ to $\tau(\theta)$, the ML estimator of $\tau(\theta)$ being $\tau(\hat{\theta})$—cf. **8.10**. Thus the LR statistic is invariant under re-parametrisation.

23.2   For the LR method to be useful in the construction of similar tests, i.e. tests based on similar critical regions, the distribution of l should be free of nuisance parameters, and it is a fact that for many common statistical problems it is so. The next two examples illustrate the method in cases where it does and does not lead to a similar test.

Example 23.1
For the normal distribution

$$dF(x) = (2\pi\sigma^2)^{-1/2} \exp\left\{ -\frac{1}{2}\left(\frac{x-\mu}{\sigma}\right)^2 \right\} dx,$$

we wish to test

$$H_0: \mu = \mu_0.$$

Here

$$L(x \mid \mu, \sigma^2) = (2\pi\sigma^2)^{-n/2} \exp\left\{ -\frac{1}{2}\sum\left(\frac{x-\mu}{\sigma}\right)^2 \right\}.$$

Using Example 18.11, we have for the unconditional ML estimators

$$\hat{\mu} = \bar{x},$$

$$\hat{\sigma}^2 = \frac{1}{n}\sum(x-\bar{x})^2 = s^2,$$

so that

$$L(x \mid \hat{\mu}, \hat{\sigma}^2) = (2\pi s^2)^{-n/2} \exp\left(-\tfrac{1}{2}n\right). \tag{23.8}$$

When H_0 holds, the ML estimator is (cf. Example 18.8)

$$\hat{\sigma}^2 = \frac{1}{n} \sum (x - \mu_0)^2 = s^2 + (\bar{x} - \mu_0)^2,$$

so that

$$L(x \,|\, \mu_0, \hat{\sigma}^2) = [2\pi\{s^2 + (\bar{x} - \mu_0)^2\}]^{-n/2} \exp\left(-\tfrac{1}{2}n\right). \qquad (23.9)$$

From (23.4), (23.8) and (23.9), we find

$$l = \left\{ \frac{s^2}{s^2 + (\bar{x} - \mu_0)^2} \right\}^{n/2}$$

or

$$l^{2/n} = \frac{1}{1 + \dfrac{t^2}{n-1}},$$

where t is Student's t-statistic with $(n-1)$ degrees of freedom. Thus l is a monotone decreasing function of t^2. Hence we may use the known exact distribution of t^2 as equivalent to that of l, rejecting the 100α per cent largest values of t^2, which correspond to the 100α per cent smallest values of l. We thus obtain an equal-tails test based on the distribution of Student's t, half of the critical region consisting of extreme positive values, and half of extreme negative values, of t. This is a very reasonable test: we have seen that it is UMPU for H_0 in Example 22.14.

Example 23.2
Consider again the problem of two means, extensively discussed in Chapters 20 and 22. We have samples of sizes n_1, n_2 from normal distributions with means and variances (μ_1, σ_1^2), (μ_2, σ_2^2) and wish to test $H_0 : \mu_1 = \mu_2$, which we may re-parametrise (cf. **22.2**) as $H_0 : \theta \equiv \mu_1 - \mu_2 = 0$. We call the common unknown value of the means μ. We have

$$L(x \,|\, \mu_1, \mu_2, \sigma_1^2, \sigma_2^2) = (2\pi)^{-(n_1 + n_2)/2} \sigma_1^{-n_1} \sigma_2^{-n_2}$$

$$\times \exp\left\{ -\frac{1}{2}\left(\sum_{j=1}^{n_1} \frac{(x_{1j} - \mu_1)^2}{\sigma_1^2} + \sum_{j=1}^{n_2} \frac{(x_{2j} - \mu_2)^2}{\sigma_2^2} \right) \right\}.$$

The unconditional ML estimators are

$$\hat{\mu}_1 = \bar{x}_1, \quad \hat{\mu}_2 = \bar{x}_2, \quad \hat{\sigma}_1^2 = s_1^2, \quad \hat{\sigma}_2^2 = s_2^2,$$

so that

$$L(x \,|\, \hat{\mu}_1, \hat{\mu}_2, \hat{\sigma}_1^2, \hat{\sigma}_2^2) = (2\pi)^{-(n_1 + n_2)/2} s_1^{-n_1} s_2^{-n_2} \exp\left\{ -\tfrac{1}{2}(n_1 + n_2) \right\}.$$

When H_0 holds, the ML estimators are roots of the set of three equations

$$\left. \begin{array}{r} \dfrac{n_1(\bar{x}_1 - \mu)}{\sigma_1^2} + \dfrac{n_2(\bar{x}_2 - \mu)}{\sigma_2^2} = 0, \\[2mm] \sigma_1^2 = \dfrac{1}{n_1} \sum\limits_{j=1}^{n_1} (x_{1j} - \mu)^2 = s_1^2 + (\bar{x}_1 - \mu)^2, \\[2mm] \sigma_2^2 = \dfrac{1}{n_2} \sum\limits_{j=1}^{n_2} (x_{2j} - \mu)^2 = s_2^2 + (\bar{x}_2 - \mu)^2. \end{array} \right\} \tag{23.10}$$

When the solutions of (23.10) are substituted into the LF, we get

$$L(x \mid \hat{\hat{\mu}}, \hat{\hat{\sigma}}_1^2, \hat{\hat{\sigma}}_2^2) = (2\pi)^{-(n_1 + n_2)/2} \hat{\hat{\sigma}}_1^{-n_1} \hat{\hat{\sigma}}_2^{-n_2} \exp\{-\tfrac{1}{2}(n_1 + n_2)\},$$

and the likelihood ratio is

$$l = \left(\frac{s_1}{\hat{\hat{\sigma}}_1}\right)^{n_1} \left(\frac{s_2}{\hat{\hat{\sigma}}_2}\right)^{n_2} = \left\{\frac{s_1^2}{s_1^2 + (\bar{x}_1 - \hat{\hat{\mu}})^2}\right\}^{n_1/2} \left\{\frac{s_2^2}{s_2^2 + (\bar{x}_2 - \hat{\hat{\mu}})^2}\right\}^{n_2/2} \tag{23.11}$$

We need then only to determine $\hat{\hat{\mu}}$ to be able to use (23.11). Now by (23.10), we see that $\hat{\hat{\mu}}$ is a solution of a cubic equation in μ whose coefficients are functions of the n_i and of the sums and sums of squares of the two sets of observations. We cannot therefore write down $\hat{\hat{\mu}}$ as an explicit function, though we can solve for it numerically in any given case. Its distribution is, in any case, not independent of the ratio σ_1^2/σ_2^2, for $\hat{\hat{\mu}}$ is a function of both s_1^2 and s_2^2, and l is therefore of the form

$$l = g(s_1^2, s_2^2) h(s_1^2, s_2^2).$$

Thus the LR method fails in this case to give us a similar test.

23.3 If, as in Example 23.1, we find that the LR test statistic is a one-to-one function of some statistic whose distribution is either known exactly (as in that Example) or can be found, there is no difficulty in constructing a valid test of H_0, though we shall have shortly to consider what desirable properties LR tests as a class possess. However, it frequently occurs that the LR method is not so convenient, when the test statistic is a more or less complicated function of the observations whose exact distribution cannot be obtained, as in Example 23.2. In such a case, we have to resort to approximations to its distribution.

Since l is distributed on the interval $(0, 1)$, we see that for any fixed constant $c > 0$, $w = -2c \log l$ will be distributed on the interval $(0, \infty)$. It is therefore natural to seek an approximation to its distribution by means of a χ^2 variate, which is also on the interval $(0, \infty)$, adjusting c to make the approximation as close as possible. The inclination to use such an approximation is increased by the fact, to be proved in **23.7**, that as n increases, the distribution of $-2 \log l$ when H_0 holds tends to a chi-squared distribution with r degrees of freedom. In fact, we shall be able to find the asymptotic distribution of $-2 \log l$ when H_1 holds also, but in order to do this we must introduce a generalisation of the chi-squared distribution.

The non-central chi-squared distribution

23.4   We have seen in **16.2–3** that the sum of squares of n independent standardised normal variates is distributed in the χ^2 form with n degrees of freedom, (16.1), and c.f. given by (16.3). We now consider the distribution of the statistic

$$z = \sum_{i=1}^{n} x_i^2$$

where the x_i are still independent normal variates with unit variance, but where their means can differ from zero and

$$E(x_i) = \mu_i, \qquad \sum \mu_i^2 = \lambda. \tag{23.12}$$

We write the joint distribution of the x_i as

$$dF \propto \exp\left\{-\tfrac{1}{2}(\mathbf{x}-\boldsymbol{\mu})'(\mathbf{x}-\boldsymbol{\mu})\right\} \prod_i dx_i,$$

and make an orthogonal transformation to a new set of independent normal variates with variances unity,

$$\mathbf{y} = \mathbf{B}'\mathbf{x}.$$

Since

$$E(\mathbf{x}) = \boldsymbol{\mu},$$

we have

$$\boldsymbol{\theta} = E(\mathbf{y}) = \mathbf{B}'\boldsymbol{\mu},$$

so that

$$\boldsymbol{\theta}'\boldsymbol{\theta} = \boldsymbol{\mu}'\boldsymbol{\mu} = \lambda, \tag{23.13}$$

since $\mathbf{BB}' = \mathbf{I}$. We now choose the first $(n-1)$ components of $\boldsymbol{\theta}$ equal to zero. Then by (23.13),

$$\theta_n^2 = \lambda.$$

Thus

$$z = \mathbf{x}'\mathbf{x} = \mathbf{y}'\mathbf{y}$$

is a sum of squares of n independent normal variates, the first $(n-1)$ of which are standardised, and the last of which has mean $\lambda^{1/2}$ and variance 1. We write

$$u = \sum_{i=1}^{n-1} y_i^2, \qquad v = y_n^2$$

and we know that u is distributed like χ^2 with $(n-1)$ degrees of freedom. The distribution of y_n is

$$dF \propto \exp\left\{-\tfrac{1}{2}(y_n - \lambda^{1/2})^2\right\} dy_n,$$

so that the distribution of v is

$$f_1(v) \, dv \propto \frac{dv}{2v^{1/2}} [\exp\{-\tfrac{1}{2}(v^{1/2} - \lambda^{1/2})^2\} + \exp\{-\tfrac{1}{2}(-v^{1/2} - \lambda^{1/2})^2\}]$$

$$\propto v^{-1/2} \exp\{-\tfrac{1}{2}(v + \lambda)\} \sum_{r=0}^{\infty} \frac{(v\lambda)^r}{(2r)!} \, dv. \tag{23.14}$$

The joint distribution of v and u is

$$dG \propto f_1(v) f_2(u) \, dv \, du, \tag{23.15}$$

where f_2 is the χ^2 distribution with $(n-1)$ degrees of freedom

$$f_2(u) \, du \propto e^{-u/2} u^{(n-3)/2} \, du. \tag{23.16}$$

We put (23.14) and (23.16) into (23.15) and make the transformation

$$\left. \begin{aligned} z &= u + v, \\ w &= \frac{u}{u+v}, \end{aligned} \right\}$$

with Jacobian equal to z. We find for the joint distribution of z and w

$$dG(z, w) \propto e^{-(z+\lambda)/2} z^{(n-2)/2} w^{(n-3)/2} (1-w)^{-1/2} \sum_{r=0}^{\infty} \frac{\lambda^r z^r}{(2r)!} (1-w)^r \, dw \, dz.$$

We now integrate out w over its range from 0 to 1, getting for the marginal distribution of z

$$dH(z) \propto e^{-(z+\lambda)/2} z^{(n-2)/2} \sum_{r=0}^{\infty} \frac{\lambda^r z^r}{(2r)!} B\{\tfrac{1}{2}(n-1), \tfrac{1}{2}+r\} \, dz. \tag{23.17}$$

To obtain the constant in (23.17), we recall that it does not depend on λ, and put $\lambda = 0$. (23.17) should then reduce to (16.1), which is the ordinary χ^2 distribution with n degrees of freedom. The non-constant factors agree, but whereas (16.1) has a constant term $1/[2^{n/2}\Gamma(\tfrac{1}{2}n)]$, (23.17) has

$$B\{\tfrac{1}{2}(n-1), \tfrac{1}{2}\} = \frac{\Gamma\{\tfrac{1}{2}(n-1)\}\Gamma(\tfrac{1}{2})}{\Gamma(\tfrac{1}{2}n)}.$$

We must therefore divide (23.17) by the factor $2^{n/2}\Gamma\{\tfrac{1}{2}(n-1)\}\Gamma(\tfrac{1}{2})$ so that $B\{\tfrac{1}{2}(n-1), \tfrac{1}{2}+r\}$ becomes $\Gamma(\tfrac{1}{2}+r)/\{2^{n/2}\Gamma(\tfrac{1}{2})\Gamma(\tfrac{1}{2}n+r)\}$. Since by (3.66)

$$\Gamma(\tfrac{1}{2}+r)/\{\Gamma(\tfrac{1}{2})(2r)!\} = 1/\{2^{2r}r!\},$$

we have for any λ, writing ν for n,

$$dH(z) = \frac{e^{-(z+\lambda)/2} z^{(\nu-2)/2}}{2^{\nu/2}} \sum_{r=0}^{\infty} \frac{\lambda^r z^r}{2^{2r} r! \Gamma(\tfrac{1}{2}\nu + r)} \, dz. \tag{23.18}$$

Guenther (1964) gives a simple geometric derivation of (23.18) with references to other geometric proofs. McNolty (1962) obtains it by inverting its c.f., which the reader is asked to find (and to invert more simply) in Exercise 23.1.

As for central χ^2 in **16.2**, we see that (23.18) is valid for any $\nu > 0$ even if not an integer, and $\lambda \geqslant 0$; indeed, we can here (as we could not in **16.2**) allow $\nu = 0$ also if $\lambda > 0$—see Exercise 23.19, which expresses (23.18) as a Poisson mixture of central χ^2 variables: for $\nu = 0$, (23.18) holds for $z > 0$, with a discrete probability $e^{-\lambda/2}$ at $z = 0$.

23.5 The distribution (23.18) is called the non-central χ^2 distribution with ν degrees of freedom and non-central parameter* λ, and written $\chi'^2(\nu, \lambda)$. It was first given by Fisher (1928a), and has been studied by Wishart (1932), Patnaik (1949), Tiku (1965a) and Han (1975).

> The d.f. of z is discussed in Exercise 23.26. Johnson and Pearson (1969) give the 0.5, 1, 2.5, 5, 95, 97.5, 99 and 99.5 percentage points of the distribution of $\sqrt{z}$ for $\sqrt{\lambda} = 0.2$ (0.2) 6.0 and $\nu = 1$ (1) 12, 15, 20 to 4 significant figures, and also approximations for $\sqrt{\lambda} = 8$, 10. Their tables are reproduced in the *Biometrika Tables*, Vol. II.

Since the first two cumulants of χ'^2 are (cf. Exercise 23.1)

$$\left.\begin{array}{l} \kappa_1 = \nu + \lambda, \\ \kappa_2 = 2(\nu + 2\lambda), \end{array}\right\} \tag{23.19}$$

it can be approximated by a (central) χ^2 distribution as follows. The first two cumulants of a χ^2 with ν^* degrees of freedom are (putting $\lambda = 0$, $\nu = \nu^*$ in (23.19))

$$\kappa_1 = \nu^*, \qquad \kappa_2 = 2\nu^*. \tag{23.20}$$

If we equate the first two cumulants of χ'^2 with those of $\rho\chi^2$, where ρ is a constant to be determined, we have, from (23.19) and (23.20),

$$\left.\begin{array}{l} \nu + \lambda = \rho\nu^*, \\ 2(\nu + 2\lambda) = 2\rho^2\nu^*, \end{array}\right\}$$

so that χ'^2/ρ is approximately a central χ^2 variate with

$$\left.\begin{array}{l} \rho = \dfrac{\nu + 2\lambda}{\nu + \lambda} = 1 + \dfrac{\lambda}{\nu + \lambda}, \\[2mm] \nu^* = \dfrac{(\nu + \lambda)^2}{\nu + 2\lambda} = \nu + \dfrac{\lambda^2}{\nu + 2\lambda}, \end{array}\right\} \tag{23.21}$$

ν^* in general being fractional. If $\nu \to \infty$, $\rho \to 1$ and $\nu^* \sim \nu$, and it is easily seen in Exercise 23.1 that the c.f. of the approximation tends to the exact result; but if $\lambda \to \infty$, $\rho \to 2$ and $\nu^* \sim \frac{1}{2}\lambda$.

Patnaik (1949) shows that this approximation to the d.f. of χ'^2 is adequate for many purposes, but he also gives better approximations obtained from Edgeworth series expansions.

---

* In the literature, $\frac{1}{2}\lambda$ is sometimes used as the parameter, and occasionally λ^2 written for λ, but our notation is now the standard one.

If ν^* is large, we may make the approximation simpler by approximating the χ^2 approximating distribution itself, for (cf. **16.6**) $(2\chi'^2/\rho)^{1/2}$ tends to normality with mean $(2\nu^*-1)^{1/2}$ and variance 1, while, more slowly, χ'^2/ρ becomes normal with mean ν^* and variance $2\nu^*$.

E. S. Pearson (1959) gives a more accurate central χ^2 approximation by equating three moments. Johnson and Pearson (1969) fit a Pearson Type I distribution using four moments. Jensen and Solomon (1972) use the Wilson–Hilferty–Haldane method of **16.7** to find the fractional power that best normalises the distribution of χ'^2—it is

$$h = \frac{1}{3}\left\{1 + 2\bigg/\left(\frac{\nu}{\lambda}+2\right)^2\right\},$$

reducing to $h = \frac{1}{3}$ as in **16.7** when $\lambda = 0$.

Just as we saw in **15.11** that a weighted sum of central χ^2 variates may be approximated by a single central χ^2, Muller and Barton (1989) show that a positively weighted sum of χ'^2 variates may be approximated by a single χ'^2.

23.6 We may now generalize our derivation of **23.4**. Suppose that $\mathbf{x}$ is a vector of n multinormal variates with mean $\boldsymbol{\mu}$ and non-singular covariance matrix $\mathbf{V}$. We can find an orthogonal transformation $\mathbf{x} = \mathbf{B}\mathbf{y}$ that reduces the quadratic form $\mathbf{x}'\mathbf{V}^{-1}\mathbf{x}$ to the diagonal form $\mathbf{y}'\mathbf{B}'\mathbf{V}^{-1}\mathbf{B}\mathbf{y} = \mathbf{y}'\mathbf{C}\mathbf{y}$, the elements of the diagonal of $\mathbf{C}$ being the latent roots of $\mathbf{V}^{-1}$. To $\mathbf{y}'\mathbf{C}\mathbf{y}$ we apply a further scaling transformation $\mathbf{y} = \mathbf{D}\mathbf{z}$, where the leading diagonal elements of the diagonal matrix $\mathbf{D}$ are the reciprocals of the square roots of the corresponding elements of $\mathbf{C}$, so that $\mathbf{D}^2 = \mathbf{C}^{-1}$. Thus $\mathbf{x}'\mathbf{V}^{-1}\mathbf{x} = \mathbf{y}'\mathbf{C}\mathbf{y} = \mathbf{z}'\mathbf{z}$, and $\mathbf{z}$ is a vector of n independent normal variates with unit variances and mean vector $\boldsymbol{\theta}$ satisfying $\boldsymbol{\mu} = \mathbf{B}\mathbf{D}\boldsymbol{\theta}$, where $\lambda = \boldsymbol{\theta}'\boldsymbol{\theta} = \boldsymbol{\mu}'\mathbf{V}^{-1}\boldsymbol{\mu}$. We have now reduced our problem to that considered in **23.4**. We see that the distribution of $\mathbf{x}'\mathbf{V}^{-1}\mathbf{x}$, where $\mathbf{x}$ is a multinormal vector with covariance matrix $\mathbf{V}$ and mean vector $\boldsymbol{\mu}$, is a non-central χ^2 distribution with n degrees of freedom and non-central parameter $\boldsymbol{\mu}'\mathbf{V}^{-1}\boldsymbol{\mu}$. This generalises the result of **15.10** for multinormal variates with zero means.

Graybill and Marsaglia (1957) have generalised the theorems on the distribution of quadratic forms in normal variates, discussed in **15.10–21** and Exercises 15.13–18, to the case where $\mathbf{x}$ has mean $\boldsymbol{\mu} \neq \mathbf{0}$. Idempotency of a matrix is then a necessary and sufficient condition that its quadratic form is distributed in a non-central χ^2 distribution, and all the theorems of Chapter 15 hold with this modification, as we shall see in Chapter 29.

The asymptotic distribution of the LR statistic
23.7 We saw in **18.16** that under regularity conditions the ML estimator (temporarily written t) of a single parameter θ attains the MVB asymptotically. It follows from **17.17** that the LF is asymptotically of the form

$$\frac{\partial \log L}{\partial \theta} = -E\left(\frac{\partial^2 \log L}{\partial \theta^2}\right)(t-\theta), \tag{23.22}$$

which is the leading term (of order $n^{1/2}$) obtained by differentiating the logarithm of

$$L \propto \exp\left\{\tfrac{1}{2}E\left(\frac{\partial^2 \log L}{\partial \theta^2}\right)(t-\theta)^2\right\}, \tag{23.23}$$

showing that the LF reduces to the normal distribution of the 'asymptotically sufficient' statistic t.

For a k-component vector of parameters $\boldsymbol{\theta}$, the matrix analogue of (23.22) is

$$\frac{\partial \log L}{\partial \boldsymbol{\theta}} = (\mathbf{t} - \boldsymbol{\theta})'\mathbf{V}^{-1}, \tag{23.24}$$

where $\mathbf{V}^{-1}$ is defined by (cf. **18.26**)

$$V_{ij}^{-1} = -E\left(\frac{\partial^2 \log L}{\partial \theta_i \, \partial \theta_j}\right).$$

When integrated, (23.24) gives the analogue of (23.23)

$$L \propto \exp\{-\tfrac{1}{2}(\mathbf{t} - \boldsymbol{\theta})'\mathbf{V}^{-1}(\mathbf{t} - \boldsymbol{\theta})\}. \tag{23.25}$$

We saw in **18.26** that under regularity conditions the vector of ML estimators $\mathbf{t}$ is asymptotically multinormally distributed with covariance matrix $\mathbf{V}$. Thus the LF reduces to the multinormal distribution of $\mathbf{t}$. This result was rigorously proved by Wald (1943a).

We may now easily establish the asymptotic distribution of the LR statistic l defined at (23.4). From (23.25), we may reduce the problem to considering the ratio of the maximum of the right-hand side of (23.25) given H_0 to its maximum given H_1. When H_1 holds, the maximum of (23.25) is when $\boldsymbol{\theta} = \hat{\boldsymbol{\theta}} = \mathbf{t}$, so that every component of $(\mathbf{t} - \boldsymbol{\theta})$ is equal to zero and we have

$$L(x|\hat{\boldsymbol{\theta}}_r, \hat{\boldsymbol{\theta}}_s) \propto 1. \tag{23.26}$$

When H_0 holds, the s components of $(\mathbf{t} - \boldsymbol{\theta})$ corresponding to $\boldsymbol{\theta}_s$ will still be zero, for the maximum of (23.25) occurs when $\boldsymbol{\theta}_s = \hat{\hat{\boldsymbol{\theta}}}_s = \mathbf{t}_s$, say. (23.25) may now be written

$$L(x|\boldsymbol{\theta}_{r0}, \hat{\hat{\boldsymbol{\theta}}}_s) \propto \exp\{-\tfrac{1}{2}(\mathbf{t}_r - \boldsymbol{\theta}_{r0})'\mathbf{V}_r^{-1}(\mathbf{t}_r - \boldsymbol{\theta}_{r0})\}, \tag{23.27}$$

the suffix r signifying that we are now confined to an r-dimensional distribution. Thus, from (23.26) and (23.27),

$$l = \frac{L(x|\boldsymbol{\theta}_{r0}, \hat{\hat{\boldsymbol{\theta}}}_s)}{L(x|\hat{\boldsymbol{\theta}}_r, \hat{\boldsymbol{\theta}}_s)} = \exp\{-\tfrac{1}{2}(\mathbf{t}_r - \boldsymbol{\theta}_{r0})'\mathbf{V}_r^{-1}(\mathbf{t}_r - \boldsymbol{\theta}_{r0})\}.$$

Thus

$$-2 \log l = (\mathbf{t}_r - \boldsymbol{\theta}_{r0})'\mathbf{V}_r^{-1}(\mathbf{t}_r - \boldsymbol{\theta}_{r0}).$$

Now we have seen that $\mathbf{t}_r$ is multinormal with covariance matrix $\mathbf{V}_r$ and mean vector $\boldsymbol{\theta}_{r0}$. Thus, by the result of **23.6**, $-2 \log l$ for a hypothesis imposing r constraints is asymptotically distributed in the non-central χ^2 distribution with r degrees of freedom and non-central parameter

$$\lambda = (\boldsymbol{\theta}_r - \boldsymbol{\theta}_{r0})'\mathbf{V}_r^{-1}(\boldsymbol{\theta}_r - \boldsymbol{\theta}_{r0}), \tag{23.28}$$

a result due to Wald (1943a). If λ is to be bounded away from infinity, we must, since $\mathbf{V}_r^{-1}$ is of order n, restrict $(\boldsymbol{\theta}_r - \boldsymbol{\theta}_{r0})$ to be of order $n^{-1/2}$—for the case of fixed alternatives, see Stroud (1972). When H_0 holds, $\lambda = 0$ and this reduces to a central χ^2 distribution with r degrees of freedom, a result originally due to Wilks (1938a).

> A simple rigorous proof of the H_0 result is given by K. P. Roy (1957). Stroud (1973) proves the non-central result when sampling is from a member of the exponential family. The singular case is discussed by D. S. Moore (1977).
> Woodroofe (1978) studies the tail of the H_0 central approximation, and shows that it is accurate for deviations of $o(n)$.

Using the LR test is therefore asymptotically equivalent to basing a test on the ML estimators of the parameters tested. It should be emphasised that these results only hold if the conditions for the asymptotic normality and efficiency of the ML estimators are satisfied.

> See also the generalisations by Chernoff (1954) and Feder (1968, 1975b). Chant (1974) considers the asymptotic equivalence of other tests. Self and Liang (1987) obtain the asymptotic distribution of ML and LR statistics when the parameter has a boundary value. Foutz and Srivastava (1977) and Kent (1982) study the asymptotic distribution of l when the form of the f.f. is mis-specified.

The asymptotic power of LR tests

23.8 The result of **23.7** makes it possible to calculate the asymptotic power function of the LR test in any case satisfying the conditions for that result to be valid. We have first to evaluate the matrix $\mathbf{V}_r^{-1}$, and then to evaluate the integral

$$P = \int_{\chi_\alpha'^2(\nu,0)}^{\infty} d\chi'^2(\nu, \lambda) \tag{23.29}$$

where $\chi_\alpha'^2(\nu, 0)$ is the $100(1 - \alpha)$ per cent point of the central χ^2 distribution. P is the power of the test, and its size when $\lambda = 0$.

> Harter and Owen (1970) give tables, due to G. E. Hayman, Z. Govindarajulu and F. C. Leone, of P to 4 d.p. for $\alpha = 0.1, 0.05, 0.025, 0.01, 0.005, 0.001$; $\lambda = 0$ (0.1) 1 (0.2) 3 (0.5) 5 (1) 40 (2) 50 (5) 100; and $\nu = 1$ (1) 30 (2) 50 (5) 100; they also give to 3 d.p., for the same values of α and ν, inverse tables of the values of λ required to attain power $P = 0.1$ (0.02) 0.7 (0.01) 0.99. The *Biometrika Tables*, Vol. II, give 3 d.p. inverse tables, compiled from the same source, of the values of λ for $\alpha = 0.05, 0.01, P$ (their β) $= 0.25, 0.50, 0.60, 0.70$ (0.05) 0.95, 0.97, 0.99 and ν as above. Fix (1949) gave inverse tables of λ for $\nu = 1$ (1) 20 (2) 40 (5) 60 (10) 100, $\alpha = 0.05, 0.01$ and P (her β) $= 0.1$ (0.1) 0.9.
> Peers (1971) obtains an asymptotic series expansion for the power function of l against local alternatives when H_0 is simple, and compares it with some asymptotically equivalent tests.

If we use the approximation of **23.5** for the non-central distribution, (23.29) becomes, using (23.21) with $\nu = r$,

$$P = \int_{[(r+\lambda)/(r+2\lambda)]\chi_\alpha^2(r)}^{\infty} d\chi^2\left(r + \frac{\lambda^2}{r+2\lambda}\right), \tag{23.30}$$

where $\chi^2(r)$ is the central χ^2 distribution with r degrees of freedom and $\chi_\alpha^2(r)$ its $100(1 - \alpha)$ per cent point. Putting $\lambda = 0$ in (23.30) gives the size of the test.

The degrees of freedom in (23.30) are usually fractional, and interpolation in the tables of χ^2 is necessary.

From the fact that the non-central parameter λ defined by (23.28) is, under the regularity conditions assumed, a quadratic form with the elements of V_r^{-1} as coefficients, it follows, since the variances and covariances are of the order n^{-1}, that λ will have a factor n and hence that the power (23.29) tends to 1 as n increases—this is evident from the approximation (23.30). On the other hand, Das Gupta and Perlman (1974) show that for fixed λ, (23.29) is a decreasing function of ν—cf. Exercise 23.3.

Example 23.3

To test $H_0: \sigma^2 = \sigma_0^2$ for the normal distribution of Example 23.1. The unconditional ML estimators are as given there, so that (23.8) remains the unconditional maximum of the LF. Given our present H_0, the ML estimator of μ is $\hat{\hat{\mu}} = \bar{x}$ (Example 18.2). Thus

$$L(x \mid \hat{\hat{\mu}}, \sigma_0^2) = (2\pi\sigma_0^2)^{-n/2} \exp\left\{ -\frac{1}{2} \frac{\sum (x - \bar{x})^2}{\sigma_0^2} \right\}. \tag{23.31}$$

The ratio of (23.31) to (23.8) gives

$$l = \left(\frac{s^2}{\sigma_0^2} \right)^{n/2} \exp\left[-\tfrac{1}{2}n\left\{ \frac{s^2}{\sigma_0^2} - 1 \right\} \right],$$

so that

$$z = e^{-1} l^{2/n} = \frac{t}{n} e^{-t/n}, \tag{23.32}$$

where $t = ns^2/\sigma_0^2$. z is a monotone function of l, but is not a monotone function of t/n, its derivative being

$$\frac{dz}{dt} = \frac{1}{n}\left(1 - \frac{t}{n} \right) e^{-t/n},$$

so that z increases steadily for $t < n$ to a maximum at $t = n$ and then decreases steadily. Putting $l \leqslant c_\alpha$ is therefore equivalent to putting

$$t \leqslant a_\alpha, \qquad t \geqslant b_\alpha,$$

where a_α, b_α are determined, using (23.32), by

$$\left. \begin{array}{c} P\{t \leqslant a_\alpha\} + P\{t \geqslant b_\alpha\} = \alpha, \\ a_\alpha\, e^{-a_\alpha/n} = b_\alpha\, e^{-b_\alpha/n}. \end{array} \right\} \tag{23.33}$$

Since the statistic t has a χ^2 distribution with $(n-1)$ d.fr. when H_0 holds, we can use tables of that distribution to satisfy (23.33).

Now consider the approximate distribution of

$$-2 \log l = (t - n) - n \log (t/n).$$

Since $E(t) = n - 1$, var $t = 2(n - 1)$, we may write

$$-2 \log l = (t - n) - n \log \left\{ 1 + \frac{t - n}{n} \right\}$$

$$= (t - n) - n \sum_{r=1}^{\infty} (-1)^{r-1} \left(\frac{t - n}{n} \right)^r / r$$

$$= (t - n) - n \left\{ \frac{t - n}{n} - \frac{(t - n)^2}{2n^2} + o(n^{-2}) \right\}$$

$$= \frac{(t - n)^2}{2n} + o(n^{-1}). \tag{23.34}$$

We have seen (16.3) that, as $n \to \infty$, a χ^2 distribution with $(n - 1)$ degrees of freedom is asymptotically normally distributed with mean $(n - 1)$ and variance $2(n - 1)$; or equivalently, that $(t - n)/(2n)^{1/2}$ tends to a standardised normal variate. Its square, the first term on the right of (23.34), is therefore a χ^2 variate with 1 degree of freedom. This is precisely the distribution of $-2 \log l$ given by the general result of 23.7 when H_0 holds. This result also tells us that when H_0 is false, $-2 \log l$ has a non-central χ^2 distribution with 1 degree of freedom and non-central parameter, by (23.28),

$$\lambda = -E \left\{ \frac{\partial^2 \log L}{\partial (\sigma^2)^2} \right\} (\sigma^2 - \sigma_0^2)^2 = \frac{n}{2\sigma^4} (\sigma^2 - \sigma_0^2)^2 = \frac{n}{2} \left(1 - \frac{\sigma_0^2}{\sigma^2} \right)^2.$$

Thus the expression (23.30) for the approximate power of the LR test in this case, where $r = 1$, is

$$P = \int_{[(1+\lambda)/(1+2\lambda)]\chi_\alpha^2(1)}^{\infty} d\chi^2 \left(1 + \frac{\lambda^2}{1 + 2\lambda} \right). \tag{23.35}$$

For illustrative purposes we shall evaluate P for one value of λ and of n conveniently chosen. Choose $\chi_\alpha^2(1) = 3.84$ to give a test of size 0.05. Consider the alternative $\sigma^2 = \sigma_1^2 = 1.25\sigma_0^2$. We then have $\lambda = 0.02n$, and we choose $n = 50$ to give $\lambda = 1$. (23.35) is then

$$P = \int_{2.56}^{\infty} d\chi^2 \left(\frac{4}{3} \right),$$

and from the *Biometrika Tables* we find by simple interpolation between 1 and 2 degrees of freedom that $P = 0.166$ approximately. The exact power may be obtained from the normal d.f.: it is the power of an equal-tails size-α test against an alternative with mean $\lambda^{1/2} = 1$ standard deviations distant from the mean on H_0, i.e. the proportion of the alternative distribution lying outside the interval $(-2.96, +0.96)$ standard deviations from its mean. The normal tables give the value $P = 0.170$. The approximation to the power function is thus quite accurate enough.

Closer approximations to the distribution of the LR statistic

23.9 Confining ourselves now to the distribution of l, when H_0 holds, we may seek closer approximations than the asymptotic result of 23.7. As indicated in 23.3, if we wish to find χ^2 approximations to the distribution of a function of l, we can gain some flexibility by considering the distribution of $w = -2c \log l$ and adjusting c to improve the approximation.

The simplest way of doing this would be to find the expected value of w and adjust c so that

$$E(w) = r,$$

the expectation of a χ^2 variate with r degrees of freedom. An approximation of this kind was first given by Bartlett (1937), and a general method for deriving the value of c has been given by Lawley (1956), who uses essentially the methods of **20.18** to investigate the moments of $-2 \log l$. If

$$E(-2 \log l) = r\left\{1 + \frac{a}{n} + O\left(\frac{1}{n^2}\right)\right\}, \tag{23.36}$$

Lawley shows that by putting either

$$\left.\begin{array}{c} w_1 = -2\left(\dfrac{1}{1+\dfrac{a}{n}}\right) \log l \\[3em] \text{or} \\[1em] w_2 = -2\left(1-\dfrac{a}{n}\right) \log l \end{array}\right\} \tag{23.37}$$

we not only have

$$E(w) = r + O\left(\frac{1}{n^2}\right),$$

which follows immediately from (23.36) and (23.37), but also that *all* the cumulants of w conform, to order n^{-1}, with those of a χ^2 distribution with r degrees of freedom. In the continuous case, this simple scaling correction which adjusts the mean of w to the correct value is an unequivocal improvement. Barndorff-Nielsen and Cox (1984) demonstrate this by a quite different argument, using the saddlepoint method of **11.13–16**. See also Jensen (1986), Barndorff-Nielsen and Hall (1988), and McCullagh and Cox (1986), who show that the correction can be expressed in terms of the cumulants of the first- and second-order derivatives of the log-LF. Ross (1987) obtains related results. Frydenberg and Jensen (1989) show that in the discrete case, the improvement is less consistent.

> If even closer approximations are required, they can be obtained in a large class of situations by methods due to G. E. P. Box (1949), who gives improved χ^2 approximations, shows how to derive a function of $-2 \log l$ which is distributed in the variance-ratio distribution, and also derives an asymptotic expansion for its distribution function in terms of Incomplete Gamma functions. See also asymptotic expansions of the distribution of l under local alternatives to H_0, given for simple H_0 by Peers (1971) and for composite H_0 by Hayakawa (1975, 1977), and related work by Harris (1985, 1986) and Cordeiro (1987).
>
> Cordeiro (1983, 1987) obtains the expected value of $-2 \log l$ for members of the natural exponential family.

Example 23.4

k independent samples of sizes n_i $(i = 1, 2, \ldots, k; \; n_i \geq 2)$ are taken from different normal populations with means μ_i and variances σ_i^2. To test

$$H_0: \sigma_1^2 = \sigma_2^2 = \cdots = \sigma_k^2,$$

a composite hypothesis imposing the $r = k - 1$ constraints

$$\frac{\sigma_2^2}{\sigma_1^2} = \frac{\sigma_3^2}{\sigma_1^2} = \cdots = \frac{\sigma_k^2}{\sigma_1^2} = 1.$$

Call the common unknown value of the variances σ^2.

The unconditional maximum of the LF is obtained, just as in Example 23.1, by putting

$$\left. \begin{aligned} \hat{\mu}_i &= \bar{x}_i, \\ \hat{\sigma}_i^2 &= \frac{1}{n_i} \sum_{j=1}^{n_i} (x_{ij} - \bar{x}_i)^2 = s_i^2, \end{aligned} \right\}$$

giving

$$L(x \mid \hat{\mu}_1, \ldots, \hat{\mu}_k, \hat{\sigma}_1^2, \ldots, \hat{\sigma}_k^2) = (2\pi)^{-n/2} \prod_{i=1}^{k} (s_i^2)^{-n_i/2} \, e^{-n/2}, \qquad (23.38)$$

where

$$n = \sum_{i=1}^{k} n_i.$$

Given H_0, the ML estimators of the means and the common variance σ^2 are

$$\left. \begin{aligned} \hat{\hat{\mu}}_i &= \bar{x}_i, \\ \hat{\hat{\sigma}}^2 &= \frac{1}{n} \sum_{i=1}^{k} n_i s_i^2 = s^2, \end{aligned} \right\}$$

so that

$$L(x \mid \hat{\hat{\mu}}_1, \ldots, \hat{\hat{\mu}}_k, \hat{\hat{\sigma}}^2) = (2\pi)^{-n/2} (s^2)^{-n/2} \, e^{-n/2}. \qquad (23.39)$$

From (23.4), (23.38) and (23.39),

$$l = \prod_{i=1}^{k} \left(\frac{s_i^2}{s^2} \right)^{n_i/2}, \qquad (23.40)$$

so that

$$-2 \log l = n \log (s^2) - \sum_{i=1}^{k} n_i \log (s_i^2). \qquad (23.41)$$

Now when H_0 holds, each of the statistics $(n_i s_i^2 / 2\sigma^2)$ has a gamma distribution with parameter $\frac{1}{2}(n_i - 1)$, and their sum $ns^2 / 2\sigma^2$ has the same distribution with parameter

$\sum_{i=1}^{k} \frac{1}{2}(n_i - 1) = \frac{1}{2}(n - k)$. For a gamma variate x with parameter p, we have

$$E\{\log (ax)\} = \frac{1}{\Gamma(p)} \int_0^\infty \log (ax)\, e^{-x} x^{p-1}\, dx$$

$$= \log a + \frac{d}{dp} \log \Gamma(p),$$

which, using Stirling's series (3.63), becomes

$$E\{\log (ax)\} = \log a + \log p - \frac{1}{2p} - \frac{1}{12p^2} + O\left(\frac{1}{p^3}\right). \tag{23.42}$$

Using (23.42) in (23.41), we have

$$E\{-2 \log l\} = n \left\{\log \left(\frac{2\sigma^2}{n}\right) + \log \{\tfrac{1}{2}(n - k)\} - \frac{1}{(n-k)} - \frac{1}{3(n-k)^2} + O\left(\frac{1}{n^3}\right)\right\}$$

$$- \sum_{i=1}^{k} n_i \left\{\log \left(\frac{2\sigma^2}{n_i}\right) + \log \{\tfrac{1}{2}(n_i - 1)\} - \frac{1}{(n_i-1)} - \frac{1}{3(n_i-1)^2} + O\left(\frac{1}{n_i^3}\right)\right\}$$

$$= n \left\{\log \left(1 - \frac{k}{n}\right) - \frac{1}{(n-k)} - \frac{1}{3(n-k)^2} + O\left(\frac{1}{n^3}\right)\right\}$$

$$- \sum_{i=1}^{k} n_i \left\{\log \left(1 - \frac{1}{n_i}\right) - \frac{1}{(n_i-1)} - \frac{1}{3(n_i-1)^2} + O\left(\frac{1}{n_i^3}\right)\right\}$$

$$= (k-1) + \left[\left(\sum_{i=1}^{k} \frac{1}{n_i - 1} - \frac{k}{n-k}\right) + \frac{1}{2}\left(\sum_{i=1}^{k} \frac{1}{n_i} - \frac{k^2}{n}\right)\right.$$

$$\left. + \frac{1}{3}\left\{\sum_{i=1}^{k} \frac{n_i}{(n_i-1)^2} - \frac{n}{(n-k)^2}\right\}\right] + O\left(\frac{1}{N^3}\right), \tag{23.43}$$

where we now write N indifferently for n_i and n. We could now improve the χ^2 approximation, in accordance with (23.37), with the expression in square brackets in (23.43) as $(k-1)(a/n)$.

Now consider Bartlett's (1937) modification of the LR statistic (23.40) in which n_i is replaced throughout by the degrees of freedom $n_i - 1 = \nu_i$, so that n is replaced by $\nu = \sum_{i=1}^{k} (n_i - 1) = n - k$. We write this

$$l^* = \prod_{i=1}^{k} \left(\frac{s_i^2}{s^2}\right)^{\nu_i/2},$$

where now

$$\left.\begin{array}{l} s_i^2 = \dfrac{1}{\nu_i} \sum_{j=1}^{n_i} (x_{ij} - \bar{x}_i)^2, \\[2mm] s^2 = \dfrac{1}{\nu} \sum_{i=1}^{k} \nu_i s_i^2. \end{array}\right\}$$

Thus

$$-2 \log l^* = \nu \log s^2 - \sum_{i=1}^{k} \nu_i \log s_i^2. \tag{23.44}$$

We shall see in Example 23.6 that l^* has the advantage over l that it gives an unbiased test for any values of the n_i. If we retrace the passage from (23.42) to (23.43), we find that

$$E(-2 \log l^*) = -\nu \left\{ \frac{1}{\nu} + \frac{1}{3\nu^2} + O\left(\frac{1}{\nu^3}\right) \right\} + \sum_{i=1}^{k} \nu_i \left\{ \frac{1}{\nu_i} + \frac{1}{3\nu_i^2} + O\left(\frac{1}{\nu_i^3}\right) \right\}$$

$$= (k-1) + \frac{1}{3} \left(\sum_{i=1}^{k} \frac{1}{\nu_i} - \frac{1}{\nu} \right) + O\left(\frac{1}{\nu_i^3}\right). \tag{23.45}$$

From (23.37) and (23.45) it follows that $-2 \log l^*$ defined at (23.44) should be divided by the scaling constant

$$1 + \frac{1}{3(k-1)} \left(\sum_{i=1}^{k} \frac{1}{\nu_i} - \frac{1}{\nu} \right)$$

to give a closer approximation to a χ^2 distribution with $(k-1)$ degrees of freedom.

> For $k = 3$ (1) 10 and all $\nu_i = 4$ (1) 11, 14 (5) 29, 49, 99, Glaser (1976) gives exact critical values of l^* for $\alpha = 0.01, 0.05, 0.10$.
>
> For $k = 3$ (1) 12 and all $\nu_i = 1$ (1) 10, Harsaae (1969) gives exact 3 d.p. critical values of $-2 \log l^*$ for α and $1 - \alpha = 0.001, 0.01, 0.05, 0.10$; Glaser (1976) gives 4 d.p. values of $(l^*)^{2/\nu}$ for $k = 3$ (1) 10, all $\nu_i = 4$ (1) 11, 14 (5) 29, 49, 99 and $\alpha = 0.01, 0.05, 0.10$, as do Dyer and Keating (1980) for $k = 2$ (1) 10, all $\nu_i = 2$ (1) 29 (10) 59 (20) 99 and $\alpha = 0.01, 0.05, 0.10, 0.25$. For unequal ν_i Chao and Glaser (1975) obtained the exact probability of l^* in usable form, and Glaser (1980) gave a computable algorithm; Nagarsenker (1984) obtained a simpler representation of the distribution in terms of Incomplete Beta functions, yielding an asymptotic expansion that gives a close approximation; Dyer and Keating (1980) give an excellent approximation based on a weighted average of the equal-ν_i critical values. Approximate critical values are given in the *Biometrika Tables* for $k = 3(1)15$, $(\sum_{i=1}^{k} \nu_i^{-1} - \nu^{-1}) = 0$ (0.5) 5 (1) 10 (2) 14 and $\alpha = 0.01, 0.05$.
>
> Rivest (1986) shows that the test size assuming normality is an understatement for various classes of distribution when all ν_i are equal.

LR tests when the range depends upon the parameter

23.10   The asymptotic distribution of the LR statistic, given in **23.7**, depends essentially on the regularity conditions necessary to establish the asymptotic normality of ML estimators. We have seen in Example 18.5 that these conditions break down where the range of the underlying distribution is a function of the parameter θ. What can be said about the distribution of the LR statistic in such cases? It is a remarkable fact that, as Hogg (1956) showed, for certain hypotheses concerning such distributions the statistic $-2 \log l$ is distributed *exactly* as χ^2, but with $2r$ degrees of freedom, i.e. twice as many as there are constraints imposed by the hypothesis.

23.11   We first derive some preliminary results concerning uniform distributions. If k variables z_i are independently distributed as

$$dF = dz_i, \qquad 0 \leqslant z_i \leqslant 1, \tag{23.46}$$

the distribution of

$$t_i = -2 \log z_i$$

is at once seen to be

$$dG = \tfrac{1}{2} \exp\left(-\tfrac{1}{2}t_i\right) dt_i, \qquad 0 \le t_i < \infty,$$

a χ^2 distribution with 2 degrees of freedom, so that the sum of k such independent variates

$$t = \sum_{i=1}^{k} t_i = -2 \sum_{i=1}^{k} \log z_i = -2 \log \prod_{i=1}^{k} z_i$$

has a χ^2 distribution with $2k$ degrees of freedom.

It follows from (14.1) that the distribution of $y_{(n_i)}$, the largest among n_i independent observations from a uniform distribution on the interval $(0, 1)$, is

$$dH = n_i y_{(n_i)}^{n_i - 1} \, dy_{(n_i)}, \qquad 0 \le y_{(n_i)} \le 1, \tag{23.47}$$

and hence that $y_{(n_i)}^{n_i} = z_i$ is uniformly distributed as in (23.46). Hence for k independent samples of size n_i, $t = -2 \log \prod_{i=1}^{k} y_{(n_i)}^{n_i}$ has a χ^2 distribution with $2k$ degrees of freedom.

Now consider the distribution of the largest of the k largest values $y_{(n_i)}$. Since all the observations are independent, this is simply the largest of $n = \sum_{i=1}^{k} n_i$ observations from the original uniform distribution. If we denote this largest value by $y_{(n)}$ the distribution of $-2 \log y_{(n)}^n$ will, by the argument above, be a χ^2 distribution with 2 degrees of freedom. We now show that the statistics $y_{(n)}$ and $u = \prod_{i=1}^{k} y_{(n_i)}^{n_i} / y_{(n)}^n$ are independently distributed. Introduce the parameter θ, so that the original uniform distribution is on the interval $(0, \theta)$. The joint density of the $y_{(n_i)}$ then becomes, from (23.47),

$$f = \prod_{i=1}^{k} \{ n_i y_{(n_i)}^{n_i - 1} / \theta^{n_i} \} = \frac{1}{\theta^n} \prod_{i=1}^{k} n_i y_{(n_i)}^{n_i}.$$

By **17.40**, $y_{(n)}$ is sufficient for θ, and by **22.12** its distribution is complete. Thus by Exercise 22.7 we need only observe that the distribution of u is free of the parameter θ to establish the result that u is distributed independently of the complete sufficient statistic $y_{(n)}$. We see that

$$(-2 \log u) + (-2 \log y_{(n)}^n) = -2 \log \prod_{i=1}^{k} y_{(n_i)}^{n_i},$$

and since $y_{(n)}$ and u are independent, $y_{(n)}^n$ and u are. The c.f. of a sum of independent variates is the product of their c.f.s (cf. **7.28**), so that if $\phi(t)$ is the c.f. of $-2 \log u$, we have

$$\phi(t) \cdot (1 - 2it)^{-1} = (1 - 2it)^{-k},$$

using the χ_2^2 and χ_{2k}^2 results just derived, whence

$$\phi(t) = (1 - 2it)^{-(k-1)},$$

so that $-2 \log u$ has a χ^2 distribution with $2(k-1)$ degrees of freedom.

Collecting our results finally, we have established that if we have k variates $y_{(n_i)}$ independently distributed as in (23.47), then $-2 \log \prod_{i=1}^{k} y_{(n_i)}^{n_i}$ has a χ^2 distribution with $2k$ degrees of freedom, while, if $y_{(n)}$ is the largest of the $y_{(n_i)}$,

$$-2 \log \left\{ \prod_{i=1}^{k} y_{(n_i)}^{n_i} / y_{(n)}^n \right\}$$

has a χ^2 distribution with $2(k-1)$ degrees of freedom.

23.12 We now consider in turn the two classes of situation in which a single sufficient statistic exists for θ when the range is a function of θ, taking first the case when only one terminal (say, the upper) depends on θ. We then have, from **17.40**, the necessary form for the frequency function

$$f(x \mid \theta) = g(x)/h(\theta), \qquad a \leqslant x \leqslant \theta. \tag{23.48}$$

Now suppose we have k ($\geqslant 1$) separate populations of this form, $f(x_i \mid \theta_i)$, and wish to test

$$H_0: \theta_1 = \theta_2 = \cdots = \theta_k = \theta_0,$$

a simple hypothesis imposing k constraints, on the basis of samples of sizes n_i ($i = 1, 2, \ldots, k$). We now find the LR criterion for H_0. The unconditional ML estimator of θ_i is the largest observation $x_{(n_i)}$ (cf. Example 18.1). Thus

$$L(x \mid \hat{\theta}_1, \ldots, \hat{\theta}_k) = \prod_{i=1}^{k} \prod_{j=1}^{n_i} \{g(x_{ij})/[h(x_{(n_i)})]^{n_i}\}. \tag{23.49}$$

Since H_0 is simple, the LF when it holds is determined and no ML estimation is necessary. We have

$$L(x \mid \theta_0, \theta_0, \ldots, \theta_0) = \prod_{i=1}^{k} \prod_{j=1}^{n_i} \{g(x_{ij})\}/[h(\theta_0)]^n.$$

Hence the LR statistic is

$$l = \frac{L(x \mid \theta_0, \ldots, \theta_0)}{L(x \mid \hat{\theta}_1, \ldots, \hat{\theta}_k)} = \prod_{i=1}^{k} \left[\frac{h(x_{(n_i)})}{h(\theta_0)} \right]^{n_i}. \tag{23.50}$$

When H_0 holds, $y_i = h(x_{(n_i)})/h(\theta_0)$ is the probability that an observation falls below or at $x_{(n_i)}$ and is itself a random variable with distribution obtained from that of $x_{(n_i)}$ as

$$dF = n_i y_i^{n_i - 1} dy_i, \qquad 0 \leqslant y_i \leqslant 1,$$

of the form (23.47). Thus, from the result of the previous section,

$$-2 \log \prod_{i=1}^{k} y_i^{n_i} = -2 \log l$$

has a χ^2 distribution with $2k$ degrees of freedom.

23.13   We now investigate the composite hypothesis, for $k \geqslant 2$ populations,

$$H_0: \theta_1 = \theta_2 = \cdots = \theta_k$$

which imposes $(k-1)$ constraints, leaving the common value of θ unspecified. The unconditional maximum of the LF is given by (23.49) as before. The maximum under our present H_0 is $L(x \mid \hat{\theta}, \hat{\theta}, \ldots, \hat{\theta})$, where $\hat{\theta}$ is the ML estimator for the pooled samples, which is $x_{(n)}$. Thus we have the LR statistic

$$l = \frac{L(x \mid \hat{\theta}, \ldots, \hat{\theta})}{L(x \mid \hat{\theta}_1, \ldots, \hat{\theta}_k)} = \prod_{i=1}^{k} \frac{[h(x_{(n_i)})]^{n_i}}{[h(x_{(n)})]^{n}}. \tag{23.51}$$

By writing this as

$$l = \prod_{i=1}^{k} \left[\frac{h(x_{(n_i)})}{h(\theta)} \right]^{n_i} \Big/ \left[\frac{h(x_{(n)})}{h(\theta)} \right]^{n},$$

where θ is the common unspecified value of the θ_i, we see that in the notation of the previous section,

$$l = \left[\prod_{i=1}^{k} y_{(n_i)}^{n_i} \right] \Big/ y_{(n)}^{n},$$

so that by **23.11** we have that in this case $-2 \log l$ is distributed like χ^2 with $2(k-1)$ degrees of freedom.

23.14   When both terminals of the range are functions of θ, we have from **17.41** that if there is a single sufficient statistic for θ, then

$$f(x \mid \theta) = g(x)/h(\theta), \qquad \theta \leqslant x \leqslant b(\theta), \tag{23.52}$$

where $b(\theta)$ must be a monotone decreasing function of θ. For $k \geqslant 1$ such populations $f(x_i \mid \theta_i)$, we again test the simple

$$H_0: \theta_1 = \theta_2 = \cdots = \theta_k = \theta_0$$

on the basis of samples of sizes n_i. The unconditional ML estimator of θ_i is the sufficient statistic

$$t_i = \min \{x_{(1i)}, b^{-1}(x_{(ni)})\},$$

where $x_{(1i)}, x_{(ni)}$ are respectively the smallest and largest observations in the ith sample. When H_0 holds, the LF is specified by $L(x \mid \theta_0, \ldots, \theta_0)$. Thus the LR statistic

$$l = \frac{L(x \mid \theta_0, \ldots, \theta_0)}{L(x \mid \hat{\theta}_1, \ldots, \hat{\theta}_k)} = \prod_{i=1}^{k} \left[\frac{h(t_i)}{h(\theta_0)} \right]^{n_i}. \tag{23.53}$$

Just as for (23.50), we see that

$$l = \prod_{i=1}^{k} y_i^{n_i},$$

where the y_i are distributed in the form (23.47), and hence $-2 \log l$ again has a χ^2 distribution with $2k$ degrees of freedom.

Similarly for the composite hypothesis with $(k-1)$ constraints $(k \geqslant 2)$

$$H_0 : \theta_1 = \theta_2 = \cdots = \theta_k,$$

we find, just as in **23.13**, that the LR statistic is

$$l = \frac{L(x \mid \hat{\hat{\theta}}, \ldots, \hat{\hat{\theta}})}{L(x \mid \hat{\theta}_1, \ldots, \hat{\theta}_k)} = \prod_{i=1}^{k} [h(t_i)]^{n_i} / [h(t)]^n$$

where $t = \min \{t_i\}$ is the combined ML estimator $\hat{\hat{\theta}}$, so that by writing

$$l = \prod_{i=1}^{k} \left[\frac{h(t_i)}{h(\theta)} \right]^{n_i} / \left[\frac{h(t)}{h(\theta)} \right]^n$$

we again reduce l to the form required in **23.11** for $-2 \log l$ to be distributed like χ^2 with $2(k-1)$ degrees of freedom.

23.15   We have thus obtained exact χ^2 distributions for two classes of hypotheses concerning distributions whose terminals depend upon the parameter being tested. Exercises 23.8 and 23.9 give further examples, one exact and one asymptotic, of LR tests for which $-2 \log l$ has a χ^2 distribution with twice as many degrees of freedom as there are constraints imposed by the hypothesis tested. It will have been noted that these χ^2 forms spring not from any tendency to multinormality on the part of the ML estimators, as did the limiting results of **23.7** for 'regular' situations, but from the intimate connection between the uniform and χ^2 distributions explored in **23.11**. One effect of this difference is that the power functions of the tests take a quite different form from that obtained by use of the non-central χ^2 distribution in **23.8**.

Barr (1966) finds the power function of the test of the simple hypothesis based on the LR statistic (23.50), and shows that the test is unbiased. When $k = 1$, the test is shown to be UMP (cf. Example 21.6 and the condition (21.37)), but there is no UMP test for $k > 1$, and the LR test is not even UMPU. For the composite hypothesis of **23.13** with $k = 2$, the power function of the LR test statistic (23.51) is used to show that the LR test is UMPU.

The properties of LR tests
23.16   So far, we have been concerned entirely with the problems of determining the distribution of the LR statistic, or a function of it. We now have to inquire into the properties of LR tests, in particular the question of their unbiasedness and whether they are optimum tests in any sense. First, however, we consider a weaker property, that of consistency, which we now define for the first time.

Test consistency
23.17   A test of a hypothesis H_0 against a class of alternatives H_1 is said to be consistent if, when any member of H_1 holds, the probability of rejecting H_0 tends to

1 as sample size(s) tend to infinity. If w is the critical region, and $\mathbf{x}$ the sample point, we write this

$$\lim_{n \to \infty} P\{\mathbf{x} \in w \,|\, H_1\} = 1. \qquad (23.54)$$

The idea of test consistency, which is a simple and natural one, was first introduced by Wald and Wolfowitz (1940). It seems perfectly reasonable to require that, as the number of observations increases, any test worth considering should reject a false hypothesis with increasing certainty, and in the limit with complete certainty. Test consistency is as intrinsically acceptable as is consistency in estimation (**17.7**), of which it is in one sense a generalisation. For if a test concerning the value of θ is based on a statistic that is a consistent estimator of θ, it is immediately obvious that the test will be consistent too. But an inconsistent estimator may still provide a consistent test. For example, if t tends in probability to $a\theta$, t will give a consistent test of hypotheses about θ. In general, it is clear that it is sufficient for test consistency that the test statistic, when regarded as an estimator, should tend in probability to some one-to-one function of θ.

Since the condition that a size-α test be unbiased is (cf. (22.53)) that

$$P\{\mathbf{x} \in w \,|\, H_1\} \geq \alpha, \qquad (23.55)$$

it is clear from (23.54) and (23.55) that a consistent test will lose its bias, if any, as $n \to \infty$. However, an unbiased test need not be consistent.*

The consistency and unbiasedness of LR tests
23.18   We saw in **18.10**, **18.16** and **18.26** that under certain conditions, the ML estimator $\hat{\theta}$ of a parameter-vector θ is consistent, though in other circumstances it need not be. If we take it that we are dealing with a situation in which all the ML estimators are consistent, we see from the definition of the LR statistic at (23.4) that, as sample sizes increase,

$$l \to \frac{L(x \,|\, \theta_{r0}, \theta_s)}{L(x \,|\, \theta_r, \theta_s)}, \qquad (23.56)$$

where θ_r, θ_s are the true values of those parameters, and θ_{r0} is the hypothetical value of θ_r being tested. Thus, when H_0 holds, $l \to 1$ in probability, and the critical region (23.6) will therefore have its boundary c_α approaching 1. When H_0 does not hold, the limiting value of l in (23.56) will be some constant k satisfying (cf. (18.20))

$$0 \leq k < 1$$

and thus we have

$$P\{l \leq c_\alpha\} \to 1 \qquad (23.57)$$

and the LR test is consistent.

In **23.8** we confirmed from the approximate power function that LR tests are consistent under regularity conditions, in conformity with our present discussion.

---

* Cf. the remark in **17.9** on consistent and asymptotically unbiased estimators.

23.19   When we turn to the question of unbiasedness, we recall the penultimate sentence of **23.17** which, coupled with the result of **23.18**, ensures that most LR estimators are asymptotically unbiased. Of itself, this is not very comforting (though it would be so if it could be shown under reasonable restrictions that the maximum extent of the bias is always small), for the criterion of unbiasedness in tests is intuitively attractive enough to impose itself as a necessity for all sample sizes. Example 23.5 shows that an important LR test is biased.

Example 23.5

Consider again the hypothesis H_0 of Example 23.3. The LR test uses as its critical region the tails of the χ^2_{n-1} distribution of $t = ns^2/\sigma_0^2$ determined by (23.33). Now in Examples 22.12 and 22.14 we saw that the unbiased (actually UMPU) test of H_0 was determined from the distribution of t by the relations

$$\left. \begin{array}{l} P\{t \leqslant a_\alpha\} + P\{t \geqslant b_\alpha\} = \alpha, \\ a_\alpha^{(n-1)/2} \exp\left(-a_\alpha/2\right) = b_\alpha^{(n-1)/2} \exp\left(-b_\alpha/2\right), \end{array} \right\} \tag{23.58}$$

the second of which is (22.58) in the notation of Example 23.3. It is clear on comparison of (23.58) with (23.33) that they would only give the same result if $a_\alpha = b_\alpha$, which cannot hold except in the trivial case $a_\alpha - b_\alpha = 0$, $\alpha = 1$. In all other cases, the tests have different critical regions, the LR test having higher values of a_α and b_α than the unbiased test, i.e. a larger fraction of α concentrated in the lower tail. It is easy to see that for alternative values of σ^2 just larger than σ_0^2, for which the distribution of t is slightly displaced towards higher values of t compared to its H_0 distribution, the probability content lost to the critical region of the LR test through its larger value of b_α will exceed the gain due to the larger value of a_α; and thus the LR test will be biased.

It will be seen by reference to Example 22.12 that whereas the LR test has values of a_α, b_α too large for unbiasedness, the equal-tails test there discussed has values too small for unbiasedness. Thus the two more or less intuitively acceptable critical regions 'bracket' the unbiased critical region.

If in (23.33) we replace n by $n - 1$, it becomes precisely equivalent to the unbiased (23.58), confirming the general fact that the LR test loses its bias asymptotically. It is suggestive to trace this bias to its source. If, in constructing the LR statistic in Example 23.3, we had adjusted the unconditional ML estimator of σ^2 to be unbiased, s^2 would have been replaced by $[n/(n-1)]s^2$, and the adjusted LR test would have been unbiased: the estimation bias of the ML estimator lies behind the test bias of the LR test.

> The test corresponding to the physically shortest confidence intervals based on s^2, given in Exercise 20.15, replaces n in (23.58) by $(n+2)$, so its critical values will be above even those of the LR test.
>
> Spjøtvoll (1972b) showed that LR tests—he discussed the corresponding integrals—are unbiased under certain regularity conditions if there is no nuisance parameter. Peers (1971) showed in this simple H_0 case that l is unbiased against local alternatives, and compared it with asymptotically equivalent tests, but Hayakawa (1975, 1977) (see also a correction by

Harris (1986)) and Harris and Peers (1980) showed that the result does not hold generally for composite H_0.

Unbiased invariant tests for location and scale parameters

23.20   Example 23.5 suggests that a good principle in constructing an LR test is to adjust all the ML estimators used in the process so that they are unbiased. A further confirmation of this principle is contained in Example 23.4, where we stated that the adjusted LR statistic l^* gives an unbiased test. We now prove this, developing a method due to Pitman (1939b) for this purpose.

If the hypothesis being tested concerns a set of k location parameters θ_i $(i = 1, 2, \ldots, k)$, we write the joint distribution of the variates as

$$dF = f(x_1 - \theta_1, x_2 - \theta_2, \ldots, x_k - \theta_k) \, dx_1 \cdots dx_k. \tag{23.59}$$

We wish to test

$$H_0: \theta_1 = \theta_2 = \cdots = \theta_k. \tag{23.60}$$

Any test statistic t, to be satisfactory intuitively, must satisfy the invariance condition

$$t(x_1, x_2, \ldots, x_k) = t(x_1 - \lambda, x_2 - \lambda, \ldots, x_k - \lambda), \tag{23.61}$$

for a change in the origin of measurement should not affect the test. We may therefore without loss of generality take the common value of the θ_i in (23.60) to be zero. We suppose that $t > 0$, and that w_0, the size-α critical region based on the distribution of t, is defined by

$$t \le c_\alpha; \tag{23.62}$$

if either or both of these statements were not true, we could transform to a function of t for which they were.

Because of its invariance property (23.61), t must be constant in the k-dimensional sample space W on any line L parallel to the equiangular vector V defined by $x_1 = x_2 = \cdots = x_k$. When H_0 holds, the content of w_0 is its size

$$\alpha = \int_{w_0} dF(x_1, x_2, \ldots, x_k), \tag{23.63}$$

and when H_0 is not true the content of w_0 is its power

$$1 - \beta = \int_{w_0} dF(x_1 - \theta_1, x_2 - \theta_2, \ldots, x_k - \theta_k) = \int_{w_1} dF(x_1, x_2, \ldots, x_k), \tag{23.64}$$

where w_1 is derived from w_0 by translation in W without rotation. We define the integral, on any line L parallel to V,

$$P(L) = \int_L f(x_1, x_2, \ldots, x_k) \, d\bar{x}; \tag{23.65}$$

the variation along any L being the same for each co-ordinate x_j, it can be summarised in the variation of the mean co-ordinate $\bar{x}$. Since the aggregate of lines L is the whole of W, we have

$$\int P(L)\, \mathrm{d}L = \int \left\{ \int_L f\, \mathrm{d}\bar{x} \right\} \mathrm{d}L = \int \cdots \int f\, \mathrm{d}x_1 \cdots \mathrm{d}x_k = 1. \qquad (23.66)$$

We now determine w_0 as the aggregate of all lines L for which the statistic $P(L) \leqslant$ some constant h. Then $P(L)$ will exceed h on any L which is in w_1 but not in w_0. Hence, from (23.63), (23.64) and (23.66),

$$\alpha = \int_{w_0} \mathrm{d}F \leqslant \int_{w_1} \mathrm{d}F = 1 - \beta,$$

so that the test is unbiased. We therefore define the test statistic t so that at any point on a line L, parallel to V, it is equal to $P(L)$. Now using the invariance property (23.61) with $\lambda = \bar{x}$ we have from (23.65)

$$t(x) = P(L) = \int_L f(x_1 - \bar{x}, x_2 - \bar{x}, \ldots, x_k - \bar{x})\, \mathrm{d}\bar{x},$$

and replacing $\bar{x}$ by u, this is

$$t(x) = \int_{-\infty}^{\infty} f(x_1 - u, x_2 - u, \ldots, x_k - u)\, \mathrm{d}u, \qquad (23.67)$$

the unbiased size-α region being defined by (23.62). It will be seen that the unbiased test thus obtained is unique. An example of the use of (23.67) is given in Exercise 23.15.

23.21   Turning now to tests concerning scale parameters, which are more to our present purpose, suppose that the joint distribution of k variates is

$$\mathrm{d}G = g\left(\frac{y_1}{\phi_1}, \frac{y_2}{\phi_2}, \ldots, \frac{y_k}{\phi_k} \right) \frac{\mathrm{d}y_1}{\phi_1} \cdot \frac{\mathrm{d}y_2}{\phi_2} \cdots \frac{\mathrm{d}y_k}{\phi_k}, \qquad (23.68)$$

where all the scale parameters ϕ_i are positive. We make the transformation

$$x_i = \log |y_i|, \qquad \theta_i = \log \phi_i$$

and find for the distribution of the x_i

$$\mathrm{d}F = g\{\exp(x_1 - \theta_1), \exp(x_2 - \theta_2), \ldots, \exp(x_k - \theta_k)\}$$
$$\times \exp\left\{ \sum_{i=1}^{k} (x_i - \theta_i) \right\} \mathrm{d}x_1 \cdots \mathrm{d}x_k. \qquad (23.69)$$

(23.69) is of the form (23.59) that we have already discussed. To test

$$H_0': \phi_1 = \phi_2 = \cdots = \phi_k \qquad (23.70)$$

is the same as to test H_0 of (23.60). The statistic (23.67) becomes

$$t(x) = \int_{-\infty}^{\infty} g\{\exp(x_1 - u), \exp(x_2 - u), \ldots, \exp(x_k - u)\} \exp\left\{\sum_{i=1}^{k} x_i - ku\right\} du,$$

which when expressed in terms of the y_i becomes

$$t(y) = \prod_{i=1}^{k} |y_i| \int_0^{\infty} g\left(\frac{y_1}{v}, \frac{y_2}{v}, \ldots, \frac{y_k}{v}\right) \frac{dv}{v^{k+1}}. \tag{23.71}$$

23.22 Now consider the special case of k independently distributed gamma variates y_i/ϕ_i with parameters m_i. Their joint distribution is

$$dG = \prod_{i=1}^{k} \left\{\frac{1}{\Gamma(m_i)}\left(\frac{y_i}{\phi_i}\right)^{m_i-1}\right\} \exp\left(-\sum_{i=1}^{k} \frac{y_i}{\phi_i}\right) \prod_{i=1}^{k} \frac{dy_i}{\phi_i}. \tag{23.72}$$

To test H_0' of (23.70), we use (23.71) and obtain

$$t(y) = \prod_{i=1}^{k} \left\{\frac{y_i^{m_i}}{\Gamma(m_i)}\right\} \int_0^{\infty} \exp\left(-\sum_{i=1}^{k} y_i/v\right) \frac{dv}{v^{m+1}}, \tag{23.73}$$

where $m = \sum_{i=1}^{k} m_i$. On substituting $u = \sum_{i=1}^{k} y_i/v$ in (23.73), we find

$$t(y) = \left[\frac{\Gamma(m)}{\prod_i \Gamma(m_i)}\right] \frac{\prod_i y_i^{m_i}}{(\sum_i y_i)^m}. \tag{23.74}$$

We now neglect the constant factor in square brackets in (23.74). From the remainder, T, the maximum attainable value of t, occurs when $y_i/\sum_i y_i = m_i/m$, when

$$T = \prod m_i^{m_i}/m^m. \tag{23.75}$$

We now write

$$t^* = -\log\left(\frac{t}{T}\right) = m \log\left(\frac{\sum_i y_i}{m}\right) - \sum_i m_i \log\left(\frac{y_i}{m_i}\right). \tag{23.76}$$

t^* will be unbiased for H_0', and will range from 0 to ∞, large values being rejected.

Example 23.6
We may now apply (23.76) to the problem of testing the equality of k normal variances, discussed in Example 23.4. For each of the quantities $\sum_j (x_{ij} - \bar{x}_i)^2/(2\sigma^2)$ is, when H_0 holds, a gamma variate with parameter $\frac{1}{2}(n_i - 1)$. We thus have to substitute in (23.76)

$$\left.\begin{array}{l} y_i = \sum (x_{ij} - \bar{x}_i)^2 = v_i s_i^2, \\ m_i = \frac{1}{2}(n_i - 1) = \frac{1}{2}v_i, \\ m = \sum_i m_i = \frac{1}{2}(n - k) = \frac{1}{2}v, \end{array}\right\} \tag{23.77}$$

and we find for the unbiased test statistic

$$2t^* = v \log\left(\frac{\sum_i v_i s_i^2}{v}\right) - \sum v_i \log s_i^2. \tag{23.78}$$

(23.78) is identical with (23.44), so that $2t^*$ is simply $-2 \log l^*$ which we discussed there. Thus the l^* test is unbiased, as stated in Example 23.4. From this, it is fairly evident that the unadjusted LR test statistic l of (23.40), which employs another weighting system, cannot also be unbiased in general. When all sample sizes are equal, the two tests are equivalent, as Exercise 23.7 shows. Even in the case $k = 2$, the unadjusted LR test is biased when $n_1 \neq n_2$: this is left to the reader to prove in Exercise 23.14.

A quite different proof is given by A. Cohen and Strawderman (1971).

23.23 Paulson (1941) investigated the bias of a number of LR tests for exponential distributions—some of his results are given in Exercises 23.16 and 23.18.

Other properties of LR tests

23.24 Apart from questions of consistency and unbiasedness, what can be said in general concerning the properties of LR tests? In the first place, we know that unique ML estimators are functions of the sufficient statistics (cf. **18.4**) so that the LR statistic (23.4) may be rewritten

$$l = \frac{L(x \mid \theta_{r0}, t_s)}{L(x \mid T_{r+s})} \tag{23.79}$$

where t_s is the statistic minimal sufficient for θ_s when H_0 holds and T_{r+s} is the statistic sufficient for all the parameters when H_0 does not hold. As we have seen in **17.38**, it is not true in general that the components of T_{r+s} include the components of t_s—the sufficient statistic for θ_s when H_0 holds may no longer form part of the sufficient set when H_0 does not hold, and even when it does may not then be separately sufficient for θ_s, merely forming part of T_{r+s} which is sufficient for (θ_r, θ_s). Thus all that we can say of l is that it is *some* function of the two sets of sufficient statistics involved. There is, in general, no reason to suppose that it will be the right function of them.

Exercise 23.25 shows that the LR method does not necessarily produce a UMP test when one exists in the case of testing a simple H_0 against a simple H_1.

If we are seeking a UMPU test, the LR method is handicapped by its own general biasedness, but we have seen that a simple bias adjustment will sometimes remove this difficulty. The adjustment takes the form of a reweighting of the test statistic by substituting unbiased estimators for the ordinary ML estimators (Examples 23.4 and 23.6, Example 23.5), or sometimes equivalently of an adjustment of the critical region of the statistic to which the LR method leads (Exercise 23.14). Exercise 23.16 shows that two UMP similar tests derived in Exercises 22.25–26 for an exponential and a uniform distribution are produced by the LR method, while the UMPU test for an exponential distribution given in Exercise 22.24 is not equivalent to the LR test, which is biased.

Wald (1943a) shows that the LR test *asymptotically* has a number of optimum power properties under regularity conditions—but see **25.4** and Example 25.1.

The LR principle is an intuitively appealing one when there is no 'optimum' test. It is of particular value in tests of linear hypotheses (which we shall discuss in the second part of this chapter) for which, in general, no UMPU test exists. But it is as well to be reminded of the possible fallibility of the LR method in exceptional circumstances, and the following example, adapted from one due to C. Stein and given by Lehmann (1950), is a salutary warning against using the method without investigation of its properties in the particular situation concerned.

Example 23.7
A discrete random variable x is defined at the values $0, \pm1, \pm2$, and the probabilities at these points given a hypothesis H_1 are:

$$x: \quad 0 \qquad\qquad \pm1 \qquad\quad +2 \qquad -2$$
$$P|H_1: \quad \alpha\left(\frac{1-\theta_1}{1-\alpha}\right) \quad (\tfrac{1}{2}-\alpha)\left(\frac{1-\theta_1}{1-\alpha}\right) \quad \theta_1\theta_2 \quad \theta_1(1-\theta_2) \tag{23.80}$$

The parameters θ_1, θ_2, are restricted by the inequalities

$$0 \leqslant \theta_1 \leqslant \alpha < \tfrac{1}{2}, \qquad 0 \leqslant \theta_2 \leqslant 1,$$

where α is a known constant. We wish to test the simple

$$H_0: \theta_1 = \alpha, \qquad \theta_2 = \tfrac{1}{2},$$

H_1 being the general alternative (23.80), on the evidence of a single observation. The probabilities on H_0 are:

$$x: \quad 0 \quad \pm1 \quad +2 \quad -2$$
$$P|H_0: \quad \alpha \quad \tfrac{1}{2}-\alpha \quad \tfrac{1}{2}\alpha \quad \tfrac{1}{2}\alpha \tag{23.81}$$

The LF (23.80) is independent of θ_2 when $x = 0, \pm1$, and is maximised unconditionally by making θ_1 as small as possible, i.e. putting $\hat{\theta}_1 = 0$. The LR statistic is therefore

$$l = \frac{L(x|H_0)}{L(x|\hat{\theta}_1, \hat{\theta}_2)} = 1 - \alpha, \qquad x = 0, \pm1. \tag{23.82}$$

When $x = +2$ or -2, the LF is maximised unconditionally by choosing θ_2 respectively as large or as small as possible, i.e. $\hat{\theta}_2 = 1, 0$, respectively; and by choosing θ_1 as large as possible, i.e. $\hat{\theta}_1 = \alpha$. The maximum value of the LF is therefore α and the LR statistic is

$$l = \tfrac{1}{2}, \qquad x = \pm2. \tag{23.83}$$

Since $\alpha < \tfrac{1}{2}$, it follows from (23.82) and (23.83) that the LR test consists of rejecting H_0 when $x = \pm2$. From (23.81) this test is seen to be of size α. But from (23.80) its power is seen to be θ_1 exactly, so for any value of θ_1 in

$$0 \leqslant \theta_1 < \alpha \tag{23.84}$$

the LR test will be biased for all θ_2, while for $\theta_1 = \alpha$ the test will have power equal to its size α for all θ_2. In this latter extreme case the test is useless, but in the former case it is worse than useless, for we can get a test of size and power α by using a table of random numbers as the basis for our decision concerning H_0. Furthermore, a useful test exists, for if we reject H_0 when $x = 0$, we still have a size-α test by (23.81) and its power, from (23.80), is $\alpha(1 - \theta_1)/(1 - \alpha)$ which exceeds α when (23.84) holds and equals α when $\theta_1 = \alpha$.

Apart from the fact that the random variable is discrete, the noteworthy feature of this cautionary example is that the range of one of the parameters is determined by α, the size of the test.

> D. R. Cox (1961, 1962) considers the distribution of LR test statistics when H_0 and H_1 are entirely separate families of composite hypotheses (so that (23.5) no longer holds) and obtains some large-sample results. See also Feder (1968), Atkinson (1970), Dyer (1973, 1974) and Kent (1986). Pereira (1977) gives an annotated bibliography. These tests are not of controlled size: Loh (1985) proposes a method based on the bootstrap of **17.10** to remedy this deficiency.

The general linear hypothesis and its canonical form

23.25 We are now in a position to discuss the problem of testing hypotheses in the linear model of Chapter 19. As at (19.8), we write

$$\mathbf{y} = \mathbf{X}\boldsymbol{\theta} + \boldsymbol{\varepsilon}, \tag{23.85}$$

where the ε_i have zero means, equal variances σ^2 and are uncorrelated. For the moment, we make no further assumptions about the form of their distribution. We take $\mathbf{X}'\mathbf{X}$ to be non-singular: if it were not, we could make it so by augmentation, as in **19.13–15**.

Suppose that we wish to test the hypothesis

$$H_0 : \mathbf{A}\boldsymbol{\theta} = \mathbf{c}_0, \tag{23.86}$$

where $\mathbf{A}$ is an $(r \times k)$ matrix and $\mathbf{c}_0$ an $(r \times 1)$ vector, each of known constants. (23.86) imposes $r \ (\leq k)$ constraints, which we take to be functionally independent, so that $\mathbf{A}$ is of rank r. The alternative H_1 is simply the negation of H_0. When $r = k$, $\mathbf{A}'\mathbf{A}$ is non-singular and (23.86) is equivalent to $H_0 : \boldsymbol{\theta} = (\mathbf{A}'\mathbf{A})^{-1}\mathbf{A}'\mathbf{c}_0$. If $\mathbf{A}$ is the first r rows of the $(n \times k)$ matrix $\mathbf{X}$, we also have a particularly direct H_0, in which we are testing the means of the first $r \ y_i$.

Consider the $(n \times 1)$ vector

$$\mathbf{z} = \mathbf{C}(\mathbf{X}'\mathbf{X})^{-1}\mathbf{X}'\mathbf{y} = \mathbf{C}\hat{\boldsymbol{\theta}}, \tag{23.87}$$

where $\mathbf{C}$ is an $(n \times k)$ matrix and $\hat{\boldsymbol{\theta}}$ is the least squares (LS) estimator of $\boldsymbol{\theta}$ given at (19.12). Then, from (23.87) and (23.85),

$$\mathbf{z} = \mathbf{C}\boldsymbol{\theta} + \mathbf{C}(\mathbf{X}'\mathbf{X})^{-1}\mathbf{X}'\boldsymbol{\varepsilon},$$

so that

$$\boldsymbol{\mu} = E(\mathbf{z}) = \mathbf{C}\boldsymbol{\theta} \tag{23.88}$$

and the covariance matrix of $\mathbf{z}$ is, as in **19.6**,

$$\mathbf{V} = \sigma^2 \mathbf{C}(\mathbf{X}'\mathbf{X})^{-1}\mathbf{C}'. \tag{23.89}$$

Let us now choose C so that the components of z are all uncorrelated, i.e. so that

$$V = \sigma^2 I.$$

From (23.89), this requires that

$$C(X'X)^{-1}C' = I$$

or, if $C'C$ is non-singular, that

$$C'C = X'X. \tag{23.90}$$

(23.90) is the condition that the z_i be uncorrelated, and implies, with (23.87) and (23.88), that

$$(z-\mu)'(z-\mu) = \varepsilon'X(X'X)^{-1}X'\varepsilon = \{X(\hat{\theta}-\theta)\}'\{X(\hat{\theta}-\theta)\}. \tag{23.91}$$

23.26   We now write

$$C = \begin{pmatrix} A \\ D \\ F \end{pmatrix},$$

where A is the $(r \times k)$ matrix in (23.86), D is a $((k-r) \times k)$ matrix and F is an $((n-k) \times k)$ matrix satisfying

$$F\theta = 0. \tag{23.92}$$

Since A is of rank r, we can choose D so that the $(k \times k)$ matrix $\begin{pmatrix} A \\ D \end{pmatrix}$ is non-singular, and thus C is of rank k. $C'C$ is then also of rank k, and hence non-singular as required above (23.90).

From (23.88), we have

$$\mu = E(z) = \begin{pmatrix} A \\ D \\ F \end{pmatrix} \theta. \tag{23.93}$$

Thus the means of the first r z_i are precisely the left-hand side of (23.86), so that H_0 is equivalent to testing

$$H_0: \mu_i = E(z_i) = c_{0i}, \qquad i = 1, 2, \ldots, r, \tag{23.94}$$

a composite hypothesis imposing r constraints upon the parameters. Since, by (23.92), the last $(n-k)$ of the μ_i are zero, there are k non-zero parameters μ_i, which together with σ^2 make up the total of $(k+1)$ parameters.

23.27   Thus we have reduced our problem to the following terms: we have a set of n mutually uncorrelated variates z_i with equal variances σ^2. $(n-k)$ of the z_i have zero means, and the others non-zero means. The hypothesis to be tested is that r of these k variates have specified means. This is called the canonical form of the general linear hypothesis.

In order to make progress with the hypothesis-testing problem, we need to make assumptions about the distribution of the errors in the linear model (23.85): specifically, we take each ε_i to be normal and hence, since they are uncorrelated, independent. The z_i, being linear functions of them, will also be normally distributed and, being uncorrelated, independently normally distributed. Their joint distribution therefore gives the LF

$$L(z|\mu, \sigma^2) = (2\pi\sigma^2)^{-n/2} \exp\left\{-\frac{1}{2\sigma^2}(z-\mu)'(z-\mu)\right\}$$

$$= (2\pi\sigma^2)^{-n/2} \exp\left[-\frac{1}{2\sigma^2}\left\{(z_r-\mu_r)'(z_r-\mu_r)\right.\right.$$

$$\left.\left. + (z_{k-r}-\mu_{k-r})'(z_{k-r}-\mu_{k-r}) + z'_{n-k}z_{n-k}\right\}\right], \tag{23.95}$$

where suffixes to vectors denote the number of components in the vector. Our hypothesis is

$$H_0: \mu_r = c_0 \tag{23.96}$$

and H_1 is its negation.

We saw in Example 22.14 that if we have only one constraint ($r = 1$), there is a UMPU test of H_0 against H_1, as is otherwise obvious in our present application from the fact that we are then testing the mean of a single normal population with unknown variance: the UMPU test is, as we saw in Example 22.14, the ordinary equal-tails Student's t-test for this hypothesis.

Kolodzieczyk (1935), to whom the first general results concerning the linear hypothesis are due, demonstrated the impossibility of a UMP test with more than one constraint, and showed that there is a pair of one-sided UMP similar tests when $r = 1$: these are the one-sided Student's t-tests (cf. Example 22.7). We have just seen that there is a two-sided Student's t-test which is UMPU for $r = 1$, but the critical region of this test is different according to which of the μ_i is being tested: thus there is no common UMPU critical region for $r > 1$.

Since there is no 'optimum' test in any sense that we have so far discussed, we are tempted to use the LR method to give an intuitively reasonable test.

23.28 The derivation of the LR statistic is simple enough. The unconditional maximum of (23.95) is obtained by solving the set of equations

$$\left.\begin{array}{ll} \dfrac{\partial \log L}{\partial \mu_i} = 0, & i = 1, 2, \ldots, k, \\[3mm] \dfrac{\partial \log L}{\partial (\sigma^2)} = 0. & \end{array}\right\}$$

The ML estimators thus obtained are

$$\hat{\mu}_i = z_i, \qquad i = 1, 2, \ldots, k,$$
$$\hat{\sigma}^2 = \frac{1}{n} \sum_{i=k+1}^{n} z_i^2,$$

whence

$$(\mathbf{z} - \hat{\mathbf{\mu}})'(\mathbf{z} - \hat{\mathbf{\mu}}) = n\hat{\sigma}^2.$$

Thus the unconditional maximum of the LF is

$$L(\mathbf{z} \,|\, \hat{\mathbf{\mu}}, \hat{\sigma}^2) = (2\pi\hat{\sigma}^2 \, e)^{-n/2} = \left(\frac{2\pi \, e}{n} \sum_{i=k+1}^{n} z_i^2 \right)^{-n/2}. \tag{23.97}$$

When the hypothesis (23.96) holds, the ML estimators of the unspecified parameters are

$$\hat{\hat{\mu}}_i = z_i, \qquad i = r+1, r+2, \ldots, k,$$
$$\hat{\hat{\sigma}}^2 = \frac{1}{n} \left\{ \sum_{i=k+1}^{n} z_i^2 + \sum_{i=1}^{r} (z - c_{0i})^2 \right\},$$

whence

$$(\mathbf{z} - \hat{\hat{\mathbf{\mu}}})'(\mathbf{z} - \hat{\hat{\mathbf{\mu}}}) = n\hat{\hat{\sigma}}^2,$$

so that the conditional maximum of the LF is

$$L(\mathbf{z} \,|\, \mathbf{c}_0, \hat{\hat{\mathbf{\mu}}}, \hat{\hat{\sigma}}^2) = (2\pi\hat{\hat{\sigma}}^2 \, e)^{-n/2} = \left[\frac{2\pi \, e}{n} \left\{ \sum_{i=k+1}^{n} z_i^2 + \sum_{i=1}^{r} (z_i - c_{0i})^2 \right\} \right]^{-n/2}. \tag{23.98}$$

From (23.97) and (23.98) the LR statistic l is given by

$$l^{2/n} = \frac{\hat{\sigma}^2}{\hat{\hat{\sigma}}^2} = \frac{1}{1 + W}, \tag{23.99}$$

where

$$W = (\mathbf{z}_r - \mathbf{c}_0)'(\mathbf{z}_r - \mathbf{c}_0)/\mathbf{z}_{n-k}'\mathbf{z}_{n-k}$$
$$= \frac{\sum_{i=1}^{r} (z_i - c_{0i})^2}{\sum_{i=k+1}^{n} z_i^2} = \frac{\hat{\hat{\sigma}}^2 - \hat{\sigma}^2}{\hat{\sigma}^2}. \tag{23.100}$$

It will be observed from above that $n\hat{\sigma}^2$, $n\hat{\hat{\sigma}}^2$ are respectively the minima of $(\mathbf{z} - \mathbf{\mu})'(\mathbf{z} - \mathbf{\mu})$ with respect to $\mathbf{\mu}$ under H_0 and H_1. By (23.91), these are the same as the minima of

$$R = \{\mathbf{X}(\hat{\mathbf{\theta}} - \mathbf{\theta})\}'\{\mathbf{X}(\hat{\mathbf{\theta}} - \mathbf{\theta})\}$$

with respect to $\mathbf{\theta}$. The identity in $\mathbf{\theta}$

$$S = \mathbf{\varepsilon}'\mathbf{\varepsilon} = (\mathbf{y} - \mathbf{X}\mathbf{\theta})'(\mathbf{y} - \mathbf{X}\mathbf{\theta}) \equiv (\mathbf{y} - \mathbf{X}\hat{\mathbf{\theta}})'(\mathbf{y} - \mathbf{X}\hat{\mathbf{\theta}}) + R$$

is easily verified by direct expansion, and the term on its right

$$(\mathbf{y} - \mathbf{X}\hat{\boldsymbol{\theta}})'(\mathbf{y} - \mathbf{X}\hat{\boldsymbol{\theta}})$$

does not depend on $\boldsymbol{\theta}$. Minimisation of R with respect to $\boldsymbol{\theta}$ is therefore equivalent to minimisation of S. But the process of minimising S for $\boldsymbol{\theta}$ is precisely the means by which we arrived at the least squares solution in **19.4**. To obtain $\hat{\hat{\sigma}}^2$, $\hat{\sigma}^2$ in (23.100), therefore, we minimise S in the original model under H_0 and H_1 respectively.

Since l is a monotone decreasing function of W, the LR test is equivalent to rejecting H_0 when W is large. If we divide the numerator and denominator of W by σ^2, we see that when H_0 holds, W is the ratio of the sum of squares of r independent normal variates to an independent sum of squares of $(n - k)$ such variates, i.e. is the ratio of two independent χ^2 variates with r and $(n - k)$ degrees of freedom. Thus, when H_0 holds, $F = [(n - k)/r]\, W$ is distributed in the variance-ratio distribution (cf. **16.15**) with $(r, n - k)$ degrees of freedom and the LR test is carried out in terms of F, large values forming the critical region.

Many of the standard tests in statistics may be reduced to tests of a linear hypothesis, and we shall be encountering them frequently in later chapters.

Example 23.8

As a special case of particular importance, consider the hypothesis

$$H_0 : \boldsymbol{\theta}_r = \mathbf{0},$$

where $\boldsymbol{\theta}_r$ is an $(r \times 1)$ subvector of $\boldsymbol{\theta}$ in (23.85). We may therefore rewrite (23.85) as

$$\mathbf{y} = (\mathbf{X}_1\ \ \mathbf{X}_2)\binom{\boldsymbol{\theta}_r}{\boldsymbol{\theta}_{k-r}} + \boldsymbol{\varepsilon}$$

where $\mathbf{X}_1$ is of order $(n \times r)$ and $\mathbf{X}_2$ is of order $(n \times (k - r))$. Then H_0 becomes equivalent to specifying

$$\mathbf{y} = \mathbf{X}_2 \boldsymbol{\theta}_{k-r} + \boldsymbol{\varepsilon}.$$

In accordance with **23.28**, we find the minima of $S = \boldsymbol{\varepsilon}'\boldsymbol{\varepsilon}$. Since we are here estimating all the parameters of a linear model both on H_0 and on H_1, we may use the result of **19.9**. We have at once that the minimum under H_0 is

$$n\hat{\hat{\sigma}}^2 = \mathbf{y}'\{\mathbf{I} - \mathbf{X}_2(\mathbf{X}_2'\mathbf{X}_2)^{-1}\mathbf{X}_2'\}\mathbf{y}$$

while under H_1 it is

$$n\hat{\sigma}^2 = \mathbf{y}'\{\mathbf{I} - \mathbf{X}(\mathbf{X}'\mathbf{X})^{-1}\mathbf{X}'\}\mathbf{y}$$

where $\mathbf{X} = (\mathbf{X}_1\ \ \mathbf{X}_2)$. The statistic

$$F = \frac{n - k}{r}\left(\frac{\hat{\hat{\sigma}}^2 - \hat{\sigma}^2}{\hat{\sigma}^2}\right)$$

is distributed in the variance-ratio distribution with $(r, n - k)$ degrees of freedom, the critical region for H_0 being the 100α per cent largest values of F.

23.29 In **23.28** we saw that the LR test is based on the statistic (23.100) which may be rewritten

$$W = \frac{\sum_{i=1}^{r}(z_i - c_{0i})^2/\sigma^2}{\sum_{i=k+1}^{n} z_i^2/\sigma^2}.$$

Whether H_0 holds or not, the denominator of W is distributed like χ^2 with $(n-k)$ degrees of freedom. When H_0 holds, as we saw, the numerator is also a χ^2 variate with r degrees of freedom, but when H_0 does not hold this is no longer so: in fact, the numerator will always be a non-central χ^2 variate (cf. **23.4**) with r degrees of freedom and non-central parameter

$$\lambda = \sum_{i=1}^{r}(c_{0i} - \mu_i)^2/\sigma^2 = (\boldsymbol{\mu}_r - \mathbf{c}_0)'(\boldsymbol{\mu}_r - \mathbf{c}_0)/\sigma^2, \qquad (23.101)$$

where μ_i is the true mean of z_i. Only when H_0 holds is λ equal to zero, giving the central χ^2 distribution of **23.28**. Since we wish to investigate the distribution of W (or equivalently of F) when H_0 is not true, so that we can evaluate the power of the LR test, we are led to the study of the ratio of a non-central to a central χ^2 variate.

The non-central F distribution

23.30 Consider first the ratio of two variates z_1, z_2, independently distributed in the non-central χ^2 form (23.18) with degrees of freedom ν_1, ν_2 and non-central parameters λ_1, λ_2 respectively. Using (11.74) the distribution of $u = z_1/z_2$ is given by

$$\mathrm{d}H(u) = \mathrm{d}u \int_0^\infty \frac{e^{-(uv+\lambda_1)/2}(uv)^{(\nu_1/2)-1}}{2^{\nu_1/2}} \sum_{r=0}^\infty \frac{\lambda_1^r(uv)^r}{2^{2r}r!\,\Gamma(\tfrac{1}{2}\nu_1+r)}$$

$$\times \frac{e^{-(v+\lambda_2)/2} v^{(\nu_2/2)-1}}{2^{\nu_2/2}} \sum_{s=0}^\infty \frac{\lambda_2^s v^s}{2^{2s}s!\,\Gamma(\tfrac{1}{2}\nu_2+s)}\, v\, \mathrm{d}v. \qquad (23.102)$$

If we write $\lambda = \lambda_1 + \lambda_2$, $\nu = \nu_1 + \nu_2$, and simplify, this becomes

$$\mathrm{d}H(u) = \frac{e^{-\lambda/2}}{2^{\nu/2}} \sum_{r=0}^\infty \sum_{s=0}^\infty \frac{\lambda_1^r \lambda_2^s}{2^{2r+2s} r!\, s!\, \Gamma(\tfrac{1}{2}\nu_1+r)\Gamma(\tfrac{1}{2}\nu_2+s)}$$

$$\times \left\{ \int_0^\infty e^{-v(1+u)/2} v^{(\nu/2)+r+s-1}\,\mathrm{d}v \right\} u^{(\nu_1/2)+r-1}\,\mathrm{d}u. \qquad (23.103)$$

The integral in (23.103) is equal to $\Gamma(\tfrac{1}{2}\nu+r+s)/((1+u)/2)^{(\nu/2)+r+s}$. Thus

$$\mathrm{d}H(u) = e^{-\lambda/2} \sum_{r=0}^\infty \sum_{s=0}^\infty \frac{(\tfrac{1}{2}\lambda_1)^r}{r!} \frac{(\tfrac{1}{2}\lambda_2)^s}{s!} u^{(\nu_1/2)+r-1} \left(\frac{1}{1+u}\right)^{(\nu/2)+r+s} \frac{\mathrm{d}u}{B(\tfrac{1}{2}\nu_1+r, \tfrac{1}{2}\nu_2+s)}, \qquad (23.104)$$

a result obtained by Tang (1938) and studied by Price (1964). If we put

$$F'' = \frac{z_1/\nu_1}{z_2/\nu_2} = \frac{\nu_2}{\nu_1} u$$

in (23.104), we obtain the doubly non-central F-distribution, the computation of whose d.f. is considered by Bulgren (1971). Harter and Owen (1974) give tables of the d.f. due to M. L. Tiku.

23.31 If now we put $\lambda_2 = 0$ and $\lambda_1 = \lambda$, F'' is only singly non-central, and we write it F'. From (23.104), its distribution is

$$dG(F') = e^{-\lambda/2} \sum_{r=0}^{\infty} \frac{\frac{(\frac{1}{2}\lambda)^r}{r!} \left(\frac{\nu_1}{\nu_2}\right)^{(\nu_1/2)+r}}{B(\frac{1}{2}\nu_1 + r, \frac{1}{2}\nu_2)} \frac{(F')^{(\nu_1/2)+r-1}}{\left(1 + \frac{\nu_1}{\nu_2}F'\right)^{r+(\nu_1+\nu_2)/2}} dF'. \quad (23.105)$$

(23.105) is a generalisation of the variance-ratio (F) distribution (16.24), to which it reduces when $\lambda = 0$. It is called the non-central F distribution with degrees of freedom ν_1, ν_2 and non-central parameter λ. We sometimes write it $F'(\nu_1, \nu_2, \lambda)$. Like (23.18), it was first discussed by Fisher (1928a), and it has been studied by Wishart (1932), Tang (1938), Patnaik (1949), Price (1964), and Tiku (1965). Formulae for its moments are given in Exercise 23.21.

> With $\lambda_2 = 0$, (23.104) is a Poisson mixture of beta distributions of the second kind (6.16)—cf. Exercise 23.19 for the parallel result for (23.18). Thus the c.f. of F' is the same Poisson mixture of central F c.f.s—see **16.17** and Phillips (1982).
>
> The 50, 75, 90 and 95 percentage points of F' are tabulated by Wallace and Toro-Vizcarrondo (1969) for ν_1 and $\nu_2 = 1$ (1) 30, 40, 60, 120, 200, 400, 1000 and $\lambda = 1$ only (they use $\frac{1}{2}\lambda$ as parameter).
>
> As $\nu_2 \to \infty$, $\nu_1 F'(\nu_1, \nu_2, \lambda) \to \chi'^2(\nu_1, \lambda)$ defined at (23.18)—this result when $\lambda = 0$ has already been noted at **16.22** (6). Cf. Exercise 23.20.

The power function of the LR test of the general linear hypothesis

23.32 It follows at once from **23.28-9** and **23.31** that the power function of the LR test of the general linear hypothesis is

$$P = \int_{F'_\alpha(\nu_1, \nu_2, 0)}^{\infty} dG\{F'(\nu_1, \nu_2, \lambda)\}, \quad (23.106)$$

where F'_α is the $100(1 - \alpha)$ per cent point of the distribution, $\nu_1 = r$, $\nu_2 = n - k$, and λ is defined at (23.101).

> Numerous tables and charts of P have been constructed:
> (1) Tang (1938) gives $1 - P$ (i.e. the Type II error β) to 3 d.p. for test sizes $\alpha = 0.01, 0.05$; ν_1 (his f_1) = 1 (1) 8; ν_2 (his f_2) = 2 (2) 6 (1) 30, 60, ∞; and $\phi = \{\lambda/(\nu_1 + 1)\}^{1/2} = 1$ (0.5) 3 (1) 8. These tables are reproduced in Mann (1949) and in Kempthorne (1952).
> (2) Tiku (1967a, 1972) extends Tang's tables (1) to 4 d.p. for $\alpha = 0.005, 0.01, 0.025, 0.05, 0.10$; $\nu_1 = 1$ (1) 10, 12; $\nu_2 = 2$ (2) 30, 40, 60, 120, ∞; and $\phi = 0.5, 1.0$ (0.2) 2.2 (0.4) 3.0.
> (3) Lehmer (1944) gives inverse tables of ϕ for $\alpha = 0.01, 0.05$; $\nu_1 = 1$ (1) 10, 12, 15, 20, 24, 30, 40, 60, 80, 120, ∞; $\nu_2 = 2$ (2) 20, 24, 30, 40, 60, 80, 120, 240, ∞; and P (her β) = 0.7, 0.8.
> (4) Dasgupta (1968) gives tables of $\frac{1}{2}\lambda$ (his δ) for ν_1 (his M) = 1 (1) 10; ν_2 (his N) = 10 (5) 50 (10) 100, ∞; $\alpha = 0.01, 0.05$ and P (his β) = 0.1 (0.1) 0.9.
> (5) Kastenbaum et al. (1970a) give tables of $\tau = \{2\lambda(\nu_1 + 1)/(\nu_1 + \nu_2 + 1)\}^{1/2}$ to 3 d.p. for $\alpha = 0.01, 0.05, 0.1, 0.2$; $\beta = 1 - P = 0.005, 0.01, 0.05, 0.1, 0.2, 0.3$; ν_1 (their $k-1$) = 1 (1) 5; and $(\nu_1 + \nu_2 + 1)/(\nu_1 + 1)$ (their N) = 2 (1) 8 (2) 30, 40 (20) 100, 200, 500, 1000. See also their related smaller tables (1970b) and Bowman (1972).
> (6) E. S. Pearson and Hartley (1951) give eight charts of the power function, one for each value of ν_1 from 1 to 8. Each chart shows the power for $\nu_2 = 6$ (1) 10, 12, 15, 20, 30, 60,

∞; $\alpha = 0.01$, 0.05; and ϕ ranging from 1 (except when $\nu_1 = 1$, when ϕ ranges from 2 for $\alpha = 0.01$, 1.2 for $\alpha = 0.05$) to a value large enough for the power to be at least 0.98. The charts are reproduced in the *Biometrika Tables*, Vol. II, with two more for $\nu_1 = 12$, 24.

(7) M. Fox (1956) gives inverse charts, one for each of the combinations of $\alpha = 0.01$, 0.05, with power P (his β) = 0.5, 0.7, 0.8, 0.9. Each chart shows, for $\nu_1 = 3$ (1) 10 (2) 20 (20) 100, 200, ∞; $\nu_2 = 4$ (1) 10 (2) 20 (20) 100, 200, ∞, the contours of constant ϕ. He also gives a nomogram for each α to facilitate interpolation in β.

(8) A. J. Duncan (1957) gives two charts, one for $\alpha = 0.01$ and one for $\alpha = 0.05$. Each shows, for $\nu_2 = 6$ (1) 10, 12, 15, 20 (10) 40, 60, ∞, the values of ν_1 (ranging from 1 to 8) and ϕ required to attain power $P = 1 - \beta = 0.50$ and 0.90.

Approximation to the power function of the LR test

23.33 As will be seen from the form of (23.105), the computation of the exact power function (23.106) is a laborious matter, and even now its tabulation is far from complete. However, we may obtain a simple approximation to the power function in the manner of, and using the results of, our approximation to the non-central χ^2 distribution in **23.5**. If z_1 is a non-central χ^2 variate with ν_1 degrees of freedom and non-central parameter λ, we have from (23.21) that $z_1/((\nu_1 + 2\lambda)/(\nu_1 + \lambda))$ is approximately a central χ^2 variate with degrees of freedom $\nu^* = (\nu_1 + \lambda)^2/(\nu_1 + 2\lambda)$. Thus

$$z_1 \Big/ \left\{ \nu^* \left(\frac{\nu_1 + 2\lambda}{\nu_1 + \lambda} \right) \right\} = z_1/(\nu_1 + \lambda)$$

is approximately a central χ^2 variate divided by its degrees of freedom ν^*. Hence z_1/ν_1 is approximately a multiple $(\nu_1 + \lambda)/\nu_1$ of such a variate. If we now define the non-central F'-variate

$$F' = \frac{z_1/\nu_1}{z_2/\nu_2},$$

where z_2 is a central χ^2 variate with ν_2 degrees of freedom, it follows at once that approximately

$$F' = \frac{\nu_1 + \lambda}{\nu_1} F, \tag{23.107}$$

where F is a central F-variate with degrees of freedom $\nu^* = (\nu_1 + \lambda)^2/(\nu_1 + 2\lambda)$ and ν_2. The first two moments of (23.107) are exact, because those of the χ'^2 approximation in **23.5** are, through the exact moment relationship given in Exercise 23.21.

The simple approximation (23.107) is surprisingly effective. By making comparisons with Tang's (1938) exact tables, Patnaik (1949) shows that the power function calculated by use of (23.107) is generally accurate to two significant figures; it will therefore suffice for all practical purposes.

To calculate the power of the LR test of the linear hypothesis, we therefore replace (23.106) by the approximate central F-integral

$$P = \int_{(\nu_1/(\nu_1 + \lambda))F_\alpha(\nu_1, \nu_2)}^{\infty} dG \left\{ F \left(\frac{(\nu_1 + \lambda)^2}{(\nu_1 + 2\lambda)}, \nu_2 \right) \right\}, \tag{23.108}$$

the size of the test being determined by putting $\lambda = 0$. $(\nu_1 + \lambda)^2 / (\nu_1 + 2\lambda)$ is generally fractional, and interpolation is necessary. Even the central F distribution, however, is not yet so very well tabulated as to make the accurate evaluation of (23.108) easy—see the list of tables in **16.19**.

A central F-approximation due to Tiku (1965, 1966) obtained by equating three moments is even more accurate—see also Pearson and Tiku (1970).

Dar (1962) gives a simple normal approximation to the distribution of the ratio of two independent identical non-central F-variables.

The non-central t distribution

23.34 When $\nu_1 = 1$, the non-central F distribution (23.105) reduces to the non-central t^2 distribution, just as for the central distributions (cf. **16.15**), and as before we write t' instead of t to indicate non-centrality. Evidently, from the derivation of non-central χ^2 as the sum of non-central squared normal variates, we may write

$$t' = (z + \delta) / w^{1/2}, \tag{23.109}$$

where z is a standardised normal variate and w is independently distributed like χ^2 / f with f degrees of freedom (we write f instead of ν_2 in (23.105), and $\delta^2 = \lambda$, in this case and sometimes write the variate as $t'(f, \delta)$). Our discussion of the F' distribution covers the t'^2 distribution, but the t' distribution has received special attention because of its importance in applications. It is derived in Exercise 23.23, and its d.f. is given in Exercise 23.24.

Johnson and Welch (1939) studied the distribution and gave tables for finding $100(1 - \alpha)$ per cent points of the distribution of t' for α or $1 - \alpha = 0.005, 0.01, 0.025, 0.05, 0.1 (0.1) 0.5$, $f = 4 (1) 9, 16, 36, 144, \infty$, and any δ; and conversely for finding δ for given values of t'. Resnikoff (1962) gives additional tables.

Resnikoff and Lieberman (1957) have given tables of the density and the d.f. of t' to 4 d.p., at intervals of 0.05 for $t/f^{1/2}$, for $f = 2 (1) 24 (5) 49$, and for the values of δ defined by

$$\int_{\delta/(f+1)^{1/2}}^{\infty} (2\pi)^{-1/2} \exp\left(-\tfrac{1}{2}x^2\right) \mathrm{d}x = \alpha,$$

$\alpha = 0.001, 0.0025, 0.004, 0.01, 0.025, 0.04, 0.065, 0.10, 0.15, 0.25$. They, and also Scheuer and Spurgeon (1963), give some percentage points of the distributions. Locks *et al.* (1963) give similar tables at intervals of 0.2 for t', with $f = 1 (1) 20 (5) 40$ and δ defined by $\delta(f+1)^{-1/2}$ or $\delta(f+2)^{-1/2} = 0 (0.25) 3$. Owen (1963) gives very extensive tables of percentage points from which the *Biometrika Tables*, Vol. II, provides tables (a) for 7 values of α, to compute critical values of t' given δ and f; and (b) for 4 values of α, inverse tables to compute δ in terms of f and the critical value t'. Hogben *et al.* (1961) give a method of obtaining the moments, with tables (reproduced in the *Biometrika Tables*, Vol. II) for the first four—simple formulae are given in Exercise 23.22. Amos (1964) studies series approximations of the distribution.

Krishnan (1967, 1968) and Bulgren and Amos (1968) study and tabulate the doubly non-central t distribution analogous to F'' of **23.30**. Extensive tables due to Bulgren are given in Harter and Owen (1974).

23.35 A particularly important application of the t' distribution is in evaluating the power of a Student's t-test for which the critical region is in one tail only (the equal-tails case, of course, corresponds to the t'^2 distribution). The test is that $\delta = 0$ in (23.109), the critical region being determined from the central t distribution. Its power is evidently just the integral of the non-central t distribution over the critical region. It has been specially tabulated by Neyman *et al.* (1935), who give, for $\alpha = 0.05$, 0.01, f (their n) $= 1$ (1) 30, ∞ and δ (their p) $= 1$ (1) 10, tables and charts of the complement $1 - P$ of the power of the test, together with the values of δ for which $P = 1 - \alpha$. Owen (1965) gives 5 d.p. tables of δ for $\alpha = 0.05$, 0.025, 0.01 and 0.005; $f = 1$ (1) 30 (5) 100 (10) 200, ∞ and $1 - P = 0.01$, 0.05, 0.10 (0.10) 0.90. The Pearson–Hartley chart of **23.32** (6) for $\nu_1 = 1$ applies to equal-tails tests on t' and is reproduced in each volume of the *Biometrika Tables*. Hodges and Lehmann (1967) give an asymptotic series for the power and use it (1968) to construct an extremely compact table.

Optimum properties of the LR test of the general linear hypothesis

23.36 We saw in **23.27** that, apart from the case $r = 1$, there is no UMPU test of the general linear hypothesis. Nevertheless, the LR test of that hypothesis has certain optimum properties which we now proceed to develop, making use of simplified proofs due to Wolfowitz (1949) and Lehmann (1950).

In **23.28** we derived the ML estimators of the $(k - r + 1)$ unspecified parameters when H_0 holds. They are the components of

$$\mathbf{t} = (\hat{\hat{\mu}}, \hat{\hat{\sigma}}^2),$$

which are defined above (23.98). When H_0 holds, the components of $\mathbf{t}$ are a set of $(k - r + 1)$ sufficient statistics for the unspecified parameters. By **22.10**, their distribution is complete. Thus, by **22.19**, every similar size-α critical region w for H_0 will consist of a fraction α of every surface $\mathbf{t} = \text{constant}$. Here every component of $\mathbf{t}$ is to be constant, and in particular the component $\hat{\hat{\sigma}}^2$. Let

$$n\hat{\hat{\sigma}}^2 = \sum_{i=k+1}^{n} z_i^2 + \sum_{i=1}^{r} (z_i - c_{0i})^2 = a^2, \tag{23.110}$$

where a is a constant.

Now consider a fixed value of λ, defined at (23.101), say $\lambda = d^2 > 0$. The power of any similar region on this surface will consist of the aggregate of its power on (23.110) for all a. For fixed a, the power on the surface $\lambda = d^2$ is

$$P(w \mid \lambda, a) = \int_{\lambda = d^2} L(\mathbf{z} \mid \boldsymbol{\mu}, \sigma^2) \, d\mathbf{z}, \tag{23.111}$$

where L is the LF defined at (23.95). We may write this out fully as

$$P(w \mid \lambda, a) = (2\pi\sigma^2)^{-n/2} \int_{\lambda = d^2} \exp\left\{ -\frac{1}{2\sigma^2} [\{(\mathbf{z}_r - \mathbf{c}_0) - (\boldsymbol{\mu}_r - \mathbf{c}_0)\}'\{(\mathbf{z}_r - \mathbf{c}_0) - (\boldsymbol{\mu}_r - \mathbf{c}_0)\} \right.$$
$$\left. + (\mathbf{z}_{k-r} - \boldsymbol{\mu}_{k-r})'(\mathbf{z}_{k-r} - \boldsymbol{\mu}_{k-r}) + \mathbf{z}'_{n-k}\mathbf{z}_{n-k}] \right\} d\mathbf{z}. \tag{23.112}$$

Using (23.110) and (23.101), (23.112) becomes

$$P(w \mid \lambda, a) = (2\pi\sigma^2)^{-(n-k+r)/2} \exp\left\{ -\frac{1}{2}\left(d^2 + \frac{a^2}{\sigma^2}\right) \right\} \int_{\lambda = d^2} \exp\left\{ (\mathbf{z}_r - \mathbf{c}_0)'(\boldsymbol{\mu}_r - \mathbf{c}_0) \right\} d\mathbf{z}_r, \tag{23.113}$$

the vector $\mathbf{z}_{k-r}$ having been integrated out over its whole range since its distribution is free of λ and independent of a. The only non-constant factor in (23.113) is the integral, which is to be maximised to obtain the critical region w with maximum P. The integral is over the surface $\lambda = d^2$ or $(\boldsymbol{\mu}_r - \mathbf{c}_0)'(\boldsymbol{\mu}_r - \mathbf{c}_0) = $ constant. It is clearly a monotone increasing function of $|\mathbf{z}_r - \mathbf{c}_0|$, i.e. of

$$(\mathbf{z}_r - \mathbf{c}_0)'(\mathbf{z}_r - \mathbf{c}_0) = \sum_{i=1}^{r} (z_i - c_{0i})^2.$$

Now if $\sum_{i=1}^{r} (z_i - c_{0i})^2$ is maximised for fixed a in (23.110), W defined at (23.100) is also maximised. Thus for any fixed λ and a, the maximum value of $P(w|\lambda, a)$ is attained when w consists of large values of W. Since this holds for each a, it holds when the restriction that a be fixed is removed. We have therefore established that on any surface $\lambda = d^2 > 0$, the LR test, which consists of rejecting large values of W, has maximum power, a result due to Wald (1942).

An immediate consequence is P. L. Hsu's (1941) result, that the LR test is UMP among all tests whose power is a function of λ only.

Invariant tests

23.37 In developing unbiased tests for location parameters in **23.20**, we found it quite natural to introduce the invariance condition (23.61) as a necessary condition that any acceptable test must satisfy. Similarly for scale parameters in **23.21**, the logarithmic transformation from (23.68) to (23.69) requires implicitly that the test statistic t satisfies

$$t(y_1, y_2, \ldots, y_n) = t(cy_1, cy_2, \ldots, cy_n), \qquad c > 0. \tag{23.114}$$

Frequently, it is reasonable to restrict the class of tests considered to those that are invariant under transformations that leave the hypothesis to be tested invariant; if this is not done, e.g. in the problem of testing the equality of location (or scale) parameters, it means that a change of origin (or unit) of measurement affects the conclusions reached by the test. The relationship between invariance and sufficiency principles in general is discussed by W. J. Hall *et al.* (1965), with a theorem due to C. Stein which gives conditions under which it does not matter in which order the principles are applied.

If we examine the canonical form of the general linear hypothesis in **23.27** from this point of view, we see at once that the problem is invariant under:
(a) any orthogonal transformation of $(\mathbf{z}_r - \mathbf{c}_0)$ (this leaves $(\mathbf{z}_r - \mathbf{c}_0)'(\mathbf{z}_r - \mathbf{c}_0)$ unchanged);
(b) any orthogonal transformation of $\mathbf{z}_{n-k}$ (this leaves $\mathbf{z}'_{n-k}\mathbf{z}_{n-k}$ unchanged);
(c) the addition of any constant a to each component of $\mathbf{z}_{k-r}$ (the mean vector of which is arbitrary);
(d) the multiplication of all the variables by $c > 0$ (which affects only the common variance σ^2).

It is easily seen that a statistic t is invariant under all the operations (a) to (d) if, and only if, it is a function of $W = (\mathbf{z}_r - \mathbf{c}_0)'(\mathbf{z}_r - \mathbf{c}_0)/\mathbf{z}'_{n-k}\mathbf{z}_{n-k}$ alone. Clearly if t is a function of W alone, its power function, like that of W, will depend only on λ. By the last sentence of **23.36**, therefore, the LR test, rejecting large values of W, is UMP among invariant tests of the general linear hypothesis.

EXERCISES

23.1 Show that the c.f. of the non-central χ^2 distribution (23.18) is

$$\phi(t) = (1 - 2it)^{-\nu/2} \exp\left\{\frac{\lambda it}{1 - 2it}\right\},$$

giving cumulants $\kappa_r = (\nu + r\lambda)2^{r-1}(r-1)!$. In particular,

$$\kappa_1 = \nu + \lambda, \qquad \kappa_2 = 2(\nu + 2\lambda),$$
$$\kappa_3 = 8(\nu + 3\lambda), \qquad \kappa_4 = 48(\nu + 4\lambda).$$

Hence show that the sum of two independent non-central χ^2 variates is another such, with both degrees of freedom and non-central parameter equal to the sum of those of the component distributions.

Show that the c.f. of the central χ^2 approximation at (23.21) tends to the exact result as $\nu \to \infty$.

By writing $\frac{1}{2}\lambda\{(1 - 2it)^{-1} - 1\}$ for $\lambda it(1 - 2it)^{-1}$, use the inversion theorem for c.f.s to deduce (23.18) from the central χ^2 density.

23.2 Show that if the non-central normal variates x_i of **23.4** are subjected to k orthogonal linear constraints

$$\sum_{i=1}^{n} a_{ij} x_i = b_j \qquad j = 1, 2, \ldots, k,$$

where

$$\sum_{i=1}^{n} a_{ij}^2 = 1, \quad \sum_{i=1}^{n} a_{ij} a_{il} = 0, \qquad j \neq l,$$

then

$$y^2 = \sum_{i=1}^{n} x_i^2 - \sum_{j=1}^{k} b_j^2$$

has the non-central χ^2 distribution with $(n - k)$ degrees of freedom and non-central parameter

$$\lambda = \sum_{i=1}^{n} \mu_i^2 - \sum_{j=1}^{k} \left(\sum_{i=1}^{n} a_{ij} \mu_i\right)^2.$$

(Patnaik (1949). Cf. also Bateman (1949).)

23.3 Show that for any fixed r, the first r moments of a non-central χ^2 distribution with fixed λ tend, as degrees of freedom increase, to the corresponding moments of the central χ^2 distribution with the same degrees of freedom. Hence show that, in testing a hypothesis H_0, if the test statistic has a non-central χ^2 distribution with degrees of freedom an increasing function of sample size n, and non-central parameter λ a non-increasing function of n such that $\lambda = 0$ when H_0 holds, then the test will become ineffective as $n \to \infty$, i.e. its power will tend to its size α.

23.4 Show that the LR statistic l defined by (23.40) for testing the equality of k normal variances has moments about zero

$$\mu_r' = \frac{n^{rm/2}\Gamma\{\frac{1}{2}(n-k)\}}{\Gamma\{\frac{1}{2}[(r+1)n-k]\}} \prod_{i=1}^{k} \frac{\Gamma\{\frac{1}{2}[(r+1)n_i - 1]\}}{n_i^{rm_i/2}\Gamma\{\frac{1}{2}(n_i - 1)\}}.$$

(Neyman and Pearson, 1931)

23.5 For testing the hypothesis H_0 that k normal distributions are identical in mean and variance, show that the LR statistic is, for sample sizes $n_i \geqslant 2$,

$$l_0 = \prod_{i=1}^{k} \left(\frac{s_i^2}{s_0^2} \right)^{n_i/2}$$

where

$$s_i^2 = \frac{1}{n_i} \sum_{j=1}^{n_i} (x_{ij} - \bar{x}_i)^2, \qquad \bar{x} = \frac{1}{n} \sum_{i=1}^{k} n_i \bar{x}_i$$

and

$$s_0^2 = \frac{1}{n} \sum_{i=1}^{k} n_i \{ s_i^2 + (\bar{x}_i - \bar{x})^2 \},$$

and that its moments about zero are

$$\mu_r' = \frac{n^{rn/2} \Gamma\{\frac{1}{2}(n-1)\}}{\Gamma\{\frac{1}{2}[(r+1)n-1]\}} \prod_{i=1}^{k} \frac{\Gamma\{\frac{1}{2}[(r+1)n_i - 1]\}}{n_i^{rn_i/2} \Gamma\{\frac{1}{2}(n_i - 1)\}}.$$

(Neyman and Pearson, 1931)

23.6 For testing the hypothesis H_2 that k normal distributions with the same variance have equal means, show that the LR statistic (with sample sizes $n_i \geqslant 2$) is

$$l_2 = l_0 / l$$

where l and l_0 are as defined for Exercises 23.4 and 23.5, and that the exact distribution of $z = 1 - l_2^{2/n}$ when H_2 holds is

$$dF \propto z^{(k-3)/2}(1-z)^{(n-k-2)/2} \, dz, \qquad 0 \leqslant z \leqslant 1.$$

Find the moments of l_2 and hence show that when the hypothesis H_0 of Exercise 23.5 holds, l and l_2 are independently distributed.

(Neyman and Pearson, 1931; Hogg (1961). See also Hogg (1962) for a test of H_2.)

23.7 Show that when all the sample sizes n_i are equal, the LR statistic l of (23.40) and its modified form l^* of (23.44) are connected by the relation

$$n \log l^* = (n - k) \log l,$$

so that in this case the tests based on l and l^* are equivalent.

23.8 For samples from k distributions of form (23.48) or (23.52), show that if l is the LR statistic for testing the hypothesis

$$H_0 : \theta_1 = \theta_2 = \cdots = \theta_{p_1}; \; \theta_{p_1+1} = \theta_{p_1+2} = \cdots = \theta_{p_2}; \; \theta_{p_2+1} = \cdots = \theta_{p_3}$$
$$\cdots; \; \theta_{p_{r-1}+1} = \cdots = \theta_{p_r},$$

so that the θ_i fall into r distinct groups (not necessarily of equal size) within which they are equal, then $-2 \log l$ is distributed exactly like χ^2 with $2(n-r)$ degrees of freedom.

(Hogg, 1956)

23.9 In a sample of n observations from

$$dF = dx/2\theta, \qquad \mu - \theta \leqslant x \leqslant \mu + \theta,$$

show that the LR statistic for testing $H_0 : \mu = 0$ is

$$l = \left(\frac{x_{(n)} - x_{(1)}}{2z} \right)^n = \left(\frac{R}{2z} \right)^n$$

where $z = \max\{-x_{(1)}, x_{(n)}\}$. Using Exercise 22.7, show that l and z are independently distributed, so that we have the factorisation of c.f.s

$$E[\exp\{(-2\log R^n)it\}] = E[\exp\{(-2\log l)it\}]E[\exp\{[-2\log (2z)^n]it\}].$$

Hence show that the c.f. of $-2\log l$ is

$$\phi(t) = \frac{(n-1)}{n(1-2it)-1}$$

so that, as $n \to \infty$, $-2\log l$ is distributed as χ^2 with 2 degrees of freedom.

(Hogg and Craig, 1956)

23.10 In **23.6**, show that a quadratic form $\mathbf{x}'\mathbf{Ax}$ has a non-central χ^2 distribution if and only if $\mathbf{AV}$ is idempotent, and that if the distribution has n degrees of freedom this implies $\mathbf{A} = \mathbf{V}^{-1}$.

23.11 k independent samples, of sizes $n_i \geq 2$, $\sum_{i=1}^{k} n_i = n$, are taken from exponential populations

$$dF_i(x) = \exp\left\{-\left(\frac{x-\theta_i}{\sigma_i}\right)\right\} dx \bigg/ \sigma_i, \qquad \theta_i \leq x < \infty.$$

Show that the LR statistic for testing

$$H_0: \theta_1 = \theta_2 = \cdots = \theta_k; \sigma_1 = \sigma_2 = \cdots = \sigma_k$$

is

$$l_0 = \prod_{i=1}^{k} d_i^{n_i}/d^n$$

where $d_i = \bar{x}_i - x_{(1)i}$ the difference between the mean and smallest observation in the ith sample, and d is the same function of the combined samples, i.e.

$$d = \bar{x} - x_{(1)}.$$

Show that the moments of $l_0^{1/n}$ are

$$\mu'_p = \frac{n^p \Gamma(n-1)}{\Gamma(n+p-1)} \prod_{i=1}^{k} \frac{\Gamma\{(n_i-1)+pn_i/n\}}{n_i^{pn_i/n}\Gamma(n_i-1)}.$$

(P. V. Sukhatme, 1936)

23.12 In Exercise 23.11, show that for testing

$$H_1: \sigma_1 = \sigma_2 = \cdots = \sigma_k,$$

the θ_i being unspecified, the LR statistic is

$$l_1 = \frac{\prod_{i=1}^{k} d_i^{n_i}}{\left(\frac{1}{n}\sum_{i=1}^{k} n_i d_i\right)^n},$$

and that the moments of $l_1^{1/n}$ are

$$\mu'_p = \frac{n^p \Gamma(n-k)}{\Gamma(n-k+p)} \prod_{i=1}^{k} \frac{\Gamma\{(n_i-1)+pn_i/n\}}{n_i^{pn_i/n}\Gamma(n_i-1)}.$$

Given that the θ_i are all known to be zero, show that the LR statistic for H_1 is l_1 with d_i replaced by $\bar{x}_i$ and that the moments μ'_p have (n_i-1) replaced by n_i and $(n-k)$ replaced by n. Show that when $k = 2$, the LR test of H_1 is equivalent to a variance-ratio (F) test with $2n_1$ and $2n_2$ d.fr.

(Nagarsenker (1980) gives 5 per cent and 1 per cent points when each $n_i = 4(1)10, 20, 40, 80, 100$ and $k = 3(1)6$.)

23.13 In Exercise 23.11, show that if it is known that the σ_i are all equal, the LR statistic for testing

$$H_2: \theta_1 = \theta_2 = \cdots = \theta_k$$

is

$$l_2 = l_0/l_1,$$

where l_0 and l_1 are defined in Exercises 23.11-12. Show that the exact distribution of $l_2^{1/n} = u$ is

$$dF = \frac{1}{B(n-k, k-1)} u^{n-k-1}(1-u)^{k-2} \, du, \qquad 0 \le u \le 1,$$

and find the moments of u. Show that when H_0 of Exercise 23.11 holds, l_1 and l_2 are independently distributed.

(P. V. Sukhatme, 1936; Hogg (1961). Cf. also Hogg and Tanis (1963) for other tests of these hypotheses.)

23.14 Show by comparison with the unbiased test of Exercise 22.17 that the LR test for the hypothesis that two normal populations have equal variances is biased for unequal sample sizes n_1, n_2.

23.15 Show by using (23.67) that an unbiased similar size-α test of the hypothesis H_0 that k independent observations x_i $(i = 1, 2, \ldots, k)$ from normal populations with unit variance have equal means is given by the critical region

$$\sum_{i=1}^{k} (x_i - \bar{x})^2 \ge c_\alpha,$$

where c_α is the $100(1 - \alpha)$ per cent point of the distribution of χ^2 with $(n - 1)$ degrees of freedom. Show that this is also the LR test.

23.16 Show that the three test statistics of Exercises 22.24-26 are equivalent to the LR statistics in the situations given; that the critical region of the LR test in Exercise 22.24 is not the UMPU region and is in fact biased; but that in the other two Exercises the LR test coincides with the UMP similar test.

23.17 Extending the results of **22.10-13**, show that if a distribution is of form

$$f(x \mid \theta_1, \theta_2, \ldots, \theta_k) = Q(\boldsymbol{\theta}) M(x) \exp\left\{ \sum_{j=3}^{k} B_j(x) A_j(\theta_3, \theta_4, \ldots, \theta_k) \right\},$$

$$a(\theta_1, \theta_2) \le x \le b(\theta_1, \theta_2)$$

(the terminals of the distribution depending only on the two parameters not entering into the exponential term), the statistics $t_1 = x_{(1)}$, $t_2 = x_{(n)}$, $t_j = \sum_{i=1}^{n} B_j(x_i)$ are jointly sufficient for $\boldsymbol{\theta}$ in a sample of n observations, and that their distribution is complete.

(Hogg and Craig, 1956)

23.18 Using the result of Exercise 23.17, show that in independent samples of sizes n_1, n_2 from two distributions

$$dF = \exp\left\{ -\frac{(x_i - \theta_i)}{\sigma} \right\} \frac{dx_i}{\sigma}, \qquad \sigma > 0; \; x_i \ge \theta_i; \; i = 1, 2,$$

the statistics

$$z_1 = \min \{x_{i(1)}\},$$

$$z_2 = \sum_{j=1}^{n_1} x_{1j} + \sum_{j=1}^{n_2} x_{2j},$$

are sufficient for θ_1 and θ_2 and complete.

Show that the LR statistic for $H_0: \theta_1 = \theta_2$ is

$$l = \left\{ \frac{z_2 - (n_1 x_{1(1)} + n_2 x_{2(1)})}{z_2 - (n_1 + n_2) z_1} \right\}^{n_1 + n_2}$$

and that l is distributed independently of z_1, z_2 and hence of its denominator. Show that l gives an unbiased test of H_0.

(Paulson, 1941)

23.19 Show that (23.18) may be written as a mixture (cf. **5.22**) of central χ^2 distributions with $(\nu + 2r)$ d.fr., with Poisson frequencies as the mixing distribution. By representing these χ^2 distributions as the sum of a $\chi^2(\nu)$ and $r \chi^2(2)$ distributions, all independent, use Exercise 5.22 to establish the c.f. of (23.18) given in Exercise 23.1; show that if for $\lambda > 0$ we put $\nu = 0$, the c.f., cumulants and additive property there remain valid, while the f.f. has discrete probability $e^{-\lambda/2}$ at $z = 0$ and (23.18) holds for $z > 0$.

Show further that (23.18) satisfies the relations

$$\chi'^2(\nu, \lambda) = \left(1 + 2 \frac{d}{d\lambda} \right) \chi'^2(\nu - 2, \lambda)$$

$$= \left(1 + 2 \frac{d}{dz} \right) \chi'^2(\nu + 2, \lambda),$$

this last result generalising the first one of Exercise 16.7 in the central case.

(cf. Siegel (1979) and J. D. Cohen (1988).)

23.20 For the LR test of the general linear hypothesis based on (23.99), show that the asymptotic non-central χ^2 distribution of $-2 \log l$, given in **23.7**, agrees with the asymptotic result for $\nu_1 F'(\nu_1, \nu_2, \lambda)$ at the end of **23.31**.

23.21 As in **23.31**, write the non-central F variate in the form

$$F'(\nu_1, \nu_2, \lambda) = \frac{z_1 / \nu_1}{z_2 / \nu_2},$$

where z_1 is a non-central $\chi'^2(\nu_1, \lambda)$ variate and z_2 an independent central $\chi^2(\nu_2)$ variate. Let z_3 be a central $\chi^2(1)$ variate, independent of z_1 and z_2. Show that

$$E\{(F')^r\} = E\left\{ \left(\frac{z_3}{z_2 / \nu_2} \right)^r \right\} E(z_1^r) / \{\nu_1^r E(z_3^r)\}$$

whence symbolically

$$\mu_r'\{F'(\nu_1, \nu_2, \lambda)\} = \frac{\mu_r'\{F(1, \nu_2)\} \mu_r'\{\chi'^2(\nu_1, \lambda)\}}{\nu_1^r \mu_r'\{\chi^2(1)\}}$$

if $2r < \nu_2$, so that the central F moments exist by Exercise 16.1. Hence, using (16.4–5), (16.28) and (23.19), show that $F'(\nu_1, \nu_2, \lambda)$ has mean and variance given by

$$\mu'_1 = (\nu_1 + \lambda)\nu_2 / \{\nu_1(\nu_2 - 2)\}, \qquad \nu_2 > 2,$$

$$\mu_2 = \frac{2\nu_2^2}{\nu_1^3(\nu_2 - 2)^2(\nu_2 - 4)} \{(\nu_1 + 2\lambda)(\nu_1 + \nu_2 - 2) + \lambda^2\}, \qquad \nu_2 > 4.$$

(Cf. Bain, 1969; Pearson and Tiku (1970) give μ_3 and μ_4).

23.22   Apply the method of Exercise 23.21 to the non-central t' variate defined at (23.109) by multiplying and dividing $E\{(t')^r\}$ by $E\{z_3^{r/2}\}$, and show that

$$\mu'_r\{t'(f, \delta)\} = \frac{\mu'_r\{N(\delta, 1)\}\mu'_{r/2}\{F(1, f)\}}{\mu'_{r/2}\{\chi^2(1)\}}, \qquad r < f,$$

where $N(\delta, 1)$ is a normal variate with mean δ and variance 1. If r is even, say $= 2s$, show that this may be written

$$\mu'_{2s}\{t'(f, \delta)\} = \mu'_{2s}\{N(\delta, 1)\}\mu_{2s}\{t(f)\}/\mu_{2s}\{N(0, 1)\},$$

where $t(f)$ is the central t-distribution with f degrees of freedom.

(Cf. Bain, 1969)

23.23   In (23.109), write the distribution of the numerator $u = z + \delta$ as

$$p_\delta(u) = \exp\left(-\tfrac{1}{2}\delta^2 + \delta u\right)p_0(u),$$

and hence show that the non-central t' distribution is

$$h(t') = \{\Gamma(\tfrac{1}{2})\Gamma(\tfrac{1}{2}f)f^{1/2} \exp(\tfrac{1}{2}\delta^2)\}^{-1} \sum_{s=0}^{\infty} \frac{\{(2/f)^{1/2}\delta t'\}^s}{s!} \frac{\Gamma\{\tfrac{1}{2}(f + s + 1)\}}{\left(1 + \dfrac{t'^2}{f}\right)^{(f+s+1)/2}},$$

reducing to the central Student's t distribution (16.15) when $\delta = 0$. Show that the distribution of t'^2 obtained from $h(t')$ agrees with (23.105) with $\nu_1 = 1$, $\nu_2 = f$, $\lambda = \delta^2$ and $F = t'^2$.

23.24   Using the transformation $\xi = (1 + t'^2/f)^{-1}$ as in **16.11** for the central case, show that the d.f. of the non-central t' distribution in Exercise 23.23 is given by

$$F(t') = 1 - \tfrac{1}{2}\exp\left(-\tfrac{1}{2}\delta^2\right) \sum_{s=0}^{\infty} \frac{(\delta/2^{1/2})^s}{\Gamma(\tfrac{1}{2}s + 1)} I_\xi(\tfrac{1}{2}f, \tfrac{1}{2}(s + 1)),$$

reducing to (16.17) when $\delta = 0$.
    Show from (23.109) that $t' > 0$ if and only if $z + \delta > 0$, so that $1 - F(0)$ is $G(\delta)$ where G is the standard normal d.f. Hence deduce the result of Exercise 5.18 for the normal d.f.

(Hawkins, 1975)

23.25   In testing a simple H_0: $L = L_0$ against a simple H_1: $L = L_1$, show that the LR statistic (23.4) is $l = L_0/\max(L_0, L_1) = \min(1, L_0/L_1)$ with critical region (23.6), as against the BCR (21.6). Show that the two tests coincide if their size

$$\alpha \le \text{Prob}(L_1 > L_0)$$

but that the LR test cannot achieve the BCR if $\text{Prob}(L_0 \ge L_1) > 1 - \alpha > 0$.

23.26   Using Exercise 23.19 and Exercise 16.7, show that the d.f. of (23.18) may be written, for even ν,

$$H(z) = \text{Prob}\{u - v \ge \tfrac{1}{2}\nu\},$$

where u and v are independent Poisson variates with parameters $\frac{1}{2}z$ and $\frac{1}{2}\lambda$ respectively; and that for any ν we have

$$H(z) = \sum_{r=0}^{\infty} e^{-\lambda/2} \frac{(\frac{1}{2}\lambda)^r}{r!} \{F_\nu(z) - 2 \sum_{p=1}^{r} f_{\nu+2p}(z)\}$$

$$= F_\nu(z) - 2 \sum_{p=1}^{\infty} f_{\nu+2p}(z) F_{2p}(\lambda),$$

where $F_\nu(z)$, $f_\nu(z)$ are the d.f. and f.f. of a central χ_ν^2 variable. Show that, regarded as a function of λ for fixed ν and z, $1 - H(z) = G(\lambda)$, say, is a non-decreasing function of λ. Regarding this as a d.f., show that its m.g.f. is

$$M_\lambda(t) = (1 - 2t)^{(\nu/2)-1} \exp\left(\frac{zt}{1-2t}\right) F_\nu\left(\frac{z}{1-2t}\right)$$

and that near $t = 0$, if $1 - F_\nu(z)$ is negligible, we may approximate it by the proper m.g.f.

$$M_\lambda^*(t) = (1 - 2t)^{(\nu/2)-1} \exp\left(\frac{zt}{1-2t}\right)$$

with cumulants (cf. Exercise 23.1) $\kappa_r^* = \{rz - (\nu - 2)\}2^{r-1}(r-1)!$

(Cf. Johnson 1959a), and Venables (1975), who uses the approximate cumulants in a Cornish–Fisher expansion (6.25-6) to find percentiles of $G(\lambda)$ for setting confidence intervals, and extends the analysis to λ in the F'-distribution (23.105). See also Winterbottom (1979).)

23.27 For the inverse Gaussian distribution of Exercises 18.41 and 11.28-9, show that if λ is known the LR test of $H_0: \mu = \mu_0$ against $H_1: \mu \neq \mu_0$ rejects H_0 when

$$l_1 = n\lambda (\bar{x} - \mu_0)^2 / (\bar{x}\mu_0^2)$$

exceeds the $100(1 - \alpha)\%$ point of its χ_1^2 distribution. Show that if λ is unknown, the LR test rejects H_0 when

$$l_2 = l_1 / \{\lambda v / (n-1)\}$$

exceeds the $100(1 - \alpha)\%$ point of its distribution, which is Student's t^2 with $(n-1)$ d.fr. in consequence of the results of Exercise 18.41.

(Chhikara and Folks (1976) show that these tests are UMPU).

CHAPTER 24

SEQUENTIAL METHODS

24.1 When considering sampling problems in the foregoing chapters we have usually assumed that the sample size n was fixed. This may be because we chose it beforehand; but it may be that n was not at our choice, as for example when we are presented with the results of a finished experiment; or it may be that the sample size was determined by some other criterion, as when we decide to observe for a given period of time. For example, in setting a standard error to an estimate, we were making probability statements within a field of samples all of size n. If n is determined in some way that is independent of the values of the observations, such a conditional argument seems reasonable, but it is as well to realize that there is nothing automatic about accepting it. We hope we will not be thought quite cynical if we add that, in our view, the reason why so many statistical arguments are made conditionally upon an observed value of n is that this procedure is very convenient mathematically—cf. **21.32** above.

Even if n is fixed, we are sometimes able to improve upon even the most powerful test procedures discussed in earlier chapters. For example, if π is the parameter of a binomial distribution, Exercise 21.2 showed that the UMP test of $H_0: \pi = \frac{1}{2}$ against $H_1: \pi > \frac{1}{2}$ rejects H_0 when the number of successes in the sample, x, is large enough. Using the normal approximation to the distribution of x, and assuming a test size 0.05 based on $n = 100$ observations, H_0 is rejected when

$$x > n\pi_0 + 1.645\{n\pi_0(1 - \pi_0)\}^{1/2} = \tfrac{1}{2}n + 0.822n^{1/2},$$

i.e. $x > 58.22$. Now suppose that we had paused after collecting three-quarters of the sample, and noticed that we already had $x = 65$. We are clearly bound to reject H_0 for $n = 100$, and indeed for any n in $75 \le n \le 100$, since $an + bn^{1/2}$ is an increasing function of n. Thus it is clearly wasteful to collect the last 25 observations at all. It evidently pays, in this example at least, to make the sample size a function of the values so far observed.

Sequential procedures

24.2 Occasionally, the sample size is a random variable explicitly dependent upon the values of the observations. One of the simplest cases is one we have already touched upon in Example 9.15. Suppose we are sampling human beings one by one to discover what proportion belongs to a rare blood-group. Instead of sampling, say, 1000 individuals and counting the number of members of that blood-group we may go on sampling until 20 such members have occurred. We shall see later why this may be a preferable procedure; for the moment we take for granted that it is worth considering. In successive trials of such an inquiry we should find that for a fixed number of successes, say 20, the sample size n required to achieve them varied considerably. It must be at least 20 but it might be infinite (although the probability of going on indefinitely is zero, so that we are almost certain to stop sooner or later).

24.3 Procedures like this are called *sequential.* Their typical feature is a *sampling scheme*, which lays down a *stopping rule* under which we decide after each selection for the sample whether to stop or to continue sampling. In our present example the rule is very simple: if we draw a failure, continue; if we draw a success, continue unless 19 successes have previously occurred, in which event, stop. The decision at any point is, in general, dependent on the observations made up to that point. Thus, for a sequence of values $x_1, x_2, \ldots, x_n$, the sample size at which we stop is not independent of the xs. It is this fact that gives sequential analysis its characteristic features.

Sequential methods were first developed during the Second World War, principally by Wald (whose work is summarised in his book, Wald (1947)) in U.S.A., and simultaneously in England by G. A. Barnard (1946).

24.4 If the probability is one that the procedure will terminate, the scheme is said to be *closed.* If there is a non-zero probability that sampling can continue indefinitely the scheme is called *open.* Open schemes are obviously of little practical use compared to closed schemes, and we usually have to reduce them to closed form by putting an upper limit to the size of the sample. Such truncation often makes their properties difficult to determine exactly.

Usage in this matter is not entirely uniform in the literature of the subject. 'Closed' sometimes means 'truncated', that is to say, applies to the case where the stopping rule puts an upper limit to the sample size, and 'open' then means 'non-truncated'.

The case where we fix a sample size beforehand may be regarded as a very special case of a sequential scheme with stopping rule: go on until you have obtained n members, irrespective of what values arise.

Example 24.1
As an example of a sequential scheme let us consider sequential sampling for attributes (Example 9.15), where we proceed until k successes are observed and then stop. It scarcely needs proof that such a scheme is closed. The probability that in an infinite sequence we do not observe k successes is zero.

From (9.15),

$$\binom{n-1}{k-1} \pi^k (1-\pi)^{n-k}, \qquad n = k, k+1, \ldots, \tag{24.1}$$

is the distribution of n. The frequency-generating function of n (with the origin at zero) is given by (5.45) as, in our present notation,

$$[\pi t/\{1-(1-\pi)t\}]^k, \tag{24.2}$$

while for the cumulant-generating function we have from (5.47)

$$\psi(t) = k \log\left[\pi e'/\{1-(1-\pi) e'\}\right].$$

From (5.48), we find

$$\kappa_1 = k/\pi, \tag{24.3}$$
$$\kappa_2 = k(1-\pi)/\pi^2. \tag{24.4}$$

From (9.16), an unbiased estimator of π is

$$p = (k-1)/(n-1). \tag{24.5}$$

For $k > 1$, the variance of p is expressible as a series. We have

$$E(p^2) = (k-1)\pi^{k-1}(1-\pi)^{1-k} \sum_{n=k}^{\infty} \binom{n-2}{k-2} \frac{(1-\pi)^{n-1}}{n-1}$$

$$= (k-1)\pi^k(1-\pi)^{1-k} \int_0^{(1-\pi)} \sum_{n=k}^{\infty} \binom{n-2}{k-2} t^{n-2} \, dt$$

$$= (k-1)\pi^k(1-\pi)^{1-k} \int_0^{(1-\pi)} t^{k-2}(1-t)^{1-k} \, dt. \tag{24.6}$$

Putting $u = \pi t/\{(1-\pi)(1-t)\}$ we find

$$E(p^2) = (k-1)\pi^2 \int_0^1 \frac{u^{k-2} \, du}{\pi + (1-\pi)u} \tag{24.7}$$

$$= (k-1)\pi^2 \int_0^1 u^{k-2} \left\{ \sum_{j=0}^{\infty} (1-\pi)^j (1-u)^j \right\} du$$

$$= (k-1)\pi^2 \sum_{j=0}^{\infty} (1-\pi)^j B(k-1, j+1)$$

$$= \pi^2 \sum_{j=0}^{\infty} \binom{k+j-1}{j}^{-1} (1-\pi)^j. \tag{24.8}$$

Hence, subtracting π^2, we have

$$\operatorname{var} p = \pi^2 \sum_{j=1}^{\infty} \binom{k+j-1}{j}^{-1} (1-\pi)^j. \tag{24.9}$$

As $\pi \to 0$ in (24.7), if $k > 2$, $E(p^2)/\pi^2 \to (k-1)/(k-2)$, and the coefficient of variation of p, $(\operatorname{var} p)^{1/2}/E(p) \to (k-2)^{-1/2}$.

Prasad and Sahai (1982) give close bounds for var p—see Exercise 24.23.
Best (1974) gives a finite series for var p, and finds its m.s.e. numerically—it is usually preferable to that of the biased ML estimator k/n.

We may obtain an unbiased estimator of var p in a simple closed form. In the same manner that we obtained the unbiasedness of p in Example 9.15, we have

$$E \frac{(k-1)(k-2)}{(n-1)(n-2)} = \pi^2.$$

Hence

$$E\left\{\left(\frac{k-1}{n-1}\right)^2 - \frac{(k-1)(k-2)}{(n-1)(n-2)}\right\} = E\left(\frac{k-1}{n-1}\right)^2 - \pi^2 = \text{var } p.$$

Thus the expression in braces on the left is the estimator of var p; it simplifies to

$$\frac{(k-1)(n-k)}{(n-1)^2(n-2)} = \frac{p(1-p)}{n-2}. \tag{24.10}$$

For large n this is asymptotically equal to the corresponding result for fixed sample size n.

Since $n-2 = (k-1-p)/p$, (24.10) is equal to $p^2(1-p)/(k-1-p)$, and for small p is approximately $p^2/(k-1)$. Thus the coefficient of variation of p is (biasedly) estimated by $\{(1-p)/(k-1-p)\}^{1/2}$ and for small p this is approximately $(k-1)^{-1/2}$. For this sequential procedure, the estimated sampling error of p is thus approximately constant for small p.

> This sequential scheme is often called *inverse* binomial sampling—it was first discussed by Haldane (1945) and Finney (1949b). W. Knight (1965) unifies its theory for binomial, Poisson, hypergeometric and exponential distributions.

24.5 The sampling of attributes plays such a large part in sequential analysis that we may, before proceeding to more general considerations, discuss a useful diagrammatic method of representing the procedure.

Take a grid such as that of Fig. 24.1 and measure number of failures along the abscissa, number of successes along the ordinate. The sequential drawing of a sample

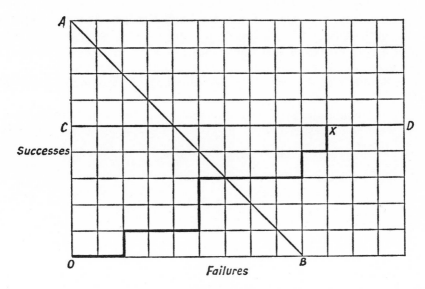

Fig. 24.1

may be represented on this grid by a path from the origin, moving one step to the right for a failure F and one step upwards for a success S. The path OX corresponds, for example, to the sequence $FFSFFFFSSFFFFFSFS$. A stopping rule is equivalent to some sort of barrier on the diagram. For example, the line AB is such that $S + F = 9$ and thus corresponds to the case of fixed sample size $n = 9$. The line CD corresponds to $S = 5$ and is thus of the type we considered in Example 24.1 with $k = 5$. The path OX, involving a sample of 15, is then one sample that would terminate at X. If X is the point whose co-ordinates are (x, y) the number of different paths from O to X is the number of ways in which x can be selected from $(x + y)$. The probability of arriving at X is this number times the probability of x Ss and y Fs, namely

$$\binom{x+y}{x} \pi^x (1 - \pi)^y.$$

Example 24.2. Gambler's ruin

One of the oldest problems in the theory of probability concerns a sequential procedure. Consider two players, A and B, playing a series of games at each of which A's chance of success is π and B's is $1 - \pi$. The loser at each game pays the winner one unit. If A starts with a units and B with b units what are their chances of ruin (a player being ruined when he has lost his last unit)?

A series of games like this is a sequential set representable on a diagram like Fig. 24.1. We may take A's winning as a success. The game continues so long as both A and B have any money left but stops when A or B has $a + b$ (when the other player has lost all his initial stake). The boundaries of the scheme are therefore the lines $y - x = -a$ and $y - x = b$.

Figure 24.2 shows the situation for the case $a = 5$, $b = 3$. The lines AB, CD are at 45° to the axes and go through $F = 0$, $S = 3$ and $F = 5$, $S = 0$ respectively. For any point between these lines $S - F$ is less than 3 and $F - S$ is less than 5. On AB, $S - F$ is 3, and if a path arrives at that line B has lost three more games than A and is ruined; similarly, if the path arrives at CD, A is ruined. The stopping rule is, then: if the point lies between the lines, continue sampling; if it reaches AB, stop with the ruin of B; if it reaches CD, stop with the ruin of A.

The actual probabilities are easily obtained. Let u_x be the probability that A will be ruined when he possesses x units. By considering a single game we see that

$$u_x = \pi u_{x+1} + (1 - \pi) u_{x-1}, \tag{24.11}$$

with boundary conditions

$$u_0 = 1, \qquad u_{a+b} = 0. \tag{24.12}$$

The general solution of (24.11) is

$$u_x = A t_1^x + B t_2^x$$

where t_1 and t_2 are the roots of

$$\pi t^2 - t + (1 - \pi) = 0,$$

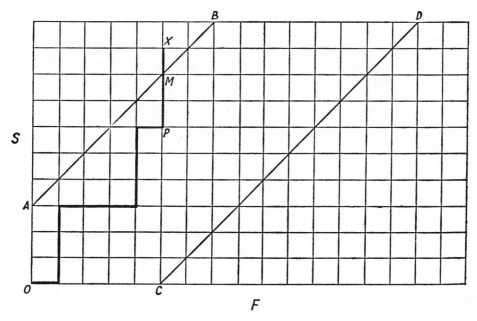

Fig. 24.2

namely

$$t = 1 \quad \text{and} \quad t = (1 - \pi)/\pi.$$

Provided that $1 - \pi \neq \pi$, the solution is then found to be, on using (24.12),

$$u_x = \frac{\left(\dfrac{1 - \pi}{\pi}\right)^{a+b} - \left(\dfrac{1 - \pi}{\pi}\right)^{x}}{\left(\dfrac{1 - \pi}{\pi}\right)^{a+b} - 1}, \qquad \pi \neq \tfrac{1}{2}. \tag{24.13}$$

If, however, $\pi = \tfrac{1}{2}$, the solution is

$$u_x = \frac{a + b - x}{a + b}. \tag{24.14}$$

In particular, at the start of the game, for $\pi = \tfrac{1}{2}$, $x = a$,

$$u_a = \frac{b}{a + b}. \tag{24.15}$$

24.6   We can obviously generalise this kind of situation in many ways and, in particular, can set up various types of boundary. A closed scheme is one for which it is virtually certain that the boundary will be reached.

Suppose, in particular, that the scheme specifies that if A loses he pays one unit but if B loses he pays k units. The path on Fig. 24.2 representing a series then consists of unit steps parallel to the abscissa and k-unit steps parallel to the ordinate. And this enables us to emphasise a point that constantly bedevils the mathematics of sequential schemes: a path may not end exactly on a boundary, but may cross it. For example, with $k = 3$ such a path might be OX in Fig. 24.2. After two successes and five failures we arrive at P. Another success would take us to X, crossing the boundary at M. We stop, of course, at this stage, whether the boundary is reached or crossed. The point of the example is that the exact probability of reaching the boundary at M is zero—in fact, this point is inaccessible. As we shall see, such discontinuities sometimes make it difficult to put forward exact and concise statements about the probabilities of what we are doing. We refer to such situations as 'end-effects'. In most practical circumstances they may be neglected.

Sequential tests of hypotheses

24.7 Let us apply the ideas of sequential analysis to testing hypotheses and, in the first instance, to choosing between H_0 and H_1. We suppose that these hypotheses concern a parameter θ that may take values θ_0 and θ_1 respectively; i.e. H_0 and H_1 are simple. We seek a sampling scheme that divides the sample space into three mutually exclusive domains: (a) domain ω_a, such that if the sample point falls within it we accept H_0 (and reject H_1); (b) domain ω_r, such that if the sample point falls within it we accept H_1 (and reject H_0); (c) the remainder of the sampling space, ω_c—if a point falls here we continue sampling. In Fig. 24.2, taking A's ruin as H_0, B's ruin as H_1, the region ω_a is the region to the right of CD, including the line itself; ω_r is the region above AB, including the line itself; ω_c is the region between the lines.

The operating characteristic

24.8 The probability of accepting H_0 when H_1 is true is a function of θ_1 which we shall denote by $K(\theta_1)$. If the scheme is closed, the probability of rejecting H_0 when H_1 is true is then $1 - K(\theta_1)$. Considered as a function of θ_1 this is simply the power function. As in our previous work we could, of course, work in terms of power; but in sequential analysis it has become customary to work with $K(\theta_1)$ itself.

$K(\theta)$ considered as a function of θ is called the operating characteristic (OC) of the scheme. Graphed as ordinate against θ as abscissa it gives us the OC curve, the complement (to unity) of the power function.

The average sample number

24.9 A second function which is used to describe the performance of a sequential test is the average sample number (ASN). This is the expectation of the sample size n required to reach a decision to accept H_0 or H_1 and therefore to discontinue sampling. The OC for H_0 and H_1 does not depend on the sample number, but only on constants determined initially by the sampling scheme. The ASN is the expected amount of sampling that we have to do to implement the scheme.

Example 24.3

Consider sampling from a (large) population of attributes of which a proportion π are successes, and let π be small. We are interested in the possibility that π is less than some given value π_0. This is, for example, a frequent situation where a manufacturer wishes to guarantee that the proportion of rejects in a batch of articles is below some declared figure. Consider first of all the alternative $\pi_1 > \pi_0$.

We will take a very simple scheme. If no success appears we proceed to sample until a pre-assigned sample number n_0 has appeared and accept π_0. If, however, a success appears we accept π_1 and stop sampling.

If the true probability of success is π, the probability that we accept the hypothesis is then $(1 - \pi)^{n_0}$. This is the OC. It is a J-shaped curve decreasing monotonically from $\pi = 0$ to $\pi = 1$. For two particular values we merely take the ordinates at π_0 and π_1.

The common sense of the situation requires that we should accept the smaller of π_0 and π_1 if no success appears, and the larger if a success does appear. Let π_0 be the smaller; then the probability α of a Type I error equals $1 - (1 - \pi_0)^{n_0}$ and that of an error of Type II, β, equals $(1 - \pi_1)^{n_0}$. If we were to interchange π_0 and π_1, the α-error would be $1 - (1 - \pi_1)^{n_0}$ and the β-error $(1 - \pi_0)^{n_0}$, both of which are greater than in the former case.

We can use the OC in this particular case to provide a test of the composite hypothesis $H_0: \pi \le \pi_0$ against $H_1: \pi > \pi_0$. In fact, if $\pi < \pi_0$ the chance of an α-error is less than $1 - (1 - \pi_0)^{n_0}$ and the chance of a β-error is less than $(1 - \pi_0)^{n_0}$.

The ASN is found by ascertaining the expectation of m, the sample number at which we terminate. For any given π this is clearly

$$\sum_{m=1}^{n_0-1} m\pi(1-\pi)^{m-1} + n_0(1-\pi)^{n_0-1} = -\pi \frac{\partial}{\partial \pi} \sum_0^{n_0-1} (1-\pi)^m + n_0(1-\pi)^{n_0-1}$$

$$= \frac{1-(1-\pi)^{n_0}}{\pi}. \tag{24.16}$$

The ASN in this case is a decreasing function of π since it equals

$$\frac{1-(1-\pi)^{n_0}}{1-(1-\pi)} = 1 + (1-\pi) + (1-\pi)^2 + \cdots + (1-\pi)^{n_0-1}.$$

We observe that the ASN will differ according to whether π_0 or π_1 is the true value.

A comparison of the results of the sequential procedure with those of an ordinary fixed sample-size is not easy to make for discrete distributions, especially as we have to compare two kinds of error. Consider, however, $\pi_0 = 0.1$ and $n = 30$. From tables of the binomial (e.g. *Biometrika Tables*, Table 37) we see that the probability of 5 successes or more is about 0.18. Thus on a fixed sample-size basis we may reject $\pi = 0.1$ in a sample of 30 with a Type I error of 0.18. For the alternative $\pi = 0.2$ the probability of 4 or fewer successes is 0.26, which is then the Type II error.

With the sequential test, for a sample of n_0 the Type I error is $1 - (1 - \pi_0)^{n_0}$ and the Type II error is $(1 - \pi_1)^{n_0}$. For a sample of 2 the Type I error is 0.19 and the

Type II error 0.64. For a sample of 6 the errors are 0.47 and 0.26 respectively. We clearly cannot make both types of errors correspond in this simple case, but it is evident that samples of smaller size are needed in the sequential case to fix either type of error at a given level. With more flexible sequential schemes, both types of error can be fixed at given levels with smaller ASN than the fixed-size sample number. In fact, their economy in sample number is one of their principal recommendations— cf. Example 24.10.

Wald's sequential probability-ratio test

24.10 Suppose we take a sample of n values in succession, $x_1, x_2, \ldots, x_n$, from a population $f(x, \theta)$. At any stage the ratio of the probabilities of the sample on hypotheses $H_0(\theta = \theta_0)$ and $H_1(\theta = \theta_1)$ is

$$L_n = \prod_{i=1}^{n} f(x_i, \theta_1) \Big/ \prod_{i=1}^{n} f(x_i, \theta_0). \qquad (24.17)$$

We select two numbers A and B, related to the desired α- and β-errors in a manner to be described later, and set up a sequential test as follows: so long as $B < L_n < A$ we continue sampling; at the first occasion when $L_n \geq A$ we accept H_1; at the first occasion when $L_n \leq B$ we accept H_0.

An equivalent but more convenient form for computation is the logarithm of L_n, the critical inequality then being

$$\log B < \sum_{i=1}^{n} \log f(x_i, \theta_1) - \sum_{i=1}^{n} \log f(x_i, \theta_0) < \log A. \qquad (24.18)$$

We shall refer to this family of tests as sequential probability-ratio tests (SPR tests).

24.11 We shall often find it convenient to write

$$z_i = \log \{ f(x_i, \theta_1) / f(x_i, \theta_0) \}, \qquad (24.19)$$

and the critical inequality (24.18) is then equivalent to a statement concerning the cumulative sums of z_is. Let us first of all prove that an SPR test terminates with probability one, i.e. is closed.

The sampling terminates if either

$$\sum z_i \geq \log A$$

or

$$\sum z_i \leq \log B.$$

The z_is are independent random variables with variance, say $\sigma^2 > 0$. $\sum_{i=1}^{n} z_i$ then has variance $n\sigma^2$. As n increases, the dispersion becomes greater and the probability that a value of $\sum z_i$ remains within the finite limits $\log B$ and $\log A$ tends to zero. More precisely, the mean $\bar{z}$ tends under the central limit effect to a (normal) distribution with variance σ^2/n, and hence the probability that it falls between $(\log B)/n$ and $(\log A)/n$ tends to zero.

It was shown by Stein (1946) that $E(e^{nt})$ exists for any complex number t whose real part is less than some $t_0 > 0$. It follows that the random variable n has moments of all orders.

Example 24.4
Consider again the binomial distribution, the probability of success being π. If there are k successes in the first n trials the SPR criterion is given by

$$\log L_n = k \log \frac{\pi_1}{\pi_0} + (n-k) \log \frac{1-\pi_1}{1-\pi_0}. \tag{24.20}$$

This quantity is computed as we go along, the sampling continuing until we reach the boundary values $\log B$ or $\log A$. How we decide upon A and B will appear in a moment.

24.12　It is a remarkable fact that the numbers A and B can be derived very simply (at least to an acceptable degree of approximation) from the probabilities of errors of the first and second kinds, α and β, without knowledge of the population. This does not mean that the sequential process is unaffected by the underlying distribution. All that is happening is that our knowledge of the frequency distribution is put into the criterion L_n of (24.17) and we work with this ratio of probabilities directly. It will not, then, come as a surprise to find that SPR tests have certain optimum properties; for they use all the available information, including the order in which the sample values occur.

Consider a sample for which L_n lies between A and B for the first $n-1$ trials and then becomes $\geqslant A$ at the nth trial, so that we accept H_1 (and reject H_0). By definition, the probability of getting such a sample is at least A times as large under H_1 as under H_0. This, being true for any one sample, is true for the aggregate of all possible samples resulting in the acceptance of H_1. The probability of accepting H_1 when H_0 is true is α, and that of accepting H_1 when H_1 is true is $1-\beta$. Hence

$$A \leqslant \frac{1-\beta}{\alpha}. \tag{24.21}$$

In like manner we see from the cases in which we accept H_0 that

$$B \geqslant \frac{\beta}{1-\alpha}. \tag{24.22}$$

24.13　If our boundaries were such that A and B were exactly attained when attained at all, i.e. if there were no end-effects, we could write

$$A = \frac{1-\beta}{\alpha}, \qquad B = \frac{\beta}{1-\alpha}. \tag{24.23}$$

Wald (1947) showed that for all practical purposes these equalities may be assumed to hold. Suppose that we have exactly

$$a = \frac{1-\beta}{\alpha}, \qquad b = \frac{\beta}{1-\alpha} \tag{24.24}$$

and that the true errors of first and second kind for the limits a and b are α', β'. We then have, from (24.21),

$$\frac{\alpha'}{1-\beta'} \leq \frac{1}{a} = \frac{\alpha}{1-\beta}, \tag{24.25}$$

and from (24.22)

$$\frac{\beta'}{1-\alpha'} \leq b = \frac{\beta}{1-\alpha}. \tag{24.26}$$

Hence

$$\alpha' \leq \frac{\alpha(1-\beta')}{1-\beta} \leq \frac{\alpha}{1-\beta}, \tag{24.27}$$

$$\beta' \leq \frac{\beta(1-\alpha')}{1-\alpha} \leq \frac{\beta}{1-\alpha}. \tag{24.28}$$

Furthermore, from (24.25) and (24.26),

$$\alpha'(1-\beta) + \beta'(1-\alpha) \leq \alpha(1-\beta') + \beta(1-\alpha')$$

or

$$\alpha' + \beta' \leq \alpha + \beta. \tag{24.29}$$

Now in practice α and β are small, often conventionally 0.01 or 0.05. It follows from (24.27) and (24.28) that the amount by which α' can exceed α, or β' exceed β, is negligible. Moreover, from (24.29) we see that either $\alpha' \leq \alpha$ or $\beta' \leq \beta$. Hence by using a and b in place of A and B, the worst we can do is to increase one of the errors, and then only by a very small amount. Such a procedure, then, will always be on the safe side in the sense that for all practical purposes it will not increase the errors of wrong decision. To avoid tedious repetition we shall henceforward use the equalities (24.23) except where the contrary is specified.

The inequalities for A and B were also derived for the critical value $1/k_\alpha$ of the fixed sample size probability-ratio test in Exercise 21.14. The fact that A and B practically attain their limits implies that the two sequential critical values enclose the single fixed-n critical value, as is intuitively acceptable.

Example 24.5
Consider again the binomial distribution of Example 24.4 with $\alpha = 0.01$, $\beta = 0.10$, $\pi_0 = 0.01$ and $\pi_1 = 0.03$. We have, for k successes and $n-k$ failures (taking logarithms to base 10),

$$\log \frac{\beta}{1-\alpha} \leq (n-k)\log \frac{1-\pi_1}{1-\pi_0} + k\log \frac{\pi_1}{\pi_0} \leq \log \frac{1-\beta}{\alpha}$$

or

$$\log \frac{10}{99} \leq (n-k)\log \frac{97}{99} + k\log 3 \leq \log 90$$

or

$$-0.995\ 635 \leqslant -0.008\ 863\ 5(n - k) + 0.477\ 121\ k \leqslant 1.954\ 243.$$

Dividing through by 0.008 863 5 we find, to the nearest integer,

$$-112 \leqslant 54k - n \leqslant 220. \tag{24.30}$$

For a test of this kind, for example, if no success occurred in the first 112 drawings we should accept π_0. If one occurred at the 100th drawing and another at the 200th, we could not accept before the 220th (i.e. $112 + (2 \times 54)$) drawing. And if, by the 200th drawing, there had occurred 6 successes, say at the 50th, 100th, 125th, 150th, 175th, 200th, we could not reject, $54k - n$ being 124 at the 200th drawing; but if that experience was then repeated, the quantity $54k - n$ would exceed 220 and we should accept π_1.

The OC of the SPR test
24.14 Consider the function

$$L^h = \left\{ \frac{f(x, \theta_1)}{f(x, \theta_0)} \right\}^h, \tag{24.31}$$

where h is a function of θ. $L^h f(x, \theta)$, say $g(x, \theta)$, is a frequency function for any value of θ provided that

$$E(L^h) = \int \left\{ \frac{f(x, \theta_1)}{f(x, \theta_0)} \right\}^h f(x, \theta)\, dx = 1. \tag{24.32}$$

It may be shown (cf. Exercise 24.4) that there is at most one non-zero value of h satisfying this equation. Consider the rule: accept H_0, continue sampling, or accept H_1 according to the inequality

$$B^h \leqslant \frac{\Pi\{L^h f(x, \theta)\}}{\Pi\{f(x, \theta)\}} \leqslant A^h. \tag{24.33}$$

This is evidently equal to the ordinary rule of (24.18) provided that $h > 0$. Consider testing H: that the true distribution is $f(x, \theta)$, against G: that the true distribution is $g(x, \theta)$. If α', β' are the two errors, the probability ratio is the one appearing in (24.33), and we then have, using (24.23),

$$A^h = \frac{1 - \beta'}{\alpha'}, \qquad B^h = \frac{\beta'}{1 - \alpha'}, \tag{24.34}$$

and hence

$$\alpha' = \frac{1 - B^h}{A^h - B^h},$$

and since α' is the power function when H_1 holds, its complement, the OC, is given by

$$1 - \alpha' = K(\theta) = \frac{A^h - 1}{A^h - B^h}. \tag{24.35}$$

The same formula holds if $h < 0$.

We can now find the OC of the test. When $h(\theta) = 1$ we have the performance at $\theta = \theta_0$. When $h(\theta) = -1$ we have the performance at $\theta = \theta_1$. For other values we have, in effect, to solve (24.32) for θ and then substitute in (24.35). But this is, in fact, not necessary in order to plot the OC curve of $K(\theta)$ against θ. We can take $h(\theta)$ itself as a parameter and plot (24.35) against it.

Example 24.6

Consider once again the binomial distribution of previous examples. We may write for the discrete values 1 (success) and 0 (failure)

$$f(1, \pi) = \pi,$$
$$f(0, \pi) = 1 - \pi.$$

Then (24.32) becomes

$$\pi \left(\frac{\pi_1}{\pi_0} \right)^h + (1 - \pi) \left(\frac{1 - \pi_1}{1 - \pi_0} \right)^h = 1,$$

or

$$\pi = \frac{1 - \left(\dfrac{1 - \pi_0}{1 - \pi_1} \right)^h}{\left(\dfrac{\pi_1}{\pi_0} \right)^h - \left(\dfrac{1 - \pi_1}{1 - \pi_0} \right)^h}. \tag{24.36}$$

For $A = (1 - \beta)/\alpha$, $B = \beta/(1 - \alpha)$ we then have from (24.35)

$$K(\pi) = \frac{\left(\dfrac{1 - \beta}{\alpha} \right)^h - 1}{\left(\dfrac{1 - \beta}{\alpha} \right)^h - \left(\dfrac{\beta}{1 - \alpha} \right)^h}. \tag{24.37}$$

We can now plot $K(\pi)$ against π by using (24.36) and (24.37) as parametric equations in h.

The ASN of the SPR test

24.15 Consider a sequence of n random variables z_i. If n were a fixed number we should have

$$E\left(\sum_{i=1}^{n} z_i\right) = nE(z).$$

This is not true for sequential sampling, but we have instead the result

$$E\left(\sum_{i=1}^{n} z_i\right) = E(n)E(z), \tag{24.38}$$

which is not quite as obvious as it looks. The result is due to Wald and to Blackwell (1946), the following proof being due to Johnson (1959b).

Let each z_i have mean value μ, $E|z_i| \leq C < \infty$, and let the probability that n takes the value k be P_k ($k = 1, 2, \ldots$). Consider the 'marker' variable y_i which is unity if z_i is observed (i.e. if $n \geq i$) and zero otherwise. Then

$$P(y_i = 1) = P(n \geq i) = \sum_{j=i}^{\infty} P_j. \tag{24.39}$$

Now let $Z_n = \sum_{i=1}^{n} z_i$. Then

$$Z_n = \sum_{i=1}^{\infty} y_i z_i,$$

$$E(Z_n) = E \sum_{i=1}^{\infty} (y_i z_i) = \sum_{i=1}^{\infty} E(y_i z_i), \tag{24.40}$$

which will be finite if $E(n)$ is finite. Furthermore, since y_i depends only on $z_1, z_2, \ldots, z_{i-1}$ and not on z_i, we have

$$E(y_i z_i) = E(y_i)E(z_i). \tag{24.41}$$

Hence

$$E(Z_n) = \sum E(y_i)E(z_i) = \mu \sum E(y_i)$$

$$= \mu \sum_{i=1}^{\infty} \{P_i + P_{i+1} + \cdots\}$$

$$= \mu \sum_{i=1}^{\infty} iP_i$$

$$= \mu E(n),$$

which is (24.38).
   We then have

$$E(n) = \frac{E(Z_n)}{E(z)}. \tag{24.42}$$

But, to our usual approximation, Z_n can take only two values for the sampling to terminate, $\log A$ with probability $1 - K(\theta)$ and $\log B$ with probability $K(\theta)$. Thus

$$E(n) = \frac{K \log B + (1 - K) \log A}{E(z)},\tag{24.43}$$

which is the approximate formula for the average sample number.

Siegmund (1975) improves on the approximations (24.35) and (24.43).

Example 24.7
For the binomial we find

$$E(z) = E \log \left(\frac{f(x, \pi_1)}{f(x, \pi_0)} \right)$$

$$= \pi \log \frac{\pi_1}{\pi_0} + (1 - \pi) \log \frac{1 - \pi_1}{1 - \pi_0}.\tag{24.44}$$

The ASN can then be calculated from (24.43) when π_1, π_2, A and B (or α and β) are given. It is, of course, a function of π.

24.16 For practical application, sequential testing for attributes is often expressed in such a way that the calculations are in terms of integers. Equation (24.30) is a case in point. We may rewrite it as

$$332 \geqslant 220 + (n - k) - 53k \geqslant 0.$$

We may imagine a game in which we start with a score of 220. If a failure occurs we add one to the score; if a success occurs we lose 53 units. The game stops as soon as the score falls to zero or rises to 332, corresponding to acceptance of the values π_0 and π_1 respectively.

24.17 In such a scheme, suppose that we start with a score S_2. For every failure we gain one unit, but for every success we lose b units. If the score rises by S_1 so as to reach $S_1 + S_2$ $(=2S$, say) we accept one hypothesis; if it falls to zero we accept the other. Let the score at any point be x and the probability be u_x that it will ultimately reach $2S$ without in the meantime falling to zero. Consider the outcome of the next trial. A failure increases the score by unity to $x + 1$, a success diminishes it by b to $x - b$. Thus

$$u_x = (1 - \pi) u_{x+1} + \pi u_{x-b},\tag{24.45}$$

with initial conditions

$$u_0 = u_{-1} = u_{-2} = \cdots = u_{-b+1} = 0,\tag{24.46}$$

$$u_{2S} = 1.\tag{24.47}$$

For $b = 1$ this equation is easy to solve, as in Example 24.2. For integer $b > 1$ the solution is more cumbrous. We quote without proof the solution obtained by Burman (1946)

$$u_x = \frac{F(x)}{F(2S)}\tag{24.48}$$

where

$$F(x) = (1-\pi)^{-x}\left\{1 - \binom{x-b-1}{1}\pi(1-\pi)^b + \binom{x-2b-1}{2}[\pi(1-\pi)^b]^2 \right.$$
$$\left. + \binom{x-3b-1}{3}[\pi(1-\pi)^b]^3 + \cdots\right\}, \qquad x > 0, \qquad (24.49)$$

$$=0, \qquad\qquad x \leqslant 0.$$

Here the series continues as long as $x - kb - 1$ is positive. Burman also gave expressions for the ASN and the variance of the sample number.

24.18   Anscombe (1949a) tabulated functions of this kind. Putting

$$R_1 = \frac{S_1}{b+1}, \qquad R_2 = \frac{S_2}{b+1}, \qquad\qquad (24.50)$$

Anscombe tabulates R_1, R_2 for certain values of the errors α, β (actually $1-\alpha$ and β) and the ratio S_1/S_2, the values for $\pi(b+1)$ being also provided.

Given π_0, π_1, α, β we can find R_1 and R_2. There remains an element of choice according to how we fix the ratio S_1/S_2.

Thus, for $\pi_0 = 0.01$, $\pi_1 = 0.03$, $S_2 = 2S_1$, $\alpha = 0.01$, $\beta = 0.10$ we find $R_2 = 4$, $R_1 = 2$ approximately. Also $\pi(b+1) = 0.571$ or $b = 56$. We then find, from (24.50), $S_1 = 114$, $S_2 = 228$. The agreement with Example 24.5 ($S_1 = 112$, $S_2 = 220$, $b = 53$) is very fair. The ASN for $\pi = 0.01$ is 253 and that for $\pi = 0.03$ is 306.

24.19   It is instructive to consider what happens in the limit when the units 1 and b are small compared to the total score. We can imagine this, in Fig. 24.1, as a shrinkage of the mesh so that the routes approach a continuous random path of a particle subject to infinitesimal disturbances in two perpendicular directions. From this viewpoint the subject links up with the theory of Brownian motion and diffusion. If Δ is the difference operator defined by

$$\Delta u_x = u_{x+1} - u_x$$

we may write equation (24.45) in the form

$$\{(1-\pi)(1+\Delta) + \pi(1+\Delta)^{-b} - 1\}u_x = 0. \qquad (24.51)$$

For small b this is nearly equivalent to

$$\{(1-\pi) + \pi - 1 + (1-\pi)\Delta - b\pi\Delta + \tfrac{1}{2}b(b+1)\pi\Delta^2\}u_x = 0,$$

namely, to

$$\{(1-\pi-b\pi)\Delta + \tfrac{1}{2}b(b+1)\pi\Delta^2\}u_x = 0. \qquad (24.52)$$

In the limit this becomes

$$\frac{d^2u}{dx^2} + \lambda\frac{du}{dx} = 0, \qquad\qquad (24.53)$$

where

$$\lambda = \frac{2(1 - \pi - b\pi)}{\pi b(b+1)}. \tag{24.54}$$

The general solution of (24.53) is

$$u_x = k_1 + k_2 e^{-\lambda x},$$

and since the boundary conditions are

$$u_{2S} = 1, \qquad u_0 = 0,$$

we have

$$u_x = \frac{1 - \exp(-\lambda x)}{1 - \exp\{-\lambda(S_1 + S_2)\}}. \tag{24.55}$$

Thus for $x = S_2$ the probability of acceptance is

$$u_{S_2} = \frac{\exp(\lambda S_2) - 1}{\exp(\lambda S_2) - \exp(-\lambda S_1)}. \tag{24.56}$$

24.20   As before, write

$$R_1(b+1) = S_1, \qquad R_2(b+1) = S_2,$$

and let π tend to zero so that $\pi(b+1) = \gamma$, say, remains finite. From (24.54) we see that λ tends to zero, but

$$b\lambda \to \frac{2\{1 - \pi(b+1)\}}{\pi(b+1)} = \frac{2(1 - \gamma)}{\gamma} = \delta, \quad \text{say.} \tag{24.57}$$

Then λS_1 tends to $S_1\delta/b$, i.e. to $R_1\delta$, provided that $\gamma \neq 1$, and (24.56) becomes

$$u_{S_2} = \frac{\exp(R_2\delta) - 1}{\exp(R_2\delta) - \exp(-R_1\delta)}, \qquad \gamma \neq 1. \tag{24.58}$$

If δ tends to zero we find

$$u_{S_2} = \frac{R_2}{R_2 + R_1}, \qquad \gamma = 1. \tag{24.59}$$

(24.58) may be compared to (24.37) which, for small h, can be written

$$u_{S_2} = \frac{\exp\left(\frac{1 - \beta}{\alpha} h\right) - 1}{\exp\left(\frac{1 - \beta}{\alpha} h\right) - \exp\left(-\frac{1 - \alpha}{\beta} h\right)}. \tag{24.60}$$

24.21   The use of sequential methods in the control of the quality of manufactured products has led to considerable developments of the kind of results mentioned in

24.17–18. We shall not have the space here to discuss the subject in detail and the reader who is interested is referred to some of the textbooks on quality control. We will merely mention some of the extensions of the foregoing theory by way of illustrating the scope of the subject.

(a) *Stopping rules.* Even for a closed scheme it may be desirable to call a halt in the sampling at some stage. For instance, circumstances may prevent sampling beyond a certain point of time; or, in clinical trials, medical ethics may require a change of treatment to a new drug which looks promising even before its value is fully established. Sequential schemes may be truncated in various ways, the simplest being to require stopping either after a given sample size has been reached or when a given time has elapsed. In such cases our general notions about performance characteristics and average sample numbers remain unchanged, but the actual mathematics and arithmetic are usually far more troublesome. Armitage (1957) considers sequential sampling under various restrictions. Madsen (1974) gives a procedure for truncating SPR tests that also improves on the approximations (24.23) and (24.43) in the untruncated case for the normal mean, which are used in Examples 24.8 and 24.10 below.

Harter and Owen (1974) give tables, due to Blyth and Hutchinson, of a truncated test.

(b) *Rectifying inspection.* In the schemes we have considered the hypotheses were that the batch or population under inquiry should be accepted or rejected as having a specified proportion of an attribute. If the attribute is 'defectiveness' we may prefer not to reject a batch *in toto* but to inspect every member of it and to replace the defective ones. This does not of itself affect the general character of the scheme—the decision to reject is replaced by a decision to rectify—but it does, of course, affect the proportion of rejects in the whole aggregate of batches to which the sampling plan is applied—what is known as the average outgoing quality level (AOQL); and hence it affects the values of the parameters which we put into the plan. The theory was examined by Bartky (1943).

(c) *Double sampling.* As an extension of this idea, we may find it economical to proceed in stages. For example, we may decide to have four possible decisions: to accept outright; to reject outright; to continue sampling; to suspend judgement but to inspect fully. There is evidently a wide variety of possible choice here. An excellent example is the double sampling procedure of Dodge and Romig (1944). We shall encounter the idea again later in **24.36**.

Example 24.8
Consider testing $H_0(\mu = \mu_0)$ against $H_1(\mu = \mu_1)$ for the mean of a normal distribution with unit variance. With z defined as at (24.19) we have

$$z_i = -\tfrac{1}{2}(x_i - \mu_1)^2 + \tfrac{1}{2}(x_i - \mu_0)^2$$
$$= (\mu_1 - \mu_0)x_i - \tfrac{1}{2}(\mu_1^2 - \mu_0^2).$$

$$Z_n = \sum_{i=1}^{n} z_i = n(\mu_1 - \mu_0)\bar{x} - \tfrac{1}{2}n(\mu_1^2 - \mu_0^2). \tag{24.61}$$

We accept H_0 or H_1 according as (24.61) $\leqslant \log B$ or $\geqslant \log A$. For the appropriate OC curve we have, from (24.35),

$$K(\mu) = \frac{A^h - 1}{A^h - B^h},\qquad(24.62)$$

where h is given by

$$\frac{1}{\sqrt{(2\pi)}}\int_{-\infty}^{\infty} \exp\left[h\{(\mu_1 - \mu_0)x - \tfrac{1}{2}(\mu_1^2 - \mu_0^2)\}\right]\exp\left\{-\tfrac{1}{2}(x - \mu)^2\right\}dx = 1,\quad(24.63)$$

which is easily seen to be equivalent to

$$\exp\{\mu^2 - h\mu_1^2 + h\mu_0^2 - (\mu - h\mu_1 + h\mu_0)^2\} = 1$$

or to

$$h = \frac{\mu_1 + \mu_0 - 2\mu}{\mu_1 - \mu_0},\qquad \mu_1 \neq \mu_0,\quad h \neq 0.\qquad(26.64)$$

We can then draw the OC curve for a range of values of μ by calculating h from (24.64) and substituting in (24.62).

Likewise for the ASN we have

$$E(z) = \frac{1}{\sqrt{(2\pi)}}\int_{-\infty}^{\infty}\exp\left\{-\tfrac{1}{2}(x - \mu)^2\right\}\{(\mu_1 - \mu_0)x - \tfrac{1}{2}(\mu_1^2 - \mu_0^2)\}\,dx$$

$$= (\mu_1 - \mu_0)\mu - \tfrac{1}{2}(\mu_1^2 - \mu_0^2).\qquad(24.65)$$

Again, for a range of μ the ASN can be determined from this equation in conjunction with (24.62) and

$$E(n) = \frac{K \log B + (1 - K)\log A}{E(z)}.\qquad(24.66)$$

Manly (1970a) gives charts that permit choice of A and B for specified α, β, and also give the value of $E(n)$.

Example 24.9
Suppose that the mean of a normal distribution is known to be μ. To test a hypothesis concerning its variance, $H_0: \sigma^2 = \sigma_0^2$ against $H_1: \sigma^2 = \sigma_1^2$, we have

$$Z_n = \sum z_i = -n \log \sigma_1 - \frac{1}{2\sigma_1^2}\sum (x - \mu)^2 + n \log \sigma_0 + \frac{1}{2\sigma_0^2}\sum (x - \mu)^2.\quad(24.67)$$

This lies between the limits $\log\{\beta/(1 - \alpha)\}$ and $\log\{(1 - \beta)/\alpha\}$ if

$$\log\frac{\beta}{1 - \alpha} < -n \log\frac{\sigma_1}{\sigma_0} - \frac{1}{2}\left(\frac{1}{\sigma_1^2} - \frac{1}{\sigma_0^2}\right)\sum (x - \mu)^2 < \log\frac{1 - \beta}{\alpha}\qquad(24.68)$$

or

$$\frac{2\log\dfrac{\beta}{1-\alpha}+n\log\dfrac{\sigma_1^2}{\sigma_0^2}}{\dfrac{1}{\sigma_0^2}-\dfrac{1}{\sigma_1^2}}\leqslant\sum(x-\mu)^2\leqslant\frac{2\log\dfrac{1-\beta}{\alpha}+n\log\dfrac{\sigma_1^2}{\sigma_0^2}}{\dfrac{1}{\sigma_0^2}-\dfrac{1}{\sigma_1^2}}. \tag{24.69}$$

The OC and ASN are given in Exercises 24.18 and 24.19. See also Exercise 24.20.

The efficiency of a sequential test

24.22   In general, many different tests may be derived for given α and β, θ_0 and θ_1. There is no point in comparing their power for given sample sizes because they are arranged so as to have the same β-error. We may, however, define efficiency in terms of sample size or ASN. The test with the smaller ASN may reasonably be said to be the more efficient. Following Wald (1947) we shall prove that when end-effects are negligible the SPR test is a most efficient test. More precisely, if S' is an SPR test and S is some other test based on the sum of logarithms of identically distributed variables,

$$E_i(n\,|\,S)\geqslant E_i(n\,|\,S'),\qquad i=0,1, \tag{24.70}$$

where E_i denotes the expected value of n on hypothesis H_i.

Note first of all that if u is any random variable,

$$\exp\{u-E(u)\}\geqslant 1+\{u-E(u)\}.$$

On taking expectations we have

$$E[\exp\{u-E(u)\}]\geqslant 1, \tag{24.71}$$

which gives

$$E(u)\leqslant\log E\{\exp(u)\}. \tag{24.72}$$

We also have, from (24.42), for *any* closed sequential test based on the sums of type Z_n

$$E_i(n\,|\,S)=\frac{E_i(\log L_n\,|\,S)}{E_i(z)}. \tag{24.73}$$

If E^* denotes the conditional expectation when H_0 is true, and E^{**} the conditional expectation when H_1 is true, we have, as at (24.22), neglecting end-effects,

$$E^*(L_n\,|\,S)=\frac{\beta}{1-\alpha}, \tag{24.74}$$

and similarly, as at (24.21),

$$E^{**}(L_n\,|\,S)=\frac{1-\beta}{\alpha}. \tag{24.75}$$

Hence

$$E_0(n\,|\,S) = \frac{1}{E_0(z)}\{(1-\alpha)E^*(\log L_n\,|\,S) + \alpha E^{**}(\log L_n\,|\,S)\}. \qquad (24.76)$$

In virtue of (24.72), (24.74) and (24.75) we then have $E_0(z) < 0$ and

$$E_0(n\,|\,S) \geqslant \frac{1}{E_0(z)}\left\{(1-\alpha)\log\frac{\beta}{1-\alpha} + \alpha\log\frac{1-\beta}{\alpha}\right\}, \qquad (24.77)$$

and interchanging H_0 and H_1, α and β in (24.77) gives, with $E_1(z) > 0$,

$$E_1(n\,|\,S) \geqslant \frac{1}{E_1(z)}\left\{\beta\log\frac{\beta}{1-\alpha} + (1-\beta)\log\frac{1-\beta}{\alpha}\right\}. \qquad (24.78)$$

When $S = S'$ these inequalities, as at (24.43), are replaced (neglecting end-effects) by equalities. Hence (24.70).

Example 24.10
One of the hoped-for properties of the sequential method (see **24.1**) is that for a given (α, β), it requires a smaller sample on the average than the method employing a fixed sample size. General formulae comparing the two would be difficult to derive, but we may illustrate the point on the testing of a mean in normal variation (Example 24.8).
    For fixed n and α the test consists of finding a value d such that

$$\text{Prob}\{\mu_0 - d \leqslant \bar{x} \leqslant \mu_0 + d \,|\, H_0\} = 1 - \alpha,$$
$$\text{Prob}\{\mu_0 - d \leqslant \bar{x} \leqslant \mu_0 + d \,|\, H_1\} = \beta,$$

and, putting

$$\lambda_0 = \sqrt{n}(d - \mu_0),$$
$$\lambda_1 = \sqrt{n}(d - \mu_1),$$

we have

$$n = \frac{(\lambda_1 - \lambda_0)^2}{(\mu_0 - \mu_1)^2}. \qquad (24.79)$$

Using (24.79), let us compare it with the ASN of an SPR test. Taking the approximate formula (24.43), which is

$$E_i(n) = \frac{1}{E_i(z)}[K(\mu)\log B + \{1 - K(\mu)\}\log A], \qquad (24.80)$$

we find, since from (24.65)

$$E_1(z) = \tfrac{1}{2}(\mu_0 - \mu_1)^2$$
$$E_0(z) = -\tfrac{1}{2}(\mu_0 - \mu_1)^2,$$
$$\frac{E_1(n)}{n} = \frac{2}{(\lambda_1 - \lambda_0)^2}\{\beta\log B + (1 - \beta)\log A\}. \qquad (24.81)$$

Likewise we find

$$\frac{E_0(n)}{n} = -\frac{2}{(\lambda_1 - \lambda_0)^2}\{(1-\alpha)\log B + \alpha \log A\}. \tag{24.82}$$

Thus, for $\alpha = 0.01$, $\beta = 0.03$, $A = 97$, $B = 3/99$ and we find $\lambda_0 = 2.5758$, $\lambda_1 = -1.8808$. The ratio $E_0(n)/n$ is then 0.43 and $E_1(n)/n = 0.55$. We thus require in the sequential case, on the average, either 43 or 55 per cent of the fixed sample size needed to attain the same performance.

> It should be emphasised that a reduced ASN will only be obtained by using an SPR test when one of H_0 or H_1 is true, and not necessarily for other parameter values, not even for intermediate values. Kiefer and Weiss (1957) provided a theoretical basis for improving upon the SPR test in this respect. The normal case was studied by T. W. Anderson (1960), Madsen (1974), Weiss (1962) and by D. Freeman and Weiss (1964); the latter two papers also solve the binomial case explicitly—see also Alling (1966) and Breslow (1970). D. G. Hoel and Mazumdar (1969) treat the scale parameter of an exponential distribution.
>
> Even when comparing the SPR test with the fixed sample size test, the former is not always more efficient, though usually so. Bechhofer (1960) studies the case of the normal mean (Example 24.10); when $\beta = c\alpha \to 0$, $c > 0$, the ratio of the ASN to the fixed n is $|(\mu_1 - \mu_0)/\{4(\mu_0 + \mu_1 - 2\mu)\}|$, which $= \frac{1}{4}$ if $\mu = \mu_0$ or μ_1 and $\to 0$ as $\mu \to \pm\infty$ but is very large near $\mu = \frac{1}{2}(\mu_0 + \mu_1)$, as is intuitively reasonable. Berk (1975, 1976, 1978) devises general methods of computing efficiency in such comparisons (one of which is an analogue of ARE in Chapter 25) and gives references. See also Holm (1977). Lai (1978) obtained a direct sequential equivalent of asymptotic efficiency, and showed that under general conditions it takes the same value as in the fixed-sample-size case. Berk and Brown (1978) extended another efficiency criterion to the sequential case.

Composite sequential hypotheses

24.23 Although we have considered the test of a simple H_0 against a simple H_1, the OC and ASN functions are, in effect, calculated against a range of alternatives and therefore give us the performance of the test for a simple H_0 against a composite H_1. We now consider the case of a composite H_0. Suppose that θ may vary in some domain Ω. We require to test that it lies in some subdomain ω_a against the alternatives that it lies either in a rejection subdomain ω_r, or in a region of indifference $\Omega - \omega_a - \omega_r$ (which may be empty). We shall require that the probability that an error of the first kind, $\alpha(\theta)$, which in general varies with θ, shall not exceed some fixed number α for all θ in ω_a; and that the probability than an error of the second kind, $\beta(\theta)$, shall not exceed β for all θ in ω_r. Wherever our parameter point θ really lies, then, we shall have upper limits to the errors, given by α and β.

24.24 Such requirements, however, are not very efficient. We are always on the safe side, but may be so far on the safe side in particular cases as to lose a good deal of efficiency. Wald (1947) suggested that it might be better instead to consider the average of $\alpha(\theta)$ over ω_a and of $\beta(\theta)$ over ω_r. This raises the question of what sort of averages should be used. Wald defines two weighting functions, $w_a(\theta)$ and $w_r(\theta)$, such that

$$\int_{\omega_a} w_a(\theta)\, d\theta = 1, \qquad \int_{\omega_r} w_r(\theta)\, d\theta = 1, \tag{24.83}$$

and we then define

$$\int_{\omega_a} w_a(\theta) \alpha(\theta) \, d\theta = \alpha, \tag{24.84}$$

$$\int_{\omega_r} w_r(\theta) \beta(\theta) \, d\theta = \beta. \tag{24.85}$$

By these means we reduce the problem to one of testing simple hypotheses. In fact, if we let

$$L_{0n} = \int_{\omega_a} f(x_1, \theta) f(x_2, \theta) \cdots f(x_n, \theta) \alpha(\theta) \, d\theta, \tag{24.86}$$

$$L_{1n} = \int_{\omega_r} f(x_1, \theta) f(x_2, \theta) \cdots f(x_n, \theta) \beta(\theta) \, d\theta, \tag{24.87}$$

the probability ratio L_{0n}/L_{1n} can be used in the ordinary way with errors α and β. We may, if we like, regard (24.86) and (24.87) as the posterior probabilities of the sample when θ itself has prior probabilities $w_a(\theta)$ and $w_r(\theta)$.

24.25 This procedure, of course, throws the problem into the form of finding or choosing the weight functions $w_a(\theta)$ and $w_r(\theta)$. We are in the same position as the probabilist wishing to apply Bayes' theorem. We may resolve it by some form of Bayes' postulate, e.g. by assuming that $w_a(\theta) = 1$ everywhere in ω_a. Another possibility is to choose $w_a(\theta)$ and $w_r(\theta)$ so as to optimise some properties of the test.

For example, the choice of the test is made when we select α and β (or, to our approximation, A and B) and the weight functions. Among all such tests there will be maximum values of $\alpha(\theta)$ and $\beta(\theta)$. If we choose the weight functions so as to minimise (max α, max β), we have a test which, for given A and B, has the lowest possible bound to the average errors. If it is not possible to minimise the maxima of α and β simultaneously we may, perhaps, minimise the maximum of some function such as their sum.

> Wald (1947) proposed a two-sided SPR test for the normal mean μ with σ^2 known (cf. Examples 24.8, 24.10), giving weight $\frac{1}{2}$ to each of the two alternative hypothesis values of μ. Billard (1972) defined a more general procedure whose α and ASN can be closely approximated, and evaluated them for Wald's test. See also Borgan (1979).

A sequential *t*-test
24.26 A test proposed by Wald (1947) and, in a modified form, by other writers sets out to test the mean μ of a normal distribution when the variance is unknown. It is known as the sequential *t*-test because it deals with the same problem as Student's *t* in the fixed-sample-size case; but it does not free itself from the scale parameter σ in the same way and the name is, perhaps, somewhat misleading.

Specifically, we wish to test that, compared to some value μ_0, the deviation $(\mu - \mu_0)/\sigma$ is small, say $< \delta$. The three subdomains of **24.23** are then as follows:

ω_a consists of (μ_0, σ) for all σ;

ω_r consists of values for which $|\mu - \mu_0| \geq \sigma \delta$, for all σ;

$\Omega - \omega_a - \omega_r$ consists of values for which $0 \leq |\mu - \mu_0| < \sigma \delta$, for all σ.

We define weight functions for σ:

$$\left. \begin{array}{ll} v_{ac} = \dfrac{1}{c}, & 0 \leq \sigma \leq c, \\[2mm] = 0 & \text{elsewhere.} \end{array} \right\} \tag{24.88}$$

$$\left. \begin{array}{ll} v_{rc} = \dfrac{1}{2c}, & 0 \leq \sigma \leq c, \quad \mu = \mu_0 + \delta\sigma \\[2mm] = 0 & \text{elsewhere.} \end{array} \right\} \tag{24.89}$$

Then

$$L_{1n} = \int_0^c v_{rc} \frac{1}{(2\pi)^{n/2} \sigma^n} \exp\left\{ -\frac{1}{2\sigma^2} \Sigma (x_i - \mu)^2 \right\} d\sigma$$

$$= \frac{1}{(2\pi)^{n/2}} \cdot \frac{1}{2c} \int_0^c \left[\frac{1}{\sigma^n} \exp\left\{ -\frac{1}{2\sigma^2} \Sigma (x_i - \mu_0 - \delta\sigma)^2 \right\} \right.$$

$$\left. + \frac{1}{\sigma^n} \exp\left\{ -\frac{1}{2\sigma^2} \Sigma (x_i - \mu_0 + \delta\sigma)^2 \right\} \right] d\sigma. \tag{24.90}$$

$$L_{0n} = \frac{1}{(2\pi)^{n/2}} \frac{1}{c} \int_0^c \frac{1}{\sigma^n} \exp\left\{ -\frac{1}{2\sigma^2} \Sigma (x_i - \mu_0)^2 \right\} d\sigma. \tag{24.91}$$

The limit of the ratio L_{1n}/L_{0n} as c tends to infinity then becomes

$$\lim L_{1n}/L_{0n} = \frac{\dfrac{1}{2} \int_0^\infty \dfrac{d\sigma}{\sigma^n} \left[\exp\left\{ -\dfrac{1}{2\sigma^2} \Sigma (x_i - \mu_0 - \delta\sigma)^2 \right\} + \exp\left\{ -\dfrac{1}{2\sigma^2} \Sigma (x_i - \mu_0 + \delta\sigma)^2 \right\} \right]}{\displaystyle\int_0^\infty \dfrac{d\sigma}{\sigma^n} \exp\left\{ -\dfrac{1}{2\sigma^2} \Sigma (x_i - \mu_0)^2 \right\}}. \tag{24.92}$$

This depends on the xs, which are observed, and on μ_0 and δ, which are given, but not on σ, which we have integrated out of the problem by the weight functions (24.88) and (24.89). If we can evaluate the integrals in (24.92) we can apply this ratio to give a sequential test.

24.27 The rather arbitrary-looking weight functions are, in fact, such as to optimise the test. To prove this we first of all establish (a) that $\alpha(\mu, \sigma)$ is constant in ω_a; (b) that $\beta(\mu, \sigma)$ is a function of $|(\mu - \mu_0)/\sigma|$ alone: and (c) that $\beta(\mu, \sigma)$ is monotonically decreasing in $|(\mu - \mu_0)/\sigma|$.

If $\bar{x}$ is the sample mean and S^2 is the sum $\sum (x_i - \bar{x})^2$, the distribution of the ratio $(\bar{x} - \mu_0)/S$ depends only on $(\mu - \mu_0)/\sigma$. If then we can show that (24.92) is a single-valued function of $(\bar{x} - \mu_0)/S$, the properties (a) and (b) will follow, for $(\mu - \mu_0)/\sigma$ is zero in ω_a and $\beta(\mu, \sigma)$ depends only on the distribution of (24.92). Now the numerator and denominator of (24.92) are both homogeneous functions of $(x_i - \mu_0)$ of degree $n-1$, as may be verified by putting $x_i = \lambda x_i$, $\mu_0 = \lambda \mu_0$, $\sigma = \lambda \sigma$. Thus the ratio of (24.92) is of degree zero. Further, it is a function of $\sum (x - \mu_0)^2$ and $\sum (x - \mu_0)$ only, and hence we do not change it by putting $(x_i - \mu_0)/\sqrt{\sum (x_i - \mu_0)^2}$ for $x_i - \mu_0$. The ratio is, then, a function only of $\sum (x_i - \mu_0)/\sqrt{\sum (x_i - \mu_0)^2}$, and is, in actual fact, a function of the square of that quantity, namely of

$$\frac{(\bar{x} - \mu_0)^2}{\sum (x_i - \mu_0)^2} = \frac{(\bar{x} - \mu_0)^2}{n(\bar{x} - \mu_0)^2 + S^2}.$$

It is therefore a single-valued function of $(\bar{x} - \mu_0)/S$.

To show that $\beta(\mu, \sigma)$ is monotonically decreasing in $|(\mu - \mu_0)/\sigma|$ it is sufficient to show that the ratio (24.92) is a strictly increasing function of $|(\bar{x} - \mu_0)/S|$, or equivalently of $(\bar{x} - \mu_0)^2/\sum (x_i - \mu_0)^2$. Now for fixed $\sum (x_i - \mu_0)^2$ the denominator of (24.92) is fixed and the numerator is an increasing function of $(\bar{x} - \mu_0)^2$. Thus the whole ratio is increasing in $(\bar{x} - \mu_0)^2$ for fixed $\sum (x_i - \mu_0)^2$ and the required result follows.

24.28   Under these conditions we can prove that the sequential t-test is optimal. In fact, any test is optimal if (i) $\alpha(\theta)$ is constant in ω_a; (ii) $\beta(\theta)$ is constant over the boundary of ω_r; and (iii) $\beta(\theta)$ does not exceed its boundary value for any θ inside ω_r.

To prove this, let v_a and v_r be two weight functions obeying these conditions and w_a, w_r two other weight functions. Let α, β be the errors for the first set, α^*, β^* those for the second. Then we have

$$\frac{1 - \beta}{\alpha} = A, \qquad \frac{\beta}{1 - \alpha} = B,$$

and hence

$$\int_{\omega_a} \alpha^*(\theta) w_a(\theta)\, d\theta = \alpha = \frac{1 - B}{A - B}, \tag{24.93}$$

$$\int_{\omega_r} \beta^*(\theta) w_r(\theta)\, d\theta = \beta = \frac{B(A - 1)}{A - B}. \tag{24.94}$$

Thus, in ω_a the maximum of $\alpha^*(\theta)$ is greater than $(1 - B)/(A - B)$, for the integral of $w_a(\theta)$ over that region is unity. But if v_a has constant α over that domain, its maximum is equal to $(1 - B)/(A - B)$. Hence

$$\max \alpha^*(\theta) \geqslant \max \alpha(\theta) \text{ in } \omega_a.$$

Likewise, in ω_r the maximum of $\beta^*(\theta)$ is attained somewhere outside ω_r and cannot exceed $B(A - 1)/(A - B)$, whereas for $\beta(\theta)$ the maximum value must be attained somewhere. Hence

$$\max \beta^*(\theta) \geqslant \max \beta(\theta) \text{ in } \omega_r.$$

The result follows. The conditions we have considered are sufficient but by no means necessary.

> Lai (1975) showed that the sequential t-test and a sequential F-test are closed in an exponentially strong sense, as are Wald's and other SPR tests. See also Wijsman (1977), Perng (1978) and Rootzén and Simons (1977), who show that this is so for invariant SPR tests when H_0 or H_1 holds.

24.29 Some tables for the use of this test were provided by Armitage (1947). The integrals occurring in (24.92) are, in fact, expressible in terms of the confluent hypergeometric function and, in turn, in terms of the distribution of non-central t. *Tables to Facilitate Sequential t-Tests* (U.S.A. National Bureau of Standards, Applied Mathematics Series 7, 1951) is a co-operative work with an introduction by K. J. Arnold. See also later work by Armitage and others, in particular Myers *et al.* (1966).

An alternative method of attack is given by D. R. Cox (1952a, b) for the case where the distribution of a set of sufficient statistics factorises as in (31.14) below. An SPR test of θ_l can then be based on T_s, free of the nuisance (often scale) parameters θ_k, in the ordinary way. This approach also requires an invariance condition—cf. W. J. Hall *et al.* (1965). See also Chand (1974). Hajnal (1961) develops a two-sample sequential t-test along these lines.

Bartlett (1946) gave a large-sample sequential test for any composite hypothesis, based on ML theory—cf. Exercise 24.21.

> Weiss and Wolfowitz (1972) give a simple class of sequential tests for the normal mean with variance unknown, all of which are asymptotically efficient.
> Köllerström and Wetherill (1979) compare SPR tests for the bivariate normal correlation parameter.

Sequential estimation; the moments and distribution of n

24.30 In testing hypotheses we usually fix the errors in advance and proceed with the sampling until acceptance or rejection is reached. We may also use the sequential process for estimation, but our problems are then more difficult to solve and may even be difficult to formulate. We draw a sequence of observations with the object of estimating some parameter of the distribution; but in general it is not easy to decide what is the appropriate estimator, what biases are present, what are the sampling errors or what should be the rules determining the ending of the sampling process. The basic difficulty is that the sample size has a complicated distribution. A secondary nuisance is the end-effect to which we have already referred.*

24.31 We derived at (24.42) the mean of n, and now rewrite it in the form

$$E\{Z_n - nE(z)\} = 0. \tag{24.95}$$

---

* Lehmann and Stein (1950) considered the concept of completeness (cf. **22.9**) in the sequential case, but general criteria are not easy to apply even in attribute sampling—cf. de Groot (1959).

Let us assume that absolute second moments exist, that the variance of each z_i is equal to σ^2, and that $E(n^2)$ exists. The variance of $\{Z_n - nE(z)\}$ may then be derived (the proof is left to the reader as Exercise 24.5) as

$$E\{Z_n - nE(z)\}^2 = \sigma^2 E(n) \tag{24.96}$$

and it follows that, with $E(z) = \mu$ as before,

$$\mu^2 E(n^2) = \sigma^2 E(n) + 2\mu E(nZ_n) - E(Z_n^2).$$

If Z_n and n are uncorrelated, this simplifies to

$$\mu^2 \operatorname{var} n = \sigma^2 E(n) - \operatorname{var} Z_n. \tag{24.97}$$

Similar results for higher moments have been obtained by Wolfowitz (1947)—cf. Exercise 24.6. See also Chow *et al.* (1965).

Approximate expressions for the first four moments of n are obtained by B. K. Ghosh (1969) by differentiating its c.f., given in Exercise 24.11 below. He applies his results to the cases of the normal mean, normal variance, the binomial, the Poisson and the exponential scale parameter. The distribution of n is also discussed—Exercise 24.13 below treats the case of the normal mean. See also Manly (1970b) and Chanda (1971).

24.32   Now let

$$Y_n = \sum_{i=1}^{n} \frac{\partial}{\partial \theta} \log f(x_i, \theta). \tag{24.98}$$

Then, under regularity conditions, $E(\partial \log f / \partial \theta) = 0$ as at (17.18) and we have

$$E(Y_n) = 0, \tag{24.99}$$

and

$$\operatorname{var} Y_n = E\left(\sum \frac{\partial \log f}{\partial \theta}\right)^2 = E(n)E\left\{\left(\frac{\partial \log f}{\partial \theta}\right)^2\right\}. \tag{24.100}$$

as in **24.15**. If t is an estimator of θ with bias $b(\theta)$, i.e. is such that

$$E(t) = \theta + b(\theta)$$

we have, differentiating this equation,

$$\operatorname{cov}\left(t, \sum \frac{\partial \log f}{\partial \theta}\right) = E\left(t \sum \frac{\partial \log f}{\partial \theta}\right) = 1 + b'(\theta). \tag{24.101}$$

Then, by the Cauchy-Schwarz inequality

$$(\operatorname{var} t)E\left(\sum \frac{\partial \log f}{\partial \theta}\right)^2 \geq \{1 + b'(\theta)\}^2,$$

and hence, by (24.100),

$$\operatorname{var} t \geq \frac{\{1 + b'(\theta)\}^2}{E(n)E\left\{\left(\frac{\partial \log f}{\partial \theta}\right)^2\right\}}, \tag{24.102}$$

which is Wolfowitz's form of the lower bound to the variance in sequential estimation. It consists simply of putting $E(n)$ for n in the MVB (17.22). Wolfowitz (1947) also gives an extension of the result to the simultaneous estimation of several parameters.

Example 24.11
Consider the binomial with unit index

$$f(x, \pi) = \pi^x (1 - \pi)^{1-x}, \qquad x = 0, 1.$$

We have

$$\frac{\partial \log f}{\partial \pi} = \frac{x}{\pi} - \frac{1-x}{1-\pi}, \qquad E\left\{ \left(\frac{\partial \log f}{\partial \pi} \right)^2 \right\} = \frac{1}{\pi(1-\pi)}.$$

If p is any unbiased estimator of π in a sample from this distribution, we then have

$$\text{var } p \geqslant \frac{\pi(1-\pi)}{E(n)}.$$

From (24.3), $E(n) = k/\pi$, so

$$\text{var } p \geqslant \pi^2 (1 - \pi)/k. \tag{24.103}$$

Comparing this with (24.9) for the estimator (24.5), we see that the first term in (24.9) is the bound (24.103), the other terms being of lower order in k.

24.33 If n is large the theory of sequential estimation is very much simplified in virtue of a general result due to Anscombe (1949b, 1952, 1953). Simply stated, this amounts to saying that for statistics where a central limit effect is present, the formulae for standard errors are the same for sequential samples as for samples of fixed size. We might argue this heuristically from (24.102). n varies about its mean n_0 with standard deviation of order $n_0^{-1/2}$ and thus formulae accurate to order n^{-1} remain accurate to that order if we use n_0 instead of n. More formally:

Let $\{Y_n\}$, $n = 1, 2, \ldots$ be a sequence of random variables. Let there exist a real number θ, a sequence of positive numbers $\{w_n\}$, and a distribution function $F(x)$ such that

(a) Y_n converges to θ in the scale of w_n, namely

$$P\left\{ \frac{Y_n - \theta}{w_n} \leqslant x \right\} \to F(x) \quad \text{as } n \to \infty; \tag{24.104}$$

(b) $\{Y_n\}$ is uniformly continuous in probability, namely given (small) positive ε and η,

$$P\left\{ \left| \frac{Y_{n'} - Y_n}{w_n} \right| < \varepsilon \quad \text{for all } n, n' \text{ such that } |n' - n| < \varepsilon n \right\} > 1 - \eta. \tag{24.105}$$

Let $\{n_r\}$ be an increasing sequence of positive integers tending to infinity and $\{N_r\}$ be a sequence of random variables taking positive integral values such that $N_r/n_r \to 1$ in probability as $r \to \infty$. Then

$$P\left\{\frac{Y_{N_r} - \theta}{w_{N_r}} \le x\right\} \to F(x) \quad \text{as } r \to \infty \tag{24.106}$$

at all continuity points of $F(x)$.

The complexity of the enunciation and the proof are due to the features we have already noticed: end-effects (represented by the relation between N_r and n_r) and the variation in n_r.

In fact, let (24.105) be satisfied with ν large enough so that for any $n_r > \nu$

$$P\{|N_r - n_r| < cn_r\} > 1 - \eta. \tag{24.107}$$

Consider the event $E : |N_r - n_r| < cn_r$ and $|Y_{N_r} - Y_{n_r}| < \varepsilon w_{N_r}$, and the events

$$A : |Y_{n'} - Y_n| < \varepsilon w_n, \quad \text{all } n' \text{ such that } |n' - n| < \varepsilon n,$$

$$B : |N_r - n_r| < cn_r.$$

Then

$$P(E) \ge P\{A \text{ and } B\} = P(A) - P\{A \text{ and not-}B\}$$
$$\ge P(A) - P(\text{not-}B)$$
$$\ge 1 - 2\eta. \tag{24.108}$$

Also

$$P\{Y_{N_r} - \theta \le x w_{N_r}\} = P\{Y_{N_r} - \theta \le x w_{n_r} \text{ and } E\}$$
$$+ P\{Y_{N_r} - \theta \le x w_{n_r} \text{ and not-}E\}.$$

Thus, in virtue of the definition of E we find

$$P\{Y_{n_r} - \theta \le (x - \varepsilon)w_{n_r}\} - 2\eta < P\{Y_{n_r} - \theta \le x w_{n_r}\}$$
$$< P\{Y_{n_r} - \theta \le (x + \varepsilon)w_{n_r}\} + 2\eta,$$

and (24.106) follows. It is to be noted that the proof does not assume N_r and Y_n to be independent.

24.34 To apply this result to sequential estimation, let $x_1, x_2, \ldots$ be a sequence of observations and Y_n an estimator of a parameter θ, D_n an estimator of the scale w_n of Y_n. The samping rule is: given some constant k, sample until the first occurring $D_n \le k$ and then calculate Y_n. We show that Y_n is an estimator of θ with scale asymptotically equal to k if k is small.

Let conditions (24.104) and (24.105) be satisfied and $\{k_r\}$ be a sequence of positive numbers tending to zero. Let $\{N_r\}$ be the sequence of random variables such that N_r is the least integer n for which $D_n \le k_r$; and let $\{n_r\}$ be the sequence such that n_r is the least n for which $w_n \le k_r$. We require two further conditions:

(c) $\{w_n\}$ converges monotonically to zero and $w_n/w_{n+1} \to 1$ as $n \to \infty$;

(d) N_r is a random variable for all r and $N_r/n_r \to 1$ in probability as $r \to \infty$.

Condition (c) implies that $w_{n_r}/k_r \to 1$ as $n \to \infty$. It then follows from our previous result that

$$P\left\{\frac{Y_{N_r} - \theta}{k_r} \leq x\right\} \to F(x) \quad \text{as } r \to \infty. \tag{24.109}$$

24.35 It may also be shown that if the xs are independently and identically distributed, the conditions (a) and (c)—which are easily verifiable—together imply condition (b) and the distribution of their sum tends to a distribution function. In particular, these conditions are satisfied for maximum likelihood estimators, for estimators based on means of some functions of the observations, and for quantiles.

Siegmund (1978) discussed setting confidence intervals following sequential tests, with special reference to the case of the normal mean with variance known or unknown. Whitehead (1986) considers the bias of the ML estimator in these circumstances, and gives a method of reducing the bias and approximating the standard error.

Example 24.12

Consider the estimation of the mean μ of a normal distribution with unknown variance σ^2. We require of the estimator a (small) variance k^2.

The obvious statistic is $Y_n = \bar{x}_n$. For fixed n this has variance σ^2/n estimated as

$$D_n^2 = \frac{1}{n(n-1)} \sum (x_i - \bar{x})^2. \tag{24.110}$$

Conditions (a) and (c) are obviously satisfied and in virtue of the result quoted in **24.35** this entails the satisfaction of condition (b). To show that (d) holds, transform by Helmert's transformation

$$\xi_i = \left(x_{i+1} - \frac{1}{i}\sum_{j=1}^{i} x_j\right)\sqrt{\frac{i}{i+1}}.$$

Then

$$D_n^2 = \frac{1}{n(n-1)} \sum_{i=1}^{n-1} \xi_i^2.$$

By the strong law of large numbers, given ε, η, there is a ν such that

$$P\left\{\left|\frac{1}{n-1}\sum_{i=1}^{n-1} \xi_i^2 - \sigma^2\right| < \varepsilon \quad \text{for all } n > \nu\right\} > 1 - \eta. \tag{24.111}$$

If k is small enough, the probability exceeds $1 - \eta$ that $D_n \leq k$ for any n in the range $2 \leq n \leq \nu$. Thus, given $N > \nu$, (24.111) implies that

$$\left|\frac{N}{\sigma^2/k^2} - 1\right| < \frac{\varepsilon}{\sigma^2}$$

with probability exceeding $1 - \eta$. Hence, as k tends to zero, condition (d) holds.

The rule is, then, that we select k and proceed until $D_n \leq k$. The mean $\bar{x}$ then has variance approximately equal to k^2.

Example 24.13

Consider the Poisson distribution with parameter equal to λ. If we proceed until the variance of the mean, estimated as $\bar{x}/n$, is less than k^2, we have an estimator $\bar{x}$ of λ with variance k^2. This is equivalent to proceeding until the sum of the xs falls below $k^2 n^2$. But we should not use this result for small n.

On the other hand, suppose we wanted to specify in advance not the variance but the coefficient of variation, say l. The method would then fail. It would propose that we proceed until $\bar{x}/\sqrt{(\bar{x}/n)}$ is less than l, i.e. until $n\bar{x} \leqslant l^2$ or the sum of observations falls below l^2. But the sum must ultimately exceed any finite number. This is related to the result noted in Example 24.1 where we saw that for sequential sampling of rare attributes the coefficient of variation is approximately constant.

> For the Poisson mean, see Weiler (1972). Starr and Woodroofe (1972) treat the exponential distribution mean, and Binns (1975) the negative binomial mean.

Stein's double sampling method

24.36 At the end of Example 22.7, we observed that for fixed n no similar test of the mean μ of a normal population with unknown variance σ^2 could have power independent of σ^2. This implied (Example 20.2) that no confidence interval of pre-assigned length can be found for μ. However, if we use a sequential method, these statements are no longer true, as Stein (1945) pointed out.

24.37 We consider a normal population with mean μ and variance σ^2 and require to estimate μ with confidence coefficient $1 - \alpha$, the length of the confidence interval being l. We choose first of all a sample of fixed size n_0, and then a further sample $n - n_0$ where n now depends on the observations in the first sample.

Take a Student's t-variable with $n_0 - 1$ degrees of freedom, and let the probability that it lies in the range $-t_\alpha$ to t_α be $1 - \alpha$. Define

$$\sqrt{z} = \frac{1}{2t_\alpha}. \tag{24.112}$$

Let s^2 be the estimated variance of the sample of n_0 values, i.e.

$$s^2 = \frac{1}{n_0 - 1} \sum_{i=1}^{n_0} (x_i - \bar{x})^2. \tag{24.113}$$

We determine n by

$$n = \max \{n_0, 1 + [s^2/z]\}, \tag{24.114}$$

where $[s^2/z]$ means the greatest integer less than s^2/z.

Consider the n observations altogether, and let them have mean Y_n. Then Y_n is distributed independently of s and consequently $(Y_n - \mu)\sqrt{n}$ is independent of s; and hence $(Y_n - \mu)\sqrt{n}/s$ is distributed as t with $n_0 - 1$ d.fr. Hence

$$P\left\{ \left| \frac{(Y_n - \mu)\sqrt{n}}{s} \right| < t_\alpha \right\} = 1 - \alpha,$$

or

$$P\left\{Y_n - \frac{st_\alpha}{\sqrt{n}} \le \mu \le Y_n + \frac{st_\alpha}{\sqrt{n}}\right\} = 1 - \alpha,$$

or

$$P\{Y_n - \tfrac{1}{2}l \le \mu \le Y_n + \tfrac{1}{2}l\} \ge 1 - \alpha. \tag{24.115}$$

The appearance of the inequality in (24.115) is due to the end-effect that s^2/z may not be integral, which in general is small, so that the limits given by $Y_n \pm \tfrac{1}{2}l$ are close to the exact limits for confidence coefficient $1 - \alpha$. We can, by a device suggested by Stein, obtain exact limits, though the procedure entails rejecting observations and is probably not worth while in practice.

Seelbinder (1953) and Moshman (1958) discuss the optimum choice of first sample size in Stein's method—it is clearly not efficient to make it too small, since only the first sample is needed to estimate σ^2. P. Hall (1981) inserts an intermediate re-estimation of n in a triple sampling scheme. Bhattacharjee (1965) shows that Stein's procedure is more sensitive to non-normality than Student's t-test and that, as we should expect, non-normality re-introduces the dependence of the interval length (and corresponding test power) upon σ^2.

24.38 Chapman (1950) and B. K. Ghosh (1975) extend Stein's method to testing the means of two normal variables, the tests being independent of both variances. They depend on the distribution of the difference of two t-variables, for which see Exercise 16.26. D. R. Cox (1952c) considered the problem of estimation in double sampling, obtaining a number of asymptotic results, and considered corrections to the single and double sampling results to improve the approximations of asymptotic theory—see Exercises 24.15–17. A. Birnbaum and Healy (1960) discuss a general class of double sampling procedures to attain prescribed variance. Graybill and Connell (1964) give a double sampling procedure for estimating a normal variance within a fixed interval. Goldman and Zeigler (1966) compare different methods in estimating a normal mean or variance: Stein's is best for the mean. S. Banerjee (1967) generalises Stein's method to obtain pre-assigned-length intervals for the problem of two means (cf. **20.22–33**).

Finster (1985) discusses the sequential estimation of the parameters of the linear model.

Distribution-free tests

24.39 By the use of order-statistics we can reduce many procedures to the binomial case. Consider, for example, the testing of the hypothesis that the mean of a normal distribution is greater than μ_0 (a one-sided test). Replace the mean by the median and variate values by a score of, say, $+$ if the sample value falls above it and $-$ in the opposite case. On the hypothesis $H_0: \mu = \mu_0$ these signs will be distributed binomially with $\pi = \tfrac{1}{2}$. On the hypothesis $H_1: \mu = \mu_0 + k\sigma$ the probability of a positive sign is

$$\pi_1 = \frac{1}{\sqrt{(2\pi)}} \int_{-k}^{\infty} \exp\left(-\tfrac{1}{2}x^2\right) \mathrm{d}x. \tag{24.116}$$

We may then set up an SPR test of π_0 against π_1 in the usual manner. This will have a type I error α and a type II error β of accepting H_0 when H_1 is true; and this type II error will be $\leqslant \beta$ when $\mu - \mu_0 > k\sigma$. This is a sequential form of the sign test, to be discussed in a later volume—see Exercises 25.1 and 25.2.

Tests of this kind are often remarkably efficient, and the sacrifice of efficiency may be well worth while for the simplicity of application. Armitage (1947) compared this particular test with Wald's t-test and came to the conclusion that, as judged by sample number, the optimum test is not markedly superior to the sign test.

24.40 Jackson (1960) has provided a bibliography on sequential analysis, classified by topic. Johnson (1961) gives a useful review of the subject.

24.1 In Example 24.1, show by use of Exercise 9.13 that (24.3) implies the biasedness of k/n for π.

24.2 Referring to Example 24.6, sketch the OC curve for a binomial with $\alpha = 0.01$, $\beta = 0.03$, $\pi_1 = 0.1$, $\pi_2 = 0.2$. (The curve is half a bell-shaped curve with a maximum at $\pi = 0$ and zero at $\pi = 1$. Six points are enough to give its general shape.) Similarly, sketch the ASN curve for the same binomial.

24.3 Two samples, each of size n, are drawn from populations, P_1 and P_2, with proportions π_1 and π_2 of an attribute. They are paired off in order of occurrence. t_1 is the number of pairs in which there is a success from P_1 and a failure from P_2; t_2 is the number of pairs in which there is a failure from P_1 and a success from P_2. Show that in the (conditional) set of such pairs the probability of a member of t_1 is

$$\pi = (1 - \pi_1)\pi_2 / \{\pi_1(1 - \pi_2) + \pi_2(1 - \pi_1)\}.$$

Considering this as an ordinary binomial in the set of $t = t_1 + t_2$ values, show how to test the hypothesis that $\pi_1 \geqslant \pi_2$ by testing $\pi = \frac{1}{2}$. Hence derive a sequential test for $\pi_1 \geqslant \pi_2$.

If

$$u = \frac{\pi_2(1 - \pi_1)}{\pi_1(1 - \pi_2)},$$

show that $\pi = u/(1 + u)$ and hence derive the following acceptance and rejection numbers:

$$a_t = \frac{\log \dfrac{\beta}{1 - \alpha}}{\log u_1 - \log u_0} + t \frac{\log \dfrac{1 + u_1}{1 + u_0}}{\log u_1 - \log u_0},$$

$$r_t = \frac{\log \dfrac{1 - \beta}{\alpha}}{\log u_1 - \log u_0} + t \frac{\log \dfrac{1 + u_1}{1 + u_0}}{\log u_1 - \log u_0},$$

where u_i is the value of u corresponding to H_i ($i = 0, 1$). (Wald, 1947)

24.4 Referring to the function $h \neq 0$ of **24.14** show that if z is a random variable such that $E(z)$ exists and is not zero; if there exists a $\delta > 0$ such that $P(e^z < 1 - \delta) > 0$ and $P(e^z > 1 + \delta) > 0$; and if for any real h, $E(\exp hz) = g(h)$ exists, then

$$\lim_{h \to \infty} g(h) = \infty = \lim_{h \to -\infty} g(h)$$

and that $g''(h) > 0$ for all real values of h. Hence show that $g(h)$ is strictly decreasing over the interval $(-\infty, h^*)$ and strictly increasing over (h^*, ∞), where h^* is the value for which $g(h)$ is a minimum. Hence show that there exists at most one $h \neq 0$ for which $E(\exp hz) = 1$. (Wald, 1947)

24.5 In **24.31**, deduce the expressions (24.96-7). (Cf. Johnson, 1959b)

24.6 In Exercise 24.5, show that the third moment of $Z_n - n\mu$ is

$$E(Z_n - n\mu)^3 = \mu_3 E(n) - 3\sigma^2 E\{n(Z_n - n\mu)\},$$

where μ_3 is the third moment of z. (Wolfowitz, 1947)

24.7 If z is defined as at (24.19), let t be a complex variable such that $E(\exp zt) = \phi(t)$ exists in a certain part of the complex plane. Show that

$$E[\{\exp(tZ_n)\}\{\phi(t)\}^{-n}] = 1$$

for any point where $|\phi(t)| \geqslant 1$. (Wald, 1947)

24.8   Putting $t = h$ in the previous exercise show that, if E_b refers to expectation under the restriction that $Z_n \leq -b$ and E_a to the restriction $Z_n \geq a$, then

$$K(h) E_b \exp(h Z_n) + \{1 - K(h)\} E_a \exp(h Z_n) = 1,$$

where K is the OC. Hence, neglecting end-effects, show that

$$K(h) = \frac{e^{h(a+b)} - e^{hb}}{e^{h(a+b)} - 1}, \qquad h \neq 0,$$

$$= \frac{a}{a+b}, \qquad h = 0. \qquad \text{(Girshick, 1946)}$$

24.9   Differentiating the identity of Exercise 24.7 with respect to t and putting $t = 0$, show that

$$E(n) = \frac{a\{1 - K(h)\} - b K(h)}{E(z)}$$

and hence derive equation (24.43).   (Girshick, 1946)

24.10   Assuming, as in the previous exercise, that the identity is differentiable, derive the results of Exercises 24.7 and 24.8.

24.11   In the identity of Exercise 24.7, put

$$-\log \phi(t) = \tau$$

where τ is purely imaginary. Show that if $\phi(t)$ is not singular at $t = 0$ and $t = h$, this equation has two roots $t_1(\tau)$ and $t_2(\tau)$ for sufficiently small values of τ. In the manner of Exercise 24.8, show that the characteristic function of n is given asymptotically by

$$E(e^{n\tau}) = \frac{A^{t_2} - A^{t_1} + B^{t_1} - B^{t_2}}{B^{t_1} A^{t_2} - A^{t_1} B^{t_2}}. \qquad \text{(Wald, 1947)}$$

24.12   In the case when z is normal with mean μ and variance σ^2, show that t_1 and t_2 in Exercise 24.11 are

$$t_1 = -\frac{\mu}{\sigma^2} + \frac{1}{\sigma^2} (\mu^2 - 2\sigma^2 \tau)^{1/2},$$

$$t_2 = -\frac{\mu}{\sigma^2} - \frac{1}{\sigma^2} (\mu^2 - 2\sigma^2 \tau)^{1/2},$$

where the sign of the radical is determined so that the real part of $\mu^2 - 2\sigma^2 \tau$ is positive.
      In the limiting case $B = 0$, A finite (when of necessity $\mu > 0$ if $E(n)$ is to exist), show that the c.f. is A^{-t_1}, and in the case B finite, $A = \infty$ (when $\mu < 0$), show that the c.f. is B^{-t_1}.
                                                                       (Wald, 1947)

24.13   In the first of the two limiting cases of the previous exercise, show that the distribution of $m = \mu^2 n / 2\sigma^2$ is given by

$$dF(m) = \frac{c}{2\Gamma(\frac{1}{2}) m^{3/2}} \exp\left(-\frac{c^2}{4m} - m + c\right) dm, \qquad 0 \leq m < \infty,$$

where $c = \mu \log A / \sigma^2 > 0$. This is a special case of the inverse Gaussian distribution of Exercise 11.28 with $c^2/2 = \lambda = 2\mu^2$ there.
      For large c show that $2m/c$ is approximately normal with unit mean and variance $1/c$.
                                (Wald, 1947, who also shows that when A, B are finite the distribution of n is the weighted sum of a number of variables of the above type.)

24.14   Values of u are observed from the exponential distribution

$$dF = e^{-\lambda u} \lambda \, du, \qquad 0 \leq u < \infty.$$

Show that a sequential test of $\lambda = \lambda_0$ against $\lambda = \lambda_1$ is given by

$$k_1 + (\lambda_1 - \lambda_0) \sum_{j=1}^{n} u_j \leqslant n \log(\lambda_1/\lambda_0) \leqslant k_2 + (\lambda_1 - \lambda_0) \sum_{j=1}^{n} u_j,$$

where k_1 and k_2 are constants.

Compare this with the test of Exercise 24.3 in the limiting case when π_1 and π_2 tend to zero so that $\pi_1 t = \lambda_0$ and $\pi_2 t = \lambda_1$ remain finite. (Anscombe and Page, 1954)

24.15 It is required to estimate a parameter θ with a small variance $a(\theta)/\lambda$ when λ tends to infinity. Given that t_m is an unbiased estimator in samples of fixed size m with variance $v(\theta)/m$; that $\gamma_1(t_m) = O(m^{-1/2})$ and $\gamma_2(t_m) = O(m^{-1})$; and that $a(t_m)$ and $b(t_m)$ can be expanded in series to give asymptotic means and standard errors, consider the double sampling rule:
(a) Take a sample of size $N\lambda$ and let t_1 be the estimate of θ from it.
(b) Take a second sample of size max $\{0, [\{n_0(t_1) - N\}\lambda]\}$ where $n_0(t_1) = v(t_1)/a(t_1)$. Let t_2 be the estimate of θ from the second sample.
(c) Let $t = [Nt_1 + \{n_0(t_1) - N\}t_2]/n_0(t_1)$ if $n_0(t_1) \geqslant N$.
(d) Assume that $N < n_0(\theta)$ and the distribution of $m_0(t_1) = 1/n_0(t_1)$ is such that the event $n_0(t_1) < N$ may be ignored.
Show that under this rule

$$E(t) = \theta + O(\lambda^{-1}),$$
$$\text{var } t = a(\theta)\lambda^{-1}\{1 + O(\lambda^{-1})\}. \qquad \text{(D. R. Cox, 1952c)}$$

24.16 In the previous exercise, take the same procedure except that $n_0(t_1)$ is replaced by

$$n(t_1) = n_0(t_1)\left\{1 + \frac{b(t_1)}{\lambda}\right\}.$$

Show that

$$E(t) = \theta + m_0'(\theta)v(\theta)\lambda^{-1} + O(\lambda^{-2}).$$

Put

$$t' = t - m_0'(t)v(t)\lambda^{-1} \quad \text{if } N \leqslant n(t_1)$$
$$= 0 \quad \text{otherwise,}$$

and hence show that t' has bias $O(\lambda^{-2})$.
Show further that if we put

$$b(\theta) = n_0(\theta)v(\theta)\{2m_0(\theta)m_0'(\theta)\gamma_1(\theta)v^{-1/2}(\theta) + m_0'^2(\theta) + 2m_0(\theta)m_0''(\theta) + m_0''(\theta)/(2N)\},$$

then

$$\text{var } t' = a(\theta)\lambda^{-1} + O(\lambda^{-3}). \qquad \text{(D. R. Cox, 1952c)}$$

24.17 Applying Exercise 24.15 to the binomial distribution, with

$$a(\pi) = a\pi^2, \qquad v(\pi) = \pi(1-\pi), \qquad \gamma_1(\pi) = \frac{(1-2\pi)}{\{\pi(1-\pi)\}^{1/2}},$$

show that the total sample size is

$$n(t_1) = \frac{1-t_1}{at_1} + \frac{3}{t_1(t-t_1)} + \frac{1}{aNt_1}$$

and the estimator $t' = t - (at/(1-t))$.
Thus N should be chosen as large as possible, provided that it does not exceed $(1-\pi)/(a\pi)$. (D. R. Cox, 1952c)

24.18   Referring to Example 24.9, show that

$$K(\sigma) = \left\{ \left(\frac{1-\beta}{\alpha}\right)^h - 1 \right\} \bigg/ \left\{ \left(\frac{1-\beta}{\alpha}\right)^h - \left(\frac{\beta}{1-\alpha}\right)^h \right\}$$

where h is given by

$$\sigma\left(\frac{\sigma_1}{\sigma_0}\right)^h = \left\{ \frac{h}{\sigma_1^2} - \frac{h}{\sigma_0^2} + \frac{1}{\sigma^2} \right\}^{-1/2},$$

provided that the expression in brackets on the right is positive. Hence show how to draw the OC curve.
(Wald, 1947)

24.19   In the previous exercise derive the expression for the ASN

$$\frac{K(\sigma)\{h_0 - h_1\} + h_1}{\sigma^2 - \gamma} \qquad \text{where} \quad \gamma = \log\left(\sigma_1^2/\sigma_0^2\right)\bigg/\left(\frac{1}{\sigma_0^2} - \frac{1}{\sigma_1^2}\right).$$
(Wald, 1947)

24.20   In Example 24.9, obtain a test based on $\sum (x - \bar{x})^2$ of normal variances when the population mean is unknown.
(Girshick, 1946)

24.21   In **24.10**, $f(x, \theta)$ is replaced by $f(x, \theta, \phi)$ where ϕ is a nuisance parameter, so that H_0 and H_1 are composite. $\hat{\theta}_n$ is the ML estimator of θ after n observations, and $\hat{\phi}_j$ is then the ML estimator of ϕ, given H_j $(j = 0, 1)$. If θ_0 and θ_1 differ from the true value of θ by amounts of order $n^{-1/2}$, show that an SPR test based on

$$t_n = \log L_n(x, \theta_1, \hat{\phi}_1) - \log L_n(x, \theta_0, \hat{\phi}_0)$$

is asymptotically equivalent to using

$$t_n \sim (\theta_1 - \theta_0)[(L_\theta - L_\phi L_{\theta\phi}/L_{\phi\phi}) + \{\hat{\theta}_n - \tfrac{1}{2}(\theta_0 + \theta_1)\}(L_{\theta\theta} - L_{\theta\phi}^2/L_{\phi\phi})],$$

where $L_a = \partial \log L/\partial a$, $L_{ab} = -\partial^2 \log L/\partial a\, \partial b$ $(a, b = \theta, \phi)$, with mean and variance given (writing $E(L_{ab}) = I_{ab}$) by

$$E(t_n) \sim (\theta_1 - \theta_0)\{\theta - \tfrac{1}{2}(\theta_0 + \theta_1)\}(I_{\theta\theta} - I_{\theta\phi}^2/I_{\phi\phi}),$$
$$\text{var } t_n \sim (\theta_1 - \theta_0)^2(I_{\theta\theta} - I_{\theta\phi}^2/I_{\phi\phi}).$$

> (Bartlett, 1946; Joanes (1972) shows that this statistic is superior to that of D. R. Cox (1963), based on the joint ML estimation of (θ, ϕ), for testing a normal mean with variance unknown (the one-sided sequential t-test).)

24.22   For a sample from the distribution $f(x|\theta) = g(x)/h(\theta)$, $a \leqslant x \leqslant \theta$, show that the SPR test of $H_0: \theta = \theta_0$ against $H_1: \theta = \theta_1$ $(0 < \theta_0 < \theta_1)$ has $\alpha = 0$, and has the form 'accept H_1 if $x_n > \theta_0$, $n = 1, 2, \ldots$; accept H_0 if $x_n \leqslant \theta_0$ and $(\theta_0/\theta_1)^n \leqslant \beta$; continue sampling otherwise'.
    Show directly that when H_0 holds, the sample size is constant at $n_0 = \log \beta/\log (\theta_0/\theta_1)$, and that when H_1 holds the ASN is exactly $(1 - \beta)/(1 - (\theta_0/\theta_1))$ if n_0 is an integer. Verify these formulae for the ASN from (24.43).

24.23   In Example 24.1, consider a random variable t with density

$$f(t) = ct^{k-2}(1-t)^{1-k}, \qquad 0 \leqslant t \leqslant 1 - \pi < 1; \quad k > 1,$$

where c is therefore the integral in (24.6). Show that

$$E(1-t) < E\{t(1-t)\}E(t^{-1})$$

for any variable on this range, and that if V is the coefficient of variation of p at (24.5) and

$$a = (1 + V^2)/(k - 1),$$

then

$$aE(t) = V^2,$$
$$aE(t^2) = \{kV^2 - (1 - \pi)\}/2,$$
$$aE(t^{-1}) = \{(k-2)(1 - \pi)\}^{-1}.$$

Hence show that

$$\operatorname{var} p < \pi^2\{4(k-2)(1-\pi)\}^{-1}[(k-1)\{k-1-2(1-\pi)\}+4(1-\pi)$$
$$-(k-1)\{\{k-1-2(1-\pi)\}^2+8\pi(1-\pi)\}^{1/2}].$$

(Prasad and Sahai, 1982)

CHAPTER 25

THE COMPARISON OF TESTS

25.1 In Chapters 21–23 we were concerned with the problems of finding 'optimum' tests, i.e. of selecting the test with the 'best' properties in a given situation, where 'best' means the possession by the test of some desirable property such as being UMP, UMPU, etc. We have not so far considered the question of comparing two or more tests for a given situation with the aim of evaluating their relative efficiencies. Some investigation of this subject is necessary to permit us to evelute the loss of efficiency incurred in using any other test than the optimum one. It may happen, for example, that a UMP test is only very slightly more powerful than another test, which is perhaps much simpler to compute; in such circumstances we might well decide to use the less efficient test in routine testing. Before we can decide an issue such as this, we must make some quantitative comparison between the tests.

We discussed the analogous problem in the theory of estimation in **17.29**, where we derived a measure of estimating efficiency. The reader will perhaps ask how it comes about that, whereas in the theory of estimation the measurement of efficiency was discussed almost as soon as the concept of efficiency had been defined, we have left over the question of measuring test efficiency to the end of our general discussion of the theory of tests. The answer is partly that the concept of test efficiency turns out to be more complicated than that of estimating efficiency, and therefore could not be so shortly treated. For the most part, however, we are simply following the historical development of the subject: it was not until, from about 1935 onwards, the attention of statisticians turned to the computationally simple tests to be discussed in a later volume that the need arose to measure test efficiency. Even the idea of test consistency, which we encountered in **23.17**, was not developed by Wald and Wolfowitz (1940) until nearly twenty years after the first definition of a consistent estimator by Fisher (1921a); only when 'inefficient' tests became of practical interest was it necessary to investigate the weaker properties of tests.

The comparison of power functions

25.2 In testing a given hypothesis against a given alternative for fixed sample size, the simplest way of comparing two tests is by direct examination of their power functions. If sample size is at our disposal (e.g. in the planning of a series of observations), it is natural to seek a definition of test efficiency of the same form as that used for estimating efficiency in **17.29**. If an 'efficient' test (i.e. the most powerful in the class considered) of size α requires to be based on n_1 observations to attain a certain power, and a second size-α test requires n_2 observations to attain the same power against the same alternative, we may define the *relative efficiency* of the second test in attaining that power against that alternative as n_1/n_2. This measure is, as in the case of estimation, the reciprocal of the ratio of sample sizes required for a given performance, but it will be noticed that our definition of relative efficiency is not asymptotic, and that it imposes no restriction upon the forms of the sampling distributions of the test statistics being compared. We can compare any two tests in this way because the power functions of the tests, from which the relative efficiency is calculated, take comprehensive account of the distributions of the test statistics; the power functions contain all the information relevant to our comparison.

Asymptotic comparisons

25.3 The concept of relative efficiency, although comprehensive, is not concise. Like the power functions on which it is based, it is a function of three arguments—the size α of the tests, the distance (in terms of some parameter θ) between the hypothesis tested and the alternative, and the sample size (n_1) required by the efficient test. Even if we may confine ourselves to a few typical values of α, a table of double entry is still required for the comparison of tests by this measure. It would be much more convenient if we could find a single summary measure of efficiency, and it is clear that we can only hope to achieve this by imposing some limiting process. We have thus been brought back to the necessity for restriction to asymptotic results.

25.4 An approach that suggests itself is that we let sample sizes tend to infinity, as in **17.29**, and take the ratio of the powers of the tests as our measure of test efficiency. If we consider this suggestion we immediately encounter a difficulty. If the tests we are considering are both size-α consistent tests against the class of alternative hypotheses in the problem (and henceforth we shall always assume this to be so), it follows by definition that the power function of each tends to 1 as sample size increases. If we compare the tests against some fixed alternative value of θ, it follows that the efficiency thus defined will always tend to 1 as sample size increases. The suggested measure of test efficiency is therefore quite useless.

More generally, it is easy to see that consideration of the power functions of consistent tests asymptotically in n is of limited value. For instance, Wald (1941) defined an asymptotically most powerful test as one whose power function cannot be bettered as sample size tends to infinity, i.e. which is UMP asymptotically. The following example, due to Lehmann (1949), shows that one asymptotically UMP test may in fact be decidedly inferior to another such test, even asymptotically.

Example 25.1

Consider again the problem, discussed in Examples 21.1 and 21.2, of testing the mean θ of a normal distribution with known variance, taken to be equal to 1. We wish to test $H_0: \theta = \theta_0$ against the one-sided alternative $H_1: \theta = \theta_1 > \theta_0$. In **21.17**, we saw that a UMP test of H_0 against H_1 is given by the critical region $\bar{x} \geqslant \theta_0 + d_\alpha/n^{1/2}$, and in Example 21.3 that its power function is

$$P_1 = G\{\Delta n^{1/2} - d_\alpha\} = 1 - G\{d_\alpha - \Delta n^{1/2}\}, \tag{25.1}$$

where $\Delta = \theta_1 - \theta_0$ and the fixed value d_α defines the size α of the test as at (21.16).

We now construct a two-tailed size-α test, rejecting H_0 when

$$\bar{x} \geq \theta_0 + d_{\alpha_2}/n^{1/2} \quad \text{or} \quad \bar{x} \leq \theta_0 - d_{\alpha_1}/n^{1/2},$$

where d_{α_1} and d_{α_2}, functions of n, may be chosen arbitrarily subject to the condition $\alpha_1 + \alpha_2 = \alpha$, which implies that d_{α_1} and d_{α_2} both exceed d_α. (22.48) shows that the power function of this second test is

$$P_2 = G\{\Delta n^{1/2} - d_{\alpha_2}\} + G\{-\Delta n^{1/2} - d_{\alpha_1}\}, \tag{25.2}$$

and since G is always positive, it follows that

$$P_2 > G\{\Delta n^{1/2} - d_{\alpha_2}\} = 1 - G\{d_{\alpha_2} - \Delta n^{1/2}\}. \tag{25.3}$$

Since the first test is UMP, we have, from (25.1) and (25.3),

$$G\{d_{\alpha_2} - \Delta n^{1/2}\} - G\{d_\alpha - \Delta n^{1/2}\} > P_1 - P_2 \geq 0. \tag{25.4}$$

It is easily seen that the difference between $G\{x\}$ and $G\{y\}$ for fixed $(x - y)$ is maximised when x and y are symmetrically placed about zero, i.e. when $x = -y$, i.e. that

$$G\{\tfrac{1}{2}(x - y)\} - G\{-\tfrac{1}{2}(x - y)\} \geq G\{x\} - G\{y\}. \tag{25.5}$$

Applying (25.5) to (25.4), we have

$$G\{\tfrac{1}{2}(d_{\alpha_2} - d_\alpha)\} - G\{-\tfrac{1}{2}(d_{\alpha_2} - d_\alpha)\} > P_1 - P_2 \geq 0. \tag{25.6}$$

Thus if we choose d_{α_2} for each n so that

$$\lim_{n \to \infty} d_{\alpha_2} = d_\alpha, \tag{25.7}$$

the left-hand side of (25.6) will tend to zero, whence $P_1 - P_2$ will tend to zero uniformly in Δ. The two-tailed test will therefore be asymptotically UMP.

Now consider the ratio of Type II errors of the tests. From (25.1) and (25.2), we have

$$\frac{1 - P_2}{1 - P_1} = \frac{G\{d_{\alpha_2} - \Delta n^{1/2}\} - G\{-d_{\alpha_1} - \Delta n^{1/2}\}}{G\{d_\alpha - \Delta n^{1/2}\}}. \tag{25.8}$$

As $n^{1/2} \to \infty$, numerator and denominator of (25.8) tend to zero. Using l'Hôpital's rule, we find, using a prime to denote differentiation with respect to $n^{1/2}$ and writing g for the normal f.f.,

$$\lim_{n^{1/2} \to \infty} \frac{1 - P_2}{1 - P_1} = \lim_{n^{1/2} \to \infty} \left[\frac{(d'_{\alpha_2} - \Delta)g\{d_{\alpha_2} - \Delta n^{1/2}\}}{-\Delta g\{d_\alpha - \Delta n^{1/2}\}} + \frac{(d'_{\alpha_1} + \Delta)g\{-d_{\alpha_1} - \Delta n^{1/2}\}}{-\Delta g\{d_\alpha - \Delta n^{1/2}\}} \right]. \tag{25.9}$$

Now (25.7) implies that $d_{\alpha_1} \to \infty$ with n, and therefore that the second term on the right of (25.9) tends to zero: (25.7) also implies that the first term on the right of (25.9) will tend to infinity if

$$\lim_{n \to \infty} \frac{-d'_{\alpha_2} g\{d_{\alpha_2} - \Delta n^{1/2}\}}{g\{d_\alpha - \Delta n^{1/2}\}} = \lim_{n \to \infty} -d'_{\alpha_2} \frac{\exp\{-\tfrac{1}{2}(d_{\alpha_2} - \Delta n^{1/2})^2\}}{\exp\{-\tfrac{1}{2}(d_\alpha - \Delta n^{1/2})^2\}}$$

$$= \lim_{n \to \infty} -d'_{\alpha_2} \exp\{-\tfrac{1}{2}(d_{\alpha_2}^2 - d_\alpha^2) + \Delta n^{1/2}(d_{\alpha_2} - d_\alpha)\} \tag{25.10}$$

does so. By (25.7), the first term in the exponent on the right of (25.10) tends to zero. If we put

$$d_{\alpha_2} = d_\alpha + n^{-\delta}, \qquad 0 < \delta < \tfrac{1}{2}, \tag{25.11}$$

(25.7) is satisfied and (25.10) tends to infinity with n. Thus, although both tests are asymptotically UMP, the ratio of Type II errors (25.8) tends to infinity with n. It is clear, therefore, that the criterion of being asymptotically UMP is not a very selective one.

Asymptotic relative efficiency

25.5   In order to obtain a useful asymptotic measure of test efficiency from the relative efficiency, we consider the limiting relative efficiency of tests against a sequence of alternative hypotheses in which θ approaches the value tested, θ_0, as n increases. We do this in order to avoid forcing the powers of tests to be nearly 1, as in **25.4**. This type of alternative was first investigated by Pitman (1948), whose work was generalised by Noether (1955). Other types of limiting process on relative efficiency are considered by Dixon (1953) and Hodges and Lehmann (1956).

Let t_1 and t_2 be consistent test statistics for the hypothesis $H_0 : \theta = \theta_0$ against the one-sided alternative $H_1 : \theta > \theta_0$. We assume for the moment that t_1 and t_2 are asymptotically normally distributed whatever the value of θ—we shall relax this restriction in **25.14–15**. For brevity, we shall write

$$\left. \begin{aligned}
E(t_i \mid H_j) &= E_{ij}, \\
\mathrm{var}\,(t_i \mid H_j) &= D_{ij}^2, \\
E_i^{(r)}(\theta) &= \frac{\partial^r}{\partial\theta^r}\, E_{i1}, \\
D_{i1}^{(r)} &= \frac{\partial^r}{\partial\theta^r}\, D_{i1}, \qquad D_{i0}^{(r)} = D_{i1}^{(r)}(\theta_0),
\end{aligned} \right\} \qquad i = 1, 2; \quad j = 0, 1.$$

Large-sample size-α tests are defined by the critical regions

$$t_i > E_{i0} + \lambda_\alpha D_{i0} \tag{25.12}$$

(the sign of t_i being changed if necessary to make the region of this form), where λ_α is the normal deviate defined by $G\{-\lambda_\alpha\} = \alpha$, G being the standardised normal d.f. as before. Just as in Example 21.3, the asymptotic power function of t_i is

$$P_i(\theta) = G\{[E_{i1} - (E_{i0} + \lambda_\alpha D_{i0})] / D_{i1}\}. \tag{25.13}$$

Writing $u_i(\theta, \lambda_\alpha)$ for the argument of G in (25.13), we expand $(E_{i1} - E_{i0})$ in a Taylor series, obtaining

$$u_i(\theta, \lambda_\alpha) = \left[E_i^{(m_i)}(\theta_i^*) \frac{(\theta - \theta_0)^{m_i}}{m_i!} - \lambda_\alpha D_{i0} \right] \bigg/ D_{i1}, \tag{25.14}$$

where $\theta_0 < \theta_i^* < \theta$ and m_i is the first non-zero derivative at θ_0, i.e. m_i is defined by

$$\left. \begin{aligned}
E_i^{(r)}(\theta_0) &= 0, \qquad r = 1, 2, \ldots, m_i - 1, \\
E_i^{(m_i)}(\theta_0) &\neq 0.
\end{aligned} \right\} \tag{25.15}$$

In order to define the alternative hypothesis, we assume that, as $n \to \infty$,

$$R_i = [E_i^{(m_i)}(\theta_0)/D_{i0}] \sim c_i n^{m_i \delta_i}. \tag{25.16}$$

(25.16) defines the constants $\delta_i > 0$ and c_i. Now consider the sequences of alternatives, approaching θ_0 as $n \to \infty$,

$$\theta = \theta_0 + \frac{k_i}{n^{\delta_i}}, \tag{25.17}$$

where k_i is an arbitrary positive constant. If the regularity conditions

$$\lim_{n \to \infty} \frac{E_i^{(m_i)}(\theta)}{E_i^{(m_i)}(\theta_0)} = 1, \qquad \lim_{n \to \infty} \frac{D_{i1}}{D_{i0}} = 1 \tag{25.18}$$

are satisfied, (25.16), (25.17) and (25.18) reduce (25.14) to

$$u_i(\theta, \lambda_\alpha) = \frac{c_i k_i^{m_i}}{m_i!} - \lambda_\alpha, \tag{25.19}$$

and the asymptotic powers of the tests are $G\{u_i\}$ from (25.13).

25.6 If the two tests are to have equal power against the same sequence of alternatives for any fixed α, we must have, from (25.17) and (25.19),

$$\frac{k_1}{n_1^{\delta_1}} = \frac{k_2}{n_2^{\delta_2}} \tag{25.20}$$

and

$$\frac{c_1 k_1^{m_1}}{m_1!} = \frac{c_2 k_2^{m_2}}{m_2!}, \tag{25.21}$$

where n_1 and n_2 are the sample sizes upon which t_1 and t_2 are based. We combine (25.20) and (25.21) into

$$\frac{n_1^{\delta_1}}{n_2^{\delta_2}} = \left(\frac{c_2}{c_1} \frac{m_1!}{m_2!} k_2^{m_2-m_1} \right)^{1/m_1}. \tag{25.22}$$

The right-hand side of (25.22) is a positive constant. Thus if we let $n_1, n_2 \to \infty$, the ratio n_1/n_2 will tend to a constant if and only if $\delta_1 = \delta_2$. If $\delta_1 > \delta_2$, we must have $n_1/n_2 \to 0$, while if $\delta_1 < \delta_2$ we have $n_1/n_2 \to \infty$. If we define the *asymptotic relative efficiency* (ARE) of t_2 compared to t_1 as

$$A_{21} = \lim \frac{n_1}{n_2}, \tag{25.23}$$

we therefore have the result

$$A_{21} = 0, \qquad \delta_1 > \delta_2. \tag{25.24}$$

Thus to compare two tests by the criterion of ARE, we first compare their values of δ: if one has a smaller δ than the other, it has ARE of zero compared to the other. The value of δ plays the same role here as the order of magnitude of the variance plays in measuring efficiency of estimation (cf. **17.29**).

We may now confine ourselves to the case $\delta_1 = \delta_2 = \delta$. (25.22) and (25.23) then give

$$A_{21} = \lim \frac{n_1}{n_2} = \left(\frac{c_2}{c_1} \frac{m_1!}{m_2!} k_2^{m_2 - m_1} \right)^{1/(m_1 \delta)} \tag{25.25}$$

If, in addition,

$$m_1 = m_2 = m, \tag{25.26}$$

(25.25) reduces to

$$A_{21} = \left(\frac{c_2}{c_1} \right)^{1/(m\delta)},$$

which on using (25.16) becomes

$$A_{21} = \lim_{n \to \infty} \left\{ \frac{E_2^{(m)}(\theta_0)/D_{20}}{E_1^{(m)}(\theta_0)/D_{10}} \right\}^{1/(m\delta)}. \tag{25.27}$$

(25.27) is simple to evaluate in most cases. Most commonly, $\delta = \frac{1}{2}$ (corresponding to an estimation variance of order n^{-1}) and $m = 1$. For an interpretation of the value of m, see **25.10** below.

In passing, we may note that if $m_2 \neq m_1$, (25.25) is indeterminate, depending as it does on the arbitrary constant k_2. We therefore see that tests with equal values of δ do not have the same ARE against all sequences of alternatives (25.17) unless they also have equal values of m. We shall be commenting on the reasons for this in **25.10**.

25.7 If we wish to test H_0 against the two-sided $H_1 : \theta \neq \theta_0$, our results for the ARE are unaffected if we use 'equal-tails' critical regions of the form

$$t_i > E_{i0} + \lambda_{\alpha/2} D_{i0} \quad \text{or} \quad t_i < E_{i0} - \lambda_{\alpha/2} D_{i0},$$

for the asymptotic power functions (25.13) are replaced by

$$Q_i(\theta) = G\{u_i(\theta, \lambda_{\alpha/2})\} + 1 - G\{u_i(\theta, -\lambda_{\alpha/2})\}, \tag{25.28}$$

and $Q_1 = Q_2$ against the alternative (25.17) (where k_i need no longer be positive) if (25.20) and (25.21) hold, as before. Konijn (1956) gives a more general treatment of two-sided tests, which need not necessarily be 'equal-tails' tests.

Example 25.2
Let us compare the sample median $\tilde{x}$ with the UMP sample mean $\bar{x}$ in testing the mean θ of a normal distribution with known variance σ^2. Both statistics are asymptotically normally distributed. We know that

$$E(\bar{x}) = \theta, \qquad D^2(\bar{x} \mid \theta) = \sigma^2/n$$

and $\tilde{x}$ is a consistent estimator of θ, symmetrically distributed about θ, with

$$E(\tilde{x}) = \theta, \qquad D^2(\tilde{x} \mid \theta) \sim \pi\sigma^2/(2n)$$

(cf. Example 10.7). Thus we have

$$E'(\theta_0) = 1$$

for both tests, so that $m_1 = m_2 = 1$, while from (25.16), $\delta_1 = \delta_2 = \frac{1}{2}$. Thus, from (25.27),

$$A_{\tilde{x},\bar{x}} = \lim_{n\to\infty} \left\{ \frac{1/(\pi\sigma^2/2n)^{1/2}}{1/(\sigma^2/n)^{1/2}} \right\}^2 = \frac{2}{\pi}.$$

This is precisely the result we obtained in Example 17.13 for the efficiency of $\tilde{x}$ in estimating θ. We shall see in **25.13** that this is a special case of a general relationship between estimating efficiency and ARE for tests.

ARE and derivatives of power functions
25.8   The nature of the sequence of alternative hypotheses (25.17), which approaches θ_0 as $n \to \infty$, makes it clear that the ARE is in some way related to the behaviour, near θ_0, of the power functions of the tests being compared. We shall make this relationship more precise by showing that, under certain conditions, the ARE is a simple function of the ratio of derivatives of the power functions.

We first treat the case of the one-sided H_1 discussed in **25.5–6**, where the power functions of the tests are asymptotically given by (25.13), which we write, as before,

$$P_i(\theta) = G\{u_i(\theta, \lambda_\alpha)\}. \tag{25.29}$$

Differentiating with respect to θ, we have

$$P_i'(\theta) = g\{u_i\}u_i'(\theta, \lambda_\alpha), \tag{25.30}$$

where g is the normal frequency function. From (25.13) we find

$$u_i'(\theta, \lambda_\alpha) = \frac{E_{i1}'}{D_{i1}} - \frac{D_{i1}'}{D_{i1}^2}(E_{i1} - E_{i0} - \lambda_\alpha D_{i0}). \tag{25.31}$$

As $n \to \infty$, we find, using (25.18), and the further regularity conditions

$$\lim_{n\to\infty} \frac{D_{i1}'}{D_{i0}'} = 1, \qquad \lim_{n\to\infty} \frac{E_{i1}}{E_{i0}} = 1,$$

that (25.31) becomes

$$u_i'(\theta, \lambda_\alpha) = \frac{E_i'(\theta_0)}{D_{i0}} + \frac{D_{i0}'}{D_{i0}}\lambda_\alpha, \tag{25.32}$$

so that if $m_i = 1$ in (25.15) and if

$$\lim_{n\to\infty} \frac{D_{i0}'}{E_i'(\theta_0)} = 0, \tag{25.33}$$

(25.32) reduces at θ_0 to

$$u_i'(\theta_0, \lambda_\alpha) \sim E_i'(\theta_0)/D_{i0}. \tag{25.34}$$

Since, from (25.13),

$$g\{u_i(\theta_0, \lambda_\alpha)\} = g\{-\lambda_\alpha\}, \tag{25.35}$$

(25.30) becomes, on substituting (25.34) and (25.35),

$$P_i'(\theta_0) = P_i'(\theta_0, \lambda_\alpha) \sim g\{-\lambda_\alpha\}E_i'(\theta_0)/D_{i0}. \tag{25.36}$$

Remembering that $m_i = 1$, we therefore have from (25.36) and (25.27)

$$\lim_{n \to \infty} \frac{P_2'(\theta_0)}{P_1'(\theta_0)} = A_{21}^\delta, \qquad m_1 = m_2 = 1, \tag{25.37}$$

so that the asymptotic ratio of the first derivatives of the power functions of the tests at θ_0 is simply the ARE raised to the power δ (commonly $\frac{1}{2}$). Thus if we were to use this ratio as a criterion of asymptotic efficiency of tests, we should get precisely the same results as by using the ARE. This criterion was, in fact, proposed (under the name 'asymptotic local efficiency') by Blomqvist (1950).

25.9 If $m_i > 1$, i.e. $E_i'(\theta_0) = 0$, (25.36) is zero to our order of approximation and the result of **25.8** is of no use. The differentiation process has to be taken further to yield useful results.

From (25.30), we obtain

$$P_i''(\theta) = \frac{\partial g\{u_i\}}{\partial u_i}[u_i'(\theta, \lambda_\alpha)]^2 + g\{u_i\}u_i''(\theta, \lambda_\alpha). \tag{25.38}$$

From (25.31),

$$u_i''(\theta, \lambda_\alpha) = \frac{E_{i1}''}{D_{i1}} - \frac{2E_{i1}'D_{i1}'}{D_{i1}^2} - (E_{i1} - E_{i0} - \lambda_\alpha D_{i0})\left[\frac{D_{i1}''}{D_{i1}^2} - \frac{2(D_{i1}')^2}{D_{i1}^3}\right]. \tag{25.39}$$

If (25.18) holds with $m_i = 2$ and also the regularity conditions below (25.31) and

$$\lim_{n \to \infty} E_{i1}' = E_i'(\theta_0) = 0, \qquad \lim_{n \to \infty} \frac{D_{i1}'}{D_{i0}'} = 1, \qquad \lim_{n \to \infty} \frac{D_{i1}''}{D_{i0}''} = 1, \qquad \lim_{n \to \infty} \frac{E_{i1}}{E_{i0}} = 1, \tag{25.40}$$

(25.39) gives

$$u_i''(\theta_0, \lambda_\alpha) = \frac{E_i''(\theta_0)}{D_{i0}} + \lambda_\alpha\left[\frac{D_{i0}''}{D_{i0}} - 2\left(\frac{D_{i0}'}{D_{i0}}\right)^2\right]. \tag{25.41}$$

Instead of (25.33), we now assume the conditions

$$\lim_{n \to \infty} \frac{D_{i0}''}{E_i''(\theta_0)} = 0, \qquad \lim_{n \to \infty} \frac{(D_{i0}')^2}{D_{i0}E_i''(\theta_0)} = 0. \tag{25.42}$$

(25.42) reduces (25.41) to

$$u_i''(\theta_0, \lambda_\alpha) \sim E_i''(\theta_0)/D_{i0}. \tag{25.43}$$

Returning now to (25.38), we see that since

$$\frac{\partial g\{u_i\}}{\partial u_i} = -u_i g\{u_i\},$$

we have, using (25.32), (25.35) and (25.43) in (25.38),

$$P_i''(\theta_0) \sim g\{-\lambda_\alpha\}\left\{\lambda_\alpha\left[\frac{E_i'(\theta_0)}{D_{i0}} + \frac{D_{i0}'}{D_{i0}}\lambda_\alpha\right]^2 + \frac{E_i''(\theta_0)}{D_{i0}}\right\}. \tag{25.44}$$

Since we are considering the case $m_i = 2$ here, the term in $E_i'(\theta_0)$ is zero, and from the second condition of (25.42), (25.44) may finally be written

$$P_i''(\theta_0) \sim g\{-\lambda_\alpha\}E_i''(\theta_0)/D_{i0}, \tag{25.45}$$

whence, with $m = 2$, (25.27) gives

$$\lim_{n\to\infty} \frac{P_2''(\theta_0)}{P_1''(\theta_0)} = A_{21}^{2\delta} \tag{25.46}$$

for the limiting ratio of the second derivatives.

(25.37) and (25.46) may be expressed concisely by the statement that for $m = 1, 2$, the ratio of the mth derivatives of the power functions of one-sided tests is asymptotically equal to the ARE raised to the power $m\delta$.

If, instead of (25.33) and (25.42), we had imposed the stronger conditions

$$\lim_{n\to\infty} D_{i0}'/D_{i0} = 0, \qquad \lim_{n\to\infty} D_{i0}''/D_{i0} = 0, \tag{25.47}$$

which with (25.16) imply (25.33) and (25.42), (25.34) and (25.43) would have followed from (25.32) and (25.41) as before. (25.47) may be easier to verify in particular cases.

The interpretation of the value of m

25.10 We now discuss the general conditions under which m will take the value 1 or 2. Consider again the asymptotic power function (25.13) for a one-sided alternative $H_1 : \theta > \theta_0$ and a one-tailed test (25.12). For brevity, we drop the suffix i in this section. If $\theta \to \theta_0$, and $D_1 \to D_0$ by (25.18), (25.13) becomes

$$P(\theta) = G\left\{\frac{E_1 - E_0}{D_0} - \lambda_\alpha\right\},$$

a monotone increasing function of $(E_1 - E_0)$.

If $(E_1 - E_0)$ is an increasing function of $(\theta - \theta_0)$, $P(\theta) \to 0$ as $\theta \to -\infty$ (which implies that the other 'tail' of the distribution of the test statistic would be used as a critical region if $\theta < \theta_0$). If $E'(\theta_0)$ exists, it is non-zero and $m = 1$, and $P'(\theta_0) \neq 0$ also, by (25.36).

If, on the other hand, $(E_1 - E_0)$ is an even function of $(\theta - \theta_0)$ (which implies that the same 'tail' would be used as critical region whatever the sign of $(\theta - \theta_0)$), and an increasing function of $|\theta - \theta_0|$, and $E'(\theta_0)$ exists, it must under regularity conditions equal zero, and $m > 1$—in practice, we find $m = 2$. By (25.36), $P'(\theta_0) = 0$ also to this order of approximation.

We are now in a position to see why, as remarked at the end of **25.6**, the ARE is not useful in comparing tests with differing values of m, which in practice are 1 and 2. For we are then comparing tests whose power functions behave essentially differently at θ_0, one having a regular minimum there and the other not. The indeterminacy of (25.25) in such circumstances is not really surprising.

Example 25.3
Consider the problem of testing $H_0: \theta = \theta_0$ for a normal distribution with mean θ and variance 1. The pair of one-tailed tests based on the sample mean $\bar{x}$ are UMP (cf. **21.17**), the upper or lower tail being selected according to whether H_1 is $\theta > \theta_0$ or $\theta < \theta_0$. From Example 25.2, $\delta = \frac{1}{2}$ and $m = 1$ for $\bar{x}$.
We could also use as a test statistic

$$S = \sum_{i=1}^{n} (x_i - \theta_0)^2.$$

S has a non-central chi-squared distribution with n degrees of freedom and non-central parameter $n(\theta - \theta_0)^2$, so that (cf. Exercise 23.1)

$$E(S|\theta) = n\{1 + (\theta - \theta_0)^2\},$$
$$D^2(S|\theta_0) = 2n,$$

and as $n \to \infty$, S is asymptotically normally distributed. We have $E'(\theta) = 2n(\theta - \theta_0)$, $E'(\theta_0) = 0$, $E''(\theta) = 2n = E''(\theta_0)$, so that $m = 2$ and

$$\frac{E''(\theta_0)}{D_0} = \frac{2n}{(2n)^{1/2}} = (2n)^{1/2}.$$

From (25.16), since $m = 2$, $\delta = \frac{1}{4}$. Since $\delta = \frac{1}{2}$ for $\bar{x}$, the ARE of S compared to $\bar{x}$ is zero by (25.24). The critical region for S consists of the upper tail, whatever the value of θ.

25.11 We now turn to the case of the two-sided alternative $H_1: \theta \neq \theta_0$. The power function of the 'equal-tails' test is given asymptotically by (25.28). Its derivative at θ_0 is

$$Q_i'(\theta_0) = P_i'(\theta_0, \lambda_{\alpha/2}) - P_i'(\theta_0, -\lambda_{\alpha/2}), \qquad (25.48)$$

where P_i' is given by (25.36) if $m_i = 1$ and (25.33) or (25.47) holds. Since $g\{-\lambda_\alpha\}$ in (25.36) is an even function of λ_α, (25.48) immediately gives the asymptotic result

$$Q_i'(\theta_0) \sim 0$$

so that the slope of the power function at θ_0 is asymptotically zero. This result is also implied (under regularity conditions) by the remark in **23.17** concerning the asymptotic unbiasedness of consistent tests.
The second derivative of the power function (25.28) is

$$Q_i''(\theta_0) = P_i''(\theta_0, \lambda_{\alpha/2}) - P_i''(\theta_0, -\lambda_{\alpha/2}). \qquad (25.49)$$

We have evaluated P_i'' at (25.44) where we had $m_i = 2$. (25.44) still holds for $m_i = 1$ if we strengthen the first condition in (25.47) to

$$D_{i0}'/D_{i0} = o(n^{-\delta}), \tag{25.50}$$

for then by (25.16) the second term on the right of (25.39) may be neglected and we obtain (25.44) as before. Substituted into (25.49), it gives

$$Q_i''(\theta_0) \sim 2\lambda_{\alpha/2} g\{-\lambda_{\alpha/2}\}\left\{\left(\frac{E_i'(\theta_0)}{D_{i0}}\right)^2 + \left(\frac{D_{i0}'}{D_{i0}}\lambda_\alpha\right)^2\right\},$$

and (25.50) reduces this to

$$Q_i''(\theta_0) \sim 2\lambda_{\alpha/2} g\{-\lambda_{\alpha/2}\}\left(\frac{E_i'(\theta_0)}{D_{i0}}\right)^2. \tag{25.51}$$

In this case, therefore, (25.27) and (25.51) give

$$\frac{Q_2''(\theta_0)}{Q_1''(\theta_0)} = A_{21}^{2\delta}. \tag{25.52}$$

Thus for $m = 1$, the asymptotic ratio of second derivatives of the power functions of two-sided tests is exactly that given by (25.46) for one-sided tests when $m = 2$, and exactly the square of the one-sided test result for $m = 1$ at (25.37).

The case $m = 2$ does not seem of much importance for two-tailed tests: the remarks in **25.10** suggest that where $m = 2$ a one-tailed test would often be used even against a two-sided H_1.

Example 25.4
Reverting to Example 25.2, we saw that both tests have $\delta = \frac{1}{2}$, $m = 1$ and $E'(\theta_0) = 1$. Since the variance of each statistic is independent of θ, at least asymptotically, we see that (25.33) and (25.50) are satisfied and, the regularity conditions being satisfied, it follows from (25.37) that for one-sided tests

$$\lim_{n\to\infty} \frac{P_{\tilde{x}}'(\theta_0)}{P_{\bar{x}}'(\theta_0)} = A_{\tilde{x},\bar{x}}^{1/2} = \left(\frac{2}{\pi}\right)^{1/2},$$

while for two-sided tests, from (25.52),

$$\lim_{n\to\infty} \frac{Q_{\tilde{x}}''(\theta_0)}{Q_{\bar{x}}''(\theta_0)} = A_{\tilde{x},\bar{x}} = \frac{2}{\pi}.$$

The maximum power loss and the ARE
25.12 Although the ARE of tests essentially reflects their power properties in the neighbourhood of θ_0, it does have some implications for the asymptotic power function as a whole, at least for the case $m = 1$, to which we now confine ourselves.

The power function $P_i(\theta)$ of a one-sided test is $G\{u_i(\theta)\}$, where $u_i(\theta)$, given at (25.14), is asymptotically equal, under regularity conditions (25.18), to

$$u_i(\theta) = \frac{E_i'(\theta_0)}{D_{i0}}(\theta - \theta_0) - \lambda_\alpha, \qquad (25.53)$$

when $m_i = 1$. Thus $u_i(\theta)$ is asymptotically linear in θ. If we write $R_i = E_i'(\theta_0)/D_{i0}$ as at (25.16), we may write the difference between two such power functions as

$$d(\theta) = P_2(\theta) - P_1(\theta) = G\{(\theta - \theta_0)R_2 - \lambda_\alpha\} - G\left\{(\theta - \theta_0)R_2\frac{R_1}{R_2} - \lambda_\alpha\right\}, \quad (25.54)$$

where we assume $R_2 > R_1$ without loss of generality. Consider the behaviour of $d(\theta)$ as a function of θ. When $\theta = \theta_0$, $d = 0$, and again as θ tends to infinity P_1 and P_2 both tend to 1 and d to zero. The maximum value of $d(\theta)$ depends only on the ratio R_1/R_2, for although R_2 appears in the right-hand side of (25.54) it is always the coefficient of $(\theta - \theta_0)$, which is being varied from 0 to ∞, so that $R_2(\theta - \theta_0)$ also goes from 0 to ∞ whatever the value of R_2. We therefore write $\Delta = R_2(\theta - \theta_0)$ in (25.54), obtaining

$$d(\Delta) = G\{\Delta - \lambda_\alpha\} - G\left\{\Delta\frac{R_1}{R_2} - \lambda_\alpha\right\}. \qquad (25.55)$$

The first derivative of (25.55) with respect to Δ is

$$d'(\Delta) = g\{\Delta - \lambda_\alpha\} - \frac{R_1}{R_2}g\left\{\Delta\frac{R_1}{R_2} - \lambda_\alpha\right\},$$

and if this is equated to zero, we have

$$\frac{R_1}{R_2} = \frac{g\{\Delta - \lambda_\alpha\}}{g\left\{\Delta\dfrac{R_1}{R_2} - \lambda_\alpha\right\}} = \exp\left\{-\tfrac{1}{2}(\Delta - \lambda_\alpha)^2 + \frac{1}{2}\left(\Delta\frac{R_1}{R_2} - \lambda_\alpha\right)^2\right\}$$

$$= \exp\left\{-\tfrac{1}{2}\Delta^2\left(1 - \frac{R_1^2}{R_2^2}\right) + \lambda_\alpha\Delta\left(1 - \frac{R_1}{R_2}\right)\right\}. \qquad (25.56)$$

(25.56) is a quadratic equation in Δ, whose only positive root is

$$\Delta = \frac{\lambda_\alpha + \left\{\lambda_\alpha^2 + 2\left[\left(1 + \dfrac{R_1}{R_2}\right) \middle/ \left(1 - \dfrac{R_1}{R_2}\right)\right]\log\dfrac{R_2}{R_1}\right\}^{1/2}}{\left(1 + \dfrac{R_1}{R_2}\right)}. \qquad (25.57)$$

This is the value at which (25.55) is maximised. Consider, for example, the case $\alpha = 0.05$ ($\lambda_\alpha = 1.645$) and $R_1/R_2 = 0.5$. (25.57) gives

$$\Delta = \frac{1.645 + \{1.645^2 + 6\log_e 2\}^{1/2}}{1.5} = 2.85.$$

(25.55) then gives, using tables of the normal d.f.,

$$P_2 = G\{2.85 - 1.64\} = G\{1.21\} = 0.89,$$
$$P_1 = G\{1.42 - 1.64\} = G\{-0.22\} = 0.41.$$

D. R. Cox and Stuart (1955) gave values of P_2 and P_1 at the point of maximum difference, obtained by the graphical equivalent of the above method, for a range of values of α and R_1/R_2. Their table is reproduced as Table 25.1, our worked example above corresponding to one of the entries.

Table 25.1   **Asymptotic powers per cent. of tests at the point of greatest difference**
(D. R. Cox and Stuart, 1955)

α	0.10		0.05		0.01		0.001	
R_1/R_2	P_1	P_2	P_1	P_2	P_1	P_2	P_1	P_2
0.9	67	73	63	71	49	60	54	67
0.8	61	74	56	72	49	71	43	72
0.7	59	80	51	77	42	77	39	83
0.6	54	84	47	84	39	86	29	87
0.5	48	88	41	89	30	90	20	93
0.3	35	96	27	96	14	97	7	99

It will be seen from the table that as α decreases for fixed R_1/R_2, the maximum difference between the asymptotic power functions increases steadily—it can, in fact, be made as near to 1 as desired by taking α small enough. Similarly, for fixed α, the maximum difference increases steadily as R_1/R_2 falls.

The practical consequence of the table is that if R_1/R_2 is 0.9 or more, the loss of power *along the whole course* of the asymptotic power function will not exceed 0.08 for $\alpha = 0.05$, 0.11 for $\alpha = 0.01$, and 0.13 for $\alpha = 0.001$, the most commonly used test sizes. Since R_1/R_2 is, from (25.36), the ratio of first derivatives of the power functions, we have from (25.37) that $(R_1/R_2)^{1/\delta} = A_{12}$, where δ is commonly $\frac{1}{2}$, and thus the ARE needs to be $(0.9)^{1/\delta}$ for the statements above to be true.

ARE and estimating efficiency

25.13   There is a simple connection between the ARE and estimating efficiency. If we have two consistent test statistics t_i as before, we define functions f_i, independent of n, such that the statistics

$$T_i = f_i(t_i) \tag{25.58}$$

are consistent *estimators* of θ. If we write

$$\theta = f_i(\tau_i), \tag{25.59}$$

it follows from (25.58) that since $T_i \to \theta$ in probability, $t_i \to \tau_i$ and $E(t_i)$ if it exists also tends to τ_i. Expanding (25.58) about τ_i by Taylor's theorem, we have, using

(25.59),

$$T_i = \theta + (t_i - \tau_i) \left[\frac{\partial f(t_i)}{\partial \tau_i} \right]_{t_i = t_i^*} \tag{25.60}$$

where t_i^*, intermediate in value between t_i and τ_i, tends to τ_i as n increases. Thus (25.60) may be written

$$T_i - \theta \sim (t_i - \tau_i) \left[\frac{\partial \theta}{\partial E(t_i)} \right]$$

whence

$$\operatorname{var} T_i \sim \operatorname{var} t_i \Big/ \left(\frac{\partial E(t_i)}{\partial \theta} \right)^2 . \tag{25.61}$$

If 2δ is the order of magnitude in n of the variances of the T_i, the estimating efficiency of T_2 compared to T_1 is, by (17.65) and (25.61),

$$\lim_{n \to \infty} \left(\frac{\operatorname{var} T_1}{\operatorname{var} T_2} \right)^{1/(2\delta)} = \left[\frac{\{\partial E(t_2)/\partial \theta\}^2 / \operatorname{var} t_2}{\{\partial E(t_1)/\partial \theta\}^2 / \operatorname{var} t_1} \right]^{1/(2\delta)} . \tag{25.62}$$

At θ_0, (25.62) is precisely equal to the ARE (25.27) when $m_i = 1$. Thus the ARE essentially gives the relative estimating efficiencies of transformations of the test statistics that are consistent estimators of the parameter concerned. But this correspondence is a local one: in **21.15** we saw that the connection between estimating efficiency and power is not strong in general. It follows at once that tests based upon efficient estimators have maximum ARE and (from **25.8-11**) that the derivatives of their power functions at θ_0 are maximised. (25.62) and (17.61) also imply that if T_1 is efficient, $A_{21} = \{\rho(T_1, T_2)\}^{1/\delta}$. A more general result in terms of $\rho(t_1, t_2)$ is given in Exercise 25.9.

Example 25.5
The result we have just obtained explains the fact, noted in Example 25.2, that the ARE of the sample median, compared to the sample mean, in testing the mean of a normal distribution has exactly the same value as its estimating efficiency for that parameter.

Non-normal cases
25.14 From **25.5** onwards, we have confined ourselves to the case of asymptotically normally distributed test statistics. However, examination of **25.5-7** will show that in deriving the ARE we made no specific use of the normality assumption. We were concerned to establish the conditions under which the arguments u_i of the power functions $G\{u_i\}$ in (25.19) would be equal against the sequence of alternatives (25.17). G played no role in the discussion other than of ensuring that the asymptotic power functions were of the same form, and we need only require that G is a regularly behaved d.f.

It follows that if two tests have asymptotic power functions of any two-parameter form G, only one of whose parameters is a function of θ, the results of **25.5-7** will hold, for (25.17) will fix this parameter and u_i in (25.19) then determines the other. Given the form G, the critical region for one-tailed tests can always be put in the form (25.12), where λ_α is more generally interpreted as the multiple of D_{i0} required to make (25.12) a size-α critical region.

25.15 The only important limiting distributions other than the normal are the non-central χ^2 distributions whose properties were discussed in **23.4-5**. Suppose that for the hypothesis $H_0: \theta = \theta_0$ we have two test statistics t_i with such distributions, the degrees of freedom being ν_i (independent of θ) and the non-central parameters $\lambda_i(\theta)$, where $\lambda_i(\theta_0) = 0$, so that the χ^2 distributions are central when H_0 holds. We have (cf. Exercise 23.1)

$$E_{i1} = \nu_i + \lambda_i(\theta), \qquad D_{i0}^2 = 2\nu_i. \tag{25.63}$$

All the results of **25.5-6** for one-sided tests therefore hold for the comparison of test statistics distributed in the non-central χ^2 form (central when H_0 holds) with degrees of freedom independent of θ. In particular, when $\delta_1 = \delta_2 = \delta$ and $m_1 = m_2 = m$, (25.63) substituted into (25.27) gives

$$A_{21} = \lim_{n \to \infty} \left\{ \frac{\lambda_2^{(m)}(\theta_0)/\nu_2^{1/2}}{\lambda_1^{(m)}(\theta_0)/\nu_1^{1/2}} \right\}^{1/(m\delta)} \tag{25.64}$$

A different derivation of this result is given by E. J. Hannan (1956).

Rothe (1981) gives a generalisation.

Other measures of test efficiency
25.16 Although the relative efficiency and the ARE are the principal measures of test efficiency, we conclude this chapter by discussing other methods that have been proposed.

Walsh (1946) proposed the comparison of two tests for fixed size α by a measure that takes into account the performance of the tests for all alternative hypothesis values of the parameter θ. If the tests t_i are based on sample sizes n_i and have power functions $P_i(\theta, n_i)$, the efficiency of t_2 compared to t_1 is $n_1/n_2 = e_{12}$ where

$$\int [P_1(\theta, n_1) - P_2(\theta, n_2)] \, d\theta = 0. \tag{25.65}$$

Thus, given one of the sample sizes (say, n_2), we choose n_1 so that the algebraic sum of the areas between the power functions is zero, and measure efficiency by n_1/n_2.

This measure removes the effect of θ from the table of triple entry required to compare two power functions, and does so in a reasonable way. However, e_{12} is still a function of α and, more important, of n_2. Moreover, the calculation of n_1/n_2 so that (25.65) is satisfied is inevitably tedious and probably accounts for the fact that this measure has rarely been used. As an asymptotic measure, however, it is equivalent to the use of the ARE, at least for asymptotically normally distributed test statistics with $m_i = 1$ in (25.15). For we then have, as in **25.12**,

$$P_i(\theta, n_i) = G\{(\theta - \theta_0)R_i - \lambda_\alpha\},$$

where $R_i = E_i'(\theta_0)/D_{i0}$ as at (25.16), and (25.65) then becomes

$$\int [G\{(\theta - \theta_0)R_1 - \lambda_\alpha\} - G\{(\theta - \theta_0)R_2 - \lambda_\alpha\}] \, d\theta = 0. \tag{25.66}$$

Clearly, (25.66) holds asymptotically only when $R_1 = R_2$, or, from (25.16),

$$\frac{R_1}{R_2} \sim \frac{c_1}{c_2}\left(\frac{n_1}{n_2}\right)^\delta = 1$$

whence

$$\lim \frac{n_1}{n_2} = \left(\frac{c_2}{c_1}\right)^{1/\delta} = A_{21},$$

exactly as at (25.27) with $m = 1$.

25.17 A quite different approach to the problem of measuring asymptotic efficiency for tests is due to Chernoff (1952). For a variate x with moment-generating function $M_x(t) = E(e^{xt})$, we define

$$m(a) = \inf_t M_{x-a}(t), \tag{25.67}$$

the absolute minimum value of the m.g.f. of $(x - a)$. If $E(x|H_i) = \mu_i$ for simple hypotheses H_0, H_1, we further define

$$\rho = \inf_{\mu_0 \leqslant a \leqslant \mu_1} \max\{m_0(a), m_1(a)\}, \tag{25.68}$$

where the suffix to m, defined at (25.67), indicates the hypothesis. For a one-sided test of H_0 against H_1 based on a sum of n identically distributed x_i, with size α and power $1 - \beta$, Chernoff shows that if *any* linear function $l(\alpha, \beta)$ of the probabilities of error α and β is minimised, its minimum value behaves as $n \to \infty$ like ρ^n, where ρ is defined by (25.68). Consider two such tests t_i, based on samples of size n_i. If they have equal minima for $l(\alpha, \beta)$, we therefore have

$$\lim \frac{\rho_1^{n_1}}{\rho_2^{n_2}} = 1$$

or

$$\lim \frac{n_1}{n_2} = \frac{\log \rho_2}{\log \rho_1}. \tag{25.69}$$

Thus the right-hand side of (25.69) is a measure of the asymptotic efficiency of t_2 compared to t_1. Its use is restricted to test statistics based on sums of independent observations, and the computation required may be considerable.

25.18 Bahadur (1967) reviews results (largely his own) on the comparison of tests by means of the rate of convergence to zero of the minimum size of critical region that includes the observed value of the test statistic. Under conditions given by Lambert and Hall (1982), the distribution of this 'tail area' is asymptotically lognormal. We sketch only the simple case when the standardised test statistic $z = n^{1/2}(t_n - b)$ has a limiting standardised normal d.f. $F(z)$. By (5.68),

$$1 - F(z) \sim F'(z)/z$$

with

$$F'(z) = (2\pi)^{-1/2} \exp\left(-\tfrac{1}{2}z^2\right).$$

Consider

$$\begin{aligned}
u_n &= \log\left[n^{1/2}\{1 - F(n^{1/2}t_n)\}\right] \\
&\doteqdot \log\left\{F'(n^{1/2}t_n)/t_n\right\} \\
&\doteqdot -\tfrac{1}{2}nt_n^2 - \log t_n - \tfrac{1}{2}\log(2\pi) \doteqdot -\tfrac{1}{2}nt_n^2 .
\end{aligned}$$

Now nt_n^2 tends to a $\chi'^2(1, nb^2)$, so from Exercise 23.1 it has mean $1 + nb^2$ and variance $2(1 + 2nb^2)$, and it tends to normality by **23.5** as n increases. Thus

$$n^{-1/2}u_n \doteqdot -\tfrac{1}{2}n^{1/2}t_n^2 \sim N(-\tfrac{1}{2}n^{1/2}b^2, b^2),$$

and finally we see that the tail area $\{1 - F(n^{1/2}t_n)\}$ asymptotically behaves as a multiple of a lognormal variable.

Dempster and Schatzoff (1965) use the expected value of this tail area as a criterion for choosing between tests. See also Joiner (1969), who compares these with other methods including that of Exercise 25.10, and Sievers (1969). Groeneboom and Oosterhoff (1981) show that Bahadur's efficiency measure is essentially the limit of the relative efficiency n_1/n_2 in **25.2** as size $\alpha \to 0$ with H_1 fixed (instead of keeping α fixed and letting $H_1 \to H_0$ as in the derivation of Pitman's ARE in **25.5**) and find it a worse approximation to relative efficiency in moderate-sized samples than ARE is. Wieand (1976) gives conditions under which the limiting approximate Bahadur efficiency, as $H_1 \to H_0$, equals the limiting ARE as size $\alpha \to 0$, as do Kallenberg and Ledwina (1987) for the exact Bahadur efficiency. L. D. Brown (1971) shows that appropriate LR tests are asymptotically optimum in Bahadur's sense. See also Fu (1973). Koziol (1978) discusses using the optimality of the LR test to prove the optimality of other tests without evaluating small tail probabilities.

Kallenberg (1983) discusses a measure of efficiency intermediate between Pitman's and Bahadur's.

Efficient score tests

25.19 The local nature of the ARE criterion suggests that it may be possible to construct local approximations to LR tests that are asymptotically efficient yet may remain analytically tractable when the original statistic is not. We begin by considering a single parameter, testing $H_0: \theta = \theta_0$ against the composite alternative $\theta \neq \theta_0$. Using a Taylor series expansion we may write, approximately,

$$\begin{aligned}\log l &= \log \{L(x|\theta_0)/L(x|\hat{\theta})\} \\ &= (\hat{\theta} - \theta_0)u(\hat{\theta}) + \tfrac{1}{2}(\hat{\theta} - \theta_0)^2 u^{(1)}(\theta^*)\end{aligned} \qquad (25.70)$$

(cf. **18.21**) where the score

$$\begin{aligned}u(\hat{\theta}) &= [\mathrm{d} \log L(x|\theta)/\mathrm{d}\theta]_{\theta=\hat{\theta}}, \\ u^{(1)}(\theta^*) &= [\mathrm{d}u(\theta)/\mathrm{d}\theta]_{\theta=\theta^*}\end{aligned}$$

and $|\theta^* - \theta_0| \leqslant |\hat{\theta} - \theta_0|$. Since the score function $u(\theta)$ is zero when $\theta = \hat{\theta}$, (25.70) reduces to

$$-2 \log l = -n(\hat{\theta} - \theta_0)^2 \{n^{-1}u^{(1)}(\theta^*)\}.$$

Since $\hat{\theta}$ and thus θ^* are both consistent estimators of θ under H_0 it follows from **18.12** and **18.16** that

$$n^{-1}u^{(1)}(\theta^*) \to n^{-1}u^{(1)}(\theta_0) \to n^{-1}I(\theta_0) = -n^{-1}E(\mathrm{d}^2 \log L/\mathrm{d}\theta^2)_{\theta=\theta_0}.$$

Then, from **18.16** and **23.7**, the test statistic

$$W_1 = (\hat{\theta} - \theta_0)I(\theta_0)(\hat{\theta} - \theta_0) \qquad (25.71)$$

is asymptotically distributed as χ^2 with one degree of freedom. Further, from (18.31), we know that, asymptotically,

$$\hat{\theta} - \theta_0 = [I(\theta_0)]^{-1}u(\theta_0)$$

so that the statistic

$$W_2 = u(\theta_0)[I(\theta_0)]^{-1}u(\theta_0) \qquad (25.72)$$

is also asymptotically equivalent to W_1 and the LR statistic.

25.20 The statistics (25.71) and (25.72) extend immediately to cover the case where H_0 is simple and $\boldsymbol{\theta}$ is a vector, upon replacing the scalar quantities by the appropriate vector and matrix forms. When H_0 is composite with $\boldsymbol{\theta} = (\boldsymbol{\theta}_1, \boldsymbol{\theta}_2)$ and $H_0: \boldsymbol{\theta}_1 = \boldsymbol{\theta}_{10}$, it follows that

$$W_1 = (\hat{\boldsymbol{\theta}}_1 - \boldsymbol{\theta}_{10})'[I^{11}(\hat{\boldsymbol{\theta}})]^{-1}(\hat{\boldsymbol{\theta}}_1 - \boldsymbol{\theta}_{10}),$$

where $I^{11}(\hat{\boldsymbol{\theta}})$ is the top left element of the inverse of

$$I(\hat{\boldsymbol{\theta}}) = -E\left(\frac{\partial^2 \log L}{\partial \boldsymbol{\theta} \, \partial \boldsymbol{\theta}'}\right)_{\boldsymbol{\theta}=\hat{\boldsymbol{\theta}}} = \begin{bmatrix} I_{11}(\hat{\boldsymbol{\theta}}) & I_{12}(\hat{\boldsymbol{\theta}}) \\ I_{21}(\hat{\boldsymbol{\theta}}) & I_{22}(\hat{\boldsymbol{\theta}}) \end{bmatrix},$$

or

$$[I^{11}(\hat{\boldsymbol{\theta}})]^{-1} = I_{11} - I_{12}I_{22}^{-1}I_{21}. \qquad (25.73)$$

Further, if the score vector is partitioned as (u_1, u_2), the corresponding version of W_2 becomes, using (25.73),

$$W_2 = \tilde{\mathbf{u}}_1' \mathbf{I}^{11}(\boldsymbol{\theta}_0) \tilde{\mathbf{u}}_1 \tag{25.74}$$

where

$$\tilde{\mathbf{u}}_1 = u_1(\hat{\boldsymbol{\theta}}_0) - I_{12}(\hat{\boldsymbol{\theta}}_0)\{I_{22}(\hat{\boldsymbol{\theta}}_0)\}^{-1} u_2(\hat{\boldsymbol{\theta}}_0) \tag{25.75}$$

and $\hat{\boldsymbol{\theta}}_0 = (\boldsymbol{\theta}_{10}, \hat{\boldsymbol{\theta}}_{20})$, where $\hat{\boldsymbol{\theta}}_{20}$ is the ML estimator for $\boldsymbol{\theta}_2$ under H_0. In each case, the statistic will be distributed as χ^2 when H_0 is true, with s degrees of freedom where s is the dimensionality of $\boldsymbol{\theta}_2$. The form of (25.75) is clearly simplified if $\boldsymbol{\theta}_2$ can be chosen so that $\mathbf{I}_{12}(\hat{\boldsymbol{\theta}}_0) = \mathbf{0}$.

25.21 Tests based upon the score function have an ARE of one, an equivalence originally established by Wald (1943a); the tests are also known as $C(\alpha)$ tests following the work of Neyman (1959); see also Moran (1970) and Cox and Hinkley (1974, Chapter 9). The value of such tests is that they will often provide simple criteria in otherwise complex situations.

Example 25.6
Suppose $x_1, \ldots, x_n$ are observations from Poisson distributions with parameters $\lambda_j = \exp(\theta_2 + \theta_1 z_j)$, where $\sum z_j = 0$ and θ_2 is known. We have that

$$u(\theta_1) = \frac{d \log L}{d\theta_1} = \sum x_j z_j - \sum z_j \exp(\theta_2 + \theta_1 z_j),$$

and

$$\frac{d^2 \log L}{d\theta_1^2} = -\sum z_j^2 \exp(\theta_2 + \theta_1 z_j).$$

Thus

$$u(\theta_{10}) = \sum x_j z_j, \qquad I(\theta_{10}) = (\sum z_j^2) \exp(\theta_2)$$

and from (25.74),

$$W_2 \propto \sum x_j z_j / \sum z_j^2. \tag{25.76}$$

When θ_2 is unknown,

$$u(\theta_2) = \sum x_j - \sum \exp(\theta_2 + \theta_1 z_j)$$

so that $\exp(\hat{\theta}_{20}) = \bar{x}$. Also, since $\sum z_j = 0$,

$$I(\hat{\boldsymbol{\theta}}_0) = \exp(\theta_2) \begin{bmatrix} \sum z_j^2 & 0 \\ 0 & n \end{bmatrix}$$

and W_2 is of the same form as before.

The statistic (25.78) is of a simple form because the ML estimators under H_0 are easy to find. Conversely, W_1 is useful when the full set of ML estimators is easy to find but the restricted set under H_0 is not.

25.22   A similar class of tests, also asymptotically equivalent to the LR procedures, are the Lagrange multiplier (LM) tests developed by Aitchison and Silvey (1958) and Silvey (1959). Assume that H_0 can be represented by the set of restrictions

$$h(\boldsymbol{\theta}) = \mathbf{0}$$

and write

$$h(\hat{\boldsymbol{\theta}}) = h(\boldsymbol{\theta}) + H'(\boldsymbol{\theta})(\hat{\boldsymbol{\theta}} - \boldsymbol{\theta})$$

where

$$H(\boldsymbol{\theta}) = \{\partial h_j(\boldsymbol{\theta})/\partial \theta_i\}.$$

Then, under H_0, $n^{1/2}h(\hat{\boldsymbol{\theta}})$ is asymptotically normally distributed with mean zero and variance

$$V(\boldsymbol{\theta}) = H'(\boldsymbol{\theta})\{I(\boldsymbol{\theta})\}^{-1}H(\boldsymbol{\theta}).$$

We may test H_0 using the statistic

$$W_1^* = h'(\hat{\boldsymbol{\theta}})\{V(\hat{\boldsymbol{\theta}})\}^{-1}h(\hat{\boldsymbol{\theta}}),$$

which is approximately distributed as χ^2 with s degrees of freedom. When the ML estimators are evaluated under the restrictions in H_0, the LM test reduces to W_2.

EXERCISES

25.1 The sign test for the hypothesis H_0 that a population median takes a specified value θ_0 consists of counting the number of sample observations exceeding θ_0 and rejecting H_0 when this number is too large. Show that for a normal population this test has ARE $2/\pi$ compared to the Student's t-test for H_0, and connect this with the result of Example 25.2.

(Cochran, 1937)

25.2 Generalising the result of Exercise 25.1, show that for any continuous frequency function f with variance σ^2, the ARE of the sign test compared to the t-test is $4\sigma^2\{f(\theta_0)\}^2$.

(Pitman, 1948)

25.3 The difference between the means of two normal populations with equal variances is tested from two independent samples by comparing every observation y_j in the second sample with every observation x_i in the first sample, and counting the number of times a y_j exceeds an x_i. Show that the ARE of this test compared to the two-sample Student's t-test is $3/\pi$.

(Pitman, 1948)

25.4 Generalising Exercise 25.3, show that if any two continuous densities $f(x)$, $f(x-\theta)$, differ only by a location parameter θ, and have variance σ^2, the ARE of the Wilcoxon test compared to the t-test is

$$12\sigma^2\left\{\int_{-\infty}^{\infty}\{f(x)\}^2\,dx\right\}^2.$$

(Pitman, 1948)

25.5 If x is normally distributed with mean μ_i and variance σ_i^2, given H_i $(i=0,1; \mu_0<\mu_1)$, show that (25.68) has the value

$$\rho=\exp\{-\tfrac{1}{2}[(\mu_1-\mu_0)/(\sigma_1+\sigma_0)]^2\}.$$

(Chernoff, 1952)

25.6 If x/σ_i^2 has a chi-squared distribution with r degrees of freedom, given H_i $(i=0,1; \sigma_0^2/\sigma_1^2=\tau<1)$, show that ρ in (25.68) satisfies

$$\log\rho=-\tfrac{1}{2}\tau(\delta-1-\log\delta)$$

where

$$\delta=(\log\tau)/(\tau-1).$$

(Chernoff, 1952)

25.7 t_1 and t_2 are unbiased estimators of θ, jointly normally distributed in large samples with variances σ^2, σ^2/e respectively $(0<e\leqslant 1)$. Using the results of 16.23 and 17.29, show that

$$E(t_1\,|\,t_2)=\theta(1-e)+t_2 e,$$

and hence that if t_2 is observed to differ from θ by a multiple d of its standard deviation we expect t_1 to differ from θ by a multiple $de^{1/2}$ of its standard deviation.

(D. R. Cox, 1956)

25.8 Using Exercise 25.7, show that if t_2 is used to test $H_0: \theta=\theta_0$, we may calculate the 'expected result' of a test based on the more efficient statistic t_1. In particular, show that if a one-tail test of size 0.01, using t_2, rejects H_0, we should expect a one-tail test of size 0.05, using t_1, to do so if $e>0.50$; while if an equal-tails size-0.01 test on t_2 rejects H_0, we should expect an equal-tails size-0.05 test on t_1 to do so if $e>0.58$.

25.9   Let t_1 be a statistic with maximum ARE and t_2 any other test statistic for the same problem with δ and m as for t_1. By considering

$$t_3 = a\frac{t_1}{D_{10}} + (1-a)\frac{t_2}{D_{20}},$$

show that, in (25.16),
$$R_3 = \{aR_1 + (1-a)R_2\}/\{a^2 + (1-a)^2 + 2a(1-a)\rho\}^{1/2},$$

where ρ is the asymptotic correlation coefficient of t_1 and t_2. Hence show that $\rho = R_2/R_1$, the $(m\delta)$th power of the ARE (25.27). Cf. (17.61) for estimators.

(Cf. van Eeden, 1963)

25.10   In **25.5**, let $\delta_1 = \delta_2 = \delta$ and $m_1 = m_2 = m$. It is proposed to measure the efficiency of the tests t_1, t_2 by the reciprocal of the ratio of the distances $(\theta - \theta_0)$ which they require for $E(t\,|\,\theta)$ to fall on the boundary of the critical region (25.12). Show that this is approximately the same as using the δth power of the ARE.

(This *average critical value* method is due to R. C. Geary—cf. Stuart (1967).)

CHAPTER 26

STATISTICAL RELATIONSHIP:
LINEAR REGRESSION AND CORRELATION

26.1 In the next three chapters, we consider relationships between two or more variables. We have already discussed bivariate and multivariate distributions and their properties; in particular, the normal case was examined in Chapters 15 and 16. However, a systematic discussion of the relationships between variables was deferred until the theory of estimation and testing hypotheses had been explored. Even so, the area that we are about to study is a very large one, and many more complex topics are deferred until the discussion of multivariate analysis in a later volume. We begin with a general overview.

26.2 Most of our work stems from an interest in the joint distribution of a pair of random variables: we may describe this as the problem of *statistical relationship*. There is a quite distinct field of interest concerning relationships of a strictly functional kind between variables, such as those of classical physics; this subject is of statistical interest because the functionally related variables are subject to observational or instrumental errors. We call this the problem of *functional relationship*, and discuss it in a later volume. In these three chapters, we are concerned with the problem of statistical relationship alone, where the variables are not (except in degenerate cases) functionally related, although they may also be subject to observational or instrumental errors.

Within the field of statistical relationship there is a further useful distinction to be made. We may be interested either in the *interdependence* between a number (not necessarily all) of our variables or in the *dependence* of one or more variables upon others. For example, we may be interested in whether there is a relationship between length of arm and length of leg in men; put this way, it is a problem of interdependence. But if we are interested in using leg-length measurements to convey information about arm-length, we are considering the dependence of the latter upon the former. This is a case in which either interdependence or dependence may be of interest. On the other hand, there are situations when only dependence is of interest. The relationship of crop-yields and rainfall is an example where non-statistical considerations make it clear that there is an essential asymmetry in the situation: we say, loosely, that rainfall 'causes' crop yield to vary, and we are quite certain that crop yields do not affect the rainfall, so we measure the dependence of yield upon rainfall. We refer to rainfall as an *input*, or *regressor*, variable.

There is no clear-cut distinction in statistical terminology for the techniques appropriate to these essentially different types of problem. For example, we shall see in Chapter 27 that if we are interested in the interdependence of two variables with the effects of other variables eliminated, we use the method called 'partial correlation', whereas if we are interested in the dependence of a single variable upon a group of others, we use 'multiple correlation'. Nevertheless, it is true in the main that the study of *interdependence* leads to the theory of correlation, whereas the study of *dependence* leads to the theory of regression. We shall examine the theory for two random variables in Chapter 26 and that for $k \geqslant 2$ variables in Chapter 27. Chapter 28 considers the general linear model.

26.3   Before proceeding to the exposition of the theory of correlation (largely developed around the beginning of this century by Karl Pearson and by Yule), we make one final general point. A statistical relationship, however strong and however suggestive, can never *establish* a causal connection: our ideas on causation must come from outside statistics, ultimately from some theory or other. Even in the simple example of crop yield and rainfall discussed in **26.2**, we had no *statistical* reason for dismissing the idea of dependence of rainfall upon crop yield: the dismissal is based on quite different considerations. Even if rainfall and crop yields were in perfect functional correspondence, we should not dream of reversing the 'obvious' causal connection. We need not enter into the philosophical implications of this; for our purposes, we need only reiterate that statistical relationship, of whatever kind, cannot logically imply causation.

G. B. Shaw made this point brilliantly in his Preface to *The Doctor's Dilemma* (1906):

> Even trained statisticians often fail to appreciate the extent to which statistics are vitiated by the unrecorded assumptions of their interpreters... It is easy to prove that the wearing of tall hats and the carrying of umbrellas enlarges the chest, prolongs life, and confers comparative immunity from disease.... A university degree, a daily bath, the owning of thirty pairs of trousers, a knowledge of Wagner's music, a pew in church, anything, in short, that implies more means and better nurture ... can be statistically palmed off as a magic-spell conferring all sorts of privileges..... The mathematician whose correlations would fill a Newton with admiration, may, in collecting and accepting data and drawing conclusions from them, fall into quite crude errors by just such popular oversights as I have been describing.

Although Shaw was on this occasion supporting a characteristically doubtful cause, his logic was valid. In the first flush of enthusiasm for correlation techniques, it was easy for early followers of Karl Pearson and Yule to be incautious. It was not until twenty years after Shaw wrote that Yule (1926) frightened statisticians by adducing cases of very high correlations which were obviously not causal: e.g. the annual suicide rate was highly correlated with the membership of the Church of England. Most of these 'nonsense' correlations operate through concomitant variation in time, and they had the salutary effect of bringing home to the statistician that causation cannot be deduced from any observed co-variation, however close.

Although simple correlation measures are now rarely computed as an end in themselves, much of multivariate statistical analysis starts from a matrix of correlation coefficients. Thus, the topic of correlation remains one of central importance.

Causality in regression

26.4   The issue of causality cannot, however, be overlooked. It behoves statisticians to indicate when, if at all, the results of a statistical investigation may be admissible as evidence in support of a causal relationship. Holland (1986) provides a useful survey of this important area; the ensuring discussion and list of references are also particularly useful.

One step we can take is to distinguish between controlled and observational investigations. In controlled, or experimental, studies, the input (or regressor) variables can be manipulated by the investigator and other sources of variation effectively controlled. It is then reasonable to argue that changes in the *dependent* (or *response*) variables are caused by the changes in the input variables. Ultimately, this depends upon the investigator's success in achieving the stated experimental conditions. *Observational* studies, where designed experiments are infeasible, will rarely provide evidence of causation.

When the variables have a clear time-ordering, such as a sales increase following a price cut, there is again some basis for arguing causality (Granger, 1969). However, the examples in **26.3** and the possibility of other intervening variables cannot be ignored.

26.5   In Chapter 1 (Tables 1.15, 1.23 and 1.24) we gave some examples of bivariate distributions arising in practice. Tables 26.1 and 26.2 give two further examples that will be used for illustrative purposes.

For the moment, we treat these data as populations, leaving aside the question of sampling until later in the chapter.

For univariate distributions, we constructed summary measures such as the mean and variance. We now seek measures of interdependence.

Denote the two variables by x and y. For any given value of x, x_0 say, we may consider the conditional frequency function of y, $f(y|x_0)$. The conditional expectation of y, given $x = x_0$, is

$$E(y|x_0) = \sum yf(y|x_0) \tag{26.1}$$

or, for continuous random variables,

$$E(y|x_0) = \int yf(y|x_0)\,dy$$

$$= \left\{ \int yf(x_0, y)\,dy \right\} \Big/ f_x(x_0), \tag{26.2}$$

where $f_x(x_0) = \int f(x_0, y)\,dy$ and the x-subscript is used to indicate the marginal distribution of x. All the integrals are taken over $(-\infty, \infty)$ and (26.2) exists provided

Table 26.1   Distribution of weight and stature for 4995 women in Great Britain, 1951
Reproduced, by permission, from *Women's Measurements and Sizes*, London, HMSO, 1957

Weight (y): central values of groups, in pounds	Stature (x): central values of groups in inches											Total
	54	56	58	60	62	64	66	68	70	72	74	
278.5					1							1
272.5												—
266.5					1							1
260.5						1						1
254.5												—
248.5					1	1						2
242.5							1					1
236.5							1					1
230.5					2				1			3
224.5					1	2	1					4
218.5			1		2	1		1				5
212.5				2	1	6		1	1			11
206.5				2	2	3	2		1			10
200.5			4	2	6	2						14
194.5				1	3	7	7	4	1			23
188.5			1	5	14	8	12	3	1	2		46
182.5			1	7	12	26	9	5		1	2	63
176.5			5	8	18	21	15	11	7		2	87
170.5			2	11	17	44	21	13	3	1		112
164.5		1	3	12	35	48	30	15	5	3		152
158.5			8	17	52	42	36	21	9			185
152.5		1	7	30	81	71	58	21	2	2		273
146.5		2	13	36	76	91	82	36	8	1		345
140.5		1	6	55	101	138	89	50	8			448
134.5			15	64	95	175	122	45	5			521
128.5		1	19	73	155	207	101	25	3			584
122.5		3	34	91	168	200	81	12	1	1		591
116.5		3	24	108	184	184	50	8				561
110.5		5	33	119	165	124	22	4				472
104.5	1	3	33	87	95	35	6					260
98.5	2	5	29	59	45	16	3					159
92.5		6	10	21	9							46
86.5		1	5	3								9
80.5	2	1	1									4
Total	5	33	254	813	1340	1454	750	275	56	11	4	4995

$f_x(x_0) \neq 0$ and the integral in the numerator is finite. When the conditional expectation is defined for all x-values, the function $E(y|x)$ is known as the *regression curve*, or simply, the *regression* of y on x. The regression of x on y is similarly defined as $E(x|y)$.

Figures 26.1 and 26.2 show, for the data of Tables 26.1 and 26.2, the conditional means of y (marked by crosses) and those of x (marked by circles). Lines CC' and

Table 26.2   Distribution of bust girth and stature for 4995 women in Great Britain, 1951
Data from same source as those of Table 26.1

Bust girth (y) central values of groups in inches	Stature (x): central values of groups, in inches											Total
	54	56	58	60	62	64	66	68	70	72	74	
56					1							1
54					1	2						3
52		1			3	4	1		1			10
50		1		3	5	4	1					14
48	1	3		9	7	6	3	1				30
46		4		11	17	17	7		1			57
44		2	11	26	50	45	17	10	1			162
42		2	11	42	85	73	31	12	3	2		261
40		2	20	76	132	131	71	31	9	4	3	479
38		2	36	98	158	203	126	65	17	3	1	709
36		6	48	188	317	410	263	89	15	1		1337
34	1	9	67	210	376	427	196	59	8			1353
32	3	5	39	131	163	122	31	8	1	1		504
30	1	4	11	18	25	10	2					71
28			2	1			1					4
Total	5	33	254	813	1340	1454	750	275	56	11	4	4995

RR' show the closest straight line fits to these values, given by the method of least squares—cf. **26.11**.

Tukey (1962) refers to the plot of conditional means as a *regressogram*; it forms the basis for a nonparametric estimator of the regression function based upon grouped data. By convention, the conditional mean of y is plotted at the midpoint of the x-values for which it is evaluated.

Covariance
26.6   It is natural to consider using the product-moment μ_{11} as the basis of a measure of dependence; a quantity we have encountered several times in earlier chapters. μ_{11}, which is known as the *covariance* of x and y, is defined (cf. **3.27**) as

$$\mu_{11} = \int_{-\infty}^{\infty} \int_{-\infty}^{\infty} (x - \mu_x)(y - \mu_y)\, dF(x, y)$$

$$= E\{(x - \mu_x)(y - \mu_y)\} \equiv E(xy) - E(x)E(y). \tag{26.3}$$

In turn, this may be expressed as

$$\mu_{11} = E_x[x\{E(y|x) - E(y)\}]$$

so that the covariance represents a weighted average of the conditional means, measured from the unconditional mean. If the variates x, y are independent,

$$\mu_{11} = \kappa_{11} = 0, \tag{26.4}$$

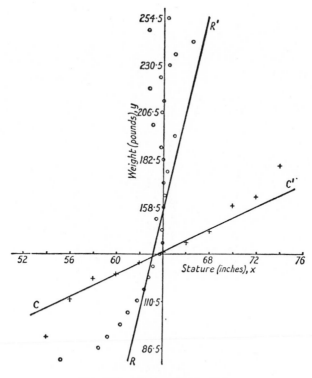

**Fig. 26.1　Regressions for data
of Table 26.1**

as we saw in Example 12.7. By that Example, too, the converse is not generally true: (26.4) does not generally imply independence, which requires

$$\kappa_{rs} = 0 \quad \text{for all } r \times s \neq 0. \tag{26.5}$$

For a *binormal* distribution, however, we know that $\kappa_{rs} = 0$ for all $r + s > 2$, so that κ_{11} is the only non-zero product-cumulant. Thus (26.4) implies (26.5) and independence for normal variables. It may also do so for other specified distributions, but it does not do so in general. Example 26.1 gives a non-normal distribution for which $\kappa_{11} = 0$ implies independence; Example 26.2 gives one where it does not.

Example 26.1
If x and y are binormal standardised variables, the joint characteristic function of x^2 and y^2 is, using (16.45),

$$\int_{-\infty}^{\infty} (2\pi)^{-1/2} \exp\left(-\tfrac{1}{2}x^2 + x^2 it\right)$$

$$\times \left[\int_{-\infty}^{\infty} \{2\pi(1-\rho^2)\}^{-1/2} \exp\{-\tfrac{1}{2}[(y-\rho x)^2/(1-\rho^2)] + y^2 iu\} \, dy\right] dx.$$

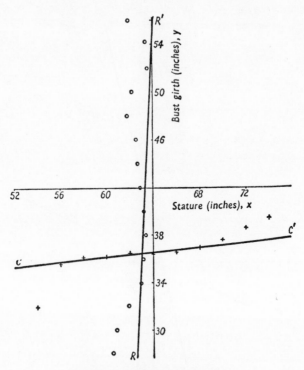

**Fig. 26.2   Regressions for data
of Table 26.2**

The inner integral is the c.f. of $(1-\rho^2)$ times a $\chi'^2(1, \rho^2 x^2/(1-\rho^2))$ variate, which by Exercise 23.1 is

$$\{1-2iu(1-\rho^2)\}^{-1/2} \exp\left\{\frac{\rho^2 x^2 iu}{1-2iu(1-\rho^2)}\right\}.$$

Hence

$$\phi(t, u) = \{1-2iu(1-\rho^2)\}^{-1/2}$$
$$\times \int_{-\infty}^{\infty} (2\pi)^{-1/2} \exp\{-\tfrac{1}{2}x^2 + x^2[it + \rho^2 iu/\{1-2iu(1-\rho^2)\}]\}\, dx.$$

The integral is the c.f. of a central χ_1^2 variate with it replaced by the expression in square brackets. Thus

$$\phi(t, u) = \{1-2iu(1-\rho^2)\}^{-1/2}\{1-2[it + \rho^2 iu/\{1-2iu(1-\rho^2)\}]\}^{-1/2}$$
$$= [(1-2it)(1-2iu) + 4\rho^2 tu]^{-1/2}.$$

We see that

$$\phi(t, 0) = (1-2it)^{-1/2},$$
$$\phi(0, u) = (1-2iu)^{-1/2},$$

so that the marginal distributions are chi-squares with one degree of freedom, as we know. By differentiating the logarithm of $\phi(t, u)$ we find

$$\mu_{11} = \kappa_{11} = \left[\frac{\partial^2 \log \phi(t, u)}{\partial(it)\,\partial(iu)} \right]_{t=u=0} = 2\rho^2.$$

Now when $\rho = 0$, we see that

$$\phi(t, u) = \phi(t, 0)\phi(0, u),$$

a necessary and sufficient condition for independence of x^2 and y^2 by **4.16–17**. Thus $\mu_{11} = 0$ implies independence in this case.

Example 26.2
Consider a bivariate distribution with uniform probability over a unit disk centred at the means of x and y. We have

$$dF(x, y) = dx\,dy/\pi, \qquad 0 \leqslant x^2 + y^2 \leqslant 1,$$

so that

$$\mu_{11} = \iint xy\,dF = \frac{1}{\pi} \iint xy\,dx\,dy$$

$$= \frac{1}{\pi} \int x \left[\int y\,dy \right] dx$$

$$= \frac{1}{\pi} \int x \left[\frac{1}{2} y^2 \right]_{-(1-x^2)^{1/2}}^{+(1-x^2)^{1/2}} dx = 0,$$

as is otherwise obvious. But clearly x and y are not independent, since the range of variation of each depends on the value of the other.

The correlation coefficient
26.7   A meaningful measure of interdependence should be invariant to changes in location and scale, as well as being symmetric in the random variables. These conditions are met by the *correlation coefficient*

$$\text{corr}\,(x, y) = \rho = \mu_{11}/(\sigma_1\sigma_2), \tag{26.6}$$

where σ_1^2 and σ_2^2 denote the variances of x and y, respectively.
  By the Cauchy–Schwarz inequality (**2.29**)

$$\mu_{11}^2 = \left\{ \iint (x - \mu_x)(y - \mu_y)\,dF \right\}^2$$

$$\leqslant \left\{ \iint (x - \mu_x)^2\,dF \right\} \left\{ \iint (y - \mu_y)^2\,dF \right\} = \sigma_1^2\sigma_2^2$$

so that

$$0 \leqslant \rho^2 \leqslant 1, \tag{26.7}$$

the upper equality in (26.7) holding if and only if x and y are in strict linear functional relationship.

Example 26.3
Since x^2 and y^2 in Example 26.1 are each chi-squared with one degree of freedom, it follows that their variances are both 2 and

$$\text{corr}(x^2, y^2) = 2\rho^2/(2 \cdot 2)^{1/2} = \rho^2.$$

A perfect linear relationship requires $x^2 = c^2 y^2$ or $x = \pm cy$.

ρ as a coefficient of interdependence
26.8   From **26.6** and Example 26.2 we see that although independence of x and y implies $\mu_{11} = \rho = 0$, the converse does not generally apply. It does apply for jointly normal variables, and sometimes for others (Example 26.1). In this lies the difficulty of interpreting ρ as a coefficient of interdependence in general. In fact, we have seen that ρ is essentially a coefficient of *linear* interdependence, and more complex forms of interdependence lie outside its vocabulary. In general, joint variation is too complex to be summarised by a single coefficient.

To *express* a quality, moreover, is not the same as to *measure* it. If $\rho = 0$ implies independence, we know from **26.7** that as $|\rho|$ increases, the interdependence also increases until when $|\rho| = 1$ we have the limiting case of linear functional relationship. Even so, it remains an open question which function of ρ should be used as a *measure* of interdependence: we see from (26.18) below that ρ^2 is more directly interpretable than ρ itself, being the ratio of the variance of the fitted line to the overall variance. Leaving this point aside, ρ gives us a measure in such cases, though there may be better measures.

On the other hand, if $\rho = 0$ does not imply independence, it is difficult to interpret ρ as a measure of interdependence, and perhaps wiser to use it as an indicator rather than as a precise measure. In practical work, we would recommend the use of ρ as a *measure* of interdependence only in cases of normal or near-normal variation.

Linear regression
26.9   If the regression of y on x is exactly linear, we have the expression

$$E(y|x) = \alpha_2 + \beta_2 x, \tag{26.8}$$

and we now wish to find α_2 and β_2. Note that

$$E(y) = \int\int y \, dF(x, y)$$

$$= \int\int \{y \, dF(y|x)\} \, dF(x)$$

$$= \underset{x \;\; y|x}{E\{ E(y|x)\}} \tag{26.9}$$

provided both sides exist, as we shall now assume. For a counterexample, see Exercise 26.24. Thus, taking expectations on both sides of (26.8) with respect to x, we see that

$$\mu_y = \alpha_2 + \beta_2 \mu_x. \tag{26.10}$$

If we subtract (26.10) from (26.8), multiply both sides by $(x - \mu_x)$ and take expectations again, we find

$$E\{(x - \mu_x)(y - \mu_y)\} = \beta_2 E\{(x - \mu_x)^2\},$$

or

$$\beta_2 = \mu_{11}/\sigma_1^2. \tag{26.11}$$

Similarly, we obtain

$$\beta_1 = \mu_{11}/\sigma_2^2 \tag{26.12}$$

for the coefficient in an exactly linear regression of x on y. (26.11) and (26.12) define the (linear) *regression coefficients** of y on x (β_2) and of x on y (β_1). Using (26.8), (26.10) and (26.11), we have the *linear regression equations*:

$$E(y|x) - \mu_y = \beta_2(x - \mu_x). \tag{26.13}$$

and

$$E(x|y) - \mu_x = \beta_1(y - \mu_y). \tag{26.14}$$

We have already encountered a case of exact linear regressions in our discussion of the bivariate normal distribution in **16.23**. It follows from (26.6), (26.11) and (26.12) that, when both regressions are linear,

$$\beta_1 \beta_2 = \rho^2. \tag{26.15}$$

Example 26.4
The regressions of x^2 and y^2 on each other in Example 26.1 are strictly linear. For, from **16.23**, putting $\sigma_1 = \sigma_2 = 1$ in (16.46), we have

$$\left. \begin{array}{l} E(y|x) = \rho x, \\ \operatorname{var}(y|x) = 1 - \rho^2. \end{array} \right\} \tag{26.16}$$

Thus

$$\begin{aligned} E(y^2|x) &\equiv \operatorname{var}(y|x) + \{E(y|x)\}^2 \\ &= 1 - \rho^2 + \rho^2 x^2. \end{aligned}$$

To each value of x^2 there correspond values $+x$ and $-x$ which occur with equal probability. Thus, since $E(y^2|x)$ is a function of x^2 only,

$$E(y^2|x^2) = \tfrac{1}{2}\{E(y^2|x) + E(y^2|-x)\} = E(y^2|x) = 1 - \rho^2 + \rho^2 x^2,$$

* The notation β_1, β_2 is unconnected with the symbolism for skewness and kurtosis in **3.31–2**; they are unlikely to be confused, since they arise in different contexts.

which we may rewrite, in the form (26.13),

$$E(y^2|x^2) - 1 = \rho^2(x^2 - 1),$$

and the regression of y^2 on x^2 is strictly linear. Similarly for x^2 on y^2. Since we saw in Example 26.1 that $\mu_{11} = 2\rho^2$, and we know that the variances $= 2$, we may confirm from (26.11) and (26.12) that ρ^2 is the regression coefficient in each of the linear regression equations.

Example 26.5
(a) In Example 26.2 it is easily seen that $E(x|y) = E(y|x) = 0$, so that we have linear regressions here, too, the coefficients being zero.

(b) Consider the variables x and y^2 in Example 26.4. We saw there that the regression of y^2 is linear on x^2, with coefficient ρ^2, and it is therefore not linear on x when $\rho \neq 0$. However, since $E(y|x) = \rho x$ we have

$$E(y|x^2) = \tfrac{1}{2}\{E(y|x) + E(y|-x)\} = 0,$$

so that the regression of y on x^2 is linear with regression coefficient zero.

26.10 The reader is asked in Exercise 26.13 to show that the angle between the two regression lines (26.13) and (26.14) is

$$\theta = \arctan\left\{\frac{\sigma_1\sigma_2}{\sigma_1^2 + \sigma_2^2}\left(\frac{1}{\rho} - \rho\right)\right\}, \qquad (26.17)$$

so that as ρ varies over its range from -1 to $+1$, θ increases steadily from 0 to its maximum of $\frac{1}{2}\pi$ when $\rho = 0$, and then decreases steadily to 0 again. Thus, if and only if x and y are in strict linear functional relationship, the two regression lines coincide ($\rho^2 = 1$). If and only if $\rho = 0$, when x and y are said to be *uncorrelated*, the regression lines are at right angles to each other.

Further, it may be shown that

$$\rho^2 = \mathrm{var}\,(\alpha_2 + \beta_2 x)/\sigma_2^2 = \mathrm{var}\,(\alpha_1 + \beta_1 y)/\sigma_1^2; \qquad (26.18)$$

see Exercise 26.23.

The terms 'regression', 'lines of regression' and 'correlation' were first used by Galton in 1886–88 (his 1877 synonym for 'regression' was 'reversion'); 'coefficient of correlation' was first used by Edgeworth in 1892. The term 'correlation' arose naturally from 'co-relation', but 'regression' requires some explanation. Galton found that the average stature of adult offspring increased with parents' stature, but not by as much, and he called this a 'regression to mediocrity'. The term has stuck firmly, but the apparently dramatic result is theoretically trivial, for as we have seen, if $\sigma_1 = \sigma_2$ (as we may reasonably suppose here), then $\beta_1 = \beta_2 = \rho \leq 1$. Thus the same phenomenon would be observed if we interchanged the roles of parents and offspring, although, as noted in **26.2**, there is an essential asymmetry in the variables for the geneticist.

Approximate linear regression: least squares

26.11   Examples 26.4 and 26.5 give instances where one or both regression functions are exactly linear. Suppose now that we are dealing with a joint distribution whose true regression structure is unknown. Although the regression is unlikely to be exactly linear, a linear approximation may be deemed appropriate.

When there are no sampling considerations involved, the choice of a method of fitting is essentially arbitrary. If we are fitting the regression of y on x, it is clearly desirable that in some sense the deviations of the points (y, x) from the fitted line should be small if the line is to represent them adequately. We might consider choosing the line to minimise the sum of the absolute deviations of the points from the line, but this gives rise to the usual mathematical difficulties accompanying an expression involving absolute values. Just as these difficulties lead us to prefer the standard deviation to the mean deviation as a measure of dispersion, they lead us here to propose that the sum of *squares* of the deviations of the points should be minimised.

We have still to determine how the deviations are to be taken: in the y-direction, the x-direction, or by dropping a perpendicular from each point to the line. As we are considering the dependence of y on x, it seems natural to minimise the sum of squared deviations in the y-direction. Thus we are led back to the method of least squares: we choose the 'best-fitting' regression line of y on x,

$$y = \alpha_2 + \beta_2 x,$$

so that the expected value of the squared deviations from the fitted regression line,

$$S = E\{(y - \alpha_2 - \beta_2 x)^2\} \tag{26.19}$$

is minimised. The problem is to determine α_2 and β_2. Differentiating (26.19) with respect to α_2 and β_2, we obtain the pair of equations

$$-\frac{1}{2}\frac{\partial S}{\partial \alpha_2} = E(y) - \alpha_2 - \beta_2 E(x) = 0$$

and

$$-\frac{1}{2}\frac{\partial S}{\partial \beta_2} = E(xy) - \alpha_2 E(x) - \beta_2 E(x^2) = 0.$$

These equations yield

$$\beta_2 = \mu_{11}/\sigma_1^2 \quad \text{and} \quad \alpha_2 = \mu_y - \beta_2\mu_x$$

so that the least squares equation is

$$y - \mu_y = \beta_2(x - \mu_x)$$

as in (26.13). Thus, we may conclude that the evaluation of an approximate regression line by the method of least squares gives results that are the same as those that arise when the regression function is exactly linear. It should be noted that the expectation in (26.19) is taken over both x and y; this is sometimes known as the *unconditional* case. If we wish to consider the conditional distribution of y_i given x_i $(i = 1, \ldots, n)$, we must minimise

$$S = \underset{y|x}{E} \{\sum (y_i - \alpha_2 - \beta_2 x_i)^2\}. \tag{26.20}$$

This results in the parameter values

$$\beta_2 = \frac{E\{\sum (y_i - \bar{y})(x_i - \bar{x})\}}{\sum (x - \bar{x})^2} \tag{26.21}$$

and $\alpha_2 = E(\bar{y}) - \beta_2 \bar{x}$. Natural estimators are then provided by using the observed y-values in place of the random variables, as in (26.22) below.

Sample coefficients

26.12 We now turn to the consideration of sampling problems for correlation and regression coefficients. As usual, we observe the convention that a roman letter (actually italic) represents a sample statistic, the Greek letter being the population equivalent. Thus we write

$$\left.\begin{aligned}
b_1 &= m_{11}/s_2^2 = \frac{1}{n}\sum (x - \bar{x})(y - \bar{y}) \Big/ \frac{1}{n}\sum (y - \bar{y})^2, \\
b_2 &= m_{11}/s_1^2 = \frac{1}{n}\sum (x - \bar{x})(y - \bar{y}) \Big/ \frac{1}{n}\sum (x - \bar{x})^2, \\
r &= m_{11}/(s_1 s_2) = \frac{1}{n}\sum (x - \bar{x})(y - \bar{y}) \Big/ \left\{\frac{1}{n}\sum (x - \bar{x})^2 \frac{1}{n}\sum (y - \bar{y})^2\right\}^{1/2},
\end{aligned}\right\} \tag{26.22}$$

for the sample regression coefficients and correlation coefficient, the summations now being over sample values. For computational purposes these expressions simplify to

$$\left.\begin{aligned}
b_1 &= \frac{\sum xy - (\sum x)(\sum y)/n}{\sum y^2 - (\sum y)^2/n}, \\
b_2 &= \frac{\sum xy - (\sum x)(\sum y)/n}{\sum x^2 - (\sum x)^2/n}, \\
r &= \frac{\sum xy - (\sum x)(\sum y)/n}{[\{\sum x^2 - (\sum x)^2/n\}\{\sum y^2 - (\sum y)^2/n\}]^{1/2}}.
\end{aligned}\right\} \tag{26.23}$$

Just as before, we have $-1 \le r \le +1$.

Example 26.6

For the data in Table 26.1, we find

$$\bar{x} = 63.06, \quad \bar{y} = 132.82,$$
$$s_1^2 = 7.25, \quad s_2^2 = 507.46, \quad \text{and } m_{11} = 19.52$$

so that $r = 0.322$ and the (approximate) linear regression equations are

$$x \text{ on } y: \quad x - 63.06 = 0.0385(y - 132.82) \text{ or}$$
$$x = 0.0385y + 57.95,$$
$$y \text{ on } x: \quad y - 132.82 = 2.692(x - 63.06) \text{ or}$$
$$y = 2.692x - 36.96.$$

These lines appear in Figure 26.1 (page 971) as RR' and CC', respectively.

> As is evident from Figure 26.1, the same regression lines would result if each single observation were replaced by the group mean of all observations with the same x-value. Any method of grouping the observations that is based only upon the values of the independent variables will continue to provide unbiased estimators for the regression coefficients. However, the correlation coefficient lacks such invariance and increases in absolute value as the level of aggregation increases; see Exercise 26.26. The misuse of correlations derived from aggregated data to represent the correlation for individuals is sometimes known as the 'fallacy of ecological correlation'.

Example 26.7

Table 26.3 shows the yields of wheat and of potatoes in 48 counties of England in 1936. We find that

$$\sum x = 758.0, \qquad \sum y = 291.1,$$
$$\sum x^2 = 12\,170.48, \qquad \sum y^2 = 1791.03, \qquad \sum xy = 4612.64.$$

Hence $r = 0.219$ and the (approximate) linear regression equations are:

x on y: $x - 15.792 = 0.612(y - 6.065)$
y on x: $y - 6.065 = 0.078(x - 15.792)$

The data and the regression lines are shown diagrammatically in Fig. 26.3, known as a *scatter plot* or *scatter diagram*; its use is strongly recommended, since it conveys quickly and simply an idea of the adequacy of the fitted regression lines (not very

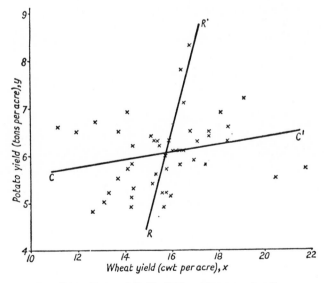

Fig. 26.3   Data of Table 26.3, with regression lines

good in our example). Indeed, a scatter diagram, plotted in advance of the analysis, will often make it clear whether the fitting of regression lines is worth while.

Standard errors

26.13 The standard errors of the coefficients (26.22) are easily obtained. In fact, we have already obtained the large-sample variance of r in Example 10.6, where we saw that it is, in general, an expression involving all the second-order and fourth-order moments of the population sampled. In the binormal case, however, we found that it simplified to

$$\operatorname{var} r = (1 - \rho^2)^2 / n, \tag{26.24}$$

though (26.24) is of little value in practice since the distribution of r tends to normality so slowly (cf. **16.29**): it is unwise to use it for $n < 500$. The difficulty is of no practical importance, since, as we saw in **16.33**, the simple transformation of r,

$$z = \tfrac{1}{2} \log \left(\frac{1+r}{1-r} \right) = ar \tanh r, \tag{26.25}$$

is for normal samples much more closely normally distributed with approximate mean

$$E(z) \doteq \tfrac{1}{2} \log \left(\frac{1+\rho}{1-\rho} \right) \tag{26.26}$$

and variance approximately

$$\operatorname{var} z \doteq \frac{1}{n-3}, \tag{26.27}$$

independent of ρ. For $n > 50$, the use of this standard error for z is adequate; closer approximations are given in **16.33**.

> Subrahmaniam *et al.* (1981) give series expansions for the distribution of z and other transformations of r.

For the sample regression coefficient of y on x,

$$b_2 = m_{11} / s_1^2,$$

the use of (10.17) gives, just as in Example 10.6 for r, the large sample approximation

$$\operatorname{var} b_2 = \left(\frac{\mu_{11}}{\sigma_1^2} \right)^2 \left\{ \frac{\operatorname{var} m_{11}}{\mu_{11}^2} + \frac{\operatorname{var} (s_1^2)}{\sigma_1^4} - \frac{2 \operatorname{cov} (m_{11}, s_1^2)}{\mu_{11} \sigma_1^2} \right\}.$$

Substituting for the variances and covariance from (10.23) and (10.24), this becomes

$$\operatorname{var} b_2 = \frac{1}{n} \left(\frac{\mu_{11}}{\sigma_1^2} \right)^2 \left\{ \frac{\mu_{22}}{\mu_{11}^2} + \frac{\mu_{40}}{\sigma_1^4} - \frac{2\mu_{31}}{\mu_{11} \sigma_1^2} \right\}. \tag{26.28}$$

Table 26.3   Yields of wheat and potatoes in 48 counties in England in 1936

County	Wheat (cwt per acre) x	Potatoes (tons per acre) y	County	Wheat (cwt per acre) x	Potatoes (tons per acre) y
Bedfordshire	16.0	5.3	Northamptonshire	14.3	4.9
Huntingdonshire	16.0	6.6	Peterborough	14.4	5.6
Cambridgeshire	16.4	6.1	Buckinghamshire	15.2	6.4
Ely	20.5	5.5	Oxfordshire	14.1	6.9
Suffolk, West	18.2	6.9	Warwickshire	15.4	5.6
Suffolk, East	16.3	6.1	Shropshire	16.5	6.1
Essex	17.7	6.4	Worcestershire	14.2	5.7
Hertfordshire	15.3	6.3	Gloucestershire	13.2	5.0
Middlesex	16.5	7.8	Wiltshire	13.8	6.5
Norfolk	16.9	8.3	Herefordshire	14.4	6.2
Lincs (Holland)	21.8	5.7	Somersetshire	13.4	5.2
Lincs (Kesteven)	15.5	6.2	Dorsetshire	11.2	6.6
Lincs (Lindsey)	15.8	6.0	Devonshire	14.4	5.8
Yorkshire (East Riding)	16.1	6.1	Cornwall	15.4	6.3
Kent	18.5	6.6	Northumberland	18.5	6.3
Surrey	12.7	4.8	Durham	16.4	5.8
Sussex, East	15.7	4.9	Yorkshire (North Riding)	17.0	5.9
Sussex, West	14.3	5.1	Yorkshire (West Riding)	16.9	6.5
Berkshire	13.8	5.5	Cumberland	17.5	5.8
Hampshire	12.8	6.7	Westmorland	15.8	5.7
Isle of Wight	12.0	6.5	Lancashire	19.2	7.2
Nottinghamshire	15.6	5.2	Cheshire	17.7	6.5
Leicestershire	15.8	5.2	Derbyshire	15.2	5.4
Rutland	16.6	7.1	Staffordshire	17.1	6.3

For a binormal population, we substitute the relations of Example 3.19 and obtain

$$\operatorname{var} b_2 = \frac{1}{n}\left(\frac{\mu_{11}}{\sigma_1^2}\right)^2 \left\{ \frac{(1+2\rho^2)\sigma_1^2\sigma_2^2}{\mu_{11}^2} + \frac{3\sigma_1^4}{\sigma_1^4} - \frac{6\rho\sigma_1^3\sigma_2}{\mu_{11}\sigma_1^2} \right\}$$

$$= \frac{1}{n}\left(\frac{\mu_{11}}{\sigma_1^2}\right)^2 \left\{ \frac{\sigma_1^2\sigma_2^2}{\mu_{11}^2} - 1 \right\}$$

$$= \frac{1}{n}\frac{\sigma_2^2}{\sigma_1^2}(1-\rho^2). \tag{26.29}$$

Similarly, for the regression coefficient of x on y,

$$\operatorname{var} b_1 = \frac{1}{n}\frac{\sigma_1^2}{\sigma_2^2}(1-\rho^2). \tag{26.30}$$

The expressions (26.29) and (26.30) are rather more useful for standard error purposes (when, of course, we substitute s_1^2, s_2^2 and r for σ_1^2, σ_2^2 and ρ in them) than (26.24), for we saw at **16.35** that the exact distribution of b_2 is symmetrical about β_2: it is left to the reader as Exercise 26.9 to show from (16.86) that (26.29) is exact when multiplied by a factor $n/(n-3)$, and that the distribution of b_2 tends to normality rapidly, its measure of kurtosis being of order $1/n$.

> As noted in **16.37**, this result may also be derived by a conditional argument applied to the linear model of **19.4**.

The estimation of ρ in normal samples

26.14   The sampling theory of the binormal distribution was developed in **16.23–36**: we may now discuss, in particular, the problem of estimating ρ from a sample, in the light of our results in the theory of estimation (Chapters 17–18).

In Example 18.14 we saw that the likelihood function contains the observations only in the form of the five statistics $\bar{x}, \bar{y}, s_1^2, s_2^2, r$. These are therefore a set of sufficient statistics for the five parameters $\mu_1, \mu_2, \sigma_1^2, \sigma_2^2, \rho$, and their distribution is complete by **22.10**. When the other four parameters are known we have the pair of sufficient statistics for ρ,

$$\left\{ \Sigma (x-\mu_1)(y-\mu_2), \frac{\Sigma (x-\mu_1)^2}{\sigma_1^2} + \frac{\Sigma (y-\mu_2)^2}{\sigma_2^2} \right\}.$$

In Example 18.14 we saw that the maximum likelihood estimator of ρ takes a different form according to which other parameters, if any, are being simultaneously estimated: the ML estimator is always a function of the set of five sufficient statistics, but it is a *different* function in different situations. When ρ alone is being estimated, the ML estimator is the root of a cubic equation (Example 18.3); when all five parameters are being estimated, the ML estimator is the sample correlation coefficient r (Example 18.14). In practice, the latter is by far the most common case, and we therefore now consider the estimation of functions of ρ by functions of r.

26.15   The exact distribution of r, which depends only upon ρ, is given by (16.61) or, more conveniently, by (16.66). Its mean value is given by (16.73). Expanding the gamma functions in (16.73) by Stirling's series (3.64) and taking the two leading terms of the hypergeometric function, we find, as stated there,

$$E(r) = \rho \left\{ 1 - \frac{(1-\rho^2)}{2n} + O(n^{-2}) \right\}.   \tag{26.31}$$

Thus r is a slightly biased estimator of ρ when $0 \neq \rho^2 \neq 1$. The bias is generally small, but it is interesting to inquire whether it can be removed.

26.16   We may approach the problem in two ways. First, we may ask: is there a function $g(r)$ such that

$$E\{g(r)\} = g(\rho)   \tag{26.32}$$

holds identically in ρ? Under certain regularity conditions Hotelling (1953) showed that if g is not dependent on n, $g(r)$ could only be a linear function of arc sin r, and Harley (1956–7) showed that in fact

$$E(\text{arc sin } r) = \text{arc sin } \rho, \tag{26.33}$$

a simple proof of Harley's result being given by Daniels and Kendall (1958). If we relax (26.32) to

$$E\{a(n)g(r) + b(n)\} = g(\rho)$$

it is satisfied by $g(r) = r^2/(1 - r^2)$; the result is given in Example 16.15 and at (27.89) below.

26.17   The second, more direct, approach is to seek a function of r unbiased for ρ itself. Since r is a function of a set of complete sufficient statistics, the unbiased function of r must be unique (cf. 22.9). Olkin and Pratt (1958) found the unbiased estimator of ρ, say r_u, to be the hypergeometric function

$$r_u = rF[\tfrac{1}{2}, \tfrac{1}{2}, \tfrac{1}{2}(n-2), (1-r^2)] \tag{26.34}$$

which, expanded as a series, gives

$$r_u = r\left\{1 + \frac{1-r^2}{2(n-2)} + \frac{9(1-r^2)^2}{8n(n-2)} + O(n^{-3})\right\}. \tag{26.35}$$

No term in the series is negative, so that

$$|r_u| \geq |r|,$$

the equality holding only if $r^2 = 0$ or 1. Since $F(\tfrac{1}{2}, \tfrac{1}{2}, \tfrac{1}{2}(n-2), 0) = 1$ and r_u is an increasing function of r, we have $0 \leq r^2 \leq r_u^2 \leq 1$.

Evidently, the first correction term in (26.35) is counteracting the downward bias of the term in $1/n$ in (26.31). Olkin and Pratt recommend the use of the two-term expansion

$$r_u^* = r\left\{1 + \frac{1-r^2}{2(n-4)}\right\}. \tag{26.36}$$

The term in braces in (26.36) gives r_u/r within 0.01 for $n \geq 8$ and within 0.001 for $n \geq 18$, uniformly in r.

Olkin and Pratt give exact tables of r_u for $n = 2(2)30$ and $|r| = 0(0.1)1$ which show that for $n \geq 14$, $|r_u|$ never exceeds $|r|$ by more than 5 per cent.

Finally, we note that as n appears only in the denominators of the hypergeometric series, $r_u \rightarrow r$ as $n \rightarrow \infty$, so that it has the same limiting distribution as r, namely a normal distribution with mean ρ and variance $(1 - \rho^2)^2/n$. A similar analysis of the estimation of ρ^2 by r^2 is to be found in 27.34 below.

Confidence intervals and tests for ρ

26.18 For testing that $\rho = 0$, the tests based on r are UMPU; this is not so when we are testing a non-zero value of ρ. However, if we confine ourselves to test statistics which are invariant under changes in location and scale, one-sided tests based on r are UMP invariant, as Lehmann (1959) shows.

Alternatively, we may make an approximate test using Fisher's z-transformation, the simplest results for which are given in **26.13**: to test a hypothetical value of ρ we compute (26.25) and test that it is normally distributed about (26.26) with variance (26.27). A one- or two-tailed test is appropriate according to whether the alternative to this simple hypothesis is one- or two-sided.

For interval estimation purposes, we may use F. N. David's (1938) charts or Odeh's tables, described in **20.34**, or Subrahmaniam and Subrahmaniam (1983), which extend up to $n = 50$. By the duality between confidence intervals and tests remarked in **21.9**, these charts and tables may also be used to read off the values of ρ to be rejected by a size-α test, i.e. all values of ρ not covered by the confidence interval for that α. F. N. David (1937) has shown that this test is slightly biased (and the confidence intervals correspondingly so). This may most easily be seen from the standpoint of the z-transformation standard-error test given in **26.13**: if the latter were exact, and z were *exactly* normal with variance independent of ρ, the test of ρ would simply be a test of the value of the mean of a normal distribution with known variance, and we know from Example 22.11 that if we use an equal-tails test for this hypothesis, it is unbiased. Since z is a one-to-one function of r, the equal-tails test on r would then also be unbiased. Thus the slight bias in the equal-tails r-test may be regarded as a reflection of the approximate nature of the z-transformation.

Exercise 26.15 shows that the LR test is based on r, but is not equal-tails except when testing $\rho = 0$.

26.19 We may use the z-transformation to test the composite hypothesis that the correlation parameters of two independently sampled bivariate normal populations are the same. For if so, the two transformed statistics z_1, z_2 will each be distributed as in **26.13**, and $(z_1 - z_2)$ will have zero mean and variance $1/(n_1 - 3) + 1/(n_2 - 3)$, where n_1 and n_2 are the sample sizes. Exercises 26.19–21 show that $(z_1 - z_2)$ is exactly the likelihood ratio statistic when $n_1 = n_2$, and approximately so when $n_1 \neq n_2$. In either case, however, the test is approximate, being a standard-error test.

The more general composite hypothesis, that the two correlation parameters ρ_1, ρ_2 differ by an amount Δ, cannot be tested in this way. For then

$$E(z_1 - z_2) = \tfrac{1}{2}\log\left(\frac{1+\rho_1}{1-\rho_1}\right) - \tfrac{1}{2}\log\left(\frac{1+\rho_2}{1-\rho_2}\right) = \tfrac{1}{2}\log\left\{\left(\frac{1+\rho_1}{1-\rho_1}\right)\left(\frac{1-\rho_2}{1+\rho_2}\right)\right\}$$

is not a function of $|\rho_1 - \rho_2|$ alone. The z-transformation could be used to test

$$H_0: \frac{1+\rho_1}{1-\rho_1} = a\left(\frac{1+\rho_2}{1-\rho_2}\right)$$

for any constant a, but this is not a hypothesis of interest. Except in the large-sample case, when we may use standard errors, there seems to be no way of testing $H_0 : \rho_1 - \rho_2 = \Delta$, for the exact distribution of the difference $r_1 - r_2$ has not been investigated.

Dunn and Clark (1969, 1971) examine methods of testing the equality of correlation parameters when the sample coefficients are not independent. See also Elston (1975).

Tests of independence and regression tests

26.20   In the particular case when we wish to test $\rho = 0$, i.e. the *independence* of the normal variables, we may use the exact result of **16.28**, that

$$t = \{(n-2)r^2/(1-r^2)\}^{1/2} \tag{26.37}$$

is distributed as Student's t-distribution with $(n-2)$ degrees of freedom. t^2 is essentially the LR test statistic for $\rho = 0$—cf. Exercise 26.15—and this is equivalent to an equal-tails test on t.

Essentially, we are testing here that μ_{11} of the population is zero, and clearly this implies that the population regression coefficients β_1, β_2 are zero. Now in **16.36**, we showed that

$$t = (b_2 - \beta_2) \left\{ \frac{s_1^2(n-2)}{s_2^2(1-r^2)} \right\}^{1/2} \tag{26.38}$$

has Student's distribution with $(n-2)$ degrees of freedom. When $\beta_2 = 0$, (26.38) is seen to be identical with (26.37). Thus the test of independence may be regarded as a test that a regression coefficient is zero, a special case of the general test (26.38) which we use for hypotheses concerning β_2. It will be noted that the exact test for any hypothetical value of β_2 is of a much simpler form than that for ρ.

Tests of independence can be made without assuming that the random variables are normally distributed, as we shall see in **26.22**.

Other measures of correlation

26.21   Hitherto, we have restricted attention to the linear correlation coefficient r defined in (26.22). However, with respect to tests for independence, a wide variety of possibilities is available. Daniels (1944) defined a class of correlation coefficients by the expression

$$r_D = \frac{\Sigma_{(2)} a_{ij} b_{ij}}{\{\Sigma_{(2)} a_{ij}^2 \Sigma_{(2)} b_{ij}^2\}^{1/2}}, \tag{26.39}$$

where $\Sigma_{(2)}$ denotes $\sum_{i=1}^{n} \sum_{j=1}^{n}$ and a_{ij}, b_{ij} depend on (x_i, x_j) and (y_i, y_j), respectively. The product moment correlation r in (26.22) is a special case of (26.39) when $a_{ij} = x_i - x_j$, $b_{ij} = y_i - y_j$. Rather than base the (a_{ij}, b_{ij}) values on the actual observations, we could use measures based on the *ranks*. That is, if the sample values are ordered so that (ignoring ties for the moment)

$$x_{(1)} < x_{(2)} < \cdots < x_{(n)}$$

with the y-values being ordered as

$$y_{(1)} < y_{(2)} < \cdots < y_{(n)},$$

we define the ranks of the pair (x_i, y_i) by their relative positions within these two sets of order statistics

$$\text{rank}\,(x_{(i)}) = i, \quad \text{rank}\,(y_{(j)}) = j$$

or

$$\text{rank}\,(x_i) = A_i, \quad \text{rank}\,(y_i) = B_i, \tag{26.40}$$

where $A_i = j$ if $x_i = x_{(j)}$ and B_i is similarly defined. Clearly, (A_i, B_i) are not restricted to the ranks as any set of coefficients might be employed in (26.40) corresponding to the pattern of dependence it is desired to detect. We leave the detailed discussion of such statistics for a later volume, providing only a few examples of principal alternatives to r in (26.22).

For a more general discussion of dependence measures, see Kent (1983).

Permutation distributions

26.22 We now restrict attention to the special case of (26.39)

$$r = \sum A_i B_i / \{\sum A_i^2 \sum B_i^2\}^{1/2}, \tag{26.41}$$

where A_i and B_i are adjusted so that $\sum A_i = \sum B_i = 0$, and all summations are taken over $i = 1, \ldots, n$. Under the hypothesis that the two random variables are independent, we may regard the $n!$ distinct assignments of $\{B_{(i_1)}, B_{(i_2)}, \ldots, B_{(i_n)}\}$ to $\{A_1, \ldots, A_n\}$ as equally likely; note that there is no loss of generality in fixing the order of one set of coefficients and then permuting the other. It follows directly that the denominator of (26.41) is invariant under these permutations so that the expected value of the numerator, taken with respect to the set of $n!$ possible assignments (denoted by E_P), is

$$E_P(\sum A_i B_i) = \sum A_i \cdot \frac{1}{n} \{B_1 + \cdots + B_n\}$$
$$= n^{-1}(\sum A_i)(\sum B_i) = 0.$$

It follows immediately that

$$E_P(r) = 0. \tag{26.42}$$

Similarly, keeping the order of the As fixed, we have that the variance of $\sum A_i B_i$ over the set of random permutations is

$$\text{var}_P\,(\sum A_i B_i) = \sum A_i^2 \,\text{var}_P\,(\tilde{B}_i) + \sum\sum_{i \neq j} A_i A_j \,\text{cov}_P\,(\tilde{B}_i, \tilde{B}_j), \tag{26.43}$$

where the random variable $\tilde{B}_i$ describes the coefficient assigned to the ith pair so that

$$P(\tilde{B}_i = B_k) = \frac{1}{n}, \qquad k = 1, \ldots, n$$

$$P(\tilde{B}_i = B_k, \tilde{B}_j = B_l) = 1/\{n(n-1)\}, \quad k, l = 1, \ldots, n; \quad k \neq l$$

and so on. It follows that

$$\text{var}_P \left(\sum A_i B_i \right) = S_{AA} S_{BB}/(n-1),$$

where $S_{AA} = \sum A_i^2$ and $S_{BB} = \sum B_i^2$. Thus, (26.43) leads to

$$\text{var}_P (r) = 1/(n-1), \qquad (26.44)$$

regardless of the actual values of the coefficients (A_i, B_i).

26.23 We have seen at (26.42) and (26.44) that the first two moments of r are independent of the actual values of (x, y) observed. By similar methods, it may be shown that

$$\left. \begin{array}{l} E_P(r^3) = \dfrac{n-2}{n(n-1)^2} \left(\dfrac{k_3}{k_2^{3/2}} \right) \left(\dfrac{k_3'}{(k_2')^{3/2}} \right) \\[3mm] E_P(r^4) = \dfrac{3}{(n^2-1)} \left\{ 1 + \dfrac{(n-2)(n-3)}{3n(n-1)^2} \left(\dfrac{k_4}{k_2^2} \right) \left(\dfrac{k_4'}{(k_2')^2} \right) \right\}, \end{array} \right\} \qquad (26.45)$$

where the ks are the k-statistics of the observed xs and the k's the k-statistics of the ys. Neglecting the differences between k-statistics and sample cumulants, we may rewrite (26.41) as

$$\left. \begin{array}{l} E(r^3) \doteqdot \dfrac{(n-2)}{n(n-1)^2} g_1 g_1' \\[3mm] E(r^4) \doteqdot \dfrac{3}{(n^2-1)} \left\{ 1 + \dfrac{(n-2)(n-3)}{3n(n-1)^2} g_2 g_2' \right\}, \end{array} \right\} \qquad (26.46)$$

where g_1, g_2 are the measures of skewness and kurtosis of the xs, and g_1', g_2' those of the ys. If these are fixed, the skewness and kurtosis measures for r follow from (25.26) as

$$\gamma_1(r) \doteqdot (g_1 g_1')/n \qquad (26.47)$$

and

$$\gamma_2(r) \doteqdot (6 + g_2 g_2')/n.$$

Thus, as $n \to \infty$, we have approximately

$$E(r^3) = 0$$

$$E(r^4) = \dfrac{3}{n^2-1}. \qquad (26.48)$$

The moments (26.42), (26.44), and (26.48) are precisely those of (16.62), the exact distribution of r in samples from a bivariate normal distribution with $\rho = 0$ as may be verified by evaluating the expected values of r^2 and r^4 for the f.f. given in (16.62). Thus, to a close approximation the permutation distribution of r is also

$$f(r) = \dfrac{1}{B\{\tfrac{1}{2}(n-2), \tfrac{1}{2}\}} (1-r^2)^{(n-4)/2}, \qquad -1 \leqslant r \leqslant 1, \qquad (26.49)$$

and we may therefore use (26.49), or equivalently the fact that $t = \{(n-2)r^2/(1-r^2)\}^{1/2}$ has a Student's distribution with $n-2$ degrees of freedom, to carry out our tests. The approximation is accurate even for small n, as we might guess from the agreement of its first two moments with those of the permutation distribution.

> The convergence of the permutation and normal-theory distributions to a common limiting normal distribution has been rigorously proved by Hoeffding (1952).

Rank correlation coefficients

26.24   Returning to the original discussion on the use of ranks in **26.21**, we may assign $A_i = i - \frac{1}{2}(n+1)$ by ordering the x-values so that (26.41) becomes

$$r_s = \sum_{i=1}^{n} R_{yi}\{i - \tfrac{1}{2}(n+1)\}/\{\tfrac{1}{12}n(n^2-1)\}, \tag{26.50}$$

where R_{yi} denotes the rank of the y-value in the ith pair and

$$\sum i = \sum R_{yi} = n(n+1)/2$$
$$S_{AA} = S_{BB} = n(n^2-1)/12$$

when ranks are used. Further, since

$$\sum_{i=1}^{n} iR_{yi} = \tfrac{1}{6}n(n+1)(2n+1) - \tfrac{1}{2}\sum_{i=1}^{n}(i-R_{yi})^2,$$

(26.50) becomes

$$r_s = 1 - \frac{6}{n(n^2-1)}\sum_{i=1}^{n}(i-R_{yi})^2. \tag{26.51}$$

This is usually known as Spearman's rank correlation coefficient, after the eminent psychologist who first introduced it in 1906. It follows from (26.42), (26.44), and (26.48) that, under independence,

$$E(r_s) = E(r_s^3) = 0$$
$$\mathrm{var}\,(r_s) = (n-1)^{-1}$$

and

$$E(r_s^4) = \frac{3}{n^2-1}\left\{1 + \frac{12(n-2)(n-3)}{25n(n-1)^2}\right\}. \tag{26.52}$$

> Kendall and Gibbons (1990) give tables of the frequency function of $\sum(i-R_{yi})^2$ for $n = 4(1)10$; the tail values are reproduced in the Biometrika tables. Owen (1973) gives tables for $n = 11$ and Otten (1973) for $n = 12(1)16$. Zar (1972) gives tables based on an intermediate (Pearson type II) approximation for $n = 12(1)50(2)100$. The approximation afforded by (26.49) is adequate for $n \geqslant 30$ and test size α down to 0.005.

26.25   The coefficient r_s is but one of many possible rank coefficients, selected primarily for its simplicity. Daniels (1948) showed that all such coefficients, defined by (26.39), are essentially coefficients of disarray, in the sense that if a pair of values of y are interchanged to bring them into the same order as the corresponding values of x, the value of any coefficient of this class will increase. We now consider the question of measuring disarray among the ranks of x and y directly.

Suppose that the ranks of x are in their natural order $1, 2, \ldots, n$ and that the corresponding ranks of y are $R_{y1}, \ldots, R_{yn}$, a permutation of $1, 2, \ldots, n$. A natural method of measuring the disarray of the y-ranks, i.e. the extent of their departure from the order $1, 2, \ldots, n$, is to count the number of inversions of order among them. For example, in the x-ranking 3214 for $n = 4$, there are 3 inversions of order, namely 3–2, 3–1, 2–1. The number of such inversions, which we shall call Q, may range from 0 to $\frac{1}{2}n(n-1)$, these limits being reached respectively when the y-ranking is $1, 2, \ldots, n$ and $n, n-1, \ldots, 1$. We therefore define a coefficient

$$t = 1 - \frac{4Q}{n(n-1)}, \qquad (26.53)$$

which is symmetrically distributed on the range $[-1, +1]$ over the $n!$ equiprobable permutations, and therefore has expectation 0 when $\rho = 0$.

The coefficient (26.53) has been discussed by several early writers (Fechner, Lipps) around the year 1900 and subsequently by several other writers, notably Lindeberg, in the 1920s (historical details are given by Kruskal (1958)), but first became widely used after a series of papers by M. G. Kendall starting in 1938 and consolidated in a monograph (Kendall, 1962; latest edition, Kendall and Gibbons, 1990). It is usually known as Kendall's rank correlation coefficient.

26.26 The statistic t may be represented in the form of (26.39) if we set

$$a_{ij} = 1 \text{ when } x_j > x_i, \quad = 0 \text{ otherwise}$$
$$b_{ij} = 1 \text{ when } y_j > y_i, \quad = 0 \text{ otherwise.}$$

Thus, the moments of t under the set of random permutations may be obtained; it follows that, under independence,

$$E(t) = 0, \quad \text{var}(t) = \frac{2(2n+5)}{9n(n-1)} \qquad (26.54)$$

and that the distribution of t tends to normality rapidly with mean and variance given by (26.54). Kendall and Gibbons (1990) give the exact distribution for $n = 4(1)10$. Beyond this point, the asymptotic normal distribution may be used with little loss of accuracy.

Intraclass correlation

26.27 On occasion, we may wish to measure the similarity between individuals that share common membership of some class, such as individuals within a family or persons in the same neighbourhood. It is natural to consider all pairs within each class, but there is no natural ordering in the pair. Consider, for example, the heights of brothers within a family; any attempt to order the individuals by age, size, or any other criterion would lead to a potentially different concept of association than the symmetric within-class measure we are seeking to define. To resolve the problem, we define *intraclass* correlation directly: suppose there are p classes with $k_1, \ldots, k_p$ members and the random variable y_{ij} refers to the jth member of the ith class. We assume that

$$E(y_{ij}) = \mu, \qquad \text{var}(y_{ij}) = \sigma^2$$

and

$$\text{cov}(y_{ij}, y_{i'j'}) = \rho_I \sigma^2, \quad i = i', j \neq j'$$
$$= 0, \qquad i \neq i', \tag{26.55}$$

ρ_I is the intraclass correlation coefficient.

26.28 When $k_i = k$, for all i, and the variates are normally distributed, it follows that the ML estimators for the parameters in (26.55) lead to

$$\hat{\mu} = \sum_i \sum_j y_{ij}/kp = \sum_i \bar{y}_{i\cdot}/p = \bar{y}_{\cdot\cdot},$$

$$S_W^2 = \sum_i \sum_j (y_{ij} - \bar{y}_{i\cdot})^2/p(k-1) \tag{26.56}$$

$$S_B^2 = k \sum_i (\bar{y}_{i\cdot} - \bar{y}_{\cdot\cdot})^2/(p-1), \tag{26.57}$$

where

$$E(S_W^2) = \sigma^2(1 - \rho_I) \tag{26.58}$$

and

$$E(S_B^2) = \sigma^2\{1 + (k-1)\rho_I\}. \tag{26.59}$$

The ratio $F = S_B^2/S_W^2$ has Fisher's F distribution (see **16.15–22**) with $\{p-1, p(k-1)\}$ d.fr. when $\rho_I = 0$; its analogues in the analysis of variance are considered in Chapter 29. For the present, we note that (26.56)–(26.59) suggest the estimator

$$r_I = (F-1)/(k-1+F). \tag{26.60}$$

It follows immediately from (26.60) that $-(k-1)^{-1} \leq r_I \leq 1$.

26.29 A confidence interval for ρ_I is readily constructed using the F distribution. Starting from the statement

$$P\{F_L \leq F^* \leq F_U\} = 1 - \alpha,$$

where $F^* = F(1 - \rho_I)/\{1 + (k-1)\rho_I\}$ we obtain the $100(1-\alpha)$ per cent confidence interval

$$\frac{F - F_U}{F + (k-1)F_U} \leq \rho_I \leq \frac{F - F_L}{F + (k-1)F_L}. \tag{26.61}$$

These results were first obtained by Fisher (1921c); see Exercise 26.14.

26.30 When the class sizes vary, the form of the estimators becomes more involved; see Shoukri and Ward (1984). However, an approximate analysis is still available using the F-statistic with

$$S_W^2 = \sum_i \sum_j (y_{ij} - \bar{y}_{i.})^2 / \sum_i (k_i - 1) \tag{26.62}$$

$$S_B^2 = \sum_i k_i (\bar{y}_{i.} - \bar{y}_{..})^2 / (p - 1) \tag{26.63}$$

and $\bar{y}_{..} = \sum_i \sum_j y_{ij} / \sum k_i$. Then $E(S_W^2)$ is unchanged, but

$$E(S_B^2) = \sigma^2 \{1 + \rho_I (g - 1)\}, \tag{26.64}$$

where

$$g = \bar{k} - \frac{\sum (k_i - \bar{k})^2}{p\bar{k}(p - 1)}, \quad p\bar{k} = \sum k_i.$$

The intraclass correlation is then defined as in (26.60) with g replacing k. Exact confidence intervals are not available for unequal k; various approximations are considered by Donner and Wells (1986). A general review of inference procedures is given by Donner (1986).

  The intraclass correlation coefficient arises naturally in the one-way random effects model in the analysis of variance, which will be discussed in a later volume. Yet, it is remarkable that the coefficient and its basic theory were developed well before that model was introduced. Shirahata (1981) considers rank-based intraclass correlations.

Tetrachoric correlation

26.31 We now discuss the estimation of ρ in a binormal population when the data are not given in full. We take first of all an extreme case exemplified by Table 26.4. This is based on the data for cows according to age and milk yield given in Table 1.24, Exercise 1.4. Suppose that, instead of being given that table we had only Table 26.4.

Table 26.4 Cows by age and milk yield

	Age 6 and over	Age 3–5 years	Total
Yield 8–18 galls.	1078	1407	2485
Yield 19 galls. and over	1546	881	2427
Total	2624	2288	4912

If we assume that the underlying distribution is binormal, how can we estimate ρ from this table? In general, consider a 2×2 table with frequencies

$$
\begin{array}{ll}
a & b \\
c & d
\end{array}
\quad
\begin{array}{l}
a + b \\
c + d
\end{array}
\tag{26.65}
$$

$$a + c \qquad b + d \qquad a + b + c + d = n$$

In (26.65) we shall always take d to be such that neither of its marginal totals contains the median value.

If this table is derived by a double dichotomy of the binormal distribution we can find h' such that

$$\int_{-\infty}^{h'} \int_{-\infty}^{\infty} f(x, y)\, dx\, dy = \frac{a+c}{n}. \tag{26.66}$$

Putting $h = (h' - \mu_1)/\sigma_1$, we find this is

$$F(h) = (2\pi)^{-1/2} \int_{-\infty}^{h} \exp\left(-\tfrac{1}{2}x^2\right) dx = \frac{a+c}{n}, \tag{26.67}$$

so that h may be found from tables of the univariate normal distribution function. Likewise k is given by

$$F(k) = (2\pi)^{-1/2} \int_{-\infty}^{k} \exp\left(-\tfrac{1}{2}y^2\right) dy = \frac{a+b}{n}. \tag{26.68}$$

The arrangement of the table in (26.65) ensures that h and k are never negative.

Having fitted univariate normal distributions to the marginal frequencies of the table in this way, we now estimate ρ using

$$\frac{d}{n} = \int_{h}^{\infty} \int_{k}^{\infty} \frac{1}{2\pi(1-\rho^2)^{1/2}} \exp\left\{ \frac{-1}{2(1-\rho^2)} (x^2 - 2\rho xy + y^2) \right\} dx\, dy. \tag{26.69}$$

After some algebra, we find that (26.69) may be expressed in terms of the tetrachoric functions which were defined at (6.44) for the purpose,

$$\frac{d}{n} = \sum_{j=0}^{\infty} \rho^j \tau_j(h) \tau_j(k), \tag{26.70}$$

where $\tau_0(x) = 1 - F(x)$. The reader is asked to derive this result as Exercise 26.27.

26.32 Formally, (26.70) provides a soluble equation for ρ, but in practice the solution by successive approximation can be very tedious. The series (26.70) always converges, but may do so slowly. It is simpler to interpolate in tables which have been prepared giving the integral d/n in terms of ρ for various values of h and k (*Tables for Statisticians and Biometricians*, Vol. 2).

The estimate of ρ derived from a sample of n in this way is known as *tetrachoric r* and is due to K. Pearson. We shall denote it by r_t.

Example 26.8

For the data of Table 26.4 we find the solution of $F(h) = 2624/4912 = 0.5342$ is $h = 0.086$, and similarly $F(k) = 2485/4912 = 0.5059$ yields $k = 0.015$. We have also for d/n the value $881/4912 = 0.1794$.

From the tables, we find for varying values of h, k and ρ the following values of d:

		$h=0$	$h=0.1$			$h=0$	$h=0.1$
$\rho = -0.30$	$k=0$	0.2015	0.1818	$\rho = -0.35$	$k=0$	0.1931	0.1735
	$k=0.1$	0.1818	0.1639		$k=0.1$	0.1735	0.1555

Linear interpolation gives us for $h = 0.086$, $k = 0.015$, the result $\rho = -0.32$ approximately. In Table 26.4, we have inverted the order of columns, and taking account of this gives us an estimate of $\rho = +0.32$. We therefore write $r_t = +0.32$. (The product-moment coefficient for Table 1.24 is $r = 0.22$.)

26.33 Tetrachoric r_t has been used mainly by psychologists, whose material is often of the 2×2 type. Karl Pearson (1913) gave an asymptotic expression for its standard error.

Since the estimation procedure equates observed frequencies with the corresponding probabilities (say, $\boldsymbol{\theta}$) in the bivariate normal distribution, it is using the ML estimator $\hat{\boldsymbol{\theta}}$. Now the ML estimator of ρ, which is a function $\rho(\boldsymbol{\theta})$, is $\rho(\hat{\boldsymbol{\theta}})$. Thus r_t is the ML estimator of ρ from a 2×2 table. Its approximate large-sample variance, obtained from **18.18** by Hamdan (1970), is

$$\left\{ n^2 \left(\frac{1}{a} + \frac{1}{b} + \frac{1}{c} + \frac{1}{d} \right) f^2(h, k \mid r_t) \right\}^{-1},$$

where $f(x, y \mid \rho)$ is the standardised bivariate normal distribution.

> Brown (1977) gives a computer algorithm for r_t and its standard error. Digby (1983) suggests various simplifying approximations.

Biserial correlation

26.34 We now consider a $2 \times q$ table for which the q x-values are ordered. Further, the y-variable is coded as $Y = 1$ or 2 according as $y <$ or $> k_0$, an unknown point of dichotomy for y. Since the regression of x on y is also linear, we have from (26.14)

$$E(x \mid y) - \mu_x = \rho \sigma_x (y - \mu_y)/\sigma_y. \tag{26.71}$$

Given two values y_1, y_2, we have that

$$\rho = \left\{ \frac{E(x \mid y_1) - E(x \mid y_2)}{\sigma_x} \right\} / \left(\frac{y_1 - y_2}{\sigma_y} \right). \tag{26.72}$$

If $Y = 2$ for n_2 out of n observations, we may estimate k_0 by k in

$$1 - F(k) = (2\pi)^{-1/2} \int_k^\infty \exp\left(-\tfrac{1}{2}u^2\right) \, du = \frac{n_2}{n_1 + n_2} = p. \tag{26.73}$$

Further, $y_2 = E(y \mid Y = 2)$ may be estimated from

$$\frac{y_2 - \mu_y}{\sigma_y} = (2\pi)^{-1/2} \int_k^\infty u \exp\left(-\tfrac{1}{2}u^2\right) \, du \left/ (2\pi)^{-1/2} \int_k^\infty \exp\left(-\tfrac{1}{2}u^2\right) \, du \right.$$

$$= (2\pi)^{-1/2} \exp\left(-\tfrac{1}{2}k^2\right) \left/ \left(\frac{n_1}{n_1 + n_2} \right) = z_k/p \right. \tag{26.74}$$

where z_k denotes the standard normal density function at k; likewise, the value corresponding to $Y = 1$ is $-z_k/q$, $q = (1-p)$. If we estimate $E(x|y_i)$ by $\bar{x}_i$, $i = 1, 2$ we arrive at the biserial estimator for ρ, from (26.72), as

$$r_b = \frac{\bar{x}_2 - \bar{x}_1}{s_x} \frac{pq}{z_k}. \tag{26.75}$$

Example 26.9 (from K. Pearson, 1909)
Table 26.5 shows the returns for 6156 candidates for the London University Matriculation Examination for 1908/9. The average ages for the two highest age-groups have been estimated.

Table 26.5

Age of candidate	Passed $Y = 2$	Failed $Y = 1$	Totals
16	583	563	1146
17	666	980	1646
18	525	868	1393
19–21	383	814	1197
22–30 (mean 25)	214	439	653
over 30 (mean 33)	40	81	121
Totals	2411	3745	6156

We have

$$\bar{x}_2 = 18.4280, \qquad \bar{x}_1 = 18.9877, \qquad s_x^2 = (3.2850)^2 \quad \text{and} \quad p = 2411/6156 = 0.3917.$$

From (26.73), $1 - F(k) = 0.3917$, so that $k = 0.275$ and $z_k = 0.384$. Hence, from (26.75),

$$r_b = -\frac{0.5597}{3.2850} \frac{0.3917 \times 0.6083}{0.384} = -0.11.$$

The estimated correlation between age and success is small.

26.35 As for r_t, the assumption of underlying normality is crucial to r_b. The distribution of biserial r_b is not known, but Soper (1914) derived the expression for its variance in normal samples

$$\text{var } r_b \sim \frac{1}{n}\left[\rho^4 + \rho^2\left\{\frac{pqk^2}{z_k^2} + (2p-1)\frac{k}{z_k} - \frac{5}{2}\right\} + \frac{pq}{z_k^2}\right], \tag{26.76}$$

and showed that (26.76) is generally well approximated by

$$\text{var } r_b \sim \frac{1}{n}\left[r_b^2 - \frac{(pq)^{1/2}}{z_k}\right]^2.$$

Maritz (1953) and Tate (1955) showed that in normal samples r_b is asymptotically normally distributed with mean ρ and variance (26.76); they also considered the maximum likelihood estimation of ρ in biserial data. It appears, as might be expected, that the variance of r_b is least, for fixed ρ, when the dichotomy is at the middle of the dichotomised variate's range $(y = \mu_y)$. When $\rho = 0$, r_b is an efficient estimator of ρ, but when $\rho^2 \to 1$ the efficiency of r_b tends to zero. Tate also tables Soper's formula (26.76) for var r_b. Cf. Exercises 26.10–12.

Terrell (1983) gives tables of percentage points for r_b. Kraemer (1981) considers a modified form that does not assume normality; its asymptotic standard error is given by Koopman (1983).

Point-biserial correlation
26.36 This is a convenient place at which to mention another coefficient, the *point-biserial correlation*, which we shall denote by ρ_{pb}, and by r_{pb} for a sample. Suppose that the dichotomy according to y is regarded, not as a section of a normal distribution, but as defined by a variable taking two values only. So far as correlations are concerned, we can take these values to be 1 or 0. For example, in Table 26.5 it is not implausible to suppose that success in the examination is a dichotomy of a normal distribution of ability to pass it. But if the y-dichotomy were according, say, to sex, this is no longer a reasonable assumption and a different approach is necessary.

Such a situation is, in fact, fundamentally different from the one we have so far considered, for we are now no longer estimating ρ in a bivariate normal population: we consider instead the product-moment of a 0–1 variable y and the variable x. If π is the true proportion of values of y with $y = 1$, we have from the binomial distribution

$$E(y) = \pi, \qquad \sigma_y^2 = \pi(1 - \pi)$$

and thus, by definition,

$$\rho_{pb} = \frac{\mu_{11}}{\sigma_x \sigma_y} = \frac{E(xy) - \pi E(x)}{\sigma_x \{\pi(1 - \pi)\}^{1/2}}.$$

We estimate $E(xy)$ by $m_{11} = \sum_{i=1}^{n_1} x_i / (n_1 + n_2)$, $E(x)$ by $\bar{x}$, σ_x by s_x, and π by $p = n_2 / (n_1 + n_2)$, obtaining

$$
\begin{aligned}
r_{pb} &= \frac{p\bar{x}_2 - p(p\bar{x}_2 + q\bar{x}_1)}{s_x (pq)^{1/2}} \\
&= \frac{(\bar{x}_2 - \bar{x}_1)(pq)^{1/2}}{s_x}.
\end{aligned}
\tag{26.77}
$$

26.37 r_{pb} in (26.77) may be compared with the biserial r_b defined at (26.75). We have

$$\frac{r_{pb}}{r_b} = \frac{z_k}{(pq)^{1/2}}.
\tag{26.78}$$

It has been shown by Tate (1953) by a consideration of Mills' ratio (cf. **5.38**) that the expression on the right of (26.78) is $\leq (2/\pi)^{1/2}$ and the values of the coefficients will thus, in general, be appreciably different.

Tate (1954) shows that r_{pb} is asymptotically normally distributed with mean ρ_{pb} and variance

$$\operatorname{var} r_{pb} \sim \frac{(1-\rho_{pb}^2)^2}{n}\left(1-\tfrac{3}{2}\rho_{pb}^2+\frac{\rho_{pb}^2}{4pq}\right), \tag{26.79}$$

which is a minimum when $p = q = \tfrac{1}{2}$.

Apart from the measurement of correlation, it is clear from (26.77) that, in effect, for a point-biserial situation, we are simply comparing the means of two samples of a variate x, the y-classification being no more than a labelling of the samples. In fact

$$\frac{r_{pb}^2}{1-r_{pb}^2} = \frac{(\bar{x}_1-\bar{x}_2)^2 \dfrac{n_1 n_2}{n_1+n_2}}{\sum (x_{1i}-\bar{x}_1)^2 + \sum (x_{2i}-\bar{x}_2)^2} = \frac{t^2}{n_1+n_2-2}, \tag{26.80}$$

where t is the usual Student's t-test used for comparing the means of two normal populations with equal variance (cf. Example 22.8). Thus if the distribution of x is normal for $y = 0, 1$, the point-biserial coefficient is a simple transformation of the t^2-statistic, which may be used to test it.

Circular correlation

26.38 Measures of correlation for random variables defined on the circle or on the sphere are considered by Stephens (1979b) and by Fisher and Lee (1986). Measures of correlation between circular and linear random variables are considered by Mardia and Sutton (1978) and by Liddell and Ord (1978).

The analytical theory of regression

26.39 We now return to consideration of the regression function, using the definition given in (26.2) which may be rewritten as

$$E(y|x) = \int_{-\infty}^{\infty} yf(x, y)\, dy \bigg/ \int_{-\infty}^{\infty} f(x, y)\, dy. \tag{26.81}$$

Any function $g(y)$ could be considered in place of y in (26.81). In particular, the rth conditional moments involve

$$g(y) = \{y - E(y|x)\}^r \tag{26.82}$$

and have attracted some interest. When $r = 2$, we have the *scedastic* curve, giving rise to the term *homoscedastic* when the scedastic function is constant for all x. The *clitic* curve and the *kurtic* curve, corresponding to standardised versions of (26.82) when $r = 3$ and 4, respectively, are of historical interest only.

26.40 Provided all the cumulants of the distribution exist, the joint density function may be represented by the bivariate Edgeworth series expansion (cf. Exercise 6.16)

$$f(x, y) = \exp\left\{\kappa_{11}D_1 D_2 + \sum_{r+s\geqslant 3} (-1)^{r+s}\kappa_{rs}\frac{D_1^r D_2^s}{r!s!}\right\}\alpha(x)\alpha(y), \tag{26.83}$$

where $\alpha(\cdot)$ denotes the standard normal density. We assume both x and y have been standardised. Also, we assume that the bivariate extensions to the conditions given in **6.22** are satisfied so that the representation (26.83) is valid. It follows that the marginal density for x, given in (6.40), is

$$g(x) = \exp\left\{\sum_{r \geq 3} (-1)^r \kappa_{r0} \frac{D^r}{r!}\right\} \alpha(x) \tag{26.84}$$

so that the conditional density of y is

$$h_x(y) = f(x, y)/g(x). \tag{26.85}$$

The regression function, $\mu'_{1x} = E(y \mid x)$, is

$$\mu'_{1x} = \left(\int yf(x, y) \, dy\right) \Big/ g(x). \tag{26.86}$$

From **6.14–18**, we have that, for $r \geq 1$,

$$\int yD^r\alpha(y) \, dy = -\int D^{r-1}\alpha(y) \, dy$$

$$= \int (-1)^r H_{r-1}(y)\alpha(y) \, dy$$

$$= -1, \quad r = 1$$
$$= 0, \quad r \geq 2,$$

where $H_r(y)$ is the rth Chebyshev–Hermite polynomial, given in (6.21). Thus, on expanding out $f(x, y)$ in (26.86) and integrating, all terms involving D_2^s, $s \geq 2$ disappear and we have

$$g(x)\mu'_{1x} = \exp\left(\sum_{r \geq 3} (-1)^r \kappa_{r0} \frac{D_1^r}{r!}\right)\left\{\sum_{r \geq 1} (-1)^r \kappa_{r1} \frac{D_1^r}{r!}\right\} \alpha(x). \tag{26.87}$$

Using (26.84), this reduces to

$$g(x)\mu'_{1x} = \left\{\sum_{r \geq 1} (-1)^r \kappa_{r1} \frac{D_1^r}{r!}\right\} g(x), \tag{26.88}$$

a result due originally to Wicksell (1934).

Example 26.10
If x is normal, so that $g(x) = \alpha(x)$, (26.88) reduces to

$$\mu'_{1x} = \sum_{r \geq 1} \frac{\kappa_{r1}}{r!} H_r(x).$$

Further, if (x, y) are standard binormal

$$\kappa_{11} = \rho, \quad \kappa_{r1} = 0, \quad r \geq 2$$

and

$$\mu'_{1x} = \rho x,$$

as expected.

The regression functions for sums of squares of normal variates are given in Exercise 26.28.

Criteria for linearity of regression

26.41 Let $\psi(t_1, t_2) = \log \phi(t_1, t_2)$ be the joint c.g.f. of x and y. We now prove: if the regression of y upon x is linear, so that

$$\mu'_{1x} = E(y|x) = \beta_0 + \beta_1 x, \tag{26.89}$$

then

$$\left[\frac{\partial \psi(t_1, t_2)}{\partial t_2}\right]_{t_2=0} = i\beta_0 + \beta_1 \frac{\partial \psi(t_1, 0)}{\partial t_1}; \tag{26.90}$$

and conversely, if a completeness condition is satisfied, (26.90) is sufficient as well as necessary for (26.89).

Write the joint c.f. as

$$\phi(t_1, t_2) = \int_{-\infty}^{\infty} \int_{-\infty}^{\infty} \exp(it_1 x + it_2 y) g(x) h_x(y) \, dy \, dx. \tag{26.91}$$

If we first differentiate with respect to t_2 and then set $t_2 = 0$, we arrive at

$$\left[\frac{\partial \phi(t_1, t_2)}{\partial t_2}\right]_{t_2=0} = i \int_{-\infty}^{\infty} \exp(it_1 x) g(x) \mu'_{1x} \, dx. \tag{26.92}$$

If (26.89) holds, this reduces to

$$\left[\frac{\partial \phi(t_1, t_2)}{\partial t_2}\right]_{t_2=0} = i \int_{-\infty}^{\infty} \exp(it_1 x) g(x)(\beta_0 + \beta_1 x) \, dx$$

$$= i\beta_0 \phi(t_1, 0) + \beta_1 \frac{\partial}{\partial t_1} \phi(t_1, 0). \tag{26.93}$$

Putting $\psi = \log \phi$ in (26.93), and dividing through by $\phi(t_1, 0)$, we obtain (26.90).

Conversely, if (26.90) holds, we rewrite it, using (26.92), in the form

$$i \int_{-\infty}^{\infty} \exp(it_1 x)(\beta_0 + \beta_1 x - \mu'_{1x}) g(x) \, dx = 0. \tag{26.94}$$

We now see that (26.94) implies

$$\beta_0 + \beta_1 x - \mu'_{1x} = 0 \tag{26.95}$$

identically in x if $\exp(it_1 x) g(x)$ is complete, and hence (26.89) follows.

26.42 If all cumulants exist, we have the definition of bivariate cumulants at (3.74)

$$\phi(t_1, t_2) = \exp\left\{ \sum_{r,s=0}^{\infty} \kappa_{rs} \frac{(it_1)^r}{r!} \frac{(it_2)^s}{s!} \right\},$$

where κ_{00} is defined to be equal to zero. Hence

$$\left[\frac{\partial\phi(t_1, t_2)}{\partial t_2}\right]_{t_2=0} = \left[i\phi(t_1, t_2) \sum_{r=0}^{\infty} \sum_{s=1}^{\infty} \kappa_{rs} \frac{(it_1)^r}{r!} \frac{(it_2)^{s-1}}{(s-1)!}\right]_{t_2=0}$$

$$= i\phi(t_1, 0) \sum_{r=0}^{\infty} \kappa_{r1} \frac{(it_1)^r}{r!}. \tag{26.96}$$

Using (26.96), (26.90) gives

$$\sum_{r=0}^{\infty} \kappa_{r1} \frac{(it_1)^r}{r!} = \beta_0 + \beta_1 \sum_{r=0}^{\infty} \kappa_{r0} \frac{(it_1)^{r-1}}{(r-1)!}. \tag{26.97}$$

Identifying coefficients of t^r in (26.97) gives

$$(r=0) \quad \kappa_{01} = \beta_0 + \beta_1 \kappa_{10}, \tag{26.98}$$

as is obvious from (26.89), and

$$(r \geq 1) \quad \kappa_{r1} = \beta_1 \kappa_{r+1,0}. \tag{26.99}$$

The condition (26.99) for linearity of regression is also due to Wicksell (1934). (26.98) and (26.99) together are sufficient, as well as necessary, for (26.90) and (given the completeness of $\exp(it_1 x)g(x)$, as before) for the linearity condition (26.89).

If we express (26.90) in terms of the c.f. ϕ, instead of its logarithm ψ, as in (26.93), and carry through the process leading to (26.99), we find for the central moments,

$$\mu_{r1} = \beta_1 \mu_{r+1,0}.$$

If the regression of x on y is also linear, of form

$$x = \beta_0' + \beta_1' y, \tag{26.100}$$

we also have

$$\kappa_{1r} = \beta_1' \kappa_{0,r+1}, \qquad r \geq 1. \tag{26.101}$$

When $r = 1$, (26.99) and (26.101) give

$$\kappa_{11} = \beta_1 \kappa_{20} = \beta_1' \kappa_{02},$$

leading to

$$\beta_1 \beta_1' = \kappa_{11}^2 / (\kappa_{20} \kappa_{02}) = \rho^2, \tag{26.102}$$

which is (26.15) again.

26.43 We now impose a further restriction on our variables: we suppose that the conditional distribution of y about its mean value (which, as before, is a function of the fixed value of x) is the same for any x, i.e. that only the mean of y changes with x. We shall refer to this restriction by saying that y has identical errors. There is thus a variate ε such that

$$y = \mu'_{1x} + \varepsilon. \tag{26.103}$$

In particular, if the regression is linear, (26.103) is

$$y = \beta_0 + \beta_1 x + \varepsilon. \tag{26.104}$$

If y has identical errors, (26.85) becomes

$$f(x, y) = g(x)h(\varepsilon) \tag{26.105}$$

where h is now the conditional distribution of ε. Conversely, (26.105) implies identical errors for y.

The c.f. for x and y is

$$\begin{aligned}
\phi(t_1, t_2) &= E\{\exp(it_1 x + it_2 y)\} \\
&= E\{\exp(it_1 x + it_2 \beta_0 + it_2 \beta_1 x + it_2 \varepsilon)\} \\
&= \phi_g(t_1 + t_2 \beta_1)\phi_h(t_2) \exp(it_2 \beta_0).
\end{aligned} \tag{26.106}$$

Note that if $\beta_1 = 0$, (26.105) shows that x and y are *independent*: linearity of regression, identical errors and a zero regression coefficient imply independence as is intuitively obvious.

A characterisation of the bivariate normal distribution

26.44 We may now prove a remarkable result: if the regressions of y on x and of x on y are both linear with identical errors, then x and y are distributed in the bivariate normal form unless (a) they are independent of each other, or (b) they are functionally related.

Given the assumptions of the theorem, we have at once, taking logarithms in (26.106)

$$\psi(t_1, t_2) = \psi_g(t_1 + t_2 \beta_1) + \psi_h(t_2) + it_2 \beta_0, \tag{26.107}$$

and similarly, from the regression of x on y,

$$\psi(t_1, t_2) = \psi_{g'}(t_2 + t_1 \beta'_1) + \psi_{h'}(t_1) + it_1 \beta'_0, \tag{26.108}$$

where primes are used as in (26.100). Equating (26.107) and (26.108), and considering successive powers of t_1 and t_2, we find, denoting the rth cumulant of g by κ_{r0}, that of g' by κ_{0r}, that of h by λ_{r0} and that of h' by λ_{0r}:

First order:

$$\kappa_{10}i(t_1 + t_2 \beta_1) + \lambda_{10}it_2 + it_2 \beta_0 = \kappa_{01}i(t_2 + t_1 \beta'_1) + \lambda_{01}it_1 + it_1 \beta'_0,$$

or, equating coefficients of t_1 and of t_2,

$$\kappa_{10} = \kappa_{01}\beta'_1 + \lambda_{01} + \beta'_0, \tag{26.109}$$

$$\kappa_{10}\beta_1 + \lambda_{10} + \beta_0 = \kappa_{01}. \tag{26.110}$$

We may assume that the errors have zero means, for, if not, the means could be absorbed into β_0 or β_0'. If we also measure x and y from their means, (26.109) and (26.110) give

$$\beta_0' = \beta_0 = 0,$$

as is obvious from general considerations.

Second order:

$$\kappa_{20}(t_1 + t_2\beta_1)^2 + \lambda_{20}t_2^2 = \kappa_{02}(t_2 + t_1\beta_1')^2 + \lambda_{02}t_1^2,$$

which, on equating coefficients of t_1^2, t_1t_2 and t_2^2 gives

$$\kappa_{20} = \kappa_{02}(\beta_1')^2 + \lambda_{02}, \quad \kappa_{20}\beta_1 = \kappa_{02}\beta_1', \quad \text{and} \quad \kappa_{20}\beta_1^2 + \lambda_{20} = \kappa_{02}.$$

These expressions give relations between g, h, g' and h'; in particular, β_1/β_1' is equal to κ_{02}/κ_{20}, the ratio of the variances.

Third order:

$$\kappa_{30}\{i(t_1 + t_2\beta_1)\}^3 + \lambda_{30}(it_2)^3 = \kappa_{03}\{i(t_2 + t_1\beta_1')\}^3 + \lambda_{03}(it_1)^3.$$

The terms in $t_1^2t_2$ and $t_1t_2^2$ give us

$$\kappa_{30}\beta_1 = \kappa_{03}(\beta_1')^2 \quad \text{and} \quad \kappa_{30}\beta_1^2 = \kappa_{03}\beta_1'.$$

Unless β_1, $\beta_1' = 0$ or $\beta_1\beta_1' = 1$, we see that these conditions imply $\kappa_{30} = \kappa_{03} = 0$. Similarly, if we take the fourth and higher powers, we find that all the higher cumulants κ_{r0}, κ_{0r} must vanish. Then it follows from equations such as those obtained from the terms in t_1^3, t_2^3 in the third-order equation, namely

$$\kappa_{30} = \kappa_{03}(\beta_1')^3 + \lambda_{03} \quad \text{and} \quad \kappa_{30}\beta_1^3 + \lambda_{30} = \kappa_{03},$$

that the cumulants after the second of h, h' also must vanish. Thus all the distributions g, h, g', h' are normal and it follows that $\psi(t_1, t_2)$ is quadratic in t_1, t_2, and hence that x, y are binormal.

In the exceptional cases we have neglected, this is no longer true. If $\beta_1\beta_1' = 1$, the correlation between x and y is ± 1 by (26.102) and x is a strict linear function of y. On the other hand, if β_1 or $\beta_1' = 0$, the variables x, y are independent.

> The first result of this kind appears to be due to Bernstein (1928). For a proof under general conditions see Féron and Fourgeaud (1952). For more recent discussions, see Kagan *et al.* (1976). These results extend to the case of several regressor variables considered in Chapter 27. See also Exercises 26.33–7.

Testing the linearity of regression

26.45 In order to assess the linearity of a regression function, we may consider the *correlation ratio* for y on x

$$\eta^2 = \text{var}\,\{E(y\,|\,x)\}/\text{var}\,(y). \tag{26.111}$$

When the regression function is linear,

$$\text{var}\,\{E(y\,|\,x)\} = E_x\{(E(y\,|\,x) - E(y))^2\} = \beta_2^2\sigma_1^2 \tag{26.112}$$

by (26.13), so that

$$\eta^2 = \beta_2^2 \sigma_1^2 / \sigma_2^2 = \mu_{11}^2 / (\sigma_1^2 \sigma_2^2) = \rho^2. \qquad (26.113)$$

However, by the Cauchy–Schwarz inequality

$$\begin{aligned}
\mu_{11}^2 &= (E_{x,y}[\{x - E(x)\}\{y - E(y)\}])^2 \\
&= (E_x[\{x - E(x)\}\{E(y|x) - E(y)\}])^2 \\
&\leqslant \text{var}(x)\,\text{var}\{E(y|x)\} \qquad\qquad (26.114)
\end{aligned}$$

so that, in general,

$$0 \leqslant \rho^2 \leqslant \eta^2 \leqslant 1 \qquad (26.115)$$

with (26.113) holding if and only if the regression is linear. The inequalities in (26.115) were established originally by M. Fréchet in 1933–35; see Kruskal (1958). It should be noted that (26.111) takes no account of the x-ordering of the conditional means so that it does not measure any particular form of non-linearity.

26.46 We now suppose that $n = n_1 + \cdots + n_k$ observations are available from a bivariate distribution with n_i replicates for x_i denoted by y_{ij}, $j = 1, \ldots, n_i$. If

$$n_i \bar{y}_i = \sum_j y_{ij} \quad \text{and} \quad n\bar{y} = \sum_i n_i \bar{y}_i$$

we can write

$$SS_Y = \sum_i \sum_j (y_{ij} - \bar{y})^2 = \sum_i \sum_j (y_{ij} - \bar{y}_i)^2 + \sum_i n_i (\bar{y}_i - \bar{y})^2 \qquad (26.116)$$

and

$$\sum_i n_i (y_i - \bar{y})^2 = \sum_i n_i \{\bar{y}_i - \bar{y} - b(x_i - \bar{x})\}^2 + b^2 \sum_i n_i (x_i - \bar{x})^2, \qquad (26.117)$$

where b is the sample regression coefficient for y on x. We now define the sample correlation ratio to be

$$e^2 = \sum n_i (\bar{y}_i - \bar{y})^2 / SS_Y \qquad (26.118)$$

and it then follows that (26.117) may be written as

$$e^2 SS_Y = (e^2 - r^2) SS_Y + r^2 SS_Y. \qquad (26.119)$$

Under the hypothesis H_0 that (x, y) is binormal (and, therefore that the regression is linear), it follows from Cochran's theorem (**15.16**) that the two sums of squares in (26.117) are independently distributed as chi-squared with $k - 2$ and $n - k$ degrees of freedom, respectively. Thus, under H_0, the test statistic

$$F = \frac{(e^2 - r^2)/(k - 2)}{(1 - e^2)/(n - k)} \qquad (26.120)$$

has an F distribution with $(k - 2, n - k)$ d.fr.

26.47 It follows that for fixed values of x_i, the results of Chapter 23 concerning the power of the LR test, based on the non-central F distribution, are applicable to this test, which is the UMP invariant test by **23.37**. However, the distribution in the binormal case, which allows the x_i to vary, will not coincide with that derived by holding the x_i fixed as above, except when the hypothesis tested is true, when the variation of the x_i is irrelevant (as we shall see in **27.29**). The power functions of the test of $\rho = 0$ are therefore different in the two cases, even though the same test is valid in each case. For large n, however, the results do coincide; we shall observe this more generally in connection with the multiple correlation coefficient (of which r^2 is a special case) in **27.29** and **27.31**.

26.48 When replicates are unavailable but the sample size is reasonably large, (26.120) may be applied after grouping the data.

Example 26.11
Consider the grouped data in Table 26.1. The set of $(n_i, \bar{y}_i)$ values is as shown in

Table 26.6

Stature (x)	Mean weight in array ($\bar{y}_i$)	n_i
54	92.5	5
56	111.4	33
58	122.0	254
60	124.4	813
62	130.2	1340
64	134.6	1454
66	140.5	750
68	146.4	275
70	157.3	56
72	163.4	11
74	179.5	4

Table 26.6, with $k = 11$ and $y = 132.8$. It follows that $e^2 = 0.1059$ whereas $r^2 = 0.1036$ so that

$$F = \frac{(0.1059 - 0.1036)/9}{(1 - 0.1059)/4984} = 1.42$$

less than the upper 5 per cent point under H_0. There appears to be no evidence of non-linearity in the regression of y on x.

26.49 The test (26.120) is an example of a *diagnostic* test used to check the assumptions of the linear regression model. We return to this theme in Chapter 28.

EXERCISES

26.1   Show that the correlation coefficient for the data of Table 26.2 is 0.072. Show that the regression lines in Fig. 26.2 are:

$$CC': y = 0.0938x + 30.56; \qquad RR': x = 0.0547y + 61.06.$$

26.2   If x_i/σ^2 $(i = 1, 2, 3)$ are independent χ^2 variates with ν_i degrees of freedom, $y_1 = x_1/x_3$ and $y_2 = x_2/x_3$, show that the joint distribution of y_1 and y_2 is

$$g(y_1, y_2) = \frac{\Gamma(\tfrac{1}{2}\nu)}{\prod_{i=1}^{3} \Gamma(\tfrac{1}{2}\nu_i)} \frac{y_1^{(\nu_1-1)/2} y_2^{(\nu_2-1)/2}}{(1+y_1+y_2)^{\nu/2}}, \qquad 0 < y_1, y_2 < \infty,$$

where $\nu = \sum_{i=1}^{3} \nu_i$. Show that the regression of y_1 on y_2, and of y_2 on y_1, is linear. Show that if $\nu_3 > 4$, their correlation coefficient is

$$\rho = [\nu_1 \nu_2 / \{(\nu_1 + \nu_3 - 2)(\nu_2 + \nu_3 - 2)\}]^{1/2}.$$

Assuming that x_1/σ^2, x_2/σ^2 are non-central $\chi'^2(\lambda_i, \nu_i)$ show, using Exercise 23.21, that this generalises to

$$\rho^2 = \prod_{i=1}^{2} \left\{ \frac{(\nu_i + \lambda_i)^2}{(\nu_i + \lambda_i)^2 + (\nu_3 - 2)(\nu_i + 2\lambda_i)} \right\}.$$

26.3   A binormal distribution is dichotomised at some value of y. The variance of x for the whole distribution is known to be σ_x^2 and that for one part of the dichotomy is σ_1^2. The correlation between x and y for the latter is c. Show that the correlation of x and y in the whole distribution may be estimated by r, where

$$r^2 = 1 - \frac{\sigma_1^2}{\sigma_x^2} (1 - c^2).$$

26.4   In the previous exercise, given that σ_y^2 is the variance of y in the whole distribution and σ_2^2 is its variance in the part of the dichotomy, show that ρ for the whole distribution may be estimated by

$$r^2 = \frac{c^2 \sigma_y^2}{\sigma_2^2 + c^2(\sigma_y^2 - \sigma_2^2)}.$$

26.5   Show that whereas tetrachoric r_t and point-biserial r_{pb} can never exceed unity in absolute value, biserial r_b may do so.

26.6   Prove that the tetrachoric series (26.70) always converges for $|\rho| < 1$.

26.7   A set of variables $x_1, x_2, \ldots, x_n$ are distributed so that the product-moment correlation of x_i and x_j is ρ_{ij}. They all have the same variance. Show that the average value of ρ_{ij} defined by

$$\bar{\rho} = \frac{1}{n(n-1)} \sum_{i=1}^{n} \sum_{j=1}^{n} \rho_{ij}, \qquad i \neq j,$$

must be not less than $-1/(n-1)$.

26.8   In the previous exercise show that $|\rho|$, the determinant of the correlation matrix, is non-negative. Hence show that

$$\rho_{12}^2 + \rho_{13}^2 + \rho_{23}^2 \leq 1 + 2\rho_{12}\rho_{13}\rho_{23}.$$

26.9 Show from (16.86) that in samples from a binormal population the sampling distribution of b_2, the regression coefficient of y on x, has exact variance

$$\operatorname{var} b_2 = \frac{1}{n-3} \frac{\sigma_2^2}{\sigma_1^2} (1-\rho^2), \qquad n \geqslant 4,$$

and that its skewness and kurtosis coefficients are

$$\gamma_1 = 0, \qquad n \geqslant 5,$$

$$\gamma_2 = \frac{6}{n-5}, \qquad n \geqslant 6.$$

Show that when $\rho = 0$, var b_2 is the expectation of the variance given for fixed xs in Example 19.6, which is in our present notation var $(b_2 | x) = \sigma_2^2 / (ns_1^2)$.

26.10 Let $\psi\{(x-\mu)/\sigma, y\}$ denote the binormal frequency with means of x and y equal to μ and 0 respectively, variances equal to σ^2 and 1 respectively, and correlation ρ. Define

$$\xi(x, \omega) = \int_{\omega}^{\infty} \psi \, dy, \qquad \eta(x, \omega) = \int_{-\infty}^{\omega} \psi \, dy.$$

Given that z_i is a random variable taking the values 0, 1 according as $y < \omega$ or $y \geqslant \omega$, show that in a biserial table the likelihood function may be written

$$L(x, y | \omega, \rho, \mu, \sigma) = \prod_{i=1}^{n} \left\{ z_i \xi \left(\frac{x_i - \mu}{\sigma}, \omega \right) + (1 - z_i) \eta \left(\frac{x_i - \mu}{\sigma}, \omega \right) \right\}.$$

If ∂^2 represents a partial differential of the second order with respect to any pair of parameters, show that

$$E(\partial^2 \log L) = n\{1 - p(x)\} E_0(\partial^2 \log \eta) + np(x) E_1(\partial^2 \log \xi)$$

where

$$p(x) = \int_{x}^{\infty} (2\pi)^{-1/2} \exp\left(-\tfrac{1}{2}t^2\right) dt,$$

and E_0, E_1 are conditional expectations with respect to x for $y < \omega$, $y \geqslant \omega$ respectively. Hence derive the inverse of the covariance matrix for the maximum likelihood estimators of the four parameters (the order of rows and columns being the same as the order of the parameters in the LF):

$$V^{-1} = \frac{n}{(1-\rho^2)} \begin{vmatrix} a_0 & \dfrac{\rho\omega a_0 - a_1}{1-\rho^2} & \dfrac{\rho a_0}{\sigma} & \dfrac{\rho a_1}{\sigma} \\[2ex] & \dfrac{a_2 - 2\rho\omega a_1 + \rho^2\omega^2 a_0}{(1-\rho^2)^2} & \dfrac{\rho^2\omega a_0 - \rho a_1}{\sigma(1-\rho^2)} & \dfrac{\rho^2\omega a_1 - \rho a_2}{\sigma(1-\rho^2)} \\[2ex] & & \dfrac{1 - \rho^2 + \rho^2 a_0}{\sigma^2} & \dfrac{\rho^2 a_1}{\sigma^2} \\[2ex] & & & \dfrac{2(1-\rho^2) + \rho^2 a_2}{\sigma^2} \end{vmatrix}$$

where

$$a_s = \int_{-\infty}^{\infty} x^s g(x, \omega, \rho) \, dx,$$

$$g(x, \omega, \rho) = (2\pi)^{-1/2} \exp\left(-\tfrac{1}{2}x^2\right) \phi \left(\frac{\omega - \rho x}{(1-\rho^2)^{1/2}} \right) \phi \left(\frac{\rho x - \omega}{(1-\rho^2)^{1/2}} \right),$$

and

$$\phi(x) = (2\pi)^{-1/2} \exp\left(-\tfrac{1}{2}x^2\right)/\{1-p(x)\}.$$

By inverting this matrix, derive the asymptotic variance of the maximum likelihood estimator $\hat{\rho}_b$ in the form

$$\text{var } \hat{\rho}_b = \frac{(1-\rho^2)^3}{n}\left\{\frac{\int_{-\infty}^{\infty} g\,dx}{\int_{-\infty}^{\infty} g\,dx \int_{-\infty}^{\infty} x^2 g\,dx - \left(\int_{-\infty}^{\infty} xg\,dx\right)^2}\right\} + \frac{\rho^2(1-\rho^2)}{n}.$$

(Tate, 1955)

26.11   In Exercise 26.10, show that when $\rho = 0$,

$$\text{var } \hat{\rho}_b = \frac{2\pi p(\omega)\{1 - p(\omega)\}}{n \exp(-k^2)}.$$

By comparing this with the large-sample formula (26.76), show that when $\rho = 0$, r_b is a fully efficient estimator.
(Tate, 1955)

26.12   In Exercise 26.10, show that $n \text{ var } \hat{\rho}_b$ tends to zero as $|\rho|$ tends to unity, and from (26.76) that $n \text{ var } r_b$ does not, and hence that r_b is of zero efficiency near $|\rho| = 1$.
(Tate, 1955; the results of Exercises 26.10–12 are extended to the multinormal distribution by J. F. Hannan and Tate (1965).)

26.13   Establish equation (26.17).

26.14   Writing l for the sample intra-class correlation coefficient (26.60) and λ for the population value, show that the exact distribution of l is given by

$$dF \propto \frac{(1-l)^{p(k-1)/2-1}\{1+(k-1)l\}^{(p-3)/2}\,dl}{\{1-\lambda+\lambda(k-1)(1-l)\}^{(kp-1)/2}},$$

reducing in the case $k = 2$ to

$$dF = \frac{\Gamma(p-\tfrac{1}{2})}{\Gamma(p-1)(2\pi)^{1/2}} \operatorname{sech}^{p-1/2}(z-\xi) \exp\{-\tfrac{1}{2}(z-\xi)\}$$

where $l = \tanh z$, $\lambda = \tanh \xi$. Hence show that, for $k = 2$, $z - \xi$ is nearly normal with mean zero and variance $1/(n-3/2)$.
(Fisher, 1921c)

26.15   Show that for testing $\rho = \rho_0$ in a bivariate normal population, the likelihood ratio statistic is given by

$$l^{1/n} = \frac{(1-r^2)^{1/2}(1-\rho_0^2)^{1/2}}{(1-r\rho_0)},$$

so that $l^{1/n} = (1-r^2)^{1/2}$ when $\rho_0 = 0$, and when $\rho_0 \neq 0$ we have

$$(1-\rho_0^2)^{-1/2}l^{1/n} = 1 + r\rho_0 + r^2(\rho_0^2 - \tfrac{1}{2}) + \cdots.$$

26.16   Show that the effect of applying Sheppard's corrections to the moments is always to increase the value of the correlation coefficient (cf. **3.18**).

26.17 Show that if x and y are respectively subject to errors of observation u, v, where u and v are uncorrelated with x, y and each other, the correlation coefficient is reduced, or attenuated, by a factor

$$\left\{ \left(1 + \frac{\sigma_u^2}{\sigma_x^2}\right)\left(1 + \frac{\sigma_v^2}{\sigma_y^2}\right)\right\}^{1/2}.$$

26.18 If x_i $(i=1,2,3)$ are mutually independent variates with means μ_i, variances σ_i^2 and coefficients of variation $V_i = \sigma_i/\mu_i$, show that the correlation between x_1/x_3 and x_2/x_3 is

$$\rho = \frac{\mu_1 \mu_2}{\left\{\sigma_1^2\left(1 + \frac{1}{V_4^2}\right) + \mu_1^2\right\}^{1/2}\left\{\sigma_2^2\left(1 + \frac{1}{V_4^2}\right) + \mu_2^2\right\}^{1/2}}$$

exactly, where V_4 is the coefficient of variation of $1/x_3$. Thus $\rho = 0$ if either μ_1 or $\mu_2 = 0$; if neither is, ρ takes the sign of their product, and

$$|\rho| = \left\{1 + V_1^2\left(1 + \frac{1}{V_4^2}\right)\right\}^{-1/2}\left\{1 + V_2^2\left(1 + \frac{1}{V_4^2}\right)\right\}^{-1/2}.$$

Show that if V_4^2 is small, we obtain approximately

$$|\rho| = V_3^2 / \{(V_1^2 + V_3^2)(V_2^2 + V_3^2)\}^{1/2},$$

since $V_3 \doteq V_4$ by the argument of **10.7** (Vol. 1).

> (The approximation was differently obtained by K. Pearson (1897) who dealt with the case where μ_1, μ_2 and therefore ρ are positive; he also treated the case where x_1, x_2 and x_3 are correlated. He called ρ a 'spurious' correlation because the original x_i are uncorrelated, but the term is inapt if one is fundamentally interested in the ratios.)

26.19 Given that two binormal populations have $\rho_1 = \rho_2 = \rho$, the other parameters being unspecified, show that the maximum likelihood estimator of ρ is

$$\hat{\rho} = \frac{n(1 + r_1 r_2) - \{n^2(1 - r_1 r_2)^2 - 4n_1 n_2(r_1 - r_2)^2\}^{1/2}}{2(n_1 r_2 + n_2 r_1)},$$

where n_i, r_i are the sample sizes and correlation coefficients $(i = 1, 2)$ and $n = n_1 + n_2$. If $n_1 = n_2 = \frac{1}{2}n$, show that if z_1, z_2 are defined by (26.25), and

$$\hat{\zeta} = \frac{1}{2}\log\left(\frac{1 + \hat{\rho}}{1 - \hat{\rho}}\right)$$

then

$$\hat{\zeta} = \frac{1}{2}(z_1 + z_2)$$

exactly.

26.20 Using the result of the previous exercise, show that the likelihood ratio test of $\rho_1 = \rho_2$ when $n_1 = n_2$ uses the statistic

$$l^{1/n} = \text{sech}\left\{\frac{1}{2}(z_1 - z_2)\right\},$$

so that it is a one-to-one function of $z_1 - z_2$, the statistic suggested in **26.19**.

(Brandner, 1933)

26.21   In Exercise 26.19, show that if $n_1 \neq n_2$, we have *approximately* for the ML estimator of ζ

$$\hat{\zeta} = \frac{1}{n}(n_1 z_1 + n_2 z_2),$$

and hence that the LR test of $\rho_1 = \rho_2$ uses the statistic

$$l = \left[\operatorname{sech}\left\{ \frac{n_1}{n}(z_1 - z_2) \right\} \right]^{n_2} \left[\operatorname{sech}\left\{ \frac{n_2}{n}(z_1 - z_2) \right\} \right]^{n_1}$$

approximately, again a one-to-one function of $(z_1 - z_2)$.   (Brandner, 1933)

26.22   To estimate a common value of ρ for two binormal populations, show that

$$\hat{\zeta}^* = \frac{(n_1 - 3)z_1 + (n_2 - 3)z_2}{n_1 + n_2 - 6}$$

is the linear combination of z_1 and z_2 with minimum variance as an estimator of ζ, but that when $n_1 \neq n_2$ this does not give the maximum likelihood estimator of ρ given in Exercise 26.19.

26.23   Show that the correlation coefficient between x and y, ρ_{xy}, satisfies

$$\rho_{xy}^2 = \frac{\operatorname{var}(\alpha_2 + \beta_2 x)}{\sigma_2^2} \equiv 1 - \frac{E\{[y - (\alpha_2 + \beta_2 x)]^2\}}{\sigma_2^2}$$

and hence establish (26.7).

26.24   Let $f(x, y) = \alpha(\alpha + 1)(x + y)^{-(\alpha + 2)}$, $1 \leq x$, $y < \infty$. Show that $E(y|x) = (\alpha + 1 + x)/\alpha$, $\alpha > 0$, whereas $E(y) = (\alpha + 1)/(\alpha - 1)$, $\alpha > 1$. (Note that $E(y|x)$ exists for all x, but $E(y)$ does not when $0 < \alpha \leq 1$.)

26.25   Show from (26.3) that the covariance

$$\mu_{11} = \int_{-\infty}^{\infty} \int_{-\infty}^{\infty} \{F(x, y) - F(x, \infty)F(\infty, y)\} \, dx \, dy.$$

26.26   Suppose that $y_{ij} = \alpha + \beta x_i + \varepsilon_{ij}$, where the ε_{ij} are independent and identically distributed with zero means and variances σ^2, and $i = 1, \ldots, m$, $j = 1, \ldots, K$. Let $z_i = \sum_j y_{ij}/K$. Show that the least squares estimators of α and β remain unbiased, but that

$$E\{\sum (z_i - \bar{z})^2\} = \beta^2 \sum (x_i - \bar{x})^2 + \frac{\sigma^2(m - 2)}{K}.$$

Hence, show that, at least in large samples, the correlation of x and z, $|r| \to 1$ as K increases.

26.27   By using the bivariate form of the inversion theorem for the characteristic function given in **4.17** show that (26.69) reduces to (26.70).

26.28   Consider the statistics $u = \sum_{i=1}^{n} x_i^2$ and $v = \sum_{i=1}^{n} y_i^2$ based on a sample of size n from a standard binormal distribution with correlation ρ. Using Example 26.1 and the results in **26.40**, derive the regression functions

$$E(v|u) = \rho^2 u + n(1 - \rho^2)$$
$$E\{\operatorname{var}(v|u)\} = 2(1 - \rho^2)\{2\rho^2 u + n(1 - \rho^2)^2\}.$$

26.29   In a standardised binormal distribution, show that on any (elliptical) contour of constant probability

$$y = \rho x \pm \{(k - x^2)(1 - \rho^2)\}^{1/2}, \quad k > 0,$$

and that y attains its maximum $k^{1/2}$ at $x = \rho k^{1/2}$. Hence, show that the ratio of the squared vertical length of the ellipse at $x = 0$ (h^2) to its maximum squared overall vertical length (H^2) is exactly $1 - \rho^2$.

Show that if all, or nearly all, of the points in a sample scatter plot are surrounded by an ellipse, and the regression of y on x is linear and homoscedastic, h^2/H^2 is approximately the ratio of the residual variance about the regression line to the total variance of y; that is, it is approximately equal to $1 - r^2$ irrespective of bivariate normality. Apply this method to the data of Fig. 26.3 and compare the resulting approximation to r^2 with the true value given in Example 26.7.   (Cf. Châtillon, 1984.)

26.30   The bivariate distribution of x and y is uniform over the region in the (x, y) plane bounded by the ellipse

$$ax^2 + 2hxy + by^2 = c, \qquad h \neq 0; \quad h^2 < ab; \quad a, b > 0.$$

Show that the regression of each variable on the other is linear and that the scedastic curves are quadratic parabolas.

26.31   The bivariate distribution of x and y is uniform over the parallelogram bounded by the lines $x = 3(y - 1)$, $x = 3(y + 1)$, $x = y + 1$, $x = y - 1$. Show that the regression of y on x is linear, but that the regression of x on y consists of sections of three straight lines joined together.

26.32   From (26.88), show that if the marginal distribution of a bivariate distribution is of the Gram–Charlier form

$$f = \alpha(x)\{1 + a_3 H_3 + a_4 H_4 + \cdots\},$$

then the regression of y on x is

$$\mu'_{1x} = \frac{\sum_{r=0}^{\infty} \sum_{s=0}^{\infty} \frac{\kappa_{r1}}{r!} a_s H_{r+s}(x)}{1 + \sum_{r=3}^{\infty} a_r H_r(x)}. \qquad \text{(Wicksell, 1917)}$$

26.33   If, for each fixed y, the conditional distribution of x is normal, show that their bivariate distribution must be of the form

$$f(x, y) = \exp\{-(a_1 x^2 + a_2 x + a_3)\}$$

where the a_i are functions of y. Show that if, in addition, the equiprobable contours of $f(x, y)$ are similar concentric ellipses, f must be binormal.   (A. Bhattacharyya, 1943)

26.34   Show that if the regression of x on y is linear, if the conditional distribution of x for each fixed y is normal and homoscedastic and if the marginal distribution of y is normal, then $f(x, y)$ must be binormal.   (A. Bhattacharyya, 1943)

26.35   Show that if the conditional distributions of x for each fixed y, and of y for each fixed x, are normal, and one of these conditional distributions is homoscedastic, then $f(x, y)$ is binormal.   (A. Bhattacharyya, 1943)

26.36   Show that if every non-degenerate linear function of x and y is normal, then $f(x, y)$ is binormal.   (A. Bhattacharyya, 1943)

26.37   Show that if the regressions of x on y and of y on x are both linear, and the conditional distribution of each for every fixed value of the other is normal, then $f(x, y)$ is either binormal or may (with a suitable choice of origin and scale) be written in the form

$$f = \exp\{-(x^2 + a^2)(y^2 + b^2)\}. \qquad \text{(A. Bhattacharyya, 1943)}$$

CHAPTER 27

PARTIAL AND MULTIPLE CORRELATION AND REGRESSION

27.1   As we saw in **26.7**, the correlation parameter ρ may be used as a measure of interdependence when the joint distribution is, at least approximately, binormal. When we come to interpret interdependence in practice, however, we often meet difficulties of the kind discussed in **26.3**: if a variable is correlated with a second variable, this may be merely incidental to the fact that both are correlated with another variable or set of variables. This consideration leads us to examine the correlations between variables when other variables are held constant, i.e. conditionally upon those other variables taking certain fixed values. These are known as the *partial correlations.*

   If we find that holding another variable fixed reduces the correlation between two variables, we infer that their interdependence arises in part through that other variable; and, if the partial correlation is zero, we infer that their interdependence is entirely attributable to that variable. Conversely, if the partial correlation is larger than the original correlation between the variables we infer that the other variable was obscuring the stronger connection or masking the correlation. But it must be remembered that we cannot assume a causal connection. As we noted in **26.4** for ordinary product-moment correlations, claims of causality must, ultimately, rely upon extra-statistical criteria.

27.2   In this branch of the subject, it is difficult at times to arrive at a notation which is unambiguous and flexible without being impossibly cumbrous. We shall rely upon matrix notation for demonstrations of the basic results, but revert to Yule's (1907) notation when discussing individual coefficients.

Partial correlation
27.3   Let $\mathbf{x}$ denote a vector of p random variables that are multinormally distributed. We exclude the singular case (cf. **15.2**) and, without loss of generality, standardise the variables so that the covariance matrix becomes the correlation matrix $\mathbf{C}$. Thus, from (15.19), the joint density function is

$$f(\mathbf{x}) = (2\pi)^{-p/2}|C|^{-1/2} \exp\{-\tfrac{1}{2}\mathbf{x}'\mathbf{C}^{-1}\mathbf{x}\}, \tag{27.1}$$

where $\mathbf{C}$ has elements

$$c_{ii} = 1$$
$$c_{ij} = \rho_{ij} = \text{corr}(x_i, x_j), \qquad i \neq j.$$

From (15.20), the c.f. is

$$\phi(\mathbf{t}) = \exp(-\tfrac{1}{2}\mathbf{t}'\mathbf{C}\mathbf{t}). \tag{27.2}$$

1010

If $\mathbf{x}$ is now partitioned into $\mathbf{x}_1$ $(k \times 1)$ and $\mathbf{x}_2$ $((p-k) \times 1)$, we may partition $\mathbf{C}$ conformably into

$$\mathbf{C} = \begin{pmatrix} \mathbf{C}_{11} & \mathbf{C}_{12} \\ \mathbf{C}_{21} & \mathbf{C}_{22} \end{pmatrix}, \tag{27.3}$$

where $\mathbf{C}_{12} = \mathbf{C}_{21}'$. It follows directly from (27.2) that the $\mathbf{x}_2$ is multinormal with mean zero and covariance matrix $\mathbf{C}_{22}$, or $\mathbf{x}_2 \sim N(\mathbf{0}, \mathbf{C}_{22})$.

27.4   The conditional distribution of $\mathbf{x}_1$ given $\mathbf{x}_2$ has density function

$$f(\mathbf{x}_1 \mid \mathbf{x}_2) = f(\mathbf{x})/f_2(\mathbf{x}_2) \tag{27.4}$$

following the usual notation. In order to evaluate (27.4) explicitly, we first note the matrix identity

$$\mathbf{G}^{-1} = \begin{pmatrix} \mathbf{A} & \mathbf{B}' \\ \mathbf{B} & \mathbf{D} \end{pmatrix}^{-1} = \begin{pmatrix} \mathbf{H}^{-1} & -\mathbf{H}^{-1}\mathbf{B}'\mathbf{D}^{-1} \\ -\mathbf{D}^{-1}\mathbf{B}\mathbf{H}^{-1} & \mathbf{D}^{-1}+\mathbf{D}^{-1}\mathbf{B}\mathbf{H}^{-1}\mathbf{B}'\mathbf{D}^{-1} \end{pmatrix}, \tag{27.5}$$

where $\mathbf{A}$ and $\mathbf{D}$ are symmetric and

$$\mathbf{H} = \mathbf{A} - \mathbf{B}'\mathbf{D}^{-1}\mathbf{B}. \tag{27.6}$$

Expression (27.5) may be demonstrated by checking that $\mathbf{G}\mathbf{G}^{-1} = \mathbf{I}$.

27.5   It follows from (27.1) that (27.4) may be written as

$$f(\mathbf{x}_1 \mid \mathbf{x}_2) = \text{const.} \times \exp\left\{-\tfrac{1}{2}(\mathbf{x}'\mathbf{C}^{-1}\mathbf{x} - \mathbf{x}_2'\mathbf{C}_{22}^{-1}\mathbf{x}_2)\right\}. \tag{27.7}$$

We now restrict attention to (-2) times the exponent in (27.7), which is

$$(\mathbf{x}_1' \quad \mathbf{x}_2')\begin{pmatrix} \mathbf{C}_{11} & \mathbf{C}_{12} \\ \mathbf{C}_{21} & \mathbf{C}_{22} \end{pmatrix}^{-1}\begin{pmatrix} \mathbf{x}_1 \\ \mathbf{x}_2 \end{pmatrix} - \mathbf{x}_2'\mathbf{C}_{22}^{-1}\mathbf{x}_2.$$

Using (27.5), this may be expressed as

$$\mathbf{x}_2'(\mathbf{C}_{22}^{-1} + \mathbf{C}_{22}^{-1}\mathbf{C}_{21}\mathbf{H}^{-1}\mathbf{C}_{12}\mathbf{C}_{22}^{-1})\mathbf{x}_2$$
$$-2\mathbf{x}_1'\mathbf{H}^{-1}\mathbf{C}_{12}\mathbf{C}_{22}^{-1}\mathbf{x}_2 + \mathbf{x}_1'\mathbf{H}^{-1}\mathbf{x}_1 - \mathbf{x}_2'\mathbf{C}_{22}^{-1}\mathbf{x}_2$$

which reduces to

$$(\mathbf{x}_1 - \mathbf{C}_{12}\mathbf{C}_{22}^{-1}\mathbf{x}_2)'\mathbf{H}^{-1}(\mathbf{x}_1 - \mathbf{C}_{12}\mathbf{C}_{22}^{-1}\mathbf{x}_2). \tag{27.8}$$

By inspection of (27.7) and (27.8), it follows that the conditional distribution of $\mathbf{x}_1$ given $\mathbf{x}_2$ is multinormal, or

$$\mathbf{x}_1 \mid \mathbf{x}_2 \sim N(\mathbf{C}_{12}\mathbf{C}_{22}^{-1}\mathbf{x}_2, \mathbf{H}), \tag{27.9}$$

where $\mathbf{H} = \mathbf{C}_{11} - \mathbf{C}_{12}\mathbf{C}_{22}^{-1}\mathbf{C}_{21}$. The constant of integration in (27.7) must be of the form

$$(2\pi)^{-k/2}|H|^{-1/2};$$

so that, from (27.1) and (27.4),

$$|H| = |C|/|C_{22}|. \tag{27.10}$$

We see from (27.9) that $\mathbf{H}$ describes the partial correlation structure for elements of $\mathbf{x}_1$, given $\mathbf{x}_2$. Further, the regression function is

$$E(\mathbf{x}_1|\mathbf{x}_2) = \mathbf{C}_{12}\mathbf{C}_{22}^{-1}\mathbf{x}_2. \tag{27.11}$$

Marsaglia (1964) shows that (27.11) and, indeed, (27.9), hold for the singular multinormal distribution provided that $\mathbf{C}_{22}^{-1}$ is replaced by the generalised inverse

$$\mathbf{C}_{22}^+ = \mathbf{T}'(\mathbf{TT}')^{-2}\mathbf{T}, \quad \text{where } \mathbf{C}_{22} = \mathbf{T}'\mathbf{T}.$$

27.6 We now set $k = 2$ so that $\mathbf{x}_1 = (x_1, x_2)'$ and examine the resulting expressions for different values of $q = p - k$. When $q = 0$, $\mathbf{H}$ yields corr $(x_1, x_2) = \rho_{12}$, as before. When $q = 1$, $\mathbf{C}_{22} \equiv 1$ and (27.6) yields

$$
\begin{aligned}
\mathbf{H} &= \begin{pmatrix} 1 & \rho_{12} \\ \rho_{12} & 1 \end{pmatrix} - \begin{pmatrix} \rho_{13} \\ \rho_{23} \end{pmatrix}(\rho_{13} \quad \rho_{23}) \\
&= \begin{pmatrix} 1 - \rho_{13}^2 & \rho_{12} - \rho_{13}\rho_{23} \\ \rho_{12} - \rho_{13}\rho_{23} & 1 - \rho_{23}^2 \end{pmatrix}
\end{aligned} \tag{27.12}
$$

so that

$$\text{corr}(x_1, x_2|x_3) = \rho_{12.3} = \frac{\rho_{12} - \rho_{13}\rho_{23}}{\{(1 - \rho_{13}^2)(1 - \rho_{23}^2)\}^{1/2}}. \tag{27.13}$$

If we revert to the use of the covariance matrix, we obtain

$$\mathbf{H} = \begin{pmatrix} \sigma_1^2 & \sigma_{12} \\ \sigma_{12} & \sigma_2^2 \end{pmatrix} - \frac{1}{\sigma_3^2}\begin{pmatrix} a^{\sigma_{13}} \\ \sigma_{23} \end{pmatrix}(\sigma_{13} \quad \sigma_{23})$$

so that

$$\sigma_{1.3}^2 = \sigma_1^2 - (\sigma_{13}^2/\sigma_3^2) \tag{27.14}$$

and

$$\sigma_{12.3}^2 = \sigma_{12} - (\sigma_{13}\sigma_{32}/\sigma_3^2). \tag{27.15}$$

27.7 If we first derive the conditional distribution of (x_1, x_2, x_3) given $\mathbf{x}_m$, where $\mathbf{x}_m$ denotes any subset of $(x_4, \ldots, x_p)$, it follows that the conditional distribution will be of form (27.1) with the elements of $\mathbf{C}$, after restandardising the variables, given by

$$C_{ij} = 1, \quad i = j$$
$$C_{ij} = \rho_{ij.m} \equiv \text{corr}(x_i, x_j|\mathbf{x}_m).$$

Henceforth, the subscript m will be used to denote any subset of conditioning variables. We may then repeat steps (27.12)–(27.13) to arrive at

$$
\begin{aligned}
\text{corr}(x_1, x_2|x_3, \mathbf{x}_m) &= \rho_{12 \cdot 3m} \\
&= \frac{\rho_{12 \cdot m} - \rho_{13 \cdot m}\rho_{23 \cdot m}}{\{(1 - \rho_{13 \cdot m}^2)(1 - \rho_{23 \cdot m}^2)\}^{1/2}}.
\end{aligned} \tag{27.16}
$$

If we redefine $\mathbf{C}$ as the covariance matrix, the extension of (27.15) is

$$\mathrm{cov}\,(x_1,\,x_2|x_3,\,\mathbf{x}_m) = \sigma_{12\cdot m}$$

$$= \sigma_{12\cdot m} - (\sigma_{13\cdot m}\sigma_{23\cdot m}/\sigma_{3\cdot m}^2). \tag{27.17}$$

The subscripts $(3, 4, \ldots, p)$ may, of course, be permuted into any order.

27.8   Although the results in **27.4–7** have been derived assuming multinormality, we shall define these to be the partial correlations whatever the underlying distribution.

Linear regression
27.9   When $k = 1$, (27.11) yields

$$E(x_1|\mathbf{x}_2) = E(x_1|x_2, \ldots, x_p) = \sum_{j=2}^{p} \beta_j x_j \tag{27.18}$$

where

$$\boldsymbol{\beta} = (\beta_2, \ldots, \beta_p)' = \mathbf{C}_{22}^{-1}\mathbf{C}_{21}. \tag{27.19}$$

The coefficients β_j, also denoted by $\beta_{1j\cdot m(j)}$, where $m(j) = \{2, 3, \ldots, j-1, j+1, \ldots, p\}$, are known as the (*partial*) *regression coefficients*. From (27.6), the conditional variance is

$$\mathrm{var}\,(x_1|\mathbf{x}_2) = \sigma_{1\cdot 2\cdots p}^2$$

$$= c_{11} - \mathbf{C}_{12}\mathbf{C}_{22}^{-1}\mathbf{C}_{21}$$

$$= c_{11} - \mathbf{C}_{12}\boldsymbol{\beta}_1 \tag{27.20}$$

Now taking $\mathbf{C}$ as the covariance matrix with elements $\{\sigma_{ij}\}$, with $\sigma_{ii} = \sigma_i^2$, (27.19) yields

$$\sigma_{1\cdot 2\cdots p}^2 = \sigma_1^2 - \sum_{j=2}^{p} \beta_j \sigma_{1j}. \tag{27.21}$$

The error variance (27.21) is independent of $(x_2, \ldots, x_p)$ if the β_j are independent of these values. The conditional distribution of x_1 is then said to be *homoscedastic*; otherwise, it is *heteroscedastic*. All conditional distributions derived from the multinormal are homoscedastic. In other cases, we must make due allowance for heteroscedasticity; the partial regression coefficients are then, perhaps, best regarded as *averages* taken over all possible values of $\mathbf{x}_2$. This interpretation is justified in **27.12** below.

27.10   When $k = 1$ and $q = 2$, it follows from (27.19) that

$$\boldsymbol{\beta}_1 = \begin{pmatrix} \sigma_{22} & \sigma_{23} \\ \sigma_{23} & \sigma_{33} \end{pmatrix}^{-1} \begin{pmatrix} \sigma_{12} \\ \sigma_{13} \end{pmatrix}$$

so that

$$\beta_{12\cdot 3} = (\sigma_{33}\sigma_{12} - \sigma_{23}\sigma_{13})/(\sigma_{22}\sigma_{33} - \sigma_{23}^2) \tag{27.22}$$

$$= \frac{\beta_{12} - \beta_{13}\beta_{32}}{1 - \beta_{23}\beta_{32}}, \tag{27.23}$$

where $\beta_{12} = \sigma_{12}/\sigma_{22}$ and so on. By a similar argument to that in **27.7**, this result extends to

$$\beta_{12 \cdot 3m} = \frac{\beta_{12 \cdot m} - \beta_{13 \cdot m}\beta_{32 \cdot m}}{1 - \beta_{23 \cdot m}\beta_{32 \cdot m}} \qquad (27.24)$$

so that the regression coefficients of a given order may be expressed in terms of the next lowest order. This recurrence relation avoids the need for matrix inversion and is particularly useful in the context of stepwise regression; see **28.26**. Finally, we note from direct use of (27.9) that

$$\beta_{12 \cdot m} = \sigma_{12 \cdot m}/\sigma_{22 \cdot m}$$

and

$$\left.\begin{array}{c} \\ \\ \end{array}\right\} \qquad (27.25)$$

$$\beta_{21 \cdot m} = \sigma_{12 \cdot m}/\sigma_{11 \cdot m}$$

so that $\rho_{12 \cdot m}^2 = \beta_{12 \cdot m}\beta_{21 \cdot m}$ for any subset m.

27.11   Since

$$f(x_1|x_2, \mathbf{x}_m) = \frac{f(x_1, x_2|\mathbf{x}_m)}{f(x_2|\mathbf{x}_m)}, \qquad (27.26)$$

it follows from (27.17) that

$$\sigma_{1 \cdot 2m}^2 = \sigma_{1 \cdot m}^2 - (\sigma_{12 \cdot m}^2/\sigma_{2 \cdot m}^2)$$
$$= \sigma_{1 \cdot m}^2(1 - \rho_{12 \cdot m}^2). \qquad (27.27)$$

Applying (27.27) to successively smaller subsets of the random variables and substituting back into the original expression, we obtain eventually

$$\sigma_{1 \cdot 2 \cdots p}^2 = \sigma_1^2(1 - \rho_{12 \cdot 3 \cdots p}^2)(1 - \rho_{13 \cdot 4 \cdots p}^2) \cdots (1 - \rho_{1,(p-1) \cdot p}^2)(1 - \rho_{1p}^2); \qquad (27.28)$$

changing the order of the subscripts yields

$$\sigma_{1 \cdot 2 \cdots p}^2 = \sigma_1^2(1 - \rho_{12}^2)(1 - \rho_{13 \cdot 2}^2) \cdots (1 - \rho_{1p \cdot 23 \cdots (p-1)}^2). \qquad (27.29)$$

From (27.29), it is evident that the conditional variance will decrease when an additional variable is included, unless x_1 and x_p are conditionally uncorrelated $(\rho_{1p \cdot 23 \cdots (p-1)} = 0)$ when it is unchanged.

Approximate linear regression
27.12   In **27.9–11**, we have assumed the regression relationship to be exactly linear, as in (27.18). We now consider the evaluation of approximate linear regression functions when the joint distribution is non-normal. By the same reasoning as in **26.11**, we are led to consider the method of least squares. Thus, we may consider the linear function

$$x_1 = \boldsymbol{\beta}'\mathbf{x} = \sum_{j=2}^{p} \beta_j x_j \qquad (27.30)$$

and choose $\boldsymbol{\beta}$ to minimise the expected sum of squares

$$S = E\{(x_1 - \boldsymbol{\beta}'\mathbf{x})^2\}. \tag{27.31}$$

The minimum value for S is achieved when

$$E(\mathbf{xx}')\boldsymbol{\beta} = E(\mathbf{x}'x_1)$$

or

$$\boldsymbol{\beta} = \mathbf{C}_{22}^{-1}\mathbf{C}_{21}, \tag{27.32}$$

where $\mathbf{C}_{22} = \mathrm{var}\,(\mathbf{x})$ and $\mathbf{C}_{21} = \mathrm{cov}\,(\mathbf{x}, x_1)$. As in **26.11**, expectations are taken over *all* $(x_1, \ldots, x_p)$, so that (27.32) represents the *unconditional* solution. If we consider the conditional distribution for $x_{i1} | (x_{i2}, \ldots, x_{ip})$, $i = 1, 2, \ldots, n$, then we are led to minimise

$$S = E\left\{\sum_{i=1}^{n} (x_{i1} - \boldsymbol{\beta}'\mathbf{x}_i)^2\right\}$$
$$= E\{(\mathbf{x}_1 - \mathbf{X}\boldsymbol{\beta})'(\mathbf{x}_1 - \mathbf{X}\boldsymbol{\beta})\}, \tag{27.33}$$

where

$$\mathbf{x}_1' = (x_{11}, x_{21}, \ldots, x_{n1})$$

and

$$\mathbf{X}' = (\mathbf{x}_2, \mathbf{x}_3, \ldots, \mathbf{x}_n) = \{x_{ij}\}'.$$

The minimum value of (27.33) is given by the expected value of (19.12), namely

$$\boldsymbol{\beta} = (\mathbf{X}'\mathbf{X})^{-1}\mathbf{X}'E(\mathbf{x}_1); \tag{27.34}$$

the least squares estimator for $\boldsymbol{\beta}$ is given by putting $\mathbf{x}_1$ in place of $E(\mathbf{x}_1)$ in (27.34).

27.13 Although there is a difference in interpretation between the linear regressions in **27.9** and **27.12**, we see that the functional form of $\boldsymbol{\beta}$ is the same and that the LS estimators derived from (27.32) or (27.34) are the same as the ML estimators we obtain for the multinormal case. It follows that the results of this chapter apply whenever the least squares argument can be invoked. In particular, the approximate linear regression derived in **27.12** is often considered to be *the* regression function, without further qualification; this may be a questionable assumption.

This equivalence is also used to justify the use of partial correlations, whatever the underlying distribution. However, it should be noted that whereas zero partial correlations imply conditional independence for the multinormal, this is *not* true in general. Korn (1984) gives limits for the partial correlations under conditional independence.

Sample coefficients

27.14 If we are using a sample of n observations and fit regressions by least squares, all the relationships we have discussed will hold between the sample coefficients. Following our usual convention, we shall use r instead of ρ, b instead of β, and s^2 instead of σ^2 to distinguish the sample coefficients from their population equivalents. The bs are determined by minimising

$$ns_{1\cdot23\cdots p}^2 = (\mathbf{x}_1 - \mathbf{X}\boldsymbol{\beta})'(\mathbf{x}_1 - \mathbf{X}\boldsymbol{\beta}) \tag{27.35}$$

and the LS estimator is, from (19.12),

$$\mathbf{b} = (\mathbf{X}'\mathbf{X})^{-1}\mathbf{X}'\mathbf{x}_1. \tag{27.36}$$

Further, the sample analogue of (27.25) is

$$r_{ij\cdot m}^2 = b_{ij\cdot m}b_{ji\cdot m},$$

while the analogues of (27.16), (27.24) and (27.29) also hold.

From (27.36), we may define the sample *residuals*

$$\mathbf{x}_{1\cdot m} = \mathbf{x}_1 - \mathbf{X}\mathbf{b} = \mathbf{x}_1 - \mathbf{X}(\mathbf{X}'\mathbf{X})^{-1}\mathbf{X}'\mathbf{x}_1 = \mathbf{M}\mathbf{x}_1, \tag{27.37}$$

where the matrix $\mathbf{M} = \mathbf{I} - \mathbf{X}(\mathbf{X}'\mathbf{X})^{-1}\mathbf{X}'$ is symmetric and *idempotent*, i.e. $\mathbf{M}' = \mathbf{M} = \mathbf{M}^2$. Thus,

$$\mathbf{x}_{1\cdot m}'\mathbf{x}_{1\cdot m} = \mathbf{x}_1'\mathbf{M}'\mathbf{M}\mathbf{x}_1 = \mathbf{x}_1'\mathbf{M}\mathbf{x}_1 = \mathbf{x}_{1\cdot m}'\mathbf{x}_1. \tag{27.38}$$

Further, if we let $\mathbf{x}_m$ comprise only $(x_{k+1}, \ldots, x_p)$ and then obtain the regression functions for x_1 on $\mathbf{x}_m$ and for x_2 on $\mathbf{x}_m$, we have, from (27.37),

$$\mathbf{x}_{i\cdot m} = \mathbf{M}\mathbf{x}_i$$

and

$$\mathbf{x}_{1\cdot m}'\mathbf{x}_{2\cdot m} = \mathbf{x}_1'\mathbf{M}\mathbf{x}_2 = \mathbf{x}_1'\mathbf{x}_{2\cdot m}. \tag{27.39}$$

Finally, since the estimators $\mathbf{b}$ are obtained by differentiating (27.35) with respect to $\boldsymbol{\beta}$, we see that the jth estimating equation becomes

$$\mathbf{x}_j'(\mathbf{x}_1 - \mathbf{X}\mathbf{b}) = 0 \tag{27.40}$$

which yields, from (27.37)

$$\mathbf{x}_j'\mathbf{M}\mathbf{x}_1 = 0, \quad j = 2, \ldots, p. \tag{27.41}$$

Relations like (27.38) and (27.39) hold for the population errors as well as the sample residuals, but we shall find them of use mainly in sampling problems, which is why we have expressed them in terms of residuals. Exercise 27.5 gives the most general rule for the omission of common secondary subscripts in summing products of residuals or of errors.

Estimation of population coefficients

27.15   As for the zero-order correlations and regressions of the previous chapter, we may use the sample coefficients as estimators of their population equivalents. If the regression concerned is linear, we know from the least squares theory in Chapter 19 that any b is an unbiased estimator of the corresponding β and that

$$\frac{n}{n-(p-1)} \, s^2_{1 \cdot 23 \cdots p}$$

is an unbiased estimator of $\sigma^2_{1 \cdot 23 \cdots p}$. However, no r is an unbiased estimator of its ρ: we saw in **26.15–17** that even for a zero-order coefficient in the normal case, r is not unbiased for ρ, but that the modification (26.34) or (26.35) is an unbiased estimator. A result to be obtained in **27.23** will enable us to estimate any partial correlation coefficient analogously in the normal case.

Geometrical interpretation of partial correlation

27.16   From our results, it is clear that the whole complex of partial regressions, correlations and variances or covariances of errors or residuals is completely determined by the variances and correlations, or by the variances and regressions, of zero order. It is interesting to consider this result from the geometrical point of view.

Suppose in fact that we have n observations on p $(<n)$ variates

$$x_{11}, \ldots, x_{1p}; \, x_{21}, \ldots, x_{2p}; \, \ldots; \, x_{n1}, \ldots, x_{np}.$$

Consider a (Euclidean) sample space of n dimensions. The observations $\mathbf{x}'_k = (x_{1k}, \ldots, x_{nk})$ on the kth variate determine one point in this space, and there are p such points, one for each variate. Call these points $Q_1, Q_2, \ldots, Q_p$. We will assume that the xs are measured about their means, and take the origin to be P.

The quantity ns_l^2 may then be interpreted as the square of the length of the vector joining the point Q_l (with co-ordinates $\mathbf{x}_l$) to P. Similarly r_{lm} may be interpreted as the cosine of the angle Q_lPQ_m, for

$$r_{lm} = \frac{\mathbf{x}'_l\mathbf{x}_m}{(\mathbf{x}'_l\mathbf{x}_l\mathbf{x}'_m\mathbf{x}_m)^{1/2}} \tag{27.42}$$

which is the cosine of the angle between PQ_l and PQ_m.

Our result may then be expressed by saying that all the relations connecting the p points in the n-space are expressible in terms of the lengths of the vectors PQ_i and of the angles between them; and the theory of partial correlation and regression is thus exhibited as formally identical with the trigonometry of an n-dimensional constellation of points.

27.17   The reader who prefers the geometrical way of looking at this branch of the subject will have no difficulty in translating the foregoing equations into trigonometrical terminology. We will indicate only the more important results required for later sampling investigations.

Note in the first place that the p points Q_i and the point P determine (except perhaps in degenerate cases) a subspace of p dimensions in the n-space. Consider the point $Q_{1.2\cdots p}$ whose co-ordinates are the n residuals $x_{1.2\cdots p}$. In virtue of (27.40) the vector $PQ_{1.2\cdots p}$ is orthogonal to each of the vectors $PQ_2, \ldots, PQ_p$ and hence to the space of $(p-1)$ dimensions spanned by $P, Q_2, \ldots, Q_p$.

Consider now the residual vectors $Q_{1.m}, Q_{2.m}$, where m represents the secondary subscripts $3, 4, \ldots, (p-1)$. The cosine of the angle between them, say θ, is $r_{12.m}$ and each is orthogonal to the space spanned by $P, Q_3, \ldots, Q_{(p-1)}$. In Fig. 27.1, let M be the foot of the perpendicular from $Q_{1.m}$ on to PQ_p and $Q'_{2.m}$ a point on $PQ_{2.m}$ such that $Q'_{2.m}M$ is also perpendicular to PQ_p. Then $MQ_{1.m}$ and $MQ_{2.m}$ are orthogonal to the space spanned by $P, Q_3, \ldots, Q_p$, and the cosine of the angle between them, say ϕ, is $r_{12.mp}$. Thus, to express $r_{12.mp}$ in terms of $r_{12.m}$ we have to express ϕ in terms of θ, or the angle between the vectors $PQ_{1.m}$ and $PQ'_{2.m}$ in terms of that between their projections on the hyperplane perpendicular to PQ_p. We now drop the prime in $Q'_{2.m}$ for convenience. By Pythagoras' theorem,

$$(Q_{1.m}Q_{2.m})^2 = PQ_{1.m}^2 + PQ_{2.m}^2 - 2PQ_{1.m} \cdot PQ_{2.m} \cos \theta$$
$$= MQ_{1.m}^2 + MQ_{2.m}^2 - 2MQ_{1.m} \cdot MQ_{2.m} \cos \phi.$$

Further,

$$PQ_{1.m}^2 = PM^2 + MQ_{1.m}^2$$

and

$$PQ_{2.m}^2 = PM^2 + MQ_{2.m}^2,$$

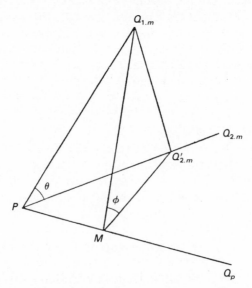

Fig. 27.1 The geometry of partial correlation

and hence we find

$$MQ_{1\cdot m}MQ_{2\cdot m}\cos\phi = PQ_{1\cdot m}PQ_{2\cdot m}\cos\theta - PM^2$$

or

$$\frac{MQ_{1\cdot m}}{PQ_{1\cdot m}}\cdot\frac{MQ_{2\cdot m}}{PQ_{2\cdot m}}\cos\phi = \cos\theta - \frac{PM}{PQ_{1\cdot m}}\cdot\frac{PM}{PQ_{2\cdot m}}. \tag{27.43}$$

Now $MQ_{1\cdot m}/PQ_{1\cdot m}$ and $PM/PQ_{1\cdot m}$ are the sine and cosine of the angle between PQ_p and $PQ_{1\cdot m}$. Since $PQ_{1\cdot m}$ is orthogonal to the space spanned by $P, Q_3, \ldots, Q_{p-1}$, its angle with PQ_p is unchanged if the latter is projected orthogonally to that space, i.e. if we replace PQ_p by $PQ_{p\cdot m}$. The cosine of the angle between $PQ_{1\cdot m}$ and $PQ_{p\cdot m}$ is $r_{1p\cdot m}$, and hence $PM/PQ_{1\cdot m} = r_{1p\cdot m}$, $MQ_{1\cdot m}/PQ_{1\cdot m} = (1 - r_{1p\cdot m}^2)^{1/2}$. The same result holds with the suffix 2 replacing 1. Thus, substituting in (27.43)

$$r_{12\cdot mp} = \frac{r_{12\cdot m} - r_{1p\cdot m}r_{2p\cdot m}}{\{(1 - r_{1p\cdot m}^2)(1 - r_{2p\cdot m}^2)\}^{1/2}}, \tag{27.44}$$

which is (27.16) again. We thus see that the expression of a partial correlation in terms of that of next lower order may be represented as the projection of an angle in the sample space on to a subspace orthogonal to the variable held fixed in the higher-order coefficient alone.

27.18   We now give an example to illustrate how partial correlations may be interpreted.

Example 27.1
In some investigations into the variation of crime in 16 large cities in the USA, Ogburn (1935) found a correlation of -0.14 between crime rate (x_1) as measured by the number of known offences per thousand of the population and church membership (x_5) as measured by the number of church members of 13 years of age or over per 100 of total population of 13 years of age or over. The obvious inference is that religious belief acts as a deterrent to crime. Let us consider this more closely.
If x_2 = percentage of males
$\quad x_3$ = percentage of total population who are male immigrants, and
$\quad x_4$ = number of children under 5 years old per 1000 married women between
$\qquad\qquad$ 15 and 44 years old,
Ogburn finds the values

$$r_{12} = +0.44, \qquad r_{24} = -0.19,$$
$$r_{13} = -0.34, \qquad r_{25} = -0.35,$$
$$r_{14} = -0.31, \qquad r_{34} = +0.44,$$
$$r_{15} = -0.14, \qquad r_{35} = +0.33,$$
$$r_{23} = +0.25, \qquad r_{45} = +0.85.$$

From these data it may be shown that the regression of x_1 on the other four variates, measured in standard form, is

$$\hat{x}_1 = 0.742x_2 - 0.512x_3 - 0.681x_4 + 0.868x_5$$
$$\phantom{\hat{x}_1 = } (3.27) \quad (-2.20) \quad (-1.75) \quad (2.20) \tag{27.45}$$

where the numbers in brackets are the t-ratios calculated from (27.50) below.

The partial correlations include

$$r_{15\cdot3} = -0.03,$$
$$r_{15\cdot4} = +0.25,$$
$$r_{15\cdot34} = +0.23.$$

From (27.45), we see that when the other factors are constant x_1 and x_5 are positively related, i.e. church membership appears to be positively associated with crime. How does this effect come to be masked so as to give a negative correlation in the coefficient of zero order r_{15}?

First of all, the partial correlation between crime and church membership when the effect of x_3, the percentage of immigrants, is eliminated, is near zero. Also, the partial correlations given x_4, the number of young children, and given both x_3 and x_4, are positive. It appears from the regression equation that a high percentage of immigrants and a high proportion of children are negatively associated with the crime-rate. Both these factors are positively correlated with church membership, which masks the positive association with crime of church membership among other members of the population.

Path analysis

27.19 It is apparent from Example 27.1 that disentangling the effects of multiple possible 'causes' is no easy matter. Sewall Wright, the geneticist, developed the method of *path analysis* (Wright, 1923, 1934) for 'working out the logical consequences of a hypothesis as to the causal relations in a system of correlated variables' (Wright, 1923, p. 254). In genetics, such well-ordered systems often exist, although the method has seen wide application in less structured areas.

To begin, we note that a path diagram for the variables $(x_1, x_2, \ldots, x_p)$ is constructed so that the pair $(x_i, x_j, i < j)$ or simply (i, j) is: unconnected, if there is no direct dependency; connected by a one-directional arrow (from j to i) if x_i depends directly on x_j; connected by a two-directional arrow if x_i and x_j are directly related but no direction has been established.

We shall suppose that the variates are decomposed into two groups $\mathbf{x}_1' = (x_1, \ldots, x_k)$ and $\mathbf{x}_2' = (x_{k+1}, \ldots, x_p)$ such that $\mathbf{x}_1$ represents the *endogenous* variables of direct interest, whereas $\mathbf{x}_2$ denotes the *exogenous* or explanatory variables. We assume that no hypothesis has been advanced about the structure of $\mathbf{x}_2$ so that the relations between its components may be connected by two-directional arrows. Finally, we assume that the variables in $\mathbf{x}_1$ may be ordered so that x_i depends only on x_j

$(j > i)$. These assumptions give rise to a *recursive* system of equations such as

$$x_1 - \beta_{12}x_2 - \beta_{13}x_3 - \cdots - \beta_{1k}x_k - \beta_{1,k+1}x_{k+1} - \cdots - \beta_{1p}x_p = u_1$$
$$x_2 - \beta_{23}x_3 - \cdots - \beta_{2k}x_k - \beta_{2,k+1}x_{k+1} - \cdots - \beta_{2p}x_p = u_2$$
$$\vdots \qquad \qquad \vdots \qquad \qquad \vdots \qquad \qquad \vdots$$
$$x_k - \beta_{k,k+1}x_{k+1} - \cdots - \beta_{kp}x_p = u_p \qquad (27.46)$$

or

$$(\mathbf{I} - \mathbf{A})\mathbf{x}_1 = \mathbf{B}\mathbf{x}_2 + \mathbf{u}, \qquad (27.47)$$

where

$$\mathbf{A} = \begin{pmatrix} 0 & \beta_{12} & \cdots & & \beta_{1k} \\ & & \ddots & & \vdots \\ & & & & \beta_{k-1,k} \\ 0 & & & & 0 \end{pmatrix} \quad \text{and} \quad \mathbf{B} = \begin{pmatrix} \beta_{1,k+1} & \cdots & \beta_{1p} \\ \vdots & & \vdots \\ \beta_{k,k+1} & & \beta_{kp} \end{pmatrix}.$$

The $\{\beta_{ij}\}$ are known as the *path coefficients*, and $\mathbf{u}$ denotes a vector of uncorrelated random errors. The system of equations in (27.47) is known as *recursive* or sometimes as a *causal chain* (Wold, 1960); note that the implication of causality comes from the prior specification of the system, and the role of statistical analysis is to examine whether or not the data support the hypothesis. In general, the hypothesis will include additional statements regarding the relationship between variables, leading to certain pairs being unconnected ($\beta_{ij} = 0$).

Example 27.2
Returning to the data of Ogburn (1935), presented in Example 27.1, we treat (x_1, x_4, x_5) as endogenous and (x_2, x_3) as exogenous. A possible structure for the variables is suggested by the following equations, which are equivalent to the path diagram shown in Fig. 27.2:

$$x_1 \qquad - \beta_{15}x_5 - \beta_{12}x_2 - \beta_{13}x_3 = u_1$$
$$x_4 - \beta_{45}x_5 \qquad - \beta_{43}x_3 = u_2$$
$$x_5 \qquad - \beta_{53}x_3 = u_3$$

That is, we consider church membership to be affected by the levels of immigration and numbers of children to be affected by both variables. Finally, crime is assumed to be directly affected by the total number of males, immigration, and church membership.

27.20   Wold (1960) showed that recursive systems like (27.47) may be estimated by the method of least squares. Further, if the u_i are taken to be independent and normally distributed, these estimators are precisely those given by ML. The estimation is usually carried out after standardising the variables to have zero mean and unit variance so that the path coefficients may be computed from the correlation matrix

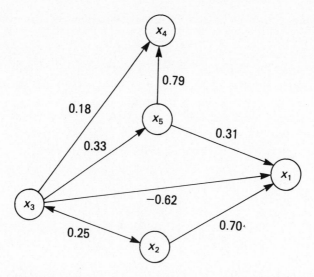

Fig. 27.2   Path diagram for data of Example 27.1, with estimated path coefficients shown

and are dimensionless, making comparisons easier. The significance of individual coefficients may be tested using the t-statistic given in (27.50) below, as was done for the original regression equation (27.45).

Example 27.3
Following Example 27.2, the estimated equations are given by

$$\hat{x}_5 = 0.33x_3$$
$$(1.37)$$
$$\hat{x}_4 = 0.179x_3 + 0.791x_5$$
$$(1.22) \qquad (5.40)$$

and

$$\hat{x}_1 = 0.703x_2 - 0.618x_3 + 0.310x_5.$$
$$(2.87) \quad (-2.54) \qquad (1.23)$$

The path coefficients are also included in Fig. 27.2. The equation for $\hat{x}_1$ should be compared with that given in (27.45). Since $n = 16$, the significance of several of the coefficients is doubtful, and the example should be viewed as being for illustrative purposes only.

27.21   When **A** in (27.47) has all upper-triangular elements present, the system is said to be *complete*. When some coefficients are preset to zero, the system is termed *incomplete*. Since the motive for studying such a system is to interpret the *correlation* structure, it is encouraging to know that when the system is complete, the covariance matrix is unrestricted and so the sample covariance matrix is the ML estimator. However, when the system is incomplete, the ML estimator for the covariance matrix does not necessarily coincide with the observed covariance matrix whenever (i, j) are connected. The heuristic estimator proposed by Wright (1934) for the incomplete case cannot, therefore, be justified in general. Conditions under which Wright's rule does correspond to ML estimation are given by Wermuth (1980).

Although the direct interpretation of the partial correlations may be lost, path diagrams are more readily interpreted by means of models like (27.47). When the system is non-recursive, more complex estimators are required (cf. Joreskög, 1981).

Sampling distributions of partial correlation and regression coefficients in the normal case

27.22   We now consider the sampling distributions of the partial correlation and regression coefficients in the normal case.

For large samples, the standard errors appropriate to zero-order coefficients (cf. **26.13**) may be used with obvious adjustments. Writing m for a set of secondary subscripts, we have, from (26.24),

$$\operatorname{var} r_{12 \cdot m} \doteqdot \frac{1}{n}(1 - \rho^2_{12 \cdot m})^2, \tag{27.48}$$

and from (26.30)

$$\operatorname{var} b_{12 \cdot m} \doteqdot \frac{1}{n}\frac{\sigma^2_{1 \cdot m}}{\sigma^2_{2 \cdot m}}(1 - \rho^2_{12 \cdot m}) = \frac{1}{n}\frac{\sigma^2_{1 \cdot 2m}}{\sigma^2_{2 \cdot m}}, \tag{27.49}$$

by (27.27). The proof of (27.48) and (27.49) by the direct methods of Chapter 10 is very tedious. They follow directly, however, from noting that the joint distribution of any two errors $x_{1 \cdot m}$ and $x_{2 \cdot m}$ is bivariate normal with correlation coefficient $\rho_{12 \cdot m}$. It follows, as Yule (1907) pointed out, that the sample correlation and regressions between the corresponding residuals have at least the large-sample distribution of a zero-order coefficient. We shall see in **27.23** that in a sample of size n, the exact distribution of $r_{12 \cdot m}$ is that of a zero-order correlation based on $(n - d)$ observations, where d is the number of secondary subscripts in m.

27.23   Consider now the geometrical representation of **27.16–17**. Suppose that we have three vectors PQ_1, PQ_2, PQ_3, representing n observations on x_1, x_2, x_3. As we saw in **27.17**, the partial correlation $r_{12 \cdot 3}$ is the cosine of the angle between PQ_1 and PQ_2 projected on to the subspace orthogonal to PQ_3, which is of dimension $(n - 1)$. If we make an orthogonal transformation (i.e. a rotation of the co-ordinate axes), the correlations, being cosines of angles, are unaffected; moreover, if the n original observations on the three variables are independent of each other, the n observations on the orthogonally transformed variables will also be. (This is a generalisation of the result of Examples 11.2 and 11.3 and of **15.27** for independent x_1, x_2, x_3, and its proof is left for the reader as Exercise 27.7; it is geometrically obvious from the radial symmetry of the standardised multinormal distribution.) If

PQ_3 is taken as one of the new co-ordinate axes in the orthogonal transformation, the distribution of $r_{12\cdot3}$ is at once seen to be the same as that of a zero-order coefficient based on $(n-1)$ independent observations. By repeated application of this argument, it follows that the distribution of a correlation coefficient of order d based on n observations is that of a zero-order coefficient based on $(n-d)$ observations: each secondary subscript involves a projection in the sample space orthogonal to that variable and a loss of one degree of freedom. The result is due to Fisher (1924a).

The results of the previous chapter are thus immediately applicable to partial correlations, with this adjustment. If d is small compared with n, the distribution of partial correlations as n increases is effectively the same as that of zero-order coefficients, confirming the approximation (27.48) to the standard error.

It also follows for partial regression coefficients that the zero-order coefficient distribution (16.86) persists when the set m of secondary subscripts is added throughout, with n replaced by $(n-d)$. In particular, the Student's distribution of (26.38) becomes, for the regression of x_1 on x_2, that of

$$t = (b_{12\cdot m} - \beta_{12\cdot m}) \left\{ \frac{s_{2\cdot m}^2(n-d-2)}{s_{1\cdot m}^2(1-r_{12\cdot m}^2)} \right\}^{1/2} \tag{27.50}$$

with $(n-d-2)$ degrees of freedom. If the set m consists of all $(p-2)$ other variates, there are $(n-p)$ degrees of freedom. Since the regression coefficients are functions of distances (variances) as well as angles in the sample space, the distribution of b_{12} itself, unlike that of r, is not directly preserved under projection with only degrees of freedom being reduced; the statistics $s_{1\cdot m}^2$, $s_{2\cdot m}^2$ in (27.50) make the necessary 'distance' adjustments for the projections.

The multiple correlation coefficient
27.24 The variance in the population of x_1 about its regression on the other variates (27.18) is $\sigma_{1\cdot2\cdots p}^2$, defined in **27.9**. We now define the *multiple correlation coefficient*[*] $R_{1(2\cdots p)}$ between x_1 and $x_2, \ldots, x_p$ by

$$1 - R_{1(2\cdots p)}^2 = \sigma_{1\cdot2\cdots p}^2 / \sigma_1^2. \tag{27.51}$$

From (27.51) and (27.29),

$$0 \leqslant R^2 \leqslant 1.$$

We shall define R as the positive square root of R^2: it is always non-negative. R is evidently not symmetric in its subscripts, and it is, indeed, a measure of the *dependence* of x_1 upon $x_2, \ldots, x_p$.

---

[*] We use a bold-face R for the population coefficient, and will later use an ordinary capital R for the corresponding sample coefficient: we are reluctant to use the Greek capital for the population coefficient, in accordance with our usual convention, because it resembles a capital P, which might be confusing.

To justify its name, we have to show that it is in fact a correlation coefficient. We have, from **27.9** and **27.14**,

$$\sigma^2_{1\cdot2\cdots p} = E(x^2_{1\cdot2\cdots p}), \tag{27.52}$$

and by the population analogue of (27.38),

$$E(x^2_{1\cdot2\cdots p}) = E(x_1 x_{1\cdot2\cdots p}). \tag{27.53}$$

(27.52) and (27.53) give, since $E(x_{1\cdot2\cdots p}) = 0$,

$$\sigma^2_{1\cdot2\cdots p} = \text{var } (x_{1\cdot2\cdots p}) = \text{cov } (x_1, x_{1\cdot2\cdots p}). \tag{27.54}$$

If we now consider the correlation between x_1 and its conditional expectation

$$E(x_1 | x_2, \ldots, x_p) = x_1 - x_{1\cdot2\cdots p},$$

we find that this is

$$\frac{\text{cov } (x_1, x_1 - x_{1\cdot2\cdots p})}{\{\text{var } x_1 \text{ var } (x_1 - x_{1\cdot2\cdots p})\}^{1/2}} = \frac{\text{var } x_1 - \text{cov } (x_1, x_{1\cdot2\cdots p})}{[\text{var } x_1\{\text{var } x_1 + \text{var } x_{1\cdot2\cdots p} - 2 \text{ cov } (x_1, x_{1\cdot2\cdots p})\}]^{1/2}},$$

and using (27.54) this is

$$\frac{\sigma^2_1 - \sigma^2_{1\cdot2\cdots p}}{\{\sigma^2_1(\sigma^2_1 - \sigma^2_{1\cdot2\cdots p})\}^{1/2}} = \left\{\frac{\sigma^2_1 - \sigma^2_{1\cdot2\cdots p}}{\sigma^2_1}\right\}^{1/2} = R_{1(2\cdots p)}, \tag{27.55}$$

by (27.51). Thus $R_{1(2\cdots p)}$ is the ordinary product-moment correlation coefficient between x_1 and the conditional expectation $E(x_1 | x_2, \ldots, x_p)$. Since the sum of squared errors (and therefore their mean $\sigma^2_{1\cdot2\cdots p}$) is minimised in finding the least squares regression, which is identical with $E(x_1 | x_2, \ldots, x_p)$ (cf. **27.12**), it follows from (27.55) that $R_{1(2\cdots p)}$ is the correlation between x_1 and the 'best-fitting' linear combination of $x_2, \ldots, x_p$. No other linear function of $x_2, \ldots, x_p$ will have greater correlation with x_1.

27.25 From (27.51) and (27.29), we have

$$1 - R^2_{1(2\cdots p)} = (1 - \rho^2_{12})(1 - \rho^2_{13\cdot2}) \cdots (1 - \rho^2_{1p\cdot23\cdots(p-1)}), \tag{27.56}$$

expressing the multiple correlation coefficient in terms of the partial correlations. Since permutation of the subscripts other than 1 is allowed in (27.29), it follows at once from (27.56) that, since each factor on the right is in the interval $[0, 1]$,

$$1 - R^2_{1(2\cdots p)} \leqslant 1 - \rho^2_{1j\cdot s}$$

where $\rho_{1j\cdot s}$ is any partial or zero-order coefficient having 1 as a primary subscript. Thus

$$R_{1(2\cdots p)} \geqslant |\rho_{1j\cdot s}|;$$

the multiple correlation coefficient is no less in value than the absolute value of any correlation coefficient with a common primary subscript. It follows that if $R_{1(2\cdots p)} = 0$, all the corresponding $\rho_{1j\cdot s} = 0$ also, so that x_1 is completely uncorrelated with all the other variables. On the other hand, if $R_{1(2\cdots p)} = 1$, at least one $\rho_{1j\cdot s}$ must be 1 also to make the right-hand side of (27.56) equal to zero (Exercise 27.21 shows that all zero-order ρ_{1j} may nevertheless be arbitrarily small). In this case, (27.51) shows that $\sigma^2_{1 \cdot 2 \cdots p} = 0$, so that all points in the distribution of x_1 lie on the regression line, and x_1 is a strict linear function of $x_2, \ldots, x_p$.

Thus $R_{1(2 \cdots p)}$ is a measure of the linear dependence of x_1 upon $x_2, \ldots, x_p$.

27.26 So far, we have considered the multiple correlation coefficient between x_1 and all the other variates, but we may evidently also consider the multiple correlation of x_1 and any subset. Thus we define

$$R^2_{1(s)} = 1 - \frac{\sigma^2_{1 \cdot s}}{\sigma^2_1} \tag{27.57}$$

for any set of subscripts s. It now follows immediately from (27.29) that

$$\sigma^2_{1 \cdot s} \leqslant \sigma^2_{1 \cdot r}, \tag{27.58}$$

where r is any subset of s: the error variance cannot be increased by the addition of a further variate. We thus have, from (27.57) and (27.58), relations of the type

$$R^2_{1(2)} \leqslant R^2_{1(23)} \leqslant R^2_{1(234)} \leqslant \cdots \leqslant R^2_{1(2 \cdots p)}, \tag{27.59}$$

expressing the fact that the multiple correlation coefficient can never be reduced by adding to the set of variables upon which the dependence of x_1 is to be measured.

In the particular case $p = 2$, we have from (27.56)

$$R^2_{1(2)} = \rho^2_{12}, \tag{27.60}$$

so that $R_{1(2)}$ is the absolute value of the ordinary correlation coefficient between x_1 and x_2.

Geometrical interpretation of multiple correlation

27.27 We may interpret $R_{1(2 \cdots p)}$ in the geometrical terms of **27.16–17**. Consider first the interpretation of the least squares regression (27.18): by **27.24**, it is that linear function of the variables $x_2, \ldots, x_p$ which minimises the sum of squares (27.31). Thus we choose the vector PV in the $(p-1)$-dimensional subspace spanned by P, $Q_2, \ldots, Q_p$, which minimises the distance $Q_1 V$, i.e. which minimises the angle between PQ_1 and PV. By (27.55), $R_{1(2 \cdots p)}$ is the cosine of this minimised angle. But this means that $R_{1(2 \cdots p)}$ is the cosine of the angle between PQ_1 and the $(p-1)$-dimensional subspace itself, for otherwise the angle would not be minimised.

If $R_{1(2 \cdots p)} = 0$, PQ_1 is orthogonal to the $(p-1)$-subspace so that x_1 is uncorrelated with $x_2, \ldots, x_p$ and with any linear function of them. If, on the other hand, $R_{1(2 \cdots p)} = 1$, PQ_1 lies in the $(p-1)$-subspace, so that x_1 is a strict linear function of $x_2, \ldots, x_p$. These are the results we obtained in **27.25**.

We shall find this geometrical interpretation helpful in deriving the distribution of the sample coefficient $R_{1(2 \cdots p)}$ in the normal case. It is a direct generalisation of the representation used in **16.24** to obtain the distribution of the ordinary product-moment correlation coefficient r which, as we observed at (27.60), is essentially the signed value of $R_{1(2)}$.

The sample multiple correlation coefficient and its conditional distribution

27.28   We now define the sample analogue of $R^2_{1(2\cdots p)}$ by

$$1 - R^2_{1(2\cdots p)} = \frac{s^2_{1\cdot 2\cdots p}}{s^2_1}, \tag{27.61}$$

and all the relations of **27.24–6** hold with the appropriate substitutions of r for ρ, and s for σ. We proceed to discuss the sampling distribution of R^2 in detail. Since, by **27.24**, it is a correlation coefficient, whose value is independent of location and scale, its distribution will be free of location and scale parameters.

First, consider the conditional distribution of R^2 when the values of $x_2, \ldots, x_p$ are fixed. From **27.14** we may write

$$\begin{aligned}
\mathbf{x}'_1\mathbf{x}_1 &\equiv \mathbf{x}'_1(\mathbf{I}-\mathbf{M})\mathbf{x}_1 + \mathbf{x}'_1\mathbf{M}\mathbf{x}_1 \\
&= \mathbf{x}'_1(\mathbf{I}-\mathbf{M})\mathbf{x}_1 + \mathbf{x}'_{1\cdot m}\mathbf{x}_{1\cdot m}.
\end{aligned} \tag{27.62}$$

Since $ns^2_1 = \mathbf{x}'_1\mathbf{x}_1$ and $ns^2_{1\cdot 2\cdots p} = \mathbf{x}'_{1\cdot m}\mathbf{x}_{1\cdot m}$, it follows from (27.61) and (27.62) that

$$\begin{aligned}
ns^2_1 &\equiv n(s^2_1 - s^2_{1\cdot 2\cdots p}) + ns^2_{1\cdot 2\cdots p}, \\
&\equiv ns^2_1 R^2_{1(2\cdots p)} + ns^2_1(1 - R^2_{1(2\cdots p)}).
\end{aligned} \tag{27.63}$$

If the observations on x_1 are independent normal variates, so that $R^2_{1(2\cdots p)} = 0$, and we standardise them, the left-hand side of (27.63) is distributed in the chi-squared form with $(n-1)$ degrees of freedom, and the quadratic forms in x_1 on the right of (27.62) may be shown to have ranks $(p-1)$ and $(n-p)$ respectively, since $\mathbf{M}$ is idempotent with rank $(p-1)$. It follows by Cochran's theorem (**15.16**) that they are independently distributed as chi-squared with these degrees of freedom and that the ratio

$$F = \frac{R^2_{1(2\cdots p)}/(p-1)}{(1 - R^2_{1(2\cdots p)})/(n-p)} \tag{27.64}$$

has the F distribution with $(p-1, n-p)$ degrees of freedom, a result first given by Fisher (1924b).

This is another example of an LR test of a linear hypothesis. We postulate that the mean of the observations of x_1 is a linear function of $(p-1)$ other variables, with $(p-1)$ coefficients and a constant term, p parameters in all. We test the hypothesis that all $(p-1)$ coefficients are zero, i.e. $H_0: R^2 = 0$. In the notation of **23.27–8**, we have $k = p$, $r = p - 1$ so that the F-test (27.64) has $(p-1, n-p)$ degrees of freedom, as we have seen. It follows immediately from **23.29–31** that when H_0 is not true, F at (27.64) has a non-central F-distribution with degrees of freedom $p-1$ and $n-p$. Its non-central parameter λ is now derived. (27.63) reflects the identity

$$x_1 \equiv (x_1 - x_{1\cdot 2\cdots p}) + x_{1\cdot 2\cdots p}.$$

With $x_2, \ldots, x_p$ fixed, the variance of x_1 is $\sigma^2_{1\cdot(2\cdots p)} = \sigma^2_1(1 - R^2_{1(2\cdots p)})$ by (27.51), and $x_{1\cdot 2\cdots p}$ has zero mean, but $(x_1 - x_{1\cdot 2\cdots p})$ has mean $\boldsymbol{\beta}'\mathbf{x}$. Thus

$$\frac{ns^2_{1\cdot 2\cdots p}}{\sigma^2_1(1 - R^2)}$$

has a central χ^2 distribution with $(n-p)$ degrees of freedom, but

$$\frac{n(s_1^2 - s_{1\cdot2\cdots p}^2)}{\sigma_1^2(1 - R^2)}$$

has a non-central χ^2 distribution with $(p-1)$ degrees of freedom, and, from **23.4**,

$$\lambda = \sum_i \beta_i' x_i x_i' \beta / \{\sigma_1^2(1 - R^2)\}$$

$$= \beta' V \beta / \{\sigma_1^2(1 - R^2)\},$$

where β is the $(p-1) \times 1$ vector of regression coefficients and V is the $(p-1) \times (p-1)$ observed covariance matrix of $x_2, \ldots, x_p$. From **23.31**, λ is also the parameter of the non-central F-distribution of (27.64), reducing to zero when $R^2 = 0$, i.e. $\beta = 0$.

The multinormal (unconditional) case

27.29 If we now allow the values of $x_2, \ldots, x_p$ to vary also, and suppose that we are sampling from a multinormal population, we find that the distribution of R^2 is unchanged if $R^2 = 0$, but quite different otherwise from that of R^2 with $x_2, \ldots, x_p$ fixed. Thus the power function of the test of $R^2 = 0$ is different in the two cases, although the same test is valid in each case. As $n \to \infty$, however, the limiting results are identical in the two cases.

We derive the multinormal result for $R^2 = 0$ geometrically, and proceed to generalise it in **27.30**.

Consider the geometrical representation of **27.27**. R is the cosine of the angle, say θ, between PQ_1 (the x_1-vector) and the vector PV, in the $(p-1)$-dimensional space S_{p-1} of the other variables, which makes the minimum angle with PQ_1. If $R = 0$, x_1 is, since the population is multinormal, independent of $x_2, \ldots, x_p$, and the vector PQ_1 will then, because of the radial symmetry of the normal distribution, be randomly directed with respect to S_{p-1}, which we may therefore regard as fixed in the subsequent argument. (We therefore see how it is that the conditional and unconditional results coincide when $R^2 = 0$.)

We have to consider the relative probabilities with which different values of θ may arise. For fixed variance s_1^2, the probability density of the sample of n observations is constant upon the $(n-2)$-dimensional surface of an $(n-1)$-dimensional hypersphere. If θ and PV are fixed, PQ_1 is constrained to lie upon a hypersphere of $(n-2) - (p-1) = (n-p-1)$ dimensions, whose content is proportional to $(\sin\theta)^{n-p-1}$ (cf. **16.24**). Now consider what happens when PV varies. PV is free to vary within S_{p-1}, where by radial symmetry it will be equiprobable on the $(p-2)$-dimensional surface of a $(p-1)$-sphere. This surface has content proportional to $(\cos\theta)^{p-2}$. For fixed θ, therefore, we have the probability element $(\sin\theta)^{n-p-1}(\cos\theta)^{p-2} d\theta$. Putting $R = \cos\theta$, and $d\theta \propto d(R^2)/\{R(1 - R^2)^{1/2}\}$, we find for the distribution of R^2 the beta distribution

$$dF \propto (R^2)^{(p-3)/2}(1 - R^2)^{(n-p-2)/2} d(R^2), \qquad 0 \leq R^2 \leq 1. \qquad (27.65)$$

The constant of integration is easily seen to be $1/B\{\tfrac{1}{2}(p-1),\tfrac{1}{2}(n-p)\}$. The transformation (27.64) applied to (27.65) then gives us exactly the same F distribution as that derived for $x_2, \ldots, x_p$ fixed in **27.28**. When $p=2$, (27.65) reduces to (16.62), which is expressed in terms of dR rather than $d(R^2)$.

27.30   We now turn to the case when $\boldsymbol{R} \neq 0$. The distribution of R in this case was first given by Fisher (1928a) by a considerable development of the geometrical argument of **27.29**. We give a much simpler derivation due to Moran (1950).
   We may write (27.56) for the sample coefficient as

$$1 - R^2_{1(2\cdots p)} = (1 - r^2_{12})(1 - T^2), \tag{27.66}$$

say, where T is the multiple correlation coefficient between $x_{1\cdot 2}$ and $x_{3\cdot 2}, x_{4\cdot 2}, \ldots, x_{p\cdot 2}$. Now $\boldsymbol{R}_{1(2\cdots p)}$ and the distribution of $R_{1(2\cdots p)}$ are unaffected if we make an orthogonal transformation of $x_2, \ldots, x_p$ so that x_2 itself is the linear function of $x_2, \ldots, x_p$ which has maximum correlation with x_1 in the population, i.e. $\rho_{12} = \boldsymbol{R}_{1(2\cdots p)}$. It then follows from (27.56) that

$$\rho^2_{13\cdot 2} = \rho^2_{14\cdot 23} = \cdots = \rho^2_{1p\cdot 23\cdots(p-1)} = 0,$$

and since subscripts other than 1 may be permuted in (27.56), it follows that *all* partial coefficients of form $\rho_{1j\cdot 2s} = 0$. Thus $x_{1\cdot 2}$ is uncorrelated with (and since the variation is normal, independent of) $x_{3\cdot 2}, x_{4\cdot 2}, \ldots, x_{p\cdot 2}$, and T in (27.66) is distributed as a multiple correlation coefficient, based on $(n-1)$ observations (since we lose one dimension by projection for the residuals), between one variate and $(p-2)$ others, with $\boldsymbol{R} = 0$. Moreover, T is distributed independently of r_{12}, for all the variates $x_{j\cdot 2}$ are orthogonal to x_2 by (27.41). Thus the two factors on the right of (27.66) are independently distributed. The distribution of r_{12}, say $f_1(r)$, is (16.60) with $\rho = \boldsymbol{R}_{1(2\cdots p)}$, integrated for β over its range, while that of T^2, say $f_2(R^2)$, is (27.65) with n and p each reduced by 1. We therefore have from (27.66) the distribution of R^2

$$dF = \int_{-R}^{R} \left\{ f_2\!\left(\frac{R^2 - r^2}{1 - r^2}\right) \right\} dF_1(r) \tag{27.67}$$

which, dropping all suffixes for convenience, is

$$
= \frac{(n-2)}{\pi} (1 - \boldsymbol{R}^2)^{(n-1)/2} \int_{-R}^{R} (1 - r^2)^{(n-4)/2} \int_{0}^{\infty} \frac{d\beta}{(\cosh \beta - \boldsymbol{R}r)^{n-1}}
$$
$$
\times \left\{ \frac{1}{B\{\tfrac{1}{2}(p-2), \tfrac{1}{2}(n-p)\}} \left(\frac{R^2 - r^2}{1 - r^2}\right)^{(p-4)/2} \left(\frac{1 - R^2}{1 - r^2}\right)^{(n-p-2)/2} \frac{dR^2}{(1 - r^2)} \right\} dr
$$
$$
= \frac{(n-2)}{\pi} \frac{(1 - \boldsymbol{R}^2)^{(n-1)/2} (1 - R^2)^{(n-p-2)/2} \, d(R^2)}{B\{\tfrac{1}{2}(p-2), \tfrac{1}{2}(n-p)\}}
$$
$$
\times \int_{-R}^{R} (R^2 - r^2)^{(p-4)/2} \left[\int_{0}^{\infty} \frac{d\beta}{(\cosh \beta - \boldsymbol{R}r)^{n-1}} \right] dr. \tag{27.68}
$$

We can substitute for the inner integral as at (16.64)–(16.65). If in (27.68) we put $r = R \cos \psi$ and write the integral with respect to β from $-\infty$ to ∞, dividing by 2 to compensate for this, we obtain Fisher's form of the distribution,

$$dF = \frac{\Gamma(\tfrac{1}{2}n)(1 - R^2)^{(n-1)/2}}{\pi \Gamma\{\tfrac{1}{2}(p-2)\}\Gamma\{\tfrac{1}{2}(n-p)\}} (R^2)^{(p-3)/2}(1 - R^2)^{(n-p-2)/2} \, d(R^2)$$

$$\times \int_0^\pi \sin^{p-3} \psi \left\{ \int_{-\infty}^\infty \frac{d\beta}{(\cosh \beta - RR \cos \psi)^{n-1}} \right\} d\psi. \qquad (27.69)$$

27.31 The distribution (27.69) may be expressed as a hypergeometric function. Expanding the integrand in a uniformly convergent series of powers of $\cos \psi$, it becomes, since odd powers of $\cos \psi$ will vanish on integration from 0 to π,

$$\sum_{j=0}^\infty \binom{n+2j-2}{2j} \frac{\sin^{p-3} \psi \cos^{2j} \psi}{(\cosh \beta)^{n-1+2j}} (RR)^{2j}$$

and since

$$\int_0^\pi \cos^{2j} \psi \sin^{p-3} \psi \, d\psi = B\{\tfrac{1}{2}(p-2), \tfrac{1}{2}(2j+1)\}$$

and

$$\int_{-\infty}^\infty \frac{d\beta}{(\cosh \beta)^{n-1+2j}} = B\{\tfrac{1}{2}, \tfrac{1}{2}(n+2j-1)\},$$

the integral in (27.69) becomes

$$\sum_{j=0}^\infty \binom{n+2j-2}{2j} B\{\tfrac{1}{2}(p-2), \tfrac{1}{2}(2j+1)\} B\{\tfrac{1}{2}, \tfrac{1}{2}(n+2j-1)\}(RR)^{2j},$$

and on writing this out in terms of gamma functions and simplifying, it becomes

$$= \frac{\pi \Gamma\{\tfrac{1}{2}(p-2)\}\Gamma\{\tfrac{1}{2}(n-1)\}}{\Gamma(\tfrac{1}{2}n)\Gamma\{\tfrac{1}{2}(p-1)\}} F\{\tfrac{1}{2}(n-1), \tfrac{1}{2}(n-1), \tfrac{1}{2}(p-1), R^2R^2\}. \qquad (27.70)$$

Substituting (27.70) for the integral in (27.69), we obtain

$$dF = \frac{(R^2)^{(p-3)/2}(1 - R^2)^{(n-p-2)/2} \, d(R^2)}{B\{\tfrac{1}{2}(p-1), \tfrac{1}{2}(n-p)\}} \cdot (1 - R^2)^{(n-1)/2}$$

$$\times F\{\tfrac{1}{2}(n-1), \tfrac{1}{2}(n-1), \tfrac{1}{2}(p-1), R^2R^2\}. \qquad (27.71)$$

This unconditional distribution should be compared with the conditional distribution of R^2 given in Exercise 27.13. Exercise 27.14 shows that as $n \to \infty$, both yield the same non-central χ^2 distribution for nR^2.

The first factor on the right of (27.71) is the distribution (27.65) when $R = 0$, the second factor then being unity. (27.71) generally converges slowly, for the first two arguments in the hypergeometric function are $\tfrac{1}{2}(n-1)$. Lee (1971) gives recurrence relations for the f.f. and d.f. of R^2 (cf. Exercise 16.14 for r) which make the computation of its distribution straightforward when p is even.

Exercises 27.22–4 derive results for the distribution of $R^2/(1-R^2)$ in the multinormal case from the conditional distribution in **27.28**.

The moments and limiting distributions of R^2

27.32   It may be shown (cf. Wishart (1931)) that the mean value of R^2 in the multinormal case is

$$E(R^2) = 1 - \frac{n-p}{n-1}(1-R^2)F\{1, 1, \tfrac{1}{2}(n+1), R^2\}, \tag{27.72}$$

$$= R^2 + \frac{p-1}{n-1}(1-R^2) - \frac{2(n-p)}{(n^2-1)} R^2(1-R^2) + O\left(\frac{1}{n^2}\right). \tag{27.73}$$

In particular, when $R^2 = 0$, (27.73) reduces to

$$E(R^2 \mid R^2 = 0) = \frac{p-1}{n-1}, \tag{27.74}$$

an exact result also obtainable directly from (27.65).

Similarly, the variance may be shown to be

$$\text{var}(R^2) = \frac{(n-p)(n-p+2)}{(n^2-1)}(1-R^2)^2 F(2, 2, \tfrac{1}{2}(n+3), R^2) - \{E(R^2) - 1\}^2 \tag{27.75}$$

$$= \frac{(n-p)}{(n^2-1)(n-1)}(1-R^2)^2$$
$$\times \left[2(p-1) + \frac{4R^2\{(n-p)(n-1)+4(p-1)\}}{(n+3)} + O\left\{\left(\frac{R^2}{n}\right)^2\right\} \right]. \tag{27.76}$$

(27.76) may be written

$$\text{var}(R^2) = \frac{4R^2(1-R^2)^2(n-p)^2}{(n^2-1)(n+3)} + O\left(\frac{1}{n^2}\right), \tag{27.77}$$

so that if $R^2 \neq 0$

$$\text{var}(R^2) \sim 4R^2(1-R^2)^2/n. \tag{27.78}$$

But if $R^2 = 0$, (27.87) is of no use, and we return to (27.76), finding

$$\text{var}(R^2) = \frac{2(n-p)(p-1)}{(n^2-1)(n-1)} \sim 2(p-1)/n^2, \tag{27.79}$$

the exact result in (27.79) being obtainable from (27.65).

27.33   The different orders of magnitude of the asymptotic variances (27.78) and (27.79) when $R \neq 0$ and $R = 0$ reflect the fundamentally different behaviour of the distribution of R^2 in the two circumstances. Although (27.74) shows that R^2 is a biased estimator of R^2, it is clearly consistent; for large n, $E(R^2) \to R^2$ and $\text{var}(R^2) \to 0$. When $R \neq 0$, the distribution of R^2 is asymptotically normal with mean R^2 and variance given by (27.78) (cf. Exercise 27.15). When $R = 0$, however, R, which is confined to the interval $[0, 1]$, is converging to the value 0 at the lower extreme of its range, and this alone is enough to show that its distribution is not normal in this case (cf. Exercises 27.14–15). It is no surprise in these circumstances that its variance is of order n^{-2}: the situation is analogous to the estimation of a terminal of a distribution, where we saw in Example 19.10 and Exercise 19.11 that variances of order n^{-2} occur.

The distribution of R behaves similarly in respect of its limiting normality to that of R^2, though we shall see that its variance is always of order $1/n$.

One direct consequence of the singularity in the distribution of R at $R^2 = 0$ should be mentioned. It follows from (27.78) that

$$\text{var } R \sim (1 - R^2)^2 / n, \tag{27.80}$$

which is the same as the asymptotic expression for the variance of the product-moment correlation coefficient (cf. (26.24))

$$\text{var } r \sim (1 - \rho^2)^2 / n.$$

It is natural to apply the variance-stabilising z-transformation of **16.33** (cf. also Exercise 16.18) to R also, obtaining a transformed variable $z = ar \tanh R$ with variance close to $1/n$, independent of the value of R. But this will not do near $R = 0$, as Hotelling (1953) pointed out, since (27.80) breaks down there; its asymptotic variance then will be given by (27.74) as

$$\text{var } R = E(R^2) - \{E(R)\}^2 \sim (p - 1)/n, \tag{27.81}$$

agreeing with the value $1/n$ obtained from (27.80) only for $p = 2$, when $R = |r|$. Lee (1971) investigates the approximation numerically, finds it inadequate, and proposes better ones.

Unbiased estimation of R^2 in the multinormal case

27.34 Since, by (27.73), R^2 is a biased estimator of R^2, we may wish to adjust it for the bias. Olkin and Pratt (1958) show that an unbiased estimator of $R^2_{1(2\cdots p)}$ is

$$t = 1 - \frac{n-3}{n-p}(1 - R^2_{1(2\cdots p)})F(1, 1, \tfrac{1}{2}(n-p+2), 1 - R^2_{1(2\cdots p)}), \tag{27.82}$$

where $n > p \geqslant 3$. t is the unique unbiased function of R^2 since it is a function of the complete sufficient statistics. (27.82) may be expanded as the series

$$t = R^2 - \frac{p-3}{n-p}(1 - R^2) - \left\{\frac{2(n-3)}{(n-p)(n-p+2)}(1 - R^2)^2 + O\left(\frac{1}{n^2}\right)\right\}, \tag{27.83}$$

whence it follows that $t \leqslant R^2$. If $R^2 = 1$, $t = 1$ also. When R^2 is zero or small, on the other hand, t is negative, as we might expect. We cannot (cf. **17.9**) find an unbiased estimator of R^2 (i.e. an estimator whose expectation is R^2 *whatever the true value of*

R^2) which takes only non-negative values, even though we know that R^2 is non-negative. We may remove the absurdity of negative estimates by using as our estimator

$$t' = \max(t, 0) \tag{27.84}$$

but (27.84) is no longer unbiased.

27.35 Lehmann (1986) shows that for testing $R^2 = 0$ in the multinormal case, tests rejecting large values of R^2 are UMP among test statistics which are invariant under location and scale changes.

Ezekiel and Fox (1959) and Kramer (1963) give charts and tables for constructing confidence intervals for R^2 from the value of R^2. Lee (1972) gives 4 d.p. tables of the upper 5 per cent and 1 per cent points of the d.f. of R for $R = 0$ (0.1) 0.9; $p - 1 = 1$ (1) 10, 12, 15, 20, 24, 30, 40; and $(n - p)^{1/2} = 60/\nu$ where $\nu = 1$ (1) 6 (2) 20. The *Biometrika Tables*, Vol. II, give a more recent 3 d.p. table by Kramer of the lower and upper 5 and 1 per cent points, for the same values of R, and $p - 1 = 2$ (2) 12 (4) 24, 30, 34, 40; $n - p = 10$ (10) 50.

Estimation of $R^2/(1 - R^2)$

27.36 The relative complexity of the estimator (27.82) does not persist if instead we seek to estimate $\theta = R^2/(1 - R^2)$. From **27.28**, the conditional distribution of (27.64) given $\mathbf{x} = (x_2, \ldots, x_p)$ is a non-central $F'(p - 1, n - p, \lambda)$, and thus by Exercise 23.21 its conditional expectation is

$$E(F'|\mathbf{x}) = \frac{(p - 1 + \lambda)(n - p)}{(p - 1)(n - p - 2)}, \quad n - p \geqslant 3;$$

hence that of $u = R^2/(1 - R^2)$ is

$$E(u|\mathbf{x}) = \left(\frac{p - 1}{n - p}\right) E(F'|\mathbf{x}) = \frac{p - 1 + \lambda}{n - p - 2}.$$

The unconditional expectation of u is therefore

$$E(u) = (n - p - 2)^{-1}\{p - 1 + E(\lambda)\}. \tag{27.85}$$

To find $E(\lambda)$, we see from **27.28** that if $\mathbf{x}$ is allowed to vary multinormally, λ/θ is the sum of squares of n independent normal variables subject to a single constraint, and is thus distributed as χ^2 with $(n - 1)$ d.fr. Hence, $E(\lambda/\theta) = n - 1$ and (27.85) becomes

$$E(u) = (n - p - 2)^{-1}\{p - 1 + (n - 1)\theta\} \tag{27.86}$$

so that an unbiased estimator of θ is given by

$$V = (n - 1)^{-1}\{(n - p - 2)u - (p - 1)\}, \tag{27.87}$$

a linear function of the corresponding statistic u. Although it is the unique MV unbiased estimator of θ, it may be negative and is subject to the remarks concerning t in **27.34**. Muirhead (1985) shows that its m.s.e. may be substantially reduced by 'shrinking' it as in Example 17.14 to aV where

$$a = \frac{(n - 1)(n - p - 4)}{(n + 1)(n - p - 2)}, \quad n - p \geqslant 5 \tag{27.88}$$

and truncating it at zero as at (27.84). For the bivariate case $p = 2$, (27.87) gives

$$E[(n - 1)^{-1}\{(n - 4)r^2/(1 - r^2) - 1\}] = \rho^2/(1 - \rho^2) \tag{27.89}$$

which was otherwise obtained in Exercise 16.15.

EXERCISES

27.1   Show that

$$\beta_{12\cdot34\cdots(p-1)} = \frac{\beta_{12\cdot34\cdots p} + \beta_{1p\cdot23\cdots(p-1)}\beta_{p2\cdot13\cdots(p-1)}}{1 - \beta_{1p\cdot23\cdots(p-1)}\beta_{p1\cdot23\cdots(p-1)}},$$

and that

$$\rho_{12\cdot34\cdots(p-1)} = \frac{\rho_{12\cdot34\cdots p} + \rho_{1p\cdot23\cdots(p-1)}\rho_{2p\cdot13\cdots(p-1)}}{\{(1 - \rho^2_{1p\cdot23\cdots(p-1)})(1 - \rho^2_{2p\cdot13\cdots(p-1)})\}^{1/2}}. \qquad \text{(Yule, 1907)}$$

27.2   Show that for p variates there are $\binom{p}{2}$ correlation coefficients of order zero and $\binom{p-2}{s}\binom{p}{2}$ of order s. Show further that there are $\binom{p}{2}2^{p-2}$ correlation coefficients altogether and $\binom{p}{2}2^{p-1}$ regression coefficients.

27.3   If the correlations of zero order among a set of variables are all equal to ρ, show that every partial correlation of the sth order is equal to $\rho/(1 + s\rho)$.

27.4   From (27.30), define the error term

$$x_{1\cdot q(1)} = x_1 - \boldsymbol{\beta}'\mathbf{x},$$

where $\boldsymbol{\beta}$ is given by (27.32) and $q(1)$ denotes all subscripts other than 1. Let $\mathbf{C}$ be the correlation matrix and denote the elements of $\mathbf{C}^{-1}$ by $\{c^{ij}\}$. Show that

$$g_{ij} = \text{cov}\,(x_{i\cdot q(i)}, x_{j\cdot q(j)}) = \sigma_i\sigma_j/c^{ij}$$

and hence that the coefficient of x_ix_j in the exponent of the multinormal distribution of $(x_1, x_2, \ldots, x_p)$ is $1/g_{ij}$.

27.5   Show from (27.39) that in summing the product of two residuals, any or all of the secondary subscripts may be omitted from a residual *all* of whose secondary subscripts are included among those of the other residual, i.e. that

$$\sum x_{1\cdot stu}x_{2\cdot st} = \sum x_{1\cdot stu}x_{2\cdot s} = \sum x_{1\cdot stu}x_2,$$

but that

$$\sum x_{1\cdot stu}x_{2\cdot st} \neq \sum x_{1\cdot su}x_{2\cdot st},$$

where s, t, u are sets of subscripts. (The same result holds for products of errors.)

(Chandler, 1950)

27.6   Using the notation in Exercise 27.5 apply the transformation

$$y_1 = x_1,$$
$$y_2 = x_{2\cdot1},$$
$$y_3 = x_{3\cdot21},$$
$$\text{etc.,}$$

to show that the multinormal distribution has density function

$$\frac{1}{(2\pi)^{p/2}\sigma_1\sigma_{2\cdot1}\sigma_{3\cdot12}\cdots} \exp\left\{-\frac{1}{2}\left(\frac{x_1^2}{\sigma_1^2} + \frac{x_{2\cdot1}^2}{\sigma_{2\cdot1}^2} + \frac{x_{3\cdot12}^2}{\sigma_{3\cdot12}^2} + \cdots\right)\right\}$$

so that the residuals $x_1, x_{2\cdot1}, \ldots$ are independent of each other. Hence show that any two residuals $x_{j\cdot r}$ and $x_{k\cdot r}$ (where r is a set of common subscripts) are distributed in the binormal form with correlation $\rho_{jk\cdot r}$.

27.7   Show that if an orthogonal transformation is applied to a set of n independent observations on p multinormal variates, the transformed set of n observations will also be independent.

27.8   For the data of Tables 26.1 and 26.2, we saw in Example 26.6 and Exercise 26.1 that

$$r_{12} = 0.32, \qquad r_{13} = 0.07,$$

where subscripts 1, 2, 3 refer to stature, weight and bust girth respectively. Given also that

$$r_{23} = 0.86$$

show that

$$R^2_{3(12)} = 0.80,$$

indicating that bust girth is fairly well determined by a linear function of stature and weight.

27.9   Show directly that no linear function of $x_2, \ldots, x_p$, has a higher correlation with x_1 than the least squares estimate of x_1.

27.10 Establish (27.73), the expression for $E(R^2)$.   (Wishart, 1931)

27.11 Establish (27.75), the expression for var (R^2).   (Wishart, 1931)

27.12 Verify that (27.82) is an unbiased estimator of R^2.

27.13 Show from the non-central F-distribution of F at (27.64) when $R^2 \neq 0$, that the distribution of R^2 in this case, when $x_2, \ldots, x_p$ are fixed, is

$$dF = \frac{1}{B\{\frac{1}{2}(p-1), \frac{1}{2}(n-p)\}} (R^2)^{(p-3)/2} (1 - R^2)^{(n-p-2)/2} \, dR^2 \cdot \exp\{-\frac{1}{2}(n-p)R^2\}$$

$$\times \sum_{j=0}^{\infty} \frac{\Gamma\{\frac{1}{2}(n-1+2j)\}\Gamma\{\frac{1}{2}(p-1)\}}{\Gamma\{\frac{1}{2}(n-1)\}\Gamma\{\frac{1}{2}(p-1+2j)\}} \frac{\{\frac{1}{2}(n-p)R^2R^2\}^j}{j!}. \qquad \text{(Fisher, 1928a)}$$

27.14 Show from (27.71) that for $n \to \infty$, p fixed, the distribution of $nR^2 = B^2$ is

$$dF = \frac{(B^2)^{(p-3)/2}}{2^{(p-1)/2}\Gamma\{\frac{1}{2}(p-1)\}} \exp(-\frac{1}{2}\beta^2 - \frac{1}{2}B^2)$$

$$\times \left\{ 1 + \frac{\beta^2 B^2}{(p-1)\cdot 2} + \frac{(\beta^2 B^2)^2}{(p-1)(p+1)\cdot 2\cdot 4} + \cdots \right\} d(B^2),$$

where $\beta^2 = nR^2$, and hence that nR^2 is a non-central χ^2 variate of form (23.18) with $\nu = p-1$, $\lambda = nR^2$. Show that the same result holds for the conditional distribution of nR^2, from Exercise 27.13.   (Fisher, 1928a)

27.15 In Exercise 27.14, use the c.f. of a non-central χ^2 variate given in Exercise 23.1 to show that as $n \to \infty$ for fixed p, R^2 is asymptotically normally distributed when $R \neq 0$, but not when $R = 0$. Extend the result to R.

27.16 Show that the distribution function of R^2 in multinormal samples may be written, if $n - p$ is even, in the form

$$(1 - R^2)^{(n-1)/2} R^{p-1} \sum_{j=0}^{(n-p-2)/2} \frac{\Gamma\{\frac{1}{2}(p-1+2j)\}}{\Gamma\{\frac{1}{2}(p-1)\}} \frac{(1-R^2)^j}{(1-R^2R^2)^{(n-1+2j)/2}}$$

$$\times F\{-j, -\frac{1}{2}(n-p), \frac{1}{2}(p-1), R^2R^2\}.$$

                        (Fisher, 1928a)

27.17 Show that in a sample $(x_1, \ldots, x_n)$ of one observation from an n-variate multinormal population with all means μ, all variances σ^2 and all correlations equal to ρ, the statistic

$$t^2 = \frac{(\bar{x} - \mu)^2}{\sum_{i=1}^{n}(x_i - \bar{x})^2/\{n(n-1)\}} \cdot \left\{ \frac{1 - \rho}{1 + (n-1)\rho} \right\}$$

has a Student's t^2-distribution with $(n-1)$ degrees of freedom. When $\rho = 0$, this reduces to the ordinary test of a mean of n independent normal variates.   (Walsh, 1947)

27.18   If $x_0, x_1, \ldots, x_n$ are normal variates with common variance α^2, $x_1, \ldots, x_n$ being independent of each other and x_0 having zero mean and correlation λ with each of the others, show that the n variates

$$y_i = x_i - ax_0, \qquad i = 1, 2, \ldots, n,$$

are multinormally distributed with all correlations equal to

$$\rho = (a^2 - 2a\lambda)/(1 + a^2 - 2a\lambda)$$

and all variances equal to

$$\sigma^2 = \alpha^2/(1-\rho).$$ (Stuart, 1958)

27.19   Use the result of Exercise 27.18 to establish that of Exercise 27.17.   (Stuart, 1958)

27.20   Show that if each pair from $x_2, \ldots, x_p$ is uncorrelated,

$$R^2_{1(2\cdots p)} = \sum_{s=2}^{p} \rho_{1s}^2 = \sum_{s=2}^{p} (\beta_{1s \cdot q} \sigma_s)^2 / \sigma_1^2.$$

27.21   Consider variables x_1, x_2, x_3 for which

$$\rho_{13} = 0, \qquad \rho_{12} = \cos\theta, \qquad \rho_{23} = \sin\theta, \qquad 0 < \theta < \pi/2.$$

Show that $\rho_{12 \cdot 3} = 1$ and hence that $R^2_{1(23)} = 1$. By letting $\theta \to \pi/2$, show that $R^2_{1(23)} = 1$ is consistent with $\rho_{13} = 0$, $\rho_{12} = \varepsilon$ for any $\varepsilon > 0$. Interpret the result geometrically.

27.22   In **27.28**, show that when $x_2, \ldots, x_p$ vary multinormally, $\lambda(1 - R^2)/R^2$ is distributed as a χ^2 with $(n-1)$ d.fr., and hence, using the c.f. of χ'^2 in Exercise 23.1, that the distribution of $R^2/(1-R^2)$ in the multinormal case is that of the ratio of independent variables y, z, with c.f.s

$$\phi_y(t) = (1 - 2it)^{(n-p)/2} \{1 - 2it(1+\theta)\}^{-(n-1)/2},$$
$$\phi_z(t) = (1 - 2it)^{-(n-p)/2},$$

where $\theta = R^2/(1 - R^2)$, and hence that y may be represented as

$$y = \chi_{p-2}^2 + (\chi_1 + \theta^{1/2}\chi_{n-1})^2,$$

all the variables being independent. (Cf. Exercise 16.6 when $p = 2$.)
                    (Gurland, 1968; Lee, 1971)

27.23   In Exercise 27.22, show that if $n - p = 2k$ is even,

$$\phi_y(t) = \{1 - 2(1+\theta)it\}^{-(p-1)/2}(1+\theta)^{-k} \sum_{j=0}^{k} \binom{k}{j}\left(\frac{\theta}{1 - 2(1+\theta)it}\right)^j,$$

and hence that $R^2/(1 - R^2)$ has the d.f.

$$G(x) = \sum_{j=0}^{k} \binom{k}{j}\left(\frac{\theta}{1+\theta}\right)^j (1+\theta)^{j-k} H_{p-1+2j,2k}\left(\frac{x}{1+\theta}\right),$$

where $H_{a,b}(x)$ is the d.f. of a ratio of independent χ^2 variates with a and b degrees of freedom.

(Gurland, 1968, who also gives two infinite series for $G(x)$, one for odd $(n-p)$, $R^2 < \frac{1}{2}$, and one for all n, p and R. A general class of series is given by Gurland and Milton (1970).)

27.24   In Exercise 27.22, approximate the distribution of $R^2/(1 - R^2)$ by assuming that it is a multiple g of the ratio of independent central χ^2 variates with ν and $n - p$ degrees of freedom, and equate the first two moments with the exact ones, as in **20.31**. Show that $g = B/A$ and $\nu = A^2/B$, where

$$A = (n-1)\theta + p - 1 \quad \text{and} \quad B = (n-1)\theta(\theta+2) + p - 1.$$

(Gurland, 1968; the result is exact when $R^2 = \theta = 0$, and the approximation seems generally good. See also Gurland and Milton (1970).)

CHAPTER 28

REGRESSION IN THE LINEAR MODEL

28.1 Our discussion in Chapters 26 and 27 assumed that the joint distribution of all the random variables was known. While such an assumption leads to interesting theoretical conclusions, it may well be an unreasonable requirement in practice. Furthermore, in experimental statistics, it is often the case that the regressor variables are set to preassigned levels, so that even the idea of random variation among such variables is unreasonable. In such circumstances, it is preferable to carry out the analysis *conditionally*, given the values of the regressor variables. That is, we return to the *linear model* of **19.4**; (19.8) may be rewritten as

$$y = X\beta + \varepsilon, \tag{28.1}$$

where β is a $(p \times 1)$ vector of parameters, or regression coefficients, X is an $(n \times p)$ matrix of known values, and ε is an $(n \times 1)$ vector of random errors. Usually, all the elements in the first column of X are set equal to one, corresponding to the constant term in the model; thus, an individual expression for the ith observation is

$$y_i = \beta_1 x_{1i} + \beta_2 x_{2i} + \cdots + \beta_p x_{pi} + \varepsilon_i, \tag{28.2}$$

where $x_{1i} \equiv 1$ in most cases.

It should be noted that the adjective *linear* will always be understood to refer to the parameter structure and not to the regressor variables; thus,

$$y = \beta_1 + \beta_2 x + \beta_3 x^2 + \varepsilon \tag{28.3}$$

is a linear model, whereas

$$y = \beta_1 + \beta_1^2 x + \varepsilon \tag{28.4}$$

is not.

28.2 Following **19.4**, the basic assumptions underlying linear regression analysis are as follows:

  (i) the model is linear (in the parameters),
 (ii) the error structure is additive,
(iii) the random errors have zero means, equal variances, and are mutually uncorrelated, written as

$$E(\varepsilon) = 0, \qquad V(\varepsilon) = \sigma^2 I, \tag{28.5}$$

(iv) the matrix X is of full rank p, so that $X'X$ is positive definite.

Conditions (i) and (ii) are implied by (28.1), whereas the assumptions in (28.5) are necessary if we are to apply the least squares (LS) theory of **19.4-19.9**. Condition (iv) is needed to ensure that the LS solution is unique. Violations of the assumptions and the subsequent effects upon the analysis are discussed in **28.29-64**.

28.3 Given these assumptions, it follows immediately from (28.1) that

$$E(\mathbf{y}) = \mathbf{X}\boldsymbol{\beta}; \tag{28.6}$$

strictly speaking, this should be written as $E(\mathbf{y}|\mathbf{X})$ so that

$$E(\mathbf{y}) = E_{\mathbf{X}}\{E(\mathbf{y}|\mathbf{X})\} = E(\mathbf{X})\boldsymbol{\beta}. \tag{28.7}$$

However, since all the analysis in this chapter is performed conditionally upon $\mathbf{X}$ being known, we use the simpler notation of (28.6) to denote the conditional expectation whenever there is no danger of confusion. The same applies for variances and other functions expressed as expectations.

It follows immediately from **19.4** that the LS estimators for $\boldsymbol{\beta}$ are

$$\hat{\boldsymbol{\beta}} = (\mathbf{X}'\mathbf{X})^{-1}\mathbf{X}'\mathbf{y} \tag{28.8}$$

and that $\hat{\boldsymbol{\beta}}$ has expected values, from (28.6) and (28.8),

$$E(\hat{\boldsymbol{\beta}}) = (\mathbf{X}'\mathbf{X})^{-1}\mathbf{X}'E(\mathbf{y}) = \boldsymbol{\beta} \tag{28.9}$$

so that the estimators are unbiased. Further, from (19.16), the covariance matrix is

$$V(\hat{\boldsymbol{\beta}}) = \sigma^2(\mathbf{X}'\mathbf{X})^{-1}. \tag{28.10}$$

The Gauss–Markov theorem in **19.6** shows the LS estimators to be the MV unbiased linear estimators for $\boldsymbol{\beta}$; this property is sometimes stated in the form that the LS estimators are best linear unbiased (BLU).

Residuals

28.4 The expected values in (28.6) may be estimated by

$$\hat{\mathbf{y}} = \mathbf{X}\hat{\boldsymbol{\beta}}; \tag{28.11}$$

sometimes termed the *fitted* values. We then define the (regression) *residuals*, **e**, as the differences between observed and fitted values:

$$\mathbf{e} = \mathbf{y} - \hat{\mathbf{y}} = \mathbf{y} - \mathbf{X}\hat{\boldsymbol{\beta}}. \tag{28.12}$$

From (28.6) and (28.9), it is clear that

$$E(\mathbf{e}) = \mathbf{0}. \tag{28.13}$$

Further, if $x_{1i} = 1$ for all i, the sum of the residuals $\mathbf{1}'\mathbf{e}$ is zero by construction. The residuals are the basic units that will be used to develop diagnostic procedures later in the chapter.

Finally, from (19.38), we have that an unbiased estimator for σ^2 is given by s^2, where

$$(n - p)s^2 = \mathbf{e}'\mathbf{e} = (\mathbf{y} - \mathbf{X}\hat{\boldsymbol{\beta}})'(\mathbf{y} - \mathbf{X}\hat{\boldsymbol{\beta}})$$
$$= \mathbf{y}'\mathbf{y} - \hat{\boldsymbol{\beta}}'\mathbf{X}'\mathbf{y}; \tag{28.14}$$

$\mathbf{e}'\mathbf{e}$ is known as the residual sum of squares (RSS) and s^2 is the estimated mean squared error (MSE).

We have already considered this model in the unconditional case in **26.11** and **27.14**; taking expectations over **X** recovers the properties for the unconditional situation.

Tests of hypotheses

28.5 It follows from **19.11** that the total sum of squares

$$\text{TSS} = \sum_{i=1}^{n} (y_i - \bar{y})^2 = \mathbf{y'y} - n\bar{y}^2 \tag{28.15}$$

partitions into two orthogonal components

$$(\mathbf{y} - \hat{\mathbf{y}})'(\mathbf{y} - \hat{\mathbf{y}}) + (\hat{\mathbf{y}}'\hat{\mathbf{y}} - n\bar{y}^2), \tag{28.16}$$

where $\hat{\mathbf{y}} = \mathbf{X}\hat{\boldsymbol{\beta}}$ and we assume that **X** includes a column of ones (this will always be true in the future unless explicitly stated otherwise). From (28.14), we note that the first term in (28.16) is the RSS; the second term is termed the explained sum of squares, denoted by ESS. As noted in (19.45), these may be expressed as

$$\text{RSS} = \boldsymbol{\varepsilon}'\{\mathbf{I} - \mathbf{X}(\mathbf{X'X})^{-1}\mathbf{X'}\}\boldsymbol{\varepsilon} \tag{28.17a}$$

and

$$\text{ESS} = \boldsymbol{\varepsilon}'\{\mathbf{X}(\mathbf{X'X})^{-1}\mathbf{X'}\}\boldsymbol{\varepsilon} \tag{28.17b}$$

from which it is apparent that the two matrices in braces in (28.17) are idempotent with ranks $(n - p)$ and $(p - 1)$, respectively.

If, in addition to the assumptions in **28.2**, we assume that the errors are normally distributed and that $\boldsymbol{\beta} = \mathbf{0}$, it follows from Cochran's theorem (**15.16**) that RSS and ESS are independently distributed chi-squared variates with $(n - p)$ and $(p - 1)$ degrees of freedom, respectively; see **19.11**. Thus, a test of the hypothesis

$$H_0: \boldsymbol{\beta} = \mathbf{0} \text{ vs. } H_1: \boldsymbol{\beta} \neq \mathbf{0} \text{ (not all } \beta_i = 0)$$

is given by the statistic

$$F = \{\text{ESS}/(p-1)\}/\{\text{RSS}/(n-p)\} \tag{28.18}$$

which has an F distribution with $(p-1, n-p)$ d.fr. under H_0.

28.6 This development was presented from a geometric perspective in **27.16–7** and **27.27–9**. In our present notation, we may write

$$R^2 = \text{ESS}/\text{TSS}$$
$$= (p-1)F/\{(n-p) + (p-1)F\}, \tag{28.19}$$

where R is the sample multiple correlation coefficient. When H_0 is true, it follows from **27.28** that

$$E(R^2 | H_0) = (p-1)/(n-1)$$

so that the adjusted coefficient

$$\bar{R}^2 = \frac{(n-1)}{(n-p)}\left[R^2 - \frac{(p-1)}{(n-1)}\right] \tag{28.20}$$

is sometimes preferred, as

$$E(\bar{R}^2 \mid H_0) = 0, \max \bar{R}^2 = 1 \text{ and } \bar{R}^2 \leqslant R^2 \leqslant 1;$$

note that $\bar{R}^2$ may be negative. Whereas R^2 cannot decrease when another variable is added to the model, $\bar{R}^2$ may decrease. In stepwise regression, the criterion $\max(\bar{R}^2)$ is sometimes used to choose the final model, but leads to the inclusion of too many terms; see **28.27-8**.

28.7   The test procedure outlined in **28.5** is often presented in the form of an analysis of variance (AV) table; see Table 28.1. (The analysis of variance is discussed

Table 28.1   Analysis of variance for a regression model

Source	DF	Sum of squares	Mean square	F
Due to regression (explained)	$p-1$	ESS	$M_E = \text{ESS}/(p-1)$	$F = M_E/M_R$
Due to error (residual)	$n-p$	RSS	$M_R = \text{RSS}/(n-p)$	
Total	$n-1$	TSS		

in detail in Chapter 29.) If a linear model is fitted using an initial set of p variables with values $\mathbf{X}_1$ and then a further q variables with values $\mathbf{X}_2$ are added to produce the model

$$\mathbf{y} = \mathbf{X}_1\boldsymbol{\beta}_1 + \mathbf{X}_2\boldsymbol{\beta}_2 + \boldsymbol{\varepsilon}, \tag{28.21}$$

the hypothesis

$$H_0\text{: } \boldsymbol{\beta}_2 = \mathbf{0}, \qquad \text{given that } \mathbf{X}_1 \text{ is in the model}$$

may be tested against the general alternative using the statistic

$$F_2 = \{(\text{ESS}_2 - \text{ESS}_1)/q\}/\{(\text{TSS} - \text{ESS}_2)/(n-p-q)\}, \tag{28.22}$$

where ESS_2 denotes the explained sum of squares for the complete model. The schematic AV is laid out in Table 28.2.

Table 28.2   AV for a regression model with two sets of regressors

Source	DF	Sum of squares	Mean square	F
Due to $\mathbf{X}_1$	$p-1$	ESS_1	$M_1 = \text{ESS}_1/(p-1)$	$F_1 = M_1/M_R$
Due to $\mathbf{X}_2$ (after fitting $\mathbf{X}_1$)	q	$\text{ESS}_2 - \text{ESS}_1$	$M_2 = (\text{ESS}_2 - \text{ESS}_1)/q$	$F_2 = M_2/M_R$
Due to error	$n-p-q$	RSS	$M_R = \text{RSS}/(n-p-q)$	
Total	$n-1$	TSS		

28.8   When the errors are normally distributed, it follows that $\hat{\boldsymbol{\beta}}$ is also normally distributed since it is linear in $\mathbf{y}$; that is,

$$\hat{\boldsymbol{\beta}} \sim N[\boldsymbol{\beta}, \sigma^2(\mathbf{X}'\mathbf{X})^{-1}]. \tag{28.23}$$

Thus, for any vector $\mathbf{c}$, $\mathbf{c}'\hat{\boldsymbol{\beta}}$ is normal

$$\mathbf{c}'\hat{\boldsymbol{\beta}} \sim N[\mathbf{c}'\boldsymbol{\beta}, \sigma^2\mathbf{c}'(\mathbf{X}'\mathbf{X})^{-1}\mathbf{c}]. \tag{28.24}$$

To test the hypothesis H_0: $\mathbf{c}'\boldsymbol{\beta} = 0$, we know that $\mathbf{c}'\hat{\boldsymbol{\beta}}$ is independent of s^2 and that s^2 is χ^2_{n-p} so that, from (16.10),

$$t = (\mathbf{c}'\hat{\boldsymbol{\beta}} - \mathbf{c}'\boldsymbol{\beta})/[s^2\mathbf{c}'(\mathbf{X}'\mathbf{X})^{-1}\mathbf{c}]^{1/2} \tag{28.25}$$

follows a Student's t distribution with $(n-p)$ degrees of freedom.

The most common use of (28.25) is to test a hypothesis of the form $H_{0(j)}$: $\beta_j = 0$ when the statistic reduces to

$$t_j = \hat{\beta}_j/[s^2 a_{jj}]^{1/2}, \tag{28.26}$$

where a_{jj} is the (j, j)th element of $(\mathbf{X}'\mathbf{X})^{-1}$.

Example 28.1
The data in Table 28.3 relate to a household's consumption of electricity in consecutive periods in State College, Pennsylvania, during the winter of 1980/81; electricity was the sole source of heating. We consider a model for daily electrical consumption, defined as

$$y = \text{electricity consumption in period/days in period.}$$

The regressor variables are

$x_2 = $ low temperature for period

$x_3 = $ high temperature for period

$x_4 = $ school attendance $=$ days in school (two children)/days in period.

Table 28.3 Data for Example 28.1

Electricity consumption (kW)	Temperatures (°F) Low	High	Days in school	Days in period
663	22	56	6	5
1018	14	48	10	7
1407	00	40	0	8
911	02	36	0	6
1758	−06	27	10	8
1165	02	34	10	6
1136	22	44	8	7
1095	10	50	10	7
1809	10	48	16	10
715	02	52	4	4
572	30	62	10	7
704	26	52	9	7
1002	17	40	10	7
848	12	56	10	7
856	12	56	10	7

The estimated model is

$$y = 219.90 - 2.403x_2 - 1.276x_3 + 16.93x_4$$
$$\quad\quad (-3.84) \quad (-1.86) \quad\quad (1.76)$$

$$s = 16.8; \quad\quad R^2 = 0.832; \quad\quad \bar{R}^2 = 0.787.$$

The numbers in brackets under the coefficients are the t-ratios calculated from (28.26); from appendix Table 5, the 5 per cent point is 2.201. The AV table is given as Table 28.4.

Table 28.4

Source	DF	SS	MS	F
Regression	3	15418	5139	18.2
Error	11	3103	282	—
Total	14	18521	—	—

From Appendix Table 7, the 5 per cent point is 3.59. Overall, the model appears to be effective, but x_3 and x_4 are of marginal value. The signs of the coefficients for x_2 and x_3 are as expected. It might be thought that an increase in x_4 should reduce heating costs but, in fact, x_4 is partly an indicator of time spent away from the house during school holidays. We shall discuss this example further later in the chapter.

Confidence and prediction intervals

28.9   Using the distributional results in **28.8**, the central confidence interval for β_j with coefficient 100 $(1-\alpha)$ is

$$\hat{\beta}_j \pm t_{1-\alpha/2} \{s^2 a_{jj}\}^{1/2}, \tag{28.27}$$

where $t_{1-\alpha/2}$ is the $100(1 - \frac{1}{2}\alpha)$ percentage point for Student's t distribution with $(n-p)$ degrees of freedom.

28.10   Suppose that, having fitted the linear regression model, we wish to estimate the expected value of y given $\mathbf{x}_0 = (x_{01}, x_{02}, \ldots, x_{0p})'$, denoted by $E(y|\mathbf{x}_0)$. From (28.11), the MVU estimator is

$$\hat{y} = \mathbf{x}_0'\hat{\boldsymbol{\beta}}, \tag{28.28}$$

and, from (28.24), it follows that the confidence interval is

$$\hat{y} \pm t_{1-\alpha/2} \{s^2 \mathbf{x}_0'(\mathbf{X}'\mathbf{X})^{-1}\mathbf{x}_0\}^{1/2}, \tag{28.29}$$

with $\nu = n - p$ as before.

28.11   Let y_{n+1} denote a potential new observation, corresponding to $\mathbf{x}_0$. Once again, the estimator is given by (28.11) as

$$\hat{y}_{n+1} = \mathbf{x}_0'\hat{\boldsymbol{\beta}}, \tag{28.30}$$

but now we are interested in the single observation rather than its expectation. Since

$$y_{n+1} = \mathbf{x}_0'\boldsymbol{\beta} + \varepsilon_{n+1},$$

we have

$$y_{n+1} - \hat{y}_{n+1} = \mathbf{x}_0'(\boldsymbol{\beta} - \hat{\boldsymbol{\beta}}) + \varepsilon_{n+1}; \qquad (28.31)$$

since the observations are independent and $\hat{\boldsymbol{\beta}}$ does not depend upon y_{n+1}, we have

$$E(y_{n+1} - \hat{y}_{n+1}) = 0,$$

showing the estimator to be unbiased and

$$\text{var}\,(y_{n+1} - \hat{y}_{n+1}) = \mathbf{x}_0'\,\text{var}\,(\hat{\boldsymbol{\beta}})\mathbf{x}_0 + \sigma^2$$

$$= \sigma^2\{1 + \mathbf{x}_0'(\mathbf{X}'\mathbf{X})^{-1}\mathbf{x}_0\}$$

$$= \sigma^2 v_{n+1}, \text{ say.} \qquad (28.32)$$

Thus, the $100(1-\alpha)$ per cent two-sided *prediction* interval for y_{n+1} is

$$\hat{y}_{n+1} \pm t_{1-\alpha/2}\{s^2 v_{n+1}\}^{1/2}. \qquad (28.33)$$

If m observations are to be made with $\mathbf{x} = \mathbf{x}_0$, the prediction interval for the mean of these m observations is also given by (28.33), but the first term in v_{n+1} becomes $1/m$ rather than one.

Example 28.2
In the simple case

$$y_i = \beta_1 + \beta_2 x_i + \varepsilon_i, \qquad i = 1, 2, \ldots, n, \qquad (28.34)$$

we have seen in Examples 19.3, 19.6, that

$$\hat{\beta}_2 = \sum_i (y_i - \bar{y})(x_i - \bar{x})/\sum_i (x_i - \bar{x})^2,$$

$$\hat{\beta}_1 = \bar{y} - \hat{\beta}_2\bar{x},$$

$$s^2 = \frac{1}{n-2}\sum_i \{y_i - (\hat{\beta}_1 + \hat{\beta}_2 x_i)\}^2,$$

and

$$(\mathbf{X}'\mathbf{X})^{-1} = \frac{1}{\sum_i (x_i - \bar{x})^2}\begin{pmatrix} \sum x^2/n & -\bar{x} \\ -\bar{x} & 1 \end{pmatrix}. \qquad (28.35)$$

Here $\mathbf{x}_0$ is the two-component vector $(1 \; x)'$, and we may proceed to construct interval estimates for $\beta_1, \beta_2, E(y|\mathbf{x}_0)$ and $E(y_{n+1}|\mathbf{x}_0)$ using (28.27), (28.29) and (28.33); in each case we have a Student's t variate with $(n-2)$ degrees of freedom.

The confidence interval for $E(y|x_0)$ is, from (28.29)

$$(x_0)'\hat{\beta} \pm t_{1-\alpha/2}\left\{\frac{s^2}{\sum(x-\bar{x})^2}\binom{1}{x_0}'\binom{\sum x^2/n \quad -\bar{x}}{-\bar{x} \quad 1}\binom{1}{x_0}\right\}^{1/2}$$

$$= (\hat{\beta}_1 + \hat{\beta}_2 x_0) \pm t_{1-\alpha/2}\left\{s^2\left(\frac{1}{n} + \frac{(x_0-\bar{x})^2}{\sum(x-\bar{x})^2}\right)\right\}^{1/2}. \qquad (28.36)$$

If we consider this as a function of the value x_0 we see that (28.36) defines the two branches of a hyperbola of which the fitted regression ($\hat{\beta}_1 + \hat{\beta}_2 x_0$) is a diameter. The confidence interval obviously has minimum length when $x_0 = \bar{x}$, the observed mean, and its length increases steadily as $|x_0 - \bar{x}|$ increases, confirming the intuitive notion that we can estimate most accurately near the 'centre' of the observed values of x. Figure 28.1 illustrates the loci of the confidence limits given by (28.36).

Robison (1964) gives ML estimates and confidence intervals for the intersection abscissa of two polynomial regressions, and a bibliography of related work.

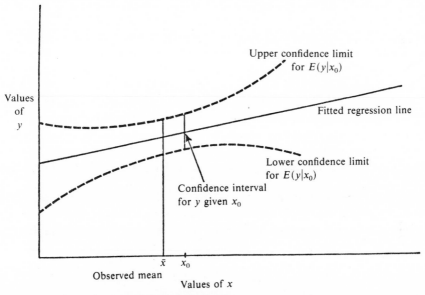

Fig. 28.1 **Hyperbolic loci of confidence limits (28.36) for an expected value of y in simple linear regression**

28.12 It should be borne in mind that these predictions are conditional upon the assumption that the linear model fitted to the original sample is valid for further observations; that is, we assume that no structural change has taken place.

Finally, we note that the confidence intervals given in **28.9–28.10** are often approximately correct because of the central limit theorem. However, we can make no such claim for the prediction interval where a non-normal error distribution may dominate the approximately normal variation in $\hat{\beta}$.

Example 28.3 (Example 28.1 continued)
In order to construct a confidence interval for $E(y|\mathbf{x}_0)$ for the data in Example 28.1, we need only the specific values in $\mathbf{x}_0$. Suppose $\mathbf{x}_0' = (5, 40, 1.429)$; it follows that $\hat{y} = 181.0$. Using (28.29), the 95 per cent confidence interval is (166.5, 195.5), whereas, from (28.33), the prediction interval is (141.3, 220.7).

Conditional and unconditional inferences
28.13   If $\mathbf{X}$ is the observed value of a set of random variables, the use of the linear model, in which $\mathbf{X}$ is a matrix of known coefficients, is conditional upon the observed $\mathbf{X}$. It is easy to see that conditionally unbiased estimators remain unbiased unconditionally, while conditional confidence intervals and tests remain valid unconditionally, since the value of any fixed conditional probability is unaffected by integrating it over the distribution of $\mathbf{X}$—the identity of (26.37) and the conditional t statistic in Example 28.2 are examples. However, the unconditional efficiency (i.e. selectivity and power) of tests and intervals generally differ from the conditional (linear model) values—cf. **27.28–31** for the multiple correlation coefficient—because the underlying statistics have different distributions. Similarly, other properties of estimators (e.g. their variances) do not generally persist unconditionally—cf. Exercise 26.9 for the regression coefficient. Sampson (1974) gives a general exposition of the subject.

Design considerations
28.14   From (28.35), it is evident that $\hat{\beta}_1$ and $\hat{\beta}_2$ will be uncorrelated if $\bar{x} = 0$; that is, the x values are measured about their mean. In general, when the matrix $(\mathbf{X}'\mathbf{X})$ is diagonal, the regressor variables are said to be *orthogonal*; this is generally desirable, as it simplifies the interpretation of the effects of each regressor on the response variable, y. Most designed experiments aim for at least some degree of orthogonality; see Chapter 29.

Further, from the example in (28.35), it is apparent that the variances of $\hat{\beta}_1$ and $\hat{\beta}_2$ will be minimised when $\bar{x} = 0$ and $\sum x^2$ is as large as possible. Both orthogonality and minimised sampling variances are therefore achieved if we choose the x_i so that (assuming n to be even)

$$x_1, x_2, \ldots, x_{n/2} = +a,$$

$$x_{n/2+1}, x_{n/2+2}, \ldots, x_n = -a,$$

and a is as large as possible. This corresponds to the intuitively obvious fact that if we are certain that the dependence of y upon x is linear with constant variance, we can most efficiently 'fix' the line at its end-points. However, if the dependence were non-linear, we should be unable to detect this if all our observations had been made at two values of x only, and it is therefore usual to spread the x-values more evenly over the range; it is always as well to be able to check the structural assumptions of our model in the course of the analysis.

> In the general case, if $\mathbf{X}$ has a column of 1s and we measure each other column about its mean, we make the other regression coefficients orthogonal to the constant term, as here.

28.15 The confidence limits for an expected value of y discussed in Example 28.3, and more generally in **28.10-12**, refer to the value of y corresponding to a particular $\mathbf{x}$; in Fig. 28.1, any particular confidence interval is given by that part of the vertical line through x_0 lying between the branches of the hyperbola. Suppose now that we require a *confidence region for an entire regression line*, i.e. a region R in the (x, y) plane (or, more generally, in the $(\mathbf{x}, y)$ space) such that there is probability $1 - \alpha$ that the true regression line $y = \mathbf{x}'\boldsymbol{\beta}$ is contained in R. This, it will be seen, is a quite distinct problem from that just discussed; we are now seeking a confidence region, not an interval, and it covers the whole line, not one point on the line. We now consider this problem, first solved in the simplest case by Working and Hotelling (1929) in a remarkable paper; our discussion follows that of P. G. Hoel (1951).

Confidence regions for a regression line
28.16 We first treat the simple case of Example 28.2 and assume σ^2 known, restrictions to be relaxed in **28.21-2**. For convenience, we measure the x_j from their mean so that $\bar{x} = 0$, and, from Example 28.2, var $\hat{\beta}_1 = \sigma^2/n$, var $\hat{\beta}_2 = \sigma^2/\sum x^2$, and $\hat{\beta}_1$ and $\hat{\beta}_2$ are normally and independently distributed. Thus

$$u = n^{1/2}(\hat{\beta}_1 - \beta_1)/\sigma, \qquad v = (\textstyle\sum x^2)^{1/2}(\hat{\beta}_2 - \beta_2)/\sigma, \qquad (28.37)$$

are independent standardised normal variates.

Let $g(u^2, v^2)$ be a single-valued even function of u and v, and let

$$g(u^2, v^2) = g_{1-\alpha}, \qquad 0 < \alpha < 1, \qquad (28.38)$$

define a family of closed curves in the (u, v) plane such that (a) whenever $g_{1-\alpha}$ decreases, the new curve is contained inside that corresponding to the larger value of $1 - \alpha$; and (b) every interior point of a curve lies on some other curve. To the implicit relation (28.38) between u and v, we assume that there corresponds an explicit relation

$$u^2 = p(v^2)$$

or

$$u = \pm h(v). \qquad (28.39)$$

We further assume that $h'(v) = dh(v)/dv$ exists for all v and is a monotone decreasing function of v taking all real values.

28.17 We see from (28.37) that for any given set of observations to which a regression has been fitted, there will correspond to the true regression line,

$$y = \beta_1 + \beta_2 x, \qquad (28.40)$$

values of u and v such that

$$\beta_1 + \beta_2 x = \left(\hat{\beta}_1 + \frac{\sigma}{n^{1/2}} u\right) + \left(\hat{\beta}_2 + \frac{\sigma}{(\sum x^2)^{1/2}} v\right) x. \qquad (28.41)$$

Substituting (28.39) into (28.41), we have two families of regression lines, with v as parameter,

$$\left(\hat{\beta}_1 \pm \frac{\sigma}{n^{1/2}} h(v)\right) + \left(\hat{\beta}_2 + \frac{\sigma}{(\sum x^2)^{1/2}} v\right) x, \tag{28.42}$$

one family corresponding to each sign in (28.39). We now find the envelopes of these families.

Differentiating (28.42) with respect to v and equating the derivative to zero, we obtain

$$x = \mp \left(\frac{\sum x^2}{n}\right)^{1/2} h'(v). \tag{28.43}$$

Substituted into (28.42), (28.43) gives the required envelopes:

$$(\hat{\beta}_1 + \hat{\beta}_2 x) \pm \frac{\sigma}{n^{1/2}} \{h(v) - vh'(v)\}, \tag{28.44}$$

where the functions of v are to be substituted for in terms of x from (28.43). The restrictions placed on $h'(v)$ below (28.39) ensure that the two envelopes in (28.44) exist for all x, are single-valued, and that all members of each family lie on one side only of its envelope. In fact, the curve given taking the upper signs in (28.44) always lies above the curve obtained by taking the lower signs in (28.44), and all members of the two families (28.42) lie between them.

28.18 Any pair of values (u, v) for which

$$g(u^2, v^2) < g_{1-\alpha} \tag{28.45}$$

will correspond to a regression line lying between the pair of envelopes (28.44), because for any fixed v, $u^2 = \{h(v)\}^2$ will be reduced, so that the constant term in (28.42) will be reduced in magnitude as a function of v, while the coefficient of x is unchanged. Thus if u and v satisfy (28.45), the true regression line will lie between the pair of envelopes (28.44). Now choose $g_{1-\alpha}$ so that the continuous random variable $g(u^2, v^2)$ satisfies

$$P\{g(u^2, v^2) < g_{1-\alpha}\} = 1 - \alpha. \tag{28.46}$$

Then we have probability $1 - \alpha$ that (28.45) holds, and the region R between the pair of envelopes (28.44) is a confidence region for the true regression line with confidence coefficient $1 - \alpha$.

28.19 In the original solution to this problem, Working and Hotelling (1929) elected to assume that $g(u^2, v^2)$ described a circle, so that

$$h(v) = (a^2 - v^2)^{1/2} \tag{28.47}$$

and

$$h'(v) = -v/h(v). \tag{28.48}$$

Their choice was motivated by the observation that, since u and v in (28.37) are independent standardised normal variates, $u^2 + v^2$ is a χ^2 variate with 2 degrees of freedom, and a^2 ($= g_{1-\alpha}$ in (28.46)) is simply the $100(1-\alpha)$ per cent point obtained from the tables of that distribution. The boundaries of the confidence region are, putting (28.47) and (28.48) into (28.44)

$$(\hat{\beta}_1 + \hat{\beta}_2 x) \pm \frac{\sigma}{n^{1/2}}\left\{h(v) + \frac{v^2}{h(v)}\right\} = (\hat{\beta}_1 + \hat{\beta}_2 x) \pm \frac{\sigma}{n^{1/2}}[\{h(v)\}^2 + v^2]^{1/2}\left[1 + \frac{v^2}{\{h(v)\}^2}\right]^{1/2}$$

$$= (\hat{\beta}_1 + \hat{\beta}_2 x) \pm \frac{\sigma}{n^{1/2}}(g_{1-\alpha})^{1/2}[1 + \{h'(v)\}^2]^{1/2}. \qquad (28.49)$$

Using (28.43), (28.49) becomes

$$(\hat{\beta}_1 + \hat{\beta}_2 x) \pm (g_{1-\alpha})^{1/2}\left\{\frac{\sigma^2}{n} + x^2 \frac{\sigma^2}{\sum x^2}\right\}^{1/2}, \qquad (28.50)$$

the terms in the braces being $\{\text{var }\hat{\beta}_1 + x^2 \text{ var }\hat{\beta}_2\}$.

If (28.50) is compared with the confidence limits (28.36) for $E(y|x_0)$ derived in Example 28.2 (where we now put $\bar{x} = 0$, as we have done here), we see that apart from the replacement of s^2 by σ^2, and of $t_{1-\alpha/2}$ by $(g_{1-\alpha})^{1/2}$, the equations are of exactly the same form. Thus the confidence region (28.50) will look exactly like the loci of the confidence limits (28.36) plotted in Fig. 28.1, being a hyperbola with the fitted line as diameter. As might be expected, for given α the branches of the hyperbola (28.50) are farther apart than those of (28.37), for we are now setting a region for the whole line where previously we had loci of limits for a single value on the line. For example, with $\alpha = 0.05$, $t_{1-\alpha/2}$ (with infinite degrees of freedom, corresponding to σ^2 known) $= 1.96$, while $g_{1-\alpha} = 5.99$ for a χ^2 distribution with two degrees of freedom, yielding the multiplier 2.45 for (28.50).

28.20 In his extension, P. G. Hoel (1951) allowed $h(v)$ to describe a family of ellipses. When $1-\alpha = 0.95$, the optimal choice is an ellipse with semi-axes 2.62 and 2.32. Hoel found that the length of the interval derived from (28.50) was less than one per cent larger than the minimum obtained from the best ellipse.

28.21 If σ^2 is unknown, only slight modifications of the argument are required. Define the variable

$$w^2 = (n-2)s^2/\sigma^2, \qquad (28.51)$$

so that w^2 is the ratio of the sum of squared residuals from the fitted regression to the true error variance, which has a χ^2 distribution with $n-2$ degrees of freedom. From (28.26) and (28.51), we see that each of the statistics

$$u^* = (n-2)^{1/2}u/w = n^{1/2}(\hat{\beta}_1 - \beta_1)/s \quad \text{and} \quad v^* = (n-2)^{1/2}v/w = n^{1/2}(\hat{\beta}_2 - \beta_2)/s$$

has a Student's t distribution with $n-2$ degrees of freedom.

Further, since u^2, v^2 and w^2 are distributed independently of one another as χ^2 with 1, 1, and $(n-2)$ degrees of freedom respectively, the ratio

$$\left(\frac{u^2+v^2}{2}\right)\bigg/\left(\frac{w^2}{n-2}\right) = \tfrac{1}{2}(u^{*2}+v^{*2})$$

has an F distribution with 2 and $n-2$ degrees of freedom. Thus if we replace σ by s in (28.50), and put $g_{1-\alpha}$ equal to twice the $100(1-\alpha)$ per cent point of this F distribution, we obtain the required confidence region. As in **28.19**, we find that the boundaries of the region are always farther apart than the loci of the confidence limits for $E(y|x_0)$.

28.22 There is no difficulty in extending our results to the case of more than one regressor, a sketch of such a generalisation having been given by Hoel (1951). With k regressors we find, generalising **28.21**, that $(u^{*2}+\sum_{i=1}^{k} v_i^{*2})/(k+1)$ has a variance-ratio distribution with $(k+1, n-k-1)$ d.fr.

> Wynn and Bloomfield (1971) give tables of the fractional multiplying factors to be applied to the Working–Hotelling region when only part of the line is to be covered for the same α. In general, the confidence region becomes narrower when this restriction is made; that is, the original Working–Hotelling region is conservative. Kanoh (1988) shows how the average width of the region can be further reduced by changing the curvature of the boundaries. Wynn (1984) extends the Working–Hotelling region to the polynomial regression case; see also Knafl et al. (1985). See also related work by Gafarian (1964), Graybill and Bowden (1967), Halperin et al. (1967), Folks and Antle (1967), Halperin and Gurian (1968), Dunn (1968), Bowden (1970), Bohrer and Francis (1972), Bohrer (1973), Hochberg and Quade (1975), Naiman (1983, 1986, 1987) and Piegorsch (1985a, b, 1987).

Stepwise regression

28.23 In experimental studies, the regressor variables are well defined, but in observational studies it may happen that a large number of potential regressors is available, with little prior evidence as to the relative merits of these competing variables. One way to reduce the dimensionality of the problem is to reduce the set of variables to a smaller set by use of either factor analysis or principal components analysis; this approach is discussed from a different perspective in **28.49** and is not pursued further here.

A second approach, which is often useful at least in exploratory analyses, is to select best subsets of variables; that is, to choose the best single regressor, then the best pair, then the best triple, and so on. This leads to two questions:

(a) how should the variables be selected to form a subset of size p?
(b) how many variables should be selected (choice of p)?

Since R^2 is used as a measure of overall goodness of fit, a natural answer to (a) is to choose those regressors that maximise R^2 for a given p. Since R^2 is non-decreasing as a function of p, question (b) may be answered by selecting a suitable cut-off value for R^2 as a function of n and p.

28.24 When the total number of regressor variables, P say, is not too large, it is feasible to consider all $\binom{P}{p}$ possible subsets for each p, i.e., $2^P - 1$ subsets in all.

Current computer programs can handle up to about $P = 20$ and say $p \leqslant 8$ in reasonable time, but larger data sets become very time-consuming. To some extent, the search process can be improved by using threshold rules such as those developed by Beale *et al.* (1967). Suppose there are s candidate variables from which we wish to choose our best $(p-1)$, and denote by j the set of $(s-1)$ variables obtained by excluding x_j from the candidates.

Define the *unconditional threshold* of x_j to be

$$T_j = R^2_{1(j)}.$$

This maximises the multiple correlation obtainable without using x_j; now if, during the search for the best subset of $(p-1)$ variables, we achieve for some $(p-1)$-subset u a value of R^2 satisfying

$$R^2_{1(u)} > T_j,$$

we can obviously only obtain a larger R^2 by including x_j in the subset to be considered. Similarly, we may define *conditional thresholds*

$$T_{jk} = R^2_{1(jk)},$$

obtained by excluding the pair (x_j, x_k). If we find an R^2 exceeding T_{jk}, we can only improve it by including one or both of x_j, x_k. By calculating such thresholds, the number of subsets considered may be substantially reduced.

28.25 For larger P, a variety of incomplete search procedures has been developed that check only some of the possible subsets. The simplest such algorithm is the *forward selection* method (cf. Efroymson, 1960) which selects the best single variable (that which maximises R^2) then *adds* the next best and so on, so that only $\frac{1}{2}P(P+1) - 1$ different subsets need be considered for all $1 \leqslant p \leqslant P$. A similar method is the *backward elimination* algorithm, which starts with all P variables in the regression and deletes them one at a time in such a way as to minimise the drop in R^2 at each stage; again only $\frac{1}{2}P(P+1) - 1$ subsets are considered. As might be expected, these two methods do not produce the same subsets of size p, nor does either procedure necessarily produce the best subset at each stage, except when $p = 1$ (forward) or $p = P - 1$ (backward). Empirical studies suggest that the backwards algorithm fares somewhat better for the choices of p of most interest since it is less likely to miss useful combinations of variables. However, the backwards procedure may be infeasible because of multicollinearity (most obviously so if $n \leqslant P$).

28.26 A natural compromise between the two methods is to consider a combined algorithm that operates as follows. Given an initial subset of $s - 1$ regressors

1. select the best variable to add to the subset to create a new subset of s variables, $x^F_{(s)}$ say;
2. select the best variable to delete from the new subset of size s, $x^B_{(s)}$ say;
3. if $x^F_{(s)} = x^B_{(s)}$, terminate the search for the best subset of size $(s-1)$ and repeat the steps for s; if $x^F_{(s)} \neq x^B_{(s)}$, return to step 1.

The exact number of subsets considered cannot be determined in advance, but it is usually of order P^2 and so fairly quick to compute. Computational speed is improved by noting that if $\mathbf{X}_p$ is the matrix of regressors for p variables

$$\mathbf{X}_{p+1} = (\mathbf{X}_p, \mathbf{x}_{p+1}) \quad \text{and} \quad \mathbf{M}_p = \mathbf{X}_p'\mathbf{X}_p,$$

then

$$\mathbf{M}_{p+1} = \mathbf{X}_{p+1}'\mathbf{X}_{p+1} = \begin{pmatrix} \mathbf{M}_p & \mathbf{m} \\ \mathbf{m}' & h \end{pmatrix}$$

has inverse

$$\begin{pmatrix} \mathbf{M}_p^{-1} + \alpha\mathbf{dd}' & -\alpha\mathbf{d} \\ -\alpha\mathbf{d}' & \alpha \end{pmatrix},$$

where $\mathbf{d} = \mathbf{M}_p^{-1}\mathbf{m}$ and $\alpha^{-1} = h - \mathbf{m}'\mathbf{M}_p^{-1}\mathbf{m}$. Thus, variables may be added or deleted without the need to invert $\mathbf{M}_p$ at each stage. For further discussions on computational efficiency, see Broerson (1986) and Ridout (1988).

28.27 Thus far, we have not commented upon the stopping rules to be used in deciding upon p. Many computer packages use percentage points of F, based on the simplifying (erroneous) assumption that only a single additional variable is being tested; others allow the user to select such an F-value by some private device.

A criterion with more solid credentials may be developed from the prediction variance given in (28.32). Consider n new observations to be recorded at each of $\mathbf{x}_1, \mathbf{x}_2, \ldots, \mathbf{x}_n$. The total prediction variance, based on p regressors, is

$$\begin{aligned} V_p &= \sigma^2 \left\{ n + \sum_{i=1}^n \mathbf{x}_i'(\mathbf{X}'\mathbf{X})^{-1}\mathbf{x}_i \right\} \\ &= \sigma^2\{n + \text{tr}\,[(\mathbf{X}'\mathbf{X})^{-1}\textstyle\sum\mathbf{x}_i\mathbf{x}_i']\} \\ &= (n+p)\sigma^2. \end{aligned} \qquad (28.52)$$

An unbiased estimator of this, assuming the selected p-regressor model to be correct, is

$$S_p = (n+p)\text{RSS}(p)/(n-p). \qquad (28.53)$$

We may then select the subset that minimises S_p over all possible subsets and $1 \leqslant p \leqslant P$. Sometimes S_p is known as the final prediction error (FPE) criterion; see Akaike (1969). A similar procedure, suggested by Mallows (1973), is to minimise the quantity

$$C_p = \frac{\text{RSS}(p)}{s^2} + 2p - n, \qquad (28.54)$$

where s^2 is the unbiased estimator of σ^2 based on all P regressors so that no misspecification error results. Additional measures include information criteria of the general form

$$\text{IC} = \log\{\text{RSS}(p)/n\} + pc(n)/n. \qquad (28.55)$$

Akaike (1969) suggested $c(n) = 2$, whereas Schwarz (1978) recommends $c(n) = \log n$ and Hannan and Quinn (1979) consider $c(n) = 2 \log (\log n)$.

Shibata (1981, 1984) shows that S_p, C_p and Akaike's IC produce equivalent results asymptotically. Breiman and Freedman (1983) demonstrate that criteria such as (28.53) are asymptotically optimal in the sense of minimising the prediction error. The Schwarz and Hannan–Quinn criteria produce consistent estimators in the sense that the probability of selecting the true model approaches one as $n \to \infty$; the other measures do not achieve this and typically include too many terms.

Example 28.4
Using the data given in Example 28.1, a stepwise analysis based on entering the variable that provides the greatest increase in R^2 proceeds as in Table 28.5.

Table 28.5

Step	Variable entered	R^2	t-value of entering variable
1	x_2	0.734	−5.99
2	x_3	0.785	−1.69
3	x_4	0.833	1.76

Many rules would, therefore, terminate the process after step 1 since the next t-value does not exceed the 5 per cent level. However, the several criteria we have just discussed all suggest the inclusion of all three variables; see Table 28.6. Stepwise regression is a useful exploratory guide, but not an infallible model-building tool!

Table 28.6   Values of selection criteria for different subsets

Subset	S_p	C_p	IC($c = 2$)	IC*($c = 2.71$)
$\{x_2\}$	6439	6.45	6.06	6.16
$\{x_3\}$	9690	15.27	6.47	6.56
$\{x_4\}$	23729	53.32	7.36	7.46
$\{x_2, x_3\}$	5967	5.10	5.98	6.12
$\{x_2, x_4\}$	6116	5.45	6.01	6.15
$\{x_3, x_4\}$	10889	22.19	6.58	6.72
$\{x_2, x_3, x_4\}$	5360	4.00	5.87	6.05

* The Schwarz value; $c \doteq 2$ for the Hannan–Quinn version.

28.28   Several attempts have been made to produce more accurate assessments of the significance levels of the tests involved in stepwise regression (cf. Draper *et al.*, 1971). Wilkinson and Dallal (1981) provide tables for tests based on the forward selection rule, whereas Butler (1984) develops bounds for the *p*-value of the *F*-test for the next best-fitting regressor variable. Miller (1984) provides a comprehensive review of the field and recommends using the test procedure developed in Spjøtvoll (1972a). Mitchell and Beauchamp (1988) consider the variable selection problem in a Bayesian context.

Narula and Wellington (1979) consider stepwise regression using least absolute deviations rather than least squares. Hoerl *et al.* (1986) show via an extensive simulation study that selection procedures based on ridge regression may be more effective than standard procedures.

Copas (1983) demonstrates that Stein-like 'shrinkage estimators' often give lower prediction mean square errors, especially when stepwise procedures have been used.

Picard and Cook (1984) show that cross-validation procedures may be used to validate tests of models developed by stepwise algorithms. In essence, the data set is divided into two parts; the model is then identified using one subset and its goodness of fit tested using the other subset.

Checking the assumptions

28.29 In **28.2**, we made some very strong assumptions in order to establish the properties of the linear model. Since regression analysis is such a widely used tool in applied statistics, it is clearly necessary to have effective procedures for checking these assumptions, insofar as this is possible. In fact, we need both *diagnostic* procedures to establish whether an assumption has been violated and *extensions* to the basic model to accommodate departures when identified. The rapid advances in statistical computing have opened the way to using a wide variety of computationally intensive diagnostics, mostly based upon the residuals, many of which are conceptually simple but were infeasible in times past.

We shall examine the assumptions in **28.2** one at a time, discussing extensions and diagnostics for each potential problem in what seems the most natural sequence. However, we should never lose sight of the fact that all violations may occur simultaneously when working with real data, so that model building is likely to remain very much an art form rather than an exact science.

28.30 Perhaps the most basic assumptions specified in **28.2** are the interwoven threesome: linearity (L), additivity (A), and, later, normality (N). As we observed in **26.44**, these assumptions arise naturally together, since $(L) + (A)$ implies (N) in some circumstances. When it is not reasonable to assume $(L + A + N)$, either the model must be completely reformulated, as for the non-linear regression models discussed in **28.84**–**7**, or transformations must be found to restore $(L + A + N)$, at least to a reasonable degree of approximation.

Even when the basic structure (28.1) is justified, a series of other assumptions must be validated, as follows:

Errors have zero means, $E(\varepsilon) = 0$. This is essentially untestable in its simplest form. If $E(\varepsilon_i) = c$ for all i, this is most likely a measurement error and the data recording process may be suspect. In more complex cases where the mean may be a function of x_i, the residuals will usually display some pattern of dependence as described below.

Errors have constant variances (are homoscedastic). The errors are said to be *heteroscedastic* if the variances differ. Such departures may be detected using tests based upon the residuals and modified estimators developed.

Errors are uncorrelated. This is the problem of *autocorrelation,* and most commonly arises for time- or space-dependent data. Again, diagnostics are based on the residuals and the patterns of dependence may be built into the extended model.

X *is of rank p.* At first sight, this is a 'technical condition' which, it might be thought, would not be violated in practice. In a sense, this is true as **X** is usually of full rank unless the investigator has unwittingly built an exact linear dependence into the set of regressor variables (see Chapter 29). In such circumstances, a suitable constraint will serve to remove the problem, as noted in **19.14**. However, there are many circumstances, particularly in observational studies, where high correlations can exist among the regressors so that the matrix **X'X** becomes *ill-conditioned*; that is, elements in the inverse become large and numerically unstable. This is the problem of *multicollinearity*, discussed briefly in **19.6**. Rather different diagnostics and modifications are required when such circumstances are suspected.

Finally, we must not forget that statistical inference requires specification of the error distribution. In general, it is assumed that the errors are normally distributed, after transformation if necessary; an exception to this approach is the *bootstrap* method, described in **28.45**.

> Pierce and Kopecky (1979) showed that the usual goodness of fit tests may be used for regression residuals, with the same asymptotic distributions. White and MacDonald (1980) and Jarque and Bera (1987) give numerical power comparisons.

We shall discuss transformations in **28.31–44** and the bootstrap in **28.45**. The issue of multicollinearity is addressed in **28.46–56**. We then examine heteroscedasticity in **28.57–60** and autocorrelation in **28.61–4**.

Transformations to the normal linear model

28.31   Following Box and Cox (1964), suppose that we are not prepared uncritically to assume that $y = X\beta + \epsilon$; rather, we seek transformations both of y and of each of the xs so that we have

$$\mathbf{y}_\lambda = \mathbf{X}_\mu \boldsymbol{\beta} + \boldsymbol{\epsilon}, \tag{28.56}$$

where the components of $\boldsymbol{\epsilon}$ are independently normal with zero means and constant variance σ^2. In (28.56), $\boldsymbol{\lambda} = (\lambda_1, \lambda_2, \dots)$ indexes the transformation of y within some selected parametric family of transformations, and similarly $\boldsymbol{\mu} = (\mu_1, \mu_2, \dots, \mu_p)$ indexes the (separate) transformations of the regressors $x_1, x_2, \dots, x_p$.

28.32   By (28.56), the LF is, in logarithmic form,

$$\log L_{\lambda,\mu}(\mathbf{y} \mid \boldsymbol{\beta}, \sigma^2) = -\tfrac{1}{2}n \log(2\pi\sigma^2)$$

$$-\frac{1}{2\sigma^2}(\mathbf{y}_\lambda - \mathbf{X}_\mu\boldsymbol{\beta})'(\mathbf{y}_\lambda - \mathbf{X}_\mu\boldsymbol{\beta}) + \log J_\lambda, \tag{28.57}$$

where J_λ is the Jacobian of the inverse transformation from $\mathbf{y}_\lambda$ (the normally distributed variable in (28.56)) to the actually observed $\mathbf{y}$. Now, when the LF (28.57) is maximised for given $\boldsymbol{\lambda}, \boldsymbol{\mu}$, with respect to $\boldsymbol{\beta}$ and σ^2, we find as in **23.28**, that the middle term becomes a constant. If we neglect constants, therefore, we have the conditional maximum for fixed $\boldsymbol{\lambda}, \boldsymbol{\mu}$,

$$\log L_{\lambda,\mu}(\mathbf{y} \mid \hat{\boldsymbol{\beta}}, \hat{\sigma}^2) = -\tfrac{1}{2}n \log \hat{\sigma}^2_{\lambda,\mu} + \log J_\lambda, \tag{28.58}$$

where $n\hat{\sigma}^2_{\lambda,\mu} = \mathbf{y}'_\lambda \mathbf{T} \mathbf{y}_\lambda$, say, is the Residual SS, again as in **23.28**.

We now need to compute the absolute maximum of the conditional maximum (28.58) over the whole range of λ, μ. This is a formidable numerical task, except when only one or two transformation indices are involved, e.g. when

(a) only the dependent variable y is transformed and λ has only one or two components; or
(b) the *same* transformation is applied to all of, or a subset of, the regressors, so that μ has only one or two components; or
(c) λ has a single component as in (a), and (b) holds with only one component in μ.

In cases (b) and (c), numerical plotting of the contours of (28.57) for all λ, μ will generally be necessary. We now confine ourselves to case (a), where only the dependent variable is being transformed. This implies that we can choose proper forms for the regressor variables before considering transformation of the dependent variable. Box and Tidwell (1962) discuss transformations of the regressors to simpler form (cf. Exercise 28.17); such transformations do not, of course, affect the normality or homoscedasticity of the errors.

28.33 In practice, the most useful transformations have been found to be the powers and the logarithm of y, possibly translated by a constant. We therefore consider the family of transformations

$$\left. \begin{aligned} y_\lambda &= (y + \lambda_2)^{\lambda_1}, & \lambda_1 \neq 0, \\ &= \log (y + \lambda_2), & \lambda_1 = 0. \end{aligned} \right\} \tag{28.59}$$

To avoid a discontinuity at $\lambda_1 = 0$, we rewrite this equivalently as

$$\left. \begin{aligned} y_\lambda &= \{(y + \lambda_2)^{\lambda_1} - 1\}/\lambda_1, & \lambda_1 \neq 0, \\ &= \log (y + \lambda_2), & \lambda_1 = 0. \end{aligned} \right\} \tag{28.60}$$

Tukey (1957) studied and charted the structural features of the family (28.59) for $\lambda_1 \leqslant 1$, and Dolby (1963) considered properties of the differential equation which it satisfies, namely $(y'_\lambda / y''_\lambda)' = (\lambda_1 - 1)^{-1}$. Healy and Taylor (1962) give tables to facilitate fractional power transformations when $\lambda_2 = 0$ and λ_1 is a multiple of 0.2.

28.34 In (28.58), we now have

$$\log J_\lambda = (\lambda_1 - 1) \sum_{i=1}^{n} \log (y_i + \lambda_2), \tag{28.61}$$

and (28.58) can be plotted for selected (λ_1, λ_2) for numerical determination of the absolute maximum. An AV must be carried out for each (λ_1, λ_2) used, to obtain the residual SS in (28.58). In the simplest case when $\lambda_2 = 0$, this can be avoided by equating to zero the first derivative of (28.58) with respect to λ_1. Using (28.60-61), this gives

$$0 = \frac{\partial \log L_{\lambda_1}(y | \hat{\boldsymbol{\beta}}, \hat{\sigma}^2)}{\partial \lambda_1} = -n \frac{\mathbf{y}'_{\lambda_1} \mathbf{T} \mathbf{u}_{\lambda_1}}{\mathbf{y}'_{\lambda_1} \mathbf{T} \mathbf{y}_{\lambda_1}} + n \lambda_1^{-1} + \sum_{i=1}^{n} \log y_i, \tag{28.62}$$

where the elements of $\mathbf{u}$ are $\{\lambda_1^{-1} y_i^{\lambda_1}, \log y_i\}$. In this case, Draper and Cox (1969) examine the precision with which λ_1 is estimated—see also Hinkley (1975) for some corrections.

Carroll (1982) shows that searching for λ over a coarse grid (the values 0, $\pm\frac{1}{2}$, ± 1, are a popular choice) may lead to a considerable increase in the prediction MSE. However, if the true value happens to be very close to a grid value, the prediction MSE may be reduced.

Schlesselman (1971) shows that (28.58) will not in general be invariant under scale changes in y unless the model (28.56) contains a general mean, which we now assume.

28.35   In order to test the hypothesis H_0: $\lambda = \lambda_0$, Andrews (1971) proposed an exact test based upon the following argument. Consider the first-order Taylor series expansion

$$\mathbf{y}_\lambda = \mathbf{y}_0 + (\lambda - \lambda_0)\mathbf{v}, \tag{28.63}$$

where $\mathbf{v} = \mathbf{v}(\mathbf{y}) = [\partial \mathbf{y}_\lambda / \partial \lambda]_{\lambda = \lambda_0}$ and y_0 denotes y evaluated at λ_0. Using (28.63), (28.56) may be expressed as

$$\mathbf{y}_0 = \mathbf{X}\boldsymbol{\beta} + (\lambda_0 - \lambda)\mathbf{v} + \boldsymbol{\varepsilon}. \tag{28.64}$$

Andrews' proposed test reduces to testing the coefficient $(\lambda_0 - \lambda)$ in (28.64) when $\mathbf{v}$ is replaced by $\hat{\mathbf{v}} = \hat{\mathbf{v}}(\hat{\mathbf{y}})$.

Atkinson (1973) suggested replacing (28.64) by

$$\mathbf{z}_0 = \mathbf{X}\boldsymbol{\beta} + (\lambda_0 - \lambda)\mathbf{w} + \boldsymbol{\varepsilon}, \tag{28.65}$$

where $\mathbf{z}_\lambda = \mathbf{y}_\lambda / \{J_\lambda\}^{1/n}$ and $\mathbf{w} = [\partial \mathbf{z}_\lambda / \partial \lambda]_{\lambda = \lambda_0}$; the test procedure is essentially the same. Simulation results (e.g. Atkinson, 1973) suggest that the revised procedure is generally more powerful, although the true probability of type I error often exceeds the nominal level. Lawrance (1987) provides a modification to the test statistic to correct this problem. It is evident that (28.64) or (28.65) may be used to provide an estimator for λ which, though inefficient, is very easy to compute, a property that will be useful in diagnostic testing, see **28.70**.

LR tests of nested hypotheses

28.36   Box and Cox (1964) present some interesting numerical examples of the application of this method of finding a transformation, and of a parallel Bayesian method of analysis which they develop. In addition, they consider the resolution of the maximised LF into three components corresponding to the normality, the homo-scedasticity, and the structure of the expectation of y_λ. Their procedure is of general applicability.

Consider sets of constraints $C_1, C_2, \ldots$, to be applied successively to a mathematical model, and let $\hat{\lambda}_{(s)}$ be the ML estimator of λ when all of $C_1, C_2, \ldots, C_s$ have been applied. $\hat{\lambda}$, without suffix, is the ML estimator when no constraint is imposed. Then, identically for any s,

$$L(y|\hat{\lambda}_{(s)}) = L(y|\hat{\lambda}) . \frac{L(y|\hat{\lambda}_{(1)})}{L(y|\hat{\lambda})} \cdot \frac{L(y|\hat{\lambda}_{(2)})}{L(y|\hat{\lambda}_{(1)})} \cdots \frac{L(y|\hat{\lambda}_{(s)})}{L(y|\hat{\lambda}_{(s-1)})}$$

$$= L(y|\hat{\lambda}) . l_1 l_2 \ldots l_s, \tag{28.66}$$

where l_p is the LR test statistic for testing the set of constraints $C_1, C_2, \ldots, C_{p-1}, C_p$ against the set $C_1, C_2, \ldots, C_{p-1}$ (cf. **23.1**). Each of the l_p lies between 0 and 1, and under regularity conditions, $-2 \log l_p$ is asymptotically a non-central χ^2 variable with d.fr. equal to the number of independent constraints upon parameters imposed by C_p (cf. **23.7**). When C_p holds, this becomes a central χ^2 variable, and thus $-2 \log l_p$ may be used to test the value of adding C_p to the already imposed $C_1, C_2, \ldots, C_{p-1}$. It should be observed that the l_p are not in general independently distributed, though in particular cases they may be independent under certain hypotheses (cf. Exercises 23.6 and 23.13, and the more general result of Exercise 28.16). The application of the resolution (28.66) to the present problem is left to the reader as Exercise 28.15, since it follows immediately from some results given in Chapter 23.

> Spitzer (1978) examined the performance of this method in samples of size 30 and 60. While the parameter-estimators were approximately normal with little bias and variance estimators of high efficiency, the t-statistics used in tests had long-tailed distributions leading to too-frequent rejection of true hypotheses. Models with $\lambda_1 > 0$ generally performed worse in bias and variance.

The purposes of transformation

28.37   The virtue of the ML approach discussed in **28.31-5** is that it requires no prior knowledge of the relationship between y and and the regressors, or of the nature of the error distribution of the untransformed y. It starts from the assumption that there exists some transformation in the family considered for which all the conditions of the linear model, including homoscedasticity and normality of the error distribution, are satisfied. In particular cases, of course, this may not be so—cf. Draper and Cox (1969), Hernandez and Johnson (1980) and Bickel and Doksum (1981); but even then, the ML procedure for choice of the transformation must presumably be an improvement on the uncritical use of y in its original form. It is a striking fact (evidenced by the numerical examples given by Box and Cox (1964)) that this ML transformation is often very close to what is suggested by non-statistical consideration of the nature of the underlying variables. Such consideration should, of course, be undertaken wherever possible as a supplement and guide to the statistical analysis itself.

J. B. Kruskal (1965) gives a method of finding the monotone transformation of the observations which minimises the Residual SS (suitably scaled) from an assumed linear model. No parametric family like (28.60) is required; nor is the normality assumption. He uses his method to re-analyse the Box and Cox examples, with several others. See also Draper and Hunter (1969).

Other approaches to transforming the data to meet the needs of the linear model have been less ambitious. They seek *either* to normalise the errors *or* to stabilise their variance *or* to remove interactions so that effects are additive; and the hope is that a transformation which effects one of these aims will at least help towards achieving the others. It is remarkable that this indeed often turns out to be the case, and we shall examine some important instances shortly, but it is over-sanguine to expect this to be always so. It is easy to construct examples where the goals of additivity and homoscedasticity conflict, for if in a two-way cross-classification the expected value of y is additive in row- and column-effects, but the errors are non-normally distributed with variance a function of $E(y)$, any transformation to remove the heteroscedasticity will destroy exact additivity, whatever may happen to the non-normality.

Hoyle (1973) gives an introductory review of the subject, with a large bibliography. We now examine these different types of transformation in turn.

Variance-stabilising transformations

28.38 Suppose that a statistic t has mean θ and variance, for fixed sample size n,

$$\text{var } t = D_n^2(\theta). \tag{28.67}$$

To eliminate this dependence of variance on the parameter θ, we seek a function $u(t)$ such that var u is a constant, c. In general, however, we are unlikely to be able to achieve this precisely, so we ask only that

$$\text{var}\{u(t)\} = c\{1 + O(R^{-1})\} \tag{28.68}$$

where R is some known constant which is large enough for R^{-1} to be negligible. In particular, we may have $R = n$, the sample size. We now assume t to be confined to a neighbourhood of its mean θ and that a Taylor expansion may be made as at (10.11), so that we have from (10.14) the approximation

$$\text{var}\{u(t)\} = \left\{\left(\frac{du(t)}{dt}\right)^2\right\}_{t=\theta} \text{var } t. \tag{28.69}$$

If (28.68) and (28.69) are equated, we have the first-order approximation

$$\left\{\left(\frac{du(t)}{dt}\right)^2\right\}_{t=\theta} = \frac{c}{D_n^2(\theta)}. \tag{28.70}$$

Since we are considering only the neighbourhood of θ, we drop the suffix '$t = \theta$', and write θ for t. Thus

$$\frac{du(\theta)}{d\theta} \propto \{D_n^2(\theta)\}^{-1/2}, \tag{28.71}$$

where we drop the constant c without loss, since this is in any case at choice, for multiplication of $u(t)$ by a constant will not affect our purpose of achieving (28.68). We now integrate the equation (28.71), again ignoring the additive constant which results from the indefinite integration without loss, since (28.68) is unaffected. We obtain

$$u(t) \propto \left\{\int \frac{d\theta}{D_n(\theta)}\right\}_{\theta=t}. \tag{28.72}$$

28.39 Although (28.72) was arrived at through approximation, we can check its validity if the theoretical distribution of t is known by computation of the theoretical variance of $u(t)$ to verify its stability as θ varies—it may be found desirable to modify $u(t)$ to improve stability. When, on the other hand, we have only observations upon t and no prior knowledge of its distribution or of the parameter θ of that distribution, we cannot even compute $D_n^2(\theta)$ precisely. In such cases, the mean and variance of t in separate groups of observations are calculated, and the latter plotted against the former to give an *estimate* of the relationship (28.67), on which the transformation (28.72) is then based. Here, the approximation is more hazardous, but nevertheless often gives satisfactory results in practice.

Example 28.5

If t has the Poisson distribution, (5.20) shows that the mean and variance are equal, so (28.67) is here simply

$$D_n^2(\theta) = \operatorname{var} t = \theta$$

and (28.72) gives

$$u(t) \propto \left\{ \int \theta^{-1/2} \, d\theta \right\}_{\theta=t} \propto t^{1/2}, \tag{28.73}$$

a simple square-root transformation. To the first order, by (28.69),

$$\operatorname{var}(t^{1/2}) = \{(\tfrac{1}{2}t^{-1/2})^2\}_{t=\theta} \operatorname{var} t = \tfrac{1}{4}, \tag{28.74}$$

verifying the variance stabilisation to this order.

Bartlett (1936) pointed out that variance stabilisation could be improved in this case by re-locating t before taking the square root. Bartlett suggested that if we define

$$u_c(t) = (t+c)^{1/2},$$

we should use $c = \tfrac{1}{2}$. Exercise 28.22 shows that $c = \tfrac{3}{8}$ is a better choice. Table 28.7

Table 28.7

θ	Variance of $u_c(t)$ as a fraction of limiting variance		
	$c = 0$	$c = \tfrac{1}{2}$	$c = \tfrac{3}{8}$
0	0	0	0
0.5	1.240	0.408	
1.0	1.608	0.640	0.717
2.0	1.560	0.856	0.924
3.0	1.360	0.928	0.983
4.0	1.224	0.960	0.999
6.0	1.104	0.980	1.002
9.0	1.052	0.988	
10.0			1.001
12.0	1.036	0.992	
15.0	1.024	0.992	
20.0			1.000

gives the variance of $u_c(t)$ as a fraction of its limiting variance as $\theta \to \infty$, for $c = 0, \frac{1}{2}$ and $\frac{3}{8}$—the calculations were made by Bartlett (1936) and Anscombe (1948).

The inadequacy of the simplest transformation with $c = 0$ is evident for small θ. For θ less than 3, the same comparison is made graphically in Fig. 28.2, adapted from Freeman and Tukey (1950), whose own variance-stabilisation proposal, $u' = t^{1/2} + (t+1)^{1/2}$, is more stable than $u_{3/8}(t)$ for $\theta \leqslant 2$, after which either is adequate. u' is within 6 per cent of stability for $\theta \geqslant 1$, and seems the best choice (cf. Exercise 28.24). Mosteller and Youtz (1961) give a table of u' for $x = 1(1)50$.

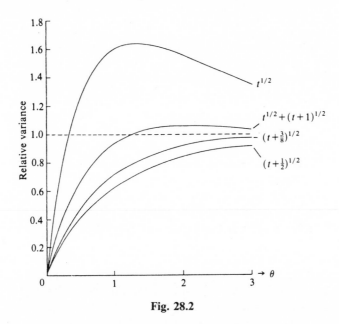

Fig. 28.2

28.40   The variance-stabilisation procedure of **28.38** can be repeated if necessary. Suppose that investigation shows the variance of $u(t)$ to be

$$\text{var}\{u(t)\} = c\left\{1 + \frac{p(\theta)}{n}\right\} + o(n^{-1}), \tag{28.75}$$

satisfying (28.68). If we now seek a second transformed variable $v(u)$ such that

$$\text{var}\{v(u)\} = d\{1 + O(n^{-2})\}, \tag{28.76}$$

we have, as at (28.69), the approximation

$$\text{var}\{v(u)\} = \left(\frac{dv(u)}{du}\right)^2_\theta \text{var } u$$

$$= \left(\frac{dv(u)}{du}\right)^2_\theta c\left\{1 + \frac{p(\theta)}{n}\right\}$$

by (28.75). Using (28.70), this is

$$\text{var}\{v(u)\} = \left(\frac{dv}{du}\right)_\theta^2 \left(\frac{du}{dt}\right)_\theta^2 D_n^2(\theta)\left\{1+\frac{p(\theta)}{n}\right\}$$

$$= \left(\frac{dv}{dt}\right)_\theta^2 D_n^2(\theta)\left\{1+\frac{p(\theta)}{n}\right\}. \tag{28.77}$$

Thus, as at (28.72),

$$v(t) \propto \left\{\int\left[D_n^2(\theta)\left\{1+\frac{p(\theta)}{n}\right\}\right]^{-1/2} d\theta\right\}_{\theta=t}. \tag{28.78}$$

We have already encountered an instance of this procedure in Hotelling's improved version of Fisher's variance-stabilising z-transformation—cf. Exercises 16.18–19.

The variance-stabilisation procedure could evidently be iterated further if this were necessary.

Exercises 28.18–20 give further applications of a single variance-stabilising transformation by the method of **28.38** to the binomial and negative binomial distributions.

Normalising transformations

28.41 In **6.25–6**, we have already examined the Cornish–Fisher method of obtaining a normalising polynomial transformation; and in **6.27–36**, we discussed Johnson's systems of functional transformations to normality.

Curtiss (1943) gives a careful mathematical discussion of the limiting normality of transformations, especially those discussed in our Examples and Exercises. As noted in **28.37**, a transformation designed to achieve one purpose (here, variance stabilisation) often also helps to achieve another (here, normalisation). For example, Exercise 28.23 treats the case dealt with in Example 28.5, where this effect occurs. However, the following example shows that this harmony of purposes is only obtainable by not pressing for optimal achievement in both directions: variance-stabilising transformations commonly normalise as a by-product, but they do not produce the *optimum* normalisation.

Example 28.6

Let z follow a gamma distribution with parameters q and θ, so that

$$f(z) = \theta^{-q} z^{q-1} e^{-z/\theta}/\Gamma(q); \tag{28.79}$$

cf. (11.25). Then

$$E(z) = q\theta \qquad \text{and} \qquad \text{var}(z) = q\theta^2.$$

If q is fixed, but θ varies between observations, we set

$$D_n^2(\theta) = q\theta^2$$

and (28.69) gives

$$u(z) \propto \left\{\int \theta^{-1} d\theta\right\}_z = \log z, \tag{28.80}$$

the logarithmic transform. However, if θ is fixed and q varies, we obtain $u(z) \propto q^{1/2}$, leading to the square root transform. Using Exercise 28.21, we find that the skewness and kurtosis measures (cf. **3.31–2**) for log z are

$$\gamma_1 \simeq -q^{-1/2}, \qquad \gamma_2 \simeq 2q^{-1}; \tag{28.81}$$

those for z are

$$\gamma_1 = 2q^{-1/2} \quad \text{and} \quad \gamma_2 = 6q^{-1} \tag{28.82}$$

so that the variance stabilisation, as a by-product, has halved skewness (with a change of sign) and reduced kurtosis by a factor of three.

When q varies, the square root transform provides effective variance stabilisation, as these values given by Bartlett (1936) show:

q	0	0.5	1.0	2.0	3.0	4.0	9.0	15.0
var $(z^{1/2})$	0	0.182	0.215	0.233	0.239	0.242	0.247	0.250

The square-root transform is equivalent to Fisher's approximation to the χ^2 distribution, treated in **16.5–6**. It has, by (16.8),

$$\gamma_1 = \tfrac{1}{2}q^{-1/2}, \qquad \gamma_2 = \tfrac{3}{16}q^{-2}, \tag{28.83}$$

distinct improvements over (28.82) for the untransformed variable (and better also than (28.81) for its logarithm). Again, the variance stabilisation has improved the normalisation here. But note that the Wilson–Hilferty cube-root transformation in **16.7** has, from (16.13), even better γ_1, of order $q^{-3/2}$, though not such good γ_2, of order q^{-1}, and (cf. **16.8**) gives the better normal approximation. However, it does not stabilise variance at all, as (16.12) shows. Hernandez and Johnson (1980) show that the cube root is also the asymptotic value of λ_1 obtained by the ML method from (28.62). Thus the best available normalising transformation sacrifices variance stability, and the square-root variance stabiliser is a better compromise transformation.

It is interesting that if both γ_1 and γ_2 are to be minimised, the appropriate transformation is $z^{5/13}$, but that in the Poisson case of Example 28.5, the power transformation is increased from $\tfrac{1}{2}$ to $\tfrac{2}{3}$ if we minimise γ_1 alone, or to $\tfrac{7}{11}$ if we minimise both γ_1 and γ_2—cf. Exercise 28.25.

28.42   The reader will see that our discussion of normalisation has been couched entirely in terms of skewness and kurtosis. Mathematically, it is taking a good deal for granted to assume that smaller values of γ_1 and γ_2 are equivalent to a closer approach to normality; but we know of no significant example where this assumption misleads us in choosing between normal approximations.

Blom (1954) seeks functional transformations $u(t)$ for which a further polynomial (Cornish–Fisher) transformation as in **6.25–6** has minimum skewness and is therefore presumably 'nearest' to symmetry and normality. This leads to a differential equation for $u(t)$ which contains all the transformations that we have encountered for the Poisson, gamma, binomial and negative binomial distributions. Haldane (1938) determined λ_1 in (28.59) when $\lambda_2 = 0$ to minimise skewness, and also in effect determined λ_1 and λ_2 to minimise skewness and kurtosis—cf. Exercise 28.25. See also Hinkley (1975).

Transformations to additivity

28.43 Although in practice it may be important to search for a scale on which effects are additive (i.e. interactions disappear) or nearly so, relatively little work has been done in this area as compared with normalisation and variance stabilisation. Some general procedures which have been proposed involve minimisation, within a class of transformations, of the value of the test statistic used for the hypothesis that interactions between regressor variables are zero. It will be recognised that the test statistic is here being used to carry out a complex estimation procedure, and nothing but intuitive justification has so far been given for this method. Additivity transformations are sometimes suggested by the analysis of residuals.

Generally, if the model appears to be additive in the original variate, but the variance is not stable, the errors should be treated as heteroscedastic; see **28.58**. Non-normality may be the result of outliers, which may be detected using the diagnostics described in **28.65–70**; alternatively, robust estimators may be used, see **28.71–4**.

Removal of transformation bias

28.44 Whatever the purpose of a transformation, it often raises problems of presentation when the analysis is complete. In particular, estimators of means or of differences which are unbiased on the transformed scale will not be so if the inverse transformation is made so that results may be presented in 'natural' terms (cf., e.g., Exercise 28.22). Adjustments of some kind must be made to remove the bias due to transformation; the jackknife method of bias-reduction was given in **17.10**. We now discuss an exact method of removing the bias.

Suppose that u is normally distributed with mean μ and variance σ^2, and that the functions of u, $(\hat{\mu}, S^2/\nu)$, are jointly sufficient statistics for these parameters, $\hat{\mu}$ being normally distributed with mean μ and variance $\lambda^2\sigma^2$, and S^2/σ^2, independent of $\hat{\mu}$, a χ^2 variate with ν d.fr. In practice, we usually have $\lambda^2 = 1/n$ and $\nu = n-1$ where n is sample size. Now consider the function $t(u)$, which in our terms is the *inverse* transformation. Neyman and Scott (1960) (cf. Schmetterer (1960)), using the approach of Exercise 28.26, showed that if $t(u)$ satisfies the second-order differential equation

$$t''(u) = A + Bt(u)$$

for constants A, B, the unique MV unbiased estimator of the mean of the untransformed variable $\theta = E(t)$ is given by

$$\hat{\theta} = \begin{cases} t(\hat{\mu}) + A(1-\lambda^2)S^2/(2\nu), & B = 0, \\ \{t(\hat{\mu}) + A/B\} \sum_{r=0}^{\infty} \dfrac{\Gamma(\frac{1}{2}\nu)}{r!\,\Gamma(\frac{1}{2}\nu + r)} \left\{ \dfrac{B(1-\lambda^2)S^2}{4} \right\}^r - \dfrac{A}{B}, & B \neq 0. \end{cases}$$

This series converges very rapidly, only a few terms usually being required for adequate accuracy.

It follows that the bias of the crude estimator $t(\hat{\mu})$, which is simply the inverse transformation of $\hat{\mu}$, is

$$E\{t(\hat{\mu}) - \theta\} = \begin{cases} -A(1-\lambda^2)\sigma^2/2, & B = 0, \\ \{\theta + (A/B)\}[\exp\{-B(1-\lambda^2)\sigma^2/2\} - 1], & B \neq 0, \end{cases}$$

and its absolute value is always a monotone decreasing function of λ^2. Since usually $\lambda^2 = 1/n$, the bias will *increase* with sample size.

The most important special cases are shown in Table 28.8. It will be seen that as $\lambda \to 0$ $(n \to \infty)$, the bias for the square-root transformation $\to -\sigma^2$. This is the result obtained directly in Exercise 28.22, where $\sigma^2 = \frac{1}{4}$ as at (28.74). The reader may also recall that the results for the logarithmic transformation with $c = 0$, $\lambda^2 = 1/n$, were contained in Exercises 18.7-8. Thöni (1969) gives a table to facilitate computation of $\hat{\theta}$ in this case. The other two transformations in the table are those of Exercises 28.18-19.

Hoyle (1968) evaluates var $\hat{\theta}$ and its MV unbiased estimator for the wider class of transformations treated in Exercise 28.26. One of his results can be obtained directly in Exercise 28.27. Land (1971) considers confidence intervals for linear functions of μ and σ^2 and (1974) approximate intervals for general functions. His tables of confidence intervals for $\theta + \frac{1}{2}\sigma^2$, useful in the logarithmic case, are given by Harter and Owen (1975).

Table 28.8

Transformation $u(t)$	Inverse transformation $t(\hat{\mu})$	A	B	Bias $E\{t(\hat{\mu})\} - \theta$	Sign of bias when $\lambda < 1$
$(t+c)^{1/2}$	$\hat{\mu}^2 - c$	2	0	$-(1-\lambda^2)\sigma^2$	Negative
$\log(t+c)$	$\exp(\hat{\mu}) - c$	c	1	$(\theta+c)[\exp\{-(1-\lambda^2)\sigma^2/2\} - 1]$	$-\mathrm{sgn}(\theta+c)$
$\arcsin(t^{1/2})$	$\sin^2(\hat{\mu})$	2	-4	$(\theta-\frac{1}{2})[\exp\{2(1-\lambda^2)\sigma^2\} - 1]$	$\mathrm{sgn}(\theta-\frac{1}{2})$
$\mathrm{arc\,sinh}(t^{1/2})$	$\sinh^2(\hat{\mu})$	2	4	$(\theta+\frac{1}{2})[\exp\{-2(1-\lambda^2)\sigma^2\} - 1]$	$-\mathrm{sgn}(\theta+\frac{1}{2})$

Bootstrap

28.45 The discussion thus far has assumed that the errors are normally distributed, after transformation if necessary. An alternative approach is to eschew any distributional assumptions and to work with the *bootstrap*; cf. **10.18**. Let

$$e_{(1)} \leq e_{(2)} \leq \cdots \leq e_{(n)}$$

denote the ordered residuals obtained from the LS fit of model (28.1), leading to the empirical d.f.

$$\tilde{F}(e) = \frac{j}{n} \qquad \text{if } e_{(j)} \leq e < e_{(j+1)}. \tag{28.84}$$

Then resample from $\tilde{F}$; that is, draw n observations with replacement, say $\tilde{e}$, and form

$$\tilde{y} = X\hat{\beta} + \tilde{e}. \tag{28.85}$$

From $(X, \tilde{y})$, compute the estimator $\tilde{\beta}$, using LS.

The resampling and estimation process is repeated B times, where B is $O(n \log n)$, to form an empirical sampling distribution for the estimator of any function of $(\boldsymbol{\beta}, \sigma^2)$ that is of interest.

The basic theory for the regression case was developed by Freedman (1981); Freedman and Peters (1984) gave some empirical results showing that the bootstrap estimates may work well and that the asymptotic standard error formulae may substantially understate the required length of the confidence interval. Some authors prefer to use the standardised residuals, $e_i^* = e_i / \{\text{var}(e_i)\}^{1/2}$, rather than e_i to define $\tilde{F}$. An extensive theoretical and empirical investigation is given in Wu (1986); see also Hinkley (1988).

Multicollinearity

28.46   As we observed in **19.6**, if one or more of the latent roots of $\mathbf{X'X}$ is near zero, $D = |\mathbf{X'X}|$ will be near zero, leading to imprecise estimators. This is the problem of multicollinearity, which sometimes plagues observational studies; experimental studies will not be so affected if the experiment is well designed. Indeed, max (D) is often used as a design criterion; see **28.14**.

When $D = 0$, this may be due to the parameterisation used, as in AV, and linear constraints can be imposed to restore a full-rank solution; see **19.13–16**. We now suppose that such constraints are not available and that D is close to zero. Further, we assume that all variables, including y, are measured about their means so that the constant term may be dropped, and that the regressor variables are expressed in some standard measure. Usually, this will mean that $\mathbf{X'X}$ becomes the correlation matrix for $(x_1, \ldots, x_p)$ but other 'standardisations' may be used, provided the resulting variables are measured on a common scale. For ease of notation, we shall assume there are p regressors after deletion of the constant.

Given this specification, it follows from (27.61) that the variance of $\hat{\beta}_j$ is

$$\text{var}(\hat{\beta}_j) = \sigma^2 / (1 - R_{(j)}^2), \tag{28.86}$$

where $R_{(j)}^2$ is the squared multiple correlation for the regression of x_j on all the regressors other than x_j. To help detect multicollinearity, we may use the *variance inflation factor* (VIF) for each variable

$$\text{VIF}(j) = 1 / (1 - R_{(j)}^2). \tag{28.87}$$

Clearly, VIF $= 1$ if x_j is orthogonal to all other variables and large values indicate potential problems.

> Stewart (1987) presents several indices for assessing multicollinearity; the ensuing discussion indicates the lively debate that persists. See also Gunst (1983).

28.47   We now consider how the effects of multicollinearity might be ameliorated. Let

$$D = \prod_{i=1}^{p} \lambda_i, \tag{28.88}$$

where $\lambda_1 \geq \lambda_2 \geq \cdots \geq \lambda_p \geq 0$ are the latent roots of $\mathbf{X'X}$. Evidently, D will be increased if we can either remove the smaller roots or increase their values; both options have been considered, as we now describe.

Ridge regression

28.48　If we add a constant c to each λ_j, we have

$$D^* = \prod_{i=1}^{p} (\lambda_i + c). \qquad (28.89)$$

In turn, this corresponds to using the biased estimator

$$\hat{\boldsymbol{\beta}}_c = (\mathbf{X'X} + c\mathbf{I})^{-1}\mathbf{X'y}, \qquad (28.90)$$

which is the ridge regression estimator described in **19.12**.

Principal components regression

28.49　It is possible to replace the original regressor variables by a set of p or fewer orthogonal variables. One such set is given by the principal components, defined by the latent roots of $\mathbf{X'X}$, denoted by $\boldsymbol{\Lambda}$:

$$(\mathbf{X'X})\mathbf{U} = \mathbf{U}\boldsymbol{\Lambda}, \qquad (28.91)$$

where $\boldsymbol{\Lambda} = \text{diag}(\lambda_1, \lambda_2, \dots, \lambda_p)$. The statistical properties of principal components will be discussed in a later volume and do not concern us here. However, since $\mathbf{U'U} = \mathbf{UU'} = \mathbf{I}$ (orthogonality), we may define the new variables

$$\mathbf{Z} = \mathbf{XU} \qquad (28.92)$$

such that the model becomes

$$\mathbf{y} = \mathbf{Z}\boldsymbol{\alpha} + \boldsymbol{\varepsilon},$$

where

$$\boldsymbol{\alpha} = \mathbf{U'}\boldsymbol{\beta} \quad \text{and} \quad \boldsymbol{\beta} = \mathbf{U}\boldsymbol{\alpha}. \qquad (28.93)$$

It follows directly that

$$\hat{\boldsymbol{\alpha}} = (\mathbf{Z'Z})^{-1}\mathbf{Z'y}$$
$$= \boldsymbol{\Lambda}^{-1}\mathbf{Z'y}$$

or

$$\alpha_j = \mathbf{z}_j'\mathbf{y}/\lambda_j, \quad j = 1, 2, \dots, p; \qquad (28.94)$$

also

$$\text{var}(\hat{\boldsymbol{\alpha}}) = \sigma^2\boldsymbol{\Lambda}^{-1}. \qquad (28.95)$$

Further, it is evident from (28.95) that the imprecise estimates arise directly from the linear combinations of the regressors for which the corresponding latent roots are near zero. If ridge regression is not favoured, we may proceed by taking only the first $k(<p)$ principal components, $\mathbf{Z}_k$ say, and employing the estimator

$$\underset{(k\times 1)}{\hat{\boldsymbol{\alpha}}_k} = \underset{(k\times k)}{\boldsymbol{\Lambda}_k^{-1}} \underset{(k\times n)}{\mathbf{Z}_k'} \underset{(n\times 1)}{\mathbf{y}},$$

where the k-subscript denotes exclusion of components $k+1,\ldots,p$. Finally, from (28.93), we estimate $\boldsymbol{\beta}$ by

$$\underset{(p\times 1)}{\hat{\boldsymbol{\beta}}_k} = \underset{(p\times k)}{\mathbf{U}_k} \underset{(k\times 1)}{\boldsymbol{\alpha}_k}, \tag{28.96}$$

where $\mathbf{U} = (\mathbf{U}_k, \mathbf{U}_k^*)$, effectively imposing the conditions

$$\alpha_{k+1} = \cdots = \alpha_p = 0 \quad \text{or} \quad (\mathbf{U}_k^*)'\boldsymbol{\beta} = \mathbf{0}. \tag{28.97}$$

As in ridge regression, the reduction in variance must be traded against the bias that is introduced.

Example 28.7 (Greenberg, 1975)
Malinvaud (1966) considers a linear regression model for French imports (Y), over the eleven years 1949–59. Imports are taken as a function of gross domestic product (x_1), stock formation (x_2) and consumption (x_3). The correlation matrix is

$$\begin{bmatrix} 1.0 & 0.169 & 0.997 \\ & 1.0 & 0.154 \\ & & 1.0 \end{bmatrix}$$

from which the variance inflation factors are

$$\text{VIF}(1) = 193.7, \text{VIF}(2) = 1.07, \text{VIF}(3) = 192.8.$$

In this case, the problem is clearly the high correlation between x_1 and x_3, although less obvious patterns will also be detected by the VIF. The latent roots and vectors, derived from the correlation matrix, are given in Table 28.9. In Table 28.10, we give estimates of the regression coefficients for the standardised variables, based on one, two and three principal components respectively. Those for $k=1$ and $k=2$ are very similar, while introduction of the third component creates chaos. A test of the linear conditions $\mathbf{u}_2'\boldsymbol{\beta} = 0$ and $\mathbf{u}_3'\boldsymbol{\beta} = 0$ fails to reject either condition at the 5 per cent level.

Table 28.9   Latent roots and vectors for the data of Example 28.7

λ_j	$1/\lambda_j$	u_j'
2.0472	0.448	$(0.692, 0.213, 0.690)$
0.9502	1.052	$(-0.143, 0.977, -0.158)$
0.0026	384.6	$(-0.708, 0.010, 0.706)$

Table 28.10   Regression derived from components for Example 28.7

	Estimate	S.E.		Estimates derived from α		
				$k=1$	$k=2$	$k=3$
$\hat{\alpha}_1$	0.6917	0.0302	$\hat{\beta}_1$	0.478	0.474	-0.648
$\hat{\alpha}_2$	0.0314	0.0430	$\hat{\beta}_2$	0.148	0.178	0.195
$\hat{\alpha}_3$	1.584	0.8516	$\hat{\beta}_3$	0.477	0.472	1.591

The results using $k=1$ and $k=2$ also seem more in accord with economic theory than the estimates given by the full set of variables.

28.50 Principal components are only one possible way of generating orthogonal regressor variables. In the particular case of polynomial regression, a different construction is desirable, which we now describe.

Polynomial regression: orthogonal polynomials

28.51 The linear model with powers of x as regressors, called the *polynomial regression* model, is

$$y_i = \beta_1 + \beta_2 x_i + \cdots + \beta_p x_i^{p-1} + \varepsilon_i, \qquad i = 1, \ldots, n \qquad (28.98)$$

leads to an $\mathbf{X}'\mathbf{X}$ matrix of the form

$$\mathbf{X}'\mathbf{X} = n\mathbf{M}_p = n \begin{bmatrix} 1 & 0 & m_2 & \cdots & m_{p-1} \\ 0 & m_2 & m_3 & \cdots & m_p \\ \vdots & \vdots & \vdots & & \vdots \\ m_{p-2} & m_{p-1} & m_p & \cdots & m_{2p-3} \\ m_{p-1} & m_p & m_{p+1} & \cdots & m_{2p-2} \end{bmatrix}, \qquad (28.99)$$

where $nm_k = \sum x_i^k$ and we set $m_1 = \bar{x} = 0$ without loss of generality. Clearly, $\mathbf{M}_p$ cannot be diagonal. However, we may choose polynomials of degree j in x, say $\phi_j(x)$, $j = 0, 1, \ldots, p-1$ that are mutually orthogonal so that (28.98) becomes

$$\mathbf{y} = \mathbf{X}\boldsymbol{\beta} + \boldsymbol{\varepsilon} = \boldsymbol{\Phi}\boldsymbol{\alpha} + \boldsymbol{\varepsilon}, \qquad (28.100)$$

where $\boldsymbol{\Phi} = \mathbf{X}\mathbf{U}$, $\boldsymbol{\alpha} = \mathbf{U}^{-1}\boldsymbol{\beta}$, and $\mathbf{U}$ is lower triangular to ensure that ϕ_j is of degree j. The requirement of orthogonality becomes

$$\boldsymbol{\Phi}'\boldsymbol{\Phi} = \mathbf{C}, \qquad (28.101)$$

where $\mathbf{C}$ is a $p \times p$ diagonal matrix and we may set $c_{00} = 1$ implying $u_{00} = 1$; the diagonal elements of $\mathbf{C}$ are arbitrary and may be set at any convenient value. It follows from (28.100) that the kth-degree polynomial is given by

$$\phi_k(x) = \begin{vmatrix} \mathbf{M}_k & \mathbf{m}_{(k)} \\ \mathbf{x}'_{(k)} & \end{vmatrix} \Big/ |\mathbf{M}_k|,$$

where

$$\mathbf{x}'_{(k)} = (1, x, \ldots, x^k) \text{ and}$$

$$\mathbf{m}'_{(k)} = (m_k, m_{k+1}, \ldots, m_{2k-1}).$$

For example,

$$\phi_1(x) = \begin{vmatrix} 1 & 0 \\ 1 & x \end{vmatrix} \bigg/ |1| = x$$

$$\phi_2(x) = \frac{\begin{vmatrix} 1 & 0 & m_2 \\ 0 & m_2 & m_3 \\ 1 & x & x^2 \end{vmatrix}}{\begin{vmatrix} 1 & 0 \\ 0 & m_2 \end{vmatrix}} = x^2 - \frac{m_3}{m_2} x - m_2$$

and so on. A simpler recursive method of obtaining the polynomials is given in Exercise 28.28.

28.52   From (28.100) and (28.101), the LS estimators

$$\hat{\boldsymbol{\alpha}} = (\boldsymbol{\Phi}'\boldsymbol{\Phi})^{-1}\boldsymbol{\Phi}'\mathbf{y}$$

reduce to

$$\hat{\alpha}_j = \sum_{i=1}^{n} y_i \phi_j(x_i) \bigg/ \sum_{i=1}^{n} \{\phi_j(x_i)\}^2 \tag{28.102}$$

with the benefit that earlier coefficients are unchanged when higher-degree polynomials are added to the model.

Equally spaced x-values

28.53   The most important applications of orthogonal polynomials in regression analysis are to situations where the regressor variable, x, takes values at equal intervals. This is often the case with observations taken at successive times, and with data grouped into classes of equal width. If we have n such equally spaced values of x, we measure from their mean and take the natural interval as unit, thus obtaining as working values of x: $-\frac{1}{2}(n-1), -\frac{1}{2}(n-3), -\frac{1}{2}(n-5), \ldots, \frac{1}{2}(n-3), \frac{1}{2}(n-1)$. For this simple case, the values of the moments in (28.99) can be explicitly calculated: in fact, apart from the mean which has been taken as origin, these are the moments of the first n natural numbers, obtainable from the cumulants given in Exercise 3.23. The odd moments are zero by symmetry; the even moments are

$$m_2 = (n^2 - 1)/12,$$
$$m_4 = m_2(3n^2 - 7)/20,$$
$$m_6 = m_2(3n^4 - 18n^2 + 31)/112,$$

and so on. Substituting these and higher moments into (28.99), we obtain for the first six polynomials

$$
\left.
\begin{aligned}
\phi_0(x) &\equiv 1, \\
\phi_1(x) &= \lambda_{1n}x, \\
\phi_2(x) &= \lambda_{2n}\{x^2 - \tfrac{1}{12}(n^2 - 1)\}, \\
\phi_3(x) &= \lambda_{3n}\{x^3 - \tfrac{1}{20}(3n^2 - 7)x\}, \\
\phi_4(x) &= \lambda_{4n}\{x^4 - \tfrac{1}{14}(3n^2 - 13)x^2 + \tfrac{3}{560}(n^2 - 1)(n^2 - 9)\}, \\
\phi_5(x) &= \lambda_{5n}\{x^5 - \tfrac{5}{18}(n^2 - 7)x^3 + \tfrac{1}{1008}(15n^4 - 230n^2 + 407)x\}, \\
\phi_6(x) &= \lambda_{6n}\{x^6 - \tfrac{5}{44}(3n^2 - 31)x^4 + \tfrac{1}{176}(5n^4 - 110n^2 + 329)x^2 \\
&\quad - \tfrac{5}{14874}(n^2 - 1)(n^2 - 9)(n^2 - 25)\}.
\end{aligned}
\right\}
\tag{28.103}
$$

Allan (1930) also gives $\phi_i(x)$ for $i = 7, 8, 9, 10$. Following Fisher (1921b), the arbitrary constants λ_{in} in (28.103) may be determined conveniently so that $\phi_i(x_j)$ is an integer for all $j = 1, 2, \ldots, n$. It will be observed that

$$
\phi_{2i}(x) = \phi_{2i}(-x) \quad \text{and} \quad \phi_{2i-1}(x) = -\phi_{2i-1}(-x);
$$

even-degree polynomials are even functions and odd-degree polynomials odd functions.

Tables of orthogonal polynomials
28.54   The *Biometrika Tables* give $\phi_i(x_j)$ for all j, $n = 3(1)52$ and $i = 1(1) \min (6, n-1)$, together with the values of λ_{in} and $\sum_{j=1}^n \phi_i^2(x_j)$.

Fisher and Yates' *Tables* give $\phi_i(x_j)$ (their ξ_i'), λ_{in} and $\sum_{j=1}^n \phi_i^2(x_j)$ for all j, $n = 3(1)75$ and $i = 1(1) \min (5, n-1)$.

The *Biometrika Tables* give references to more extensive tabulations, ranging to $i = 9$, $n = 52$, by van der Reyden, and to $i = 5$, $n = 104$, by Anderson and Houseman.

28.55   There is a large literature on orthogonal polynomials. For theoretical details, the reader should refer to the paper by Fisher (1921b) which first applied them to polynomial regression, to a paper by Allan (1930), and three papers by Aitken (1933). Rushton (1951) discussed the case of unequally spaced x-values and C. P. Cox (1958) gave a concise determinantal derivation of general orthogonal polynomials, while Guest (1954, 1956) has considered grouping problems.

Narula (1979) provides efficient numerical procedures for computing orthogonal polynomials when the data are unequally spaced. Studden (1980, 1982) develops optimal designs for polynomial regression where the aim is to provide efficient tests of the hypothesis that an rth-order polynomial is adequate against the alternative that order $s > r$ is required.

Example 28.8
The first two columns of Table 28.11 show the human population of England and Wales at the decennial Censuses from 1811 to 1931. These observations are clearly

Table 28.11

Year	Population (millions) y	$\dfrac{\text{Year}-1871}{10} = x = \phi_1(x)$	$\phi_2(x)$	$\phi_3(x)$	$\phi_4(x)$
			Polynomials with $n=13$		
1811	10.16	-6	22	-11	99
1821	12.00	-5	11	0	-66
1831	13.90	-4	2	6	-96
1841	15.91	-3	-5	8	-54
1851	17.93	-2	-10	7	11
1861	20.07	-1	-13	4	64
1871	22.71	0	-14	0	84
1881	25.97	1	-13	-4	64
1891	29.00	2	-10	-7	11
1901	32.53	3	-5	-8	-54
1911	36.07	4	2	-6	-96
1921	37.89	5	11	0	-66
1931	39.95	6	22	11	99
$\sum y$ 314.09	$\lambda_{i,13}$	1	1	1/6	7/12
	$\sum \phi_j^2(x)$	182	2002	572	68 068
	$\sum y\phi_j(x)$	474.77	123.19	-39.38	-374.30

Here $n=13$, and from the *Biometrika Tables*, Table 47, we read off the values in the last four columns of Table 28.11.

not uncorrelated, so that the regression model (28.98) is not strictly appropriate, but we carry through the fitting process for purely illustrative purposes.

From Table 28.11, we obtain

$$\hat{\alpha}_0 = 24.161, \ \hat{\alpha}_1 = 2.6086, \ \hat{\alpha}_2 = 0.061534,$$

$$\hat{\alpha}_3 = -0.068846, \text{ and } \hat{\alpha}_4 = -0.0054989.$$

The ESS for the jth degree polynomial is $\hat{\alpha}_j^2 \sum \phi_i^2$, so that the condensed AV table is as shown in Table 28.12.

The total SS is 1251.283 and the F-values relate to tests of the hypothesis $H_k: \alpha_{k+1} = 0$, given $\phi_1, \ldots, \phi_k$ in the model. We would presumably include all four terms in the model. The cubic and quartic expressions are displayed in Fig. 28.3.

Table 28.12

Component	Explained DF	Explained SS	Residual DF	Residual SS	F	R^2
Linear	1	1238.497	11	12.786	1065.5	0.9898
Quadratic	1	7.580	10	5.206	14.6	0.9958
Cubic	1	2.711	9	2.495	9.8	0.9980
Quartic	1	2.058	8	0.437	37.7	0.9997

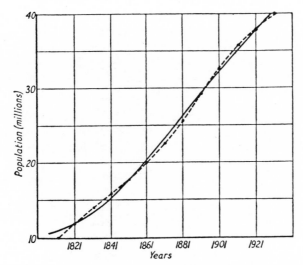

Fig. 28.3 Cubic (full line) and quartic (broken line) polynomial fitted to the data of Table 28.11

The reader should not need to be warned against the dangers of extrapolating from a fitted regression, however close, which has no theoretical basis. For example, the value predicted by the quartic regression for 1951 $(x=8)$ is 38.2, a good deal less than the Census population of 43.7 millions actually found in that year.

Distributed lags
28.56   A special case of (28.1) of particular interest arises when $x_{jt} = x_{t-j+1}$ so that

$$y_t = \sum_{j=1}^{p} \beta_j x_{t-j+1} + \varepsilon_t. \tag{28.104}$$

Often, the series x_t displays high autocorrelation, and the selected value of p is sufficiently large to create numerical difficulties if the least-squares estimators are used. To overcome this, Almon (1965) suggested that the βs be represented by an rth-order polynomial $(r+1<p)$

$$\beta_j = \sum_{j=0}^{r} j^i \delta_i, \qquad j=0, 1, \ldots, p. \tag{28.105}$$

In matrix terms, the original equation is

$$\mathbf{y} = \mathbf{X}\boldsymbol{\beta} + \boldsymbol{\varepsilon}$$

which may now be rewritten as

$$\mathbf{y} = \mathbf{Z}\boldsymbol{\delta} + \boldsymbol{\varepsilon},$$

where $\mathbf{Z} = \mathbf{XH}$ and $h_{ij} = j^i$. The reduced number of unknowns will often remove the multicollinearity while allowing explicit incorporation of the first p lags. Clearly, we need not restrict attention to polynomials, but may use any set of weights in (28.105). Such techniques are known as distributed lag methods and have proved particularly useful in econometrics; see Griliches (1967).

28.57 If the covariance matrix is not $V(\varepsilon) = \sigma^2 I$ as previously assumed, there are two principal departures to be considered:

heteroscedasticity:　　　　when $V(\varepsilon) = \sigma^2 V$, where V is diagonal and $V \neq I$
autocorrelation:　　　　　when $V(\varepsilon) = \sigma^2 V$, where the diagonal elements of V are equal, but at least some off-diagonal elements are non-zero.

We now consider these in turn.

Heteroscedasticity
28.58 If V is diagonal and known, we may use the weighted least squares (WLS) estimator

$$\hat{\beta} = (X'V^{-1}X)^{-1}X'V^{-1}y \tag{28.106}$$

whose properties are discussed in **19.19**. When V must be estimated we must either have replicates for each set of regressor variables or else constrain the elements of V to satisfy some estimable function. If we write

$$E(y_i) = \mu_i(\beta) \equiv \mu_i = x_i'\beta \tag{28.107}$$

and

$$\text{var}(y_i) = \sigma^2 v_i = \sigma^2\{g(\mu_i, z_i, \theta)\}^2 = \mu^2 g_i^2, \tag{28.108}$$

the WLS criterion would be to minimise

$$\sum_{i=1}^{n} [y_i - x_i'\beta]^2 / v_i \tag{28.109}$$

or, equivalently,

$$\sum_{i=1}^{n} [\tilde{y}_i - \tilde{x}_i'\beta]^2, \tag{28.110}$$

where $\tilde{y}_i = y_i / g_i$ and $\tilde{x}_i = x_i / g_i$. Both (28.109) and (28.110) lead directly to (28.106). It is clear from (28.109) that greater weight is assigned to the observations with smaller v_i, thereby making the WLS estimator potentially more efficient. Typical forms for g_i in (28.108) are

$$g_i = \mu_i^\theta \quad \text{or} \quad g_i = e^{\theta \mu_i}; \tag{28.111}$$

in some cases $|\mu_i|$ will be more appropriate than μ_i.

28.59 If either version of (28.111) is suspected, a natural diagnostic tool is to plot the LS residuals against μ_i or $\log(\mu_i)$, respectively. Improved plots are discussed in **28.60** below when we examine diagnostics more systematically. If the plots provide little evidence of heteroscedasticity, the LS estimators will cause only small losses of efficiency, so we now suppose that sizeable discrepancies have been identified. We may then generate the WLS estimators by an iterative process as follows:

1. using initial consistent estimators of $\boldsymbol{\beta}$ (typically the LS estimates), estimate the weights, $\hat{w}_i = 1/\hat{g}_i$;
2. generate new estimates of $\boldsymbol{\beta}$ using (28.106);

this cycle may be repeated K times. For any choice of K, the resulting estimators are asymptotically normally distributed:

$$\hat{\boldsymbol{\beta}} \sim N\{\boldsymbol{\beta}, \sigma^2(\mathbf{X}'\mathbf{V}^{-1}\mathbf{X})^{-1}\} \qquad (28.112)$$

provided the initial estimator is consistent. Although 'well known', this result was not demonstrated formally until the work of Jobson and Fuller (1980) and Carroll and Ruppert (1982).

Full iteration of steps (1) and (2) is a version of *iteratively reweighted least squares*, IRLS, (cf. Green, 1984), although the same end could, in this case, be achieved more efficiently using a Newton–Raphson method. The choice of K has been the subject of some research; see Matloff *et al.* (1984); the general view is that $K = 2$ or 3 cycles will usually suffice, although increasing K does not lead uniformly to improved estimators.

28.60 The algorithm described in **28.59** does not provide an explicit estimation procedure for the parameters in g_i. Heuristic procedures include plots of $|e_i|$ or $\log |e_i|$ against μ_i or $\log \mu_i$, where the $\{e_i\}$ are the residuals, $\mathbf{e} = \mathbf{y} - \mathbf{X}\hat{\boldsymbol{\beta}}$; Cook and Weisberg (1983) recommend the use of standardised residuals. Such procedures are reasonably satisfactory provided very small residuals are accommodated, e.g. by using $|e_i^*| = |e_i| + c$, $c > 0$. When the errors are normally distributed, ML estimators may be determined, as we now show.

Example 28.9
Let

$$\mathbf{y} \sim N(\mathbf{X}\boldsymbol{\beta}, \sigma^2\mathbf{V}),$$

where $\mathbf{V} = \text{diag}(v_i)$, $v_i = \mu_i^\theta$ and $\mu_i = \mathbf{x}_i'\boldsymbol{\beta} > 0$. The log LF is

$$l = \text{const} - \tfrac{1}{2}\log|V| - \frac{n}{2}\log \sigma^2 - \frac{1}{2\sigma^2}(\mathbf{y} - \mathbf{X}\boldsymbol{\beta})'\mathbf{V}^{-1}(\mathbf{y} - \mathbf{X}\boldsymbol{\beta}).$$

The first-order derivatives yield the estimating equations

$$(\mathbf{X}'\mathbf{V}^{-1}\mathbf{X})\boldsymbol{\beta} - \mathbf{X}'\mathbf{V}^{-1}\mathbf{y} = 0$$

$$n\sigma^2 - (\mathbf{y} - \mathbf{X}\boldsymbol{\beta})'\mathbf{V}^{-1}(\mathbf{y} - \mathbf{X}\boldsymbol{\beta}) = 0$$

and

$$-\sum \log \mu_i + \sum e_i^2 \mu_i^{-\theta} \log \mu_i = 0,$$

from which $(\hat{\boldsymbol{\beta}}, \hat{\sigma}^2, \hat{\theta})$ may be determined numerically.

For further details of estimation procedures, see Carroll and Ruppert (1988). Several tests for heteroscedasticity have been proposed; for a review and evaluation, see Kadiyala and Oberhelman (1984).

Autocorrelation

28.61 We now consider the other primary departure from the assumption of independent identical errors, where we now allow some of the correlations to be non-zero. Autocorrelation may arise through temporal or spatial patterns in the data or because the same experimental unit is used for several observations (known as *repeated measures*), as the following example illustrates.

Example 28.10

(a) When data are recorded for successive time intervals (monthly, yearly, etc.), the first-order autoregressive model is written as

$$\varepsilon_t = \phi \varepsilon_{t-1} + u_t, \tag{28.113}$$

where the u_t are taken to be independent $(0, \sigma^2)$. From (28.113), it follows that

$$\text{corr}(\varepsilon_t, \varepsilon_{t-s}) = \phi^{|s|} \tag{28.114}$$

provided that the series is *stationary* ($|\phi| < 1$ in this case). For further details of autoregressive time series, see Kendall and Ord (1990, Chapter 5).

(b) If data are recorded on a regular grid with cells (i, j), as in remote sensing for example, a possible autoregressive model is

$$\varepsilon_{ij} = \phi(\varepsilon_{i-1,j} + \varepsilon_{i+1,j} + \varepsilon_{i,j-1} + \varepsilon_{i,j+1}) + u_{ij} \tag{28.115}$$

although the bilateral nature of the spatial dependence may lead to other choices being preferred; see Cliff and Ord (1981, Chapter 6). It is not possible to develop a simple closed form for the autocorrelations.

(c) If k observations are taken on each experimental unit, we may assume that the random errors are independent between units, but not within units. For example, when there is no time ordering, we may asume that, for observations r and s on unit i,

$$E(\varepsilon_{ir}\varepsilon_{is}) = \phi\sigma^2, \qquad r \neq s \tag{28.116}$$

with $-(k-1)^{-1} < \phi < 1$ and

$$E(\varepsilon_{ir}\varepsilon_{js}) = 0, \qquad i \neq j; \tag{28.117}$$

that is, we allow intraclass correlation (cf. **26.27**). The effects of this correlation upon the estimation process are illustrated in Exercise 28.31. Further discussion of repeated measures is deferred until a later volume.

In the remainder of the discussion, we focus primarily upon the time-series case which is the most important in practice, particularly in observational studies.

28.62 The autoregressive structure (28.113) is often a plausible departure from independence and the standard test of H_0: $\phi = 0$ against H_1: $\phi \neq 0$ was developed by Durbin and Watson (1950, 1951). Their statistic may be written in terms of the residuals as

$$d = \sum_{t=2}^{n} (e_t - e_{t-1})^2 \Big/ \sum_{t=1}^{n} (e_t - \bar{e})^2. \tag{28.118}$$

When H_0 is true, d has an expected value in the neighbourhood of 2, whereas $d \to 0$ as $\phi \to 1$ and $d \to 4$ as $\phi \to -1$. The first two moments of $Z = \frac{1}{4}d$ are

$$E(Z) = S_1/(n-p) \tag{28.119}$$

$$E(Z^2) = (2S_2 + S_1^2)/\{(n-p)(n-p+2)\}, \tag{28.120}$$

where $S_j = \sum_{i=1}^{n-p} \lambda_i^j$ and $\lambda_1 \geqslant \lambda_2 \geqslant \cdots \geqslant \lambda_{n-p}$ are the non-zero latent roots of $\{I - X(X'X)^{-1}X'\}A$, and A has elements $a_{ij} = 1$ if $j = i \pm 1$, $a_{ij} = 0$ otherwise.

Evidently, even under H_0, the distribution of d depends upon X. Durbin and Watson derived upper and lower bounds, d_U and d_L, for the percentage points of d, based upon the latent roots of $X(X'X)^{-1}X'$; tables from their 1951 paper are reproduced as Appendix Tables 10 and 11. The tables are used in the following way: if $d_U < d < 4 - d_U$, H_0 is not rejected; if $d < d_L$ or $d > 4 - d_L$, reject H_0; otherwise, reserve judgement.

Example 28.11
For the model given in Example 28.1, we find that $d = 1.64$. With $n = 15$ and $p - 1 = 3$, $d_U = 1.46$ at the 1 per cent level and $d_U = 1.75$ at the 5 per cent level, so the evidence suggests rather weak autocorrelation as $\hat{\rho} = 0.18$. Inclusion of a lagged dependent variable in the model changes the other coefficients only slightly and the coefficient of the lagged term is not significantly different from zero.

Kamat and Satke (1962) show that the statistic

$$d^* = \sum |e_t - e_{t-1}| \Big/ \{\sum (e_t - \bar{e})^2\}^{1/2}$$

has an asymptotic relative efficiency of 0.766 with respect to d when the underlying population is normal and may be more efficient for non-normal alternatives. Various alternatives have been proposed to d (cf. Sims, 1975), mainly with a view to finding a statistic with simpler distributional properties, but these have usually proved to be somewhat less powerful. In a follow-up paper, Durbin and Watson (1971) suggest using

$$\tilde{d} = 2(1 - \tilde{\rho}),$$

where $\tilde{\rho}$ is the solution with absolute value less than one of the equation

$$(1 + \rho^2) \sum_{1}^{n} e_t^2 - \rho^2(e_1^2 + e_n^2) - 2\rho \sum_{2}^{n} e_t e_{t-1} = 0. \tag{28.121}$$

The distribution of $\tilde{d}$ may be approximated by a beta distribution, with the first two moments given by (28.119) and (28.120). A numerical study by Dent and Cassing (1978) suggests that $\tilde{d}$ has comparable power properties and a somewhat smaller zone of indecision.

K. F. Wallis (1972) gives an analogous procedure for testing for fourth-order serial correlation, for use with quarterly (e.g. economic) data.

28.63 These arguments may be extended to more general error structures, as follows. Consider the linear regression model

$$y_t = \beta_1 x_{1t} + \beta_2 x_{2t} + \cdots + \beta_k x_{kt} + \varepsilon_t \tag{28.122}$$

with an autoregressive error structure of order p

$$\varepsilon_t = \phi_1 \varepsilon_{t-1} + \cdots + \phi_p \varepsilon_{t-p} + u_t \tag{28.123}$$

which we may write as

$$\phi(B)\varepsilon_t = u_t, \tag{28.124}$$

where $\phi(B) = 1 - \phi_1 B - \cdots - \phi_p B^p$, using the backward shift operator, B. We use p for the AR scheme in deference to convention and switch temporarily to k for the number of regressors. Combining (28.122) and (28.124), we obtain

$$\phi(B)y_t = \sum \beta_i \phi(B)x_{it} + \phi(B)\varepsilon_t$$

or

$$y_t' = \sum \beta_i x_{it}' + u_t, \tag{28.125}$$

where $x_{it}' = \phi(B)x_{it}$ and $y_t' = \phi(B)y_t$. If the coefficients $\{\phi_j\}$ were known, we could estimate the $\{\beta_i\}$ from (28.125) by LS in the usual way. Cochrane and Orcutt (1949) suggested a recursive procedure based on fitting (28.122) by least squares, estimating the ϕs from (28.123) and re-estimating the βs from (28.125), then iterating between (28.123) and (28.125).

Durbin (1960) proposed an alternative procedure which yields asymptotically efficient estimators. Writing $\gamma_{ij} = \beta_i \phi_j$, we put (28.125) in the form

$$y_t = \sum_{i=1}^{p} \phi_j y_{t-j} + \sum_{i,j} \gamma_{ij} x_{i,t-j} + u_t. \tag{28.126}$$

If the γs were functionally independent, we could regard this as a regression of y_t on the lagged ys and the xs, and derive the least-squares estimators of ϕ and γ. If the corresponding estimators of ϕ, β, γ, are f, b, c, it follows that the quantities $f_j - \phi_j$ and $c_{ij} - \phi_j \beta_i$ are asymptotically normal with zero means and ascertainable covariance matrix. We can therefore write down their likelihood and maximise it to obtain the estimators of ϕ and β.

In certain cases, the least-squares estimators are asymptotically efficient—cf. R. L. and T. W. Anderson (1956) and Kramer (1980)—but tests of hypotheses are impaired.

28.64 Modern estimation procedures for models such as (28.122) and (28.123) are fully integrated with time-series methods. In particular, the Kalman filter-based approach of Harvey and Phillips (1979) allows efficient estimation of regression models with autoregressive moving average (ARMA) error structures. Their method uses *recursive* residuals, introduced to the statistical literature by R. L. Brown *et al.* (1975), although long familiar to engineers (cf. Kailath, 1974); our description restricts attention to the simple case of no autocorrelation.

Given r observations, the LS estimator for β is

$$\mathbf{b}_r = (\mathbf{X}_r'\mathbf{X}_r)^{-1}\mathbf{X}_r\mathbf{y}_r; \tag{28.127}$$

we may define the recursive residual for the next observation, y_{r+1}, as

$$w_{r+1} = (y_{r+1} - \mathbf{x}_{r+1}'\mathbf{b}_r)/\{1 + \mathbf{x}_{r+1}'(\mathbf{X}_r'\mathbf{X}_r)^{-1}\mathbf{x}_{r+1}\}^{1/2}, \tag{28.128}$$

which we rewrite as $w_r = z_{r+1}/\{1 + \mathbf{x}_{r+1}'\mathbf{g}_{r+1}\}$. If follows (cf. Exercise 19.20) that

$$(\mathbf{X}_{r+1}'\mathbf{X}_{r+1})^{-1} = (\mathbf{X}_r'\mathbf{X}_r)^{-1} - \mathbf{g}_{r+1}\mathbf{g}_{r+1}' \tag{28.129}$$

$$\mathbf{b}_{r+1} = \mathbf{b}_r + \mathbf{u}_r z_{r+1} \tag{28.130}$$

and the residual sum of squares is

$$SS_{r+1} = SS_r + w_{r+1}^2. \tag{28.131}$$

The recursive solutions (28.129)–(28.131) do not require any matrix inversions, so that the calculations can proceed very quickly. Under the hypothesis that the model is correct, successive w_i are independent $(0, \sigma^2)$. This property may be exploited to develop a variety of tests for model specification; see R. L. Brown *et al.* (1975).

Diagnostics

28.65 Advances in statistical computing have made computationally intensive methods of data analysis easily accessible to the model builder. This is particularly true of methods based on the residuals, used to check for departures from the assumptions made in **28.2**. In the following sections, we spell out the principles underlying these techniques, leaving the reader interested in applications to consult the monographs by Belsley *et al.* (1980), Cook and Weisberg (1982), and Atkinson (1985).

Leverage

28.66 From **28.3–4**, the fitted values are

$$\hat{\mathbf{y}} = \mathbf{X}(\mathbf{X}'\mathbf{X})^{-1}\mathbf{X}'\mathbf{y} = \mathbf{H}\mathbf{y}, \tag{28.132}$$

where $\mathbf{H} = \mathbf{X}(\mathbf{X}'\mathbf{X})^{-1}\mathbf{X}'$ is sometimes called the *hat* matrix. The vector of residuals is

$$\mathbf{r} = \mathbf{y} - \hat{\mathbf{y}} = (\mathbf{I} - \mathbf{H})\mathbf{y}. \tag{28.133}$$

Denote the (i, i)th element of $\mathbf{H}$ by

$$h_i = \mathbf{x}_i'(\mathbf{X}'\mathbf{X})^{-1}\mathbf{x}_i; \tag{28.134}$$

h_i provides a measure of the *leverage* of the observation. When h_i is near one, $\hat{y}_i$ is virtually determined by y_i and the particular observation is relatively isolated in the X-space from the remainder. Conversely, if h_i is near zero, y_i has very little impact on $\hat{y}_i$ and the ith observation is close to at least some of the others.

Note that

$$\sum_{i=1}^{n} h_i = \text{tr}\,\{\sum \mathbf{x}_i'(\mathbf{X'X})^{-1}\mathbf{x}_i\}$$
$$= \text{tr}\,\{(\mathbf{X'X})^{-1}\sum \mathbf{x}_i\mathbf{x}_i'\} = p \qquad (28.135)$$

so that the average leverage is p/n. Some authors prefer the leverage measure

$$h_i' = h_i/(1-h_i), \qquad (28.136)$$

as this tends to emphasise the highly leveraged observations.

28.67   In order to determine whether a particular observation, the ith say, appears to be in accord with the rest of the sample, a natural approach is to fit the regression model using all the observations except the ith and then to compute the residual based on the difference between y_i and the new fitted value.

Let $\mathbf{X}_{(i)}$ denote the $(n-1) \times p$ matrix after deletion of the ith row. The corresponding fitted value is

$$\hat{y}(i) = \mathbf{x}_i'\hat{\boldsymbol{\beta}}_{(i)}, \qquad (28.137)$$

where

$$\hat{\boldsymbol{\beta}}_{(i)} = (\mathbf{X}_{(i)}'\mathbf{X}_{(i)})^{-1}\mathbf{X}_{(i)}'\mathbf{y}_{(i)}. \qquad (28.138)$$

Further,

$$\text{var}\,\{\hat{y}(i)\} = \sigma^2 v_i = \sigma^2\{1 + \mathbf{x}_i'(\mathbf{X}_{(i)}'\mathbf{X}_{(i)})^{-1}\mathbf{x}_i\}. \qquad (28.139)$$

We let $s_{(i)}^2$ denote the unbiased estimator for σ^2 after removal of the ith observation. Since y_i and $\mathbf{y}_{(i)}$ are independent, we have immediately that

$$r_i^* = \frac{y_i - \hat{y}(i)}{s_{(i)}\sqrt{v_i}} \qquad (28.140)$$

follows Student's t distribution with $(n-p-1)$ degrees of freedom. Since

$$\{\mathbf{X}_{(i)}'\mathbf{X}_{(i)}\}^{-1} = (\mathbf{X'X})^{-1} + (\mathbf{X'X})^{-1}\mathbf{x}_i\mathbf{x}_i'(\mathbf{X'X})^{-1}/(1-h_i) \qquad (28.141)$$

(cf. Exercise 19.20), (28.139) reduces to

$$v_i = 1 + \{\mathbf{x}_i'(\mathbf{X'X})^{-1}\mathbf{x}_i\}^2/(1-h_i)$$

and

$$\hat{y}(i) = \mathbf{x}_i'\hat{\boldsymbol{\beta}}_{(i)} = \mathbf{x}_i'\hat{\boldsymbol{\beta}} - \frac{\mathbf{x}_i'(\mathbf{X'X})^{-1}\mathbf{x}_i r_i}{(1-h_i)}, \qquad (28.142)$$

where $r_i = y_i - \hat{y}_i$ denotes the usual residual. Thus, after some simplification,

$$r_i^* = r_i/\{s_{(i)}(1 - h_i)^{1/2}\}; \qquad (28.143)$$

the derivation is essentially that for recursive residuals given in **28.64**. Atkinson (1985) refers to $\{r_i^*\}$ as the *deletion* residuals, which seems a good description although Cook and Weisberg (1982) prefer the term *externally studentised* in contrast to the residual

$$r_i' = r_i/\{s(1 - h_i)^{1/2}\}, \qquad (28.144)$$

which they term *studentised*; r_i' is also known as the *standardised* residual. Either r_i' or r_i^* may be used to determine whether the data set contains outliers, although multiple outliers may cause *masking*; that is, the residuals are distorted by the outliers still left in the sample. In principle, this may be overcome by omitting groups of observations, but the computational effort involved is substantial.

An alternative approach, due to Atkinson (1986b), is to use a highly robust but inefficient estimator such as the least median of squares (see **28.74**) and then check for outliers in the usual way.

Miyashita and Newbold (1983) demonstrate that tests for outliers based on the largest studentised residual are very sensitive to heavy tails in the error distribution. Thus, such procedures should probably be used as diagnostic aids rather than formal inferential tools.

Influence

28.68   A key question is the effect that a particular observation has upon the location of the regression function. Cook (1977) measured this effect by the quadratic expression

$$D_i = \frac{(\hat{\boldsymbol{\beta}}_{(i)} - \hat{\boldsymbol{\beta}})'\mathbf{X}'\mathbf{X}(\hat{\boldsymbol{\beta}}_{(i)} - \hat{\boldsymbol{\beta}})}{ps^2}, \qquad (28.145)$$

known as Cook's D. If $\hat{\boldsymbol{\beta}}_{(i)}$ is close to $\hat{\boldsymbol{\beta}}$, omitting the ith observation is said to have little *influence*, reflected in a value of D_i close to zero. Conversely, the larger the value of D_i, the more influential the observation becomes.

It may be shown that

$$\hat{\boldsymbol{\beta}}_{(i)} - \hat{\boldsymbol{\beta}} = -(\mathbf{X}'\mathbf{X})^{-1}\mathbf{x}_i r_i/(1 - h_i) \qquad (28.146)$$

so that

$$D_i = r_i^2 \mathbf{x}_i'(\mathbf{X}'\mathbf{X})^{-1}\mathbf{x}_i/\{(1 - h_i)^2 ps^2\}$$
$$= r_i^2 h_i/\{(1 - h_i)^2 ps^2\}. \qquad (28.147)$$

From (28.144), this is

$$D_i = (r_i')^2 h_i'/p. \qquad (28.148)$$

Thus, the influence measure is a combination of the design effect (leverage) and the deviation of the observation from its expectation; see Fig. 28.4. Belsley *et al.* (1980, p. 15) prefer a similar measure known as DFFITS,

$$C_i = (h_i')^{1/2} r_i^*, \qquad (28.149)$$

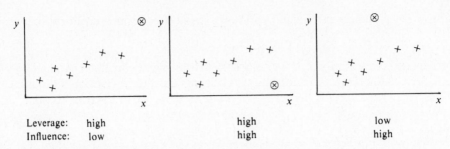

Leverage:	high	high	low
Influence:	low	high	high

Fig. 28.4 **Showing different patterns of leverage and influence for the model** $y = \beta_0 + \beta_1 x + \varepsilon$.

whereas Atkinson (1981) uses

$$C_i' = \{(n-p)/p\}^{1/2} C_i. \tag{28.150}$$

In general, plots of both D_i (or C_i, C_i') and their components should be examined.

Example 28.12
Some of the summary statistics for the data of Example 28.1 are presented in Table 28.13. In order to illustrate the performance of these coefficients, we consider the changes to the third observation shown in Table 28.14.
   The effects are given at the bottom of Table 28.13.

Table 28.13 **Summary statistics for Example 28.1**

Observation	h_i	r_i'	D_i	C_i
1	0.15	1.09	0.05	0.47
2	0.08	−0.22	0.00	−0.07
3	0.46	0.55	0.07	0.51
4	0.47	−1.50	0.45	−1.42
5	0.43	−0.10	0.00	−0.08
6	0.34	−0.41	0.02	−0.29
7	0.26	2.84	0.43	1.67
8	0.14	0.01	0.00	0.00
9	0.16	1.28	0.07	0.55
10	0.38	0.99	0.15	0.68
11	0.30	−0.78	0.07	−0.50
12	0.20	−0.80	0.04	−0.41
13	0.23	−0.60	0.03	−0.33
14	0.14	−0.39	0.01	−0.15
15	0.25	−1.57	0.18	−0.91
Revised values for third observation				
3	0.46	0.55	0.07	0.51
3a	0.53	−0.42	0.05	−0.44
3b	0.63	−1.20	0.58	−1.56
3c	0.78	−2.17	3.21	−4.15
3d	0.46	1.82	0.59	1.68
4d	0.47	−2.82	1.10	−2.67

Table 28.14

	y	x_2	x_3	x_4
Original	175.9	0	40	0
3a	175.9	−10	40	0
3b	175.9	−20	40	0
3c	175.9	−40	40	0
3d	205.9	0	40	0

For case 3*d*, the effect on the fourth observation (close to observation 3 in the *x*-values) is also listed for comparison.

28.69   The measures described so far rely upon the inclusion or total exclusion of individual observations. Cook (1986) introduced (local) influence graphs for each observation, wherein the *i*th observation is given variable weight w_i, $0 \le w_i \le 1$. Generally, these graphs will be most useful in interactive data analysis where only the plots for 'suspect' observations (e.g. those with high influence) are displayed.

Influence and transformations
28.70   Upon reflection, it is apparent that an observation may become more or less influential when transformations are used. If only the dependent variable is transformed, the leverage of the observation is unchanged but the (adjusted) residual changes. Thus, a possible outlier may become acceptable after transformation, although other observations may then become problematical. Atkinson (1983) used the scoring estimator of λ given by (28.65), in conjunction with the influence measure C' to disentangle the effects of influence and transformation; Cook and Wang (1983) developed a similar procedure using D and an approximate ML estimator. Further studies by Atkinson (1986a) indicate that the two approaches typically produce very similar results. Hinkley and Wang (1988) extend the Cook–Wang approach to cover situations where both dependent and regressor variables are transformed. Lawrance (1988) introduces perturbation methods that allow investigation of the local influence of transformations; cf. Cook (1986).

Outliers and robustness
28.71   The identification and treatment (adjustment, elimination, or whatever) of outliers has been a part of statistics since its beginnings, as illustrated in the review by Beckman and Cook (1983). The approach we have taken in **28.65–70** is to identify such observations and to decide whether to modify the model according to the several diagnostic tests applied. In other cases, of course, there may be a recording error and the observation may have to be adjusted, or even eliminated.

A different approach to the whole problem is to search for estimation procedures that are *robust*; that is, methods whose performance is not materially affected by the presence of outliers. A detailed discussion of robustness is beyond the scope of this chapter and will be the subject of a later volume; here we content ourselves with a few comments on certain aspects of that approach insofar as they impinge on our present discussions.

28.72   The oldest alternative to least squares is the method of least, or minimum absolute deviations (LAD or MAD), developed in some detail by Laplace, though originally proposed by Boscovich in 1757; see Stigler (1986, pp. 39–55) for a historical account. To find the LAD estimators, we minimise

$$\sum_{i=1}^{n} |y_i - \mathbf{x}'_i\boldsymbol{\beta}| \qquad (28.151)$$

with respect to $\boldsymbol{\beta}$, a task now easily completed by linear programming methods. The statistical and computational properties of LAD estimators are reviewed by Narula and Wellington (1982). Just as the median is much less susceptible to outliers than the sample mean, so (28.151) yields more robust estimators.

28.73   A general form for estimators for the linear model is given by solving the problem

$$\min \rho\{(\mathbf{y} - \mathbf{X}\boldsymbol{\beta})/\sigma\} \qquad (28.152)$$

which, under suitable regularity conditions, yields estimators of the form

$$\mathbf{X}'\psi\{(\mathbf{y} - \mathbf{X}\boldsymbol{\beta})/\sigma\} = \mathbf{0}, \qquad (28.153)$$

where $\psi(u) = \rho'(u)$. When $\rho(u) = u^2$ in (28.152), (28.153) yields the LS estimators, whereas $\rho(u) = |u|$ produces the LAD estimators. More general versions of $\psi(u)$ have been proposed by several authors, the best known being Huber's (1964) M-estimators, for which

$$\psi(u) = \begin{cases} |u|, & \text{if } |u| \leqslant k \\ k \, \text{sgn}\,(u), & \text{if } |u| > k, \end{cases}$$

k typically being set at around 1.5σ; clearly, a robust estimator of σ is also needed. However, subsequent research demonstrated that M-estimators were still vulnerable to highly leveraged points, which led to the development of generalised-M, or GM, estimators (Mallows, 1975) that bound the influence of such observations.

> Various other procedures have been developed, notably regression quantile methods (Koenker and Bassett, 1978) and local smoothing methods (cf. Cleveland and Devlin, 1988).

28.74   A rather different approach is to use robust methods to improve the performance of the diagnostic procedures. Carroll and Ruppert (1985, 1987) use the methods described in **28.73** for this purpose, but this can be computationally demanding for larger data sets. A simple approach that produces inefficient but highly robust estimators is the least median of squares (LMS) method developed in Rousseeuw (1984), which selects $\hat{\boldsymbol{\beta}}$ to

$$\underset{\hat{\boldsymbol{\beta}}}{\text{minimise}} \;\, \underset{i}{\text{median}} \,(r_i^2), \qquad (28.154)$$

where r_i denotes the ith residual. For problems with $p > 2$, evaluation of the LMS estimators becomes very time consuming, although Souvaine and Steele (1987) have developed a very effective algorithm. However, for diagnostic purposes, a random search procedure will often suffice. That is, we select p observations without replacement from the original n, to form the set J, say, and then evaluate

$$\hat{\boldsymbol{\beta}}_J = \mathbf{X}_J^{-1} \mathbf{y}_J, \qquad (28.155)$$

where subscript J serves to indicate the observations selected in set J. The residuals are then

$$r_{iJ} = y_i - \mathbf{x}_i' \hat{\boldsymbol{\beta}}_J \qquad (28.156)$$

and will be zero when $i \in J$. The quantity, $\text{median}_{i \notin J} (r_{iJ}^2)$, may be evaluated for the residuals not contained in set J. By replicating the process a sufficient number of times, a close approximation to (28.154) will yield approximately LMS estimators. This approach has been used by Atkinson (1986b) as the basis for generating robust estimates as inputs to diagnostic procedures (cf. **28.67**).

> Heitmann and Ord (1985) show that the usual LS estimator is a weighted average of the estimators (28.155); cf. Exercise 28.34. The weights may be used to develop a weighted LMS estimator which might be expected to improve performance.

28.75   Thus far, we have assumed that the complete set of regressor variables is available and that these values are recorded without error. We now consider situations where some of the regressor values may be unknown or recorded with error.

Calibration
28.76   In **28.9**, we considered the prediction of y given the values of the regressor variables. We are sometimes faced with the reverse problem: y is observed and we must estimate x. This arises, for example, in the calibration of laboratory instruments where x is the 'true' reading and y denotes the reading obtained from an instrument subject to measurement error. Other examples include preliminary screening tests (e.g. for blood pressure) or sample estimates of the size of a population between censuses. The more general question of measurement error in both x and y is briefly discussed in **28.81–3**.

28.77   We now restrict attention to the case where we have n pairs of observations (x_i, y_i), $i = 1, \ldots, n$ and a further y-value, y_{n+1} from which we seek to estimate x_{n+1}. Two cases emerge, which have sometimes been confused in the literature. We refer to these as the *unconditional* and *conditional* models, respectively, the terms being used in a manner that is consistent with our earlier description of regression models in these chapters. Also, we refer to the general question of estimating x_{n+1} as a *calibration* problem, rather than its alternative title *inverse* regression, as the term inverse seems open to misinterpretation.

Unconditional model. Suppose that (x, y) are binormal with parameters $(\mu_1, \mu_2; \sigma_1^2, \sigma_2^2; \rho)$. Given n pairs of observations, the joint log-likelihood for the five parameters and x_{n+1} may be partitioned as

$$l = \prod_{i=1}^{n+1} f_y(y_i | \mu_2, \sigma_2^2) f_{x|y}(x_i | y_i; \alpha_1, \beta_1, \omega^2), \tag{28.157}$$

where $\alpha_1 = \mu_1 - \beta_1 \mu_2$, $\beta_1 = \rho \sigma_1 / \sigma_2$ and $\omega_1^2 = \sigma^2 (1 - \rho^2)$. The ML estimators are

$$\hat{\mu}_2 = \bar{y}_{n+1} = \sum_{1}^{n+1} y_i / (n+1), \qquad \hat{\sigma}_2^2 = \sum_{1}^{n+1} (y_i - \bar{y}_{n+1})^2 / (n+1), \tag{28.158}$$

$$\hat{\alpha}_1 = \bar{x}_n - \hat{\beta}_1 \bar{y}_n, \tag{28.159}$$

$$\hat{\beta}_1 = \sum_{1}^{n} (x_i - \bar{x}_n)(y_i - \bar{y}_n) \Big/ \sum_{1}^{n} (y_i - \bar{y}_n)^2, \tag{28.160}$$

and

$$\hat{\omega}_1^2 = \sum_{1}^{n} e_i^2 / (n+1), \tag{28.161}$$

where $n\bar{x}_n = \sum_1^n x_i$, $n\bar{y}_n = \sum_1^n y_i$ and $e_i = (x_i - \bar{x}_n) - \hat{\beta}_1 (y_i - \bar{y}_n)$. Finally, the unconditional estimator for x_{n+1}, denoted by $\hat{x}_u$, is

$$\hat{x}_u = \hat{\alpha}_1 + \hat{\beta}_1 y_{n+1}; \tag{28.162}$$

the conditional variance of the prediction, given $\mathbf{y}$, is

$$\text{var}(\hat{x}_u | \mathbf{y}) = \omega_1^2 \left\{ 1 + 1/n + \frac{(y_{n+1} - \bar{y}_n)^2}{\sum_1^n (y_i - \bar{y}_n)^2} \right\}. \tag{28.163}$$

Further, $E(\hat{x}_u | \mathbf{y}) = x_{n+1}$ and the conditional distribution of x_u given $\mathbf{y}$ is normal.
The unconditional variance is, approximately,

$$\text{var}(\hat{x}_u | y_{n+1}) = \omega_1^2 \left\{ 1 + 1/n + \frac{(y_{n+1} - \mu_2)^2}{n\sigma_2^2} \right\}, \tag{28.164}$$

which would be estimated from (28.161) and (28.163).

Conditional model. If we regard the x_i as fixed, the conditional argument based on the distribution of y given x, $f_{y|x}(y|x, \alpha_2, \beta_2, \omega_2^2)$ leads to the usual estimators

$$\hat{\alpha}_2 = \bar{y}_n - \beta_2 \bar{x}_n, \qquad \hat{\beta}_2 = \sum_{1}^{n} (x_i - \bar{x}_n)(y_i - \bar{y}_n) \Big/ \sum_{1}^{n} (x_i - \bar{x}_n)^2 \quad \text{and} \quad n\hat{\omega}_2^2 = \sum \tilde{e}_i^2,$$

where $e_i = y_i - \bar{y}_n - \hat{\beta}_2(x_i - \bar{x}_n)$, with the usual adjustment to make $\hat{\omega}_2$ unbiased, if desired. It then follows that the conditional estimator, $\hat{x}_c$, is

$$\hat{x}_c = \bar{x}_n + \frac{(y_{n+1} - \hat{\alpha}_2)}{\hat{\beta}_2}. \tag{28.165}$$

Note that $\hat{x}_c$ is the ML estimator, but its mean and variance do not exist, since $\hat{\beta}_2$ is normally distributed and its reciprocal does not have any finite moments. For large samples we have (in the sense of **10.5**, Vol. 1)

$$E(\hat{x}_c) \doteq x_{n+1}, \text{var}(\hat{x}_c) \doteq \sigma_2^2 / \beta_2. \tag{28.166}$$

28.78 The different properties of $\hat{x}_u$ and $\hat{x}_c$ have led to some debate on their respective merits (cf. Krutchkoff, 1967; Williams, 1969). It seems to us that this debate has largely ignored the distinction between the unconditional and conditional models. Once the model is clearly stated, the choice of estimator follows directly unless, of course, one invokes criteria other than ML to argue for $\hat{x}_u$ when $\hat{x}_c$ is appropriate.

For further discussion of the issues and an extension to the multivariate case, see Brown (1982).

Missing values

28.79 When missing values relate only to the dependent variable, we are back at the prediction problem discussed in **28.9**; the multivariate analogue follows directly from the arguments presented there. When the missing values are among the regressor variables, we have a generalisation of the calibration problem. In general, non-iterative solutions may not be available and it is easier to proceed with an iterative solution using the EM algorithm (see **18.40–2**).

28.80 In this case, the M-step corresponds to carrying out the analysis on the 'complete' data, and the E-step requires that we find the expected values for the missing regressors. In order to apply this approach, we must assume that the values are *missing at random* (MAR); that is, the probability that a value is missing is functionally independent of that missing value. Further, it is usually assumed that y and $\mathbf{x}_m$, those x with missing values, are jointly normally distributed. For details of the method, see Orchard and Woodbury (1972), Beale and Little (1975), or Little and Rubin (1987, Chapter 8). Simon and Simonoff (1986) provide diagnostic plots to determine the potential effects of the missing observations without the need to assume that they are MAR.

Measurement errors

28.81 A key assumption throughout our discussion of regression analysis has been that the error terms are not related to the regressor variables. Suppose now that the true relationship is, for a single regressor,

$$y = \beta_0 + \beta_1 X + \varepsilon \tag{28.167}$$

but that the regressor is measured subject to error as

$$x = X + \delta. \tag{28.168}$$

Thus, the relationship between the observable (x, y) is

$$y = \beta_0 + \beta_1 x + (\varepsilon - \beta_1 \delta) \tag{28.169}$$

and the error and the regressor are related through (28.168); we shall assume that δ is independent of X and of ε for all observations. This framework leads to the area of *functional and structural relationships*, of which we defer discussion until a later volume. However, a few comments are in order at this stage.

Suppose we have n pairs of observations and we apply the usual LS estimators to (28.169). This leads to an estimating equation for β_1 of the form

$$\sum (x_i - \bar{x})(y_i - \bar{y}) = \hat{\beta}_1 \sum (x_i - \bar{x})^2.$$

Using (28.167) and (28.168), this becomes

$$\hat{\beta}_1 = \frac{\sum \{(X_i - \bar{X}) + (\delta_i - \bar{\delta})\}\{\beta_0 + \beta_1 X_i + \varepsilon_i\}}{\sum \{(X_i - \bar{X})^2 + 2(\delta_i - \bar{\delta})(X_i - \bar{X}) + (\delta_i - \bar{\delta})^2\}}. \tag{28.170}$$

From the independence of (ε, δ, X), it follows that for large samples, $\hat{\beta}_1$ converges to the limit

$$\beta_1 \frac{S_{XX}}{S_{XX} + \sigma_\delta^2}, \tag{28.171}$$

where $S_{XX} = \lim_{n \to \infty} \{n^{-1} \sum (X_i - \bar{X})^2\}$; that is, the estimator is inconsistent and, if its expectation exists, it is biased towards zero even asymptotically.

Instrumental variables

28.82 One resolution of the measurement error problem is to find another variable, z say, that is highly correlated with X but uncorrelated with δ, so that $n^{-1} \sum z_i \delta_i \to 0$ in probability. Then the estimator

$$\hat{\beta}_1 = \frac{\sum (z_i - \bar{z})(y_i - \bar{y})}{\sum (x_i - \bar{x})(z_i - \bar{z})} \tag{28.172}$$

is consistent. The z-variables are known as *instrumental* variables.

Various instrumental variables have been proposed over the years:

(i)   an indicator variable dividing the observations into k groups of equal size ($k = 2$, Wald, 1940; $k = 3$, Bartlett, 1949);
(ii) the ranks of the x variables;
(iii) replicate measurements on the regressor where the two measurements are uncorrelated;
(iv) lagged values of the regressors.

In order to make (i) and (ii) operational, it is necessary to assume that the measurement errors do not interfere with the identification of the correct values for the instrumental variables.

These basic ideas clearly extend to multiple regressors; see Exercise 28.33. The set of instrumental variables should be chosen as highly correlated with the regressor variables as possible, subject to being uncorrelated with the measurement errors.

28.83 A different form of measurement error arises when the observed values are rounded. Following **3.18-23**, Vol. 1, this might suggest applying Sheppard's corrections to the diagonal elements of $\mathbf{X}'\mathbf{X}$. Dempster and Rubin (1983) use a likelihood argument to show that this step may be justified in some instances, but caution that the appropriate adjustment depends critically upon the distributional assumptions. From a pragmatic viewpoint, Sheppard's corrections reduce the elements on the main diagonal of $\mathbf{X}'\mathbf{X}$ and may, therefore, worsen any multicollinearity problems.

Nonlinear regression

28.84   Thus far we have assumed that the response, y, is a linear function of the parameters, $\boldsymbol{\beta}$, or at least can be reasonably approximated by such a linear function. Such an assumption may not be plausible, and we now consider departures from linearity. We shall consider the general form:

$$y = \eta(\mathbf{x}, \boldsymbol{\beta}, \varepsilon), \tag{28.173}$$

where, as usual, ε denotes the random error term. Now examine the following four examples of η in (28.173):

$$
\left.
\begin{array}{ll}
\text{(i)} & \eta = \beta_0 + \beta_1 x + \varepsilon \\
\text{(ii)} & \eta = \beta_0 + x\, e^{\beta_1} + \varepsilon \\
\text{(iii)} & \eta = \exp(\beta_0 + \beta_1 x + \varepsilon) \\
\text{(iv)} & \eta = \beta_0/(1 + \beta_1 x) + \varepsilon.
\end{array}
\right\}
\tag{28.174}
$$

Version (i) is linear in $\boldsymbol{\beta}$ and (ii) can be made so by use of the transformation $\gamma_1 = \exp(\beta_1)$; models such as (ii) are said to display *parameter-effects* non-linearity. Model (iii) can be transformed to a linear model for $\log_e y$ and is typical of the schemes discussed in **28.31–44**. However, scheme (iv) is *intrinsically* non-linear since there is no transformation that will induce linearity. The form of the error process is critical here since

$$\eta = \beta_0/(1 + \beta_1 x + \varepsilon)$$

clearly can be transformed to a linear model for the reciprocal of y in (28.173).

28.85   We now assume that (28.173) may be rewritten as

$$y = \eta(\mathbf{x}, \boldsymbol{\beta}) + \varepsilon, \tag{28.175}$$

after transformation if necessary, and that the resulting errors satisfy the assumptions made in **28.2**. Then the parameters $\boldsymbol{\beta}$ may be estimated by non-linear least squares (NLLS); that is, by minimising

$$S = \sum_{i=1}^{n} \{y_i - \eta(\mathbf{x}_i, \boldsymbol{\beta})\}^2 \tag{28.176}$$

with respect to $\boldsymbol{\beta}$. Under suitable regularity conditions, this yields the set of first-order equations:

$$\frac{\partial S}{\partial \beta_j} = 0 = \sum \{y_i - \eta(\mathbf{x}_i, \boldsymbol{\beta})\} v_{ij}, \qquad j = 1, \dots, p, \tag{28.177}$$

where $v_{ij} = \partial \eta(\mathbf{x}_i, \boldsymbol{\beta})/\partial \beta_j$. A variety of methods is available to determine $\hat{\boldsymbol{\beta}}$, cf. **18.21**.

28.86 Again under fairly general regularity conditions (cf. Gallant, 1987, pp. 253–8), it follows that the NLLS estimators are consistent and asymptotically normally distributed with mean vector $\boldsymbol{\beta}$ and covariance matrix given by

$$E\left(\frac{\partial^2 S}{\partial\boldsymbol{\beta}\partial\boldsymbol{\beta}'}\right). \tag{28.178}$$

When the errors are normally distributed, these are, of course, the ML results. Since, in this case, the sampling distribution is exactly normal when the model is linear, we may expect 'increasing' non-linearity in the model specification to be the source of increased non-normality in the finite sampling distribution. We therefore seek suitable measures of non-linearity as sources of possible correction factors.

28.87 Most asymptotic theory (cf. Chambers, 1973) is based upon the expansion

$$\hat{\eta}_i = \eta(\mathbf{x}_i, \hat{\boldsymbol{\beta}}) = \eta(\mathbf{x}_i, \boldsymbol{\beta}) + \sum_j (\hat{\beta}_j - \beta_j)v_{ij} \tag{28.179}$$

so that a natural way to assess the intrinsic non-linearity of the model is to consider the ratio

$$\hat{N}(\boldsymbol{\beta}) = \sum_i \{\hat{\eta}_i - \eta_i - \sum_j (\hat{\beta}_j - \beta_j)v_{ij}\}^2 \Big/ \sum_i (\hat{\eta}_i - \eta_i)^2 \tag{28.180}$$

for a test set of values of $\boldsymbol{\beta}$ in the neighbourhood of $\hat{\boldsymbol{\beta}}$. Beale (1960) introduced $\hat{N}(\boldsymbol{\beta})$ as an empirical measure of non-linearity, with a second measure $N(\boldsymbol{\beta})$ defined as the limit of $\hat{N}(\boldsymbol{\beta})$ as the size of the test set goes to infinity. A third measure introduced by Beale sought the minimum of $\hat{N}(\boldsymbol{\beta})$ over all transformations $\boldsymbol{\phi} = \boldsymbol{\phi}(\boldsymbol{\beta})$ and is denoted by $\hat{N}(\boldsymbol{\phi})$, with limiting value $N(\boldsymbol{\phi})$. It is more convenient to consider $N(\boldsymbol{\phi})$ together with

$$N_1(\boldsymbol{\beta}) = N(\boldsymbol{\beta}) - N(\boldsymbol{\phi}) \tag{28.181}$$

since this quantity has a minimum value of zero when the most nearly linear transformation has been used.

Bates and Watts (1980) have shown how Beale's limiting measures may be interpreted geometrically as measures of *curvature*: $4N(\boldsymbol{\phi})$ is the (mean-square) *intrinsic* curvature and $4N_1(\boldsymbol{\beta})$ is the (mean-square) *parameter-effects* curvature. For example, in scheme (ii) of (28.174), $N_1(\boldsymbol{\beta}) > 0$ and $N(\boldsymbol{\phi}) = 0$, indicating that there is no intrinsic curvature; that is, the non-linearity could be removed by a transformation of the parameters.

The scope of the paper by Bates and Watts (1980) is much more general than we have indicated hitherto, as their measures of curvature are not restricted to the quadratic approximations based on (28.179), which may understate the extent of the non-linearity. The reader seeking further details should consult their paper, which includes an algorithm for the evaluation of the curvature measures.

Hougaard (1985, 1988) uses the Bates–Watts curvatures to develop correction factors for the asymptotic distribution of $\hat{\boldsymbol{\beta}}$ and Hamilton and Wiens (1987) provide a corrected F-test. Cook *et al.* (1986) present model diagnostics for nonlinear models and discuss the relationship between bias and curvature.

EXERCISES

28.1 Show that if (28.1) and (28.5) hold, but **X** has elements which are non-linear functions of r further parameters $\gamma_1, \ldots, \gamma_r$, making $(k+r)$ in all, the regression model can be augmented (cf. **19.13–16**) to

$$\mathbf{y} = (\mathbf{X}, \mathbf{D})\begin{pmatrix} \boldsymbol{\beta} \\ \mathbf{0} \end{pmatrix} + \boldsymbol{\varepsilon},$$

where **D** is any $(n \times r)$ matrix chosen so that $(\mathbf{X}, \mathbf{D})$ is of full rank $(k+r)$, and **0** is an $(r \times 1)$ vector of zeros. Hence obtain confidence regions for (a) the complete set of $(k+r)$ parameters, (b) the r further parameters alone.

(Halperin (1963); cf. also Hartley (1964))

28.2 Show that for interval estimation of β in the linear regression model $y_i = \beta x_i + \varepsilon_i$, the interval based on the Student's t-variate

$$t = (b - \beta)/(s^2/\textstyle\sum x_i^2)^{1/2}$$

is physically shorter *for every sample* than that based on

$$u = (\bar{y} - \beta \bar{x})/(s^2/n)^{1/2}.$$

28.3 In the regression model

$$y_i = \alpha + \beta x_i + \varepsilon_i, \qquad i = 1, 2, \ldots, n,$$

suppose that the observed mean $\bar{x} = 0$ and let x_0 satisfy $\alpha + \beta x_0 = 0$. Use the random variable $\hat{\alpha} + \hat{\beta} x_0$ to set up a confidence statement for a quadratic function, of form

$$P\{Q(x_0) \geq 0\} = 1 - \alpha.$$

Hence derive a confidence statement for x_0 itself, and show that, depending on the coefficients in the quadratic function, this may place x_0:
  (i) in a finite interval;
  (ii) outside a finite interval;
  (iii) in the infinite interval consisting of the whole real line.

(Cf. Lehmann, 1986)

28.4 To determine which of the models

$$y = \beta_0' + \beta_1' x_1 + \varepsilon', \qquad y = \beta_0'' + \beta_2' x_2 + \varepsilon'',$$

is more effective in predicting y, consider the model

$$y_i = \beta_0 + \beta_1 x_{1i} + \beta_2 x_{2i} + \varepsilon_i, \qquad i = 1, 2, \ldots, n,$$

with independent normal errors of variance σ^2, estimated by s^2 with $(n-3)$ degrees of freedom. Show that the statistics

$$z_s = \sum_i (y_i - \bar{y})(x_{si} - \bar{x}_s) \Big/ \Big\{ \sum_i (x_{si} - \bar{x}_s)^2 \Big\}^{1/2}, \qquad s = 1, 2,$$

have

$$\operatorname{var} z_1 = \operatorname{var} z_2 = \sigma^2, \qquad \operatorname{cov}(z_1, z_2) = \sigma^2 r_{12},$$

where r_{12} is the observed correlation between x_1 and x_2. Hence show that $(z_1 - z_2)$ is exactly normally distributed with mean

$$\beta_1' \Big\{ \sum_i (x_{1i} - \bar{x}_1)^2 \Big\}^{1/2} - \beta_2' \Big\{ \sum_i (x_{2i} - \bar{x}_2)^2 \Big\}^{1/2}$$

and variance $2\sigma^2(1-r_{12})$. Using the fact that

$$\sum_i (y_i - \bar{y})^2 - (\beta_s')^2 \sum_i (x_{si} - \bar{x}_s)^2$$

is the sum of squares of deviations from the regression of y on x_s alone, show that the hypothesis of equality of these two sums of squares may be tested by the statistic $t = (z_1 - z_2)/\{2s^2(1-r_{12})\}^{1/2}$, distributed in Student's form with $(n-3)$ degrees of freedom.

> (Hotelling (1940); Healy (1955). See also E. J. Williams (1959), Dunn and Clark (1969, 1971) and Choi (1977).

28.5 Show that if there are two different vectors y_1, y_2 each related to the same set of regressors **x** in a linear model, the difference between any pair of corresponding parameters in the models may be tested by applying the method of **28.8** to the differences $(y_{1i} - y_{2i})$.

> (Yates (1939b) also considers the case where the regressors are different and the **y**-vectors correlated.)

28.6 Independent samples of sizes n_i are taken from two regression models

$$y = \alpha_i + \beta_i x + \varepsilon, \qquad i = 1, 2,$$

with independently normally distributed errors. The error variance σ^2 is the same in both models. Given that b_1, b_2 are the separate least squares estimators of β_1, β_2, show that $(b_1 - b_2)$ is normally distributed with mean $(\beta_1 - \beta_2)$ and variance

$$\sigma^2\left\{\left(\sum_{i=1}^{n_1}(x_i - \bar{x}_1)^2\right)^{-1} + \left(\sum_{j=1}^{n_2}(x_j - \bar{x}_2)^2\right)^{-1}\right\},$$

and that

$$t = \{(b_1 - b_2) - (\beta_1 - \beta_2)\}\bigg/\left\{s^2\left(\frac{1}{\sum_i(x_i - \bar{x}_1)^2} + \frac{1}{\sum_j(x_j - \bar{x}_2)^2}\right)\right\}^{1/2}$$

has a Student's t-distribution with $n_1 + n_2 - 4$ degrees of freedom, where

$$s^2 = \frac{(n_1 - 2)s_1^2 + (n_2 - 2)s_2^2}{n_1 + n_2 - 4}$$

and s_1^2, s_2^2 are the separate estimators of σ^2 in the two models. Hence show that t may be used to test the hypothesis that $\beta_1 = \beta_2$ against $\beta_1 \neq \beta_2$.

> (Cf. Fisher, 1922b)

28.7 For the simple linear model $y = \beta_0 + \beta_1 x + \varepsilon$, two independent samples, of sizes m and n, have means $(\bar{y}_m, \bar{x}_m)$ and $(\bar{y}_n, \bar{x}_n)$. Show that $b_1 = (\bar{y}_m - \bar{y}_n)/(\bar{x}_m - \bar{x}_n)$ is an unbiased estimator of β_1, with variance

$$\sigma^2\left(\frac{1}{m} + \frac{1}{n}\right)\bigg/(\bar{x}_m - \bar{x}_n)^2.$$

Show that b_1 is not consistent (as $m, n \to \infty$ with m/n fixed) if the two samples were formed by random subdivision of an original sample of $(m+n)$ observations.

28.8 We are given n observations on the model

$$y = \beta_1 x_1 + \beta_2 x_2 + \varepsilon$$

with error variance σ^2, and, in addition, an extraneous unbiased estimator b_1 of β_1 together with an unbiased estimator s_1^2 of its sampling variance σ_1^2. To estimate β_2, consider the regression of $(y - b_1 x_1)$ on x_2. Show that the estimator

$$b_2 = \sum (y - b_1 x_1)x_2/\sum x_2^2$$

is unbiased, with variance

$$\text{var } b_2 = (\sigma^2 + \sigma_1^2 r^2 \sum x_1^2)/\sum x_2^2,$$

where r is the observed correlation between x_1 and x_2. Assuming that b_1 is ignored, show that the ordinary least squares estimator of β_2 has variance $\sigma^2/\{\sum x_2^2 (1 - r^2)\}$ and hence that the use of the extraneous information about β_1 increases efficiency in estimating β_2 if and only if

$$\sigma_1^2 < \frac{\sigma^2}{\sum x_1^2 (1 - r^2)},$$

i.e. if the variance of b_1 is less than that of the ordinary least squares estimator of β_1. Show that an unbiased estimator of var b_2 is given by

$$\hat{V} = \frac{1}{(n-2) \sum x_2^2} [\sum (y - b_1 x_1 - b_2 x_2)^2 + s_1^2 \sum x_1^2 \{(n-1) r^2 - 1\}],$$

but that if the errors are normally distributed this is not distributed as a multiple of a χ^2 variate.

(Durbin, 1953)

28.9 In generalisation of the situation of Exercise 28.8, let $\mathbf{b}_1$ be a vector of unbiased estimators of the h parameters $(\beta_1, \beta_2, \ldots, \beta_h)$, with covariance matrix $\mathbf{V}_1$; and let $\mathbf{b}_2$ be an independently distributed vector of unbiased estimators of the k $(>h)$ parameters $(\beta_1, \beta_2, \ldots, \beta_h, \beta_{h+1}, \ldots, \beta_k)$, with covariance matrix $\mathbf{V}_2$. Using Aitken's generalisation of Gauss's least squares theorem (**19.19**), show that the minimum variance unbiased estimators of $(\beta_1, \ldots, \beta_k)$ which are linear in the elements of $\mathbf{b}_1$ and $\mathbf{b}_2$ are the components of the vector

$$\mathbf{b} = \{(\mathbf{V}_1^{-1})^* + \mathbf{V}_2^{-1}\}^{-1} \{(\mathbf{V}_1^{-1})^* \mathbf{b}_1^* + \mathbf{V}_2^{-1} \mathbf{b}_2\},$$

with covariance matrix

$$\mathbf{V}(\mathbf{b}) = \{(\mathbf{V}_1^{-1})^* + \mathbf{V}_2^{-1}\}^{-1},$$

where an asterisk denotes the conversion of an $(h \times 1)$ vector into a $(k \times 1)$ vector or an $(h \times h)$ matrix into a $(k \times k)$ matrix by putting it into the leading position and augmenting it with zeros.

Show that $\mathbf{V}(\mathbf{b})$ reduces, in the particular case $h = 1$, to

$$\mathbf{V}(\mathbf{b}) = \sigma^2 \begin{pmatrix} \sum x_1^2 + \sigma^2/\sigma_1^2 & \sum x_1 x_2 & \cdots & \sum x_1 x_k \\ \sum x_1 x_2 & \sum x_2^2 & \cdots & \sum x_2 x_k \\ \vdots & \vdots & \ddots & \vdots \\ \sum x_1 x_k & \sum x_2 x_k & \cdots & \sum x_k^2 \end{pmatrix}^{-1},$$

differing only in its leading term from the usual least squares covariance matrix $\sigma^2 (\mathbf{X}'\mathbf{X})^{-1}$.

(Durbin, 1953)

28.10 A simple graphical procedure may be used to fit an ordinary least squares regression of y on x *without computations* when the x-values are equally spaced, say at intervals of s. Let the n observed points on the scatter diagram of (y, x) be $P_1, P_2, \ldots, P_n$ in increasing order of x. Find the point Q_2 on $P_1 P_2$ with x-coordinate $\frac{2}{3} s$ above that of P_1; find Q_3 on $Q_2 P_3$ with x-coordinate $\frac{2}{3} s$ above that of Q_2; and so on by equal steps, joining each Q-point to the next P-point and finding the next Q-point $\frac{2}{3} s$ above, until finally $Q_{n-1} P_n$ gives the last point, Q_n. Carry out the same procedure backwards, starting from $P_n P_{n-1}$ and determining Q_2', say, $\frac{2}{3} s$ below P_n in x-coordinate, and so on until Q_n' on $Q_{n-1}' P_1$ is reached, $\frac{2}{3} s$ below Q_{n-1}'. Then $Q_n Q_n'$ is the least squares line. Prove this.

(Askovitz, 1957)

28.11 In the linear model with general parameter θ_0 given in Exercise 19.1, we take independent samples of sizes n_1, n_2 with possibly different matrices $\mathbf{X}_1$, $\mathbf{X}_2$, and let S_{ml} be the sum of squared residuals in the lth sample when the LS estimators $\hat{\mathbf{\theta}}_m$ are based on the mth sample alone ($m, l = 1, 2$). Show that

$$D = S_{12} - S_{22} = (\hat{\mathbf{\theta}}_1 - \hat{\mathbf{\theta}}_2)'\mathbf{X}_2'\mathbf{X}_2(\hat{\mathbf{\theta}}_1 - \hat{\mathbf{\theta}}_2) \geq 0$$

and that $b = D/\sigma^2$ is a quadratic form in $(k+1)$ independent standardised normal variables, whose diagonal matrix contains the latent roots of $\mathbf{B} = \mathbf{I} + \mathbf{X}_2'\mathbf{X}_2(\mathbf{X}_1'\mathbf{X}_1)^{-1}$ in its leading diagonal. Using Exercise 19.3, show that $E(b) = \sigma^2 \operatorname{tr} \mathbf{B}$, var $b = 2\sigma^4 \operatorname{tr}(\mathbf{B}^2)$, and that if $r = D/S_{22}$, we have, since b is distributed independently of S_{22},

$$E(r) = \operatorname{tr}(\mathbf{B})/(n_2 - k - 3),$$

$$\operatorname{var} r = \frac{2}{(n_2 - k - 3)(n_2 - k - 5)} \left\{ \operatorname{tr}(\mathbf{B}^2) + \frac{\{\operatorname{tr}(\mathbf{B})\}^2}{n_2 - k - 3} \right\}.$$

If the sets of x-values are the same in each sample, with c times more observations at each in the second sample, show that

$$\mathbf{B} = (1+c)\mathbf{I}, \qquad \operatorname{tr}(\mathbf{B}) = (1+c)k, \qquad \operatorname{tr}(\mathbf{B}^2) = (1+c)^2 k.$$

(M. J. Gardner (1972), who also considered the case where the xs are multinormal.)

28.12 Suppose that we have available a set of n_1 observations on $(y, x_2, \ldots, x_k)$ satisfying the model

$$\mathbf{y}_1 = \mathbf{X}_1 \mathbf{\beta}_1 + \mathbf{\varepsilon}_1, \qquad \mathbf{\varepsilon}_1 \sim N(\mathbf{0}, \sigma^2 \mathbf{I}).$$

A further set of n_2 observations becomes available for which

$$\mathbf{y}_2 = \mathbf{X}_2 \mathbf{\beta}_2 + \mathbf{\varepsilon}_2, \qquad \mathbf{\varepsilon}_2 \sim N(\mathbf{0}, \sigma^2 \mathbf{I}).$$

To test $H_0: \mathbf{\beta}_1 = \mathbf{\beta}_2$ against $H_1: \mathbf{\beta}_1 \neq \mathbf{\beta}_2$, show that the LR procedure is based on the statistic

$$F = (\hat{\mathbf{\beta}}_1 - \hat{\mathbf{\beta}}_2)'[(\mathbf{X}_1'\mathbf{X}_1)^{-1} + (\mathbf{X}_2'\mathbf{X}_2)^{-1}]^{-1}(\hat{\mathbf{\beta}}_1 - \hat{\mathbf{\beta}}_2)/ks^2,$$

where $(n_1 + n_2 - 2k)s^2 = (n_1 - k)s_1^2 + (n_2 - k)s_2^2$. When H_0 is true, show that the statistic has a central F distribution with $(k, n_1 + n_2 - 2k)$ d.fr.

(Chow, 1960. A similar test is available when $V(\varepsilon_1) \neq V(\varepsilon_2)$, see Zellner, 1962; for improved approximations to the distributions in that case, see Ali and Silver, 1985.)

28.13 (*Seemingly unrelated regressions*). Suppose that

$$\begin{pmatrix} \mathbf{y}_1 \\ \mathbf{y}_2 \end{pmatrix} = \begin{pmatrix} \mathbf{X}_1 & \mathbf{0} \\ \mathbf{0} & \mathbf{X}_2 \end{pmatrix} \begin{pmatrix} \mathbf{\beta}_1 \\ \mathbf{\beta}_2 \end{pmatrix} + \begin{pmatrix} \mathbf{\varepsilon}_1 \\ \mathbf{\varepsilon}_2 \end{pmatrix},$$

where

$$E(\mathbf{\varepsilon}) = \mathbf{0} \quad \text{and} \quad V(\mathbf{\varepsilon}) = \mathbf{V} = \begin{pmatrix} \sigma_1^2 \mathbf{I} & \rho\sigma_1\sigma_2 \mathbf{I} \\ \rho\sigma_1\sigma_2 \mathbf{I} & \sigma_2^2 \mathbf{I} \end{pmatrix}.$$

Show that

$$(\mathbf{X}'\mathbf{V}^{-1}\mathbf{X})^{-1} = \begin{pmatrix} \mathbf{A}_{11} & \mathbf{A}_{12} \\ \mathbf{A}_{21} & \mathbf{A}_{22} \end{pmatrix}$$

has submatrices such as

$$\mathbf{A}_{11} = \sigma_1^2(1 - \rho^2)[\mathbf{X}_1'\mathbf{X}_1 - \rho^2 \mathbf{X}_1'\mathbf{X}_2(\mathbf{X}_2'\mathbf{X}_2)^{-1}\mathbf{X}_2'\mathbf{X}_1]^{-1}.$$

In particular, when $\mathbf{X}_1'\mathbf{X}_2 = \mathbf{0}$, show that the relative efficiency of these estimators compared to LS is $(1 - \rho^2)^{-1}$.

(Zellner, 1962; Binkley and Nelson, 1988)

28.14 In the linear model $\mathbf{y} = \mathbf{X}_1\boldsymbol{\beta}_1 + \mathbf{X}_2\boldsymbol{\beta}_2 + \boldsymbol{\varepsilon}$, show that the LS estimators may be written as

$$\hat{\boldsymbol{\beta}}_1 = (\mathbf{X}_1'\mathbf{X}_1)^{-1}\mathbf{X}_1'(\mathbf{y} - \mathbf{X}_2\hat{\boldsymbol{\beta}}_2), \qquad \hat{\boldsymbol{\beta}}_2 = (\mathbf{X}_2'\mathbf{D}\mathbf{X}_2)^{-1}\mathbf{X}_2'\mathbf{D}\mathbf{y},$$

where

$$\mathbf{D} = \mathbf{I} - \mathbf{X}_1(\mathbf{X}_1'\mathbf{X}_1)^{-1}\mathbf{X}_1'.$$

If $\boldsymbol{\beta}_1$ is first estimated from

$$\mathbf{y} = \mathbf{X}_1\boldsymbol{\beta}_1 + \boldsymbol{\varepsilon}^* \tag{A}$$

and $\boldsymbol{\beta}_2$ is then estimated, using the residuals $\mathbf{y}_r$ in (A) as though they were uncorrelated, from

$$\mathbf{y}_r = \mathbf{X}_2\boldsymbol{\beta}_2 + \boldsymbol{\eta},$$

show that the estimators obtained are

$$\boldsymbol{\beta}_1^* = (\mathbf{X}_1'\mathbf{X}_1)^{-1}\mathbf{X}_1'\mathbf{y}, \qquad \boldsymbol{\beta}_2^* = (\mathbf{X}_2'\mathbf{X}_2)^{-1}\mathbf{X}_2'\mathbf{D}\mathbf{y},$$

and that $\boldsymbol{\beta}_1^*$ and $\boldsymbol{\beta}_2^*$ are biased unless $\mathbf{X}_1'\mathbf{X}_2 = 0$ or $\boldsymbol{\beta}_2 = 0$. Given that $\boldsymbol{\beta}_2$ is a scalar parameter, show that

$$\beta_2^* = (1 - R^2)\hat{\beta}_2,$$

where R is the multiple correlation coefficient of the single variable x_2 upon all the variables in $\mathbf{X}_1$.

In the case $y_j = \beta_1 x_{1j} + \beta_2 x_{2j} + \varepsilon_j$, show that the mean-square errors of the biased two-step estimators β_1^*, β_2^* are less than the variances of the unbiased LS estimators $\hat{\beta}_1$, $\hat{\beta}_2$ if $\beta_2^2 / V(\hat{\beta}_2) < 1$.

> (Cf. Freund *et al.*, (1961), Goldberger and Jochems (1961), Goldberger (1961), Zyskind (1963) and T. D. Wallace (1964).)

28.15 There are G groups of observations, and all observations within a group are normally distributed with common mean and common variance σ_g^2, the model (28.56) holding except for the homoscedasticity condition below it. Consider the sets of constraints

$$C_1 : \text{all the } \sigma_g^2 \text{ are equal } (G-1 \text{ constraints});$$
$$C_2 : r \text{ of the } k \text{ parameters in } \boldsymbol{\beta} \text{ are zero.}$$

Working in terms of the variable $z_\lambda = y_\lambda J^{-1/n}$, so that (28.58) reduces to

$$L_\lambda(z \mid \hat{\boldsymbol{\beta}}, \hat{\sigma}^2) = \{\hat{\sigma}_\lambda^2(z)\}^{-n/2},$$

show that (28.66) gives

$$L(z \mid \hat{\lambda}_{(2)}) = L(z \mid \hat{\lambda})l_1(z)l_2(z)$$

where l_1 is the LR test statistic defined at (23.40), and

$$l_2 = \left\{ 1 + \frac{r}{n-k}F \right\}^{-n/2},$$

where F is the variance-ratio test statistic defined generally at (23.99) and for this case in Example 23.8. (This result generalises Exercise 23.6.)

> (Box and Cox, 1964)

28.16 Using Exercise 22.7, show that if in (28.66) l_p is distributed free of certain parameters for which there is a complete sufficient (vector) statistic $\mathbf{t}$, and l_q is a function of $\mathbf{t}$ alone, l_p and l_q are stochastically independent. Apply this result to establish the independence results in Exercises 23.6 and 23.13. Show in Exercise 28.15 that $l_1(z)$ and $l_2(z)$ are independent when C_1 and C_2 both hold.

> (Cf. Hogg, 1961)

28.17 Show that if a linear model contains terms $\theta_i x_i^{\mu_i}$, $\mu_i \neq 0$, we have approximately

$$x_i^{\mu_i} = x_i + (\mu_i - 1)x_i \log x_i.$$

Hence show how μ_i can be estimated. The process may be iterated.

(Box and Tidwell, 1962)

28.18 In **28.38**, show that for the binomial distribution of **5.2**, (28.72) gives the variance-stabilising transformation

$$u\left(\frac{x}{n}\right) = \arcsin\left\{\left(\frac{x}{n}\right)^{1/2}\right\},$$

where x/n is the observed proportion of 'successes'. Show that

$$2 \arcsin(y^{1/2}) = \arcsin(2y-1) + \pi/2,$$

so that $u = \arcsin(2x/n - 1)$ may be used equivalently.

(Anscombe (1948) shows that better variance stabilisation is obtained if x/n is replaced by $(x+\frac{3}{8})/(n+\frac{3}{4})$. Freeman and Tukey (1950) suggest

$$\arcsin\left\{\left(\frac{x}{n+1}\right)^{1/2}\right\} + \arcsin\left\{\left(\frac{x+1}{n+1}\right)^{1/2}\right\}$$

(cf. Example 28.5), tabulated by Mosteller and Youtz (1961). See also Laubscher (1961).)

28.19 In **28.38** show that for the negative binomial distribution of **5.16**, (28.72) gives the variance-stabilising transformation

$$u\left(\frac{x}{n}\right) = \operatorname{arc\,sinh}\left\{\left(\frac{x}{n}\right)^{1/2}\right\}$$

where x/n is the observed proportion of 'successes'.

(Anscombe (1948) shows that better variance stabilisation occurs if x/n is replaced by $(x+\frac{3}{8})/(n-\frac{3}{4})$. See also Laubscher (1961).)

28.20 In Exercise 28.18, show that the alternative transformation

$$u\left(\frac{x}{n}\right) = \log\left\{\frac{x}{n} \middle/ \left(1 - \frac{x}{n}\right)\right\}$$

stabilises the variance near $p = \frac{1}{2}$. Show that this transformation is strictly appropriate when (28.67) is $D_n^2(\theta) = c\theta^2(1-\theta)^2$.

(Cf. Bartlett, 1947)

28.21 If z has the gamma distribution given in (28.77), show that the cumulant generating function for $\log z$ is

$$\psi(w) = \log \Gamma(1+iw) + \sum_{s=1}^{q} \log(1+iw/s).$$

Using Exercise 14.4 or otherwise, show that the rth cumulant of $\log z$ is

$$\kappa_r = (-1)^r (r-1)! \sum_{s=q}^{\infty} s^{-r}, \qquad r \geq 2.$$

Finally, show that

$$\gamma_1 = \kappa_3 / \kappa_2^{3/2} \simeq -q^{-1/2}$$
$$\gamma_2 = \kappa_4 / \kappa_2^2 \simeq 2q^{-1}.$$

(Bartlett and Kendall, 1946)

28.22 In Example 28.5 expand

$$u_c(t) = (\theta + c)^{1/2}\left(1 + \frac{t - \theta}{\theta + c}\right)^{1/2}$$

in series and show that

$$E(u_c) = (\theta + c)^{1/2} - \tfrac{1}{8}\theta^{-1/2} + \frac{24c - 7}{128}\theta^{-3/2} + o(\theta^{-3/2}),$$

and

$$\text{var } u_c = \frac{1}{4}\left\{1 + \frac{3 - 8c}{8\theta} + \frac{32c^2 - 52c + 17}{32\theta^2} + o(\theta^{-2})\right\},$$

so that the choice $c = \tfrac{3}{8}$ removes the term of order θ^{-1} in the variance, reducing it to

$$\text{var } u_{3/8} \sim \frac{1}{4}\left(1 + \frac{1}{16\theta^2}\right).$$

Hence show that

$$\{E(u_c)\}^2 - c \sim \theta - \frac{1}{4} - \frac{3 - 8c}{32\theta},$$

so that if the inverse transformation is used on u_c to obtain an estimator of θ, its downward bias is nearly constant at $\tfrac{1}{4}$.

(The $c = \tfrac{3}{8}$ result is due to A. H. L. Johnson; cf. Anscombe (1948).)

28.23 In Exercise 28.22, show that the coefficients of skewness and kurtosis of $u_c(t)$ are

$$\gamma_1 = -\frac{1}{2\theta^{1/2}}\left\{1 + \frac{25 - 48c}{16\theta}\right\} + o(\theta^{-3/2}),$$

$$\gamma_2 = \frac{1}{\theta}\left\{1 + \frac{945 - 1536c}{2560}\right\} + o(\theta^{-2}),$$

compared with

$$\gamma_1 = \theta^{-1/2},$$
$$\gamma_2 = \theta^{-1},$$

for the original Poisson variable t. Thus, whatever c is chosen, γ_1 is approximately halved (with changed sign) and γ_2 is unaffected to the first order.

(Anscombe, 1948)

28.24 Using the result for var u_c in Exercise 28.22, show that the transformation

$$u_\delta' = (t + \tfrac{1}{2} + \delta)^{1/2} + (t + \tfrac{1}{2} - \delta)^{1/2}$$

has variance

$$\text{var } u_\delta' = 1 - \frac{1}{8\theta} + \frac{16\delta^2 - 1}{32\theta^2} + o(\theta^{-2}),$$

so that if we choose $\delta = \frac{1}{2}$ to give u' of Example 28.5,

$$\mathrm{var}\ u'_{1/2} = 1 - \frac{1}{80} + \frac{3}{32\theta^2} + o(\theta^{-2}).$$

28.25 x is a variate with cumulants κ_r all of order n, the sample size. Show that the moments of

$$y = \left(\frac{x}{\kappa_1}\right)^h = \left(1 + \frac{x - \kappa_1}{\kappa_1}\right)^h$$

are

$$\mu'_r(y) = 1 + \frac{(rh)^{(2)}}{2!} \frac{\kappa_2}{\kappa_1^2} + \frac{(rh)^{(3)}}{3!} \frac{\kappa_3}{\kappa_1^3} + \frac{(rh)^{(4)}}{4!} \frac{(\kappa_4 + 3\kappa_2^2)}{\kappa_1^4}$$

$$+ \frac{(rh)^{(5)}}{5!} \frac{10\kappa_3\kappa_2}{\kappa_1^5} + \frac{(rh)^{(6)}}{6!} \frac{15\kappa_2^3}{\kappa_1^6} + O(n^{-4}).$$

Hence show that

$$\mu_3(y) = \frac{h^3}{\kappa_1^4}\{\kappa_3\kappa_1 + 3(h-1)\kappa_2^2\} + \frac{h^3(h-1)}{2\kappa_1^6}$$
$$\times \{3\kappa_4\kappa_1^2 + 3(7h-10)\kappa_3\kappa_2\kappa_1 + (17h^2 - 55h + 44)\kappa_2^3\} + O(n^{-4}),$$

so that putting $h = 1 - \kappa_1\kappa_3/3\kappa_2^2$ makes $\mu_3(y)$ of order n^{-3} and $\gamma_1(y)$ of order $n^{-3/2}$. Show further that by adding a constant to x to make its mean g, and choosing

$$g = \frac{12\kappa_3\kappa_2^2}{20\kappa_3^2 - 9\kappa_4\kappa_2}, \qquad h = \frac{16\kappa_3^2 - 9\kappa_4\kappa_2}{20\kappa_3^2 - 9\kappa_4\kappa_2},$$

the variate $z = (1 + (x - \kappa_1)/g)^h$ has γ_1 of order $n^{-3/2}$ and γ_2 of order n^{-2}.

Verify in Example 28.6 that for the χ^2 distribution the first of these results gives the Wilson–Hilferty cube-root transformation, while the second gives a power $h = \frac{5}{13}$. Similarly, show in Example 28.5 for the Poisson distribution that we get $h = \frac{2}{3}$ and $h = \frac{7}{11}$ respectively, so that although $h = \frac{1}{2}$ stabilises variance in each of these Examples, normality considerations move h in opposite directions in the two Examples.

(Cf. Haldane, 1938)

28.26 In **28.44**, expand $t(u)$ in Taylor series about μ, obtaining

$$\theta = E\{t(u)\} = \sum_{m=0}^{\infty} \frac{t^{(m)}(\mu)}{m!} E(u - \mu)^m = \sum_{m=0}^{\infty} \frac{t^{(2m)}(\mu)}{m!} \left(\frac{\sigma^2}{2}\right)^m;$$

now expand $t^{(2m)}(\mu)$ in Taylor series about zero, so that

$$\theta = \sum_{m=0}^{\infty} \sum_{r=0}^{\infty} \frac{t^{(2m+r)}(0)}{m!\,r!} \mu^r \left(\frac{\sigma^2}{2}\right)^m,$$

given that these summations converge. Using the result of Exercise 17.10, show that the MV unbiased estimator of θ is

$$\hat{\theta} = \sum_{m=0}^{\infty} \sum_{r=0}^{\infty} \frac{t^{(2m+r)}(0)}{m!\,r!\,2^m} \sum_{i=0}^{[r/2]} (-1)^i \frac{r!}{i!(r-2i)!} \frac{\Gamma(\frac{1}{2}\nu)}{\Gamma\{\frac{1}{2}\nu + m + i\}} \hat{\mu}^{r-2i} \left(\frac{\lambda^2}{2}\right)^i \left(\frac{S^2}{2}\right)^{m+i}$$

Hence verify the formulae given in **28.44** for the special case $t''(u) = A + Bt(u)$.

(Neyman and Scott, 1960; Hoyle, 1968)

28.27 For the square-root transformation, show that the bias-corrected inverse transformation $\hat{\theta}$ in **28.44** has variance

$$\text{var}(\hat{\theta}) = 4\mu^2\lambda^2\sigma^2 + 2\sigma^4\{(1-\lambda^2)^2/\nu + \lambda^4\}$$

and that its MV unbiased estimator is

$$\text{est. var}(\hat{\theta}) = 4\hat{\mu}^2\lambda^2 S^2/\nu + \frac{2S^4}{\nu(\nu+2)}\{(1-\lambda^2)^2/\nu - \lambda^4\}.$$

(Cf. Hoyle, 1968)

28.28 By consideration of the case when $y_j = x_j^k$, $j = 1, 2, \ldots, n$, exactly, show that if the orthogonal polynomials defined by (28.100) are *orthonormal* (i.e. $\mathbf{C} = \mathbf{I}$ in (28.101)) then they satisfy the recurrence relation

$$\phi_k(x_j) = \frac{1}{b_k}\left\{ x_j^k - \sum_{i=0}^{k-1} \phi_i(x_j) \sum_{j=1}^{n} x_j^k \phi_i(x_j) \right\},$$

where the normalising constant b_k is defined by

$$b_k^2 = \sum_{j=1}^{n} \left\{ x_j^k - \sum_{i=0}^{k-1} \phi_i(x_j) \sum_{j=1}^{n} x_j^k \phi_i(x_j) \right\}^2.$$

Hence verify the expressions for $j = 1, 2, 3$, with appropriate adjustments.

(Robson, 1959)

28.29 In fitting orthogonal polynomials of degree k as in **28.51**, the reduction in the total SS associated with the term of degree r is $Q_r = \hat{\alpha}_r^2 \sum_{i=1}^{n} \phi_j^2(x_i)$. Show that the ratios

$$z_r = Q_{k-r+1} \bigg/ \sum_{s=k-r+2}^{k+1} Q_s, \qquad r = 1, 2, \ldots, k,$$

where $Q_{k+1} = (n-k)s^2$ is the Residual SS, are all independently distributed when the regression coefficients α_r are all zero:

(a) by using the result of Exercise 28.16; and
(b) by using the result of Exercise 22.27.

(This result indicates (cf. Hogg (1961)) that one may independently test the regression coefficients if one starts from the highest order and works downwards, 'pooling' the associated SS of those adjudged zero with the residual SS, until one is adjudged non-zero, when the process stops. All the tests are, of course, $t^2(F)$ tests, and the overall test has size $1 - (1-\alpha)^k \sim k\alpha$ if a test of size α is used at each stage. T. W. Anderson (1962a) shows under weak assumptions that this procedure maximises the probability of correctly locating a non-zero coefficient.)

28.30 Given a set of n observations with linear model

$$\mathbf{y} = \mathbf{X}\boldsymbol{\beta} + \boldsymbol{\varepsilon}, \qquad \varepsilon_i \sim \text{independent}(0, \sigma^2),$$

suppose that the data are grouped into m groups using the $(m \times n)$ matrix $\mathbf{G}$ where $g_{ij} = 1$ if the ith observation is in the jth group; $g_{ij} = 0$ otherwise. The regressors may be considered when specifying $\mathbf{G}$, but not the values of $\mathbf{y}$. Show that the resulting estimator for $\boldsymbol{\beta}$ is of WLS form (28.106), where $\mathbf{V} = \sigma^2\mathbf{G}\mathbf{G}'$ is diagonal only when the groups are non-overlapping. Further, if $m > n$, show that a WLS estimator based upon the generalised inverse may be used.

(Leech and Cowling, 1983)

28.31   Suppose that the linear model (28.1) has errors with zero means and covariance matrix

$$
\mathbf{V} = \sigma^2 \begin{bmatrix} \mathbf{G} & & & & \mathbf{0} \\ & \mathbf{G} & & & \\ & & \mathbf{G} & & \\ & & & \ddots & \\ \mathbf{0} & & & & \mathbf{G} \end{bmatrix},
$$

where $\mathbf{V}$ is $(km \times km)$ and $\mathbf{G}$ is the $(k \times k)$ matrix $(1-\phi)\mathbf{I} + \phi \mathbf{1}\mathbf{1}'$; that is, the errors have an intraclass correlation structure; cf. (28.116–17). Show that $\mathbf{G}^{-1} = (1-\phi)^{-1}\{\mathbf{I} - c\mathbf{1}\mathbf{1}'\}$, where $c = \phi/(1-\phi+k\phi)$. Partitioning $\mathbf{X}'$ into $(\mathbf{X}_1', \ldots, \mathbf{X}_m')$ and $\mathbf{y}'$ into $(\mathbf{y}_1', \ldots, \mathbf{y}_m')$ show that

$$
\hat{\boldsymbol{\beta}} = (\textstyle\sum \mathbf{X}_j'\mathbf{G}^{-1}\mathbf{X}_j)^{-1}(\textstyle\sum \mathbf{X}_j'\mathbf{G}^{-1}\mathbf{y}_j)
$$

with covariance matrix $\sigma^2(\sum \mathbf{X}_j'\mathbf{G}^{-1}\mathbf{X}_j)^{-1}$. Further, when $\mathbf{X}_j = \mathbf{1}$ for all j, show that $\hat{\beta} = \bar{y}$ as with LS, but that its variance is $\sigma^2\{1+(k-1)\phi\}/km$.

28.32   Suppose that there is a linear structural relationship between X and Y, so that

$$
Y = \alpha + \beta X.
$$

Now assume that X and Y may only be measured, subject to error, as

$$
x = X + \delta, \qquad y = Y + \varepsilon,
$$

where $\delta \sim N(0, \omega^2)$ and $\varepsilon \sim N(0, \sigma^2)$ are independent of each other and across all observations. Given that $X \sim N(0, \sigma_x^2)$ and X is independent of δ and ε, show that the LS estimator based on sample values (x_i, y_i), $i = 1, 2, \ldots, n$ yields a biased estimator for β since

$$
\lim_{n \to \infty} E(\hat{\beta}) = \frac{\beta \sigma_x^2}{\sigma_x^2 + \omega^2}.
$$

(*Note*: If the X observations are fixed, the relationship is *functional*, if they are drawn at random from some population, the relationship is *structural*.)

28.33   The instrumental variables estimators for $\boldsymbol{\beta}$ in (28.1) are given by

$$
\mathbf{Z}'(\mathbf{y} - \mathbf{X}\boldsymbol{\beta}) = \mathbf{0},
$$

where $\mathbf{Z}$ denotes the $(n \times p)$ matrix of values of the instrumental variables. Show that the large sample covariance matrix for $\hat{\boldsymbol{\beta}}$ is $V_I = \sigma^2(\mathbf{Z}'\mathbf{X})^{-1}(\mathbf{Z}'\mathbf{Z})(\mathbf{X}'\mathbf{Z})^{-1}$.

28.34   Consider estimators of the form of (28.155) based on only p observations out of n. For subset J, let

$$
w_J = |\mathbf{X}_J|^2/|\mathbf{X}'\mathbf{X}|.
$$

Show that the LS estimator is given by

$$
\hat{\boldsymbol{\beta}} = \textstyle\sum w_J \hat{\boldsymbol{\beta}}_J
$$

where the sum is taken over all $\binom{n}{p}$ subsets.

(Heitmann and Ord, 1985)

CHAPTER 29

ANALYSIS OF VARIANCE IN THE LINEAR MODEL

29.1 In developing the MV unbiased linear estimation properties of the LS estimator in the linear model $y = X\theta + \varepsilon$ at (19.8), we observed at (19.42) that the sum of squares (SS) of the observations may be written identically as the sum of two non-negative components

$$y'y \equiv (y - X\hat{\theta})'(y - X\hat{\theta}) + (X\hat{\theta})'(X\hat{\theta}) \tag{29.1}$$

of which the first is the sum of squared residuals (residual SS) from the model fitted by LS. The second component on the right of (29.1) is the reduction in the SS due to the fitted model; the greater this reduction is (i.e. the smaller the residual SS is), the more satisfactorily the fitted model represents the y–X relationship in the observations. If we rewrite (29.1) as

$$y'y \equiv (y - X\hat{\theta})'(y - X\hat{\theta}) + \hat{\theta}'(X'X)\hat{\theta} \tag{29.2}$$

and recall from (19.16) that

$$V(\hat{\theta}) = \sigma^2(X'X)^{-1} \tag{29.3}$$

we see that if the error-vector ε in the model is normally distributed, $\hat{\theta}$, being a linear function of ε, will by **15.4** be multinormally distributed with mean θ and covariance matrix (29.3). The last term on the right of (29.2) is therefore the exponent of this multinormal distribution apart from the factor $-(2\sigma^2)^{-1}$, and by **23.6**, $\hat{\theta}'X'X\hat{\theta}/\sigma^2$ is distributed in the non-central χ^2 form (23.18) with degrees of freedom $\nu = k$ and non-central parameter

$$\lambda = \theta'V^{-1}(\hat{\theta})\theta = \theta'X'X\theta/\sigma^2. \tag{29.4}$$

For brevity we write this distribution $\chi'^2(\nu, \lambda)$ as in **23.5**.

29.2 This result enables us to test the hypothesis that $\theta = \theta_0$, and in particular to test

$$H_0: \theta = 0, \tag{29.5}$$

for then λ defined by (29.4) is zero, and the distribution becomes a (central) χ^2 with k degrees of freedom (d.fr.). As we saw in **19.11**, $(y - X\hat{\theta})'(y - X\hat{\theta})/\sigma^2$ is a χ^2 with $(n - k)$ d.fr., and $y'y/\sigma^2$ is a χ^2 with n d.fr. when (29.5) holds, Cochran's theorem of **15.16** applies, and the two components on the right of (29.2) are independently distributed. Their ratio (after division by d.fr.)

$$F = \{\hat{\theta}'X'X\hat{\theta}/k\}/\{(y - X\hat{\theta})'(y - X\hat{\theta})/(n - k)\} \tag{29.6}$$

has the variance-ratio F distribution with $(k, n - k)$ d.fr. when (29.5) holds.

If we wish to investigate the power of a test of (29.5) based on (29.6), we require its distribution when (29.5) does not hold. In order to show that it is a non-central F as at (23.105), we must prove that the numerator and denominator of (29.6) remain independent when $\theta \neq 0$. If we wish to test the more general hypothesis that $\theta = \theta_0 \neq 0$, we require this distribution in order to make a test at all. Thus we need an extension of Cochran's theorem (15.16) to non-central normal variables, i.e. normal variables with means not all equal.

29.3   Apart from this particular need, the form of (29.2) is suggestive in another way. Suppose that $X'X$ is a diagonal matrix, say C, with diagonal elements c_{ii}. The last term on the right of (29.2) can then be further separated into

$$(X\hat{\theta})'(X\hat{\theta}) = \sum_{i=1}^{k} c_{ii}\,\hat{\theta}_i^2. \tag{29.7}$$

The elements c_{ii} are positive, since $X'X$ is a positive definite (non-singular) matrix. (29.7) expresses the reduction in the SS as the sum of k parts, one corresponding to each parameter. Here again, we may be interested in testing hypotheses concerning individual θ_i, and require the distribution of the components $c_{ii}\hat{\theta}_i^2$ when $\theta \neq 0$.

If $X'X$ is diagonal, so is (29.3), and the linear model is called *orthogonal* since the $\hat{\theta}_i$ are uncorrelated, and actually independent when ε is normal. We have already discussed orthogonal models in the context of regression theory in **28.51–5**, where we were concerned with the use of orthogonal polynomials. Example 28.8 illustrated the procedure of evaluating the reduction in the SS due to each further parameter, using an entirely intuitive justification. Our present discussion will be more general.

Analysis of variance

29.4   We now introduce a fundamental concept, originally developed by R. A. Fisher in the 1920s, although there had been early work in this direction by W. H. R. A. Lexis and T. N. Thiele in the late 19th century. If the SS $y'y$ can be expressed as the sum of non-negative components, each of which corresponds to a subset of the parameters of the linear model, we call this an *analysis of variance* (ANOVA, or AV) on y. (It would be more appropriate to call it an analysis of SS, but history and brevity are against this logical usage.) An AV is interpreted as a separating-out of the influences of the different subsets of parameters upon the observations y. The importance of such separations in many fields of investigation make AV the central technique of much applied statistics.

Decomposition of non-central quadratic forms

29.5   We now state a general AV problem. Suppose that y is a vector of p independent normal variates with

$$E(y) = \mu, \qquad V(y) = I, \tag{29.8}$$

and that

$$\mathbf{y'y} = \sum_{i=1}^{k} \mathbf{y'A}_i \mathbf{y}, \tag{29.9}$$

where A_i has rank r_i. Under what conditions will the k quadratic forms $Q_i = \mathbf{y'A}_i\mathbf{y}$ be independently distributed, and what will their distributions be? Since, by **23.4**, the distribution of $\mathbf{y'y} = Q$ is a $\chi'^2(p, \mathbf{\mu'\mu})$, we may expect the components to have the same distributional form.

This differs from the problem considered in **15.16** only in that there we had $\mu = 0$. We saw there that any one of the three conditions

(a) $\sum_{i=1}^{k} r_i =$ the rank of Q,
(b) each A_i is idempotent, i.e. $A_i = A_i^2$,
(c) $A_iA_j = 0$, all $i \neq j$,

implies the other two. Re-examination of the proof of this in **15.17–19** will reveal that it did not depend on the vector of means at all. Neither did the proof (in **15.13**) of Craig's theorem that Q_i and Q_j are independent if and only if $A_iA_j = 0$, which shows that (c) is equivalent to

(c′) each Q_i is independent of every other.

However, the equivalence of (b) and the statement

(b°) each Q_i is a (central) χ^2 variable with r_i d.fr.,

depended upon the result of **15.11** that if $\mu = 0$, Q_i is a χ^2 variable if and only if A_i is idempotent. It is thus (b°) that requires to be generalised through a generalisation of **15.11** to $\mu \neq 0$.

29.6 The only essential change brought about in **15.11** is that the canonically transformed variable y_i^2 in (15.43) is now a $\chi'^2(1, \mu_i^2)$ variable, by **23.4**. The c.g.f. of $\sum_{i=1}^{r} a_i y_i^2$ is therefore not (15.45), but the more general form obtained from Exercise 23.1, which yields for the cumulants of Q

$$\kappa_s = 2^{s-1}(s-1)! \sum_{i=1}^{r} a_i^s(1 + s\mu_i^2) \tag{29.10}$$

(the generalisation of (15.46)), and also shows that the cumulants of a $\chi'^2(\nu, \lambda)$ variable are

$$\kappa_s^* = 2^{s-1}(s-1)!(\nu + s\lambda). \tag{29.11}$$

For (29.10) and (29.11) to be identical, we must have

$$\left.\begin{array}{l} \sum_{i=1}^{r} a_i^s = \nu, \\[2mm] \sum_{i=1}^{r} a_i^s \mu_i^2 = \lambda, \end{array}\right\} \text{all } s. \tag{29.12}$$

(29.12) is satisfied if and only if every $a_i = 1$, so that $\nu = r$ and $\lambda = \sum_{i=1}^{r} \mu_i^2$. Since the a_i are the non-zero latent roots of $\mathbf{A}$, it follows that $\mathbf{A}$ is idempotent. We thus see that, for general μ, the statement equivalent to (b) in **29.5** is

(b') each Q_i is a $\chi'^2(r_i, \lambda_i)$ variable,

reducing to (b°) when $\mu = 0$. Moreover, if we transform orthogonally back from the canonical to the original variables, we see at once that $\lambda_i = \mu' \mathbf{A}_i \mu$, and $\sum_{i=1}^{k} \lambda_i = \mu' \mu$, the non-central parameters of the Q_i adding to that of Q (cf. Exercise 23.1).

We have thus reached the conclusion that if (29.8-9) hold, any of the conditions (a), (b), (c) implies the other two; equivalently, any of the conditions (a), (b') and (c') implies the other two of them.

29.7 The result of **29.6** is unaffected if, in (29.9), $\mathbf{y}'\mathbf{y}$ is replaced by $Q = \mathbf{y}'\mathbf{A}\mathbf{y}$, where $\mathbf{A}$ is any idempotent matrix with rank $r < p$. The argument justifying this in Chapter 15 for the case $\mu = 0$ is valid for general μ.

Even if $\mathbf{V}(\mathbf{y})$ in (29.8) is generalised so that the components of $\mathbf{y}$ have a non-singular multinormal distribution with covariance matrix $\mathbf{V}$ which is non-diagonal, the result is only slightly changed. For if

$$E(\mathbf{y}) = \mu, \qquad V(\mathbf{y}) = \mathbf{V}, \tag{29.13}$$

and

$$Q = \mathbf{y}'\mathbf{A}\mathbf{y} = \sum_{i=1}^{k} \mathbf{y}'\mathbf{A}_i\mathbf{y} = \sum_{i=1}^{k} Q_i, \tag{29.14}$$

we may write $\mathbf{V} = \mathbf{TT}'$ as in **15.21**, and the transformation $\mathbf{y} = \mathbf{Tz}$ produces independent normal variables $\mathbf{z}$, since the exponent of the multinormal distribution is

$$\mathbf{y}'\mathbf{V}^{-1}\mathbf{y} = \mathbf{z}'\mathbf{T}'(\mathbf{T}')^{-1}\mathbf{T}^{-1}\mathbf{Tz} = \mathbf{z}'\mathbf{z}.$$

We then have, from (29.14),

$$Q = \mathbf{z}'\mathbf{T}'\mathbf{A}\mathbf{Tz} = \sum_{i=1}^{k} \mathbf{z}'\mathbf{T}'\mathbf{A}_i\mathbf{Tz} = \sum_{i=1}^{k} Q_i,$$

and these are the quadratic forms with which we now deal. Condition (b) of **29.5** is now

$$\mathbf{T}'\mathbf{A}_i\mathbf{T} = \mathbf{T}'\mathbf{A}_i\mathbf{T} \cdot \mathbf{T}'\mathbf{A}_i\mathbf{T}$$

or

$$\mathbf{A}_i\mathbf{V} = \mathbf{A}_i\mathbf{V}\mathbf{A}_i\mathbf{V}$$

so that $\mathbf{A}_i\mathbf{V}$ must now be idempotent, as must $\mathbf{AV}$. Condition (c) is

$$\mathbf{T}'\mathbf{A}_i\mathbf{T} \cdot \mathbf{T}'\mathbf{A}_j\mathbf{T} = 0$$

or

$$\mathbf{A}_i\mathbf{V}\mathbf{A}_j = 0, \qquad i \neq j.$$

Condition (a) is unaffected by the transformation.

We may therefore finally state the general result:

If $\mathbf{y}$ is non-singularly multinormal with moments (29.13), and the decomposition (29.14) holds for a quadratic form Q where $\mathbf{AV}$ is idempotent, then Q is a $\chi'^2(r, \boldsymbol{\mu}'\mathbf{A}\boldsymbol{\mu})$ variable, where r is the rank of $\mathbf{A}$, and any one of the three following conditions implies the others:

(a) $\sum_{i=1}^{k} r_i = r$.

(b) Each $\mathbf{A}_i\mathbf{V}$ is idempotent; equivalently, each Q_i is $\chi'^2(r_i, \boldsymbol{\mu}'\mathbf{A}_i\boldsymbol{\mu})$, where r_i is the rank of $\mathbf{A}_i$.

(c) $\mathbf{A}_i\mathbf{V}\mathbf{A}_j = \mathbf{0}$; equivalently, each Q_i is independent of every other.

> Graybill and Marsaglia (1957) give some more general results than this. Banerjee (1964) simplifies their proofs. See also Styan (1970) for the case of singular $\mathbf{V}$. Graybill and Milliken (1969) treat the case where $\mathbf{A}_i$ has random elements.

29.8   These general results on the decomposition of quadratic forms in normal variables solve the problems in **29.2–3**, which motivated our investigation. For example, it now follows that the numerator of (29.6) is independent of the denominator, so that their ratio is duly distributed in the non-central F form (which we write, as in **23.31**, $F'(\nu_1, \nu_2, \lambda)$) where $\nu_1 = k$, $\nu_2 = n - k$ and λ is given by (29.4) in this case. Similarly, we now see that for any individual $\hat{\theta}_i$ in the orthogonal model of **29.3**, the ratio

$$F = c_{ii}\hat{\theta}_i^2/\{(\mathbf{y} - \mathbf{X}\hat{\boldsymbol{\theta}})'(\mathbf{y} - \mathbf{X}\hat{\boldsymbol{\theta}})/(n-k)\} \qquad (29.15)$$

is an $F'(1, n-k, c_{ii}\theta_i^2/\sigma^2)$ variable, and may thus be used to test hypotheses concerning θ_i.

More generally, for a hypothesis H_0 imposing $r \leqslant k$ constraints, the ratio of the SS due to H_0 and the residual SS, multiplied by $(n-k)/r$, is an $F'(r, n-k, \lambda)$ variable—cf. **23.29–31**. It should be particularly noted from **23.29** that the non-central parameter λ is always of exactly the same form as the numerator SS of the test statistic with each observation replaced by its expectation, and σ^2 as a divisor. Thus we may always obtain λ very simply from the numerator SS in the test statistic by substituting $\boldsymbol{\theta}$ for $\hat{\boldsymbol{\theta}}$ and dividing by σ^2.

These are examples of the LR test in the linear model, derived generally through the canonical form of the model in **23.25–9**. The discussion below (23.100) explicitly pointed out that the LR test of a linear hypothesis concerning any subset of r of the k parameters is based upon the reduction in the SS due to these r parameters divided by the residual SS. The canonical approach of Chapter 23 had its theoretical uses in the derivation of optimum properties of LR tests in **23.36–7**. For our present purposes, the equivalent partitioning of SS which we have been discussing is more direct and informative.

We remind the reader that exact and approximate expressions for the power function of the LR F-tests are given in **23.32–3**.

AV for classified observations
29.9   Our definition of AV in **29.4** applies to any linear model, and covers the

applications to regression theory in **28.5–8**. However, the term AV is commonly used in a narrower sense, in which it was originally developed.

We saw in **29.4** that AV is used to separate out the influence of different parameters upon y. In experimental work, the parameters are often the effects of certain 'treatments' upon the variable y. For example, in agricultural experimentation, from which this terminology derives, y might be the yield of wheat from a plot of fixed size, and the 'treatment' being investigated might be the addition of a certain fertiliser to the plot during the growing season. Naturally, the experiment would include both treated and untreated plots. The point here is that such an experiment may be brought within the scope of the general linear model by defining a 'label' variable x which is equal to 1 when the treatment is given and 0 otherwise.

It is easy to see that any pattern of treatments can be handled in this way; we need only define a label variable x for each possible ingredient of the treatments in the experiment. If there are two fertilisers in the example of the previous paragraph, we should define x_1 as the label variable for the first and x_2 as the label variable for the second fertiliser. Thus, a plot which receives both fertilisers has $x_1 = x_2 = 1$; one which receives only the first has $x_1 = 1$, $x_2 = 0$; a plot which receives only the second fertiliser has $x_1 = 0$, $x_2 = 1$; and a plot receiving no fertiliser has $x_1 = x_2 = 0$. The analysis of the linear model can now proceed without difficulty, since the elements of **X** may be any known constants.

29.10   The feature of the matrix **X** in the examples discussed in **29.9** is that all its elements are units or zeros, since they are merely labels for the presence or absence of certain ingredients in the 'treatments'. In the narrower sense, the term AV is used to describe the analysis of a linear model when this restriction holds true for all or most of the elements of **X**. For example, in the single-fertiliser experiment discussed at the beginning of **29.9**, some plots might be given a single dose, others a double dose, and others none at all of the fertiliser. We could then define $x = 2$, 1 or 0 accordingly; the analysis of this model could still be called AV. However, this formulation suffers from the fact that it implies that $E(y)$ is affected twice as much by a double as by a single dose—the model is linear in β, the 'effect' parameter expressing the dependence of y upon x. This could be overcome by defining two label variables, x_1 to denote presence or absence of a single dose, and x_2 to denote presence or absence of a double dose of the fertiliser. This alternative formulation does not (as the reader may be tempted to think) reduce the model to the two-fertiliser model at the end of **29.9**, for we cannot now have $x_1 = x_2 = 1$ for any plot—there is evidently some loss of symmetry to offset the avoidance of the implication of linearity in dose-effect.

29.11   We shall be discussing the formulation of linear models in several important AV situations, and we shall see that the simple (mainly 0–1) structure of the elements of **X** produces corresponding simplifications in the analysis itself. The simplest case is that of a classification of observations into groups, suspected to differ in their means; this is usually known as a *one-way classification*.

Example 29.1   AV in a one-way classification

Suppose that a sample of independent observations is classified into k groups, with n_i $(i = 1, 2, \ldots, k)$ observations in the ith group and $\sum_{i=1}^{k} n_i = n$. If the groups can differ only in their means, we may express this as

$$y_{iq} = \theta_i + \varepsilon_{iq}, \qquad i = 1, 2, \ldots, k; \qquad q = 1, 2, \ldots, n_i,$$

which is in the form of the general linear model

$$\mathbf{y} = \mathbf{X\theta} + \mathbf{\varepsilon},$$

where

$$\mathop{\mathbf{y}}_{(n \times 1)} = \begin{bmatrix} y_{11} \\ y_{12} \\ \vdots \\ y_{1n_1} \\ y_{21} \\ \vdots \\ y_{2n_2} \\ \vdots \\ y_{k1} \\ \vdots \\ y_{kn_k} \end{bmatrix}, \qquad \mathop{\mathbf{\theta}}_{(k \times 1)} = \begin{pmatrix} \theta_1 \\ \theta_2 \\ \vdots \\ \theta_k \end{pmatrix}$$

and

$$\mathop{\mathbf{X}}_{(n \times k)} = \begin{bmatrix} 1 & & & & & \\ \vdots & & & & & \\ 1 & & & & & \\ & 1 & & & & \\ & \vdots & & & & \\ & 1 & & & & \\ & & 1 & & & \\ & & \vdots & & & \\ & & 1 & \cdot & & \\ & & & & \cdot & \\ & & & & 1 & \\ & & & & \vdots & \\ & & & & 1 & \end{bmatrix} \begin{array}{l} \left.\vphantom{\begin{matrix}1\\ \vdots \\1\end{matrix}}\right\} n_1 \text{ rows} \\ \left.\vphantom{\begin{matrix}1\\ \vdots \\1\end{matrix}}\right\} n_2 \text{ rows} \\ \left.\vphantom{\begin{matrix}1\\ \vdots \\1\end{matrix}}\right\} n_3 \text{ rows} \\ \quad \vdots \\ \left.\vphantom{\begin{matrix}1\\ \vdots \\1\end{matrix}}\right\} n_k \text{ rows} \end{array}$$

The zero elements of $\mathbf{X}$ are omitted. We see at once that

$$\mathbf{X'X} = \begin{pmatrix} n_1 & & & 0 \\ & n_2 & & \\ & & \ddots & \\ 0 & & & n_k \end{pmatrix}$$

so that the analysis is orthogonal (cf. **29.3**) whatever the values of the n_i—in particular, they need not be equal. Also

$$\mathbf{X'y} = \begin{pmatrix} \sum_q y_{1q} \\ \sum_q y_{2q} \\ \vdots \\ \sum_q y_{kq} \end{pmatrix}$$

so the LS estimator of $\boldsymbol{\theta}$ is

$$\hat{\boldsymbol{\theta}} = (\mathbf{X'X})^{-1}\mathbf{X'y} = \begin{pmatrix} \bar{y}_{1.} \\ \bar{y}_{2.} \\ \vdots \\ \bar{y}_{k.} \end{pmatrix}$$

where $\bar{y}_{i.} = \sum_{q=1}^{n_i} y_{iq}/n_i$ is the mean of the observations in the ith group, a bar over a symbol denoting averaging over the dotted suffix(es), as it always will hereafter. The estimator is in accordance with intuition since the observations are independent. We have

$$\mathbf{X}\hat{\boldsymbol{\theta}} = \begin{bmatrix} \left.\begin{matrix} \bar{y}_{1.} \\ \vdots \\ \bar{y}_{1.} \end{matrix}\right\} & n_1 \text{ rows} \\ \left.\begin{matrix} \bar{y}_{2.} \\ \vdots \\ \bar{y}_{2.} \end{matrix}\right\} & n_2 \text{ rows} \\ \vdots & \vdots \\ \left.\begin{matrix} \bar{y}_{k.} \\ \vdots \\ \bar{y}_{k.} \end{matrix}\right\} & n_k \text{ rows} \end{bmatrix}$$

and the SS due to the fitted model as a whole is

$$S_1 = (\mathbf{X}\hat{\boldsymbol{\theta}})'(\mathbf{X}\hat{\boldsymbol{\theta}}) = \sum_{i=1}^{k} n_i \bar{y}_{i.}^2. \tag{29.16}$$

By subtraction, the residual SS is, from (29.1),

$$S_R = \mathbf{y'y} - S_1 = \sum_{i=1}^{k} \sum_{q=1}^{n_i} y_{iq}^2 - \sum_{i=1}^{k} n_i \bar{y}_{i.}^2 \equiv \sum_i \sum_q (y_{iq} - \bar{y}_{i.})^2. \tag{29.17}$$

To test the simple hypothesis imposing k constraints,

$$H_1: \boldsymbol{\theta} = \mathbf{0}, \tag{29.18}$$

we use (29.6) and obtain

$$F = \left(\frac{n-k}{k}\right)\frac{S_1}{S_R}, \tag{29.19}$$

distributed as an $F'(k, n-k, \sum_i n_i\theta_i^2/\sigma^2)$ variable, reducing to a (central) $F(k, n-k)$ variable if H_1 holds.

However, (29.18) is not the hypothesis of principal interest in most practical situations, where we usually wish to test whether the θ_i are all equal without specifying their common value. Instead of (29.18), we therefore test the composite hypothesis

$$H_2: \theta_1 - \theta_k = \theta_2 - \theta_k = \cdots = \theta_{k-1} - \theta_k = 0, \qquad (29.20)$$

which imposes only $(k-1)$ constraints. If (29.20) holds, the n observations are identically distributed with common mean

$$\theta_* \equiv \sum_{i=1}^{k} n_i\theta_i/n = \begin{pmatrix} n_1/n \\ n_2/n \\ \vdots \\ n_k/n \end{pmatrix}' \boldsymbol{\theta}.$$

The LS estimator of θ_* is then the overall sample mean

$$\hat{\theta}_* = \bar{y}_{..} = \frac{1}{n} \sum_{i=1}^{k} \sum_{q=1}^{n_i} y_{iq} = \frac{1}{n} \sum_i n_i \bar{y}_{i.}.$$

If **1** is an $(n \times 1)$ vector of units, we may rewrite the linear model temporarily in the (singular) form

$$\mathbf{y} = \mathbf{X}\boldsymbol{\theta} + \boldsymbol{\varepsilon} \equiv \mathbf{1}\theta_* + \left[\mathbf{X} - \mathbf{1} \begin{pmatrix} n_1/n \\ n_2/n \\ \vdots \\ n_k/n \end{pmatrix}' \right] \boldsymbol{\theta} + \boldsymbol{\varepsilon} \qquad (29.21)$$

and observe that the value of θ_* (a single constraint) is not involved in the hypothesis (29.20), and that the SS attributable to θ_*, namely

$$(\mathbf{1}\hat{\theta}_*)'(\mathbf{1}\hat{\theta}_*) = n\bar{y}_{..}^2,$$

must be subtracted from (29.16) (the SS due to the fitted model as a whole) to give the SS due to the other $(k-1)$ constraints. This is

$$S_2 = S_1 - n\bar{y}_{..}^2 \equiv \sum_{i=1}^{k} n_i(\bar{y}_{i.} - \bar{y}_{..})^2. \qquad (29.22)$$

The residual SS is given by S_R, defined at (29.17), as before.
29.8 now gives for the test statistic

$$F = \left(\frac{n-k}{k-1}\right) \frac{S_2}{S_R} \qquad (29.23)$$

which is distributed as an $F'(k-1, n-k, \sum_{i=1}^{k} n_i(\theta_i - \theta_*)^2/\sigma^2)$ variable, reducing to a central $F(k-1, n-k)$ when H_2 holds.

For computational purposes, S_2 and S_R are usually written as

$$
\left.\begin{aligned}
S_2 &= \sum_{i=1}^{k} \frac{(\sum_{q=1}^{n_i} y_{iq})^2}{n_i} - \frac{(\sum_i \sum_q y_{iq})^2}{n}, \\
S_R &= \sum_i \sum_q y_{iq}^2 - \sum_i \frac{(\sum_q y_{iq})^2}{n_i},
\end{aligned}\right\}
\tag{29.24}
$$

and the results assembled in tabular form, as in Table 29.1.

Table 29.1   AV for a one-way classification

Variation	D.fr.	SS	Mean square (MS) = SS/d.fr.
Between groups	$k-1$	S_2	$S_2/(k-1)$
Residual	$n-k$	S_R	$S_R/(n-k)$
	$n-1$	$S_2 + S_R = \mathbf{y}'\mathbf{y} - n\bar{y}_{..}^2$	
General mean	1	$S_1 - S_2 = n\bar{y}_{..}^2$	$n\bar{y}_{..}^2$
Total	n	$S_1 + S_R = \mathbf{y}'\mathbf{y}$	

The 'general mean' row of Table 29.1 is usually omitted as of no interest; the variance-ratio test based on the ratio of $n(\bar{y}_{..} - \theta_*)^2$ to $S_R/(n-k)$ is, of course, the ordinary Student's t^2-test for the mean, i.e. it has an $F(1, n-k)$ distribution when H_2 holds. The test (29.23) is simply the ratio of the 'between groups' MS to the residual MS, while (29.19) is obtained by adding together the 'between groups' and 'general mean' rows of the table and taking the ratio of the resulting MS to the residual MS.

Kastenbaum et al. (1970a) give, for the case where all $n_i = n/k$, tables showing how large n_i must be to achieve power of 0.7, 0.8, 0.9, 0.95, 0.99 and 0.995 when using the test (29.23) with size $\alpha = 0.01, 0.05, 0.1$ or 0.2, for $k = 2$ (1) 6 with $\max_{i,j} |\theta_i - \theta_j|/\sigma$ given. Extended tables are available from Bowman (1972). Bowman and Kastenbaum give more extensive tables in Harter and Owen (1975).

AV identities and their geometrical interpretation
29.12   The general theory of the linear model has been used in Example 29.1, but the final result can be less formally derived as follows. The identity

$$
\sum_{i=1}^{k} \sum_{q=1}^{n_i} (y_{iq} - \bar{y}_{..})^2 \equiv \sum_{i=1}^{k} \sum_{q=1}^{n_i} (y_{iq} - \bar{y}_{i.})^2 + \sum_{i=1}^{k} n_i (\bar{y}_{i.} - \bar{y}_{..})^2
\tag{29.25}
$$

splits the SS of the observations about their overall mean into an SS 'within groups' and an SS 'between groups' (i.e. between group means). If it can be verified that the two sums on the right of (29.25) are independently distributed in the χ'^2 form, the ratio of the second to the first is an intuitively acceptable criterion for testing the equality of the group means in the population. This approach leads to (29.23) as before, but it offers no direct justification for the choice of this particular test statistic, for which the general theory is necessary. In more complicated situations, the approach through algebraic identities like (29.25) is often much simpler and quicker than the direct use of linear model theory, but care is necessary in splitting the SS—ultimately, safety lies only in checking with the general theory.

29.13   The Pythagorean form of (29.25) has the virtue of drawing attention to a geometrical interpretation of the algebraic partitioning of the SS which is the essence of AV. We saw in Example 11.7 that the simpler identity (for a single group of observations)

$$\sum_i y_i^2 \equiv \sum_i (y_i - \bar{y})^2 + n\bar{y}^2 \qquad (29.26)$$

is geometrically equivalent to projecting the point $\mathbf{y} = (y_1, y_2, \ldots, y_n)$ in the n-dimensional sample space upon the equiangular vector, which it meets at $(\bar{y}, \bar{y}, \ldots, \bar{y})$, and using Pythagoras' theorem in the resulting right-angled triangle. In the more general notation that we have been using in Example 29.1 and in (29.25), (29.26) is

$$\sum_{i=1}^{k} \sum_{q=1}^{n_i} y_{iq}^2 \equiv \sum_i \sum_q (y_{iq} - \bar{y}_{..})^2 + n\bar{y}_{..}^2, \qquad (29.27)$$

and is therefore seen to be equivalent to the splitting-off of the 'general mean' row from the Total SS in Table 29.1 to give the left-hand side of (29.25). The further decomposition in (29.25) of the first term on the right of (29.27) is similarly geometrically interpretable, $\sum_i \sum_q (y_{iq} - \bar{y}_{i.})^2$ being the squared distance from $\mathbf{y}$ to the vector $\mathbf{X}\hat{\boldsymbol{\theta}}$ defined above (29.16), and $\sum_i n_i (\bar{y}_{i.} - \bar{y}_{..})^2$ being the squared distance from $\mathbf{X}\hat{\boldsymbol{\theta}}$ to the equiangular vector. From the geometrical standpoint, therefore, AV is seen to consist of a resolution of the distance from $\mathbf{y}$ to the origin into a number of components relevant to the problem in hand.

29.14   The fact that in Example 29.1 we obtained an orthogonal analysis for any classification into groups, no matter what the sizes n_i were, encourages us to investigate more complex classification systems. We shall find that orthogonality does not generally persist for unequal group sizes, but does so when the sizes are equal. We first treat the case of a two-way classification in detail, since it exhibits most of the points of general interest.

AV in a two-way cross-classification
29.15   Suppose that, instead of being simply classified into k groups as in Example 29.1, a sample of n observations is classified in a table with r rows and c columns containing frequencies

n_{11}	n_{12}	$\cdots$	n_{1c}	$n_{1.}$	
n_{21}	n_{22}	$\cdots$	n_{2c}	$n_{2.}$	
$\vdots$	$\vdots$		$\vdots$	$\vdots$	
n_{r1}	n_{r2}	$\cdots$	n_{rc}	$n_{r.}$	(29.28)
$n_{.1}$	$n_{.2}$	$\cdots$	$n_{.c}$	n	

We call this an $r \times c$ cross-classification. The marginal frequencies $n_{i.}$ and $n_{.j}$ are sums over the suffix that is replaced by a dot. The reader will see that the grand total frequency in (29.28) should strictly be written $n_{..}$, but we continue to write n instead in this one case to denote 'sample size'.

We may, of course, treat the rc cells in the body of the table (29.28) as a one-way classification (Example 29.1) with $k = rc$. However, the questions that are usually asked about the two-way cross-classification (29.28) are:

(1) Do the means of the row-classification (with frequencies $n_{1.}, n_{2.}, \ldots, n_{r.}$) differ?
(2) Do the means of the column-classification (with frequencies $n_{.1}, n_{.2}, \ldots, n_{.c}$) differ?
(3) Is there any interrelation between row- and column-means?

More rarely, we ask also

(4) Does the mean of the whole set of n observations differ from some hypothetical value?

29.16 Denote the pth observation in the ith row and jth column of the table by y_{ijp}. We then have, in our notational convention,

$$
\left.
\begin{aligned}
\bar{y}_{ij.} &= \sum_{p=1}^{n_{ij}} y_{ijp}/n_{ij}, \\
\bar{y}_{i..} &= \sum_{j=1}^{c} \sum_{p=1}^{n_{ij}} y_{ijp}/n_{i.} \quad \equiv \sum_{j=1}^{c} n_{ij} \bar{y}_{ij.}/n_{i.} \\
&\equiv \sum_{j=1}^{c} n_{ij} \bar{y}_{ij.} \bigg/ \sum_{j=1}^{c} n_{ij}, \\
\bar{y}_{.j.} &= \sum_{i=1}^{r} \sum_{p=1}^{n_{ij}} y_{ijp}/n_{.j} \quad = \sum_{i=1}^{r} n_{ij} \bar{y}_{ij.}/n_{.j} \\
&\equiv \sum_{i=1}^{r} n_{ij} \bar{y}_{ij.} \bigg/ \sum_{i=1}^{r} n_{ij}, \\
\bar{y}_{...} &= \sum_{i=1}^{r} \sum_{j=1}^{c} \sum_{p=1}^{n_{ij}} y_{ijp}/n \equiv \sum_{i=1}^{r} n_{i.} \bar{y}_{i..}/n \equiv \sum_{j=1}^{c} n_{.j} \bar{y}_{.j.}/n.
\end{aligned}
\right\} \quad (29.29)
$$

An easy way of avoiding any possible confusion in notation is to define a dummy variable n_{ijp} which is identically equal to 1 for all $p = 1, 2, \ldots, n_{ij}$. Then (29.29) becomes

$$
\left.
\begin{aligned}
\bar{y}_{ij.} &= \sum_{p} y_{ijp} \bigg/ \sum_{p} n_{ijp}, & \bar{y}_{i..} &= \sum_{j} \sum_{p} y_{ijp} \bigg/ \sum_{j} \sum_{p} n_{ijp}, \\
\bar{y}_{.j.} &= \sum_{i} \sum_{p} y_{ijp} \bigg/ \sum_{i} \sum_{p} n_{ijp}, & \bar{y}_{...} &= \sum_{i} \sum_{j} \sum_{p} y_{ijp} \bigg/ \sum_{i} \sum_{j} \sum_{p} n_{ijp},
\end{aligned}
\right\} \quad (29.30)
$$

which is easily remembered by its numerator–denominator symmetry. Here, and hereafter unless otherwise stated, i is always summed from 1 to r; j is summed from 1 to c; and p is summed from 1 to n_{ij}.

29.17   In formulating the linear model, we require one parameter for the mean of the observations in each cell of the $r \times c$ table. In order to answer the questions posed in **29.15**, however, we express this mean μ_{ij} in a cell in terms of:

μ_{**}, a mean common to all observations;

μ_{i*}, a mean common to all observations in the ith row;

μ_{*j}, a mean common to all observations in the jth column.

Since we already have the cell means μ_{ij}, common to observations in the ith row and the jth column, we now have $1 + r + c + rc$ parameters, of which only rc can be linearly independent. The singularity that we have thus introduced by our choice of parameters can easily be removed, either by the augmentation technique of **19.14–15**, or by eliminating the redundant parameters, as we shall do here.

Once μ_{**} is defined, any $(r-1)$ of the means μ_{i*} determine the other one; similarly, only $(c-1)$ of the means μ_{*j} need be considered, since with μ_{**} they will determine the other one. Once the μ_{i*} and μ_{*j} are thus determined, it is easy to see that only $(r-1)(c-1)$ of the μ_{ij} can be independently determined. We may thus confine ourselves to $(r-1)$ parameters μ_{i*} (omitting μ_{r*}, say), to $(c-1)$ parameters μ_{*j} (omitting μ_{*c}, say), and to $(r-1)(c-1)$ parameters μ_{ij} (say, $i = 1, 2, \ldots, r-1$ and $j = 1, 2, \ldots, c-1$). These, with μ_{**}, make up the rc parameters required for the model to be non-singular.

It should be noticed that we have not yet defined the parameters $\mu_{**}, \mu_{i*}, \mu_{*j}$ except to state that they are (weighted) means of the μ_{ij}.

29.18   We first define

$$\left.\begin{array}{l} \theta_{**} = \mu_{**}, \\[4pt] \theta_{i*} = \mu_{i*} - \mu_{**}, \\[4pt] \theta_{*j} = \mu_{*j} - \mu_{**}, \\[4pt] \theta_{ij} = \mu_{ij} - (\mu_{i*} + \mu_{*j}) + \mu_{**}, \end{array}\right\} \tag{29.31}$$

and write the linear model in the form

$$y_{ijp} = \theta_{**} + \theta_{i*} + \theta_{*j} + \theta_{ij} + \varepsilon_{ijp} \equiv \mu_{ij} + \varepsilon_{ijp}. \tag{29.32}$$

Evidently, $\bar{y}_{ij.}$ will be the LS estimator $\hat{\mu}_{ij}$ in (29.32).

For obvious reasons, θ_{**} is called the *general mean*, and θ_{i*}, θ_{*j} are respectively called the *i*th *row-effect* and the *j*th *column-effect*, measuring the deviation from the general mean in a particular row or column. Row-effects and column-effects together are called *main effects*.

If the deviation of the cell-mean from the general mean were exactly equal to the sum of the corresponding row-effect and column-effect, we should have

$$\mu_{ij} - \mu_{**} = (\mu_{i*} - \mu_{**}) + (\mu_{*j} - \mu_{**}),$$

which implies $\theta_{ij} = 0$. We then say, in accordance with ordinary usage, that the *i*th row and *j*th column 'act additively' or 'do not interact'. θ_{ij} as defined in (29.31) measures departures from this situation, and is called the *interaction* between the *i*th row and the *j*th column.

29.19 The $(r+c+1)$ linear relations between the parameters, discussed in **29.17**, may now be written (but we shall return to this subject in **29.26–8** below)

$$
\left.
\begin{aligned}
0 = \sum_{i=1}^{r} n_{i.}\theta_{i*} &= \sum_{j=1}^{c} n_{.j}\theta_{*j} \\
&= \sum_{i=1}^{r} n_{ij}\theta_{ij}, \qquad j = 1, 2, \ldots, c-1, \\
&= \sum_{j=1}^{c} n_{ij}\theta_{ij}, \qquad i = 1, 2, \ldots, r-1, \\
&= \sum_{i=1}^{r}\sum_{j=1}^{c} n_{ij}\theta_{ij}.
\end{aligned}
\right\}
\tag{29.33}
$$

If, as in **29.18**, we define the parameters θ in (29.31) for $i = 1, 2, \ldots, r-1$ and $j = 1, 2, \ldots, c-1$ only, the eliminated $(r+c+1)$ parameters may be expressed in terms of the others, using (29.33), as

$$
\left.
\begin{aligned}
\theta_{r*} &= -\sum_{i=1}^{r-1} n_{i.}\theta_{i*}/n_{r.}, \\
\theta_{*c} &= -\sum_{j=1}^{c-1} n_{.j}\theta_{*j}/n_{.c}, \\
\theta_{rj} &= -\sum_{i=1}^{r-1} n_{ij}\theta_{ij}/n_{rj}, \qquad j = 1, 2, \ldots, c-1, \\
\theta_{ic} &= -\sum_{j=1}^{c-1} n_{ij}\theta_{ij}/n_{ic}, \qquad i = 1, 2, \ldots, r-1, \\
\theta_{rc} &= +\sum_{i=1}^{r-1}\sum_{j=1}^{c-1} n_{ij}\theta_{ij}/n_{rc}.
\end{aligned}
\right\}
\tag{29.34}
$$

29.20   We may now write down the matrix $\mathbf{X}$ of the linear model (29.32). It is not a matrix of units and zeros only, because the expression of the eliminated parameters in terms of the others, in (29.34), involves various ratios of the ns.

To simplify the reader's verification of the elements of the matrix in (29.35), its columns are headed by the parameters to which they correspond and its rows are bordered by the frequencies in the cells to which they apply. Only non-zero elements of $\mathbf{X}$ are shown. Throughout the matrix, a vector of units $\mathbf{1}$ contains a number of components equal to the sum of the cell-frequencies (in the border of the rows) over which the vector $\mathbf{1}$ physically extends in (29.35). (See overleaf.)

Premultiplying (29.35) by its transpose, we find that

$$
\mathbf{X'X} =
\begin{array}{cc}
 & \begin{array}{cccc} 1 & (r-1) & (c-1) & (r-1)(c-1) \end{array} \\
\begin{array}{c} 1 \\ (r-1) \\ (c-1) \\ (r-1)(c-1) \end{array} &
\left(\begin{array}{c|c|c|c}
n & 0 & 0 & 0 \\ \hline
0 & A & D & 0 \\ \hline
0 & D' & B & 0 \\ \hline
0 & 0 & 0 & C
\end{array}\right)
\end{array}
\qquad (29.36)
$$

$\underset{(rc\times rc)}{\mathbf{X'X}}$

(the orders of the submatrices being indicated by the numbers bordering the matrix), where $\mathbf{A}, \mathbf{B}$ and $\mathbf{C}$ are symmetric matrices with elements above the leading diagonal given by

$$
\underset{(r-1)\times(r-1)}{\mathbf{A}} =
\begin{bmatrix}
n_{1.}+\dfrac{n_{1.}^2}{n_{r.}} & \dfrac{n_{1.}n_{2.}}{n_{r.}} & \dfrac{n_{1.}n_{3.}}{n_{r.}} & \cdots & \dfrac{n_{1.}n_{r-1,.}}{n_{r.}} \\
 & n_{2.}+\dfrac{n_{2.}^2}{n_{r.}} & \dfrac{n_{2.}n_{3.}}{n_{r.}} & \cdots & \dfrac{n_{2.}n_{r-1,.}}{n_{r.}} \\
 & & \ddots & & \vdots \\
 & & & & n_{r-1,.}+\dfrac{n_{r-1,.}^2}{n_{r.}}
\end{bmatrix},
\qquad (29.37)
$$

$$
\underset{(c-1)\times(c-1)}{\mathbf{B}} =
\begin{bmatrix}
n_{.1}+\dfrac{n_{.1}^2}{n_{.c}} & \dfrac{n_{.1}n_{.2}}{n_{.c}} & \dfrac{n_{.1}n_{.3}}{n_{.c}} & \cdots & \dfrac{n_{.1}n_{.c-1}}{n_{.c}} \\
 & n_{.2}+\dfrac{n_{.2}^2}{n_{.c}} & \dfrac{n_{.2}n_{.3}}{n_{.c}} & \cdots & \dfrac{n_{.2}n_{.c-1}}{n_{.c}} \\
 & & \ddots & & \vdots \\
 & & & & n_{.c-1}+\dfrac{n_{.c-1}^2}{n_{.c}}
\end{bmatrix}.
\qquad (29.38)
$$

The $(r-1)(c-1)\times(r-1)(c-1)$ matrix $\mathbf{C}$ is more complicated. If we label its rows and columns by the suffixes of the θ_{ij} to which they refer (so that its $\{(k-1)(c-1)+l\}$th row, $\{(m-1)(c-1)+q\}$th column is labelled as the (kl)th row, (mq)th column) then

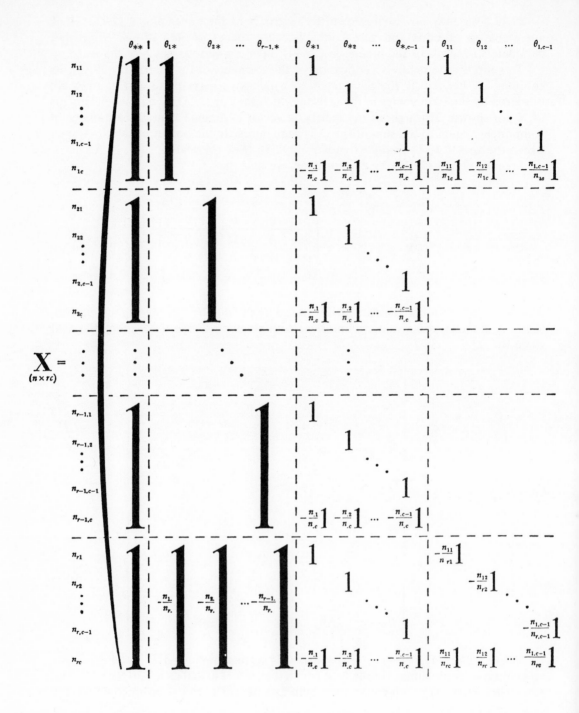

$$
\begin{array}{ccccccccccc}
\theta_{21} & \theta_{22} & \cdots & \theta_{2,c-1} & \cdots & & \theta_{r-1,1} & \theta_{r-1,2} & \cdots & \theta_{r-1,c-1} \\
\end{array}
$$

$$
\left(
\begin{array}{cccc|c|cccc}
1 & & & & & & & & \\
 & 1 & & & & & & & \\
 & & \ddots & & & & & & \\
 & & & 1 & & & & & \\
-\frac{n_{21}}{n_{2c}}1 & -\frac{n_{22}}{n_{2c}}1 & \cdots & -\frac{n_{2,c-1}}{n_{2c}}1 & & & & & \\
\hline
 & & & & \ddots & & & & \\
\hline
 & & & & & 1 & & & \\
 & & & & & & 1 & & \\
 & & & & & & & \ddots & \\
 & & & & & & & & 1 \\
 & & & & & -\frac{n_{r-1,1}}{n_{r-1,c}}1 & -\frac{n_{r-1,2}}{n_{r-1,c}}1 & \cdots & -\frac{n_{r-1,c-1}}{n_{r-1,c}}1 \\
\hline
-\frac{n_{21}}{n_{r1}}1 & & & & & -\frac{n_{r-1,1}}{n_{r1}}1 & & & \\
 & -\frac{n_{22}}{n_{r2}}1 & \ddots & & \cdots & & -\frac{n_{r-1,2}}{n_{r2}}1 & \ddots & \\
 & & & -\frac{n_{2,c-1}}{n_{r,c-1}}1 & & & & & -\frac{n_{r-1,c-1}}{n_{r,c-1}}1 \\
\frac{n_{21}}{n_{rc}}1 & \frac{n_{22}}{n_{rc}}1 & \cdots & \frac{n_{2,c-1}}{n_{rc}}1 & \cdots & \frac{n_{r-1,1}}{n_{rc}}1 & \frac{n_{r-1,2}}{n_{rc}}1 & \cdots & \frac{n_{r-1,c-1}}{n_{rc}}1 \\
\end{array}
\right.
$$

(29.35)

the element in the (kl)th row and (mq)th column of $\mathbf{C}$ is

$$C_{(kl),(mq)} = n_{kl} + n_{kl}^2 \left(\frac{1}{n_{kc}} + \frac{1}{n_{rl}} + \frac{1}{n_{rc}} \right) \quad \text{if } k = m, l = q,$$

$$= n_{kl} n_{kq} \left(\frac{1}{n_{kc}} + \frac{1}{n_{rc}} \right) \quad \text{if } k = m, l \neq q,$$

$$= n_{kl} n_{ml} \left(\frac{1}{n_{rl}} + \frac{1}{n_{rc}} \right) \quad \text{if } k \neq m, l = q,$$

$$= n_{kl} n_{mq} / n_{rc} \quad \text{if } k \neq m, l \neq q.$$

$$(29.39)$$

The only remaining non-null matrix in (29.36) is $\mathbf{D}$, of order $(r-1) \times (c-1)$, whose (i, j)th element is

$$D_{ij} = n_{ij} \left(1 - \frac{n_{ic}}{n_{.c}} \frac{n_{.j}}{n_{ij}} \right) + \frac{n_{i.} n_{.j}}{n_{r.}} \left(\frac{n_{rc}}{n_{.c}} - \frac{n_{rj}}{n_{.j}} \right). \quad (29.40)$$

29.21 In general, $\mathbf{X}'\mathbf{X}$ at (29.36) can only be inverted numerically as in the general LS procedure, but inspection reveals that if we can make $\mathbf{D} = \mathbf{0}$, the matrix will be of the form

$$\mathbf{X}'\mathbf{X} = \begin{pmatrix} n & & & \mathbf{0} \\ & \mathbf{A} & & \\ & & \mathbf{B} & \\ \mathbf{0} & & & \mathbf{C} \end{pmatrix} \quad (29.41)$$

whose inverse is simply

$$(\mathbf{X}'\mathbf{X})^{-1} = \begin{pmatrix} n^{-1} & & & \mathbf{0} \\ & \mathbf{A}^{-1} & & \\ & & \mathbf{B}^{-1} & \\ \mathbf{0} & & & \mathbf{C}^{-1} \end{pmatrix} \quad (29.42)$$

We are therefore led to examine the conditions under which $\mathbf{D} = \mathbf{0}$, i.e. every element D_{ij} defined by (29.40) is zero. The structure of D_{ij} makes it evident that this will be so if and only if

$$\frac{n_{ij}}{n_{.j}} = \frac{n_{ic}}{n_{.c}}, \quad i = 1, 2, \ldots, r-1; \quad j = 1, 2, \ldots, c-1,$$

and also

$$\frac{n_{rj}}{n_{.j}} = \frac{n_{rc}}{n_{.c}}.$$

These conditions are simply that every cell-frequency n_{ij} be proportional to its column-total frequency $n_{.j}$. It follows that every n_{ij} must then also be proportional to its row-total frequency $n_{i.}$, and that we must have

$$n_{ij} = n_{i.}n_{.j}/n, \quad \text{all } i, j \tag{29.43}$$

for **D** to be equal to the null matrix. Under this proportional frequencies condition, the analysis of the two-way classification becomes relatively simple.

The proportional-frequencies case
29.22   We first observe that the form of (29.42) implies that the LS estimator of the general mean θ_{**} is orthogonal to the estimators of all the other parameters, and similarly that the $(r-1)$ linearly independent row-effects are estimated orthogonally from all other parameters, as are the $(c-1)$ linearly independent column-effects and the $(r-1)(c-1)$ linearly independent interactions. The only non-orthogonalities occur *within* these last three groups of parameters, and have been imposed by the fact that each group has been obtained as a linearly independent subset of the larger (singular) group of parameters in which we are interested.

The reader will, perhaps, have observed that we have not yet evaluated the LS estimators themselves. The reason for this is that even when the proportional-frequencies condition (29.43) holds, the elements of **C** given at (29.39) are not such as to make its inversion simple, although of course we may evaluate $\mathbf{C}^{-1}$ numerically in any given situation. Fortunately, however, we may use the orthogonalities referred to in the preceding paragraph to obtain the LS estimators of the row- and column-effects at once, and use them later to evaluate the LS estimators of the interactions. To do this, we need only invert **A** and **B** at (29.37–8), and use (29.42) to evaluate the first $1+(r-1)+(c-1)=r+c-1$ rows of the $(rc \times 1)$ LS estimator vector.

29.23   It is easily verified by matrix multiplication that the inverse of (29.37) is

$$\mathbf{A}^{-1} = \begin{bmatrix} \dfrac{1}{n_{1.}} - \dfrac{1}{n} & -\dfrac{1}{n} & -\dfrac{1}{n} & \cdots & -\dfrac{1}{n} \\ & \dfrac{1}{n_{2.}} - \dfrac{1}{n} & -\dfrac{1}{n} & \cdots & -\dfrac{1}{n} \\ & & \ddots & & \vdots \\ & & & & \dfrac{1}{n_{r-1,.}} - \dfrac{1}{n} \end{bmatrix} \tag{29.44}$$

and similarly that (29.38) has inverse

$$\mathbf{B}^{-1} = \begin{bmatrix} \dfrac{1}{n_{.1}} - \dfrac{1}{n} & -\dfrac{1}{n} & -\dfrac{1}{n} & \cdots & -\dfrac{1}{n} \\ & \dfrac{1}{n_{.2}} - \dfrac{1}{n} & -\dfrac{1}{n} & \cdots & -\dfrac{1}{n} \\ & & \ddots & & \vdots \\ & & & & \dfrac{1}{n_{.c-1}} - \dfrac{1}{n} \end{bmatrix}. \tag{29.45}$$

We now require only the first $r+c-1$ rows of the $(rc \times 1)$ vector $\mathbf{X'y}$. From (29.35), these are seen to be, in the notation of (29.29),

$$(\mathbf{X'y})_{r+c-1} = \begin{bmatrix} n\bar{y}_{...} \\ n_{1.}(\bar{y}_{1..} - \bar{y}_{r..}) \\ n_{2.}(\bar{y}_{2..} - \bar{y}_{r..}) \\ \vdots \\ n_{r-1,.}(\bar{y}_{r-1,..} - \bar{y}_{r..}) \\ n_{.1}(\bar{y}_{.1.} - \bar{y}_{.c.}) \\ n_{.2}(\bar{y}_{.2.} - \bar{y}_{.c.}) \\ \vdots \\ n_{.c-1}(\bar{y}_{.c-1,.} - \bar{y}_{.c.}) \end{bmatrix} . \tag{29.46}$$

Using (29.42) and (29.44–6) we find, for the first $(r+c-1)$ components of the LS estimator $(\mathbf{X'X})^{-1}\mathbf{X'y}$,

$$\begin{bmatrix} \hat{\theta}_{**} \\ \hat{\theta}_{1*} \\ \vdots \\ \hat{\theta}_{r-1,*} \\ \hat{\theta}_{*1} \\ \vdots \\ \hat{\theta}_{*c-1} \end{bmatrix} = \begin{bmatrix} \bar{y}_{...} \\ \bar{y}_{1..} - \bar{y}_{...} \\ \vdots \\ \bar{y}_{r-1,..} - \bar{y}_{...} \\ \bar{y}_{.1.} - \bar{y}_{...} \\ \vdots \\ \bar{y}_{.c-1,.} - \bar{y}_{...} \end{bmatrix} . \tag{29.47}$$

Thus the LS estimator of the general mean is the overall sample mean, and the LS estimators of row- (or column-) effects are the sample differences between row- (or column-) means and the overall sample mean. It follows at once from (29.47) and the first two linear relationships in (29.34) that the same holds true for the eliminated (redundant) row- and column-effects, i.e. that

$$\hat{\theta}_{r*} = \bar{y}_{r..} - \bar{y}_{...}, \qquad \hat{\theta}_{*c} = \bar{y}_{.c.} - \bar{y}_{...}. \tag{29.48}$$

29.24 Substituting (29.47–8) into the definition of the interactions θ_{ij} in (29.31), we see that

$$\hat{\theta}_{ij} = \hat{\mu}_{ij} - \bar{y}_{i..} - \bar{y}_{.j.} + \bar{y}_{...}, \tag{29.49}$$

since θ_{ij} is a linear function of the other quantities (cf. **19.6**). Now, clearly, from the extreme right of (29.32), we must have the LS estimator

$$\hat{\mu}_{ij} = \bar{y}_{ij.},$$

and thus (29.49) becomes

$$\hat{\theta}_{ij} = \bar{y}_{ij.} - \bar{y}_{i..} - \bar{y}_{.j.} + \bar{y}_{...}. \tag{29.50}$$

Thus all the parameters are estimated, in this proportional-frequencies case, by the 'obvious' intuitive estimators.

Now that the LS estimators of all the parameters in our model are known, we may proceed, in Example 29.2, to test the various hypotheses corresponding to the questions asked in **29.15**.

Example 29.2   Two-way cross-classification with proportional frequencies
The results of our investigations so far show that the linear model (29.32) for the two-way cross-classification is representable in the non-singular form

$$\mathbf{y} = \mathbf{X}\boldsymbol{\theta} + \boldsymbol{\varepsilon}, \tag{29.51}$$

where $\mathbf{X}$ is defined by (29.35), and $\boldsymbol{\theta}$ is the vector whose transpose heads the rows of $\mathbf{X}$ in (29.35).

We first consider question (1) in **29.15**. This corresponds to asking for a test of the hypothesis

$$H_1: \theta_{1*} = \theta_{2*} = \cdots = \theta_{r-1,*} = 0, \tag{29.52}$$

imposing $(r-1)$ constraints. Question (2) in **29.15** is similarly equivalent to the $(c-1)$-constraint hypothesis

$$H_2: \theta_{*1} = \theta_{*2} = \cdots = \theta_{*c-1} = 0, \tag{29.53}$$

and question (3) in **29.15** to the $(r-1)(c-1)$-constraint hypothesis

$$H_3: \theta_{ij} = 0, \qquad i = 1, 2, \ldots, r-1; \qquad j = 1, 2, \ldots, c-1. \tag{29.54}$$

Finally, question (4) in **29.15** corresponds to the single-constraint hypothesis

$$H_4: \theta_{**} = 0. \tag{29.55}$$

All four hypotheses (29.52)–(29.55) are composite. To test any one of them, we must find the SS attributable to that hypothesis, and use the general theory that we have developed, summarised in **29.8**. Since the four hypotheses between them account for all rc parameters in the model, and have no parameter in common, we see that we have to partition the SS due to the fitted model as a whole, namely $\hat{\boldsymbol{\theta}}'\mathbf{X}'\mathbf{X}\hat{\boldsymbol{\theta}}$, into the components attributable to the four hypotheses. This is particularly straightforward here, since we have seen in **29.22** that the four groups of parameters are estimated by orthogonal sets of estimators. In fact, $\mathbf{X}'\mathbf{X}$ was given at (29.41).

We now write the LS estimators from (29.47) and (29.50) in the form

$$
\hat{\boldsymbol{\theta}} =
\begin{bmatrix}
\bar{y}_{...} \\
\\
\bar{y}_{1..} - \bar{y}_{...} \\
\vdots \\
\bar{y}_{r-1,..} - \bar{y}_{...} \\
\\
\bar{y}_{.1.} - \bar{y}_{...} \\
\vdots \\
\bar{y}_{.c-1,.} - \bar{y}_{...} \\
\\
\bar{y}_{11.} - \bar{y}_{1..} - \bar{y}_{.1.} + \bar{y}_{...} \\
\vdots \\
\bar{y}_{r-1,c-1,.} - \bar{y}_{r-1,..} - \bar{y}_{.c-1,.} + \bar{y}_{...}
\end{bmatrix}
=
\begin{bmatrix}
\hat{\theta}_{**} \\
\cdots \\
\hat{\theta}_{i*} \\
\cdots \\
\hat{\theta}_{*j} \\
\cdots \\
\hat{\theta}_{ij}
\end{bmatrix}
, \tag{29.56}
$$

where the subvectors of $\hat{\boldsymbol{\theta}}$ have $1, r-1, c-1$ and $(r-1)(c-1)$ components respectively. From (29.56) and (29.41), we have the decomposition

$$
\hat{\boldsymbol{\theta}}' \mathbf{X}' \mathbf{X} \hat{\boldsymbol{\theta}} = n\hat{\theta}_{**}^2 + \hat{\boldsymbol{\theta}}_{i*}' \mathbf{A} \hat{\boldsymbol{\theta}}_{i*} + \hat{\boldsymbol{\theta}}_{*j}' \mathbf{B} \hat{\boldsymbol{\theta}}_{*j} + \hat{\boldsymbol{\theta}}_{ij}' \mathbf{C} \hat{\boldsymbol{\theta}}_{ij}, \tag{29.57}
$$

where $\mathbf{A}, \mathbf{B}, \mathbf{C}$ are the submatrices of (29.41) defined at (29.37)–(29.39). The first term on the right of (29.57) is the SS attributable to H_4, which we write explicitly as

$$
S_4 = n\bar{y}_{...}^2. \tag{29.58}
$$

(29.37) may be written in the form

$$
\mathbf{A} = \mathbf{D}_{n_{i.}} + \mathbf{n}_r \mathbf{n}_r',
$$

where $\mathbf{D}_{n_{i.}}$ is an $(r-1) \times (r-1)$ diagonal matrix with elements $n_{i.}$ and $\mathbf{n}_r$ is the $(r-1) \times 1$ vector with elements $n_{i.}/n_{r.}^{1/2}$. The second term on the right of (29.57) is now seen to be

$$
\begin{aligned}
S_1 &= \hat{\boldsymbol{\theta}}_{i*}' \{ \mathbf{D}_{n_{i.}} + \mathbf{n}_r \mathbf{n}_r' \} \hat{\boldsymbol{\theta}}_{i*} \\
&= \hat{\boldsymbol{\theta}}_{i*}' \mathbf{D}_{n_{i.}} \hat{\boldsymbol{\theta}}_{i*} + \hat{\boldsymbol{\theta}}_{i*}' \mathbf{n}_r (\hat{\boldsymbol{\theta}}_{i*}' \mathbf{n}_r)' \\
&= \sum_{i=1}^{r-1} n_{i.} (\bar{y}_{i..} - \bar{y}_{...})^2 + \left\{ \sum_{i=1}^{r-1} \frac{n_{i.}}{n_{r.}^{1/2}} (\bar{y}_{i..} - \bar{y}_{...}) \right\}^2 \\
&= \sum_{i=1}^{r} n_{i.} (\bar{y}_{i..} - \bar{y}_{...})^2. \tag{29.59}
\end{aligned}
$$

This is the SS attributable to H_1. In an exactly similar way, using $\mathbf{B}$ at (29.38), we find for the third term on the right of (29.57)

$$
S_2 = \sum_{j=1}^{c} n_{.j} (\bar{y}_{.j.} - \bar{y}_{...})^2, \tag{29.60}
$$

the SS attributable to H_2. Finally, we find from (29.39) that the last term on the right of (29.57) is

$$S_3 = \sum_{i=1}^{r-1} \sum_{j=1}^{c-1} n_{ij}(\bar{y}_{ij.} - \bar{y}_{i..} - \bar{y}_{.j.} + \bar{y}_{...})^2 + \frac{1}{n_{rc}}\left\{ \sum_{i=1}^{r-1} \sum_{j=1}^{c-1} n_{ij}(\bar{y}_{ij.} - \bar{y}_{i..} - \bar{y}_{.j.} + \bar{y}_{...}) \right\}^2$$

$$+ \sum_{i=1}^{r-1} \frac{1}{n_{ic}}\left\{ \sum_{j=1}^{c-1} n_{ij}(\bar{y}_{ij.} - \bar{y}_{i..} - \bar{y}_{.j.} + \bar{y}_{...}) \right\}^2 + \sum_{j=1}^{c-1} \frac{1}{n_{rj}}\left\{ \sum_{i=1}^{r-1} n_{ij}(\bar{y}_{ij.} - \bar{y}_{i..} - \bar{y}_{.j.} + \bar{y}_{...}) \right\}^2$$

$$= \sum_{i=1}^{r} \sum_{j=1}^{c} n_{ij}(\bar{y}_{ij.} - \bar{y}_{i..} - \bar{y}_{.j.} + \bar{y}_{...})^2, \tag{29.61}$$

upon use of linear relationships among the estimated interactions precisely analogous to those for the interaction parameters given in (29.34). The four SS defined in (29.58–61) exhaust the SS due to the fitted model (29.51). The only other quantity we shall require is the residual SS, which here, as generally, is the difference

$$S_R = \mathbf{y}'\mathbf{y} - \hat{\boldsymbol{\theta}}'\mathbf{X}'\mathbf{X}\hat{\boldsymbol{\theta}} = \sum_i \sum_j \sum_p y_{ijp}^2 - (S_1 + S_2 + S_3 + S_4). \tag{29.62}$$

For computational purposes, the other SS are written in the forms

$$\left.\begin{aligned}
S_4 &= \left(\sum_i \sum_j \sum_p y_{ijp} \right)^2 \Big/ n, \\[2mm]
S_1 &= \sum_i \left\{ \left(\sum_j \sum_p y_{ijp} \right)^2 \Big/ n_{i.} \right\} - \left(\sum_i \sum_j \sum_p y_{ijp} \right)^2 \Big/ n, \\[2mm]
S_2 &= \sum_j \left\{ \left(\sum_i \sum_p y_{ijp} \right)^2 \Big/ n_{.j} \right\} - \left(\sum_i \sum_j \sum_p y_{ijp} \right)^2 \Big/ n, \\[2mm]
S_3 &= \sum_i \sum_j \left\{ \left(\sum_p y_{ijp} \right)^2 \Big/ n_{ij} \right\} - \sum_i \left\{ \left(\sum_j \sum_p y_{ijp} \right)^2 \Big/ n_{i.} \right\} \\[2mm]
&\quad - \sum_j \left\{ \left(\sum_i \sum_p y_{ijp} \right)^2 \Big/ n_{.j} \right\} + \left(\sum_i \sum_j \sum_p y_{ijp} \right)^2 \Big/ n,
\end{aligned}\right\} \tag{29.63}$$

which the reader should verify.

On substituting from (29.63), (29.62) becomes simply

$$S_R = \sum_i \sum_j \sum_p y_{ijp}^2 - \sum_i \sum_j \left\{ \left(\sum_p y_{ijp} \right)^2 \Big/ n_{ij} \right\}. \tag{29.64}$$

We may now, as we did in Example 29.1, assemble the results of our analysis in tabular form, as in Table 29.2.

The general theory of **29.8** tells us that the LR test of any of the hypotheses H_1 to H_4 is obtained by using the ratio of the corresponding MS in Table 29.2 to the residual MS, and rejecting the hypothesis for large values of the ratio. Each of these ratios is a non-central F-variate with d.fr. as given in the table and non-central parameter (obtained by using the general rule in **29.8**) given in the last column of

Table 29.2   AV for a two-way cross-classification with proportional frequencies

Variation	D.fr.	SS	MS	Non-central parameter λ
Between rows	$r-1$	S_1	$S_1/(r-1)$	$\sum_i n_{i.}\theta_{i*}^2/\sigma^2$
Between columns	$c-1$	S_2	$S_2/(c-1)$	$\sum_j n_{.j}\theta_{*j}^2/\sigma^2$
Interactions	$(r-1)(c-1)$	S_3	$S_3/(r-1)(c-1)$	$\sum_i\sum_j n_{ij}\theta_{ij}^2/\sigma^2$
Residual	$n-rc$	S_R	$S_R/(n-rc)$	
General mean	1	S_4	S_4	$n\theta_{**}^2/\sigma^2$
Total	n	$\mathbf{y'y}$		

the table. (As in Example 29.1, the test for the general mean is the usual Student's t^2-test.)

> Because each MS in the AV table is tested against the same residual MS, one may be tempted to merge successively each SS for which the H_0 is accepted with the residual SS to form a pooled SS, whose division by the similarly pooled d.fr. gives a pooled MS against which the remaining H_0 could be tested with increased power. Such pooling is rarely worth while for the present linear model, since the residual d.fr. are usually larger (and often much larger) than the other d.fr., so that little power is to be gained by the pooling; while the decision to pool, itself depending on a previous test, may actually be erroneous. Mead *et al.* (1975) study the problem in detail: pooling is never recommended if residual d.fr. are reasonably large, and larger than the prospectively merged d.fr., but may be worth while if the latter are considerably the larger. To control the overall test size, the preliminary test must be carried out with α fairly large. (Between 0.5 when the two d.fr. are about equal and 0.25 when the other d.fr. are more than twice the residual d.fr., if followed by a pooled test of size 0.05, gives an overall test size of approximately 0.05.) Pooling is used in other AV models where it is more likely to be advantageous.

If we may *assume* all interactions to be zero, S_3 is merged with S_R as a new pooled residual SS, with $(n-r-c+1)$ d.fr., and the other tests use this MS as denominator.

To test the comprehensive hypothesis

$$H_0: \mathbf{\theta} = \mathbf{0} \tag{29.65}$$

for all the parameters (which means that H_1, H_2, H_3 and H_4 all hold), the same theory tells us that the ratio to be used is

$$F = \frac{(S_1 + S_2 + S_3 + S_4)/rc}{S_R/(n-rc)}, \tag{29.66}$$

which is an $F'\{rc, n-rc, \sum_i\sum_j n_{ij}(\theta_{**} + \theta_{i*} + \theta_{*j} + \theta_{ij})^2/\sigma^2\}$ variable, the non-central parameter being obtained from the last term on the right of (29.64), by substituting $\mathbf{\theta}$ for $\hat{\mathbf{\theta}}$ in accordance with the general rule. This test is exactly the one mentioned in **29.15**, in which the rc cell-frequencies are treated as a one-way classification and the test (29.19) applied. Similarly, to test that H_1, H_2 and H_3 (but not H_4) hold, the numerator of (29.66) is replaced by $(S_1 + S_2 + S_3)/(rc - 1)$, this test being equivalent to (29.23) applied to the rc cell-frequencies.

The equal-frequencies (balanced) case

29.25   The most important case of the proportional-frequencies situation (29.43) arises when all cell-frequencies n_{ij} are equal, say to m. The arithmetic of the computing formulae (29.63) and (29.64) then simplifies obviously (cf. Exercise 29.1). The matrix C of (29.39) also now becomes easy to invert and the theory of **29.22–4** correspondingly more direct (cf. Exercise 29.2).

> Kastenbaum *et al.* (1970b) give tables showing how large m must be $(1 \leqslant m \leqslant 5)$ for $r = 2\,(1)$ 6   and $c = 2\,(1)$ 5, in testing H_1 with $\alpha = 0.05$, 0.01 and power 0.7, 0.8, 0.9, 0.95, 0.99, 0.995 when $\max_{ij}|\theta_{i*} - \theta_{j*}|/\sigma$ is given. Extended tables are available from Bowman (1972). Bowman and Kastenbaum give more extensive tables in Harter and Owen (1975).

Apart from these simplifications, the only new point arising in this *balanced* case occurs when $m = 1$, for then (cf. Exercise 29.1) the residual SS (29.64) is identically zero, as are its d.fr. $(n - rc)$. Since all the tests given in Example 29.2 then become nugatory, this situation clearly requires special consideration. It is not difficult to see how our problem comes about, for with $m = 1$, $n = rc$ we are put in the position of having to estimate rc parameters from the same number of observations. Not surprisingly, we can do this exactly, with no residual variation—we are in just the same position as we should be in fitting a polynomial of degree $q - 1$ (requiring q constants) to a set of q observations. Thus we can estimate all rc parameters even when $m = 1$, but only at the expense of seeing our residual SS disappear.

There is no way out of this difficulty unless we consent to reduce the number of parameters in the model, and what we shall in fact do is to discard the $(r - 1)(c - 1)$ interaction parameters θ_{ij}, leaving ourselves with $r + c - 1$ parameters to be estimated. We shall then have a new residual SS to replace (29.64), and in fact this will be seen in Example 29.3 to be precisely the former interaction SS, S_3, defined at (29.61).

It should not need to be stressed that this restricted model, without interaction parameters, is unsuitable for the analysis of data where interactions do exist. For this reason it is inadvisable to restrict ourselves voluntarily to one observation per cell of a cross-classification unless we are sure that rows and columns do not interact. However, considerations of cost or time sometimes enforce such a restriction.

Example 29.3   Two-way cross-classification with exactly one observation per cell
If the interaction parameters θ_{ij} are dropped from the linear model, we now have, with one observation per cell,

$$y_{ij} = \theta_{**} + \theta_{i*} + \theta_{*j} + \varepsilon_{ij},$$

where, to avoid singularities, we define θ_{i*} for $i = 1, 2, \ldots, r - 1$ and θ_{*j} for $j = 1, 2, \ldots, c - 1$, as previously. All the work of **29.17–19** in respect of our present parameters holds good. The matrix $\mathbf{X}$ defined at (29.35) remains valid if we use only its first $(r+c-1)$ columns, as does the leading $(r+c-1)\times(r+c-1)$ submatrix of $\mathbf{X}'\mathbf{X}$ at (29.36), in which we now still have $\mathbf{D}=\mathbf{0}$ since the proportional-frequencies condition (29.43) holds here. $\mathbf{A}$ and $\mathbf{B}$ at (29.37) and (29.38) and their inverses at (29.44) and (29.45) are unaffected, as are the vectors (29.46) and (29.47), which are now complete instead of partial vectors for the LS estimators of our parameters. We may therefore test the hypotheses H_1, H_2 and H_4 at (29.52), (29.53) and (29.55) exactly as in Example 29.2, the only difference being that what was previously the interaction SS, S_3, now becomes the new residual SS, for the four SS in the following abbreviated Table 29.3 must add to $\mathbf{y}'\mathbf{y}$, as always.

The tests of H_1, H_2 and H_4 can now be carried out with the MS $S_3/(r-1)(c-1)$ as denominator of the F-statistic.

Table 29.3   AV for a two-way cross-
classification with one observation per cell

Variation	D.fr.	SS
Between rows	$r-1$	S_1
Between columns	$c-1$	S_2
Residual	$(r-1)(c-1)$	S_3
General mean	1	S_4
Total	n	$\mathbf{y}'\mathbf{y}$

Although we have dropped the interaction parameters θ_{ij} from the model in order to obtain a residual SS, we can also use their estimators $\hat{\theta}_{ij}$ to test for zero interactions by separating off from that residual SS an appropriate component.

Consider the linear form

$$L = \sum_i \sum_j c_{ij} \hat{\theta}_{ij},$$

where $\hat{\theta}_{ij}$ is defined by (29.50) (the final suffix to the ys is now redundant) and the c_{ij} are coefficients to be determined. If the interactions θ_{ij} are all zero, but in general not otherwise, $E(L)=0$, and it is thus intuitively reasonable to use a statistic of the form L to test the hypothesis of zero interactions. If we choose the c_{ij} so that $\sum_i c_{ij} = \sum_j c_{ij} = 0$, we see from (29.50) that

$$L \equiv \sum_i \sum_j c_{ij} y_{ij},$$

and hence

$$\operatorname{var} L = C^2 \sigma^2,$$

where $C^2 = \sum_i \sum_j c_{ij}^2$, and σ^2 is the error variance as usual. Thus $L^2/(C^2\sigma^2)$ is a χ^2 variable with 1 d.fr. when the interactions are zero. Moreover, our present residual SS at (29.61) is $S_3 = \sum_i \sum_j \hat{\theta}_{ij}^2$, and $S_3 - \{L^2/(C^2\sigma^2)\}$ is independent of $L^2/(C^2\sigma^2)$, since the $\hat{\theta}_{ij}$ can be orthogonally transformed to a set of standardised independent normal variates of which one is $L/C\sigma$, and $S_3 - \{L^2/(C^2\sigma^2)\}$ will be the sum of squares of the others, distributed as χ^2 with $(r-1)(c-1) - 1 = rc - r - c$ d.fr.

It remains to choose the c_{ij}. They may be functions of the $\hat{\theta}_{i*}, \hat{\theta}_{*j}$, since the latter are distributed independently of the $\hat{\theta}_{ij}$ by (29.42), and hence the marginal distribution of L will be the same as any of its conditional distributions for fixed $\hat{\theta}_{i*}, \hat{\theta}_{*j}$, which will be as given above.

A simple choice is $c_{ij} = \hat{\theta}_{i*}\hat{\theta}_{*j}$, so that we may define

$$S_I = \left(\sum_i \sum_j \hat{\theta}_{i*}\hat{\theta}_{*j} y_{ij}\right)^2 \bigg/ \left(\sum_i \hat{\theta}_{i*}^2 \sum_j \hat{\theta}_{*j}^2\right) = \frac{\{\sum_i \sum_j (\bar{y}_{i.} - \bar{y}_{..})(\bar{y}_{.j} - \bar{y}_{..})y_{ij}\}^2}{\sum_i (\bar{y}_{i.} - \bar{y}_{..})^2 \sum_j (\bar{y}_{.j}}. \quad (29.67)$$

S_1/σ^2 is a χ^2 variable with 1 d.fr. and $(S_3 - S_1)/\sigma^2$ independently a χ^2 variable with $(rc - r - c)$ d.fr. Their ratio $S_1/(S_3 - S_1) = F$ has the variance-ratio distribution with $(1, rc - r - c)$ d.fr., and may be used to test the hypothesis that all interactions are zero. This test for complete *additivity* of effects was suggested by Tukey (1949), who generalised it further—see Scheffé (1959). M. N. Ghosh and Sharma (1963) studied its power against the alternative that there are interactions of form $\theta_{ij} = \alpha\theta_{i*}\theta_{*j}$. For the 6×6 classification, the power was found to be of the same order as that of the F-test for interactions obtained by equating adjacent pairs of the θ_{i*} and of the θ_{*j}. D. E. Johnson and Graybill (1972a, b) discuss an alternative form for the interactions and give ML estimators and LR tests for the parameters, including σ^2. Hegemann and Johnson (1976) show that Tukey's test is superior only in the presence of large main effects and significant correlations between these and the interaction effects. See also Marasinghe and Johnson (1981, 1982).

> Snee (1982) discusses the problem of discriminating between the existence of non-zero interactions and the heterogeneity of row (or column) variances, which may produce similar results in the AV table.

Choice of weights

29.26 We must now discuss a point that we deliberately passed over in formulating our linear model in **29.17-19**. We observed there that we had $(r + c + 1)$ parameters in our original model which were redundant in the sense that they were linearly dependent upon the rc other parameters, and we therefore eliminated them using the set of linear relations given in (29.33), leading to (29.34), which determined the structure of the basic matrix (29.35). It is now necessary to recognise that the set of relations given in (29.33) is essentially arbitrary—in the first relation given there, for example, we *chose* to equate to zero the particular linear function $\sum_i n_{i.}\theta_{i*}$, using as weights the marginal row frequencies $n_{i.}$. This may seem natural, but it is by no means necessary: we might have chosen instead to use equal weights, so that $\sum_i \theta_{i*} = 0$, or indeed any weights w_i, so that $\sum_i w_i\theta_{i*} = 0$.

If the complete set of n observations were a simple random sample from some population, the observed $n_{i.}/n$ would be estimates of the population relative frequencies in the row categories, and it would therefore be meaningful to define the row-effects using these weights to express their linear dependence. Similarly,

$\sum_j n_{.j} \theta_{*j} = 0$, and the other relations in (29.33) would be meaningful in the same context. We call these the *frequency weights*.

29.27 However, in many experimental contexts there is no question of the observations being a random sample from some population—the $r \times c$ cross-classification is deliberately set up to throw light on the variable (y) being studied. The use of observed frequencies as weights in the linear relations (29.33) is then no longer readily interpretable. It may even be meaningless to consider any set of weights as the 'right' ones, in the sense of reflecting an underlying population distribution; for example, if we have a 2×3 cross-classification to study the effects of two different doses of Fertiliser A and three different doses of Fertiliser B on the yield of a crop (y), one may be simply interested in the effects and interactions as such, and not as representing any population at all. There is a crucial distinction here between the 'experimental' and the 'survey' approach to data.

29.28 In experimental investigations, therefore, it is common (for lack of any known appropriate system of weights) to use equal weights throughout (29.33). For the remainder of this chapter, the *equal-weights* system signifies that (29.33) holds with all symbols $n_{i.}$, $n_{.j}$, n_{ij} suppressed, i.e. replaced by 1s. It is to be observed that, whereas in the balanced case (cf. **29.25**) the equal-weights system has in effect already been used, simply because the frequencies were equal, our general results for the proportional-frequencies case (in **29.22-4** and Example 29.2) would not hold unless the frequency weights were used.

Now that we are about to resume discussion of the general disproportional-frequencies case, which we left in **29.21**, the distinction between these two weighting systems will become acute, if only because, in this most general case, we may have a very small number of very large frequencies which tend to dominate the frequency weighting system, and perhaps distort its interpretation.

Disproportional frequencies

29.29 We first use the frequency weights (29.33) as before. The proportionality condition (29.43) does not hold, so that the matrix $\mathbf{D}$ in (29.36) is non-null, and the simplied analysis of **29.22-5** is no longer valid. It remains true even in this most general case that (29.36) may be partitioned into

$$\mathbf{X'X} = \begin{pmatrix} n & & \\ & \begin{pmatrix} \mathbf{A} & \mathbf{D} \\ \mathbf{D'} & \mathbf{B} \end{pmatrix} & \\ & & \mathbf{C} \end{pmatrix}$$

of which (29.41) was the special case $\mathbf{D} = 0$. The inverse of this can still be written down, for

$$\mathbf{E} = \begin{pmatrix} \mathbf{A} & \mathbf{D} \\ \mathbf{D'} & \mathbf{B} \end{pmatrix}^{-1} = \begin{pmatrix} (\mathbf{A} - \mathbf{D}\mathbf{B}^{-1}\mathbf{D'})^{-1} & -(\mathbf{A} - \mathbf{D}\mathbf{B}^{-1}\mathbf{D'})^{-1}\mathbf{D}\mathbf{B}^{-1} \\ -(\mathbf{B} - \mathbf{D'}\mathbf{A}^{-1}\mathbf{D})^{-1}\mathbf{D'}\mathbf{A}^{-1} & (\mathbf{B} - \mathbf{D'}\mathbf{A}^{-1}\mathbf{D})^{-1} \end{pmatrix}, \quad (29.68)$$

as may be verified by multiplication, so that

$$(\mathbf{X}'\mathbf{X})^{-1} = \begin{pmatrix} n^{-1} & & \\ & \mathbf{E} & \\ & & \mathbf{C}^{-1} \end{pmatrix}. \tag{29.69}$$

The effect of the non-nullity of $\mathbf{D}$ on the LS analysis is to change the estimator of the parameter-vector $\boldsymbol{\theta}$, for although the partial vector (29.46) is unchanged, the $(r+c-1) \times (r+c-1)$ leading diagonal submatrix of (29.42) is now replaced by that of (29.69). If we write (29.46) concisely as

$$\begin{pmatrix} n\bar{y}_{...} \\ \mathbf{v}_{r-1} \\ \mathbf{v}_{c-1} \end{pmatrix},$$

each $\mathbf{v}$ being the subvector of (29.46) with number of rows indicated by its suffix, we may generalise (29.47), using (29.68) and (29.69), to

$$((\mathbf{X}'\mathbf{X})^{-1}\mathbf{X}'\mathbf{y})_{r+c-1} = \begin{pmatrix} n^{-1} & 0 \\ 0 & \mathbf{E} \end{pmatrix} \begin{pmatrix} n\bar{y}_{...} \\ \mathbf{v}_{r-1} \\ \mathbf{v}_{c-1} \end{pmatrix}$$

$$= \begin{pmatrix} \bar{y}_{...} \\ (\mathbf{A} - \mathbf{D}\mathbf{B}^{-1}\mathbf{D}')^{-1}(\mathbf{v}_{r-1} - \mathbf{D}\mathbf{B}^{-1}\mathbf{v}_{c-1}) \\ (\mathbf{B} - \mathbf{D}'\mathbf{A}^{-1}\mathbf{D})^{-1}(\mathbf{v}_{c-1} - \mathbf{D}'\mathbf{A}^{-1}\mathbf{v}_{r-1}) \end{pmatrix}. \tag{29.70}$$

Thus the estimators $\hat{\boldsymbol{\theta}}_{i*}$, $\hat{\boldsymbol{\theta}}_{*j}$ are numerically determinable, while $\hat{\boldsymbol{\theta}}_{**} = \bar{y}_{...}$ always, as is intuitively obvious. As in **29.24**, the definition of the interactions at (29.31) then implies that their estimators satisfy

$$\hat{\theta}_{ij} = \bar{y}_{ij.} - \hat{\theta}_{i*} - \hat{\theta}_{*j} - \bar{y}_{...}, \tag{29.71}$$

so that the LS estimators of all the parameters are determined. The generalisation of the decomposition (29.57) is

$$\hat{\boldsymbol{\theta}}'\mathbf{X}'\mathbf{X}\hat{\boldsymbol{\theta}} = n\hat{\theta}_{**}^2 + \begin{pmatrix} \hat{\boldsymbol{\theta}}_{i*} \\ \hat{\boldsymbol{\theta}}_{*j} \end{pmatrix}' \begin{pmatrix} \mathbf{A} & \mathbf{D} \\ \mathbf{D}' & \mathbf{B} \end{pmatrix} \begin{pmatrix} \hat{\boldsymbol{\theta}}_{i*} \\ \hat{\boldsymbol{\theta}}_{*j} \end{pmatrix} + \hat{\boldsymbol{\theta}}'_{ij}\mathbf{C}\hat{\boldsymbol{\theta}}_{ij}. \tag{29.72}$$

Thus we have seen that the estimators of the three groups of parameters—(1) the general mean; (2) the row-effects and the column-effects; (3) the interactions—are always uncorrelated between the three groups and in general correlated within them. Proportional frequencies remove the correlations between (but not within) the sub-groups (2a): the row-effects and (2b): the column-effects.

Example 29.4   Two-way cross-classification with disproportional frequencies and frequency weights
(29.72) shows at once that H_3 and H_4 of (29.54) and (29.55) can each be tested in the manner of Example 29.2, although the SS attributable to the interactions must

now be numerically evaluated from the last term on the right of (29.72). Thus both the general mean and the interactions MS can be tested (with 1 and $(r-1)(c-1)$ d.fr. respectively) against the residual MS, since they are non-central F variables, irrespective of the row- and column-effects.*

The SS attributable to the row- and column-effects jointly is the middle term on the right of (29.72), say M. H_1 and H_2 of (29.52) and (29.53) could therefore be tested *jointly* by calculating M. However, practical interest usually lies in testing row- and column-effects separately. The SS attributable to row-effects, for example, would be obtained by calculating the reduction in the residual SS brought about by first estimating all parameters except θ_{i*} and then estimating all parameters including θ_{i*}. Similarly, θ_{*j} can have its SS evaluated. These two SS will *not* add to M, since row- and column-effects are not orthogonal in general.

If it can be assumed that all interactions are zero, the situation simplifies (cf. Exercise 29.4).

29.30 Use of the equal-weights system instead of (29.33) makes the testing of row- and of column-effects computationally a good deal simpler than with the frequency weights used in Example 29.4. We may proceed directly as follows.

Suppose that we first analyse the $r \times c$ cross-classification as if it were a one-way classification with $k = rc$. We then obtain an AV with $n - rc$ d.fr. for residual, the remaining rc d.fr. being attributable to the combined effect of row- and column-classifications. Using Example 29.1, we find the AV shown in Table 29.4.

Table 29.4   AV for any $r \times c$ cross-classification

Variation	D.fr.	SS
Due to classification as a whole	rc	$\sum_i \sum_j n_{ij} \bar{y}_{ij.}^2$
Residual	$n - rc$	$\sum_i \sum_j \sum_p (y_{ijp} - \bar{y}_{ij.})^2$
Total	n	$\sum_i \sum_j \sum_p y_{ijp}^2$

It is clear that any cell for which $n_{ij} = 1$ can contribute nothing to the residual MS (cf. **29.25** and Example 29.3 where *every* cell has $n_{ij} = 1$). To split the rc d.fr. for the classification into its component parts due to row-effects, column-effects, interactions and the general mean, we need only analyse the cell means $\bar{y}_{ij.}$.

29.31 From the model (29.32), it follows that the $\bar{y}_{ij.}$ satisfy

$$\bar{y}_{ij.} = \theta_{**} + \theta_{i*} + \theta_{*j} + \theta_{ij} + \varepsilon_{ij}, \tag{29.73}$$

---

* It should be particularly noted that if equal weights were used instead of frequency weights in (29.33), the SS attributable to interactions would not be a separate component in (29.72) but would be entangled with that for row- and column-effects just as the latter are entangled with each other.

where the errors ε_{ij} are uncorrelated, with zero means and variances σ^2/n_{ij}. Suppose now that we average the $\bar{y}_{ij.}$ over columns, that is take the *unweighted* mean $(1/c)\sum_{j=1}^{c} \bar{y}_{ij.} = \bar{y}_i$, say. It then follows from (29.73) that

$$\bar{y}_i = \theta_{**} + \theta_{i*} + \frac{1}{c}\sum_j \theta_{*j} + \frac{1}{c}\sum \theta_{ij} + \varepsilon_i, \qquad (29.74)$$

where the ε_i, uncorrelated with zero mean, have variances $(\sigma^2/c^2)\sum_j n_{ij}^{-1} = \sigma^2 V_i$, say. If we use equal weights instead of the frequency weights in (29.33), the two summations on the right of (29.74) are each equal to zero and we have simply

$$\bar{y}_i = \theta_{**} + \theta_{i*} + \varepsilon_i. \qquad (29.75)$$

(29.75) is the one-way classification model (Example 29.1) with a single observation in each group, except that the error variances are not equal. If we define $z_i = \bar{y}_i/V_i^{1/2}$, we have $E(z_i) = (\theta_{**} + \theta_{i*})/V_i^{1/2}$, and the conditions of Example 29.1 are otherwise satisfied. The effect of this on the analysis is to replace S_2 defined at (29.21), which in our present application would be

$$\sum_i \left(\bar{y}_i - \frac{1}{r}\sum_i \bar{y}_i \right)^2,$$

by the same sum with each term given the coefficient V_i^{-1}, i.e.

$$S_2 = \sum_{i=1}^{r} \left(\sum_{j=1}^{c} n_{ij}^{-1}/c^2 \right)^{-1} (\bar{y}_i - \bar{y})^2$$

where

$$\bar{y} = \sum_{i=1}^{r} \left(\sum_{j=1}^{c} n_{ij}^{-1}/c^2 \right)^{-1} \bar{y}_i \Big/ \sum_{i=1}^{r} \left(\sum_{j=1}^{c} n_{ij}^{-1}/c^2 \right)^{-1}$$

is the weighted mean of the $\bar{y}_i$, using V_i^{-1} as weights. We therefore have an AV as shown in Table 29.5.

Table 29.5   AV for $r \times c$ cross-classification using equal weights

Variation	D.fr.	SS
Due to rows	$r-1$	$\sum_{i=1}^{r} V_i^{-1}(\bar{y}_i - \bar{y})^2$
Due to remainder of classification	$r(c-1)+1$	[Obtainable as a difference]
Residual	$n-rc$	As in Table 29.4
Total	n	$\sum_i \sum_j \sum_p y_{ijp}^2$

An exactly analogous breakdown of the 'classification' SS can be made for the columns classification. We therefore have tests of row- and of column-effects in the general case. 'Rows' and 'columns' SS cannot be added, because of their non-orthogonality, so that we cannot obtain the interactions SS by differencing. However, if r or $c = 2$, a test for interactions is easily derived by this method—cf. Exercise 29.5.

29.32 The equal-weights system, whose use permitted the development of the AV (Table 29.5), is used naturally in this context, since in effect we reduce the n observations to a set of rc means and then analyse these as though they were individual observations. There is nothing, of course, to prevent a full analysis using this equal-weights system, instead of that in **29.29**, and indeed this is the customary procedure. If this is done, the results in general are different from those of **29.29**.

29.33 The method used in **29.31** is due to Yates (1934a), who called it the *method of weighted squares of means*. He also discusses other, more approximate, methods of analysis, as also do Snedecor (1946) and R. L. Anderson and Bancroft (1952). All these authors give numerical examples (cf. Exercise 29.7). Scheffé (1959) gives further theoretical details of the disproportional-frequencies analysis; in particular he allows arbitrary weight systems in (29.33). F. M. Speed (1978) set out the hypotheses tested by each of the many methods that they discuss. See also Aitkin (1978), and Searle *et al.* (1981).

Empty cells in the two-way cross-classification
29.34 Throughout our analysis of the two-way cross-classification in **29.15–33**, we made the implicit assumption that every cell in the table (29.28) contained at least one observation, i.e. $n_{ij} > 0$. In practice, it quite frequently occurs that this assumption is not fulfilled, as a result of accident, experimental failure, or other causes. We must now consider the effects that the presence of empty cells in the classification will have on the analysis of the observations. If there is at least one empty cell, the cross-classification is called *incomplete*. We have so far discussed only *complete* cross-classifications.

Incomplete classifications are an extreme form of disproportional frequencies. The analysis of **29.15–20** remains valid if some of the n_{ij} are zero, provided that no n_{rj} or $n_{ic} = 0$, i.e. we must have one row and one column in the classification without empty cells, to use as the eliminated row and column. Then the final paragraph of **29.29** will hold, and in particular the estimators of the interactions will be orthogonal to the estimators of the other parameters.

However, we clearly cannot estimate the mean μ_{ij} of an empty cell in general where the corresponding interaction θ_{ij} is non-zero, for we can get no information on θ_{ij} from other cells in the table. It follows from the definitions (29.31) and the linear relations (29.33) that none of the θ_{ij}, θ_{i*}, θ_{*j} or θ_{**} can be estimated in the general case if there are one or more empty cells in the cross-classification. But even in this case we can estimate the error variance quite easily. If we denote the number of cells containing observations by $[rc]$, we obtain the more general form (Table 29.6) of Table 29.4.

Table 29.6   AV for any $r \times c$ cross-classification*

Variation	D.fr.	SS
Due to classification	$[rc]$	$\sum_i \sum_j n_{ij} \bar{y}_{ij.}^2$
Residual	$n - [rc]$	$\sum_i \sum_j \sum_p (y_{ijp} - \bar{y}_{ij.})^2$
Total	n	$\sum_i \sum_j \sum_p y_{ijp}^2$

29.35 If the θ_{ij} in empty cells are zero, the difficulty in **29.34** disappears. Thus, if we wish to test H_3 of (29.54) (the hypothesis that *all* interactions are zero) we may proceed, as in Example 29.4, to estimate the remaining parameters, evaluate the SS due to them, and thus obtain a residual SS. The difference between the latter and the residual SS in Table 29.6 will be attributable to interactions and have $[rc] - r - c + 1$ d.fr. Similarly, if H_3 can be *postulated*, row- or column-effects can be tested as for a complete classification by the method given in Exercise 29.4. Scheffé (1959) gives further details.

Hierarchical classifications

29.36 The two-way cross-classification which we have treated at length in **29.15–35** is not the only interesting generalisation of the one-way classification in Example 29.1. Suppose that, within each of the k groups there considered, there is a further one-way classification of the observations. The n_1 observations in the first group are in l_1 subgroups, with frequencies $n_{11}, n_{12}, \ldots, n_{1l_1}$ where $\sum_{h=1}^{l_1} n_{1h} = n_1$; the second group similarly has l_2 subgroups, with frequencies $n_{21}, n_{22}, \ldots, n_{2l_2}$, $\sum_{h=1}^{l_2} n_{2h} = n_2$; and so on until in the kth group there are l_k subgroups with frequencies $n_{k1}, n_{k2}, \ldots, n_{kl_k}$, $\sum_{h=1}^{l_k} n_{kh} = n_k$. It will accord better with our notational conventions if we now replace the original group frequencies n_i of Example 29.1 by $n_{i.}$, to denote summation of the subgroup frequencies n_{ih} within the original groups. Thus we have

$$\sum_{h=1}^{l_i} n_{ih} = n_{i.}.$$

This is a two-way *hierarchical classification*† of the observations, the *separate* sub-grouping within each of the original groups contrasting with the common row-grouping of every column category in a two-way cross-classification.

---

* In Table 29.6 summations range over the $[rc]$ occupied cells only.

---

† The alternative term 'nested classification' appears to be more easily taken to imply that there is an equal number of subgroups in each original group, and we therefore do not use it, despite its appealing cosiness.

Example 29.5   AV in a two-way hierarchical classification

In Example 29.1 we have already defined k parameters, one for each group. In order to investigate variation in the means θ_{ih} of the l_i subgroups within the ith group, we use only $l_i - 1$ linearly independent parameters, for we may put

$$\sum_{h=1}^{l_i} n_{ih}\theta_{ih} = 0$$

(cf. **29.17–19** for the cross-classification) so that,* as at (29.34),

$$\theta_{il_i} = -\frac{1}{n_{il_i}}\sum_{h=1}^{l_i-1} n_{ih}\theta_{ih}. \tag{29.76}$$

We may now generalise the linear model in Example 29.1. We write $l = \sum_{i=1}^{k} l_i$, and y_{ihp} for the pth observation in the hth subgroup of the ith group. We have

$$\underset{(n\times1)}{\mathbf{y}} = \begin{bmatrix} y_{111} \\ \vdots \\ y_{11n_{11}} \\ y_{121} \\ \vdots \\ y_{12n_{12}} \\ \vdots \\ y_{1l_11} \\ \vdots \\ y_{1l_1n_{1l_1}} \\ y_{211} \\ \vdots \\ \vdots \\ y_{kl_kn_{kl_k}} \end{bmatrix} \qquad \underset{(l\times1)}{\boldsymbol{\theta}} = \begin{bmatrix} \theta_1 \\ \theta_2 \\ \vdots \\ \theta_k \\ \theta_{11} \\ \vdots \\ \theta_{1,l_1-1} \\ \theta_{21} \\ \vdots \\ \theta_{2,l_2-1} \\ \vdots \\ \theta_{k1} \\ \vdots \\ \theta_{k,l_k-1} \end{bmatrix}\text{,}$$

and

$$\underset{(n\times l)}{\mathbf{X}} = \left(\underset{(n\times k)}{\mathbf{X}_0} \middle| \begin{array}{cccc} \underset{(n_1.\times(l_1-1))}{\mathbf{X}_1} & & & \mathbf{0} \\ & \underset{(n_2.\times(l_2-1))}{\mathbf{X}_2} & & \\ & & \ddots & \\ \mathbf{0} & & & \underset{(n_k.\times(l_k-1))}{\mathbf{X}_k} \end{array} \right). \tag{29.77}$$

* These θ_{ih} are not related to the interaction parameters in the cross-classification. Interaction problems do not arise here.

The $(n \times k)$ $\mathbf{X}_0$ submatrix in (29.77) is that used in Example 29.1. Each of the other submatrices $\mathbf{X}_i$ is of the form

$$
\mathbf{X}_{(n_i \times (l_i-1))} =
\left[
\begin{array}{cccccc}
1 & & & & & \\
\vdots & & & 0 & & \\
1 & & & & & \\
0 & 1 & & & & \\
\vdots & \vdots & & 0 & & \\
0 & 1 & & & & \\
0 & 0 & 1 & & & \\
\vdots & \vdots & \vdots & 0 & & \\
0 & 0 & 1 & & & \\
 & & & \ddots & & \\
 & 0 & & & 1 & \\
 & & & & \vdots & \\
 & & & & 1 & \\
-\dfrac{n_{i1}}{n_{il_i}}\mathbf{1} & -\dfrac{n_{i2}}{n_{il_i}}\mathbf{1} & \cdots & & -\dfrac{n_{i,l_i-1}}{n_{il_i}}\mathbf{1}
\end{array}
\right]
\begin{array}{l}
\left.\rule{0pt}{22pt}\right\} n_{i1} \text{ rows} \\
\left.\rule{0pt}{22pt}\right\} n_{i2} \text{ rows} \\
\left.\rule{0pt}{22pt}\right\} n_{i3} \text{ rows} \\
\\
\left.\rule{0pt}{22pt}\right\} n_{i,l_i-1} \text{ rows} \\
\left.\rule{0pt}{8pt}\right\} n_{il_i} \text{ rows}
\end{array}
\tag{29.78}
$$

which follows at once from (29.76). (29.77) and (29.78) now give

$$
\mathbf{X}'\mathbf{X}_{(l \times 1)} =
\left[
\begin{array}{cc}
\begin{pmatrix} n_{1.} & & & 0 \\ & n_{2.} & & \\ & & \ddots & \\ 0 & & & n_{k.} \end{pmatrix} & 0 \\
 & \begin{matrix} \mathbf{A}_1 & & & \\ & \mathbf{A}_2 & & \\ 0 & & \ddots & \\ & & & \mathbf{A}_k \end{matrix}
\end{array}
\right],
\tag{29.79}
$$

where

$$
\mathbf{A}_i _{((l_i-1) \times (l_i-1))} =
\begin{pmatrix}
n_{i1} + \dfrac{n_{i1}^2}{n_{il_i}} & \dfrac{n_{i1}n_{i2}}{n_{il_i}} & \cdots & \dfrac{n_{i1}n_{i,l_i-1}}{n_{il_i}} \\
 & \ddots & & \\
 & & & n_{i,l_i-1} + \dfrac{n_{i,l_i-1}^2}{n_{il_i}}
\end{pmatrix}
$$

is of the same form as (29.37) and therefore has the inverse, of the same form as (29.44),

$$
\mathbf{A}_i^{-1} =
\begin{bmatrix}
\dfrac{1}{n_{i1}} - \dfrac{1}{n_{i.}} & -\dfrac{1}{n_{i.}} & -\dfrac{1}{n_{i.}} & \cdots & -\dfrac{1}{n_{i.}} \\
& \dfrac{1}{n_{i2}} - \dfrac{1}{n_{i.}} & -\dfrac{1}{n_{i.}} & \cdots & -\dfrac{1}{n_{i.}} \\
& & \ddots & & \vdots \\
& & & & \dfrac{1}{n_{i,l_i-1}} - \dfrac{1}{n_{i.}}
\end{bmatrix}.
\tag{29.80}
$$

Hence, from (29.79) and (29.80), the LS estimators are

$$
\hat{\boldsymbol{\theta}} = (\mathbf{X}'\mathbf{X})^{-1}\mathbf{X}'\mathbf{y} =
\begin{bmatrix}
\bar{y}_{1..} \\
\bar{y}_{2..} \\
\vdots \\
\bar{y}_{k..} \\
\bar{y}_{11.} - \bar{y}_{1..} \\
\bar{y}_{12.} - \bar{y}_{1..} \\
\vdots \\
\bar{y}_{1,l_1-1.} - \bar{y}_{1..} \\
\bar{y}_{21.} - \bar{y}_{2..} \\
\vdots \\
\bar{y}_{k,l_k-1.} - \bar{y}_{k..}
\end{bmatrix},
\tag{29.81}
$$

and thus the SS due to the fitted model is

$$
\hat{\boldsymbol{\theta}}'\mathbf{X}'\mathbf{X}\hat{\boldsymbol{\theta}} = \sum_{i=1}^{k} n_{i.}\,\bar{y}_{i..}^2 + \sum_{i=1}^{k}\sum_{h=1}^{l_i} n_{ih}(\bar{y}_{ih.} - \bar{y}_{i..})^2.
\tag{29.82}
$$

The first term on the right of (29.82) is precisely S_1 defined at (29.16) in Example 29.1. What we have now done is to partition off a further SS, the second term on the right of (29.82). Since S_1 was the SS due to the k original group parameters $\theta_1, \theta_2, \ldots, \theta_k$, the second SS is that attributable to the $\sum_{i=1}^{k}(l_i-1) = l-k$ linearly independent subgroup parameters now introduced. We may summarise our result in tabular form, as in Table 29.7.

The residual SS in Table 29.7 is obtained by subtraction. The first row of the table may be split into 'between groups' and 'general mean' components as in Example 29.1.

The ratio of the 'between subgroups within groups' MS to the residual MS is, from our general theory, a non-central F-variable with d.fr. $(l-k, n-l)$ and non-central parameter $\lambda = \sum_i \sum_h n_{ih}\theta_{ih}^2/\sigma^2$, which is zero when all subgroup means within each group are equal, giving a central F-test for this hypothesis.

Table 29.7   AV for a two-way hierarchical classification

Variation	D.fr.	SS
Due to groups	k	$\sum_i n_{i.} \bar{y}_{i..}^2$
Between subgroups within groups	$l - k$	$\sum_i \sum_h n_{ih}(\bar{y}_{ih.} - \bar{y}_{i..})^2$
Residual	$n - l$	$\sum_i \sum_h \sum_p (y_{ihp} - \bar{y}_{ih.})^2$
Total	n	$\mathbf{y'y}$

29.37   The hierarchical process can clearly be carried further, with subsubgroups and even subsubsubgroups. These would be termed three-way and four-way hierarchical classifications, and are relatively rare in practice. It should be obvious to the reader that there is no need to go again through the rather tedious algebra of LS theory to obtain the results we need here; the work of Example 29.5 essentially split the SS within each of the k groups into two components, with $(l_i - 1)$ and $(n_i - l_i)$ d.fr. respectively, and summed corresponding components over all groups to obtain the $(l - k)$ and $(n - l)$ d.fr. in Table 29.7. The same splitting-off process can now be carried out within each subgroup, and so on. The reader is asked to verify the three-way AV in Exercise 29.3. Scheffé (1959) gives theoretical details for the three-way case.

Multi-way classifications

29.38   We have just outlined the treatment of multi-way hierarchical classifications, and this leads us to a consideration of multi-way classifications in general.

We first note that, as soon as we consider three-way classifications, there is the possibility of 'mixed' classifications which are partly hierarchical and partly cross-classifications. These arise when a two-way hierarchical classification forms one (say, the row-) classification of an $r \times c$ cross-classification. In the notation of Example 29.5, we here have $r = l$. The AV is carried out in two stages. First, the cross-classification is analysed by the methods already discussed, and the total SS is resolved into the usual five components, which we represent concisely by their d.fr. in Table 29.8.

At the second stage, each of the SS involving the hierarchical (row) classifications is subdivided into two parts as indicated on the right. The first of these subdivisions is a direct application of Example 29.5 (it being remembered that the general mean component has here already been removed from the first line of Table 29.7 by the cross-classification analysis), but the simplest way of achieving both subdivisions is to merge all subgroups within the groups of the hierarchical classification and recalculate the SS for rows and interactions using the merged data—these are the required component SS, with $(k - 1)$ and $(k - 1)(c - 1)$ d.fr. respectively. The subgroups SS are then obtained as differences if the analysis is orthogonal.

Scheffé (1959) gives theoretical details for the case where there is the same number of subgroups in each group of the hierarchical classification and the same number of observations in each cell of the $l \times c$ table.

Table 29.8

SS due to		D.fr.	
General mean	1		
Rows	$l-1$	$\begin{cases}\text{Row groups} \\ \text{Row subgroups}\end{cases}$	$\begin{array}{c}k-1 \\ l-k\end{array}$
Columns	$c-1$		
Interactions	$(l-1)(c-1)$	$\begin{cases}\text{Interactions with groups} \\ \text{Interactions with subgroups}\end{cases}$	$\begin{array}{c}(k-1)(c-1) \\ (l-k)(c-1)\end{array}$
Residual	$n-lc$		
Total	n		

The three-way cross-classification

29.39   Suppose now that, instead of embedding a two-way hierarchical classification within a two-way cross-classification as in **29.38**, we carry out a new one-way classification within each cell of a cross-classification. If the *same* one-way classification is carried out in each cell, we clearly arrive at a three-way cross-classification. All the problems of formulating the linear model, discussed in **29.15–19** for the two-way case, now arise afresh, and some generalisation of our concepts is required. We now consider a sample of n observations classified into an $r \times c \times l$ table, with l 'layers' cutting across the previous $r \times c$ table (29.28).

We may clearly ask the questions of **29.15** about all three classifications in this more general situation, so that we are now interested in the general mean, in row-effects, column-effects and layer-effects (the three sets of *main effects*), and in the interactions between any pair of the row-, column-, and layer-classifications. However, there is a new feature introduced by the additional classification, for we may now wish to know whether the interactions between, say, rows and columns themselves depend upon the layer-classification. We are here concerned with a *second-order interaction*, our original interactions now being defined as of first order.

Nothing but the heavy algebra now prevents our following through in detail an analysis parallel to that already carried out in the two-way case. There is no numerical difficulty in any given case about fitting the linear model, estimating its rcl parameters and carrying out the AV. Even in the two-way case, however, worthwhile simplifications in the algebra only occurred when the frequency in each cell was proportional to the product of marginal totals as required by (29.43). Similar orthogonality conditions now require that each cell frequency should be proportional to the product of all the corresponding marginal frequencies (cf. Seber (1964)). This proportionality condition, in practice, is satisfied with equal frequencies in all the cells, i.e. in the balanced case.

The general principles of the analysis are then very simply set out. We have seen that we may regard a three-way ($r \times c \times l$) cross-classification as having been generated by imposing a new (l-fold) classification upon every cell of an existing ($r \times c$) cross-classification. It follows that the AV can be carried out in two stages, exactly as for the 'mixed' classification of **29.38**. First, we consider the $(r \times c)$ cross-classification as a one-way classification with rc cells, and carry out the AV of the $rc \times l$ two-way cross-classification in which our observations are then displayed. We obtain the schematic AV table (Table 29.9) from Table 29.2.

Table 29.9

SS due to		D.fr.	
General mean	1		
$(r \times c)$ cross-classification	$rc - 1$	rows	$r - 1$
		columns	$c - 1$
		row–column interactions	$(r-1)(c-1)$
Layer classification	$l - 1$		
Interactions of $(r \times c)$ cross-classification with l layers	$(rc-1)(l-1)$	row–layer interactions	$(r-1)(l-1)$
		column–layer interactions	$(c-1)(l-1)$
		row–column–layer second-order interactions	$(r-1)(c-1)(l-1)$
Residual	$n - rcl$		

At the second stage, each of the two SS involving the $(r \times c)$ cross-classification is subdivided into three parts as shown on the right. The simplest method (as in **29.38**) of doing this is first to merge all columns within rows and recalculate the two SS, which then are the required components with $(r-1)$ and $(r-1)(l-1)$ d.fr.; if the merging operation is now separately applied to all rows within columns, and the two SS again recalculated, we obtain the required components with $(c-1)$ and $(c-1) \times (l-1)$ d.fr. The two remaining SS (with $(r-1)(c-1)$ and $(r-1)(c-1)(l-1)$ d.fr.) are now obtainable as differences since the analysis is orthogonal.

29.40 The computation of the AV in the general disproportional-frequencies case remains formidable, even with electronic computers. Pearce (1963) reviews the situation generally. Freeman and Jeffers (1962) give a method for the non-orthogonal three-way cross-classification; Stevens (1953) used an iterative method for this case. Bradu (1965) solves the problem for the case where all interactions of all orders are assumed zero—see also Rees (1966)—and the case where all classifications but one are mutually orthogonal is treated by Freeman (1972). A general recursive AV procedure is given by Wilkinson (1970). Hemmerle (1974) gives an iterative method based on residuals. Golub and Nash (1982) give an efficient gradient algorithm. Jamshidian and Jennrich (1988) give another, which is more efficient for hierarchical models. See also Pearce *et al.* (1974) and Freeman (1975). Paik and Federer (1974)— see also their references to earlier work—discuss non-orthogonal analysis in terms of a general calculus.

Gabriel (1963) gives an expository review of the theory for analysing cell means, whose variances are inversely proportional to cell sample sizes, with special reference to the case when y is a 0–1 variable, and the cell means become proportions. An approximate method of analysing cell means is given in Exercise 29.10.

Missing observations in the linear model

29.41 The advantages of balanced arrangements in AV, namely orthogonality, ease of computation and superior robustness, are such that most designed analyses will seek to take advantage of them. Nevertheless, force of circumstances will sometimes lead to involuntary departures from the intended equality of frequencies: plants or animals may die, human subjects may prove reluctant to co-operate, or records may be lost before analysis. If this happens, we are always free to analyse the achieved unequal frequencies by the appropriate non-orthogonal methods, but, as we have seen, these are often complicated. Moreover, accidental losses of observations are rarely extreme; usually only one or a few are found to be missing. It is therefore worth investigating whether we can retain the original AV structure and correct it for the missing observations, rather than abandon it altogether. The discussion that follows holds for the linear model, of which AV situations are a special case.

29.42 Suppose, then, that m of the n intended observations are missing. Without loss of generality, we take these to be the last m components of the observation vector $\mathbf{y}$, which now become unknowns, say $u_1, \ldots, u_m$. Thus we may write

$$\mathbf{y} = \begin{pmatrix} \mathbf{z} \\ \mathbf{u} \end{pmatrix},$$

where $\mathbf{z}$ $((n-m) \times 1)$ contains the actually observed values of y and $\mathbf{u}$ $(m \times 1)$ the unknown observations. In effect, we are presented with a fresh set of unknowns to estimate, in addition to the original parameters of the model. It is natural, in these circumstances, to estimate the values in $\mathbf{u}$ by the same LS method as we use for the original parameters.

The sum of squared residuals

$$S = (\mathbf{y} - \mathbf{X}\boldsymbol{\theta})'(\mathbf{y} - \mathbf{X}\boldsymbol{\theta})$$

must therefore now be minimised not only for variation in $\boldsymbol{\theta}$ (as was done in **19.4**) but also for variation in $\mathbf{u}$. If we first minimise S with respect to $\boldsymbol{\theta}$, we shall, of course, obtain the original LS solution (i.e. the LS solution if there had been no missing observation), but the estimator and the residual SS of that solution will now both be functions of $\mathbf{u}$, say $\hat{\boldsymbol{\theta}}(\mathbf{u})$ and $S_0(\mathbf{u})$. The minimisation process could now be completed by minimising $S_0(\mathbf{u})$ for variation in $\mathbf{u}$.

However, this two-stage minimisation procedure, which was suggested by Yates (1933), is not the easiest way in general. Instead, let us minimise S first for variation in $\mathbf{u}$. Partitioning $\mathbf{X}$ into

$$\begin{pmatrix} \mathbf{X}_z \\ \mathbf{X}_u \end{pmatrix}$$

conformably with the partition

$$y = \begin{pmatrix} z \\ u \end{pmatrix},$$

we have

$$S = (z - X_z\theta)'(z - X_z\theta) + (u - X_u\theta)'(u - X_u\theta). \tag{29.83}$$

Since only the second of the two non-negative terms on the right of (29.83) depends upon u, we reduce it to zero by putting

$$u = X_u\theta; \tag{29.84}$$

thus S at (29.83) is reduced to its first term, which may then be minimised with respect to θ.

But the results of the two two-stage minimisation methods just described must be the same. Thus if we obtain $\hat{\theta}(u)$ by the first method, which gives

$$\hat{\theta}(u) = (X'X)^{-1}X'y = (X'X)^{-1}(X_z'z + X_u'u), \tag{29.85}$$

and use this in conjunction with (29.84), we have

$$u = X_u\hat{\theta}(u). \tag{29.86}$$

(29.86) states that each missing observation is to be equated to its estimated expectation in the original LS analysis.

29.43   (29.86) is a set of linear equations to be solved for u, and the solution $\hat{u}$ is then to be used in the original LS analysis. A straightforward solution was given by Tocher (1952).

First, suppose that we replace u by the null vector 0 in the original LS analysis. (29.85) becomes

$$\hat{\theta}(0) = (X'X)^{-1}X_z'z. \tag{29.87}$$

Now consider

$$\hat{u} = \{I - X_u(X'X)^{-1}X_u'\}^{-1}X_u\hat{\theta}(0). \tag{29.88}$$

Observing from (29.85) that $\hat{\theta}(u) = \hat{\theta}(0) + (X'X)^{-1}X_u'u$, we find that (29.88) reduces to

$$\{I - X_u(X'X)^{-1}X_u'\}\hat{u} = X_u\hat{\theta}(\hat{u}) - X_u(X'X)^{-1}X_u'\hat{u},$$

so that

$$\hat{u} = X_u\hat{\theta}(\hat{u}). \tag{29.89}$$

Thus $\hat{u}$ defined by (29.88) is the solution of (29.86).

29.44   We have seen, therefore, that in order to estimate the m missing observations u, so that we may preserve the computational form of the original LS analysis, we need only

(a) perform the original analysis with $\mathbf{u} = \mathbf{0}$ to obtain $\hat{\boldsymbol{\theta}}(\mathbf{0})$ at (29.87);
(b) calculate $\hat{\mathbf{u}}$ at (29.88); and
(c) again perform the original analysis using $\hat{\mathbf{u}}$ in $\mathbf{y}$.

It should be noted that the matrix in braces to be inverted in (29.88) is $(m \times m)$. Thus, if only one observation is missing, the matrix is a scalar and stage (b) above is very simple.

> The second of the four papers by Wilkinson (1957-60) on missing observations gives detailed solutions of (29.86) for many common AV situations. See also Biggers (1959) and Pearce and Jeffers (1971).

29.45 It is easy to see that the estimator $\hat{\boldsymbol{\theta}}(\hat{\mathbf{u}})$ obtained by using (29.88) in the original LS analysis is exactly the estimator that would have been obtained by using the $(n - m)$ observed values alone (generally in a non-orthogonal analysis). For, from (29.85),

$$\mathbf{X}'\mathbf{X}\hat{\boldsymbol{\theta}}(\hat{\mathbf{u}}) = \mathbf{X}'_z\mathbf{z} + \mathbf{X}'_u\hat{\mathbf{u}} = \mathbf{X}'_z\mathbf{z} + \mathbf{X}'_u\mathbf{X}_u\hat{\boldsymbol{\theta}}(\hat{\mathbf{u}})$$

using (29.89). Thus

$$\mathbf{X}'_z\mathbf{z} = (\mathbf{X}'\mathbf{X} - \mathbf{X}'_u\mathbf{X}_u)\hat{\boldsymbol{\theta}}(\hat{\mathbf{u}})$$
$$= \mathbf{X}'_z\mathbf{X}_z\hat{\boldsymbol{\theta}}(\hat{\mathbf{u}}) \qquad\qquad (29.90)$$

and (29.90) is precisely the set of equations satisfied by $\hat{\boldsymbol{\theta}}$ when $\mathbf{z}$ alone is analysed.

This result, together with the disappearance of the second term on the right of (29.83), implies at once that the residual SS obtained by using $\hat{\mathbf{u}}$ in the original LS analysis is identical with that obtained when $\mathbf{z}$ alone is analysed. However, the degrees of freedom for the residual SS must obviously be reduced, since we now have only $(n - m)$ observations. If $\mathbf{X}$ and $\mathbf{X}_z$ both have the same rank (e.g. when both have full rank) the residual SS will have its d.fr. reduced by m, the number of missing observations. More generally, the reduction will be (cf. Exercise 19.8) m minus the difference in rank between $\mathbf{X}$ and $\mathbf{X}_z$.

29.46 Although the residual SS requires no adjustment, all the other SS in the AV table will be incorrect if $\hat{\mathbf{u}}$ at (29.88) is used in the original LS analysis. This is most easily seen from the fact (cf. Example 29.4 and **29.38**) that each other SS in the AV table may be obtained as the difference between the residual SS in two linear models, one of which is a restricted form of the other. Evidently, any of these residual SS is correctly obtainable, by the argument of **29.41–45**, by using $\hat{\mathbf{u}}$ at (29.88) for that model, but of course $\hat{\mathbf{u}}$ will in general differ from that in the full model considered so far. Thus each of these residual SS will be too large if $\hat{\mathbf{u}}$ for the full model is used, since it will not be the correct (minimising) $\hat{\mathbf{u}}$. Hence the other SS in the AV table need correction by the difference between the subtractive corrections to the corresponding residual SS (or by a single subtractive correction if one of the latter is the residual SS for the full model). Correspondingly, degrees of freedom must be corrected by the difference between two adjustments of the form discussed in **29.45**, but this difference will often be zero.

In the third of his four papers, Wilkinson (1957–60) gives an explicit method of obtaining the subtractive corrections to the other residual SS, and hence the other SS in the AV table. Fortunately, as Yates (1933) pointed out, the latter corrections, being generally differences of quantities of the same sign, are often small, and the unadjusted SS may be used as approximations.

> D. B. Rubin (1976) gives a non-iterative method of finding the LS estimators that has the advantage of also producing all the correct SS in the AV table.
>
> Afifi and Elashoff (1966) extend the analysis to the case where elements of **X**, as well as of **y**, are missing—cf. Exercise 29.19. See also Beale and Little (1975).
>
> The EM algorithm (**18.40–42**) of Dempster *et al.* (1977) is widely used for ML estimation from incomplete data. Hemmerle (1979) extends his iterative method (cf. **29.40**) to the missing data case—see also Golub and Nash (1982).
>
> Tocher (1952) gives similar methods of analysis for other types of 'spoilt' experiments, namely those in which some observations are irretrievably mixed up and those in which some observations are unwittingly duplicated—Plackett (1950) also discusses the latter situation. Hoyle (1971) reviews the analysis of spoilt data, and gives a large bibliography.

Multiple comparisons

29.47   Each of the variance-ratio tests in an AV table tests a hypothesis concerning a *set* of parameters, e.g. the row-effects or the interactions between rows and columns. For practical purposes, however, it is often not enough to know, for example, that the row-effects θ_{i*} are different—we need to know which of the θ_{i*} are to be regarded as greater than the others, or more generally whether the θ_{i*} may be said to fall into distinct groups.

Now the LS estimators $\hat{\theta}_{i*}$ are, of course, the MV unbiased linear estimators of their corresponding parameters, and provide us with estimators of any of the differences $\theta_{i*} - \theta_{s*}$, but we are usually unable to nominate the differences of interest in advance, and we therefore are faced with the problem of carrying out a number of non-independent tests on the differences.

In a sense, this problem of *multiple comparisons*, as it is called, is a complex version of the problem of outlying observations, where instead of being concerned about a location-shift in one or more observations, we are now more generally asking whether the observations (here the $\hat{\theta}_{i*}$) have expected values (the θ_{i*}) that fall into distinct groups. Tukey (1953) reviews the subject.

The LSD test

29.48   For the sake of definiteness, our discussion will refer to a one-way classification as in Example 29.1, although there is no essential difference if we consider any set of effects in an AV table. In Example 29.1, the observed group means $\bar{y}_{i.}$ are the LS estimators of the k group parameters θ_i (each of which includes the general mean as a common element). If the F-test at (29.23) rejects the hypothesis (29.20) that the θ_i are all equal, we are faced with the need to decide which subsets of the θ_i may be regarded as homogeneous, and which not.

The simplest test procedure is the oldest (Student, 1908), namely to carry out an ordinary two-sample Student's t-test on every one of the $\frac{1}{2}k(k-1)$ possible pairs of y_i, y_j, $i \neq j$. If each of these tests has size α, we can say little about the size of the overall combined test, since they are non-independent.

This combined test amounts to calculating an estimated standard error of the difference between two means (using the residual SS) and comparing the observed difference with it, using the appropriate (residual SS) number of d.fr. in Student's distribution. If the number of observations in each group (n_i) is the same (equal to N, say) only a single standard error estimate is needed, of form $(2s^2/N)^{1/2}$, where s^2 is the unbiased estimator of the error variance σ^2. We thus set up a *least significant difference* (the appropriate multiple of the standard error for a two-sided test of size α) and compare each of the $\frac{1}{2}k(k-1)$ observed differences with it. In consequence, this is sometimes called the *LSD test*. One cannot say much more in general about the LSD test than that if the true group means do not differ, a proportion α of all pairs adjudged heterogeneous by the test will be wrongly so judged.

A simple modification of the LSD test, proposed by Fisher (1966), is to reduce the size of each component test from α to $\alpha/\binom{k}{2}$. This has the effect of reducing the expected number of pairs erroneously adjudged heterogeneous (when all group means are truly equal) from $\frac{1}{2}k(k-1)\alpha$ to simply α.

When using either the original or modified LSD test, it must be remembered that the expected error rates just referred to are unconditional ones, taking no account of the fact that the test is made after (and because) the overall F-test rejected the hypothesis of homogeneity.

Step-by-step and simultaneous test procedures

29.49   Like the outlier problem that it resembles, the multiple comparisons problem is often discussed in terms of sample range criteria.

The simplest of these is Tukey's (1951, 1952) studentised range test. The k group means, which we shall now write $\bar{x}_i$, $i = 1, 2, \ldots, k$, are (on the hypothesis of homogeneity) a random sample of size k from a normal distribution with variance σ^2/N, independently estimated by s^2/N, where N is the number of observations in each group as before. Suppose the group means to be ordered, and denote them by $\bar{x}_{(1)} \leqslant \bar{x}_{(2)} \cdots \leqslant \bar{x}_{(k-1)} \leqslant \bar{x}_{(k)}$. To any pair $\bar{x}_{(i)}$ and $\bar{x}_{(j)}$, $i < j$, there corresponds a difference $(\bar{x}_{(j)} - \bar{x}_{(i)})$ which is the range of a subset of $(j - i + 1)$ adjacent ordered group means. This subset is adjudged heterogeneous (and the extreme group means $\bar{x}_{(i)}$ and $\bar{x}_{(j)}$ therefore regarded as from different populations) if $(\bar{x}_{(j)} - \bar{x}_{(i)})/(s/N^{1/2})$ exceeds the $100(1-\alpha)$ per cent point, q_α, of the studentised range (i.e. the range divided by the estimated standard error) of k observations. Since no subset range can do this unless the range $(\bar{x}_{(k)} - \bar{x}_{(1)})$ does, the procedure clearly has overall size α.

In practice, $\bar{x}_{(k)}$ is successively compared with $\bar{x}_{(1)}, \bar{x}_{(2)}$, and so on until a 'homogeneous' verdict is reached. If $(\bar{x}_{(k)} - \bar{x}_{(1)})$ is homogeneous, the test ends there; otherwise all means $\bar{x}_{(1)}, \bar{x}_{(2)}, \ldots$ found to be heterogeneous from $\bar{x}_{(k)}$ are tested against $\bar{x}_{(k-1)}$, again in succession starting from $\bar{x}_{(1)}$. All means then found to be heterogeneous from $\bar{x}_{(k-1)}$ are tested against $\bar{x}_{(k-2)}$, and so on until no further 'heterogeneous' verdict can be reached. Such procedures, in which the decision on each subset depends on previous decision concerning a larger subset, are called *step-by-step* or *stepwise* procedures.

29.50   A different *simultaneous test procedure* employing a sum-of-squares, rather than a range, technique is suggested by Gabriel (1964). The 'between-groups' SS (29.22) is calculated for *every one* of the $(2^k - k - 1)$ subsets of two or more of the k groups, and tested against the *fixed* critical value obtained from the variance-ratio distribution with $(k - 1, n - k)$ d.fr. The sample sizes n_i need not now be equal. This procedure leads to transitive judgements in the sense that no subset can be adjudged heterogeneous when a larger subset containing it is not; i.e. no subset can be adjudged homogeneous unless all smaller subsets which it contains are also homogeneous (cf. Exercise 29.16). Clearly, this procedure contains the ordinary variance-ratio test as a component, when the subset is actually the whole set of k groups. This implies that the test has overall size α.

When all the n_i are equal, Tukey's test in **29.49** can be modified to be a 'simultaneous test procedure' of the same type as Gabriel's if every subset of, instead of merely every subset of *adjacent*, group means is tested by the range criterion. (This test has the additional property that if any set of more than two groups is adjudged heterogeneous, at least one subset of it would be.)

Both the Gabriel and the Tukey–Gabriel methods discussed in this section have the property that the probability of erroneously judging a subset to be heterogeneous decreases with the size of the subset. For $k = 8$ and 40 d.fr. for the residual SS, this phenomenon is more marked for the former method—Gabriel (1964) gives tables. The Tukey–Gabriel method is much simpler computationally, especially for large k.

Simultaneous confidence intervals for differences and contrasts
29.51   The studentised range test of **29.49** leads immediately to simultaneous confidence intervals for all $\frac{1}{2}k(k-1)$ differences between true group means $(\theta_i - \theta_j)$, proposed by Tukey (1951). For whatever the θ_i may be, the random variables $\bar{x}_i - \theta_i$ are identically and independently normally distributed with zero means and variances σ^2/N, and the probability is $1 - \alpha$ that their studentised range will not exceed q_α defined in **29.49**. It follows that simultaneously for all $i \neq j$,

$$\text{Prob}\left\{\left|\frac{(\bar{x}_i - \theta_i) - (\bar{x}_j - \theta_j)}{s/N^{1/2}}\right| \leq q_\alpha\right\} = 1 - \alpha \tag{29.91}$$

so that

$$(\bar{x}_i - \bar{x}_j) - q_\alpha s/N^{1/2} \leq \theta_i - \theta_j \leq (\bar{x}_i - \bar{x}_j) + q_\alpha s/N^{1/2}, \tag{29.92}$$

is simultaneously satisfied for all $i \neq j$ with probability $1 - \alpha$.

Exercise 29.11 shows that the method extends to negatively equi-correlated multi-normal $\bar{x}_i$.

29.52   The method of **29.51** enables us to make simultaneous statements about all $\frac{1}{2}k(k-1)$ differences $\theta_i - \theta_j$ with a known overall confidence coefficient $1 - \alpha$. In many applications of AV, we are interested not only in the differences but also in other linear combinations of the θ_i with constant coefficients. If its coefficients sum to zero, a linear combination is called a *contrast*, defined by

$$\psi = \sum_{i=1}^{k} c_i \theta_i, \qquad \sum_{i=1}^{k} c_i = 0. \tag{29.93}$$

The most obviously useful contrast other than the difference between any θ_i and θ_j is the difference between the average of any subset of p of the k parameters θ_i and the average of the $k - p$ others. Interactions (defined in **29.18, 29.39**) are also at once seen to be contrasts.

The method of **29.51** is easily adapted so that every contrast, and not merely every difference, of the θ_i is simultaneously covered by an interval. Since the number of contrasts is infinite, the resulting gain in generality is considerable.

29.53 Write $z_i = \bar{x}_i - \theta_i$, and let $\sum_{i=1}^{k} c_i = 0$. Consider the maximum possible value of $\sum c_i z_i$. Since $\sum c_i = 0$, the sum of the positive c_i is $\frac{1}{2} \sum_i |c_i|$ as is also the sum of the absolute values of the negative c_i. We therefore see that

$$\sum_i c_i z_i \leq \left(\frac{1}{2} \sum_i |c_i| \right) \max |z_i - z_j|,$$

i.e.

$$\sum_i c_i(\bar{x}_i - \theta_i) \leq \left(\frac{1}{2} \sum_i |c_i| \right) \max |(\bar{x}_i - \theta_i) - (\bar{x}_j - \theta_j)|. \tag{29.94}$$

Referring back to (29.91), we see that (29.94) implies that for any choice of the c_i with $\sum c_i = 0$,

$$\text{Prob} \left\{ \left| \frac{\sum_i c_i(\bar{x}_i - \theta_i)}{s/N^{1/2}} \right| \leq \left(\frac{1}{2} \sum_i |c_i| \right) q_\alpha \right\} = 1 - \alpha,$$

and hence that

$$\sum_i c_i\bar{x}_i - \left(\frac{1}{2} \sum_i |c_i| \right) q_\alpha s/N^{1/2} \leq \sum_i c_i\theta_i \leq \sum_i c_i\bar{x}_i + \left(\frac{1}{2} \sum_i |c_i| \right) q_\alpha s/N^{1/2} \tag{29.95}$$

is simultaneously satisfied for all contrasts $\psi = \sum c_i\theta_i$, with probability $1 - \alpha$. The method again generalises to negatively equi-correlated multinormal $\bar{x}_i$ (cf. Exercise 29.11). P. K. Sen (1969) extends it to unequally correlated variables.

Keselman *et al.* (1975) investigate the size and power of the test when the n_i are unequal and they are each replaced by their harmonic mean.

Spjøtvoll and Stoline (1973) (for whose method Stoline (1978) provides exact tables) and Hochberg (1975) extend the method further to give approximate simultaneous intervals, when sample sizes are unequal, for all linear combinations—when differences $\theta_i - \theta_j$ are of interest. Dunnett (1980a) recommends use of Tukey's own modification of (29.95) for the unequal sample sizes case, which replaces $N^{1/2}$ by $\{N_i^{-1} + N_j^{-1})/2\}^{-1/2}$ and is shown to be conservative by Hayter (1984)—see also Uusipaikka (1985). Hochberg (1976) further modifies the method for the case of unequal variances, for which Tamhane (1979) and Dunnett (1980b) review and compare these and other methods. Keselman and Rogan (1978) and Keselman *et al.* (1979) compare these and other methods for differences between means. For the case of only two distinct sample sizes, Genizi and Hochberg (1978) give an improved method which is extended by Felzenbaum *et al.* (1983). See also the simple method proposed by Gabriel (1978). Hochberg *et al.* (1982) discuss graphical procedures.

29.54 In **29.53**, simultaneous confidence intervals for all contrasts were obtained from intervals for all differences by the use of a rather wasteful inequality. It is not surprising, therefore, that in general these are not the most useful intervals for all contrasts. To obtain a more useful set, we make an entirely different approach.

The estimator of any contrast (29.93) is

$$\hat{\psi} = \sum c_i \hat{\theta}_i. \tag{29.96}$$

Clearly,

$$E(\hat{\psi}) = \psi,$$

and, further, if the $\hat{\theta}_i$ are normally distributed, so will $\hat{\psi}$ be. If we now consider any set of r ($\leqslant k$) estimated contrasts, which we write in the form $\hat{\boldsymbol{\psi}} = \mathbf{C}\hat{\boldsymbol{\theta}}$, it will be multinormally distributed (cf. **15.4**, Vol. 1) with mean vector equal to $\boldsymbol{\psi} = \mathbf{C}\boldsymbol{\theta}$ and covariance matrix

$$\mathbf{V} = \mathbf{V}(\hat{\boldsymbol{\psi}}) = \mathbf{C}\mathbf{V}(\hat{\boldsymbol{\theta}})\mathbf{C}'. \tag{29.97}$$

In our present discussion of the one-way classification with equal frequencies N, $\mathbf{V}(\hat{\boldsymbol{\theta}})$ is diagonal, with elements σ^2/N, so that

$$\mathbf{V} = \frac{\sigma^2}{N} \mathbf{C}\mathbf{C}'. \tag{29.98}$$

We assume that $\mathbf{V}$ is non-singular.

29.55 The result of **15.10** now implies that the quadratic form

$$Q = (\hat{\boldsymbol{\psi}} - \boldsymbol{\psi})'\mathbf{V}^{-1}(\hat{\boldsymbol{\psi}} - \boldsymbol{\psi})$$

has a χ^2 distribution with degrees of freedom equal to r, the rank of $\mathbf{V}$. Independently of Q, the residual SS (divided by σ^2) is also distributed in this form with, say, ν d.fr. Thus the ratio

$$F = (Q/r)/(s^2/\sigma^2),$$

where $s^2 = $ (residual SS)$/\nu$ as usual, has the variance-ratio distribution with (r, ν) d.fr. In the simplest case, (29.98), this gives the statistic

$$F = (\hat{\boldsymbol{\psi}} - \boldsymbol{\psi})'(\mathbf{C}\mathbf{C}')^{-1}(\hat{\boldsymbol{\psi}} - \boldsymbol{\psi})/(rs^2/N).$$

If we call the $100(1 - \alpha)$ per cent point of this F distribution $F_{\alpha, r, \nu}$, we now have

$$\text{Prob}\,\{(\hat{\boldsymbol{\psi}} - \boldsymbol{\psi})'(\mathbf{C}\mathbf{C}')^{-1}(\hat{\boldsymbol{\psi}} - \boldsymbol{\psi})/(s^2/N) \leqslant rF_{\alpha, r, \nu}\} = 1 - \alpha. \tag{29.99}$$

The corresponding general result is at once available by using (29.97) instead of (29.98).

29.56 Since $\mathbf{V}$ must be non-singular, its rank r cannot exceed q, the number of linearly independent comparisons possible among the θ_i ($k-1$ in the one-way classification—see Example 29.1), which is equal to the d.fr. for their SS in the AV table. (29.99) with $r=q$ holds for *any* set of q linearly independent contrasts we may choose, but this does not imply that it holds for *every* such set simultaneously. However, Scheffé (1953, 1959) showed by geometrical methods that it does imply for every single contrast ψ simultaneously that

$$\text{Prob}\,\{(\hat{\psi}-\psi)^2 \leqslant qF_{\alpha,q,\nu}\,\hat{V}(\hat{\psi})\} = 1-\alpha. \tag{29.100}$$

Here, $\hat{V}(\hat{\psi})$ is the estimated variance of $\hat{\psi}$, in which σ^2 is estimated by s^2 with ν d.fr. Cf. Exercise 29.12 for an analytic proof and Exercise 29.15 for an extremely simple algebraic one. See also Gabriel and Peritz (1973).

Scheffé (1953, 1959) went on to show numerically that the intervals for all contrasts yielded by (29.100) are generally shorter than those obtained from (29.95) unless the contrasts happen to be differences—for which (29.95) reduces to (29.92), designed specifically for differences—or otherwise have very few non-zero c_i. See also Spjøtvoll and Stoline (1973).

> Exercise 29.13 gives a very simple device that permits either (29.95) or (29.100) to yield simultaneous intervals for all linear combinations, and not merely all contrasts, of the θ_i.

29.57 If we now reconsider the variance-ratio test of the overall hypothesis that all the θ_i are equal ((29.23) in the one-way classification case), we see that this hypothesis (cf. (29.20)) states that q linearly independent contrasts are all zero. This implies that *all* contrasts are zero, for every contrast may be regarded as a linear combination of the q linearly independent ones. Thus the overall test is logically equivalent to testing the hypothesis that each of the infinite number of possible contrasts is zero, i.e. seeing whether at least one of the infinite number of intervals given by (29.100) does not cover the value zero. (See also Exercises 29.12 and 29.15.) This property extends at once to Gabriel's (1964) simultaneous test procedure in **29.50**. A subset will be adjudged heterogeneous by that method if and only if some contrast within the subset has interval (29.100) not covering zero.

This is the main use of Scheffé's all-contrasts method, which may be used to examine any contrasts to reveal whether they are in fact reasons for rejecting the hypothesis of homogeneity, and to calculate confidence intervals for them—they need not be nominated in advance. A natural way of seeking the contrasts which may be to blame for the rejection of overall homogeneity is to start with all $\frac{1}{2}k(k-1)$ differences. All this may be done without affecting the size of the overall test. The reader who now refers back to the original discussion of the purposes of multiple comparisons in **29.47**, will probably agree that Scheffé's all-contrasts method is very close to achieving those purposes. But of course, the intervals (29.100) are unconditional ones, and if they are used *only* when the overall homogeneity H_0 is rejected, the residual SS will tend to be small and the probability of coverage too low—cf. Olshen (1973).

For a modification mainly of theoretical interest, see Scheffé (1970). Gabriel (1969) gives a general theory of simultaneous test procedures.

> Petroudas and Gabriel (1983) investigate the use of permutation tests in simultaneous inference.

29.58 Dunn (1961) considers a procedure intermediate between setting confidence intervals for a single contrast and setting them for all contrasts. Her method requires the prior nomination of m contrasts as of special interest. The intervals obtained (based on Student's t-statistic) are shorter than those obtained from either Tukey's or Scheffé's method if k (the number of parameters) exceeds 2 and m is not too large—this advantage increases as k, or the number of d.fr. for residual, or the confidence coefficient $1 - \alpha$, increases. The very simple result that underlies this method is given in Exercise 29.14. The procedure is improved by Siotani (1964).

Bohrer (1973) gives conditions under which Scheffé's method is optimum and (1967) improves on it in the case when the contrasts are of variables known to be all non-negative. Spjøtvoll (1972c) generalises it in the one-way classification with unequal group variances. Bradu and Gabriel (1974) discuss several methods for interactions in the balanced two-way classification.

> O'Neill and Wetherill (1971) review multiple comparisons techniques and give a classified bibliography. See also R. G. Miller (1966, 1977) and the book by Hochberg and Tamhane (1987).

Selection and ranking of populations

29.59 Suppose now that we know in advance that the k groups of observations originated from different populations. A test of their homogeneity is thus superfluous. Of course we may still use **29.51–7** to construct simultaneous confidence intervals for the differences in the value of some parameter θ specifying the distinct populations, but we may now be interested in other specific questions.

For example, we may wish simply to select the 'best' population, say that with the largest mean, or that with the smallest variance, among otherwise identical populations, or (in a binomial populations problem) that with the smallest proportion of members with a particular defect. More generally, we may wish to select the s 'best' of the k populations, and possibly also to specify their rank order in θ. (If $s = k$, we would be ranking the θ-values in all k populations.) Evidently, any statistical procedure will make the correct selection with a certain probability. Our aim is then to maximise this probability for given sizes of sample from the k populations, or to choose these sample sizes to achieve a desired probability of making the correct selection. It should be particularly noticed that we are not required in the selection problem to estimate the value of the parameter in the best population(s); we merely have to identify the best population(s).

A second method of approach in the selection problem is to choose a subset (either of fixed or of random size) of the k populations that with given probability contains the best (or the s best) population(s), again possibly specifying the order of θ-values. Other types of problem also arise.

The theory of selection and ranking procedures has developed considerably in recent years, led by the contributions of R. E. Bechhofer and M. Sobel. The book by Gibbons *et al.* (1977) treats the subject fully, with many tables to assist computations in particular problems. The monographs by Bechhofer *et al.* (1968) deals with sequential methods in this field.

Analysis of covariance

29.60   A natural extension of AV arises when, in an analysis of classified data such as we have been discussing in this chapter, we have available to us not only the observations on y but also the values of one or more further variables x, known or suspected to influence the value of y. If the data were not classified, we should carry out an ordinary regression analysis of y on the xs, but what we wish to investigate now is the joint effect upon y of the classification (possibly a complex one) and of the measured variables x. There is more than one possible purpose for such an analysis.

Commonly, the effects of the xs are to be eliminated by regression methods, so that we may be free to analyse the effect of the classification upon y after discounting the influence of x—for example, when x is the value on an earlier occasion (before the treatments giving rise to the classification) of y itself. Thus, if the effect of different teaching methods upon children's performance in a school subject is to be measured, their initial levels of performance would be the values of x, and their final levels the values of y.

The purpose of the analysis is to ensure 'fair' comparisons among treatments and also, by removing unwanted variation due to x, the reduction of residual variation. It makes interpretation of results simpler if x is unaffected by the treatment yielding the classification, but the analysis may be carried out in any case.

We may also be interested in whether the regression of y on x is affected by the classification; in our example, we might ask whether the regression of final performance level upon initial performance level is the same for each method used. Our motive is now intrinsic interest in the relationship between the variables.

This branch of the subject has come to be called *analysis of covariance* (ANCOVA), because the regression calculations involve partitioning sums of products of y and x in the same way as ordinary AV involves the partitioning of SS. The variables x are usually called *concomitant variables*, implying that y is of greater interest.

Since we have already seen that regression analysis and AV can each be treated within the framework of a linear model, it is evident that analysis of covariance, a mixture of the two, can be so treated. The interpretative convenience of having the concomitant variables unaffected by the treatments is now seen to be a special case of the convenience of having different sets of regressors uncorrelated in linear models. One can therefore set up a linear model *ab initio* for any situation requiring an analysis of covariance.

However, this process can be avoided if the AV that is embedded within the analysis of covariance is of a known form; we can then extend an AV to produce an associated analysis of covariance, by extending the linear model appropriately. This process was outlined for the linear model in **19.17–8**; a more detailed discussion is given by Cox and McCullagh (1982). The resulting summary of the results is best illustrated by an example.

Example 29.6 (Fairfield Smith, 1957)
An experiment investigating the effect of different methods of curing alfalfa on the vitamin A potency of milk considered six blocks of three cows each; the resulting data are given in Table 29.10 after conversion to logarithms. (The full experimental design was more complex, but this need not concern us here.)
    The AV and analysis of covariance results are given in Table 29.11.

Table 29.10   Carotene intake (X) and vitamin A potency per pound of butter fat (Y) for three types of alfalfa forage: $x = 10^3(\log_{10} X - 2)$, $y = 10^3(\log_{10} Y - 3.8)$

Block	Field hay		Barn hay		Silage	
	x	y	x	y	x	y
1	149	136	297	138	781	231
2	164	154	393	174	766	280
3	64	111	299	131	675	219
4	90	96	279	207	659	261
5	218	76	389	66	919	263
6	193	85	360	85	979	272
Total	878	658	2017	801	4779	1526
Mean	146.3	109.7	336.2	133.5	796.5	254.3

Table 29.11   AV and ANCOVA results for data of Table 29.10

Variation	AV			ANCOVA		
	D.fr.	SS	F-ratio	D.fr.	SS	F-ratio
Between blocks	5	9 886	1.71	5	19 978	5.28
Between treatments	2	72 194	31.20	2	2 111	1.39
Residual	10	11 564		9	6 815	
Total (adjusted)	—	—	—	16	28 904	
Due to regression	—	—	—	1	64 740	(85.5)
Total	17	93 644		17	93 644	

The AV results are straightforward; block effects are slight and large treatment effects are evident. The barn hay observation in block 4 is evidently an outlier. At first sight the ANCOVA results are puzzling because now the treatment effects are insignificant and the block effects significant. The reason for this is that the covariate is strongly correlated with the treatments; indeed, the major component of the treatment differences is the variation in carotene intake, as can be seen from the F-value for the regression effect. Such a covariate is sometimes termed a *treatment-induced concomitant*. In no way does this suggest a failure of the ANCOVA analysis; rather it suggests that the nature of the covariates must be properly understood and that, especially when covariates are correlated with treatments, the significance of treatment effects may diminish as well as increase.

The analysis may be taken further to enquire, for example, whether the regression coefficients are equal for all treatments. In general, unequal slopes may make interpretation of treatment effects more awkward; little insight is gained from such extensions in this example.

29.61 Developments in statistical computing have eliminated the need for the various numerical short-cuts that used to be common in ANCOVA and details have been omitted here. Also, the EM algorithm (**18.40–2**) is now generally preferred for dealing with missing values, so we do not consider that aspect of ANCOVA analysis either. Nevertheless, the method remains conceptually valuable and is computationally useful for large designs.

> Hemmerle (1976, 1982) discusses computational procedures.

29.62 The September 1957 issue of *Biometrics* (**13** (3)) contained seven papers on ANCOVA and its uses and has become a classic reference; it was followed up by an anniversary issue in September 1982 (**38** (3)) that includes discussions of ANCOVA methods for complex designs and non-linear models as well as non-parametric and robust procedures.

Other models for AV

29.63 Throughout this chapter, we have been concerned with the application of the linear model to the analysis of observations classified into groups. Underlying the whole of the discussion was the assumption, explicitly written into the linear model, that the classification affected the observations through their mean values, but not otherwise. In the case of the linear model, therefore, analysis of variance (AV) is accurately described as an analysis of *means*, which is carried out through certain sums of squares (SS) computed from the observations.

It is a remarkable fact that we are led to very similar (and in some simple cases, even identical) computations of SS when investigating a quite different type of situation. In the early development of the subject, this similarity in analysis tended to obscure the fundamental distinctions between the underlying mathematical models, which were first set out in detail by Eisenhart (1947). AV based on the LS analysis of the linear model was there called Model I AV, a name in common use subsequently. The other well-established mathematical model, Model II AV, considers the superficially similar model

$$\underset{(n\times 1)}{\mathbf{y}} = \underset{(n\times 1)}{\mathbf{1}\theta} + \underset{(n\times p)(p\times 1)}{\mathbf{Xu}} + \underset{(n\times 1)}{\boldsymbol{\varepsilon}} \, . \qquad (29.101)$$

In (29.101), we isolate the 'general mean' θ, which here will need no subscript. As before, $\mathbf{1}$ is a vector of units and $\mathbf{X}$ a known matrix of constants, while $\boldsymbol{\varepsilon}$ is the vector of errors in the observations. The crucial change is the replacement of the parameter-vector $\boldsymbol{\theta}$ in (19.8) by a vector $\mathbf{u}$ of p *random variables*. Thus (29.101) states that y_i $(i = 1, 2, \ldots, n)$ is composed of the general mean θ, plus a linear combination of p random variables u_j, plus an error term, ε_i. There are $(p+1)$ random components of y_i, instead of only the one in (19.8).

We assume, as at (19.9) and (19.10) for the linear model, that

$$E(\boldsymbol{\varepsilon}) = \mathbf{0}; \qquad \mathbf{V}(\boldsymbol{\varepsilon}) = E(\boldsymbol{\varepsilon}\boldsymbol{\varepsilon}') = \sigma_\varepsilon^2 \mathbf{I}, \tag{29.102}$$

where we have added an identifying subscript to the error variance σ_ε^2 to distinguish it from the variances of our new random variables. We further assume that

$$E(\mathbf{u}) = \mathbf{0}. \tag{29.103}$$

which is why we isolated the general mean θ, and that

$$\left.\begin{array}{ll} \operatorname{var} u_j = \sigma_j^2; & \operatorname{cov}(u_j, u_l) = 0, \quad l \neq j; \\ \operatorname{cov}(u_j, \varepsilon_i) = 0, & \text{all } i, j. \end{array}\right\} \tag{29.104}$$

Thus all our random variables have zero means and are uncorrelated in pairs.

The reader will realise that the Model I definition (cf. **29.4, 29.9–10**) of the term 'analysis of variance' must now be broadened. We define AV generally as the study, whether by means of classified data (cf. **29.15**) or not, of the resolution of SS into component SS attributable to various factors, acting singly and in combination.

T. P. Speed (1987) proposes an alternative definition of AV based on a general class of dispersion models. His group-theoretical combinatorial arguments are remarkably general, clarifying the links between AV and other fields. See also the discussion following his paper.

The parameters of interest in the model (29.101) are the σ_j^2, the variances of the u_j, and the error variance σ_ε^2. It follows immediately that in the simplest (one-way classification) case where all $\sigma_j^2 = \sigma^2$ and $\mathbf{X}$ is as in Example 29.1, the variance of an observation is $\sigma^2 + \sigma_\varepsilon^2$, so that here the two parameters of the model are literally components of the variance of the observations. It was this that gave rise to the term 'components of variance' that is often used for the model (29.101) in general.

29.64   Both Model I and Model II are extreme models, in the sense that *all* the elements of the vector $\boldsymbol{\theta}$ in (19.8) are constants (parameters) while *all* the elements of $\mathbf{u}$ in (29.101) are random variables, uncorrelated with each other and with the error-vector $\boldsymbol{\varepsilon}$. In practice, we may meet situations where a *mixed model* is apposite, i.e. where

$$\mathbf{y} = \mathbf{1}\theta_0 + \mathbf{X}_1\boldsymbol{\theta} + \mathbf{X}_2\mathbf{u} + \boldsymbol{\varepsilon},$$

where θ_0 is now used for the general mean to distinguish it from the parameter-vector $\boldsymbol{\theta}$ of constants. Model II and mixed models will be treated in a later volume.

EXERCISES

29.1   Verify that if all $n_{ij} = m$, the SS in (29.63) and (29.64) become

$$S_4 = \left(\sum_i \sum_j \sum_p y_{ijp} \right)^2 \Big/ (rcm),$$

$$S_1 = \sum_i \left(\sum_j \sum_p y_{ijp} \right)^2 \Big/ (cm) - S_4,$$

$$S_2 = \sum_j \left(\sum_i \sum_p y_{ijp} \right)^2 \Big/ (rm) - S_4,$$

$$S_3 = \sum_i \sum_j \left(\sum_p y_{ijp} \right)^2 \Big/ m - (S_1 + S_2 + S_4),$$

$$S_R = \sum_i \sum_j \sum_p y_{ijp}^2 - (S_1 + S_2 + S_3 + S_4),$$

and show that if $m = 1$, these SS reduce (the summation over p now being redundant) to

$$S_4 = \left(\sum_i \sum_j y_{ij} \right)^2 \Big/ (rc),$$

$$S_1 = \sum_i \left(\sum_j y_{ij} \right)^2 \Big/ c - S_4,$$

$$S_2 = \sum_j \left(\sum_i y_{ij} \right)^2 \Big/ r - S_4,$$

$$S_3 = \sum \sum y_{ij}^2 - (S_1 + S_2 + S_4),$$

$$S_R \equiv 0.$$

29.2   Verify that if all $n_{ij} = m$, the matrix $\mathbf{C}$ defined at (29.39) becomes

$$\underset{(r-1)(c-1) \times (r-1)(c-1)}{\mathbf{C}} = m \begin{pmatrix} 2\mathbf{E} & \mathbf{E} & \mathbf{E} & \cdots & \mathbf{E} \\ \mathbf{E} & 2\mathbf{E} & \mathbf{E} & \cdots & \mathbf{E} \\ \vdots & & \ddots & \ddots & \mathbf{E} \\ \mathbf{E} & & \cdots & \mathbf{E} & 2\mathbf{E} \end{pmatrix}$$

where

$$\underset{(c-1) \times (c-1)}{\mathbf{E}} = \begin{pmatrix} 2 & 1 & \cdots & 1 \\ 1 & & \ddots & \vdots \\ \vdots & \ddots & & 1 \\ 1 & \cdots & 1 & 2 \end{pmatrix}.$$

Show that

$$\mathbf{C}^{-1} = \frac{1}{rm} \begin{pmatrix} (r-1)\mathbf{E}^{-1} & -\mathbf{E}^{-1} & \cdots & & -\mathbf{E}^{-1} \\ -\mathbf{E}^{-1} & & \ddots & \ddots & \vdots \\ \vdots & \ddots & & & -\mathbf{E}^{-1} \\ -\mathbf{E}^{-1} & \cdots & & -\mathbf{E}^{-1} & (r-1)\mathbf{E}^{-1} \end{pmatrix},$$

where

$$\mathbf{E}^{-1} = \frac{1}{c} \begin{pmatrix} c-1 & -1 & \cdots & -1 \\ -1 & c-1 & & \vdots \\ \vdots & & \ddots & -1 \\ -1 & \cdots & -1 & c-1 \end{pmatrix}$$

and hence verify that the LS estimator of θ_{ij} is given by (29.50).

29.3 Generalise the AV table of Example 29.5 to a three-way hierarchical classification and give the three LR tests of the hypothesis that there is no variation in (a) the group means; (b) subgroup means within groups; and (c) subsubgroup means within subgroups.

29.4 In Example 29.4, show that if it is postulated that H_3 of (29.54) holds, so that there are known to be no interactions, the SS attributable to row-effects is $M - S_2$, and the SS attributable to column-effects is $M - S_1$, where S_1 and S_2 are defined by (29.59-60). Show that the residual MS has $(n - r - c + 1)$ d.fr. in this case.

29.5 Show that if, in an $(r \times 2)$ cross-classification, the differences of cell-means $d_i = y_{i1.} - y_{i2.}$ are analysed by the method of weighted squares of means applied to the $\bar{y}_i$ in **29.31**, the SS $\sum_i W_i (d_i - \sum_i W_i d_i / \sum_i W_i)^2$, where $W_i^{-1} = n_{i1}^{-1} + n_{i2}^{-1}$, provides the test of the hypothesis that the interactions are all zero. (Yates, 1934a)

29.6 In a $2 \times 2 \times 2 \times \cdots = 2^m$ cross-classification, show that if a $2 \times 2^{m-1}$ table is formed for any one of the classifications (A, say) against all possible combinations of the others, the unweighted mean of the differences of the row-cell means provides an efficient estimator and test of the effect of A. Show that this is a generalisation of **29.31**, and that any interaction can similarly be tested. (Yates, 1934a; F. M. Speed and Monlezun, 1979)

29.7 Tables 29.12 gives Brandt's data, used by Yates (1934a), for an 8×2 cross-classification by breed and sex of 533 slaughtered pigs; cell frequencies n_{ij} and the cell totals of the variable studied (the logarithm of the percentage bacon yielded by the carcass) are shown.

The total SS, which is not obtainable from the above, was 13.0142. Using **29.31** and Exercise 29.5, show that (neglecting the 1 d.fr. for the general mean) the AV is as shown in Table 29.13.

Show that breed and sex each have effects upon bacon yield, but that they do not interact.

Table 29.12

Breed	Sex	Female		Male
	n_{i1}	$\sum_p y_{i1p}$	n_{i2}	$\sum_p y_{i2p}$
I	33	66.55	89	181.04
II	51	98.69	141	281.43
III	13	25.90	17	34.20
IV	4	7.62	9	17.58
V	8	14.64	4	8.20
VI	15	28.11	32	64.42
VII	35	66.90	47	90.52
VIII	12	23.32	23	46.70
Totals	171	331.73	362	724.09

Table 29.13

Variation	d.fr.	SS	MS
Due to breeds	7	0.6056	0.0865
Due to sexes	1	0.3032	0.3032
Breeds and sexes	8	1.0415	
Interactions	7	0.2300	0.0329
Total between classes	15	1.2715	0.0848
Residual	517	11.7427	0.0227
Total minus general mean	532	13.0142	

29.8 Given that x_i/σ^2 $(i = 0, 1, 2, \ldots, k)$ are independent χ^2 variates with ν_i d.fr., show that the joint distribution of $y_i = x_i/x_0$ $(i = 1, 2, \ldots, k)$ is

$$g(y_1, y_2, \ldots, y_k) = \frac{\Gamma(\nu/2)}{\prod_{i=0}^{k}\Gamma(\nu_i/2)} \frac{\prod_{i=1}^{k} y_i^{(\nu_i/2)-1}}{(1+\sum_{i=1}^{k} y_i)^{\nu/2}}, \qquad 0 \leqslant y_i < \infty,$$

where $\nu = \sum_{i=0}^{k} \nu_i$. (The distribution of $\nu_1 F/\nu_2$ in (16.24), Vol. 1, is the case $k = 1$. Exercise 26.2 dealt with $k = 2$.)

29.9 Show that the AV for an $(r \times c)$ cross-classification with m observations in every cell may be formally constructed from an $(r \times c \times m)$ cross-classification with exactly one observation per cell, in which the layer classification factor is 'replication', its main effect and all interactions concerned with it being defined to be identically zero.

29.10 For a cross-classification with unequal cell frequencies, show that if the cell means are analysed as single observations, their average variance may be estimated by s^2/H, where s^2 is the residual MS of the original observations and H is the harmonic mean of the cell frequencies. Hence show how an approximate AV of the cell means may be carried out.

Apply the method to the numerical data of Exercise 29.7 to show that the approximate AV for cell means is as in Table 29.14.

Table 29.14

	SS	d.fr.	MS
Between sexes	0.023	1	0.023
Between breeds	0.020	7	0.003
Interactions	0.006	7	0.001
Sub-total	0.049	15	
Residual		517	0.0017 $(=0.0227 \times 0.0759)$

Compare the values of the F-ratios in this table with the exact values in Exercise 29.7. (Cf. Scheffé, 1959; Rankin (1974) improves the procedure in the one- and two-way cases by adjusting the d.fr. for effects and interactions to reflect the unequal cell frequencies.)

29.11 If in **29.51** the k variables $(\bar{x}_i - \theta_i)$ are multinormally distributed with all variances σ^2/N and all covariances equal to $-\lambda^2\sigma^2/N$, show that the k variables $z_i = \bar{x}_i - \theta_i + x_0$ are independently normal if x_0 is a normal variable independent of the $\bar{x}_i$ with zero mean and variance

$\lambda^2\sigma^2/N$. By applying the method of **29.51** to the z_i, show that, with probability $1-\alpha$,

$$\bar{x}_i - \bar{x}_j - q_\alpha\{(1+\lambda^2)s^2/N\}^{1/2} \le \theta_i - \theta_j \le \bar{x}_i - \bar{x}_j + q_\alpha\{(1+\lambda^2)s^2/N\}^{1/2}$$

holds simultaneously for all $\frac{1}{2}k(k-1)$ differences $\theta_i - \theta_j$. Show that the result of **29.53** generalises in exactly the same way. (Scheffé, 1959)

29.12 In the one-way classification of Example 29.1, without loss of generality, take $\sum_{i=1}^{k} n_i\hat{\theta}_i$ as origin and σ as unit. Show that the value of $t^2 = (\sum_i c_i\hat{\theta}_i)^2/(\sum_i c_i^2/n_i)$ is maximised for choice of the c_i when $c_i \propto n_i\hat{\theta}_i$, so that $\sum_i c_i = 0$ and $|t|$ is the largest observed absolute ratio of a contrast to its standard error. Show further that $t^2 = S_2$, the numerator SS (defined by (29.22)) of the overall variance-ratio test defined by (29.23), so that the overall test essentially tests the largest observed contrast.

(This result holds quite generally—cf. Scheffé (1959) and Gabriel (1964).)

29.13 By defining a dummy parameter $\theta_0 \equiv 0$ with estimator $\hat{\theta}_0 \equiv 0$ also, show that all linear combinations $\psi = \sum_i c_i\theta_i$ (where $\sum_i c_i$ need not be zero) may have confidence intervals set for them by (29.100) with q increased by 1; and that the method of (29.95) may similarly be used if k is increased by 1 and $\frac{1}{2}\sum_i |c_i|$ is replaced by

$$\max\left\{\sum_i c_i, c_i > 0; \left|\sum_i c_i\right|, c_i < 0\right\}.$$

(Tukey, 1953; Scheffé, 1959)

29.14 In **29.58**, consider k non-independent events with equal probabilities P_1 of occurring. Show that P_k, the probability that they all occur, satisfies $P_k \ge 1 - k(1 - P_1)$, and let P_1 be the probability $1 - \alpha$ that a Student's t-statistic (ν d.fr.) lies in the interval $(-t_\alpha, t_\alpha)$. Show that if m linear combinations $\lambda_s = \sum_{i=1}^{k} c_{si}\theta_i$, $s = 1, 2, \ldots, m$, are estimated by $l_s = \sum_i c_{si}\hat{\theta}_i$, then $(l_s - \lambda_s)/\{\hat{V}(l_s)\}^{1/2}$ is distributed in Student's form with ν d.fr. for each s. Hence show that

$$\text{Prob}\{l_s - t_\alpha[\hat{V}(l_s)]^{1/2} \le \lambda_s \le l_s + t_\alpha[\hat{V}(l_s)]^{1/2}\} \ge 1 - \alpha. \qquad \text{(Dunn, 1961)}$$

29.15 In **29.56-7**, show by the Cauchy inequality that

$$\max_{c_i}\left\{\sum_{i=1}^{k} c_i(\hat{\theta}_i - \theta_i)\right\}^2 = \sum_{i=1}^{k} c_i^2 \sum_{i=1}^{k} (\hat{\theta}_i - \theta_i)^2,$$

and hence that the largest squared difference between an observed contrast and its expectation is distributed as $(\sum_i c_i^2)qs^2F$, where F is the test statistic for the overall hypothesis that the θ_i are equal. Hence establish (29.100).

29.16 In **29.50**, let R and P be any subsets of the k groups such that R contains P. Show that (29.22) calculated for R can never be less than the same SS calculated for P.

29.17 Show that the simultaneous test procedure of **29.50** has the property that the probability of wrongly adjudging any homogeneous subset to be heterogenous is at most α. (Gabriel, 1964)

29.18 In **29.50** define a new step-by-step procedure based on the SS (29.21), but applied in the manner of the last paragraph of **29.49**. This has the same overall size α as the simultaneous test procedure of **29.50**. Show that for such a step-by-step procedure the critical value used for the SS must increase with the size of the subset being tested, and hence that a subset can be adjudged heterogeneous by the step-by-step method only if it is by the simultaneous method. (Gabriel, 1964)

29.19   In **29.42–4**, generalise by letting

$$
\mathbf{y} = \begin{pmatrix} \mathbf{z} \\ \mathbf{u} \\ \mathbf{v} \\ \mathbf{w} \end{pmatrix},
$$

where $\mathbf{z}$ contains the observed y-values with no corresponding x-values missing, $\mathbf{u}$ the missing y-values with no corresponding x-values missing, $\mathbf{v}$ the observed y-values with some corresponding x-values missing, and $\mathbf{w}$ the missing y-values with some corresponding x-values missing.

$$
\begin{pmatrix} \mathbf{X}_z \\ \mathbf{X}_u \\ \mathbf{X}_v \\ \mathbf{X}_w \end{pmatrix}
$$

is the conformable partition of $\mathbf{X}$, and $\mathbf{X}_0$, say, is $\mathbf{X}$ when $\mathbf{X}_v = \mathbf{X}_w = \mathbf{0}$.

(a) Put $\mathbf{u} = \mathbf{w} = \mathbf{0}$ and replace $\mathbf{X}$ by $\mathbf{X}_0$. Estimate $\boldsymbol{\theta}$, obtaining $\hat{\boldsymbol{\theta}}(0)$, say;
(b) Estimate $\mathbf{u}$ by (29.88), with $\mathbf{X}_0$ instead of $\mathbf{X}$;
(c) Re-estimate $\boldsymbol{\theta}$, putting $\mathbf{u} = \hat{\mathbf{u}}$, $\mathbf{w} = \mathbf{0}$.

   Show that this is the LS estimator.   (Afifi and Elashoff, 1966)

CHAPTER 30

TESTS OF FIT

30.1 In our discussions of estimation and test procedures from Chapter 17 onwards, we have concentrated entirely on problems concerning the parameters of distributions of known form. In our classification of hypothesis-testing problems in **21.3** we did indeed define a non-parametric hypothesis, but we have not yet investigated non-parametric hypotheses or estimation problems.

In the present chapter, we confine ourselves to a particular class of procedures that stand slightly apart, and are of sufficient practical importance to justify this special treatment.

Tests of fit
30.2 Let $x_1, x_2, \ldots, x_n$ be independent observations on a random variable with distribution function $F(x)$ which is unknown. Suppose that we wish to test the hypothesis

$$H_0: F(x) = F_0(x), \tag{30.1}$$

where $F_0(x)$ is some particular d.f., which may be continuous or discrete. The problem of testing (30.1) is called a *goodness-of-fit problem.* Any test of (30.1) is called a *test of fit.*

Hypotheses of fit, like parametric hypotheses, divide naturally into simple and composite hypotheses. (30.1) is a simple hypothesis if $F_0(x)$ is completely specified; e.g. the hypothesis (a) that the n observations have come from a normal distribution with specified mean and variance is a simple hypothesis. On the other hand, we may wish to test (b) whether the observations have come from a normal distribution whose parameters are unspecified, and this would be a composite hypothesis (in this case it would often be called a 'test of normality'). Similarly, if (c) the normal distribution has its mean, but not its variance, specified, the hypothesis remains composite. This is precisely the distinction we discussed in the parametric case in **21.4**.

30.3 It is clear that (30.1) is no more than a restatement of the general problem of testing hypotheses; we have merely expressed the hypothesis in terms of the d.f. instead of the frequency function. What is the point of this? Shall we not merely be retracing our previous steps?

The reasons for the new formulation are several. The parametric hypothesis-testing methods developed earlier were necessarily concerned with hypotheses imposing one or more constraints (cf. **21.4**) in the parameter space; they afford no means whatever of testing a hypothesis like (b) in **30.2**, where no constraint is imposed upon parameters and we are testing the non-parametric hypothesis that the d.f. is a member of a specified (infinite) family of distributions. In such cases, and even in cases where the hypothesis does impose one or more parametric constraints, as in (a) or (c) of **30.2**, the reformulation of the hypothesis in the form (30.1) provides us with new methods. For we are led by intuition to expect the whole distribution of the sample observations to mimic closely that of the true d.f. $F(x)$. It is therefore natural to seek to use the whole observed distribution directly as a means of testing (30.1), and we shall find that the most important tests of fit do just this. Furthermore, the 'optimum' tests we have devised for parametric hypotheses, H_0, have been recommended by the properties of their power functions against alternative hypotheses that differ from H_0 only in the values of the parameters specified by H_0. It seems at least likely that a test based on the whole distribution of the sample will have reasonable power properties against a wider (infinite) class of alternatives, even though it may not be optimum against any one of them.

The LR and Pearson X^2 tests of fit for simple H_0

30.4 Two well-known methods of testing goodness-of-fit depend on a very simple device, which essentially reduces the problem to a parametric one. We consider it first in the case when $F_0(x)$ is completely specified, so that (30.1) is a simple hypothesis.

Suppose that the range of the variate x is arbitrarily divided into k mutually exclusive classes. (These need not be, though in practice they are usually taken as, successive intervals in the range of x.)* Then, since $F_0(x)$ is specified, we may calculate the probability of an observation falling in each class. If these are denoted by p_{0i}, $i = 1, 2, \ldots, k$, and the observed frequencies in the k classes by $n_i (\sum_{i=1}^{k} n_i = n)$, the n_i are multinomially distributed (cf. **5.49**), and from (5.123) we see that the LF is

$$L(n_1, n_2, \ldots, n_k \,|\, p_{01}, p_{02}, \ldots, p_{0k}) \propto \prod_{i=1}^{k} p_{0i}^{n_i}. \tag{30.2}$$

On the other hand, if the true distribution function is $F_1(x)$, where F_1 may be any d.f., we may denote the probabilities in the k classes by p_{1i}, $i = 1, 2, \ldots, k$, and the likelihood is

$$L(n_1, n_2, \ldots, n_k \,|\, p_{11}, p_{12}, \ldots, p_{1k}) \propto \prod_{i=1}^{k} p_{1i}^{n_i}. \tag{30.3}$$

We may now use **23.1** to find the LR test of the hypothesis (30.1), the composite alternative hypothesis being

$$H_1 : F(x) = F_1(x).$$

The likelihood (30.3) is maximised when we substitute the ML estimators for p_{1i}:

$$\hat{p}_{1i} = n_i / n.$$

---

* We discuss the choice of k and of the classes in **30.20–3**, **30.28–30** below. For the present, we allow them to be arbitrary.

The LR statistic for testing H_0 against H_1 is therefore

$$l = \frac{L(n_1, n_2, \ldots, n_k \,|\, p_{01}, p_{02}, \ldots, p_{0k})}{L(n_1, n_2, \ldots, n_k \,|\, \hat{p}_{11}, \hat{p}_{12}, \ldots, \hat{p}_{1k})}$$

$$= n^n \prod_{i=1}^{k} (p_{0i}/n_i)^{n_i}. \tag{30.4}$$

H_0 is rejected when l is small enough, as at (23.6).

The exact distribution of (30.4) is unknown. However, we know from **23.7** that as $n \to \infty$ when H_0 holds, $-2 \log l$ is asymptotically distributed in the χ^2 form, with $k-1$ degrees of freedom (since there are $r = k-1$ independent constraints on the p_{1i} because $\sum_{i=1}^{k} p_{1i} = 1$).

30.5 (30.4) is not, however, the classical *chi-squared statistic* put forward by Pearson (1900) for this situation. This procedure, which has been derived already as Example 15.3, uses the asymptotic k-variate normality of the multinomial distribution of the n_i, and the fact that, given H_0, the quadratic form in the exponent of this distribution is distributed in the χ^2 form with degrees of freedom equal to its rank, $k-1$. In our present notation, this quadratic form was found in Example 15.3 to be*

$$X^2 = \sum_{i=1}^{k} \frac{(n_i - np_{0i})^2}{np_{0i}}. \tag{30.5}$$

From (30.4) we have

$$-2 \log l = 2 \sum_{i=1}^{k} n_i \log (n_i/np_{0i}). \tag{30.6}$$

The two distinct statistics (30.5) and (30.6) thus have the same distribution asymptotically, given H_0. More than this, however, they are asymptotically equivalent statistics when H_0 holds, for if we write $\Delta_i = (n_i - np_{0i})/np_{0i}$, we have

$$-2 \log l = 2 \sum_i n_i \log (1 + \Delta_i)$$

$$= 2 \sum_i \{(n_i - np_{0i}) + np_{0i}\}\{\Delta_i - \tfrac{1}{2}\Delta_i^2 + O_p(n^{-3/2})\}$$

$$= 2 \sum_i \left\{ (n_i - np_{0i})\Delta_i + np_{0i}\Delta_i - \frac{np_{0i}}{2}\Delta_i^2 + O_p(n^{-1/2}) \right\},$$

and since $\sum p_{0i}\Delta_i = 0$, we have

$$-2 \log l = \sum_i \{np_{0i}\Delta_i^2 + O_p(n^{-1/2})\} = X^2\{1 + O_p(n^{-1/2})\}. \tag{30.7}$$

---

* Following standard practice, we write X^2 for the test statistic and reserve the symbol χ^2 for the distributional form we have so frequently discussed. Earlier writers confusingly wrote χ^2 for the statistic as well as the distribution.

For small n, the test statistics differ. Pearson's form (30.5) may alternatively be expressed as

$$X^2 = \frac{1}{n} \sum_i \frac{n_i^2}{p_{0i}} - n,$$ (30.8)

which is easier to compute; but (30.5) has the advantage over (30.8) of being a direct function of the differences between the observed frequencies n_i and their hypothetical expectations np_{0i}, differences that are themselves of obvious interest. The corresponding simplification of (30.6) is computationally inconvenient.

> M. E. Wise (1963, 1964) examines the approximations to the multinomial involved in using (30.5) as a χ^2_{k-1} variable, and shows that the error is particularly small when the np_{0i} are equal or nearly so—the latter need not then be large (cf. **30.22**, **30.30**, which also cover composite H_0). Some small-sample tabulations of the distributions of (30.5) and (30.6) in the case when each $p_{0i} = 1/k$ are given by Good *el al.* (1970); Zahn and Roberts (1971) tabulate the case $p_{0i} = 1/n$. See also Chapman (1976).
>
> Larntz (1978) shows that for moderate np_{0i} the χ^2_{k-1} approximation is better for X^2 than for the LR statistic (30.6), which rejects H_0 more often than the approximation indicates. Koehler and Larntz (1980) consider the asymptotic distributions of (30.5) and (30.6) as n and $k \to \infty$ together—they tend to different normal distributions. When the p_{0i} are all equal, X^2 has better power, but on the whole, (30.6) performs better when they are unequal. P. J. Smith *et al.* (1981) give the approximate mean and variance of (30.6) and (as in **23.9**) adjust (30.6) multiplicatively to have mean $(k-1)$, which gives a better χ^2_{k-1} approximation when all $p_{0i} = k^{-1}$ and $k \geq 4$.
>
> (30.5) and (30.6) are special cases of the general statistic
>
> $$t_\lambda = \frac{2}{\lambda(\lambda+1)} \sum_{i=1}^k n_i \left\{ \left(\frac{n_i}{np_{0i}} \right)^\lambda - 1 \right\},$$
>
> where $\lambda = 1$ gives X^2 at (30.8) while at $\lambda = 0$ l'Hôpital's rule shows that $t_0 = -2 \log l$ at (30.6). Cressie and Read (1984) study the family t_λ in detail, and show that under H_0 they are all asymptotically equivalent, as they also are for alternative hypotheses of the type in **30.27** below. Convergence to the asymptotic moments is fastest for $0.3 \leq \lambda < 2.7$. When all $p_{0i} = k^{-1}$, the χ^2 approximation to the distribution is adequate for $k \leq 6$, $n \geq 10$ and $\frac{1}{3} < \lambda < \frac{3}{2}$—for other λ it underestimates test size. The two values of λ that perform best overall are $\lambda = 1$ (i.e. X^2) and, more surprisingly, $\lambda = \frac{2}{3}$, although in accordance with **25.18**, $\lambda = 0$ (i.e. $-2 \log l$) is asymptotically best using Bahadur's measure of test efficiency. See also Read (1984). Drost *et al.* (1989) give improved approximations to the distributions.

Choice of critical region for X^2

30.6   Since H_0 is rejected for small values of l, (30.7) implies that when using (30.5) as test statistic, H_0 is to be rejected when X^2 is large. There has been some uncertainty in the literature about this, the older practice being to reject H_0 for small as well as large values of X^2, i.e. to use a two-tailed rather than an upper-tail test. For example, Cochran (1952) approves this practice on the grounds that extremely small X^2 values are likely to have resulted from numerical errors in computation, while on other occasions such values have apparently been due to the frequencies n_i having been biased, perhaps inadvertently, to bring them closer to the hypothetical expectations np_{0i}.

Now there is no doubt that computations should be checked for accuracy, but there are likely to be more direct and efficient methods of doing this than by examining the value of X^2 reached. After all, we have no assurance that a moderate and acceptable value of X^2 has been any more accurately computed than a very small one. Cochran's second consideration is a more cogent one, but it is plain that in this case we are considering a different and rarer hypothesis (that there has been voluntary or involuntary irregularity in collecting the observations) which must be precisely formulated before we can determine the best critical region to use (cf. Stuart (1954a)). Leaving such irregularities aside, we use the upper tail of the distribution of X^2 as critical region. This will be justified from the point of view of its asymptotic power in **30.27**.

30.7 The essence of the LR and Pearson tests of fit is the reduction of the problem to one concerning the multinomial distribution. The need to group the data into classes clearly involves the sacrifice of a certain amount of information, especially if the underlying variable is continuous. However, this defect also carries with it a corresponding virtue: we do not need to know the values of the individual observations, so long as we have k classes for which the hypothetical p_{0i} can be computed. In fact, there need be no underlying variable at all—we may use either of these tests of fit even if the original data refer to a simple categorisation. The point is illustrated by Example 30.1.

Example 30.1
In some classical experiments on pea-breeding, Mendel observed the frequencies of different kinds of seeds in crosses from plants with round yellow seeds and plants with wrinkled green seeds. They are given in Table 30.1, together with the theoretical probabilities on the Mendelian theory of inheritance.
(30.8) gives

$$X^2 = \frac{1}{556} \times 16 \left\{ \frac{315^2}{9} + \frac{101^2}{3} + \frac{108^2}{3} + \frac{32^2}{1} \right\} - 556$$

$$= \frac{16}{556} \times 19\,337.3 - 556 = 0.47.$$

Table 30.1

Seeds	Observed frequency n_i	Theoretical probability p_{0i}
Round and yellow	315	9/16
Wrinkled and yellow	101	3/16
Round and green	108	3/16
Wrinkled and green	32	1/16
	$n = 556$	1

For $(k-1)=3$ degrees of freedom, tables of χ^2 show that the probability of a value exceeding 0.47 lies between 0.90 and 0.95, so that the fit of the observations to the theory is very good indeed: a test of any size $\alpha \leqslant 0.90$ would not reject the hypothesis.

For the LR statistic, (30.6) gives $-2 \log l = 0.48$, very close to the value for X^2.

Composite H_0

30.8 Confining our attention now to Pearson's test statistic (30.5), we consider the situation when the hypothesis tested is composite—the LR test remains asymptotically equivalent when H_0 holds (cf. Exercise 30.11). Suppose that $F_0(x)$ is specified as to its form, but that some (or perhaps all) of the parameters are left unspecified, as in (b) or (c) of **30.2**. The specification of the form of $F_0(x)$ is essential, since this cannot be estimated from the sample without disruptive effects on the distribution of X^2, for we should be imposing an unknown number of non-linear constraints upon the agreement between the n_i and the np_{0i}. In the multinomial formulation of **30.4**, the new feature is that the hypothetical probabilities p_{0i} are not now immediately calculable, since they are functions of the s (assumed $<k-1$) unspecified parameters $\theta_1, \theta_2, \ldots, \theta_s$, which we may denote collectively by $\boldsymbol{\theta}$. Thus we must write them $p_{0i}(\boldsymbol{\theta})$. In order to make progress, we must estimate $\boldsymbol{\theta}$ by some vector of estimators $\mathbf{t}$, and use (30.5) in the form

$$X^2 = \sum_{i=1}^{k} \frac{\{n_i - np_{0i}(\mathbf{t})\}^2}{np_{0i}(\mathbf{t})}.$$

This clearly changes our distribution problem, for now the $p_{0i}(\mathbf{t})$ are themselves random variables, and it is not obvious that the asymptotic distribution of X^2 will be of the same form as in the case of a simple H_0. In fact, the term $n_i - np_{0i}(\mathbf{t})$ does not necessarily have a zero expectation. We may write X^2 identically as

$$X^2 = \sum_{i=1}^{k} \frac{1}{np_{0i}(\mathbf{t})} [\{n_i - np_{0i}(\boldsymbol{\theta})\}^2 + n^2 \{p_{0i}(\mathbf{t}) - p_{0i}(\boldsymbol{\theta})\}^2 \\ - 2n\{n_i - np_{0i}(\boldsymbol{\theta})\}\{p_{0i}(\mathbf{t}) - p_{0i}(\boldsymbol{\theta})\}]. \tag{30.9}$$

Now we know from the theory of the multinomial distribution in **5.49** that asymptotically

$$n_i - np_{0i}(\boldsymbol{\theta}) \sim cn^{1/2},$$

so that the first term in the square brackets in (30.9) is of order n. If we also have

$$p_{0i}(\mathbf{t}) - p_{0i}(\boldsymbol{\theta}) = o_p(n^{-1/2}), \tag{30.10}$$

the second and third term will be of order less than n, and relatively negligible, so that (30.9) asymptotically behaves like its first term. Even this, however, still has the random variable $np_{0i}(\mathbf{t})$ as its denominator, but to the same order of approximation we may replace this by $np_{0i}(\boldsymbol{\theta})$. We thus see that if (30.10) holds, (30.8) behaves asymptotically just as (30.5)—it is distributed in the χ^2 form with $(k-1)$ degrees of freedom. However, if the $p_{0i}(\mathbf{t})$ are 'well-behaved' functions of $\mathbf{t}$, they will differ from the $p_{0i}(\mathbf{\theta})$ by the same order of magnitude as t does from $\mathbf{\theta}$. Then for all practical purposes (30.10) requires that

$$\mathbf{t} - \mathbf{\theta} = o_p(n^{-1/2}). \tag{30.11}$$

(30.11) is not customarily satisfied, since we usually have estimators with variances and covariances of order n^{-1} and then

$$\mathbf{t} - \mathbf{\theta} = O_p(n^{-1/2}). \tag{30.12}$$

In this 'regular' case, therefore, our argument above does not hold. But it does hold in cases where estimators have variances of order n^{-2}, as we have found to be the case for estimators of parameters that locate the end-points of the range of a variable (cf. **18.19** and Exercises 19.11–12). In such cases, therefore, we may use (30.8) with no new theory required. In the more common case where (30.12) holds, we must investigate further.

30.9 It will simplify our discussion if we first give Fisher's (1922a) alternative and revealing proof of the asymptotic distribution of (30.5) for the simple hypothesis case.

Suppose that we have k independent Poisson variates, the ith having parameter np_{0i}, where $n = \sum_{i=1}^{k} n_i$ and $\sum_i p_{0i} = 1$. The probability that the first takes the value n_1, the second n_2 and so on, is

$$P(n_1, n_2, \ldots, n_k, n) = \prod_{i=1}^{k} e^{-np_{0i}} (np_{0i})^{n_i} / n_i! = e^{-n} n^n \prod_{i=1}^{k} p_{0i}^{n_i} / n_i!; \tag{30.13}$$

n appears explicitly as a variate in (30.13), although of course the resulting $(k+1)$-variate distribution is singular since $n = \sum_i n_i$. Its marginal distribution $g(n)$ is easily found, since the sum of the k independent Poisson variables is itself (cf. Example 11.17) a Poisson variable with parameter equal to $\sum_{i=1}^{k} np_{0i} = n$. Thus

$$g(n) = e^{-n} n^n / n! \tag{30.14}$$

and the conditional distribution

$$h(n_1, n_2, \ldots, n_k \mid n) = \frac{P(n_1, n_2, \ldots, n_k, n)}{g(n)}$$

$$= \frac{n!}{n_1! n_2! \ldots n_k!} p_{01}^{n_1} p_{02}^{n_2} \cdots p_{0k}^{n_k}. \tag{30.15}$$

We see at once that (30.15) is precisely the multinomial distribution of the n_i on which our test procedure is based. Thus, as an alternative to the proof of the asymptotic distribution of X^2 given in Example 15.3 (cf. **30.5**), we may obtain it by regarding the n_i as the values of k independent Poisson variables with parameters np_{0i}, conditional upon n being fixed. By Example 4.9, the standardised variable

$$x_i = \frac{n_i - np_{0i}}{(np_{0i})^{1/2}} \tag{30.16}$$

is asymptotically normal as $n \to \infty$. Hence, as $n \to \infty$,

$$X^2 = \sum_{i=1}^{k} x_i^2$$

is the sum of squares of k independent standardised normal variates, subject to the single condition $\sum_i n_i = n$, which is equivalent to $\sum_i (np_{0i})^{1/2} x_i = 0$. By Example 11.6, X^2 therefore has a χ^2 distribution asymptotically, with $(k-1)$ degrees of freedom.

30.10   The utility of this alternative proof is that, in conjunction with Example 11.6, to which it refers, it shows clearly that if s further homogeneous linear conditions are imposed on the n_i, the only effect on the asymptotic distribution of X^2 will be to reduce the degrees of freedom from $(k-1)$ to $(k-s-1)$.

We now return to the composite hypothesis of **30.8** in the case when (30.12) holds. Suppose that we choose as our set of estimators t of θ the maximum likelihood (or other asymptotically equivalent efficient) estimators, so that $t = \hat{\theta}$. Now the likelihood function L in this case is simply the multinomial (30.15) regarded as a function of the θ_j, on which the p_{0i} depend. Thus, generalising **18.20** (where $s = 1$),

$$\frac{\partial \log L}{\partial \theta_j} = \sum_{i=1}^{k} n_i \frac{\partial p_{0i}}{\partial \theta_j} \frac{1}{p_{0i}}, \qquad j = 1, 2, \dots, s, \tag{30.17}$$

and the ML estimators in this regular case are the roots of the s equations obtained by equating (30.17) to zero for each j. Clearly, each such equation is a homogeneous linear relationship among the n_i. We thus see that, in this regular case, we have s additional constraints imposed by the process of efficient estimation of θ from the multinomial distribution, so that the statistic (30.5) is asymptotically distributed in the χ^2 form with $(k-s-1)$ degrees of freedom—see **30.14** below. A more rigorous and detailed proof is given by Cramér (1946)—see also Birch (1964). We shall call $\hat{\theta}$ the *multinomial* ML estimators.

The effect of estimation on the distribution of X^2

30.11   We may now, following Watson (1959), consider the general problem of the effect of estimating the unknown parameters on the asymptotic distribution of the X^2 statistic. We confine ourselves to the regular case, when (30.12) holds, and we write for any estimator t of θ

$$t - \theta = n^{-1/2} \mathbf{A} x + o(n^{-1/2}), \tag{30.18}$$

where $\mathbf{A}$ is an arbitrary $(s \times k)$ matrix and x is the $(k \times 1)$ vector whose ith element is

$$x_i = \frac{n_i - np_{0i}(\theta)}{\{np_{0i}(\theta)\}^{1/2}}, \tag{30.19}$$

defined just as at (30.16) for the simple hypothesis case; we assume $\mathbf{A}$ to have been chosen so that

$$E(\mathbf{A}x) = 0. \tag{30.20}$$

It follows at once from (30.18) and (30.20) that the covariance matrix of $\mathbf{t}$, $\mathbf{V}(\mathbf{t})$, is of order n^{-1}. By a Taylor expansion applied to $\{p_{0i}(\mathbf{t}) - p_{0i}(\boldsymbol{\theta})\}$ in (30.9), we find that we may write

$$X^2 = \sum_{i=1}^{k} y_i^2$$

where, as $n \to \infty$,

$$y_i = x_i - n^{1/2} \sum_{j=1}^{s} (t_j - \theta_j) \frac{\partial p_{0i}(\boldsymbol{\theta})}{\partial \theta_j} \frac{1}{\{p_{0i}(\boldsymbol{\theta})\}^{1/2}} + o_p(1),$$

or, in matrix form,

$$\mathbf{y} = \mathbf{x} - n^{1/2}\mathbf{B}(\mathbf{t} - \boldsymbol{\theta}) + o_p(1), \tag{30.21}$$

where $\mathbf{B}$ is the $(k \times s)$ matrix whose (i, j)th element is

$$b_{ij} = \frac{\partial p_{0i}(\boldsymbol{\theta})}{\partial \theta_j} \frac{1}{\{p_{0i}(\boldsymbol{\theta})\}^{1/2}}. \tag{30.22}$$

Substituting (30.18) into (30.21), we find simply

$$\mathbf{y} = (\mathbf{I} - \mathbf{BA})\mathbf{x} + o_p(1). \tag{30.23}$$

30.12 Now from equation (30.19) the x_i have zero means. As $n \to \infty$, they tend to multinormality, by the multivariate central limit theorem, and their covariance matrix is, temporarily writing p_i of $p_{0i}(\boldsymbol{\theta})$,

$$\mathbf{V}(\mathbf{x}) = \begin{pmatrix} 1 - p_1 & -(p_1 p_2)^{1/2} & -(p_1 p_3)^{1/2} & \cdots & -(p_1 p_k)^{1/2} \\ -(p_2 p_1)^{1/2} & 1 - p_2 & -(p_2 p_3)^{1/2} & \cdots & -(p_2 p_k)^{1/2} \\ \vdots & \vdots & \vdots & & \vdots \\ -(p_k p_1)^{1/2} & -(p_k p_2)^{1/2} & -(p_k p_3)^{1/2} & \cdots & 1 - p_k \end{pmatrix}$$

$$= \mathbf{I} - (\mathbf{p}^{1/2})(\mathbf{p}^{1/2})' \tag{30.24}$$

where $\mathbf{p}^{1/2}$ is the $(k \times 1)$ vector with ith element $\{p_{0i}(\boldsymbol{\theta})\}^{1/2}$. It follows at once from (30.23) and (30.24) that the y_i also are asymptotically normal with zero means and covariance matrix

$$\mathbf{V}(\mathbf{y}) = (\mathbf{I} - \mathbf{BA})\{\mathbf{I} - (\mathbf{p}^{1/2})(\mathbf{p}^{1/2})'\}(\mathbf{I} - \mathbf{BA})'. \tag{30.25}$$

Thus $X^2 = \mathbf{y}'\mathbf{y}$ is asymptotically distributed as the sum of squares of k normal variates with zero means and covariance matrix (30.25). Exercise 15.18 shows that the distribution of X^2 is of the χ^2 form with r degrees of freedom if and only if $\mathbf{V}^3 = \mathbf{V}^2$ and rank $(\mathbf{V}^2) = \mathrm{tr}\,(\mathbf{V}) = r$. This implies that all latent roots are 0 or 1, and since $\mathbf{V}$ is symmetric this is equivalent to the idempotency of $\mathbf{V}$ by the argument near the end of **15.11**.

30.13 We now consider particular cases of (30.25). First, the case of a simple hypothesis, where no estimation is necessary, is formally obtainable by putting $\mathbf{A} \equiv \mathbf{0}$ in (30.18). (30.25) then becomes simply

$$\mathbf{V(y)} = \mathbf{V(x)} = \mathbf{I} - (\mathbf{p}^{1/2})(\mathbf{p}^{1/2})'. \tag{30.26}$$

Since $(\mathbf{p}^{1/2})'(\mathbf{p}^{1/2}) = \sum_{i=1}^{k} p_{0i}(\boldsymbol{\theta}) = 1$, (30.26) is seen on squaring it to be idempotent, and its trace is $(k-1)$. Thus X^2 is a χ^2_{k-1} variable in this case, as we already know from two different proofs.

30.14 The composite hypothesis case is not so straightforward. First, suppose as in **30.10** that the multinomial ML estimators $\hat{\boldsymbol{\theta}}$ are used. We seek the form of the matrix $\mathbf{A}$ in (30.18) when $\mathbf{t} = \hat{\boldsymbol{\theta}}$. Now we know from **18.26** that the elements of the inverse of the covariance matrix of $\hat{\boldsymbol{\theta}}$ are asymptotically given by

$$\{\mathbf{V}(\hat{\boldsymbol{\theta}})\}^{-1}_{jl} = -E\left\{\frac{\partial^2 \log L}{\partial \theta_j\, \partial \theta_l}\right\}, \qquad j, l = 1, 2, \ldots, s. \tag{30.27}$$

From (30.17), the multinomial ML equations give

$$\frac{\partial^2 \log L}{\partial \theta_j\, \partial \theta_l} = \sum_{i=1}^{k} \frac{n_i}{p_{0i}}\left(\frac{\partial^2 p_{0i}}{\partial \theta_j\, \partial \theta_l} - \frac{1}{p_{0i}}\frac{\partial p_{0i}}{\partial \theta_j}\frac{\partial p_{0i}}{\partial \theta_l}\right). \tag{30.28}$$

On taking expectations in (30.28), we find

$$-E\left\{\frac{\partial^2 \log L}{\partial \theta_j\, \partial \theta_l}\right\} = n\left\{\sum_{i=1}^{k} \frac{1}{p_{0i}}\frac{\partial p_{0i}}{\partial \theta_j}\frac{\partial p_{0i}}{\partial \theta_l} - \sum_{i=1}^{k}\frac{\partial^2 p_{0i}}{\partial \theta_j\, \partial \theta_l}\right\}. \tag{30.29}$$

The second term on the right of (30.29) is zero, since it is $(\partial^2/\partial\theta_j\,\partial\theta_l)\sum_{i=1}^{k} p_{0i}$. Thus, using (30.22),

$$-E\left\{\frac{\partial^2 \log L}{\partial \theta_j\, \partial \theta_l}\right\} = n\sum_{i=1}^{k} b_{ij}b_{il},$$

so that, from (30.27),

$$\{\mathbf{V}(\hat{\boldsymbol{\theta}})\}^{-1} = n\mathbf{B'B}$$

or

$$\mathbf{C} = n\mathbf{V}(\hat{\boldsymbol{\theta}}) = (\mathbf{B'B})^{-1}. \tag{30.30}$$

But from (30.18) and (30.24) we have

$$\mathbf{D} = n\mathbf{V(t)} = \mathbf{AV(x)A'} = \mathbf{A}\{\mathbf{I} - (\mathbf{p}^{1/2})(\mathbf{p}^{1/2})'\}\mathbf{A'}. \tag{30.31}$$

Here (30.30) and (30.31) are alternative expressions for the same matrix.
We choose $\mathbf{A}$ to satisfy (30.30) by noting that

$$\mathbf{B'}(\mathbf{p}^{1/2}) = \mathbf{0} \tag{30.32}$$

(since the jth element of this $(s \times 1)$ vector is $(\partial/\partial\theta_j)\sum_{i=1}^{k} p_{0i} = 0$), and hence that if $\mathbf{A} = \mathbf{GB'}$ where $\mathbf{G}$ is symmetric and non-singular, (30.31) gives $\mathbf{GB'BG'}$. If this is to be equal to (30.30), we obviously have $\mathbf{G} = (\mathbf{B'B})^{-1}$, so finally

$$\mathbf{A} = (\mathbf{B'B})^{-1}\mathbf{B'} \tag{30.33}$$

in the case of multinomial ML estimation. (30.25) then becomes, using (30.32),

$$\mathbf{V(y)} = \{\mathbf{I} - \mathbf{B}(\mathbf{B'B})^{-1}\mathbf{B'}\}^2 - (\mathbf{p}^{1/2})(\mathbf{p}^{1/2})'$$
$$= \mathbf{I} - \mathbf{B}(\mathbf{B'B})^{-1}\mathbf{B'} - (\mathbf{p}^{1/2})(\mathbf{p}^{1/2})'. \tag{30.34}$$

By squaring, this matrix is shown to be idempotent. Its rank is equal to its trace, which is given by

$$\text{tr } \mathbf{V(y)} = \text{tr } \{\mathbf{I} - (\mathbf{p}^{1/2})(\mathbf{p}^{1/2})'\} - \text{tr } \mathbf{B'} \cdot \mathbf{B}(\mathbf{B'B})^{-1},$$

and using **30.13** this is

$$\text{tr } \mathbf{V(y)} = (k-1) - s.$$

Thus the distribution of X^2 is χ^2_{k-s-1} asymptotically, as we saw in **30.10**.

30.15 Our present approach enables us to inquire further: what happens to the asymptotic distribution of X^2 if some other estimators than $\hat{\boldsymbol{\theta}}$ are used? This question was first considered in the simplest case by Fisher (1928b). Chernoff and Lehmann (1954) considered a case of particular interest, when the estimators used are the ML estimators based on the n individual observations and not the multinomial ML estimators $\hat{\boldsymbol{\theta}}$, based on the k frequencies n_i, that we have so far discussed. If we have the values of the n observations, it is clearly an efficient procedure to utilise this knowledge in estimating $\boldsymbol{\theta}$, even though we are going to use the k class-frequencies alone in carrying out the test of fit. We shall find, however, that the X^2 statistic obtained in this way no longer has an asymptotic χ^2 distribution.

30.16 Let us return to the general expression (30.25) for the covariance matrix. Multiplying it out, we rewrite it

$$\mathbf{V(y)} = \{\mathbf{I} - (\mathbf{p}^{1/2})(\mathbf{p}^{1/2})'\} - \mathbf{BA}\{\mathbf{I} - (\mathbf{p}^{1/2})(\mathbf{p}^{1/2})'\}$$
$$- \{\mathbf{I} - (\mathbf{p}^{1/2})(\mathbf{p}^{1/2})'\}\mathbf{A'B'} + \mathbf{BA}\{\mathbf{I} - (\mathbf{p}^{1/2})(\mathbf{p}^{1/2})'\}\mathbf{A'B'}. \tag{30.35}$$

Rather than find the latent roots λ_i of $\mathbf{V(y)}$, we consider those of $\mathbf{I} - \mathbf{V(y)}$, which are $1 - \lambda_i$. We use (30.31) to write this matrix in the form

$$\mathbf{I} - \mathbf{V(y)} = (\mathbf{p}^{1/2})(\mathbf{p}^{1/2})' + \mathbf{B}[\mathbf{A}\{\mathbf{I} - (\mathbf{p}^{1/2})(\mathbf{p}^{1/2})'\} - \tfrac{1}{2}\mathbf{A}\{\mathbf{I} - (\mathbf{p}^{1/2})(\mathbf{p}^{1/2})'\}\mathbf{A'B'}]$$
$$+ [\{\mathbf{I} - (\mathbf{p}^{1/2})(\mathbf{p}^{1/2})'\}\mathbf{A'} - \tfrac{1}{2}\mathbf{BA}\{\mathbf{I} - (\mathbf{p}^{1/2})(\mathbf{p}^{1/2})'\}\mathbf{A'}]\mathbf{B'}$$
$$= (\mathbf{p}^{1/2})(\mathbf{p}^{1/2})' + \mathbf{B}[\mathbf{A}\{\mathbf{I} - (\mathbf{p}^{1/2})(\mathbf{p}^{1/2})'\} - \tfrac{1}{2}\mathbf{DB'}]$$
$$+ [\{\mathbf{I} - (\mathbf{p}^{1/2})(\mathbf{p}^{1/2})'\}\mathbf{A'} - \tfrac{1}{2}\mathbf{BD}]\mathbf{B'}. \tag{30.36}$$

(30.36) may be written as the product of two partitioned matrices,

$$\mathbf{I} - \mathbf{V(y)} = \begin{pmatrix} (\mathbf{p}^{1/2}) \\ \mathbf{B} \\ \{\mathbf{I} - (\mathbf{p}^{1/2})(\mathbf{p}^{1/2})'\}\mathbf{A'} - \tfrac{1}{2}\mathbf{BD} \end{pmatrix} \begin{pmatrix} (\mathbf{p}^{1/2})' \\ \mathbf{A}\{\mathbf{I} - (\mathbf{p}^{1/2})(\mathbf{p}^{1/2})'\} - \tfrac{1}{2}\mathbf{DB'} \\ \mathbf{B'} \end{pmatrix}'.$$

The matrices on the right may be transposed without affecting the non-zero latent roots. This converts their product from a $(k \times k)$ to a $(2s+1) \times (2s+1)$ matrix, which is reduced, on using (30.30) and (30.32), to

$$
\begin{pmatrix} 1 & 0 & 0 \\ \hline 0 & & \\ 0 & & M \end{pmatrix} = \begin{pmatrix} 1 & 0_{1 \times s} & 0_{1 \times s} \\ \hline 0_{s \times 1} & AB - \frac{1}{2}DC^{-1} & D - \frac{1}{2}ABD - \frac{1}{2}DB'A' + \frac{1}{4}DC^{-1}D \\ 0_{s \times 1} & C^{-1} & B'A' - \frac{1}{2}C^{-1}D \end{pmatrix}. \tag{30.37}
$$

(30.37) has one latent root of unity and $2s$ others which are those of the matrix M partitioned off in its south-east corner. If $k \geq 2s+1$, which is almost invariably the case in applications, this implies that (30.36) has $(k-2s-1)$ zero latent roots, one of unity, and $2s$ others which are the roots of M. Thus for $V(y)$ itself, we have $(k-2s-1)$ latent roots of unity, one of zero and $2s$ which are the complements to unity of the latent roots of M.

30.17 We now consider the problem introduced in **30.15**. Suppose that, to estimate θ, we use the ML estimators based on the n individual observations, which we shall call the 'ordinary ML estimators' and denote by $\hat{\theta}^*$. We know from **18.26** that if f is the density function of the observations, we have asymptotically

$$
D = nV(\hat{\theta}^*) = -\left\{ E\left(\frac{\partial^2 \log f}{\partial \theta_j \, \partial \theta_i} \right) \right\}^{-1} \tag{30.38}
$$

and that the elements of $\hat{\theta}^*$ are the roots of

$$
\frac{\partial \log L}{\partial \theta_i} = 0, \qquad i = 1, 2, \ldots, s,
$$

where L is now the ordinary (not the multinomial) likelihood function. Thus if θ_0 is the true value, we have the Taylor expansion

$$
0 = \left[\frac{\partial \log L}{\partial \theta_i} \right]_{\theta_j = \theta_j^*} = \left[\frac{\partial \log L}{\partial \theta_i} \right]_{\theta_j = \theta_{0j}} + (\hat{\theta}_j^* - \theta_{0j}) \left[\frac{\partial^2 \log L}{\partial \theta_i \, \partial \theta_j} \right]_{\theta_j = \theta_{0j} + \varepsilon} \qquad i, j = 1, 2, \ldots, s,
$$

and as in **18.26** this gives asymptotically, using (30.38),

$$
n^{1/2}(\hat{\theta}^* - \theta) = n^{-1/2} D\left(\frac{\partial \log L}{\partial \theta} \right). \tag{30.39}
$$

In this case, we evaluate $V(y)$ directly from (30.21) with $t = \hat{\theta}^*$, where we see that asymptotically

$$
\begin{aligned} V(y) &= E\{x - n^{1/2}B(\hat{\theta}^* - \theta)\}\{x - n^{1/2}B(\hat{\theta}^* - \theta)\}' \\ &= V(x) + nBV(\hat{\theta}^*)B' - (TB' + BT') \end{aligned} \tag{30.40}
$$

where

$$
T = E\{x \cdot n^{1/2}(\hat{\theta}^* - \theta)'\} = E\left\{ x \cdot n^{-1/2} \left(\frac{\partial \log L}{\partial \theta} \right)' \right\} D
$$

asymptotically by (30.39). We write this $T = ED$.

If we write $\delta_{ir} = 1$ if the rth observation falls into the ith class, C_i, and $\delta_{ir} = 0$ otherwise, we can write (30.19) as

$$x_i = \sum_{r=1}^{n} (\delta_{ir} - p_{0i})/(np_{0i})^{1/2},$$

so that the characteristic term of E is

$$n^{-1}p_{0i}^{-1/2}E\left\{\sum_{r=1}^{n}(\delta_{ir} - p_{0i}) \cdot \frac{\partial \log L}{\partial \theta_j}\right\} = p_{0i}^{-1/2}E\left\{(\delta_{ir} - p_{0i})\frac{\partial \log L}{\partial \theta_j}\right\}$$

$$= p_{0i}^{-1/2}E\left\{\delta_{ir}\frac{\partial \log f(x_r \mid \theta)}{\partial \theta_j}\right\},$$

where we have used (17.18) and the independence of the observations. Thus we have

$$p_{0i}^{-1/2}\int_{C_i}\frac{\partial \log f}{\partial \theta_j}f\,dx = p_{0i}^{-1/2}\int_{C_i}\frac{\partial f}{\partial \theta_j}dx = p_{0i}^{-1/2}\frac{\partial}{\partial \theta_j}\int_{C_i}f\,dx = p_{0i}^{-1/2}\frac{\partial}{\partial \theta_j}p_{0i}.$$

This is precisely b_{ij} of (30.22). Thus

$$\mathbf{E} = \mathbf{B}, \qquad \mathbf{T} = \mathbf{BD}. \tag{30.41}$$

Using (30.24), (30.38) and (30.41) in (30.40), we obtain

$$\mathbf{V}(\mathbf{y}) = \mathbf{I} - (\mathbf{p}^{1/2})(\mathbf{p}^{1/2})' - \mathbf{BDB}'. \tag{30.42}$$

30.18   (30.42) is not idempotent, as may be seen by squaring it. Moreover, (30.41) shows that the non-negative covariance matrix

$$\mathbf{V}\left\{\mathbf{B}'\mathbf{x} - n^{-1/2}\left(\frac{\partial \log L}{\partial \boldsymbol{\theta}}\right)\right\} = \mathbf{B}'\mathbf{B} + \mathbf{D}^{-1} - (\mathbf{B}'\mathbf{E} + \mathbf{E}'\mathbf{B}) = \mathbf{D}^{-1} - \mathbf{B}'\mathbf{B} = \mathbf{D}^{-1} - \mathbf{C}^{-1},$$

using (30.30). Thus $\mathbf{C} - \mathbf{D}$ is non-negative and (30.42) may be written in the form

$$\mathbf{V}(\mathbf{y}) = \mathbf{I} - (\mathbf{p}^{1/2})(\mathbf{p}^{1/2})' - \mathbf{BCB}' + \mathbf{B}(\mathbf{C} - \mathbf{D})\mathbf{B}'. \tag{30.43}$$

The first two terms on the right of (30.43) are what we got at (30.26) for the case when no estimation takes place, when $\mathbf{V}(\mathbf{y})$ has $(k-1)$ latent roots unity, and one of zero. The first three terms are (30.34), when, with multinomial ML estimation, $\mathbf{V}(\mathbf{y})$ has $(k-s-1)$ latent roots unity and $(s+1)$ of zero. Because of the non-negative definiteness of all its terms, reduction of (30.43) to canonical form shows that its latent roots are bounded by the corresponding latent roots of (30.26) and (30.34). Thus (30.43) has $(k-s-1)$ latent roots of unity, one of zero, and s between zero and unity, as established by Chernoff and Lehmann (1954).

In general, the values of the s latent roots depend upon $\boldsymbol{\theta}$, but if $\boldsymbol{\theta}$ contains only a location and a scale parameter, this is not so—cf. Exercise 30.13. In any case, it follows from the fact that the two sets of ML estimators $\hat{\boldsymbol{\theta}}$ and $\hat{\boldsymbol{\theta}}^*$ draw closer together as k increases, so that $\mathbf{D} \to \mathbf{C}$, that the last s latent roots tend to zero as $k \to \infty$.

Molinari (1977) shows that if parameters are estimated by the method of moments, X^2 is no longer necessarily bounded above by a χ^2_{k-1} variable.

Chase (1972) shows that if parameters are estimated from an independent sample, there is over-recovery of d.fr., the distribution of X^2 then being bounded below by a χ^2_{k-1} variable whether the ordinary or the multinomial ML estimators are used.

30.19 What we have found, therefore, is that X^2 does not have an asymptotic χ^2 distribution when fully efficient (ordinary ML) estimators are used in estimating parameters—there is partial recovery of the s degrees of freedom lost by the multinomial ML estimators. However, the distribution of X^2 is bounded between a χ^2_{k-1} and a χ^2_{k-s-1} variable, and as k becomes large these are so close together that the difference can be ignored—this is another way of expressing the final deduction in **30.18**. But for k small, the effect of using the χ^2_{k-s-1} distribution for test purposes may lead to serious error; for the probability of exceeding any given value will be greater than we suppose. s is rarely more than 1 or 2, but it is as well to be sure, when ordinary ML estimation is being used, that the critical values of χ^2_{k-s-1} and χ^2_{k-1} are both exceeded by X^2. The tables of χ^2 show that, for a test of size $\alpha = 0.05$, the critical value for $(k-1)$ degrees of freedom exceeds that for $(k-s-1)$ degrees of freedom, if s is small, by Cs, approximately, where C declines from about 1.5 at $(k-s-1)=5$ to about 1.2 when $(k-s-1)=30$. For $\alpha = 0.01$, the corresponding values of C are about 1.7 and 1.3.

The choice of classes for the X^2 test

30.20 The whole of the asymptotic theory of the X^2 test which we have discussed so far is valid however we determine the k classes into which the observations are grouped, so long as they are determined *without reference to the observations*. The italicised condition is essential, for we have made no provision for the class boundaries themselves being random variables. However, it is common practice to determine the class boundaries, and sometimes even to fix k itself, after reference to the general picture presented by the observations. We must therefore discuss the formation of classes, and then consider how far it affects the theory we have developed.

We first consider the determination of class boundaries, leaving the choice of k until later. If there is a natural discreteness imposed by the problem (as in Example 30.1 where there are four natural groups) or if we have a sample of observations from a discrete distribution, the class-boundary problem arises only in the sense that we may decide (in order to reduce k, or in order to improve the accuracy of the asymptotic distribution of X^2 as we shall see in **30.30** below) to amalgamate some of the hypothetical frequencies at the discrete points. Indeed, if a discrete distribution has infinite range, like the Poisson, we are forced to some amalgamation of hypothetical frequencies if k is not to be infinite with most of the hypothetical frequencies very small indeed.

But the class-boundary problem arises in its most acute form only when we are sampling from a continuous distribution. There are now no natural hypothetical frequencies at all. If we suppose k to be determined in advance in some way, how are the boundaries to be determined?

In practice, arithmetical convenience is usually allowed to dictate the solution: the classes are taken to cover equal ranges of the variate, except at an extreme where the range of the variate is infinite. The range of a class is roughly determined by the dispersion of the distribution, while the location of the distribution helps to determine where the central class should fall. Thus, if we wished to form $k = 10$ classes for a sample to be tested for normality, we might roughly estimate (perhaps by eye) the mean $\bar{x}$ and the standard deviation s of the sample and take the class boundaries as $\bar{x} \pm \frac{1}{2} js$, $j = 0, 1, 2, 3, 4$. The classes would then be

$$(-\infty, \bar{x} - 2s), \qquad (\bar{x} - 2s, \bar{x} - 1.5s), \qquad (\bar{x} - 1.5s, \bar{x} - s), \qquad (\bar{x} - s, \bar{x} - 0.5s),$$
$$(\bar{x} - 0.5s, \bar{x}), \qquad (\bar{x}, \bar{x} + 0.5s), \qquad (\bar{x} + 0.5s, \bar{x} + s), \qquad (\bar{x} + s, \bar{x} + 1.5s),$$
$$(\bar{x} + 1.5s, \bar{x} + 2s), \qquad (\bar{x} + 2s, \infty).$$

30.21 Although this procedure is not very precise, it clearly makes the class boundaries random variables, and it is not obvious that the asymptotic distribution of X^2, calculated for classes formed in this way, is the same as when the classes are fixed. However, intuition suggests that since the asymptotic theory holds for *any* set of k fixed classes, it should hold also when the class boundaries are determined from the sample. That this is so when the class boundaries are determined by consistent estimation of parameters in the regular case was shown for the normal distribution by Watson (1957b) and for continuous distributions in general by A. R. Roy (1956), Watson (1958, 1959) and Chibisov (1971)—cf. also Dahiya and Gurland (1972, 1973). D. S. Moore (1971) extends the results to the multivariate case—see also Ruymgaart (1975). D. S. Moore and Spruill (1975) give a unified general treatment of the large-sample theory.

We may thus neglect the random variations of the class boundaries so far as the asymptotic distribution of X^2, when H_0 holds, is concerned. Small-sample distributions, of course, will be affected. (We discuss small-sample distributions of X^2 in the fixed-boundaries case in **30.30** below.)

The equal-probabilities method of constructing classes

30.22 We may now directly face the question of how class boundaries should be determined, in the light of the assurance of the last paragraph of **30.21**. If we now seek an optimum method of boundary determination, it ought to be in terms of the power of the test; we should choose that set of boundaries which maximises power for a test of given size. Failing this, we seek some means of avoiding the unpleasant fact that there is a multiplicity of possible sets of classes, any one of which will in general give a different result for the same data; we require a rule which is plausible and practical.

One such rule has been suggested by Mann and Wald (1942) and by Gumbel (1943); given k, choose the classes so that the hypothetical probabilities p_{0i} are all equal to $1/k$. This procedure is perfectly definite and unique. It varies arithmetically from the usual method, described in **30.20** (in which the classes are variate intervals of equal length) in that we have to use tables to ensure that the p_{0i} are equal. This requires for exactness that the data should be available ungrouped. The procedure is illustrated in Example 30.2.

Example 30.2
Quenouille (1959) gives, apart from a change in location, 1000 random values from the distribution
$$dF = \exp(-x)\,dx, \qquad 0 \leqslant x < \infty.$$
The first 50 of these, arranged in order of variate-value, are:
0.01, 0.01, 0.04, 0.17, 0.18, 0.22, 0.22, 0.25, 0.25, 0.29, 0.42, 0.46, 0.47, 0.47, 0.56, 0.59, 0.67, 0.68, 0.70, 0.72, 0.76, 0.78, 0.83, 0.85, 0.87, 0.93, 1.00, 1.01, 1.01, 1.02, 1.03, 1.05, 1.32, 1.34, 1.37, 1.47, 1.50, 1.52, 1.54, 1.59, 1.71, 1.90, 2.10, 2.35, 2.46, 2.46, 2.50, 3.73, 4.07, 6.03.

Suppose that we wished to form four classes for an X^2 test that the underlying distribution is indeed as stated. A natural grouping with equal-width intervals would be as shown in Table 30.2.

Table 30.2

Variate values	Observed frequency	Hypothetical frequency
0–0.50	14	19.7
0.51–1.00	13	11.9
1.01–1.50	10	7.2
1.51 and over	13	11.2
	50	50.0

Since $2x$ is a χ_2^2 variable, the hypothetical frequencies are obtained from the *Biometrika Tables* d.f. of a χ_2^2. We find $X^2 = 3.1$ with 3 d.fr., a value which would not reject the hypothetical distribution for any test of size less than $\alpha = 0.37$; the agreement of observation and hypothesis is therefore very satisfactory.

Let us now consider how the same data would be treated by the method of **30.22**. We first determine the values of the hypothetical variable dividing it into four equal probability classes—these are, of course, the quartiles. The *Biometrika Tables* give the values 0.288, 0.693, 1.386. We now form Table 30.3.

Table 30.3

Variate values	Observed frequency	Hypothetical frequency
0–0.28	9	12.5
0.29–0.69	9	12.5
0.70–1.38	17	12.5
1.39 and over	15	12.5
	50	50.0

X^2 is now easier to calculate, since (30.8) reduces to

$$X^2 = \frac{k}{n} \sum_{i=1}^{k} n_i^2 - n \tag{30.44}$$

since all hypothetical probabilities $p_{0i} = 1/k$. We find here that $X^2 = 3.9$, which would not lead to rejection unless the test size exceeded 0.27. The result is still very satisfactory, but the equal-probabilities test seems rather more critical of the hypothesis than the other test was.

It will be seen that there is little extra arithmetical work involved in the equal-probabilities method of carrying out the X^2 test. Instead of a constant class-length, with hypothetical frequencies to be looked up in a table (or, if necessary, to be calculated) we have irregular class-lengths determined from the tables so that the hypothetical frequencies are equal.

We have had no parameters to estimate in this example. If parameters must be estimated, we can only equalise the *estimated* hypothetical probabilities.

30.23 Apart from the virtue of removing the class-boundary decision from uncertainty, the equal-probabilities method of forming classes for the X^2 test will not necessarily increase the power of the test, for one would suspect that a goodness-of-fit hypothesis is likely to be most vulnerable at the extremes of the range of the variable, and the equal-probabilities method may well result in a loss of sensitivity at the extremes unless k is rather large. This brings us to the question of how k should be chosen, and in order to discuss this question we must consider the power of the X^2 test. First, we investigate the moments of the X^2 statistic.

The moments of the X^2 test statistic
30.24 We suppose, as before, that we have hypothetical probabilities p_{0i} when H_0 holds, so that our test statistic is, as at (30.8),

$$X^2 = \frac{1}{n} \sum_{i=1}^{k} \frac{n_i^2}{p_{0i}} - n.$$

We confine ourselves to the simple hypothesis. Suppose now that the true probabilities are p_{1i}, $i = 1, 2, \ldots, k$. The expected value of the test statistic is then

$$E(X^2) = \frac{1}{n} \sum_{i=1}^{k} \frac{1}{p_{0i}} E(n_i^2) - n.$$

From the moments of the multinomial distribution at (5.126),

$$E(n_i^2) = np_{1i}(1 - p_{1i}) + n^2 p_{1i}^2, \tag{30.45}$$

whence

$$E(X^2) = \sum_{i=1}^{k} \frac{p_{1i}(1 - p_{1i})}{p_{0i}} + n\left\{\sum_{i=1}^{k} \frac{p_{1i}^2}{p_{0i}} - 1\right\}. \tag{30.46}$$

When H_0 holds ($p_{1i} = p_{0i}$), this reduces to

$$E(X^2 | H_0) = k - 1. \tag{30.47}$$

This exact result is already known to hold asymptotically, since X^2 is then a χ^2_{k-1} variate. If we differentiate (30.46) with respect to the p_{1i}, subject to $\sum_i p_{1i} = 1$, we find that, as $n \to \infty$, (30.46) has its minimum value when $p_{1i} = p_{0i}$. For any hypothesis H_1 specifying a set of probabilities $p_{1i} \neq p_{0i}$, we therefore have asymptotically

$$E(X^2 | H_1) > k - 1. \tag{30.48}$$

(30.48), like the asymptotic argument based on the LR statistic in **30.6**, indicates that the critical region for the X^2 test consists of the upper tail, although this indication is not conclusive since the asymptotic distribution of X^2 is not of the χ^2 form when H_1 holds. This alternative distribution is, in fact, a non-central χ^2 under the conditions given in **30.27** below.

Even the variance of X^2 is a relatively complicated function of the p_{0i} and p_{1i} (cf. Exercise 30.5). However, when H_0 holds it simplifies there to

$$\operatorname{var}(X^2 | H_0) = n^{-1}\{2(n-1)(k-1) - k^2 + \sum_i p_{0i}^{-1}\}$$

$$\xrightarrow[n \to \infty]{} 2(k-1), \tag{30.49}$$

this limit being the asymptotic χ^2_{k-1} variance. (30.49) is exactly equal to $2(k-1) \times (n-1)/n$ when all $p_{0i} = k^{-1}$, but may be much larger than its limiting value $2(k-1)$ if some p_{0i} are very small.

From (30.46) and Exercise 30.3, we also have in the equal-probabilities case $(p_{0i} = k^{-1})$

$$\left.\begin{array}{l} E(X^2 | H_1) = (k-1) + (n-1)(k \sum_i p_{1i}^2 - 1) > k - 1, \\[2mm] \operatorname{var}(X^2 | H_1) \sim 4(n-1)k^2\{\sum_i p_{1i}^3 - (\sum_i p_{1i}^2)^2\}. \end{array}\right\} \tag{30.50}$$

Consistency and unbiasedness of the X^2 test

30.25 Equations (30.50) are sufficient to demonstrate the consistency of the equal-probabilities X^2 test. For the test consists in comparing the value of X^2 with a fixed critical value, say c_α, in the upper tail of its distribution when H_0 holds. Now when H_1 holds, the mean value and variance of X^2 are each of order n. By Chebyshev's inequality (3.95),

$$P\{|X^2 - E(X^2)| \geq \lambda[\operatorname{var}(X^2)]^{1/2}\} \leq \frac{1}{\lambda^2}. \tag{30.51}$$

Since c_α is fixed, it differs from $E(X^2)$ by a quantity of order n, so that if we require the probability that X^2 differs from its mean sufficiently to fall below c_α, the multiplier λ on the left of (30.51) must be of order $n^{1/2}$, and the right-hand side is therefore of order n^{-1}. Thus

$$\lim_{n \to \infty} P\{X^2 < c_\alpha\} = 0,$$

and the test is consistent for any H_1 specifying unequal class-probabilities.

The general X^2 test, with unequal p_{0i}, is also consistent against alternatives specifying at least one $p_{1i} \neq p_{0i}$, as is intuitively reasonable. A proof is given by Neyman (1949).

30.26 Although the X^2 test is consistent (and therefore asymptotically unbiased), it is not unbiased in general for fixed n, as may be seen for the simplest case $k = 2$ in Exercise 30.12. For, from (30.46), we may have $E(X^2 | H_1) < k - 1$ in finite samples, and in view of the upper-tail critical region, this indicates the possibility that the test is biased against H_1 for some values of α. Haberman (1988) discusses this point in detail and recommends care when k is large and many p_{0i} are small and variable.

However, A. Cohen and Sackrowitz (1975) show that the X^2 test, and also the LR test, are strictly unbiased if we use equal probabilities, and that X^2 minimises $|\partial^2 P / \partial \theta_i \, \partial \theta_j|$ $(i, j = 1, 2, \ldots, k - 1)$ at H_0. R. Thompson (1979) shows that every unbiased test against one-sided alternatives has a monotone power function.

The limiting power function of the X^2 test

30.27 Suppose that, as in our discussion of ARE in Chapter 25, we allow H_1 to approach H_0 as n increases, at a rate sufficient to keep the power bounded away from 1. In fact, let $p_{1i} - p_{0i} = c_i n^{-1/2}$ where the c_i are constants not dependent on n. Then the distribution of X^2 is asymptotically a non-central χ^2 with degrees of freedom $k - s - 1$ (where s parameters are estimated by the multinomial ML estimators) and non-central parameter

$$\lambda = \sum_{i=1}^{k} \frac{c_i^2}{p_{0i}} = n \sum_{i=1}^{k} \frac{(p_{1i} - p_{0i})^2}{p_{0i}}. \tag{30.52}$$

This result, first announced by Eisenhart (1938), follows at once from the representation of **30.9–10**; its proof is left to the reader as Exercise 30.4. We now see that the second term on the right of (30.46) is simply λ of (30.52), in accordance with κ_1 in Exercise 23.1. It now follows by using the Neyman–Pearson lemma at (21.6) on (23.18) that the best critical region for testing $\lambda = 0$ consists of the upper tail of the distribution of X^2.

The tables described in **23.8** may be used to evaluate the asymptotic power of the X^2 test, given by the integral (23.29). The approximation (23.30) may also be used.

> For *fixed* p_{1i}, the limiting distribution as $n \to \infty$ of $\{X^2 - E(X^2)\}/\{\mathrm{var}\,(X^2)\}^{1/2}$ is standardised normal; when power is large, this gives a better power approximation for simple H_0 than (30.52) does—see Broffitt and Randles (1977).

Example 30.3
We may illustrate the use of the limiting power function by returning to the problem of Example 30.2 and examining the effect on the power of the equal-probabilities procedure of doubling k. To facilitate use of the *Biometrika Tables*, we actually take four classes with slightly unequal probabilities (Table 30.4).

In the table, the p_{0i} are obtained from the gamma distribution with parameter 1, as before, and the p_{1i} from the gamma distribution with parameter 1.5. For these 4

Table 30.4

Values	p_{0i}	p_{1i}	$(p_{1i}-p_{0i})^2$	$(p_{1i}-p_{0i})^2/p_{0i}$
0–0.3	0.259	0.104	0.0240	0.0927
0.3–0.7	0.244	0.190	0.0029	0.0119
0.7–1.4	0.250	0.282	0.0010	0.0040
1.4 and over	0.247	0.424	0.0313	0.1267
				$\overline{0.2353} = \lambda/n$

classes, and $n = 50$ as in Example 30.2, we evaluate the non-central parameter of (30.52) as $\lambda = 0.2353 \times 50 = 11.8$. With 3 degrees of freedom for X^2, this gives a power when $\alpha = 0.05$ of 0.83, from Patnaik's table.

Suppose now that we form eight classes by splitting each of the above classes into two, with the new p_{0i} as equal as is convenient for use of the tables. We obtain Table 30.5.

Table 30.5

Values	p_{0i}	p_{1i}	$(p_{1i}-p_{0i})^2$	$(p_{1i}-p_{0i})^2/p_{0i}$
0–0.15	0.139	0.040	0.0098	0.0705
0.15–0.3	0.120	0.064	0.0031	0.0258
0.3–0.45	0.103	0.071	0.0010	0.0097
0.45–0.7	0.141	0.119	0.0005	0.0035
0.7–1.0	0.129	0.134	0.0000	0.0002
1.0–1.4	0.121	0.148	0.0007	0.0058
1.4–2.1	0.125	0.183	0.0034	0.0272
2.1 and over	0.122	0.241	0.0142	0.1163
				$\overline{0.2590} = \lambda/n$

For $n = 50$, we now have $\lambda = 13.0$ with 7 degrees of freedom. The approximate power for $\alpha = 0.05$ is now about 0.75 from Patnaik's table. The doubling of k has increased λ, but only slightly. The power is actually reduced, because for given λ the central and non-central χ^2 distributions draw closer together as degrees of freedom increase (cf. Exercise 23.3) and here this effect is stronger than the increase in λ. However, n is too small here for us to place any reliance on the exact values of the power obtained from the limiting power function, and we should perhaps conclude that the doubling of k has affected the power very little.

The choice of k with equal probabilities
30.28 With the aid of the asymptotic power function of **30.27**, we can get a heuristic indication of how to choose k in the equal-probabilities case. The non-central parameter (30.52) is then, putting $\theta_i = p_{1i} - k^{-1}$,

$$\lambda = nk \sum_{i=1}^{k} \theta_i^2. \tag{30.53}$$

We now assume that $|\theta_i| \leqslant 1/k$, all i, and consider what happens as k becomes large. $\theta = \sum_{i=1}^{k} \theta_i^2$, as a function of k, will then be of the same order of magnitude as a sum of squares in the interval $(-1/k, 1/k)$, i.e.

$$\theta \sim a \int_{-1/k}^{1/k} u^2 \, du = 2a \int_{0}^{1/k} u^2 \, du. \tag{30.54}$$

The asymptotic power of the test is a function $P\{k, \lambda\}$ which therefore is $P\{k, \lambda(k)\}$; it is a monotone increasing function of λ, and has its stationary values when $d\lambda(k)/dk = 0$. We thus put, using (30.53) and (30.54),

$$0 = \frac{1}{n} \frac{d\lambda}{dk} \sim \theta + k \cdot 2a \left(\frac{1}{k}\right)^2 \left(-\frac{1}{k^2}\right) = \theta - 2ak^{-3}$$

giving

$$k^{-3} \sim \theta/(2a). \tag{30.55}$$

We cannot let $k \to \infty$ without restriction since all θ_i then $\to 0$, but we assume k large enough so that both the H_0 and H_1 distribution of X^2 are near normality, and the approximate power function of the test is (cf. (25.53)) therefore

$$P = G\left\{\frac{\left[\frac{\partial}{\partial \theta} E(X^2 | H_1)\right]_{\theta=0}}{[\text{var}\,(X^2 | H_0)]^{1/2}} \cdot \theta - \lambda_\alpha\right\}, \tag{30.56}$$

where

$$G(-\lambda_\alpha) = \alpha \tag{30.57}$$

determines the size of the test. From (30.49) and (30.50)

$$\left[\frac{d}{d\theta} E(X^2 | H_1)\right]_{\theta=0} = (n-1)k, \tag{30.58}$$

$$\text{var}\,(X^2 | H_0) = 2(k-1)(n-1)/n, \tag{30.59}$$

and if we insert these values and also (30.55) into (30.56), we obtain approximately

$$P = G\{2^{1/2} a(n-1)k^{-5/2} - \lambda_\alpha\}. \tag{30.60}$$

This is the approximate power function at the point where power is maximised for choice of k. If we choose a value P_0 at which we wish the maximisation to occur, we have, on inverting (30.60),

$$G^{-1}\{P_0\} = 2^{1/2} a(n-1)k^{-5/2} - \lambda_\alpha, \tag{30.61}$$

or

$$k = b \left\{\frac{2^{1/2}(n-1)}{\lambda_\alpha + G^{-1}\{P_0\}}\right\}^{2/5}, \tag{30.62}$$

where $b = a^{2/5}$.

30.29   In the special case $P_0 = \frac{1}{2}$ (where we wish to choose k to maximise power when it is 0.5), $G^{-1}(0.5) = 0$ and (30.62) simplifies. In this case, Mann and Wald (1942) obtained (30.62) by a much more sophisticated and rigorous argument—they found $b = 4$ in the case of the simple hypothesis. Our own derivation suggests that the same essential argument applies for the composite hypothesis, but b may be different in this case.

We conclude that k should be increased in the equal-probabilities case in proportion to $n^{2/5}$, and that k should be smaller if we are interested in the region of high power (when $G^{-1}\{P_0\}$ is large) than if we are interested in the 'neighbouring' region of low power (when $G^{-1}\{P_0\}$ approaches $-\lambda_\alpha$, from above since the test is unbiased).

With $b = 4$ and $P_0 = \frac{1}{2}$, (30.62) leads to much larger values of k than are commonly used. k will be doubled when n increases by a factor of $4\sqrt{2}$. When $n = 200$, $k = 31$ for $\alpha = 0.05$ and $k = 27$ for $\alpha = 0.01$—these are about the lowest values of k for which the approximate normality assumed in our argument (and also in Mann and Wald's) is at all accurate. In this case, Mann and Wald recommend the use of (30.62) when $n \geq 450$ for $\alpha = 0.05$ and $n \geq 300$ for $\alpha = 0.01$. It will be seen that n/k, the hypothetical expectation in each class, increases as $n^{3/5}$, and is equal to about 6 and 8 respectively when $n = 200$, $\alpha = 0.05$ and 0.01.

C. A. Williams (1950) reported that k can be halved from the Mann–Wald optimum without serious loss of power at the 0.50 point. But it should be remembered that n and k must be substantial before (30.62) produces good results. Example 30.4 illustrates the point, which is also borne out by calculations made by Hamdan (1963) for tests of a normal mean. Hamdan (1968) verifies in the bivariate normal case that the use of (30.62) can result in loss of power compared with the use of equal-length classes. For a restricted class of alternatives, Quine and Robinson (1985) show that asymptotic efficiency is best if not too many classes are used—see also Kallenberg *et al.* (1985) and Kallenberg (1985). The rate of increase of k as $n \to \infty$ is crucial, essentially for the reason discussed in Example 30.3.

Example 30.4

Consider again the problem of Example 30.3. We there found that we were at around the 0.8 value for power. From a table of the normal distribution $G^{-1}(0.8) = 0.84$. With $b = 4$, $\alpha = 0.05$, $\lambda_a = 1.64$, (30.62) gives for the optimum k around this point

$$k = 4 \left\{ \frac{2^{1/2}(n-1)}{2.48} \right\}^{2/5} = 3.2(n-1)^{2/5}.$$

For $n = 50$, this gives $k = 15$ approximately.

Suppose now that we use the *Biometrika Tables* to construct a 15-class grouping with probabilities p_{0i} as nearly equal as is convenient. We obtain Table 30.6.

Here $\lambda = 13.7$ and Patnaik's table gives a power of 0.64 for 14 degrees of freedom. λ has again been increased, but power reduced because of the increase in k. We are not at the optimum here. With large k (and hence large n), the effect of increasing degrees of freedom would not offset the increase of λ in this way.

Table 30.6

Values	p_{0i}	p_{1i}	$(p_{1i} - p_{0i})^2/p_{0i}$
0–0.05	0.049	0.008	0.034
0.05–0.15	0.090	0.032	0.037
0.15–0.20	0.042	0.020	0.012
0.20–0.30	0.078	0.044	0.015
0.30–0.40	0.071	0.047	0.008
0.40–0.50	0.063	0.048	0.004
0.50–0.65	0.085	0.072	0.002
0.65–0.75	0.050	0.047	0.000
0.75–0.90	0.065	0.067	0.000
0.90–1.1	0.074	0.083	0.000
1.1–1.3	0.060	0.075	0.004
1.3–1.6	0.071	0.095	0.008
1.6–2.0	0.067	0.101	0.018
2.0–2.7	0.068	0.116	0.034
2.7 and over	0.067	0.145	0.098
			$0.274 = \lambda/n$

30.30   We must not make k too large, since the multinormal approximation to the multinomial distribution cannot be expected to be satisfactory if the np_{0i} are very small. If the H_0 distribution is unimodal, and equal-length classes are used in the conventional manner, small hypothetical frequencies will occur only at the tails. Yarnold (1970), generalising earlier recommendations by Cochran (1952, 1954), concludes from a detailed theoretical investigation of the simple H_0 case that for $\alpha = 0.05$ or 0.01 the *minimum* hypothetical frequency may be as small as $5f$, where f is the proportion of the $k \geqslant 3$ classes with expectations less than 5—Lawal and Upton (1980) suggest a lognormal approximation for ν d.fr. when this minimum exceeds $fk\nu^{-3/2}$, which is usually considerably smaller than $5f$. Roscoe and Byars (1971) found in extensive sampling experiments that if $k > 2$, the approximation is good for an *average* hypothetical frequency $\geqslant 2$ for test size $\alpha = 0.05$ and $\geqslant 4$ for $\alpha = 0.01$.

In the equal-probabilities case, all the hypothetical frequencies will be equal. Slakter (1966, 1968) shows that even for fractional equal hypothetical frequencies, the approximation remains good, but that power is approximately 80 per cent of its nominal value—see also Roscoe and Byars (1971), Kempthorne (1967) and **30.5** above for the simple H_0.

The Mann–Wald procedure of **30.29** leads to hypothetical frequencies always greater than 5 for $n \geqslant 200$. It is interesting to note that in Examples 30.3–4, the application of the 5 limit would have ruled out the 15-class procedure, and that the more powerful 8-class procedure, with hypothetical frequencies ranging from 5 to 7, would have been acceptable.

Finally, we remark that the large-sample nature of the distribution theory of X^2 is not a disadvantage in practice, for we do not usually wish to test goodness-of-fit except in large samples.

Recommendations for the X^2 test of fit

30.31 We summarise the above discussion with a few practical recommendations:

(1) If the distribution being tested has been tabulated, use classes with equal, or nearly equal, probabilities.

(2) Determine the number of classes when n exceeds 200 approximately by (30.62) with b between 2 and 4.

(3) If parameters are to be estimated, use the ordinary ML estimators in the interest of efficiency, but recall that there is partial recovery of degrees of freedom (**30.19**) so that critical values should be adjusted upwards; if the multinomial ML estimators are used, no such adjustment is necessary.

None of the theory above will hold if the *form* (instead of the parameters alone) of $F_0(x)$ is estimated from the data used to test goodness of fit.

X^2 tests of independence

30.32 The general theory of the X^2 test may be used to test the fit of hypotheses of independence for categorised variables, where we are given only the frequencies falling into various categories. For the independence implies a set of hypothetical frequencies, and we test the fit of the observed frequencies to these in the usual way.

Example 30.5   Large-sample test of independence in a 2×2 table
Suppose the members of a population (assumed infinite) are categorised by the presence or absence of attributes A and B. The four possible outcomes will have probabilities

$$
\begin{array}{c|cc|c}
 & B & \text{not-B} & \text{Totals} \\
\hline
A & p_{11} & p_{12} & p_{1\cdot} \\
\text{not-A} & p_{21} & p_{22} & p_{2\cdot} \\
\hline
\text{Totals} & p_{\cdot 1} & p_{\cdot 2} & 1
\end{array}
\tag{30.63}
$$

where a dot suffix denotes summation over that suffix. This is a simple multinomial as in (5.124) with 4 classes. A random sample of size n from the population has corresponding frequencies arranged in a *2×2 table*:

$$
\begin{array}{cc|c}
n_{11} & n_{12} & n_{1\cdot} \\
n_{21} & n_{22} & n_{2\cdot} \\
\hline
n_{\cdot 1} & n_{\cdot 2} & n
\end{array}
\tag{30.64}
$$

The ML estimates of the p_{ij} are obtained by maximising $\log L = \text{constant} + \sum_{i=1}^{2} \sum_{j=1}^{2} n_{ij} \log p_{ij}$, subject to $\sum_i \sum_j p_{ij} = 1$. This results in

$$
\hat{p}_{ij} = n_{ij}/n,
\tag{30.65}
$$

the generalisation of the binomial result in Example 17.9 and **18.5**. Hence we have

$$
\left.
\begin{aligned}
\hat{p}_{i\cdot} &= n_{i\cdot}/n, \\
\hat{p}_{\cdot j} &= n_{\cdot j}/n.
\end{aligned}
\right\}
\tag{30.66}
$$

Now consider the hypothesis that the two attributes are independently distributed over the population, i.e. that there is the same probability of an A among the B's as among the not-B's. This is

$$H_0: p_{11}/p_{\cdot 1} = p_{12}/p_{\cdot 2} \tag{30.67}$$

which implies

$$p_{11}/p_{\cdot 1} = p_{1 \cdot}/1,$$

i.e.

$$p_{11} = p_{1 \cdot} p_{\cdot 1},$$

and similarly any

$$p_{ij} = p_{i \cdot} p_{\cdot j}. \tag{30.68}$$

We require the ML estimators of the p_{ij} under the constraint imposed by H_0. We can save ourselves the routine algebra by observing that the marginal probabilities $p_{i \cdot}$ and $p_{\cdot j}$ are quite unaffected by the hypothesis of independence, which concerns only the relations between the probabilities p_{ij} in the body of the table. Thus the ML estimators under H_0 of $p_{i \cdot}$ and $p_{\cdot j}$ will remain as at (30.66). In consequence, the estimators of the individual p_{ij} under H_0 will, by (30.68), be

$$\hat{p}_{ij} = \frac{n_{i \cdot}}{n} \frac{n_{\cdot j}}{n}. \tag{30.69}$$

The hypothetical frequencies corresponding to the n_{ij} will therefore be

$$e_{ij} = n\hat{p}_{ij} = n_{i \cdot} n_{\cdot j}/n \tag{30.70}$$

and we test the fit by

$$X^2 = \sum_{i=1}^{2} \sum_{j=1}^{2} \frac{(n_{ij} - e_{ij})^2}{e_{ij}}. \tag{30.71}$$

The asymptotic equivalence of X^2 to the LR test $-2 \log l$ and the fact that by **23.7** the latter is asymptotically a χ^2 variable with d.fr. equal to the number of constraints imposed by H_0 show that X^2 is asymptotically a χ^2 variable with one degree of freedom, since the single constraint (30.67) implies the set of four hypothetical frequencies.

(30.71) may be simplified in this example, for clearly each $(n_{ij} - e_{ij})^2$ takes the same value, say D^2. Thus

$$X^2 = D^2 \sum_i \sum_j \frac{1}{e_{ij}} = n\left(n_{11} - \frac{n_{1 \cdot} n_{\cdot 1}}{n}\right)^2 \left(\frac{1}{n_{1 \cdot} n_{\cdot 1}} + \frac{1}{n_{1 \cdot} n_{\cdot 2}} + \frac{1}{n_{2 \cdot} n_{\cdot 1}} + \frac{1}{n_{2 \cdot} n_{\cdot 2}}\right)$$

$$= \frac{n(n_{11} n_{22} - n_{12} n_{21})^2}{n_{1 \cdot} n_{\cdot 1} n_{2 \cdot} n_{\cdot 2}}, \tag{30.72}$$

a convenient form for computation.

Yates (1934b) proposed a correction to improve the continuous asymptotic approximation to the discrete distribution of X^2. This is

$$X_c^2 = \frac{n(|n_{11}n_{22} - n_{12}n_{21}| - \frac{1}{2}n)^2}{n_1. n_{.1} n_2. n_{.2}},$$
(30.73)

equivalent to decreasing (increasing) n_{11} and n_{22} and increasing (decreasing) n_{12} and n_{21} by $\frac{1}{2}$ each when $n_{11}n_{22} > (<)n_{12}n_{21}$.

> Cochran (1954) recommended use of (30.73) for $n \geqslant 40$, and if all $e_{ij} \geqslant 5$ for $n \geqslant 20$ also. On the basis of large-scale sampling experiments, Roscoe and Byars (1971) recommend use of the uncorrected (30.72) if all $e_{ij} \geqslant 7.5$. In general, (30.73) is not necessarily an improvement. Loh (1989) shows that, whether or not the continuity correction is used, the actual size of the test can be many times larger than the intended α as $n \to \infty$, but is closer to it as the minimum e_{ij} increases. Yates (1984) critically reviews methods for 2×2 tables—see also the ensuing discussion.

Example 30.6   Independence in an $r \times c$ table
The generalisation of Example 30.5 to the case where there are r ($\geqslant 2$) categories for the attribute A and s ($\geqslant 2$) categories for the attribute B is quite straightforward. The probabilities are now

p_{11}	p_{12}	$\cdots$	p_{1c}	$p_1.$
p_{21}	p_{22}	$\cdots$	p_{2c}	$p_2.$
$\vdots$	$\vdots$		$\vdots$	$\vdots$
p_{r1}	p_{r2}		p_{rc}	$p_r.$
$p_{.1}$	$p_{.2}$		$p_{.c}$	1

(30.74)

instead of (30.63), while (30.64) is replaced by the $r \times c$ table

n_{11}	n_{12}	$\cdots$	n_{1c}	$n_1.$
n_{21}	n_{22}	$\cdots$	n_{2c}	$n_2.$
$\vdots$	$\vdots$		$\vdots$	$\vdots$
n_{r1}	n_{r2}		n_{rc}	$n_r.$
$n_{.1}$	$n_{.2}$		$n_{.c}$	n

(30.75)

As before, the unconstrained ML estimations are given by (30.65–6). The hypothesis of independence of A and B now states that there is the same set of probabilities in each of the c columns, i.e.

$$H_0: \frac{p_{i1}}{p_{.1}} = \frac{p_{i2}}{p_{.2}} = \cdots = \frac{p_{ic}}{p_{.c}}, \qquad i = 1, 2, \ldots, r.$$
(30.76)

For each i, H_0 imposes $c - 1$ constraints, and (30.76) for any $(r - 1)$ values of i implies (30.76) for the other value of i. Thus there are $(r - 1)(c - 1)$ constraints imposed by H_0.

As before, the ML estimators of the marginal probabilities are unaffected by the hypothesis of independence. Since (30.76) implies (30.68) as before, we have (30.69) once again for the constrained ML estimators of the individual probabilities. X^2 will now have $(r-1)(c-1)$ d.fr. in its limiting χ^2 distribution, and may be written, from the extension of (30.71) to r rows, c columns, as

$$X^2 = n\left\{ \sum_{i=1}^{r} \sum_{j=1}^{c} \frac{n_{ij}^2}{n_{i.}n_{.j}} - 1 \right\}. \tag{30.77}$$

(30.77) is a form of (30.8).

> Roscoe and Byars (1971) found the asymptotic approximation to the distribution of X^2 is adequate if the average e_{ij} at (30.70) ≥ 4 for test size $\alpha = 0.05$ and ≥ 6 for $\alpha = 0.01$. See also Cochran (1954). A decrease in d.fr. improves the closeness of the test's size to the intended α, and so does an increase in the minimum e_{ij}—see Loh (1989).

30.33 Apart from the difficulties we have already discussed in connection with X^2 tests, which are not very serious, they have been criticised on two counts. In each case, the criticism is of the power of the test. Firstly, the fact that the essential underlying device is the reduction of the problem to a multinomial distribution problem itself implies the necessity for grouping the observations into classes. In a broad general sense, we must lose information by grouping in this way, and we suspect that the loss will be greatest when we are testing the fit of a continuous distribution. Secondly, the fact that the X^2 statistic is based on the *squares* of the deviations of observed from hypothetical frequencies implies that the X^2 test will be insensitive to the patterns of signs of these deviations, which is clearly informative. The first of these criticisms is the more radical, since it must clearly lead to the search for other test statistics to replace X^2, and we postpone discussion of such tests until after we have discussed the second criticism.

The signs of deviations

30.34 Let us consider how we should expect the pattern of deviations (of observed from hypothetical frequencies) to behave in some simple cases. Suppose that a simple hypothesis specifies a continuous unimodal distribution with location and scale parameters, say equal to mean and standard deviation; and suppose that the hypothetical mean is too high. For any set of k classes, the p_{0i} will be too small for low values of the variate, and too large thereafter. Since in large samples the observed proportions will converge stochastically to the true probabilities, the pattern of signs of observed deviations will be a series of positive deviations followed by a series of negative deviations. If the hypothetical mean is too low this pattern is reversed.

Suppose now that the hypothetical value of the scale parameter is too low. The pattern of deviations in large samples is now a series of positives, followed by a series of negatives, followed by positives again. If the hypothetical scale parameter is too high, all these signs are reversed.

Now of course we do not knowingly use the X^2 test for changes in location or scale alone, since we can then find more powerful test statistics. However, when there is error in both location and scale parameters, the situation is essentially unchanged; we shall still have three (or in more complicated cases, somewhat more) 'runs' of signs of deviations. More generally, whenever the parameters have true values differing from their hypothetical values, or when the true distributional form is one differing 'smoothly' from the hypothetical form, we expect the signs of deviations to cluster in this way instead of being distributed randomly, as they should be if the hypothetical frequencies were the true ones.

This observation suggests that we supplement the X^2 test with a test of the number of runs of signs among the deviations, small numbers forming the critical region. The elementary theory of runs necessary for this purpose is given as Exercise 30.8. Before we can use it in any precise way, however, we must investigate the relationship between the 'runs' test and the X^2 test. F. N. David (1947), Seal (1948) and Fraser (1950) showed that when H_0 holds the tests are asymptotically independent (cf. Exercise 30.7) and that for testing the simple hypothesis all patterns of signs are equiprobable, so that the distribution theory of Exercise 30.8 can be combined with the X^2 test as indicated in Exercise 30.9.

The supplementation by the 'runs' test is likely to be valuable in increasing sensitivity when testing a simple hypothesis, as in the illustrative discussion above. For the composite hypothesis, of particular interest where tests of fit are concerned, when all parameters are to be estimated from the sample, it is of no practical value, since the patterns of signs of deviations, although independent of X^2, are not equiprobable as in the simple hypothesis case, and the distribution theory of Exercise 30.8 is therefore of no use (cf. Fraser, 1950).

Other tests of fit

30.35 We now turn to the discussion of alternative tests of fit. Since these have striven to avoid the loss of information due to grouping suffered by the X^2 test, they cannot avail themselves of multinomial simplicities, and we must expect their theory to be more difficult. Before we discuss the most important test, we remark on a feature they have in common.

It will have been noticed that, when using X^2 to test a simple hypothesis, its distribution is asymptotically χ^2_{k-1} *whatever the simple hypothesis may be*, although its exact distribution does depend on the hypothetical distribution specified. It is clear that this result is achieved because of the intervention of the multinomial distribution and its tendency to joint normality. Moreover, the same is true of the composite hypothesis situation if multinomial ML estimators are used—in this case $X^2 \to \chi^2_{k-s-1}$ *whatever the composite hypothesis may be*, though its exact distribution is even more clearly seen to be dependent on the composite hypothesis concerned. When other estimators are used (even when fully efficient ordinary ML estimators are used) these pleasant asymptotic properties do not hold in general: even the asymptotic distribution of X^2 now depends on the latent roots of the matrix (30.37), which are in general functions both of the hypothetical distribution and of the values of the parameters θ.

We express these results by saying that, in the first two instances above, the distribution of X^2 is asymptotically *distribution-free* (i.e. free of the influence of the hypothetical distribution's form and parameters), whereas in the third instance it is not asymptotically distribution-free or even *parameter-free* (i.e. free of the influence of the parameters of F_0 without being distribution-free).

30.36 The most important alternative tests of fit all make use, directly or indirectly, of the *probability-integral transformation,* which we have encountered on various occasions (e.g. **1.27**) as a means of transforming any known continuous distribution to the uniform distribution on the interval $(0, 1)$. In our present notation, if we have a simple hypothesis of fit specifying a d.f. $F_0(x)$, to which a density $f_0(x)$ corresponds, then the variable $y = \int_{-\infty}^{x} f_0(u)\,du = F_0(x)$ is uniformly distributed on $(0, 1)$. Thus if we have a set of n observations x_i and transform them to a new set y_i by the probability-integral transformation for a known $F_0(x)$, and use a function of the y_i to test the departure of the y_i from uniformity, the distribution of the test statistic will be distribution-free, not merely asymptotically but for any n.

When the hypothetical distribution is composite, say $F_0(x\,|\,\theta_1, \theta_2, \ldots, \theta_s)$ with the s parameters θ to be estimated, we must select s functions $t_1, \ldots, t_s$ of the x_i for this purpose. The transformed variables are now

$$y_i = \int_{-\infty}^{x_i} f_0(u\,|\,t_1, t_2, \ldots, t_s)\,du,$$

but they are neither independent nor uniformly distributed, and their distribution will depend in general both on the hypothetical distribution F_0 and on the true values of its parameters, as F. N. David and Johnson (1948) showed in detail. However (cf. Exercise 30.10), if F has only parameters of location and scale, suitably invariantly estimated, the distribution of the y_i will depend on the form of F but not on its parameters. It follows that for finite n, no test statistic based on the y_i can be distribution-free for a composite hypothesis of fit (although it may be parameter-free if only location and scale parameters are involved). Of course, such a test statistic may still be asymptotically distribution-free.

Tests of fit based on the sample distribution function: Kolmogorov's D_n

30.37 The remaining general tests of fit are all functions of the cumulative distribution of the sample, or *sample distribution function,* defined by

$$S_n(x) = \begin{cases} 0, & x < x_{(1)}, \\[2mm] \dfrac{r}{n}, & x_{(r)} \leqslant x < x_{(r+1)}, \\[2mm] 1, & x_{(n)} \leqslant x. \end{cases} \tag{30.78}$$

The $x_{(r)}$ are the order-statistics, i.e. the observations arranged so that

$$x_{(1)} \leqslant x_{(2)} \leqslant \cdots \leqslant x_{(n)}.$$

$S_n(x)$ is simply the proportion of the observations not exceeding x. If $F_0(x)$ is the true d.f., fully specified, from which the observations come, we have, for each value of x, from the strong law of large numbers,

$$\lim_{n \to \infty} P\{S_n(x) = F_0(x)\} = 1, \tag{30.79}$$

and in fact stronger results are available concerning the convergence of the sample d.f. to the true d.f.

In a sense, (30.79) is the fundamental relationship on which all statistical theory is based. If something like it did not hold, there would be no point in random sampling. In our present context, it is clear that a test of fit can be based on any measure of divergence of $S_n(x)$ and $F_0(x)$.

Durbin (1972) surveys the field of tests based on the sample distribution function. The most important is based on deviations of the sample d.f. $S_n(x)$ from the completely specified continuous hypothetical d.f. $F_0(x)$. The measure of deviation used, however, is simple, being the maximum absolute difference between $S_n(x)$ and $F_0(x)$. Thus we define

$$D_n = \sup_x |S_n(x) - F_0(x)|. \qquad (30.80)$$

The appearance of the modulus in the definition (30.80) leads us to expect difficulties in the investigation of the distribution of D_n, but remarkably enough, the asymptotic distribution was obtained by Kolmogorov (1933) when he first proposed the statistic.

$$\lim_{n \to \infty} P\{D_n > z n^{-1/2}\} = 2 \sum_{r=1}^{\infty} (-1)^{r-1} \exp\{-2r^2 z^2\} \qquad (30.81)$$

Smirnov (1948) tabulates (30.81) (actually its complement) for $z = 0.28 \ (0.01) \ 2.50 \ (0.05)$ 3.00 to 6 d.p. or more. This is the whole effective range of the limiting distribution.

As well as deriving the limiting result (30.81), Kolmogorov (1933) gave recurrence relations for finite n, which have since been used to tabulate the distribution of D_n. Z. W. Birnbaum (1952) gives tables of $P\{D_n < c/n\}$ to 5 d.p., for $n = 1 \ (1) \ 100$ and $c = 1 \ (1) \ 15$, and inverse tables of the values of D_n for which this probability is 0.95 for $n = 2 \ (1) \ 5 \ (5) \ 30 \ (10) \ 100$ and for which the probability is 0.99 for $n = 2 \ (1) \ 5 \ (5) \ 30 \ (10) \ 80$. L. H. Miller (1956) gives inverse tables for $n = 1 \ (1) \ 100$ and probabilities 0.90, 0.95, 0.98, 0.99. Massey (1950, 1951) had previously given $P\{D_n < c/n\}$ for $n = 5 \ (5) \ 80$ and selected values of $c \leqslant 9$, and also inverse tables for $n = 1 \ (1) \ 20 \ (5) \ 35$ and probabilities 0.80, 0.85, 0.90, 0.95, 0.99.

It emerges that the critical values of the asymptotic distribution are $1.3581 n^{-1/2}$ for $\alpha = 0.05$ and $1.6276 n^{-1/2}$ for $\alpha = 0.01$ and that these are always greater than the exact values for finite n. The approximation for these values of α is satisfactory at $n = 80$.

Confidence limits for distribution functions

30.38 Because the distribution of D_n is distribution-free and adequately known for all n, and because it uses as its measure of divergence the maximum absolute deviation between $S_n(x)$ and $F_0(x)$, we may reverse the procedure of testing for fit and use D_n to set confidence limits for a (continuous) distribution function *as a whole*. For, *whatever the true $F(x)$*, we have, if d_α is the critical value of D_n for test size α,

$$P\{D_n = \sup_x |S_n(x) - F(x)| > d_\alpha\} = \alpha.$$

We may invert this into the confidence statement

$$P\{S_n(x) - d_\alpha \leqslant F(x) \leqslant S_n(x) + d_\alpha, \quad \text{all } x\} = 1 - \alpha. \qquad (30.82)$$

Thus we simply set up a band of width $\pm d_\alpha$ around the sample d.f. $S_n(x)$, and there is probability $1 - \alpha$ that the true $F(x)$ lies *entirely* within this band. This is a remarkably simple and direct method of estimating a distribution function. No other test of fit permits this inversion of test into confidence limits since none uses so direct and simply interpretable a measure of divergence as D_n.

One can draw useful conclusions from this confidence limits technique as to the sample size necessary to approximate a d.f. closely. For example, from the critical values given at the end of **30.37**, it follows that a sample of 100 observations would have probability 0.95 of having its sample d.f. everywhere within 0.13581 of the true d.f. To be within 0.05 of the true d.f. everywhere, with probability 0.99, would require a sample size of $(1.6276/0.05)^2$, i.e. more than 1000.

> Noether (1963) shows that the left side of (30.82) holds with probability $\geqslant 1 - \alpha$ for discrete distributions. Thus the D_n test is then also conservative. Conover (1972) and Gleser (1985) give exact results for the discrete case and Wood and Altavela (1978) some asymptotic results.
>
> Exercise 30.15 shows how to set somewhat analogous conservative confidence intervals, using X^2, for all the theoretical probabilities p_{0i} simultaneously.
>
> Using the fact that symmetry of a d.f. about a known point θ implies
>
> $$F(x) = 1 - F(2\theta - x),$$
>
> Schuster (1973) gives a Kolmogorov-type test for this case that exactly halves the width of the confidence band (30.82). P. V. Rao *et al.* (1975) give an estimator of the centre of symmetry θ based on D_n, and Schuster (1975) an estimator of the d.f. See also Doksum *et al.* (1977), Koziol (1980), Randles *et al.* (1980) and Breth (1982).

30.39 Because it is a modular quantity, D_n does not permit us to set one-sided confidence limits for $F(x)$, but we may consider positive deviations only and define

$$D_n^+ = \sup_x \{S_n(x) - F_0(x)\} \tag{30.83}$$

as was done by Wald and Wolfowitz (1939) and Smirnov (1939a).

The variable $y = 2n(D_n^+)^2$ is asymptotically distributed in the negative exponential form

$$\mathrm{d}F(y) = \exp(-y)\,\mathrm{d}y, \qquad 0 \leqslant y < \infty.$$

Alternatively, we may express this by saying that $2y = 4n(D_n^+)^2$ is asymptotically a χ^2 variate with 2 degrees of freedom. Evidently, exactly the same theory will hold if we consider only negative deviations and a statistic D_n^-.

> Z. W. Birnbaum and Tingey (1951) give an expression for the exact distribution of D_n^+, and tabulate the values it exceeds with probabilities 0.10, 0.05, 0.01, 0.001, for $n = 5, 8, 10, 20, 40, 50$. As for D_n, the asymptotic values exceed the exact values, and the differences are small for $n = 50$.
>
> We may evidently use D_n^+ to obtain one-sided confidence regions which take the form $P\{S_n(x) - d_\alpha^+ \leqslant F(x)\} = 1 - \alpha$, where d_α^+ is the critical value of D_n^+.
>
> Stephens (1970, 1974) gives simple modifications of the statistics D_n, D_n^+ which permit very compact tables of critical values (reproduced in the *Biometrika Tables*, Vol. II) and makes some power comparisons.

Comparison of D_n and X^2 tests

30.40 Using calculations made by C. A. Williams (1950), Massey (1951) compared the values of Δ for which the large-sample powers of the X^2 and the D_n tests are at least 0.5. For test size $\alpha = 0.05$, the D_n test can detect with power 0.5 a Δ about half the magnitude of that which the X^2 test can detect with this power; even with $n = 200$, the ratio of Δs is 0.6, and it declines steadily in favour of D_n as n increases. For $\alpha = 0.01$ the relative performances are very similar. Since this comparison is based on a poor lower bound to the power of D_n, we must conclude that D_n is a very much more sensitive test for the fit of a continuous distribution.

Kac *et al.* (1955) point out that if the Mann–Wald equal-probabilities procedure of **30.28-9** is used, the X^2 test requires Δ to be of order $n^{-2/5}$ to attain power $\frac{1}{2}$, whereas D_n requires Δ to be of order $n^{-1/2}$. Thus D_n asymptotically requires sample size to be of order $n^{4/5}$ compared to n for the X^2 test, and is asymptotically very much more efficient—in fact the relative efficiency of X^2 will tend to zero as n increases.

Computation of D_n

30.41 If we are setting confidence limits for the unknown $F(x)$, no computations are required beyond the simple calculation of $S_n(x)$ and the setting of bounds distant $\pm d_\alpha$ from it. In using D_n for testing, however, we have to face the possibility of calculating $F_0(x)$ for every observed value of x, a procedure that is tedious even when $F_0(x)$ is well tabulated. However, because the test criterion is the maximum deviation between $S_n(x)$ and $F_0(x)$, it is often possible by preliminary examination of the data to locate the intervals in which the deviations are likely to be large. If initial calculations are made only for these values, computations may be stopped as soon as a single deviation exceeding d_α is found. (This abbreviation of the calculations is not possible for statistics that depend on *all* deviations.)

A further considerable saving of labour may be effected as in the following example, due to Z. W. Birnbaum (1952).

Example 30.7
A sample of 40 observations is to hand, where values are arranged in order:
0.0475, 0.2153, 0.2287, 0.2824, 0.3743, 0.3868, 0.4421, 0.5033, 0.5945, 0.6004, 0.6255, 0.6331, 0.6478, 0.7867, 0.8878, 0.8930, 0.9335, 0.9602, 1.0448, 1.0556, 1.0894, 1.0999, 1.1765, 1.2036, 1.2344, 1.2543, 1.2712, 1.3507, 1.3515, 1.3528, 1.3774, 1.4209, 1.4304, 1.5137, 1.5288, 1.5291, 1.5677, 1.7238, 1.7919, 1.8794.
We wish to test, with $\alpha = 0.05$, whether the true $F_0(x)$ is normal with mean 1 and variance 1/6. From Z. W. Birnbaum's (1952) tables we find for $n = 40$, $\alpha = 0.05$ that $d_\alpha = 0.2101$. Consider the smallest observation, $x_{(1)}$. To be acceptable, $F_0(x_{(1)})$ should lie between 0 and d_α, i.e. in the interval $(0, 0.2101)$. The observed value of $x_{(1)}$ is 0.0475, and from tables of the normal d.f. we find $F_0(x_{(1)}) = 0.0098$, within the above interval, so the hypothesis is not rejected by this observation. Further, it cannot possibly be rejected by the next higher observations until we reach an $x_{(i)}$ for which *either* (a) $i/40 - 0.2101 > 0.0098$, i.e. $i > 8.796$, *or* (b) $F_0(x_{(i)}) > 0.2101 + 1/40$, i.e. $x_{(i)} > 0.7052$ (from the tables again). The 1/40 is added on the right of (b) because we know that $S_n(x_{(i)}) \geqslant 1/40$ for $i > 1$. Now from the data, $x_{(i)} > 0.7052$ for $i \geqslant 14$. We next need, therefore, to examine $i = 9$ (from the inequality (a)). We find there the

acceptance interval for $F_0(x_{(9)})$

$$(S_9(x) - d_\alpha, S_8(x) + d_\alpha) = (9/40 - 0.2101, 8/40 + 0.2101) = (0.0149, 0.4101).$$

We find from the tables $F_0(x_{(9)}) = F_0(0.5945) = 0.1603$, which is acceptable. To reject H_0, we now require *either*

$$i/40 - 0.2101 > 0.1603, \quad \text{i.e.} \quad i > 14.82$$

or

$$F_0(x_{(i)}) > 0.4101 + 1/40, \quad \text{i.e.} \quad x_{(i)} > 0.9052, \quad \text{i.e.} \quad i \geqslant 17.$$

We therefore proceed to $i = 15$, and so on. The reader should verify that only the 6 values $i = 1, 9, 15, 21, 27, 34$ require computations in this case. The hypothesis is accepted because in every one of these six cases the value of F_0 lies in the confidence interval; it would have been rejected, and computations ceased, if any one value had lain outside the interval.

30.42 For the composite hypothesis of fit, with unspecified parameters, Kolmogorov-type tests were investigated by Durbin (1975). By **30.36**, the tests are not now distribution-free, but will be parameter-free if the parameters are those of location and scale.

The most important special cases are the normal distribution, discussed in **30.43** below, and the exponential distribution with unknown scale parameter. For the latter, Durbin (1975) tabulated the percentage points of D_n^+, D_n^- and D_n up to $n = 100$—Iman (1982) gives them in graphical form for $\alpha = 0.01, 0.05, 0.10$ and $n = 5, 10, 20, 30, 50, 100$; Stephens' (1970) approximations for D_n in the composite case are adequate for $n \geqslant 5$, being $1.09n^{-1/2}$ and $1.31n^{-1/2}$ asymptotically for $\alpha = 0.05, 0.01$, or about four-fifths of the simple H_0 values in **30.37**. See also Margolin and Maurer (1976). Stephens (1976) obtains asymptotic percentage points.

> Stephens (1977) and Chandra *et al.* (1981) give tests for the extreme-value distribution of Exercise 18.6 and the Weibull distribution of **5.33**, and Stephens (1979a) for the logistic distribution.

Tests of normality

30.43 To conclude this chapter, we refer briefly to the problem of testing normality, i.e. the problem of testing whether the underlying d.f. is normal, the parameters being unspecified. Of course, any general test of fit for the composite hypothesis may be employed to test normality. However, it is common to test the observed moment-ratios $\sqrt{b_1}$ and b_2 against their distributions given the hypothesis of normality (cf. **12.18** and Exercises 12.9–10 and the percentage points given in the *Biometrika Tables*) and these are sometimes called 'tests of normality', although they are better called tests of skewness and kurtosis respectively.

D. S. Moore (1971) gives values of X^2 for testing normality with $\alpha = 0.001, 0.005, 0.01, 0.05, 0.10, 0.25$ and $k = 5, 7, 9, 11, 15, 21$ when fully efficient ML estimators are used to determine boundaries with estimated equal probabilities in classes.

D. S. Moore (1982) shows that if the observations are normal but positively dependent, the X^2 test tends to reject the hypothesis of normality. This holds also for non-normal distributions—see Gleser and Moore (1983).

Kac *et al.* (1955) discuss the distributions of D_n in testing normality when the two parameters (μ, σ^2) are estimated from the sample by $(\bar{x}, s^2)$. The limiting distributions are parameter-free (because μ and σ are location and scale parameters—cf. **30.36**) but are not obtained explicitly. Lilliefors (1967) gave critical values of D_n in testing normality, re-computed to much greater accuracy by Dallal and Wilkinson (1986) for $\alpha = 0.001$, 0.01, 0.05 (0.05) 0.20 and $n = 4$ (1) 20 (5) 30, 40, 100, 400, 900, with an analytic approximation—Iman (1982) gives graphs for $\alpha = 0.01$, 0.05, 0.10 and $n = 5$ (5) 20, 30, 50, 100. These critical values are roughly two-thirds of the simple H_0 values in **30.37**—e.g. for $\alpha = 0.05$, the value is $0.886n^{-1/2}$ and for $\alpha = 0.01$ it is $1.031n^{-1/2}$.

Shapiro and Wilk (1965) give a criterion for testing normality based on the regression of the order-statistics upon their expected values, using the theory of **19.20–1** and extensive sampling experiments to establish its distribution. It is defined by $W = (\sum_{i=1}^{n} a_i x_{(i)})^2 / \sum_{i=1}^{n} (x_i - \bar{x})^2$ where the a_i are tabulated coefficients. Small values of W are critical. Leslie *et al.* (1986) study its asymptotic distribution and establish its consistency. See also Verrill and Johnson (1987, 1988), who give tables and large-sample theory for censored samples. Tables of W are given in the *Biometrika Tables*, Vol. II. Shapiro and Francia (1972) give a simplified approximate version, W', for large samples which is proved consistent by Sarkadi (1975)—see also Weisberg (1974).

Shapiro *et al.* (1968) and Stephens (1974) make power comparisons from extensive sampling experiments and show that W is usually somewhat superior to the other tests given in this chapter for testing normality, although $\sqrt{b_1}$ and b_2 together are sensitive to non-normality. In sampling experiments using many different populations, E. S. Pearson *et al.* (1977) compare the powers of W, W' and tests based on $\sqrt{b_1}$ and b_2 with those of 'omnibus' tests combining the latter. W does well for symmetric populations with $\beta_2 < 3$, but one-tailed b_2 tests are better; for $\beta_2 > 3$, W performs poorly. Similarly, one-tailed $\sqrt{b_1}$ is recommended for skew populations, though W does well compared with other omnibus tests. Saniga and Miles (1979) also recommend b_2 and $\sqrt{b_1}$ against asymmetric stable alternatives.

A simple test of normality against asymmetric alternatives, based on the characteristic independence of sample mean and variance (cf. Exercise 12.21) is proposed by Lin and Mudholkar (1980); its power compares well with those of the other tests above.

Doksum *et al.* (1977) discuss plots and tests for assessing symmetry, some of which compared favourably in power with W above.

The *Biometrika Tables*, Vol. II, give simple modifications for using D_n and other statistics to test the composite H_0 of normality or exponentiality—cf. Stephens (1976).

EXERCISES

30.1 Show that if, in testing a composite hypothesis, an inconsistent set of estimators **t** is used, the statistic $X^2 \to \infty$ as $n \to \infty$.

(Cf. Fisher, 1924c)

30.2 Using (30.33), show that the matrix **M** defined at (30.37) reduces, when the vector of multinomial ML estimators $\hat{\boldsymbol{\theta}}$ is used, to

$$\mathbf{M} = \begin{pmatrix} \tfrac{1}{2}\mathbf{I} & \tfrac{1}{4}\mathbf{C} \\ \mathbf{C}^{-1} & \tfrac{1}{2}\mathbf{I} \end{pmatrix},$$

and that **M** is idempotent with tr $\mathbf{M} = s$. Hence confirm the result of **30.10** and **30.14** that X^2 is asymptotically distributed like χ^2_{k-s-1} when $\hat{\boldsymbol{\theta}}$ is used.

(Watson, 1959)

30.3 Show from the limiting joint normality of the n_i that as $n \to \infty$, the variance of the simple-hypothesis X^2 statistic in the equal-probabilities case ($p_{0i} = 1/k$) is

$$\operatorname{var}(X^2) = \lim_{n \to \infty} \left[2k^2 \left\{ \sum_i p_{1i}^2 - \left(\sum_i p_{1i}^2 \right)^2 \right\} + 4(n-1)k^2 \left\{ \sum_i p_{1i}^3 - \left(\sum_i p_{1i}^2 \right)^2 \right\} \right]$$

where p_{1i}, $i = 1, 2, \dots, k$ are the true class probabilities. Verify that this reduces to the correct value $2(k-1)$ when

$$p_{1i} = p_{0i} = 1/k.$$

(Mann and Wald, 1942)

30.4 Establish the non-central χ^2 result of **30.27** for the alternative hypothesis distribution of the X^2 test statistic.

(Cf. Cochran, 1952)

30.5 Show from the moments of the multinomial distribution (cf. (5.126)) that the exact variance of the simple-hypothesis X^2 statistic is given by

$$n \operatorname{var}(X^2) = 2(n-1) \left\{ 2(n-2) \sum_i \frac{p_{1i}^3}{p_{0i}^2} - (2n-3) \left(\sum_i \frac{p_{1i}^2}{p_{0i}} \right)^2 - 2 \left(\sum_i \frac{p_{1i}^2}{p_{0i}} \right) \left(\sum_i \frac{p_{1i}}{p_{0i}} \right) + 3 \sum_i \frac{p_{1i}^2}{p_{0i}^2} \right\}$$
$$- \left(\sum_i \frac{p_{1i}}{p_{0i}} \right)^2 + \sum_i \frac{p_{1i}}{p_{0i}^2}.$$

Show that the limiting results of Exercise 30.3 follow when all $p_{0i} = k^{-1}$. When H_0 holds ($p_{1i} = p_{0i}$), show that

$$n \operatorname{var} X^2 = 2(n-1)(k-1) - k^2 + \sum_i p_{0i}^{-1}$$

exactly, reducing to $n \operatorname{var} X^2 = 2(n-1)(k-1)$ if all $p_{0i} = k^{-1}$.

(Cf. Patnaik, 1949)

30.6 For the same alternative hypothesis as in Example 30.3, namely the gamma distribution with parameter 1.5, use the *Biometrika Tables* to obtain the p_{1i} for the unequal-probabilities four-class grouping in Example 30.2. Calculate the non-central parameter (30.52) for this case, and show by comparison with Example 30.3 that the unequal-probabilities grouping would require about a 25 per cent larger sample than the equal-probabilities grouping in order to attain the same power against this alternative.

30.7 k independent standardised normal variables x_j are subject to c homogeneous linear constraints. Show that $S = \sum_{j=1}^{k} x_j^2$ is distributed independently of the signs of the x_j. Given that $c = 1$, and that the constraint is $\sum_{j=1}^{k} x_j = 0$, show that all sequences of signs are equiprobable (except all signs positive, or all signs negative, which cannot occur), but that this is not so generally for $c > 1$. Hence show that any test based on the sequence of signs of the deviations of observed from hypothetical frequencies $(n_i - np_{0i})$ is asymptotically independent of the X^2 test when H_0 holds.

(F. N. David, 1947; Seal, 1948; Fraser, 1950)

30.8 M elements of one kind and N of another are arranged in a sequence at random $(M, N > 0)$. A *run* is defined as a subsequence of elements of one kind immediately preceded and succeeded by elements of the other kind. Let R be the number of runs in the whole sequence $(2 \le R \le M + N)$. Show that

$$P\{R = 2s\} = 2\binom{M-1}{s-1}\binom{N-1}{s-1} \bigg/ \binom{M+N}{M},$$

$$P\{R = 2s - 1\} = \left\{\binom{M-1}{s-2}\binom{N-1}{s-1} + \binom{M-1}{s-1}\binom{N-1}{s-2}\right\} \bigg/ \binom{M+N}{M},$$

and that

$$E(R) = 1 + \frac{2MN}{M+N}$$

$$\text{var } R = \frac{2MN(2MN - M - N)}{(M+N)^2(M+N-1)}.$$

(Stevens (1939); Wald and Wolfowitz (1940). Swed and Eisenhart (1943) tabulate the distribution of R for $M \le N \le 20$.)

30.9   From Exercises 30.7 and 30.8, show that if there are M positive and N negative deviations $(n_i - np_{0i})$, we may use the runs test to supplement the X^2 test for the simple hypothesis. From Exercise 16.4, show that if P_1 is the probability of a value of X^2 not less than that observed and P_2 is the probability of a value of R not greater than that observed, then $U = -2(\log P_1 + \log P_2)$ is asymptotically distributed like χ^2 with 4 degrees of freedom, large values of U forming the critical region for the combined test.

(F. N. David, 1947)

30.10 $x_1, x_2, \ldots, x_n$ are independent random variables with the same distribution $f(x | \theta_1, \theta_2)$. θ_1 and θ_2 are estimated by statistics $t_1(x_1, x_2, \ldots, x_n)$, $t_2(x_1, x_2, \ldots, x_n)$. Show that the random variables

$$y_i = \int_{-\infty}^{x_i} f(u | t_1, t_2) \, du$$

are not independent and that they have a distribution depending in general on f, θ_1 and θ_2; but that if f is of form $\theta_2^{-1} f\{(x - \theta_1)/\theta_2\}$, $\theta_2 > 0$, and t_1, t_2 respectively satisfy $t_1(\mathbf{x} + \alpha) = t_1(\mathbf{x}) + \alpha$, $t_2(\beta \mathbf{x}) = \beta t_2(\mathbf{x})$, $\beta > 0$, the distribution of y_i is not dependent on θ_1 and θ_2, but on the form of f alone.

(F. N. David and Johnson, 1948)

30.11   Show that for testing a composite hypothesis the X^2 test statistic using multinomial ML estimators is asymptotically equivalent to the LR test statistic when H_0 holds.

30.12   Show that when there are $k = 2$ classes and $n = 3$ observations, of which n_1 fall into the first class which has hypothetical probability $p_{01} = \frac{1}{4}$, X^2 has its critical region made up cumulatively of the points $n_1 = 3$, $n_1 = 2$, $n_1 = 0$, $n_1 = 1$ in that order, according to the value of the test size α. Show that if $\alpha < 1$, the critical region is biased.

(Cf. A. Cohen and Sackrowitz, 1975)

30.13   Show that if the distribution tested is of the form

$$\frac{1}{\theta_2} f\left(\frac{x-\theta_1}{\theta_2}\right), \quad \theta_2 > 0,$$

the matrix **BDB'** in (30.42) does not depend upon θ_1, θ_2, so that its latent roots also do not.

(Cf. Watson, 1958)

30.14   Show that in the equal-probabilities case, X^2 defined at (30.44) varies by multiples of $2k/n$ and hence that we may expect the χ^2 approximation to the distribution of X^2 to be improved by a continuity correction of $-k/n$. Show also that the minimum attainable value of X^2 is

$$\frac{n - k\left\lfloor\dfrac{n}{k}\right\rfloor}{n}\left\{k\left(\left\lfloor\dfrac{n}{k}\right\rfloor + 1\right) - n\right\},$$

where $[z]$ is the integral part of z, and hence is zero only when n is an integral multiple of k.

30.15   If X^2 for the simple or composite H_0 is asymptotically distributed in the χ^2 form with f degrees of freedom, its $100(1-\alpha)$ percentile being $\chi^2_{f,\alpha}$, show that if $\max |(n_i - np_{0i})/n| = \Delta \leqslant 0.5$, the minimum possible value of X^2 for a fixed Δ, whatever the n_i and the p_{0i}, is $4n\Delta^2$. Putting $n \geqslant n_0 = \chi^2_{f,\alpha}/(4\Delta^2)$, show that

$$\text{Prob}\,\{X^2 \leqslant 4n\Delta^2\} \geqslant 1 - \alpha$$

for any set of p_{0i} whatever, and that for sufficiently large n, a conservative set of $100(1-\alpha)$ per cent confidence intervals for all the true p_{0i} simultaneously is given by

$$\left\{\frac{n_i}{n} \pm 0.5(\chi^2_{f,\alpha}/n)^{1/2}\right\}, \quad i = 1, 2, \ldots, k.$$

(Naddeo, 1968)

30.16   Let $\boldsymbol{\theta}$ be the unspecified parameters in testing a composite hypothesis of fit for n observations **x**. Suppose that **t**, with less than n components, is minimal sufficient for $\boldsymbol{\theta}$, and that we can make a $1-1$ transformation from **x** to $(\mathbf{t}, \mathbf{u})$, where **u** is distributed independently of **t**. Show that if the value of **t** is discarded, and is replaced by a random observation **t'** from its distribution with a known value of $\boldsymbol{\theta}$, then the set of observations **x'** obtained by the inverse transformation from $(\mathbf{t'}, \mathbf{u})$ is distributed independently of $\boldsymbol{\theta}$, so that the hypothesis of fit becomes simple.

(Durbin, 1961)

30.17   A random sample of n observations u_r is taken from the uniform distribution on the interval $(0, 1)$, dividing that interval into $(n+1)$ lengths c_j, where $c_j \geqslant 0$ and $\sum_{j=1}^{n+1} c_j = 1$. The c_j are ordered so that $c_{(1)} \leqslant c_{(2)} \leqslant \cdots \leqslant c_{(n+1)}$. Show that the non-negative variables

$$\begin{cases} g_1 = (n+1)c_{(1)}, \\ g_j = (n+2-j)(c_{(j)} - c_{(j-1)}), \quad j = 2, 3, \ldots, n+1; \ \sum_{j=1}^{n+1} g_j = 1, \end{cases}$$

have the distribution

$$\mathrm{d}F_g = n!\,\mathrm{d}g_1 \cdots \mathrm{d}g_n,$$

and that the unordered c_j also have this distribution, so that

$$\mathrm{d}F_c = n!\,\mathrm{d}c_1 \cdots \mathrm{d}c_n.$$

(the $(n+1)$th variable being omitted in each case to remove the singularity of the distribution). Hence show that the variables

$$w_r = \sum_{j=1}^{r} g_j, \quad r = 1, 2, \ldots, n,$$

are distributed exactly as the order-statistics of the original sample, $u_{(r)}$, $r = 1, 2, \ldots, n$. Thus any test of fit based on the probability-integral transformation may be applied to the w_r, as well as to the $u_{(r)}$ obtained from the transformation.

> (Durbin (1961), who finds from sampling experiments that a one-sided Kolmogorov test (D_n^-) applied to the w_r has better power properties than the ordinary two-sided D_n test for detecting changes in distributional form.)

CHAPTER 31

COMPARATIVE STATISTICAL INFERENCE

31.1 Statistical inference is an inductive process from sample to population. In thinking of a hypothesis (H) and observational data, or evidence (E), there is no problem in making probabilistic statements of the form $P(E|H)$; indeed, these are justified by deductive logic once the axioms of probability are specified, and such statements have been used repeatedly throughout these volumes. However, the very existence of inductive statements of the form $P(H|E)$ has been questioned and many philosophers, notably Sir Karl Popper (1968, 1969), have concluded that such probabilities do not exist. Such probabilities are, of course, the posterior probabilities of the Bayesian approach, so the debate is of vital interest to statisticians. Indeed, it shows no signs of abating, and the interested reader should consult Popper and Miller (1987), Good (1988), Gemes (1989), and Miller (1990) for recent developments.

Any inferential procedure must be based on some more or less rational set of rules, but the rationality of any such given system and the apparent value of the conclusions it allows remain open to debate.

In these volumes, we have adopted the *frequentist* paradigm, sometimes known as the *classical* approach, which has been the dominant school of statistical thought for most of this century. However, the *Bayesian* viewpoint has grown in popularity since the 1950s and several other approaches to inference, some more complete than others, have been developed in recent years.

31.2 In this chapter, we attempt to outline both the areas of agreement and the principal differences between the major schools; it is not our intent to develop each approach in detail. For example, the companion volume by O'Hagan (to be published) provides a comprehensive development of inference from the Bayesian viewpoint. General discussions of comparative inference are given by Barnett (1982), Dawid (1983), and the volume edited by Godambe and Sprott (1971). A more philosophical discussion, supporting the subjective Bayesian approach, appears in Howson and Urbach (1989).

Since our discussion is a rather short appraisal of a large and complex literature, we shall emphasise only the principal points at issue. Thus, we examine 'standard' positions within each school and do not emphasise debates within a school (e.g., choice of axioms for subjective probability). We hope that these broad brush strokes are seen to produce portraits and not caricatures!

A framework for inference

31.3 In general terms, the inferential process contains the following ingredients:

A *measurable (vector) random variable*, X, taking on values in the sample space, $\mathcal{X}$.

The *unknown parameter(s)*, θ, which may be partitioned into parameters of direct interest and nuisance parameters, then denoted by θ and ϕ, respectively. The set of possible θ values are defined by the parameter space Ω.

The *population of interest* which we assume is representable in terms of one of a family of probability distributions $\{F(x, \theta)\}$, indexed by θ. We use

$$F \equiv F(\theta) \equiv F(x, \theta) = P(X \leq x \mid \theta)$$

interchangeably where no ambiguity arises. The functional form of F may be completely specified or a member of some class of distributions, $\mathscr{F}$.

A *statistical experiment* that produces a set of observations, described by the random vector $X = (X_1, X_2, \ldots, X_n)$, with a particular realisation, the *sample data*, denoted by $x = \{x_1, x_2, \ldots, x_n\}$. The experimental procedure specifies the mode of sampling and the form of the stopping rule, S, whether or not such information is required.

Our notation will not distinguish vectors and scalars, unless the discussion explicitly requires that the distinction be made.

In addition, there may be historical (or prior) information regarding θ of either a personal or objective nature which we summarise by some function $p(\theta)$. Since the specification, use and even existence of such information is a matter for considerable debate, we defer further discussion of this topic until **31.35**. The general form of the inference problem is to use the available information

$$I = \{\mathscr{X}, \Omega, \mathscr{F}, x, S, p\} \tag{31.1}$$

to make inductive statements about θ. We now examine the various approaches to this problem, beginning with an overview of the frequentist approach that we have adopted hitherto. This discussion leads us to consider such issues as ancillarity, conditionality, and sufficiency which, in turn, lead to an appraisal of the likelihood approach. Then, after an examination of the fiducial approach, we turn our attention to Bayesian inference and so to decision theory. The chapter concludes with an evaluation of the different approaches and a discussion of attempts at reconciliation between these schools of thought.

The frequentist approach

31.4 The frequency theory of probability, introduced in **7.11**, assumes that it is possible to consider an infinite sequence of independent replications of the same statistical experiment.

We now confine attention primarily to point estimation, but include comments on interval estimation and testing hypotheses as appropriate; we compared Bayesian and fiducial intervals with confidence intervals in **20.45–6**. We may consider a statistic or estimator, $T(X)$, as a summary of the information about θ; for simplicity, we will often restrict attention to a single parameter. In Chapter 17, we identified certain desirable properties for T, such as *consistency* (**17.7**) and *unbiasedness* (**17.9**). Since there is often a multiplicity of estimators satisfying these requirements, we sought measures of *efficiency* (**17.28–9**) and identified desirable estimators as MVU, minimum variance unbiased (**17.13–27**). The broader criterion of MMSE or minimum mean square error (**17.30**) is sometimes felt to be more appropriate and applicable.

Although these criteria may be deemed desirable in themselves, they lack a method for constructing suitable statistics, T. Within the exponential family, the set of sufficient statistics may be identified (**17.31–41**) which leads to the MVU estimator of θ if it exists (**17.35**). More generally, in Chapter 18, we established that the maximum likelihood estimator, (MLE), obtained as

$$\hat{\theta} = \max_{\theta \in \Omega} L(\theta \mid x), \qquad (31.2)$$

where

$$L(\theta \mid x) = \prod_{i=1}^{n} f(x_i \mid \theta) \qquad (31.3)$$

is consistent and asymptotically unbiased under mild regularity conditions when the observations are independent and from the same distribution. Furthermore, the MLE is a function of the sufficient statistics and is asymptotically MVU.

31.5   Even at this stage, we find some parting of the ways, which the large-sample properties of the MLE tend to obscure. If T is an unbiased estimator for θ, then $g(T)$ is not unbiased for $\phi = g(\theta)$, whereas the MLE is functionally invariant so that

$$\hat{\theta} = T \Leftrightarrow \hat{\phi} = g(T). \qquad (31.4)$$

Example 31.1
Given a random sample of n observations, X_i, from a normal population with mean θ and variance 1, we have that

$$\hat{\theta} = \bar{X} \quad \text{and} \quad \hat{\phi} = (\bar{X})^2$$

when $\phi = \theta^2$. However, the MVU for ϕ is

$$T = (\bar{X})^2 - (1/n). \qquad (31.5)$$

Although $E(T) = \phi$ and $\phi \geq 0$, it may happen that the observed value of T is negative. Common sense suggests replacing negative values of T by zero, even though this violates the property of unbiasedness. In general, such adjustments produce estimators with smaller mean square error, so that different criteria may lead to different estimators.

> *Ad hoc* estimators obtained by solving $T = g(\hat{\theta})$, where $E(T) = g(\theta)$, are widely used and justified by appeal to the unbiasedness for $g(\theta)$, even though these estimators are biased for θ unless $g(\theta)$ is a linear function.

Stopping rules
31.6   Another component of the information set that affects the unbiasedness property is the stopping rule. Suppose that the random variable describes the presence or absence of an attribute, with $\pi = P(\text{present})$. We know from **9.32** that, if X denotes the number of successes in a sample of fixed size n, the estimator

$$p = X/n \quad \text{has} \quad E(p) = \pi$$

and is MVU. However, if we sample until the kth success, observing X failures along the way, the MVU estimator becomes

$$(k-1)/(k+X-1);$$

see Example 9.15. More generally, as we saw in **24.30–35**, it is difficult to obtain exact sampling distributions for sequential procedures.

Censoring mechanisms

31.7 Unbiasedness is clearly a property of the sample space since it depends upon taking the expectation over $\mathscr{X}$. Suppose, for example, that $\mathscr{X} \equiv [0, \infty)$, but the exact value of x is recorded only in the range $[0, u]$, otherwise the observation is censored. Any unbiasedness property possessed by an estimator would be destroyed by the censoring.

Example 31.2

Suppose that X is exponentially distributed with mean θ. Given a sample of size n, clearly $E(X) = \theta$, but if we now censor the observation at u, it follows that

$$E(X) = g(\theta) = \theta\{1 - \exp(-u/\theta)\} \tag{31.6}$$

so that the solution of $\bar{X} = g(\hat{\theta})$ now yields an *ad hoc* estimator for θ which is consistent but only asymptotically unbiased.

This led to a famous discourse in Pratt (1962) which we summarise briefly:

> Suppose that the researcher carries out the experiment as described above with observations censored at u since the measuring instrument will only record in the range $[0, u]$. Then the statistician recommends use of the estimator based on (31.6) even if *no observations were actually censored.*
>
> The researcher then reports that, had censoring occurred, another machine was available that would allow recording up to $u' > u$; the statistician duly modifies the estimator derived from (31.6), with u' in place of u. Finally, the researcher reports that the second machine was broken and the statistician reverts to his first estimator.

As Pratt notes, none of these machinations affect what was actually observed and the dependence of the estimator on what was not observed appears unreasonable. This criticism of frequentist methods has been raised by many others over the years (e.g., Jeffreys, 1961).

31.8   In part, the response to this issue relates to the nature of the estimator used. Suppose that n observations are taken with censoring at u and that j observations are actually censored. The likelihood function for $(\mathbf{x}, j)$ may be written as

$$L(\mathbf{x}, j \mid \theta) = L(\mathbf{x} \mid j, \theta) L(j \mid \theta)$$

$$= \prod_{i=1}^{n-j} \left\{ \frac{f(x_i \mid \theta)}{1 - Q} \right\} \binom{n}{j} Q^j (1 - Q)^{n-j}, \tag{31.7}$$

where

$$Q = Q(\theta) = P(X > u). \tag{31.8}$$

The ML estimator for θ based on the conditional likelihood $L(\mathbf{x}|j, \theta)$ is, for the exponential, precisely that suggested by (31.6) earlier. From (31.7), we can see that the statistician has landed in the predicament described by Pratt through use of an *inefficient* estimator; the information contained in $L(j|\theta)$ has been ignored. When the full likelihood in (31.7) is used and $j=0$, we obtain the original uncensored LF, as can be seen by inspection. Alternatively, one might argue that censoring should only be taken into account when it actually occurs, an application of the conditionality principle discussed in **31.15** below.

31.9   The critic of the frequentist approach might justifiably claim that we have saved the 'statistician' only by rejecting the frequency approach. Our response is that the estimator based on (31.6) is not wrong, but *inefficient*. A user of statistical methods must decide upon the properties considered desirable in an estimator and an overly rigid insistence upon unbiasedness may lead to difficulties.

Nevertheless, the notion of unbiasedness has considerable intuitive appeal and many would be reluctant to abandon it. Therefore, let us take a step backwards and consider the problem in a more general setting. It is evident that many estimation procedures involve finding the global maximum (or minimum) of some function $G(\theta|x)$. Under suitable regularity conditions, this value is given by setting the derivative to zero so that $\hat{\theta}$ is given by

$$G' \equiv G'(\hat{\theta}|x) = 0; \tag{31.9}$$

this is true for both ML and LS estimators, for example. Since $E(G')=0$ directly from (31.9), we term (31.9) an *unbiased estimating equation* (Godambe, 1960, 1976). This weaker concept of unbiasedness often produces the usual estimators based on fixed size sampling rules; however, the estimators are unaffected by the stopping rule.

Example 31.3
The exponential family equation in **17.19** has f.f.

$$f(x|\theta) = \exp\{A(\theta)B(x) + C(x) + D(\theta)\}. \tag{31.10}$$

The ML estimators are given by

$$G' = A(\theta)\Sigma B(x_i) + nD'(\theta) = 0. \tag{31.11}$$

From **17.17**, it is evident that the estimating equation gives the UMVU estimator when it exists.

Example 31.4
The LS estimating equation for the regression model is, from **19.4** and (31.9),

$$\mathbf{X'X\theta} - \mathbf{X'y} = \mathbf{0}. \tag{31.12}$$

When $\mathbf{X}$ is of full rank and fixed, the standard LS estimator results.

Example 31.5
Given x successes in n trials for a Bernoulli process with probability of success θ,

the likelihood function is

$$L(x|\theta) = \text{const} - x \log \theta - (n-x) \log (1-\theta)$$

so that $G' = 0$ yields $\hat{\theta} = x/n$. The unbiased estimators for fixed and sequential sampling are, respectively,

$$x/n \quad \text{and} \quad (x-1)/(n-1);$$

see **9.28** and **9.37**, Vol. 1.

31.10   The estimators generated by (31.9) will be optimal when ML is used if there are no nuisance parameters or if it is possible to condition upon statistics that are complete and sufficient for the nuisance parameters (Godambe, 1976). When this is not feasible, it is still possible to consider the partial likelihood (cf. **18.44**). Lindsay (1982) develops a weaker optimality criterion that is met by such estimators.

It might be argued that this revised concept of unbiasedness gives away too much, but since users of unbiased estimators often use them to estimate nonlinear functions of the parameters (when unbiasedness is lost), we feel that the estimating equation approach has considerable merit.

31.11   Our discussion has touched on a variety of issues, including the use of conditional arguments and heavy reliance upon the likelihood function. Conditional estimators are most effective when the statistics on which we condition contain no information about θ. This leads to the notion of *ancillary* statistics and paves the way for discussions of other approaches to inference.

Ancillary statistics

31.12   Consider a set of $r+s$ $(r \geq 1, s \geq 0)$ statistics, written (T_r, T_s), that are minimal sufficient for $k+l$ $(k \geq 1, l \geq 0)$ parameters, which we shall write (θ_k, θ_l). Suppose now that the subset T_s has a distribution free of θ_k. (This is possible only if the distribution of (T_r, T_s) is not boundedly complete—cf. Exercise 22.7.) We then have the factorisation of the likelihood function into

$$L(\mathbf{x}|\theta_k, \theta_l) = g(T_r, T_s | \theta_k, \theta_l) h(\mathbf{x})$$

$$= g_1(T_r | T_s, \theta_k, \theta_l) g_2(T_s | \theta_l) h(\mathbf{x}). \tag{31.13}$$

Fisher (e.g., 1956) calls T_s an *ancillary statistic*, while Bartlett (e.g., 1939) calls the conditional statistic $(T_r | T_s)$ a *quasi-sufficient statistic* for θ_k, the term arising from the resemblance of (31.13) when θ_l is known to the factorisation of (17.84) that characterises a sufficient statistic.

It should be noted that some authors do not require (T_r, T_s) to be minimally sufficient. In general, this increases the problems associated with the non-uniqueness of the ancillary statistics, and we shall restrict attention to the minimally sufficient case.

The interest in ancillary statistics derives from the idea that, since the distribution of T_s does not depend on θ_k, we should base inferences about θ_k on the conditional distribution of $T_r | T_s$.

If T_s is sufficient for θ_l when θ_k is known, it follows from (31.13) that

$$L(\mathbf{x}|\theta_k, \theta_l) = g_1(T_r|T_s, \theta_k)g_2(T_s|\theta_l)h(\mathbf{x}); \qquad (31.14)$$

thus, we may confine ourselves to functions of $(T_r|T_s)$ in making inferences about θ_k.

However, the real question is whether we should confine ourselves to the conditional statistic when T_s is *not* sufficient for θ_l. The difficulty is essentially that only the *marginal* distribution of T_s is free of θ_k; T_s remains a component of the set of minimal sufficient statistics for all the parameters, whose *joint* distribution depends on θ_k.

31.13   If we denote the information matrix (17.88) for θ_k by $I(\mathbf{x})$ and that for inferences based on $T_r|T_s$ by $I(T_r|T_s)$, we know that when (31.14) holds

$$E_{T_s}\{I(T_r|T_s)\} = I(\mathbf{x}) \qquad (31.15)$$

so that use of T_s allows complete recovery of the information. In any particular case, the conditional information may be more or less than the expected information. This compares with Example 31.2 where conditioning on j results in a reduction in the expected information.

Example 31.6 (Cox and Hinkley, 1974)
An experiment involves drawing a sample from an $N(\theta, \sigma^2)$ population where σ^2 is known. The sample size, A, is either n or kn (k an integer, $k > 1$) and is selected by spinning a fair coin. The conditional information is n/σ^2, given $A = n$ and kn/σ^2, given $A = kn$; and $\hat{\theta} = \bar{X}$ in both cases. The UMP test, given A, is of the usual form but power considerations based upon the unconditional experiment lead to a different and less intuitively appealing test procedure.

Example 31.7
When (X, Y) are binormal with joint f.f.

$$f(x, y) = f_1(x)f_2(y|x),$$

where $E(X) = \mu$, $V(X) = \sigma^2$, $E(Y|x) = \beta_0 + \beta_1 x$, and $V(Y|x) = \omega^2$, we may partition the parameters into

$$\theta = (\beta_0, \beta_1, \omega^2), \qquad \phi = (\mu, \sigma^2)$$

and the minimal sufficient statistics into

$$T = \{\bar{Y}, \hat{\beta}_1, \Sigma(Y - \hat{Y})^2\}, \qquad A = \{\bar{X}, \Sigma(X - \bar{X})^2\}.$$

It is easily verified that (31.14) holds. As we saw in Chapters 19 and 26, inferences for θ are usually made conditionally and the conditional information matrix is

$$\omega^{-2}\begin{bmatrix} n & \Sigma x \\ \Sigma x & \Sigma x^2 \end{bmatrix}.$$

These examples suggest that conditioning is desirable, but the argument is not one-sided. For example, Welch (1939) gave an example (Exercise 22.23) which showed that the conditional test based on $(T_r | T_s)$ may be uniformly less powerful than an alternative (unconditional) test. Furthermore, the ancillary statistics may not be uniquely determined; see Exercise 31.1 and also D. Basu (1964), Cox (1971), and Dawid (1975).

31.14   Cox (1971) suggested that the usefulness of an ancillary statistic should be measured by the variation in the conditional information, $I(T|A)$; he recommends choosing A to maximise some measure of the variability in $I(T|A)$ such as the variance; see also Becker and Gordon (1983). Barnard and Sprott (1971) argue that ancillary statistics should be chosen to preserve the invariance properties of the likelihood function.

The conditionality principle
31.15   The preceding discussion and Example 31.2 in particular suggest that when some random mechanism independent of θ is used to select a statistical experiment from a set of possible procedures, only the experiment actually performed is relevant to subsequent inferences. That is, the selection mechanism is ancillary, so that we should condition upon the outcome of that initial selection procedure.

This principle is often stated as a slightly weaker form, due to Basu (1975) but implicit in Cox (1958). Denote the jth experiment by

$$E_j = (X_j, \theta, \{f_j\}), \qquad j = 1, 2,$$

where $f_j \equiv f_j(x_j | \theta)$, and the mixed experiment by

$$E^* = (X^*, \theta, f^*),$$

wherein $J = 1$ or 2 is observed with $P(J = 1) = P(J = 2) = \frac{1}{2}$; further, let $X^* = (J, X_J)$ and $f^* \equiv f^*(\{j, x_j\}) = \frac{1}{2} f(x_j | \theta)$. Only θ need be common to E_1 and E_2. Finally, the *evidence* obtained from the experiment (the term is used to indicate potentially greater generality than *information* as previously defined) is denoted by $Ev(E, x)$ for experiment E.

The *weak conditionality principle* (WCP) may be stated as follows (cf. Birnbaum, 1962, p. 172):

Suppose there are two experiments, E_1 and E_2, with common θ, and a mixed experiment E^* containing E_1 and E_2 as its two constituents. Then, for each outcome (E_j, x_j) of E^*, we have

$$Ev(E^*, x^*) = Ev(E_j, x_j); \tag{31.16}$$

that is, the evidence from E^* is exactly that from E_j when E_j is actually performed.

The CP is sometimes paraphrased as stating 'the irrelevance of unobserved outcomes'.

The WCP seems undemanding, yet its acceptance runs counter to the purer frequentist approaches as we would have to rule out some of the inferential procedures based upon randomisation arguments; see Godambe (1982) and Genest and Schervish (1985).

31.16   Since conditional arguments often arise in hypothesis testing, we now examine this aspect in greater detail. First, we consider a familiar example from this viewpoint.

Example 31.8
We have seen (Example 17.17) that in normal samples the pair $(\bar{x}, s^2)$ is jointly sufficient for (μ, σ^2), and we know that the distribution of s^2 does not depend on μ. Thus we have

$$L(\mathbf{x} | \mu, \sigma^2) = g_1(\bar{x} | s^2, \mu, \sigma^2) g_2(s^2 | \sigma^2) h(\mathbf{x}),$$

a case of (31.13) with $k = l = r = s = 1$. The conditionality principle states that the statistic $\bar{x} | s^2$ is to be used in drawing inferences about μ. (It happens that $\bar{x}$ is actually independent of s^2 in this case, but this is merely a simplification irrelevant to the general argument.) But s^2 is not a sufficient statistic for the nuisance parameter σ^2, so that the distribution of $\bar{x} | s^2$ is not free of σ^2. If we have no prior distribution given for σ^2 we can only make progress by integrating out σ^2 in some more or less arbitrary way. If we are prepared to use its fiducial distribution and integrate over that, we arrive back at the discussion of Example 20.10, where we found that this gives the same result as that obtained from the standpoint of maximising power in Examples 22.7 and 22.14, namely that Student's t distribution should be used.

Another conditional test principle
31.17   Another principle may be invoked (cf. D. R. Cox (1958)) to suggest the use of $(T_r | T_s)$ whenever $l \geqslant 1$ and T_s is sufficient for θ_l when θ_k is known, irrespective of whether its distribution depends on θ_k, for then we have

$$L(\mathbf{x} | \theta_k, \theta_l) = g_1(T_r | T_s, \theta_k) g_2(T_s | \theta_k, \theta_l) h(\mathbf{x}), \tag{31.17}$$

so that the conditional statistic is distributed independently of the nuisance parameter θ_l. Here again, we have no obvious reason to suppose that the procedure is optimal in any sense.

The justification of conditional tests
31.18   The results of **22.30–2** enable us to see that, if the distribution of the sufficient statistics (T_r, T_s) is of the exponential form (22.73), then the use of the conditional distribution of T_r for given T_s will give UMPU tests, for in our previous notation the statistic T_r is $s(\mathbf{x})$ and T_s is $t(\mathbf{x})$, and we have seen that the UMPU tests are always based on the distribution of T_r for given T_s. If the sufficient statistics are not distributed in the form (22.73) (e.g. in the case of a distribution with range depending on the parameters) this justification is no longer valid. However, following Lindley, we may derive a further justification of the conditional statistic $T_r | T_s$ in **31.17** provided that $g_2(T_s | \theta_k, \theta_l)$ in (31.17) is boundedly complete when θ_k is known. For then, by **22.19**, every size-α critical region similar with respect to θ_l will consist of a fraction α of all surfaces of constant T_s. Thus any similar test on θ_k will be a conditional test based on $T_r | T_s$, and any optimum conditional test will be an optimum similar test.

Welch's (1939) counter-example, which is given in Exercise 22.23, falls within the scope of neither of our justifications of the use of conditional test statistics. There, the range of $f(x)$ depends on the only parameter θ, and $l = 0$. The two-component minimal sufficient statistic (M, R), a 1-1 function of $(x_{(1)}, x_{(n)})$, is not boundedly complete, since

$$E\left(R - \frac{n-1}{n+1}\right) = 0,$$

so that Exercise 22.7 does not preclude R from being an ancillary statistic.

> D. Basu (1977) gives a general review of methods of eliminating nuisance parameters. Kiefer (1977) discusses and extends the literature on conditional confidence interval and test procedures, where it is decided in advance to make certain types of inference given that certain random events occur, with confidence coefficients that are conditional upon those events.
>
> Given a vector of p sufficient statistics for a scalar parameter, θ, McCullagh (1984) shows that this may be partitioned into a locally sufficient statistic and $(p-1)$ locally ancillary statistics, but that these are not unique. For a recent discussion of conditioning and frequentist inference, see Berger (1985).

The sufficiency principle

31.19   A second principle that is widely accepted is based upon the property of *sufficiency* that has been often used throughout this volume. A *weak* version of the *sufficiency principle* (WSP) may be stated as follows (Dawid, 1977):

Consider an experiment E and suppose that the statistic $T(X)$ is sufficient for θ. Then, if $T(x_1) = T(x_2)$,

$$Ev(E, x_1) = Ev(E, x_2). \tag{31.18}$$

Example 31.9

Two sets of five observations from $N(\theta, 1)$ populations result in the observations $(-1.5, -0.8, 0, 0.8, 1.5)$ and $(-2, -2, -2, -2, 8)$. In each case, the WSP leads to $\hat{\theta} = 0$. Although the second sample appears to have come from a possibly skewed population with a much larger variance, model criticism is precluded.

31.20   The *strong sufficiency principle* (SSP) requires that (31.18) should hold even when two different experiments (e.g., with different stopping rules) are performed.

Example 31.10

If a series of attribute trials produces x successes and $(n-x)$ failures, the likelihood function is proportional to

$$\theta^x (1-\theta)^{n-x}$$

and the stopping rule does not affect the inferences; cf. Example 31.5. Note that the use of unbiased estimating equations (31.9) is compatible with the SSP.

It is clear that discussions of the conditionality and sufficiency principles rely heavily on the likelihood function; we now consider its role in inference in more detail.

Barnard and Godambe (1982) present criticisms of the sufficiency principle, reporting that it ignores relevant features of an experiment.

The likelihood principle

31.21 In **18.43**, we considered Fisher's recommendation that the LF be used as an information summary. However, it is possible to take this line of reasoning further and to argue that all inferential procedures should be based solely upon the LF. This view may be stated formally as the *likelihood principle* (LP), which also comes in weak and strong forms. The weak principle (WLP) states that all the information about θ obtained from statistical experiment, E, is contained in the LF, $L(x|\theta)$. If two replications, yielding observations x_1 and x_2, lead to proportional likelihoods:

$$L(x_1|\theta) = c(x_1, x_2)L(x_2|\theta),$$

where the function c is independent of θ, x_1 and x_2 provide the same information about θ, or

$$Ev(E, x_1) = Ev(E, x_2). \tag{31.19}$$

The strong form (SLP) extends the principle to include two different experiments, E_1 and E_2, so that

$$Ev(E_1, x_1) = Ev(E_2, x_2).$$

In particular, the SLP implies the irrelevance of the stopping rule, as in Example 31.10. The likelihood approach to inference was explored in detail by Edwards (1972) and, more recently, by Berger and Wolpert (1988); Edwards (1974) traces the history of the LP.

Example 31.11
Consider a random sample of size n from an $N(\theta, 1)$ population. The standard test of H_0: $\theta = 0$ against the two-sided alternative yields the critical region $\{|\bar{X}| > zn^{-1/2}\}$. The LP leads to the rejection of H_0 when $|\bar{X}| > (2k)n^{-1/2}$ if we employ the rule

$$\text{reject } H_0 \text{ when } \lambda = \log\{L(\theta = 0|\mathbf{x})/(L(\hat{\theta}|\mathbf{x})\} < k;$$

however, there is no basis for linking the value of k to that of z. In general, k will be determined by the ordinate of the distribution rather than the tail area. When σ^2 is unknown, one possibility is to eliminate the nuisance parameter by using the LR statistic as in Example 23.1. Again, the ordinate of the t distribution would be used in the decision rule.

An alternative approach is the use of a pivotal method, wherein a probability statement can be made for the random variable $Z = g(t, \theta)$ that does not involve the nuisance parameters. This statement may then be used as the basis for tests of hypotheses or to generate interval estimates, as in Example 20.2 where we developed a confidence interval for the mean using Student's distribution. By contrast, the Bayesian approach is to integrate out the nuisance parameter; cf. Examples 20.13 and 20.14.

One of the conclusions to be drawn from this example is that different approaches will often produce similar numerical results, but the basis upon which the inferences are drawn is quite different.

31.22   The unrestrained use of the LP lacks appeal for many statisticians who find it intuitively unacceptable to ignore the sample space when making inferences, but a result of Birnbaum (1962) makes serious consideration of the LP necessary. Although it has undergone several subsequent modifications, the essence of Birnbaum's result is that

$$\text{WCP} + \text{WSP} \Leftrightarrow \text{WLP}. \tag{31.20}$$

Our demonstration of the result assumes that the sample space is discrete; the proof follows Berger and Wolpert (1988, pp. 27–8).

Proof of $\Rightarrow$.   The WCP implies that

$$Ev(E^*, \{j, x_j\}) = Ev(E_j, x_j). \tag{31.21}$$

Experiment E^* is defined as for the WCP in **31.15** and has random outcome $X_j^* = (J, X_J)$. Consider the statistic

$$T(X_j^*) = \begin{cases} (1, x_{10}) & \text{if } J = 2, X_2 = x_{20}, \\ (J, X_J) & \text{otherwise} \end{cases}$$

where x_{10} and x_{20} are particular outcomes for E_1 and E_2. Clearly, $(1, x_{10})$ and $(2, x_{20})$ result in the same value of T. It follows that T is sufficient for θ since

$$P(X^* = (j, x_j) \mid T = t \neq (1, x_{10})) = \begin{cases} 1 & \text{if } (j, x_j) = t \\ 0, & \text{otherwise} \end{cases}$$

and

$$P(X^* = (1, x_{10}) \mid T = (1, x_{10})) = 1 - P(X^* = (2, x_{20}) \mid T = (1, x_{10}))$$

$$= \frac{\frac{1}{2}f_1(x_{10} \mid \theta)}{\frac{1}{2}f_1(x_{10} \mid \theta) + \frac{1}{2}f_2(x_{20} \mid \theta)}$$

which is independent of θ from (31.19). Thus, the WSP implies that

$$Ev(E^*, (1, x_1^*)) = Ev(E^*, (2, x_2^*)). \tag{31.22}$$

(31.21) and (31.22) combine to imply that

$$Ev(E_1, x_1) = Ev(E_2, x_2),$$

so the WLP holds.

Proof of $\Leftarrow$.   To establish that LP $\Rightarrow$ WCP, we note that, for E^*,

$$L(\theta \mid (j, x_j)) = \tfrac{1}{2}f_j(x_j \mid \theta)$$

which is clearly proportional to $L(\theta|x_j)$, the likelihood for E_j when x_j is observed. Hence,

$$Ev(E^*, (j, x_j)) = Ev(E_j, x_j).$$

Finally, if the sufficient statistic T yields $T(x_1) = T(x_2)$ for realisations x_1 and x_2, it follows that the likelihoods must be proportional so that WLP⇒WSP. If the stopping rule is ignored so that the SCP and SSP may be assumed, (31.20) may be restated as an equivalence with the SLP. Birnbaum (1972) produced another proof wherein the SP was replaced by a logically weaker axiom of *mathematical equivalence*.

31.23   When we turn to continuous sample spaces, the principles must be restated to allow possible exceptions of measure zero; a relative likelihood principle may then be established; see Berger and Wolpert (1988, pp. 28–36) for details.

31.24   This result has led to a considerable discussion. Durbin (1970) pointed out that the theorem does not hold if we first reduce the problem to consideration of the *minimal* sufficient statistics. However, Birnbaum (1970), Savage (1970), and Berger and Wolpert (1988, pp. 45–6) claim that this restriction is inappropriate in general. Kalbfleisch (1975) pointed out that the proof fails if sufficiency is held to be inapplicable to simple mixture experiments; see also the discussion following his paper.

Akaike (1982) demonstrated that, in some cases, the notion of mathematical equivalence is equivalent to the LP. In the discussion on Birnbaum (1962), Kempthorne asks whether it is not the case that CP⇒SP so that, in fact, CP⇔LP; Birnbaum concurred. This has recently been demonstrated rigorously by Evans, Fraser, and Monette (1985, 1986); these authors go on to argue forcibly against the LP, providing examples that suggest information may be suppressed when SP and CP are applied uncritically. Joshi (1989) provides a counter-example for the case of a finite sample space.

31.25   It is interesting to note that in his later writings, Birnbaum concluded that the LP was inadequate as a basis for inference since it conflicted with the *confidence principle*, which he formulated (Birnbaum, 1977, p. 24) as follows:

a concept of statistical evidence is not plausible unless it finds 'strong evidence' for H_2 as against H_1 with small probability (α) when H_1 is true and with much larger probability ($1-\beta$) when H_2 is true.

This principle is, of course, an excellent foundation for frequentist inference. For further discussion, see the special volume of *Synthèse*, **36**(1), 1977, devoted to the foundations of statistical inference.

Fiducial inference
31.26   The construction of fiducial intervals for one- and two-sample problems was examined at some length in **20.41–4**. Therefore, our present discussion concentrates on the underlying assumptions rather than detailing further examples.

As noted in **20.43**, if t is the (minimal) sufficient statistic for the single parameter θ, with d.f. $F(t|\theta)$, the fiducial distribution of θ given t has probability density function

$$g(\theta|t) = \frac{\partial G(\theta|t)}{\partial \theta} = -\frac{\partial F(t|\theta)}{\partial \theta}, \tag{31.23}$$

provided F is monotone decreasing in θ. Some of the difficulties with this approach are how to proceed in the absence of a sufficient statistic, lack of uniqueness (in multiparameter cases), and the lack of a frequency interpretation.

31.27 Let us consider the last concern first. Fisher's writings in this area were evidently influenced by Keynes (1921), Carnap (1962), and others, who sought to develop an epistemic view of probability that would measure the 'degree of rational credibility' of a hypothesis H relative to data or evidence E. Thus, although the initial development of fiducial probability was confused, the ultimate aim was clear: to make probabilistic statements of the form $P(H|E)$ or, in our present context, to develop a d.f. $G(\theta|t)$. By construction, and intent, G is designed to provide statements about θ *for a single trial*, so the absence of a frequency interpretation is hardly surprising.

It is clear that the fiducial approach is seeking to deliver a quite different inductive statement than is available from the frequency viewpoint. Although it is possible, as Barnard (1950) has shown, to justify the Fisher–Behrens solution of the two-means problem from a different frequency standpoint, as he himself goes on to argue, the idea of a fixed 'reference set', in terms of which frequencies are to be interpreted, is really foreign to the fiducial approach. Thus, the statistician must choose between confidence intervals, which make precise frequency-interpretable statements and which may on exceptional occasions be trivial, and other methods, which forgo frequency interpretations in the interests of what are, perhaps intuitively, felt to be more relevant inferences.

For the present, we now accept the declared aims of the fiducial approach and examine the methods in greater detail.

Paradoxes and restrictions in fiducial theory
31.28 The principal difficulties with the fiducial approach may be illustrated in the following way. Assume that (t_1, t_2) are jointly sufficient for (θ_1, θ_2), and write the alternative factorisations

$$L(x|\theta_1, \theta_2) \propto g(t_1, t_2|\theta_1, \theta_2) = g_1(t_1|t_2, \theta_1, \theta_2)g_2(t_2|\theta_1, \theta_2)$$

$$= g_3(t_2|t_1, \theta_1, \theta_2)g_4(t_1|\theta_1, \theta_2).$$

We may distinguish two special structures of the sufficient statistics:

(a) One of the statistics depends on only one parameter. The factorisation becomes either

or

$$\left. \begin{array}{l} L \propto g_1(t_1|t_2, \theta_1, \theta_2)g_2(t_2|\theta_2) \\ L \propto g_3(t_2|t_1, \theta_1, \theta_2)g_4(t_1|\theta_1) \end{array} \right\}. \tag{31.24}$$

(b) One of the conditional distributions depends on only one parameter, giving either

$$
\left. \begin{array}{c}
L \propto g_1(t_1 \,|\, t_2, \theta_1) g_2(t_2 \,|\, \theta_1, \theta_2) \\
L \propto g_3(t_2 \,|\, t_1, \theta_2) g_4(t_1 \,|\, \theta_1, \theta_2)
\end{array} \right\}.
\tag{31.25}
$$

or

If the first line of (31.25) holds, t_2 is singly sufficient for θ_2 when θ_1 is known; if the second line holds, t_1 is singly sufficient for θ_1 when θ_2 is known; note that (31.24) and (31.25) correspond to (31.13) and (31.17) respectively, in a slightly different notation.

Either line of (31.24), or of (31.25), permits a joint fiducial distribution to be constructed by first obtaining the fiducial distribution of one parameter from the factor in which it appears alone, and then obtaining the conditional fiducial distribution of the other parameter (the value of the first parameter being fixed) from the factor in which both parameters appear. The product of these distributions is taken as the joint fiducial distribution, on the analogy of the multiplication theorem for probabilities. (31.24) and (31.25) were used in this way by Fisher (1956) and Quenouille (1958) respectively.

Referring back to the discussion for the means of one and two samples, in Examples 20.10–11, it will be seen that both (31.24) and (31.25) hold, because the sample mean (or difference of means) t_1 is distributed independently of the sample variance(s) t_2. In general, however, even these special sufficiency structures are not enough to guarantee the uniqueness of the joint fiducial distribution, as Tukey (1957) and Brillinger (1962) showed by counter-examples. The non-uniqueness arises precisely because *both* lines of (31.24), or of (31.25), can hold simultaneously, and the joint fiducial distribution may depend on which line we use to construct it. See also Mauldon (1955) and Dempster (1963).

31.29 One critical difficulty of fiducial theory is exemplified by the derivation of Student's distribution in fiducial form given in Example 20.10. It appears to us that this particular matter has been widely misunderstood, except by Jeffreys. Since the Student distribution gives the same result for fiducial theory as for confidence theory, whereas the two methods differ on the problem of two means, both sides seem to have sought their basic differences in the second, not in the first. But in our view *c'est le premier test qui coûte*. If the logic of this is agreed, the more general Fisher-Behrens result follows by a very simple extension. This is also evident from the Bayes-Jeffreys approach, in which Example 20.14 is an obvious extension of Example 20.13 for two independent samples.

The question, as noted in Example 20.10, is whether, given the joint distribution of $\bar{x}$ and s (which are independent in the ordinary sense), we can replace $d\bar{x}\,ds$ by $d\mu\,d\sigma/\sigma$. It appears to us that this is not obvious and, indeed, requires a new postulate, just as (20.163) is a new postulate. On this point, the paper by Yates (1939a) is explicit.

For further discussions of the fiducial approach, see Dempster (1964), Wilkinson (1977), Pedersen (1978), Seidenfeld (1979), Dawid and Stone (1982), and Barnard and Sprott (1983).

Structural inference

31.30 Fraser (1968) gives a general theory of *structural inference*, essentially fiducial in character, which is more simply illustrated and discussed in Fraser (1976) and the comments following it.

The approach involves the specification of a structural equation that relates the measurement, X, to the physical quantity of interest, θ. For example, we may consider

$$X = \theta + \varepsilon,$$

where ε denotes the random error, which is $N(0, \sigma^2)$.

Then, given n observations $(x_1, x_2, \ldots, x_n)$, we can transform to

$$u_1 = \theta + \varepsilon_1, \quad u_j = \varepsilon_j - \varepsilon_{j-1} = x_j - x_{j-1}, \quad j \geq 2.$$

The variates u_j $(j \geq 2)$ have distributions independent of θ and form ancillary statistics. We are thus led to the *reduced* distribution for $\bar{\varepsilon} = \bar{X} - \theta$; $\bar{\varepsilon}$ is then used as a pivotal function to provide inferences about θ. In this case, the structural argument is very similar to the fiducial, but it may be extended to include general transformations of the form $X = T(\theta)\varepsilon$.

The LR as a credibility measure

31.31 The difficulties with the fiducial approach are seen to stem from a lack of uniqueness once we move from the special case of a single parameter and a single sufficient statistic. Since the aim of the fiducial approach is to condition upon the ancillary statistics and then to produce a measure that may be interpreted as a degree of credibility, a somewhat similar end could be reached by considering the ML estimator(s) which are at least 'asymptotically sufficient' in the sense noted in **18.16**. If the parameter set (θ, ϕ) contains nuisance parameters ϕ, these may be replaced by their ML estimators as functions of θ, $\hat{\phi}(\theta)$ say. That is, we might consider the LR in the form

$$g(\theta \mid x) = L\{x \mid \theta, \hat{\phi}(\theta)\} / L(x \mid \hat{\theta}, \hat{\phi}). \tag{31.26}$$

Birnbaum (1962) refers to

$$\alpha(\theta) = 1 / \{1 + g(\theta \mid x)\} \tag{31.27}$$

as the *intrinsic significance level* associated with θ, implying that it could be used for testing hypotheses in this way. Since α is a monotone function of g, the suggestion is comparable to the use of the LR test, although the critical regions will differ since (31.27) bases the test on the ordinate of g rather than a tail area.

31.32 If we are willing to define $g(\theta \mid x)$ for all $\theta \in \Omega$, the integral over θ, suitably scaled, could be used as a measure of credibility; of course, the denominator in (31.26) is then redundant. Compared to the fiducial argument, this approach has the benefit that there is no problem with reference sets since the use of g is in accordance with the conditionality principle.

This approach is developed more formally in Barndorff-Nielsen (1983), Cox and Reid (1987) and Fraser and Reid (1989).

Example 31.12

Consider again the case of a sample of size n from a normal population with mean μ and variance σ^2 unknown. The likelihood reduces to

$$L(x|\mu, \sigma^2) \propto \sigma^{-n} \exp\{-\tfrac{1}{2}\Sigma(x_i - \mu)^2/\sigma^2\};$$

since

$$n\hat{\sigma}^2(\mu) = \Sigma(x_i - \bar{x})^2 + n(\bar{x} - \mu)^2 = ns^2 + n(\bar{x} - \mu)^2$$

(31.27) yields

$$g(\mu|\bar{x}) \propto \left\{1 + \frac{(\bar{x} - \mu)^2}{s^2}\right\}^{-n/2}. \tag{31.28}$$

(31.28) differs from the Student's t form by having n in place of $(n-1)$. The exact Student's t form could be obtained by using the marginal likelihood for

$$t = (\bar{x} - \mu)/s.$$

In general, the likelihood-based argument produces a measure of credibility that is equal (at least asymptotically) to the fiducial d.f. and avoids the construction difficulties noted earlier. In Exercise 31.2, the reader is asked to confirm that (31.26) recovers the Fisher–Behrens solution to the two-means problem in Example 20.11 with n_i in place of $(n_i - 1)$.

31.33 In order to make this approach more operational, we may use Taylor's theorem to observe that, when $l(\theta) = \log L(\theta)$,

$$l(\theta) \doteq l(\hat{\theta}) + (\theta - \hat{\theta})\left(\frac{\partial l}{\partial \theta}\right)_{\hat{\theta}} + \tfrac{1}{2}(\theta - \hat{\theta})^2\left(\frac{\partial^2 l}{\partial \theta^2}\right)_{\hat{\theta}}. \tag{31.29}$$

Since $(\partial l/\partial \theta)_{\hat{\theta}} = 0$, by construction, it follows that

$$L(\theta)/L(\hat{\theta}) \doteq \exp\left\{\tfrac{1}{2}(\theta - \hat{\theta})^2\left(\frac{\partial^2 l}{\partial \theta^2}\right)_{\hat{\theta}}\right\}. \tag{31.30}$$

Since $I(\hat{\theta}) = (\partial^2 l/\partial \theta^2)_{\hat{\theta}}$ is independent of θ, it follows directly from (31.30) that, for large samples, the LR plotted in the parameter space is shaped like the normal f.f., where $z = (\theta - \hat{\theta})\{I(\hat{\theta})\}^{1/2}$ is $N(0, 1)$. The quadratic approximation in (31.29) may be improved using transformations, cf. **28.37–42**.

31.34 This argument extends readily to cover several parameters and the elimination of nuisance parameters when (31.30) becomes the usual LR criterion, l, and the resulting distribution of $-2 \log l$ is chi-squared; cf. **23.7**. Using the d.f. thus constructed, we have an LR-based significance test that would reject the hypothesis H_0: $\theta = \theta_0$ if θ_0 did not fall in the interval $[\theta_1, \theta_2]$, where

$$P^*[\theta_1 \leqslant \theta \leqslant \theta_2] = 1 - \alpha; \tag{31.31}$$

P^* denotes the measure of credibility derived from (31.26). Interestingly, this has elements of a pure significance test introduced by Fisher before the Neyman–Pearson theory was developed, yet anchored by the ML estimator.

The difficulties with such an approach are, as before, the lack of a frequency interpretation for P^* or, indeed, any direct interpretation of the function. Here, as elsewhere, the statistician must decide whether he or she is willing to make the logical leap in order to justify inferential statements that relate to single experiments.

Bayesian inference

31.35 The Bayesian approach to the problem of induction is to suppose that a *prior* distribution may be specified for parameter θ, $p(\theta)$ say, defined on the parameter space $\theta \in \Omega$. Given the likelihood function, $L(x|\theta)$, it follows from an application of Bayes theorem (cf. **8.2**, Vol. 1) that the *posterior* distribution for θ is

$$P(\theta|x) \propto p(\theta)L(x|\theta). \tag{31.32}$$

Several examples are given in Chapter 8 and the accompanying exercises. Once the notion of specifying a prior distribution for θ is accepted, the framework of Bayesian inference may be developed deductively from one of several systems of axioms (e.g., Ramsey, 1926; Good, 1950; Savage, 1962; DeGroot, 1970); for a detailed evaluation, see Fishburn (1986). We leave a detailed exposition of this subject to the companion volume by O'Hagan (to be published).

31.36 Thus, the key question is how to specify the prior distribution. Three possible approaches may be considered:
 (i) as a frequency distribution, based on past experience;
 (ii) as an objective representation of rational initial beliefs regarding the parameter;
 (iii) as a subjective statement about what You (a specific individual) believe before the data are collected.

31.37 We shall consider alternative (i) only briefly. In keeping with the frequency approach, we would need to assume an underlying process that generates the parameter values that is stable, or at least predictable. Examples include industrial production runs where a prior distribution for the proportion of defectives, say, may be assessed from past records. More generally, state-space models in time series assume that the parameter (state) develops over time according to a state equation such as

$$\theta_t = \theta_{t-1} + \delta_t, \tag{31.33}$$

where δ_t represents a random disturbance at time t. See Kendall and Ord (1990, Chapter 9) for further discussion.

In some ways, this may seem to be mixing oil and water and the counter-claim could be made that prior information is not allowable in the frequentist scheme. Such a claim is indeed made by critics of the frequency approach, but seems to represent an overly literal interpretation of that viewpoint. Indeed, it should be noted that, even though the prior is specified in frequentist terms, (31.33) still requires that we are willing to consider the posterior distribution for θ.

Objective probability

31.38   Objective, or logical, probability was developed, notably by Jeffreys (1961, revised version of his 1939 book) and others, to provide a substantive measure of the weight of evidence favouring a given hypothesis in light of the data. That is, an agreed prior distribution was sought that would allow posterior probability statements to be made on the basis of a particular trial.

Much of Jeffreys' work focused on the specification of a prior distribution in situations where nothing is known about the parameter before the statistical experiment takes place. Interestingly, most subjective Bayesians, such as Lindley (1971), would now argue that there is always *some* information available and that the specification of prior ignorance is a non-issue. When the number of values of θ in Ω is finite, it is feasible to make use of *Bayes' postulate* (also known as the *principle of insufficient reason* or the *principle of indifference*) and to assign equal prior probabilities to each possible value. This requires that a satisfactory base of possible parameter values can be established, not always a trivial task.

Example 31.13
An urn contains an unknown number of balls of equal size and weight that are made of the same material. What is the prior probability that a white ball will be selected on the first drawing when you are told that the urn contains balls that are
   (a) white or not-white
   (b) white, red or blue?
The principle of insufficient reason leads us to conclude $p = \frac{1}{2}$ in case (a), but $p = \frac{1}{3}$ in case (b).

Despite this example, the principle may often serve as a reasonable starting point. One implication of the principle is *Laplace's law of succession* (Example 8.3, Vol. 1) which shows that if we start from

$$\Omega = \{0, 1/N, 2/N, \ldots, (N-1)/N, 1\} \qquad (31.34)$$

and assign equal prior probabilities $1/(N+1)$ to each state, then

$$P\{(m+1)\text{th trial is a success}\,|\,\text{first } m \text{ trials are successes}\} = \frac{m+1}{m+2} \qquad (31.35)$$

for any m and $N \geq 1$.

Example 31.14
If a coin is tossed m times and comes down heads each time, would we accept that the probability of the next spin yielding a head was given by (31.35)?

The answer is probably not, because we are drawing on a lot of past experience that tells us coins have a head and a tail and that either side is 'equally likely' to fall face uppermost. However, this does not violate the principle, rather it tells us that assigning equal probabilities to the values in (31.34) was not an accurate statement of prior beliefs. Conversely, if there are three coins: one with two heads, one standard, and one with two tails, specifying equal probabilities on (31.34) with $N = 2$ would be very plausible. Note that we do not require that a coin be selected at random, but only that we are ignorant of the selection process.

31.39 Now suppose that Ω is a continuum; even if the prior for θ is uniform over some finite interval, that for any nonlinear transform of $g(\theta)$ will not be. This led Jeffreys to propose the use of the prior

$$p(\theta) \propto \{I(\theta)\}^{1/2}, \tag{31.36}$$

where $I(\theta) = -E(\partial^2 \log L/\partial\theta^2)$. He arrived at (31.36) by selecting that form of $g(\theta)$ for which $p\{g(\theta)\}$ is uniform, even if improper in some cases; the function of $g(\theta)$ then corresponds to a location parameter for the distribution, at least locally. Jeffreys termed priors given by (31.36) *invariant*. The resulting functions $g(\theta)$ are precisely those we obtained when considering variance-stabilising transforms in **28.38–40**:

distribution	$g(\theta)$	$p(\theta)$
normal, $N(\theta, 1)$	θ	1
normal, $N(0, \theta)$	$\log \theta$	θ^{-1}
Poisson, $P(\theta)$	$\theta^{1/2}$	$\theta^{-1/2}$
binomial, $B(n, \theta)$	$\sin^{-1}(\theta^{1/2})$	$\{\theta(1-\theta)\}^{-1/2}$
negative binomial, $NB(k, \theta)$	$\sinh^{-1}(\theta^{1/2})$	$\theta^{-1}(1-\theta)^{-1/2}$.

The two binomials have different priors for the same parameter, a property that would appear to be in violation of the strong likelihood principle. Nevertheless, several authors have recommended vague priors that depend on the stopping rule (e.g. Jaynes, 1968; Box and Tiao, 1973, Zellner, 1977, Bernardo, 1979), and claim that such a step is indeed desirable.

When the number of parameters increases, so does the difficulty of specifying a vague prior and we must introduce new postulates such as the independence of prior beliefs about different parameters; see Examples 20.13–14.

31.40 Although the concept that Jeffreys was trying to make operational is an attractive one, it does not seem possible to develop it in a consistent fashion; see Barnett (1982, Chapter 6) and Howson and Urbach (1989, Chapter 9) for recent critiques. It is interesting to speculate whether Jeffreys would have adopted (31.36) if its results had not matched existing ones.

> Exercise 31.5 shows that a paradox may arise if an improper prior is used; the problem disappears once a proper prior is used.

Subjective probabilities

31.41 We now leave the objectivist viewpoint and accept that prior probabilities are necessarily personal and based upon our own experience. In order to make such a scheme operational, it is necessary that

(a) You have beliefs about the parameter of interest, which can be expressed in the form of probabilities;

(b) Your probabilities may be compared one with another (though they need not be comparable with anyone else's);

(c) Your probabilities can be assessed by some scheme of hypothetical bets.

If Your betting behaviour is internally consistent, it follows that Your probabilities satisfy the standard rules of probability and You are said to be *coherent*; otherwise, You are *incoherent* and a Bayesian could make bets with You in such a way that You would be bound to lose money. This is the *principle of coherence*, which states that Your system of bets should be internally consistent. Presumably coherence was used to avoid confusion with Fisher's use of consistency in estimation and testing. Clearly, non-Bayesians do not have a monopoly of virtuous keywords!

31.42　　The key requirement is now the assessment of the prior distribution. Most subjectivists (e.g. Ramsey, 1931; Savage, 1954) use some method of assessing fair bets either directly for the phenomenon under study or by comparison with some standardised experiment (e.g. an urn scheme). It is assumed that such assessments can be made directly for the probabilities, uncontaminated by the relative utilities of different outcomes.

31.43　　Once You have established Your prior distributions, subjective Bayesian analysis proceeds straightforwardly, although it will often be desirable to use conjugate priors (see **8.16**, Vol. 1) to simplify the algebra. If the set of parameters is (θ, ϕ), where ϕ denotes nuisance parameter(s), the standard approach is to examine the *marginal posterior*:

$$P(\theta \mid x) = \int P(\theta, \phi \mid x) \, d\phi$$

$$= \int L(x \mid \theta, \phi) p(\theta, \phi) \, d\phi. \tag{31.37}$$

Explicit evaluation of (31.37) can prove very difficult for higher-dimensional problems. However, the innovative numerical integration procedures developed by Naylor and Smith (1988) among others have greatly contributed to the feasibility of this approach.

For more general updating rules, see Diaconis and Zabell (1982).

Bayesian estimation

31.44　　Point estimation is usually based upon either the mode or the mean of the posterior distribution. The *posterior mode* is given by $\tilde{\theta}$, where

$$P(\tilde{\theta} \mid x) = \max_{\theta} P(\theta \mid x); \tag{31.38}$$

when the prior distribution is uniform, $\tilde{\theta}$ will be equivalent to the ML estimator ($\hat{\theta}$).

The *posterior mean*, given by

$$\bar{\theta} = E(\theta \mid x), \tag{31.39}$$

will be equal to the ML estimator, if at all, only for distribution-specific choices of the prior.

Example 31.15

Let π denote the probability of success in a Bernoulli trial with prior f.f.

$$p(\pi) \propto \pi^{a-1}(1-\pi)^{b-1}.$$

Given n trials with x successes, the posterior is

$$P(\pi|x) \propto \pi^{a+x-1}(1-\pi)^{b+n-x-1},$$

whence

$$\tilde{\theta} = \frac{(a+x-1)}{(n+a+b-2)} \quad \text{and} \quad \bar{\theta} = \frac{(a+x)}{(n+a+b)},$$

compared to $\hat{\theta} = x/n$. Upon inspection, $\tilde{\theta} = \hat{\theta}$ for the uniform prior $(a = b = 1)$, whereas $\bar{\theta} = \hat{\theta}$ when $a = b = 0$, a degenerate choice that is not feasible.

31.45 Interval estimates may be obtained from the posterior distribution directly; Bayesian inference allows the statement 'with probability $1 - \alpha$, θ lies between the values θ_1 and θ_2' or

$$P(\theta_1 \leq \theta \leq \theta_2) = P(t_2|x) - P(t_1|x) = 1 - \alpha. \tag{31.40}$$

The interval $[\theta_1, \theta_2]$ is known as a $100(1 - \alpha)$ per cent *credible* region. Parallel to the notion of the physically shortest interval (**20.13**), we may choose the set of θ-values, Ω_1, such that (31.40) is satisfied and

$$\left\{ \theta \in \Omega_1 : \frac{\partial P(\theta)}{\partial \theta} \geq c \right\}. \tag{31.41}$$

Such an interval (or region) is known as the *highest posterior density* (HPD) credible region.

Example 31.16

For a random sample of size n from a normal population with known variance, $N(\mu, \sigma^2)$ say, consider the prior distribution $N(\phi, \tau^2)$. From **8.15**, Vol. 1, the posterior distribution for μ is $N(\mu_P, \sigma_P^2)$, where

$$\mu_P = \frac{\phi\sigma^2 + n\bar{x}\tau^2}{\sigma^2 + n\tau^2} \quad \text{and} \quad \sigma_P^2 = \frac{\sigma^2\tau^2}{(\sigma^2 + n\tau^2)}.$$

The HPD credible region for μ is

$$\mu_P \pm z_{1-\alpha/2}\sigma_P,$$

where z represents the percentage points of $N(0, 1)$. In this example, $\tilde{\theta} = \bar{\theta}$ and these will be equal to $\hat{\theta}$ for the uniform, improper prior given by letting $\tau \to \infty$; the credible and confidence intervals will then be identical (numerically speaking!).

Bayesian tests
31.46   The two one-sided hypotheses

$$H_0: \theta \leq \theta_0 \qquad \text{and} \qquad H_1: \theta > \theta_0$$

are readily compared by computing their posterior probabilities

$$P(H_0) = P(\theta_0 | x), \qquad P(H_1) = 1 - P(\theta_0 | x). \qquad (31.42)$$

However, the comparison of

$$H_0: \theta = \theta_0 \qquad \text{and} \qquad H_1: \theta \neq \theta_0$$

raises some difficulties. Jeffreys (1961, Chapter 5) argued that the value of θ_0 is distinguished from all other θ values and so a prior probability may be assigned to that point:

$$p_0 = p(\theta_0) > 0.$$

The posterior odds in favour of H_0 are then

$$P(\theta_0 | x) \bigg/ \int_{\Omega - \theta_0} \mathrm{d}P(\theta_0 | x). \qquad (31.43)$$

Such an assumption is clearly plausible in some cases, such as testing whether a regression coefficient is zero, but depends heavily on the value of p_0 selected. The frequentist view would be that the null hypothesis often is deserving of special attention, but that there is no reasonable way of arriving at an appropriate value of p_0.

A similar device is to expand H_0 to

$$H_0: \theta_0 - a \leq \theta \leq \theta_0 + b; \; a, b > 0 \qquad (31.44)$$

and then consider the posterior odds as before; unfortunately, this depends heavily on the selection of a and b.

31.47   Bernardo (1980) examined the structure of Bayesian tests and concluded that there are no problems when H_0 and H_1 have the same dimensionality. In other cases, it appears that the conclusions to be drawn from such tests are clearly interpretable only when $p(\theta_0)$ depends on the overall prior $p(\theta)$, $\theta \neq \theta_0$.

In general, tests of hypotheses now receive less attention from Bayesians, who tend to favour the use of decision theory; see **31.52-7**.

The relationship between Bayesian and fiducial approaches
31.48   Lindley (1958) obtained a simple yet far-reaching result which not only illuminates the relationship between fiducial and Bayesian arguments, but also limits the claims of fiducial theory to provide a general method of inference, consistent with and combinable with Bayesian methods. In fact, Lindley shows that the fiducial argument is consistent with Bayesian methods if and only if it is applied to a random variable x and a parameter θ which may be (separately) transformed to u and τ

respectively so that τ is a location parameter for u; and in this case, it is equivalent to a Bayesian argument with a uniform prior distribution for τ. The criticism applies equally to 'confidence distributions' defined at the end of **20.6**, in so far as they coincide with fiducial distributions.

31.49 Using (31.23), we write for the fiducial distribution of θ

$$g_x(\theta) = -\frac{\partial}{\partial\theta} F(x\,|\,\theta), \tag{31.45}$$

while the posterior distribution for θ given a prior distribution $p(\theta)$ is, by Bayes' theorem,

$$P(\theta\,|\,x) = p(\theta)f(x\,|\,\theta) \bigg/ \int p(\theta)f(x\,|\,\theta)\,d\theta, \tag{31.46}$$

where $f(x\,|\,\theta) = \partial F(x\,|\,\theta)/\partial x$, the frequency function. Writing $r(x)$ for the denominator on the right of (31.46), we thus have

$$P(\theta\,|\,x) = \frac{p(\theta)}{r(x)} \frac{\partial F(x\,|\,\theta)}{\partial x}. \tag{31.47}$$

If there is some prior distribution $p(\theta)$ for which the fiducial distribution is equivalent to a Bayes posterior distribution, (31.45) and (31.47) will be equal, or

$$\frac{-\dfrac{\partial}{\partial\theta} F(x\,|\,\theta)}{\dfrac{\partial}{\partial x} F(x\,|\,\theta)} = \frac{p(\theta)}{r(x)}. \tag{31.48}$$

(31.48) shows that the ratio on its left-hand side must be a product of a function of θ and a function of x. We rewrite it

$$\frac{1}{r(x)} \frac{\partial F}{\partial x} + \frac{1}{p(\theta)} \frac{\partial F}{\partial\theta} = 0. \tag{31.49}$$

For given $p(\theta)$ and $r(x)$, we solve (31.49) for F. The only non-constant solution is

$$F = G\{R(x) - P(\theta)\}, \tag{31.50}$$

where G is an arbitrary function and R, P are respectively the integrals of r, p with respect to their arguments. If we write $u = R(x)$, $\tau = P(\theta)$, (31.50) becomes

$$F = G\{u - \tau\}, \tag{31.51}$$

so that τ is a location parameter for u. Conversely, if (31.51) holds, (31.48) is satisfied with u and τ for x and θ and $p(\tau)$ a uniform distribution. Thus (31.51) is a necessary and sufficient condition for (31.48) to hold, i.e. for the fiducial distribution to be equivalent to some Bayes posterior distribution.

31.50　Now consider the situation where we have two independent samples, summarised by sufficient statistics x, y, from which to make an inference about θ. We can do this in two ways:

(a) we may consider the combined evidence of the two samples simultaneously, and derive the fiducial distribution $g_{x,y}(\theta)$;

(b) we may derive the fiducial distribution $g_x(\theta)$ from the first sample above, and use this as the prior distribution in a Bayesian argument on the second sample, to produce a posterior distribution $P(\theta|x, y)$.

Now if the fiducial argument is consistent with Bayesian arguments, (a) and (b) are logically equivalent and we should have $g_{x,y}(\theta) = P(\theta|x, y)$.

Take the simplest case, where x and y have the same distribution. Since it admits a single sufficient statistic for θ, the frequency function is of the form (17.83), from which we may assume (cf. Exercise 17.14) that the distribution of x itself is of form

$$f(x|\theta) = f(x)h(\theta) \exp(x\theta), \tag{31.52}$$

and similarly for y in the other sample. Moreover, in the combined samples, $x + y$ is evidently sufficient for θ, and thus the combined fiducial distribution $g_{x,y}(\theta)$ is a function of $(x + y)$ and θ only. We now ask for the conditions under which $P(\theta|x, y)$ is also a function of $(x + y)$ and θ only. Since by Bayes' theorem

$$P(\theta|x, y) = \frac{g_x(\theta)f(y|\theta)}{\int g_x(\theta)f(y|\theta)\,d\theta},$$

if $P(\theta|x, y)$ is a function of $(x + y)$ and θ only, so also will be the ratio for two different values of θ

$$\frac{P(\theta|x, y)}{P(\theta'|x, y)} = \frac{g_x(\theta)f(y|\theta)}{g_x(\theta')f(y|\theta')}. \tag{31.53}$$

Thus (31.53) must be invariant under interchange of x and y. Using (31.52) in (31.53), we therefore have

$$\frac{g_x(\theta)h(\theta)}{g_x(\theta')h(\theta')} \exp\{y(\theta - \theta')\} = \frac{g_y(\theta)h(\theta)}{g_y(\theta')h(\theta')} \exp\{x(\theta - \theta')\},$$

so that

$$\frac{g_x(\theta)}{g_x(\theta')} \cdot \frac{g_y(\theta')}{g_y(\theta)} = \exp\{(x - y)(\theta - \theta')\}$$

or

$$g_x(\theta) = \frac{g_x(\theta')\,e^{-x\theta'} \cdot g_y(\theta)\,e^{-y\theta}}{g_y(\theta')\,e^{-y\theta'}}\,e^{x\theta}, \tag{31.54}$$

and if we regard θ' and y as constants, we may write (31.54) as

$$g_x(\theta) = A(x) \cdot B(\theta)\,e^{x\theta}, \tag{31.55}$$

where A and B are arbitrary functions. Using (31.45), (31.52) and (31.55), we have

$$\frac{-\dfrac{\partial}{\partial \theta} F(x \mid \theta)}{\dfrac{\partial}{\partial x} F(x \mid \theta)} = \frac{g_x(\theta)}{f(x \mid \theta)} = \frac{A(x)B(\theta)}{f(x)h(\theta)}. \tag{31.56}$$

But (31.56) is precisely the condition (31.48), for which we saw (31.51) to be necessary and sufficient. Thus we can have $g_{x,y}(\theta) = P(\theta \mid x, y)$ if and only if x and θ are transformable to (31.51) with τ a location parameter for u, and $p(\tau)$ a uniform distribution. Thus the fiducial argument is consistent with Bayes' theorem if and only if the problem is transformable into a location parameter problem, the prior distribution of the parameter then being uniform. An example where this is not so is given as Exercise 31.3.

Lindley goes on to show that in the exponential family of distributions (17.83), the normal and the gamma distributions are the only ones obeying the condition of transformability to (31.51): this explains the identity of the results obtained by fiducial and Bayesian methods in these cases (cf. Example 20.12). Sprott (1960, 1961) shows that these remain the only such distributions if x and y are differently distributed.

Welch and Peers (1963), Welch (1965), and Peers (1965) examine the problem of correspondence of Bayesian and confidence intervals with special reference to asymptotic solutions. Thatcher (1964) examines this correspondence for binomial predictions. Geisser and Cornfield (1963) and Fraser (1964) display further difficulties with fiducial distributions in the multivariate case. See also the I.S.I. Symposium (*Bulletin of the International Statistical Institute*, 1964, **40**(2), pp. 833-939).

Fraser (1962) proposes a modification of the fiducial method which extends the range of its consistency with Bayesian methods.

Empirical Bayes methods

31.51 An interesting variation on the Bayesian approach is the empirical Bayes scheme developed by Robbins (1956, 1964); see Maritz and Lwin (1989) for a detailed exposition. Suppose that a sample of n observations is available with f.f. $f(x_i \mid \theta_i)$, where θ_i represents a random drawing from prior $p(\theta \mid \phi)$ and ϕ represents the parameters for the prior distribution. We may then consider the marginal distribution (or mixture, cf. **5.20-4**, Vol. 1):

$$f(x \mid \phi) = \int f(x \mid \theta) p(\theta \mid \phi) \, d\theta \tag{31.57}$$

and use ML methods to estimate ϕ. The posterior distribution for θ_i is then approximated by

$$P(\theta_i \mid x_i) \propto f(x_i \mid \theta_i) p(\theta_i \mid \hat{\phi}). \tag{31.58}$$

In particular cases (e.g. with conjugate priors), explicit determination of (31.57) may be possible; otherwise, numerical procedures must be used.

This approach is something of an amalgam of Bayesian and frequentist ideas and has had a mixed reception. For example, Neyman (1962) hailed it as a breakthrough, whereas Lindley (1971) regards it as involving no new point of principle.

Decision theory

31.52   Abraham Wald's work on sequential analysis (cf. Chapter 24) led also to the development of a general theory of decision making. Consider a situation where, given the data, it is necessary to make a decision; further, assume that the consequences of these decisions are known and that they can be evaluated numerically. These are not trivial assumptions; for example, in their development of hypothesis tests Neyman and Pearson concluded that such information was *unlikely* to be available. Given the necessary background, the problem is to decide on optimum decision rules with reference to some performance measure. We now proceed to outline the basics of such a theory; for more detailed expositions, see Wald (1950), Blackwell and Girshick (1954), Ferguson (1967), and DeGroot (1970) among others.

31.53   Suppose we can specify a set of possible actions $A = \{a\}$ and a *decision rule* $d(x)$ that specifies the action to be taken when x is observed. The consequence of taking that action is to incur a *loss* $L[d(x), \theta]$ when the parameter value is θ. Some authors use a utility function rather than a loss function; for most purposes, loss can be viewed as negative utility, although utility may be deemed to be bounded whereas loss functions are often allowed to be unbounded.

The expected loss is known as the *risk function*:

$$R(d, \theta) = \int L[d(x), \theta] f(x \mid \theta) \, dx. \qquad (31.59)$$

A decision rule, d, is *admissible* if there is no rule d' such that

$$R(d', \theta) \leqslant R(d, \theta) \quad \text{for all } \theta \qquad (31.60)$$

with strict inequality for at least some θ. Quite generally, nothing is lost by restricting attention to the class of admissible decision rules, although this class may be large.

31.54   In order to select a particular decision rule, we may use a criterion such as *minimax*; that is, we choose the rule $d(x)$ that minimises risk taken over all θ:

$$\min_{d} \max_{\theta} R(d, \theta). \qquad (31.61)$$

Example 31.17

Suppose we use the squared-error-loss function

$$L[d(x), \theta] = \{d(x) - \theta\}^2$$

and consider five estimators for the location parameter of a normal distribution given a sample of size n:

(i) sample mean, $d_1(x) = \bar{x}$
(ii) sample median, $d_2(x) = m$
(iii) the first observation, $d_3(x) = x_1$
(iv) the number 631, $d_4(x) = 631$
(v) the weighted mean, $d_5(x) = \Sigma w_i x_i, \Sigma w_i = 1$.

It follows that the risk functions are, with $R_i = R(d_i, \theta)$,

$$R_1 = \sigma^2/n, \ R_2 \doteq 1.57\sigma^2/n, \ R_3 = \sigma^2$$

$$R_4 = (631 - \theta)^2 \quad \text{and} \quad R_5 = \sigma^2 \left\{ \frac{1}{n} + \Sigma(w_i - \bar{w})^2 \right\}.$$

By inspection, d_2, d_3, and d_5 are inadmissible; d_4 would be ruled inferior to all others by the minimax criterion even though it is admissible since R_4 is smallest when θ is near 631.

31.55   The partial ordering induced by admissibility or selection by a criterion such as minimax are as much as can be achieved in the absence of a prior distribution for θ. Hence, most modern research in decision theory follows the Bayesian path and assumes the existence of a prior. Thus, Bayesian decision theory is a logical extension of the Bayesian approach.

Given $p(\theta)$, we may compute the expected risk:

$$E(R) = \int R(d, \theta) p(\theta) \, d\theta$$

$$= \int_\theta \int_x L[d(x), \theta] f(x|\theta) p(\theta) \, dx \, d\theta; \quad (31.62)$$

(31.62) may be re-expressed as

$$E(R) = \int_x \left\{ \int_\theta L[d(x), \theta] P(\theta|x) \, d\theta \right\} f(x) \, dx \quad (31.63)$$

provided that the order of integration may be reversed. From (31.63), it is apparent that we may minimise $E(R)$ by choosing $d(x)$ to minimise the inner integral, for each x. The resulting decision rule is known as the *Bayes rule* and its expected risk is termed the *Bayes risk*. Any admissible rule is a Bayes rule for *some* prior distribution (a result due to Wald).

Example 31.18
When squared-error-loss is used, it follows that, provided the inner integral in (31.63) is bounded, it is minimised by selecting the posterior mean, whatever the distribution.

The James–Stein estimator

31.56 Suppose now that we have a sample of size n from each of K normal populations, $x_i \sim N(\mu_i, \sigma^2)$; we take σ^2 to be known and assume that the μ_i have common prior $N(\phi, \tau^2)$. Let the decision rule $d(t)$, $t = (t_1, \ldots, t_K)$, assign estimate t_i to parameter μ_i, $i = 1, \ldots, K$, and let the loss function be

$$L[d(t), \mu] = (t_1 - \mu_1)^2 + \cdots + (t_K - \mu_K)^2. \tag{31.64}$$

The obvious (and ML) estimator is $t_{ML} = (\bar{x}_1, \ldots, \bar{x}_K)$ for which

$$E(R) = K\sigma^2/n. \tag{31.65}$$

However, from Example 31.16, the Bayes rule is given by the posterior means:

$$t_i^* = \frac{\phi\sigma^2 + n\bar{x}_i\tau^2}{\sigma^2 + n\tau^2}$$

for which

$$E(R) = \frac{K\tau^2\sigma^2}{\sigma^2 + n\tau^2},$$

which is clearly less than the expected risk in (31.65). Stein (1956) showed that t_{ML} is inadmissible when $K \geqslant 3$; James and Stein (1961) produced the improved, but also inadmissible, estimator $t_{JS} = (\tilde{t}_1, \ldots, \tilde{t}_K)$, where

$$\tilde{t}_i = \left\{ 1 - \frac{(K-2)\sigma^2}{n\Sigma(\bar{x}_j - \phi)^2} \right\} \bar{x}_i + \frac{(K-2)\sigma^2\phi}{n\Sigma(\bar{x}_j - \phi)^2}. \tag{31.66}$$

For an excellent discussion of these shrinkage estimators and simple proofs of the admissibility results, see Stigler (1990).

31.57 These results have been used by some to cast doubt upon the value of ML estimators. Alternatively, they might be used to sound a strong note of caution about using decision theory unless one is very sure of the loss function and the prior distribution.

For a review of the historical development of decision theory and an extensive bibliography, see Fishburn (1989).

Discussion

31.58 There has been so much controversy about the various methods of estimation we have described that, at this point, we shall leave our customary objective standpoint. The remainder of this chapter is an expression of personal views. We think that it is the correct viewpoint; and it represents the result of many years' reflection on the issues involved, a serious attempt to understand what the protagonists say and to divine what they mean.

31.59 We have, then, to examine six major approaches, although some are more closely related than others: frequency, likelihood, fiducial, objective and subjective Bayesian, and decision theory. We must not be misled by, though we may derive some comfort from, the similarity of the results to which they lead in certain simple cases. We shall, however, develop the thesis that, where they differ, the basic reason is not that one or more are wrong, but that they are consciously or unconsciously either answering different questions or resting on different postulates.

31.60 In setting out the differences, it is useful to adopt Lakatos' (1974) concept of competing research programs, and to establish the *hard core* of assumptions underlying each theory. Supporting each theory is a *protective layer* of auxiliary assumptions, so the conclusions that may be drawn then follow deductively from these foundations. We are not concerned to debate at length which assumptions are major and which auxiliary, but rather to use this as a framework for our discussions.

31.61 The hard core underlying the frequency theory may be summarised as follows:

(a) the Kolmogorov axioms;
(b) well-defined random sampling procedures, that include specification of the sample space and stopping rule;
(c) the frequency interpretation of probability;
(d) a version of the *repeated sampling principle* (Cox and Hinkley, 1974, p. 45) which states that statistical procedures are to be assessed by their behaviour in hypothetical repetitions under the same conditions. This is Cox and Hinkley's *strong* version; the *weak* version requires only that we should not follow procedures that would be misleading for some parameter combinations (most of the time, in hypothetical repetitions). This principle is essentially the same as Birnbaum's *confidence principle*, introduced in **31.25**. As noted earlier, it is the conflict between this principle and the likelihood principle that is at the root of the debate between frequentists and Bayesians.

The protective belt then includes such concepts as consistency, unbiasedness, efficiency, sufficiency, power, etc. discussed in earlier chapters of this volume. In his discussion of maximum likelihood and decision theory, Efron (1982, p. 343) refers to these concepts as 'ingenious evasions' used by Fisher to avoid a decision-theoretic approach. However, it should be noted that Fisher, like Neyman and Pearson, was at pains to avoid strong assumptions regarding the existence and form of loss functions and prior distributions; the notions are certainly ingenious but form part of an alternative paradigm, not an evasion.

31.62 The frequency approach leads then to point and interval estimates and tests of hypotheses that are keyed to an interpretation of performance *in the long run*. The other approaches we have described constitute several attempts to develop an additional, or alternative, notion of probability that enables the investigator to make inferential statements conditionally upon the data recorded in a *particular* statistical experiment.

31.63   The protective belt of any theory evolves over time as, for example, when the Neyman–Pearson approach to testing hypotheses supplanted Fisher's pure tests of significance; note that we are not claiming such changes are either instantaneous or free from controversy! Another possible modification would be the use of unbiased estimating equations in place of unbiasedness (cf. **31.9**).

Two of the difficulties facing the frequency approach in practice are the specification of the sample space and need to ensure random sampling. Johnstone (1989) argues that it is not necessary that we know that the sample was drawn at random; '[A]ll that is necessary logically is that we *not have knowledge to the contrary*' (his italics). Following Fisher, Johnstone terms this a *postulate of ignorance* which is distinct from Bayes' postulate in that it applies to the sample space rather than the parameter space; hence, it is also distinct from our earlier argument in **9.9**, Vol. 1, which presumed random sampling.

Johnstone's ideas are clearly open to abuse but, used carefully, have considerable merit. For example, the distribution of the error term in a regression equation applied to some macroeconomic aggregate has a much more plausible interpretation when Johnstone's postulate is used.

31.64   The frequency approach is quite general in that it may be applied to any sampling situation once the sampling process is fully specified. However, there may be difficulties in execution. For example, when no single sufficient statistic exists, the confidence intervals may not be real or may be otherwise nugatory (cf. **20.12** and Example 20.4). Thus sufficiency is desirable, although it is not required. Perhaps it would be better to say that problems of interpretation may exist when nested and simply connected intervals cannot be obtained.

The principal argument in favour of the frequency theory of probability is that it does without any assumptions concerning prior distributions such as are essential to the Bayesian approach. This, in our opinion, is undeniable. But it is fair to ask whether it achieves this economy of basic assumption without losing something which the Bayesian theory possesses. Our view is that it does lose something on occasion, and that this something may be important for the purposes of estimation.

Prior information

31.65   Consider the case where we are estimating the mean μ of a normal population with known variance, and suppose that we *know* that μ lies between 0 and 1. According to Bayes' postulate, we should have

$$P(\mu \mid \bar{x}) = \frac{\exp\left\{-\dfrac{n}{2}(\mu - \bar{x})^2\right\}}{\displaystyle\int_0^1 \exp\left\{-\dfrac{n}{2}(\mu - \bar{x})^2\right\} d\mu}, \tag{31.67}$$

and the problem of setting limits to μ, though not free from mathematical complexity, is determinate. What has confidence-interval theory to say on this point? It can do no more than reiterate statements like

$$P\{\bar{x} - 1.96/\sqrt{n} \le \mu \le \bar{x} + 1.96/\sqrt{n}\} = 0.95.$$

These are still true in the required proportion of cases, but the statements take no account of our prior knowledge about the range of μ and may occasionally be idle. It may be true, but is absurd, to assert $-1 \le \mu \le 2$ if we know already that $0 \le \mu \le 1$. Of course, we may truncate our interval to accord with the prior information. In our example, we could assert only that $0 \le \mu \le 1$: the observations would have added nothing to our knowledge.

Thus, it appears that the frequency theory has the defect of its principal virtue: it attains its generality at the price of being unable to incorporate prior knowledge into its statements. When we make our final judgement about μ, we have to synthesise the information obtained from the observations with our prior knowledge. Bayes' theorem attempts this synthesis at the outset. Frequency theory leaves it until the end (and, we feel bound to remark, in most current expositions ignores the point completely).

31.66 Fiducial theory, as we have remarked, has been confined by Fisher to the case where sufficient statistics are used, or quite generally, to cases where all the information in the likelihood function can be utilised. No systematic exposition has been given of the procedure to be followed when prior information is available, but there seems no reason why a similar method to that exemplified by equation (31.67) should not be used. That is, if we derive the fiducial distribution $f(\mu)$ over a general range but have the supplementary information that the parameter must lie in the range μ_0 to μ_1 (within that general range), we modify the fiducial distribution by truncation to

$$f(\mu) \Big/ \int_{\mu_0}^{\mu_1} f(\mu)\, \mathrm{d}\mu.$$

Falsificationism
31.67 One final observation is relevant concerning the frequentist approach. Its development was paralleled by the development of *falsificationism* in the philosophy of science, spearheaded by Sir Karl Popper (cf. Popper, 1963). The basis of Popper's scheme is that evidence may or may not refute a theory but does not sustain it; that is, science progresses by performing experiments that challenge theories. This view of science does not allow the results of an experiment to provide explicit corroboration for a theory; it is symbolised by the stricture that we speak of 'not rejecting H_0' rather than 'accepting H_0'. More fundamentally, the frequentist approach does not seek to provide measures of corroboration, and it is the search for such measures that has, in part, fuelled the development of alternative paradigms for statistical inference. Indeed, all the other approaches described in this chapter allow corroborative statements to be made on the basis of the experiment just performed and conditionally upon the observations.

Likelihood-based inference

31.68 The likelihood function is recognised by all schools of thought as being a comprehensive summary of the data. Indeed, as we saw in **18.43**, Fisher (1956) suggested plotting the LF against θ; others (e.g. Efron, 1982) also strongly support the use of the LF as an effective *summary*. Edwards (1972), arguing that the LF described the relative *support* for different values of θ, went further and suggested that inferences be made on the basis of these support values. Clearly, such an approach is consistent with the strong likelihood principle (SLP), although it is incomplete unless supplemented with some procedure for handling nuisance parameters, such as the use of partial likelihood (cf. **18.44**) or the approaches discussed in **31.31–4**. Such methods have the advantage that prior information may be incorporated through a prior likelihood function.

Bayesian procedures are always consistent with the SLP, although empirical Bayes methods need not be. Fiducial inference may violate the SLP, although such violations tend to be uncommon.

Probability as a degree of belief

31.69 In the Bayesian and fiducial arguments, we must first assume the existence of a different concept of probability that measures the degree of belief or credibility in a hypothesis or theory. Carnap (1962) termed this probability$_1$, as distinct from the frequency concept, probability$_2$. Viewed in this light, the failure (?) of the frequency approach to deliver statements on the credibility of a hypothesis is almost axiomatic, since frequentists are unwilling to accept any probability$_1$ concepts that do not have a frequency interpretation.

The fiducialist argument rests on the assumption that probability$_2$ can be converted into probability$_1$ by means of a pivoting operation. Following the discussion in **31.26–9** we know that the process is possible; the key question is whether the resulting probability measure is meaningful.

31.70 The hard core for Bayesian inference is an axiomatic development that provides the framework for specifying prior probabilities and updating such probabilities by Bayes' theorem. For the objectivist, this means that there should be an agreed process by which a prior may be generated that is acceptable to all. Such a rule is necessarily mechanistic, since subjective interpretation is not admissible; yet, if the performance of the rule cannot be judged by either frequency or subjective criteria, its meaning remains rather obscure. Indeed, we are asked to accept that the prior be flat on $(-\infty, \infty)$ but inversely proportional to θ on $(0, \infty)$. Sophisticated arguments concerning the distinction somehow fail to impress us as touching the root of the problem. Further, it is found that working with uninformative priors can lead to some theoretical difficulties; see Stone (1976) and Exercise 31.5.

31.71 The subjectivists' hard core requires the individual to be willing to bet on anything, but in a logical fashion, as noted in **31.41**. Given this framework, You can certainly begin with Your prior and derive Your posterior probability statement regarding the plausibility of a hypothesis. Let us begin by considering the process of specifying the prior.

If the prior distribution is specified in conjugate form, such as the normal mean being $N(\phi, \tau^2)$, then we are faced with a potential infinite regress in specifying the prior for (ϕ, τ^2) and so on. This is resolved only by claiming knowledge of the (hyper) parameters at some stage (cf. Lindley and Smith, 1972). If the prior distribution is determined within a framework of bets, You must be able to specify Your utility function. Once this is available, then an axiomatic development such as that of Savage (1954) shows that coherent behaviour leads to degrees of belief that satisfy the axioms of probability.

Once the prior is available, You may proceed to make inferences in a manner that is consistent with the SLP (and, therefore, possibly inconsistent with the confidence principle). Whether this is a source of strength or weakness lies in the eye of the beholder, but the fact remains that all the inferences made are subjective, Your own assessments.

Whether such individual statements are acceptable is problematic. When making a decision in a context that lacks opportunities for replication, the use of Your probabilities seems reasonable when You are responsible for the decision. However. we believe that many, if not most, statistical analyses cannot be reasonably fitted into a decision-theoretic framework. Furthermore, an expression of personal beliefs has not proved acceptable as a way of reporting the results of an investigation.

Reconciliation?

31.72   As might be expected, there have been a number of attempts to reconcile the different approaches to statistical inference; we shall review some of these briefly. We begin by noting that, in large samples, all methods are consistent with the strong likelihood principle.

31.73   We saw in **8.8**, Vol. 1, that the use of Bayes' theorem with a uniform prior distribution gives a posterior mode that is equal to the ML estimator. Even if a non-uniform prior distribution is used, the methods are *asymptotically* equivalent. Equation (8.6) may be written in our present notation as

$$P(\theta \mid x) \propto p(\theta) L(x \mid \theta). \qquad (31.68)$$

To maximise this for choice of θ is equivalent to maximising its logarithm,

$$\log p(\theta) + \log L(x \mid \theta) = \sum_{i=1}^{n} \{\log f(x_i \mid \theta) + (1/n) \log p(\theta)\}. \qquad (31.69)$$

As $n \to \infty$, the second term in braces on the right is negligible, and we are effectively maximising $\log L(x \mid \theta)$ to obtain the ML estimator. We may express this by saying that, given enough observations, the prior distribution becomes irrelevant; this is known as the principle of *stable estimation*. For small n, however, there may be wide differences between the ML and Bayesian estimates—cf. Exercise 31.4.

Diaconis and Freedman (1986a, b) show that when the parameter space is high-dimensional (or infinite dimensional as in some non-parametric problems), the prior may swamp the data no matter how many observations are available. In this sense, Bayesian estimators may lack consistency; the discussion following their 1986a paper should also be consulted.

31.74   One aspect of the Bayesian approach is, as we have suggested on occasion, that it demands too much. For example, we need to be able to specify the functional form of the LF and to list all the variables of interest. Yet, much of the appeal of procedures such as cross-validation and bootstrapping derives from their application in circumstances where it may not be possible to specify the LF precisely. Likewise, randomisation in experimental design safeguards against factors that may not have been recognised.

Following this theme, Durbin (1988) points out that the overall complexity of many models makes specification of the LF, and therefore application of the likelihood principle, impractical. Nevertheless, simple diagnostic tests often guide the model builder well, and Durbin suggests that the practical effects of philosophical differences are often small compared to the need for effective statistical modelling.

31.75   Box (1980) identifies two components in statistical modelling: *criticism* and *estimation*. Starting from (31.32), Box would use the posterior distribution for θ for estimation, but the predictive distribution

$$f(x) = \int p(\theta) L(x \mid \theta) \, d\theta \qquad (31.70)$$

for model criticism. Although $f(x)$ is derived on the assumption that the prior $p(\theta)$ is available, Box recommends frequency procedures for the criticism part of the modelling process. This is similar in spirit to Durbin's comments in **31.74** and also to our earlier discussion in **31.31-4**, save that there θ was removed by maximisation rather than integration.

31.76   Giere (1977) distinguishes between *testing* and *information* in statistical inference, suggesting that the information criterion allows a direct measure of evidence for a hypothesis so that the Bayesian approach may be invoked. In the testing framework, no such measure exists, as noted by many frequentist writers from Neyman and Pearson onwards. Giere goes on to argue for probability as a measure of propensity that would allow statements to be made for single experiments.

31.77   I. J. Good (cf. 1976, 1983, 1988) calls for a Bayesian–non-Bayesian compromise from a different viewpoint. For Good, frequentist methods often represent a collection of *ad hoc* procedures, and he would accept frequentist procedures whenever they match up sufficiently well with the Bayesian solution. While such an approach may serve to reduce contention, 'compromise' is perhaps an inappropriate descriptor.

31.78 Putting these several considerations together, we see that good statistical practice can often emerge from different paradigms and that, indeed, different notions of probability may be appropriate in different circumstances. Nevertheless, the frequentist approach remains firmly rooted in the Popperian tradition of falsification-ism, and any attempt to go beyond that requires recognition of some other concept of probability.

It may be tempting to think in terms of Kuhn's (1962) notion of a scientific revolution whereby the current (frequentist) paradigm is challenged by the newcomer (Bayesian), from which a new orthodoxy will emerge. However, this view is somewhat inappropriate; rather we should recognise that the Bayesian approach seeks to deliver more but, in order to do so, requires stronger assumptions.

In conclusion, it is fitting to quote some words written long ago (Kendall, 1949):

> The frequentist seeks for objectivity in defining his probabilities by reference to frequencies; but he has to use a primitive idea of randomness or equiprobability in order to calculate the probability in any given practical case. The non-frequentist begins by taking probabilities as a primitive idea but he has to assume that the values which his calculations give to a probability reflect, in some way, the behaviour of events .... *Neither party can avoid using the ideas of the other in order to set up and justify a comprehensive theory.*

EXERCISES

31.1   In the multinomial distribution of Example 18.10, show that the four-component minimal sufficient statistic for θ can be written either as $(a+b, c+d, b, d)$ or as $(a+c, b+d, c, d)$, and that the first two components in each case form an ancillary pair of statistics.

31.2   Consider the LF for the two-means problem when the variates are $N(\mu_i, \sigma_i^2)$, $i = 1, 2$. Use (31.26) to show that the LR credibility measure is algebraically the same as the Fisher–Behrens distribution, with n_i in place of $(n_i - 1)$.

31.3   Show that if the distribution of a sufficient statistic x is

$$f(x \mid \theta) = \frac{\theta^2}{\theta+1}(x+1)\,e^{-x\theta}, \qquad x > 0, \quad \theta \geqslant 0,$$

the fiducial distribution of θ for combined samples with sufficient statistics x, y, is

$$g_{x,y}(\theta) = \frac{e^{-z\theta}}{(\theta+1)^3}[\theta^3(2z^2 + \tfrac{4}{3}z^3 + \tfrac{1}{6}z^4) + \theta^4(z^2 + z^3 + \tfrac{1}{6}z^4)]$$

(where $z = x + y$), while that for a single sample is

$$g_x(\theta) = \frac{\theta x\, e^{-x\theta}}{(\theta+1)^2}[1 + (1+\theta)(1+x)].$$

(Note that the minus sign in (31.23) is unnecessary here, since $F(x \mid \theta)$ is an increasing function of θ.) Hence show that the Bayes posterior distribution from the second sample, using $g_x(\theta)$ as prior distribution, is

$$P(\theta \mid y; x) \propto e^{-z\theta}\left(\frac{\theta}{\theta+1}\right)^3 x(1+y)[1 + (1+\theta)(1+x)],$$

so that

$$P(\theta \mid y; x) \neq g_{x,y}(\theta). \text{ Note that } P(\theta \mid y; x) \neq P(\theta \mid x; y) \text{ also.}$$

(Lindley, 1958)

31.4   If the prior distribution of $\nu\theta^{-1}$ is a χ^2 with ν d.fr., and an observation x is distributed normally with mean zero and variance θ, show (cf. Exercise 16.25) that the posterior distribution of θ is

$$P(\theta \mid x) \propto \exp\{-(\nu + x^2)/(2\theta)\}\theta^{-(\nu+3)/2}.$$

Hence show that the Bayes estimate of θ must *always* exceed the preassigned constant $k_\nu = \nu/(\nu+3)$, however small $|x|$ is, although the prior probability that $\theta < k_\nu$ tends to 0.5 as $\nu \to \infty$, exceeding 0.2 at $\nu = 10$. Show that the ML estimator of θ is $\hat{\theta} = x^2$.

31.5   An amount x is deposited into a box, B_1, and $2x$ into a second box B_2. The boxes appear indistinguishable from the outside to a contestant, C, who is allowed to choose one box. Having made a choice, C is then offered the opportunity to switch boxes. C reckons that if the selected box contains y, the other box is equally likely to contain $y/2$ or $2y$ so that the switch is justified. Having made the switch, C repeats the argument and switches again, and so on.

Let Y_S, Y_N denote the expected amount from opening the selected and nonselected boxes, respectively. Show that

$$E(Y_N \mid Y_S = y) = y \frac{\{\frac{1}{2}p(y/2) + 2p(y)\}}{\{p(y/2) + p(y)\}}, \tag{*}$$

where $p(x)$ is the prior density for C for the amount placed in B_1. Also show that $E(Y_N) = E(Y_S) = 3E(x)/2$ when $E(x)$ exists, whatever the form of $p(x)$. Note that if the improper prior:

$$p(x) = \text{constant}, \qquad 0 < x < \infty$$

is used in (*), C appears justified in switching.

31.6 Consider the linear model $\mathbf{y} = \mathbf{X}\boldsymbol{\beta} + \boldsymbol{\varepsilon}$ where $\boldsymbol{\varepsilon}$ has a multinormal distribution with mean $\mathbf{0}$ and covariance matrix $\sigma^2\mathbf{I}$, and $\mathbf{X}$ is of rank $k < p$.

Given that $\boldsymbol{\beta}$ has a p-dimensional multinormal prior distribution with mean $\boldsymbol{\gamma}$ and $V(\boldsymbol{\beta}) = \mathbf{C}^{-1}$, show that $\boldsymbol{\beta}$ is estimable provided $(\mathbf{X}'\mathbf{X} + \sigma^2\mathbf{C})$ is of rank p. Setting $\mathbf{C} = \sigma^{-2}\mathbf{B}'\mathbf{B}$ and $\mathbf{a} = \mathbf{B}\boldsymbol{\gamma}$, compare this estimator with that given in **19.14**.

APPENDIX TABLES

Appendix Table 1   Density function of the normal distribution $y = \dfrac{1}{\sqrt{(2\pi)}} e^{-\frac{1}{2}x^2}$ **with first and second differences**

x	y	$\Delta^1(-)$	Δ^2	x	y	$\Delta^1(-)$	Δ^2
0.0	0.39894	199	−392	2.5	0.01753	395	+79
0.1	0.39695	591	−374	2.6	0.01358	316	+66
0.2	0.39104	965	−347	2.7	0.01042	250	+53
0.3	0.38139	1312	−308	2.8	0.00792	197	+45
0.4	0.36827	1620	−265	2.9	0.00595	152	+36
0.5	0.35207	1885	−212	3.0	0.00443	116	+27
0.6	0.33322	2097	−159	3.1	0.00327	89	+23
0.7	0.31225	2256	−104	3.2	0.00238	66	+17
0.8	0.28969	2360	−52	3.3	0.00172	49	+13
0.9	0.26609	2412	0	3.4	0.00123	36	+10
1.0	0.24197	2412	+46	3.5	0.00087	26	+7
1.1	0.21785	2366	+84	3.6	0.00061	19	+6
1.2	0.19419	2282	+118	3.7	0.00042	13	+4
1.3	0.17137	2164	+143	3.8	0.00029	9	+2
1.4	0.14973	2021	+161	3.9	0.00020	7	+3
1.5	0.12952	1860	+173	4.0	0.00013	4	—
1.6	0.11092	1687	+177	4.1	0.00009	3	—
1.7	0.09405	1510	+177	4.2	0.00006	2	—
1.8	0.07895	1333	+170	4.3	0.00004	2	—
1.9	0.06562	1163	+162	4.4	0.00002	—	—
2.0	0.05399	1001	+150	4.5	0.00002	—	—
2.1	0.04398	851	+137	4.6	0.00001	—	—
2.2	0.03547	714	+120	4.7	0.00001	—	—
2.3	0.02833	594	+108	4.8	0.00000	—	—
2.4	0.02239	486	+91				

Appendix Table 2   Distribution function of the normal distribution

The table shows the area under the curve $y = (2\pi)^{-\frac{1}{2}} e^{-\frac{1}{2}x^2}$ lying to the left of specified deviates x; e.g. the area corresponding to a deviate 1.86 ($= 1.5 + 0.36$) is 0.9686.

Deviate	0.0+	0.5+	1.0+	1.5+	2.0+	2.5+	3.0+	3.5+
0.00	5000	6915	8413	9332	9772	9^2379	9^2865	9^377
0.01	5040	6950	8438	9345	9778	9^2496	9^2869	9^378
0.02	5080	6985	8461	9357	9783	9^2413	9^2874	9^378
0.03	5120	7019	8485	9370	9788	9^2430	9^2878	9^379
0.04	5160	7054	8508	9382	9793	9^2446	9^2882	9^380
0.05	5199	7088	8531	9394	9798	9^2461	9^2886	9^381
0.06	5239	7123	8554	9406	9803	9^2477	9^2889	9^381
0.07	5279	7157	8577	9418	9808	9^2492	9^2893	9^382
0.08	5319	7190	8599	9429	9812	9^2506	9^2897	9^383
0.09	5359	7224	8621	9441	9817	9^2520	9^2900	9^383
0.10	5398	7257	8643	9452	9821	9^2534	9^303	9^384
0.11	5438	7291	8665	9463	9826	9^2547	9^306	9^385
0.12	5478	7324	8686	9474	9830	9^2560	9^310	9^385
0.13	5517	7357	8708	9484	9834	9^2573	9^313	9^386
0.14	5557	7389	8729	9495	9838	9^2585	9^316	9^386
0.15	5596	7422	8749	9505	9842	9^2598	9^318	9^387
0.16	5636	7454	8770	9515	9846	9^2609	9^321	9^387
0.17	5675	7486	8790	9525	9850	9^2621	9^324	9^388
0.18	5714	7517	8810	9535	9854	9^2632	9^326	9^388
0.19	5753	7549	8830	9545	9857	9^2643	9^329	9^389
0.20	5793	7580	8849	9554	9861	9^2653	9^331	9^389
0.21	5832	7611	8869	9564	9864	9^2664	9^334	9^390
0.22	5871	7642	8888	9573	9868	9^2674	9^336	9^390
0.23	5910	7673	8907	9582	9871	9^2683	9^338	9^404
0.24	5948	7704	8925	9591	9875	9^2693	9^340	9^408
0.25	5987	7738	8944	9599	9878	9^2702	9^342	9^412
0.26	6026	7764	8962	9608	9881	9^2711	9^344	9^415
0.27	6064	7794	8980	9616	9884	9^2720	9^346	9^418
0.28	6103	7823	8997	9625	9887	9^2728	9^348	9^422
0.29	6141	7852	9015	9633	9890	9^2736	9^350	9^425
0.30	6179	7881	9032	9641	9893	9^2744	9^352	9^428
0.31	6217	7910	9049	9649	9896	9^2752	9^353	9^431
0.32	6255	7939	9066	9656	9898	9^2760	9^355	9^433
0.33	6293	7967	9082	9664	9901	9^2767	9^357	9^436
0.34	6331	7995	9099	9671	9904	9^2774	9^358	9^439
0.35	6368	8023	9115	9678	9906	9^2781	9^360	9^441
0.36	6406	8051	9131	9686	9909	9^2788	9^361	9^443
0.37	6443	8078	9147	9693	9911	9^2795	9^362	9^446
0.38	6480	8106	9162	9699	9913	9^2801	9^364	9^448
0.39	6517	8133	9177	9706	9916	9^2807	9^365	9^450
0.40	6554	8159	9192	9713	9918	9^2813	9^366	9^452
0.41	6591	8186	9207	9719	9920	9^2819	9^368	9^454
0.42	6628	8212	9222	9726	9922	9^2825	9^369	9^456
0.43	6664	8238	9236	9732	9925	9^2831	9^370	9^458
0.44	6700	8264	9251	9738	9927	9^2836	9^371	9^459
0.45	6736	8289	9265	9744	9929	9^2841	9^372	9^461
0.46	6772	8315	9279	9750	9931	9^2846	9^373	9^463
0.47	6808	8340	9292	9756	9932	9^2851	9^374	9^464
0.48	6844	8365	9306	9761	9934	9^2856	9^375	9^466
0.49	6879	8389	9319	9767	9936	9^2861	9^376	9^467

Note—Decimal points in the body of the table are omitted. Repeated 9's are indicated by powers, e.g. 9^371 stands for 0.99971.

Appendix Table 3   Quantiles of the d.f. of χ^2

(Reproduced from Table III of Sir Ronald Fisher's *Statistical Methods for Research Workers*, Oliver and Boyd Ltd., Edinburgh, by kind permission of the author and publishers)

$P = 1 - F$	0.99	0.98	0.95	0.90	0.80	0.70	0.50	0.30	0.20	0.10	0.05	0.02	0.01
$\nu = 1$	0.0^3157	0.0^3628	0.0^2393	0.0158	0.0642	0.148	0.455	1.074	1.642	2.706	3.841	5.412	6.635
2	0.0201	0.0404	0.103	0.211	0.446	0.713	1.386	2.408	3.219	4.605	5.991	7.824	9.210
3	0.115	0.185	0.352	0.584	1.005	1.424	2.366	3.665	4.642	6.251	7.815	9.837	11.345
4	0.297	0.429	0.711	1.064	1.649	2.195	3.357	4.878	5.989	7.779	9.488	11.668	13.277
5	0.554	0.752	1.145	1.610	2.343	3.000	4.351	6.064	7.289	9.236	11.070	13.388	15.086
6	0.872	1.134	1.635	2.204	3.070	3.828	5.348	7.231	8.558	10.645	12.592	15.033	16.812
7	1.239	1.564	2.167	2.833	3.822	4.671	6.346	8.383	9.803	12.017	14.067	16.622	18.475
8	1.646	2.032	2.733	3.490	4.594	5.527	7.344	9.524	11.030	13.362	15.507	18.168	20.090
9	2.088	2.532	3.325	4.168	5.380	6.393	8.343	10.656	12.242	14.684	16.919	19.679	21.666
10	2.358	3.059	3.940	4.865	6.179	7.267	9.342	11.781	13.442	15.987	18.307	21.161	23.209
11	3.053	3.609	4.575	5.578	6.989	8.148	10.341	12.899	14.631	17.275	19.675	22.618	24.725
12	3.571	4.178	5.226	6.304	7.807	9.034	11.340	14.011	15.821	18.549	21.026	24.054	26.217
13	4.107	4.765	5.892	7.042	8.634	9.926	12.340	15.119	16.985	19.812	22.362	25.472	27.688
14	4.660	5.368	6.571	7.790	9.467	10.821	13.339	16.222	18.151	21.064	23.685	26.873	29.141
15	5.229	5.985	7.261	8.547	10.307	11.721	14.339	17.322	19.311	22.307	24.996	28.259	30.578
16	5.812	6.614	7.962	9.312	11.152	12.624	15.338	18.418	20.465	23.542	26.296	29.633	32.000
17	6.408	7.255	8.672	10.085	12.002	13.531	16.338	19.511	21.615	24.769	27.587	30.995	33.409
18	7.015	7.906	9.390	10.865	12.857	14.440	17.338	20.601	22.760	25.989	28.869	32.346	34.805
19	7.633	8.567	10.117	11.651	13.716	15.352	18.338	21.689	23.900	27.204	30.144	33.687	36.191
20	8.260	9.237	10.851	12.443	14.578	16.266	19.337	22.775	25.038	28.412	31.410	35.020	37.566
21	8.897	9.915	11.591	13.240	15.445	17.182	20.337	23.858	26.171	29.615	32.671	36.343	38.932
22	9.542	10.600	12.338	14.041	16.314	18.101	21.337	24.939	27.301	30.813	33.924	37.659	40.289
23	10.196	11.293	13.091	14.848	17.187	19.021	22.337	26.018	28.429	32.007	35.172	38.968	41.638
24	10.856	11.992	13.848	15.659	18.062	19.943	23.337	27.096	29.553	33.196	36.415	40.270	42.980
25	11.524	12.697	14.611	16.473	18.940	20.867	24.337	28.172	30.675	34.382	37.652	41.566	44.314
26	12.198	13.409	15.379	17.292	19.820	21.792	25.336	29.246	31.795	35.563	38.885	42.856	45.642
27	12.879	14.125	16.151	18.114	20.703	22.719	26.336	30.319	32.912	36.741	40.113	44.140	46.963
28	13.565	14.847	16.928	18.939	21.588	23.647	27.336	31.391	34.027	37.916	41.337	45.419	48.278
29	14.256	15.574	17.708	19.768	22.475	24.577	28.336	32.461	35.139	39.087	42.557	46.693	49.588
30	14.953	16.306	18.493	20.599	23.364	25.508	29.336	33.530	36.250	40.256	43.773	47.962	50.892

Note—For values of ν greater than 30 the quantity $\sqrt{(2\chi^2)}$ may be taken to be distributed normally about mean $\sqrt{(2\nu - 1)}$ with unit variance.

Appendix Table 4a Distribution function of χ^2 for one degree of freedom for values $\chi^2 = 0$ to $\chi^2 = 1$ by steps of 0.01

χ^2	$P = 1 - F$	Δ	χ^2	$P = 1 - F$	Δ
0	1.00000	7966	0.50	0.47950	436
0.01	0.92034	3280	0.51	0.47514	430
0.02	0.88754	2505	0.52	0.47084	423
0.03	0.86249	2101	0.53	0.46661	418
0.04	0.84148	1842	0.54	0.46243	411
0.05	0.82306	1656	0.55	0.45832	406
0.06	0.80650	1516	0.56	0.45426	400
0.07	0.79134	1404	0.57	0.45026	395
0.08	0.77730	1312	0.58	0.44631	389
0.09	0.76418	1235	0.59	0.44242	384
0.10	0.75183	1169	0.60	0.43858	379
0.11	0.74014	1111	0.61	0.43479	374
0.12	0.72903	1060	0.62	0.43105	369
0.13	0.71843	1015	0.63	0.42736	365
0.14	0.70828	974	0.64	0.42371	360
0.15	0.69854	938	0.65	0.42011	355
0.16	0.68916	905	0.66	0.41656	351
0.17	0.68011	874	0.67	0.41305	346
0.18	0.67137	845	0.68	0.40959	343
0.19	0.66292	820	0.69	0.40616	338
0.20	0.65472	795	0.70	0.40278	334
0.21	0.64677	773	0.71	0.39944	330
0.22	0.63904	752	0.72	0.39614	326
0.23	0.63152	731	0.73	0.39288	322
0.24	0.62421	713	0.74	0.38966	318
0.25	0.61708	696	0.75	0.38648	315
0.26	0.61012	679	0.76	0.38333	311
0.27	0.60333	663	0.77	0.38022	308
0.28	0.59670	648	0.78	0.37714	304
0.29	0.59022	634	0.79	0.37410	301
0.30	0.58388	620	0.80	0.37109	297
0.31	0.57768	607	0.81	0.36812	294
0.32	0.57161	595	0.82	0.36518	291
0.33	0.56566	583	0.83	0.36227	287
0.34	0.55983	572	0.84	0.35940	285
0.35	0.55411	560	0.85	0.35655	281
0.36	0.54851	551	0.86	0.35374	278
0.37	0.54300	540	0.87	0.35096	276
0.38	0.53760	530	0.88	0.34820	272
0.39	0.53230	521	0.89	0.34548	270
0.40	0.52709	512	0.90	0.34278	267
0.41	0.52197	503	0.91	0.34011	264
0.42	0.51694	495	0.92	0.33747	261
0.43	0.51199	487	0.93	0.33486	258
0.44	0.50712	479	0.94	0.33228	256
0.45	0.50233	471	0.95	0.32972	253
0.46	0.49762	463	0.96	0.32719	251
0.47	0.49299	457	0.97	0.32468	248
0.48	0.48842	449	0.98	0.32220	246
0.49	0.48393	443	0.99	0.31974	243
0.50	0.47950	436	1.00	0.31731	241

Appendix Table 4b   Distribution function of χ^2 for one degree of freedom for values of χ^2 from 1 to 10 by steps of 0.1

χ^2	$P = 1 - F$	Δ	χ^2	$P = 1 - F$	Δ
1.0	0.31731	2304	5.5	0.01902	106
1.1	0.29427	2095	5.6	0.01796	99
1.2	0.27332	1911	5.7	0.01697	94
1.3	0.25421	1749	5.8	0.01603	89
1.4	0.23672	1605	5.9	0.01514	83
1.5	0.22067	1477	6.0	0.01431	79
1.6	0.20590	1361	6.1	0.01352	74
1.7	0.19229	1258	6.2	0.01278	71
1.8	0.17971	1163	6.3	0.01207	66
1.9	0.16808	1078	6.4	0.01141	62
2.0	0.15730	1000	6.5	0.01079	59
2.1	0.14730	929	6.6	0.01020	56
2.2	0.13801	864	6.7	0.00964	52
2.3	0.12937	803	6.8	0.00912	50
2.4	0.12134	749	6.9	0.00862	47
2.5	0.11385	699	7.0	0.00815	44
2.6	0.10686	651	7.1	0.00771	42
2.7	0.10035	609	7.2	0.00729	39
2.8	0.09426	568	7.3	0.00690	38
2.9	0.08858	532	7.4	0.00652	35
3.0	0.08326	497	7.5	0.00617	33
3.1	0.07829	465	7.6	0.00584	32
3.2	0.07364	436	7.7	0.00552	30
3.3	0.06928	408	7.8	0.00522	28
3.4	0.06520	383	7.9	0.00494	26
3.5	0.06137	359	8.0	0.00468	25
3.6	0.05778	337	8.1	0.00443	24
3.7	0.05441	316	8.2	0.00419	23
3.8	0.05125	296	8.3	0.00396	21
3.9	0.04829	279	8.4	0.00375	20
4.0	0.04550	262	8.5	0.00355	19
4.1	0.04288	246	8.6	0.00336	18
4.2	0.04042	231	8.7	0.00318	17
4.3	0.03811	217	8.8	0.00301	16
4.4	0.03594	205	8.9	0.00285	15
4.5	0.03389	192	9.0	0.00270	14
4.6	0.03197	181	9.1	0.00256	14
4.7	0.03016	170	9.2	0.00242	13
4.8	0.02846	160	9.3	0.00229	12
4.9	0.02686	151	9.4	0.00217	12
5.0	0.02535	142	9.5	0.00205	10
5.1	0.02393	134	9.6	0.00195	11
5.2	0.02259	126	9.7	0.00184	10
5.3	0.02133	119	9.8	0.00174	9
5.4	0.02014	112	9.9	0.00165	8
5.5	0.01902	106	10.0	0.00157	8

Appendix Table 5   Quantiles of the d.f. of t

(Reproduced from Sir Ronald Fisher and Dr F. Yates: *Statistical Tables for Biological, Medical and Agricultural Research*, Oliver and Boyd Ltd., Edunburgh, by kind permission of the authors and publishers)

$P=2$ $(1-F)$ $\nu =$	0.9	0.8	0.7	0.6	0.5	0.4	0.3	0.2	0.1	0.05	0.02	0.01	0.001
1	0.158	0.325	0.510	0.727	1.000	1.376	1.963	3.078	6.314	12.706	31.821	63. 657	636.619
2	0.142	0.289	0.445	0.617	0.816	1.061	1.386	1.886	2.920	4.303	6.965	9.925	31.598
3	0.137	0.277	0.424	0.584	0.765	0.978	1.250	1.638	2.353	3.182	4.541	5.841	12.924
4	0.134	0.271	0.414	0.569	0.741	0.941	1.190	1.533	2.132	2.776	3.747	4.604	8.610
5	0.132	0.267	0.408	0.559	0.727	0.920	1.156	1.476	2.015	2.571	3.365	4.032	6.869
6	0.131	0.265	0.404	0.553	0.718	0.906	1.134	1.440	1.943	2.447	3.143	3.707	5.959
7	0.130	0.263	0.402	0.549	0.711	0.896	1.119	1.415	1.895	2.365	2.998	3.499	5.408
8	0.130	0.262	0.399	0.546	0.706	0.889	1.108	1.397	1.860	2.306	2.896	3.355	5.041
9	0.129	0.261	0.398	0.543	0.703	0.883	1.100	1.383	1.833	2.262	2.821	3.250	4.781
10	0.129	0.260	0.397	0.542	0.700	0.879	1.093	1.372	1.812	2.228	2.764	3.169	4.587
11	0.129	0.260	0.396	0.540	0.697	0.876	1.088	1.363	1.796	2.201	2.718	3.106	4.437
11	0.128	0.259	0.395	0.539	0.695	0.873	1.083	1.356	1.782	2.179	2.681	3.055	4.318
13	0.128	0.259	0.394	0.538	0.694	0.870	1.079	1.350	1.771	2.160	2.650	3.012	4.221
14	0.128	0.258	0.393	0.537	0.692	0.868	1.076	1.345	1.761	2.145	2.624	2.977	4.140
15	0.128	0.258	0.393	0.536	0.691	0.866	1.074	1.341	1.753	2.131	2.602	2.947	4.073
16	0.128	0.258	0.392	0.535	0.690	0.865	1.071	1.337	1.746	2.120	2.583	2.921	4.015
17	0.128	0.257	0.392	0.534	0.689	0.863	1.069	1.333	1.740	2.110	2.567	2.898	3.965
18	0.127	0.257	0.392	0.534	0.688	0.862	1.067	1.330	1.734	2.101	2.552	2.878	3.922
19	0.127	0.257	0.391	0.533	0.688	0.861	1.066	1.328	1.729	2.093	2.539	2.861	3.883
20	0.127	0.257	0.391	0.533	0.687	0.860	1.064	1.325	1.725	2.086	2.528	2.845	3.850
21	0.127	0.257	0.391	0.532	0.686	0.859	1.063	1.323	1.721	2.080	2.518	2.831	3.819
22	0.127	0.256	0.390	0.532	0.686	0.858	1.061	1.321	1.717	2.074	2.508	2.819	3.792
23	0.127	0.256	0.390	0.532	0.685	0.858	1.060	1.319	1.714	2.069	2.500	2.807	3.767
24	0.127	0.256	0.390	0.531	0.685	0.857	1.059	1.318	1.711	2.064	2.492	2.797	3.745
25	0.127	0.256	0.390	0.531	0.684	0.856	1.058	1.316	1.708	2.060	2.485	2.787	3.725
26	0.127	0.256	0.390	0.531	0.684	0.856	1.058	1.315	1.706	2.056	2.479	2.779	3.707
27	0.127	0.256	0.389	0.531	0.684	0.855	1.057	1.314	1.703	2.052	2.473	2.771	3.690
28	0.127	0.256	0.389	0.530	0.683	0.855	1.056	1.313	1.701	2.048	2.467	2.763	3.674
29	0.127	0.256	0.389	0.530	0.683	0.854	1.055	1.311	1.699	2.045	2.462	2.756	3.659
30	0.127	0.256	0.389	0.530	0.683	0.854	1.055	1.310	1.697	2.042	2.457	2.750	3.646
40	0.126	0.255	0.388	0.529	0.681	0.851	1.050	1.303	1.684	2.021	2.423	2.704	3.551
60	0.126	0.254	0.387	0.527	0.679	0.848	1.046	1.296	1.671	2.000	2.390	2.660	3.460
120	0.126	0.254	0.386	0.526	0.677	0.845	1.041	1.289	1.658	1.980	2.358	2.617	3.373
∞	0.126	0.253	0.385	0.524	0.674	0.842	1.036	1.282	1.645	1.960	2.326	2.576	3.291

Appendix Table 6   5 per cent points of the distribution of z
(values at which the d.f. = 0.95)

(Reprinted from Table VI of Sir Ronald Fisher's *Statistical Methods for Research Workers*, Oliver and Boyd Ltd., Edinburgh, by kind permission of the author and publishers)

		Values of ν_1								
	1	2	3	4	5	6	8	12	24	∞
1	2.5421	2.6479	2.6870	2.7071	2.7194	2.7276	2.7380	2.7484	2.7588	2.7693
2	1.4592	1.4722	1.4765	1.4787	1.4800	1.4808	1.4819	1.4830	1.4840	1.4851
3	1.1577	1.1284	1.1137	1.1051	1.0994	1.0953	1.0899	1.0842	1.0781	1.0716
4	1.0212	0.9690	0.9429	0.9272	0.9168	0.9093	0.8993	0.8885	0.8767	0.8639
5	0.9441	0.8777	0.8441	0.8236	0.8097	0.7997	0.7862	0.7714	0.7550	0.7368
6	0.8948	0.8188	0.7798	0.7558	0.7394	0.7274	0.7112	0.6931	0.6729	0.6409
7	0.8606	0.7777	0.7347	0.7080	0.6896	0.6761	0.6576	0.6369	0.6134	0.5862
8	0.8355	0.7475	0.7014	0.6725	0.6525	0.6378	0.6175	0.5945	0.5682	0.5371
9	0.8163	0.7242	0.6757	0.6450	0.6238	0.6080	0.5862	0.5613	0.5324	0.4979
10	0.8012	0.7058	0.6553	0.6232	0.6009	0.5843	0.5611	0.5346	0.5035	0.4657
11	0.7889	0.6909	0.6387	0.6055	0.5822	0.5648	0.5406	0.5126	0.4795	0.4387
12	0.7788	0.6786	0.6250	0.5907	0.5666	0.5487	0.5234	0.4941	0.4592	0.4156
13	0.7703	0.6682	0.6134	0.5783	0.5535	0.5350	0.5089	0.4785	0.4419	0.3957
14	0.7630	0.6594	0.6036	0.5677	0.5423	0.5233	0.4964	0.4649	0.4269	0.3782
15	0.7568	0.6518	0.5950	0.5585	0.5326	0.5131	0.4855	0.4532	0.4138	0.3628
16	0.7514	0.6451	0.5876	0.5505	0.5241	0.5042	0.4760	0.4428	0.4022	0.3490
17	0.7466	0.6393	0.5811	0.5434	0.5166	0.4964	0.4676	0.4337	0.3919	0.3366
18	0.7424	0.6341	0.5753	0.5371	0.5099	.04894	0.4602	0.4255	0.3827	0.3253
19	0.7386	0.6295	0.5701	0.5315	0.5040	0.4832	0.4535	0.4182	0.3743	0.3151
20	0.7352	0.6254	0.5654	0.5265	0.4986	0.4776	0.4474	0.4116	0.3668	0.3057
21	0.7322	0.6216	0.5612	0.5219	0.4938	0.4725	0.4420	0.4055	0.3599	0.2971
22	0.7294	0.6182	0.5574	0.5178	0.4894	0.4679	0.4370	0.4001	0.3536	0.2892
23	0.7269	0.6151	0.5540	0.5140	0.4854	0.4636	0.4325	0.3950	0.3478	0.2818
24	0.7246	0.6123	0.5508	0.5106	0.4817	0.4598	0.4283	0.3904	0.3425	0.2749
25	0.7225	0.6097	0.5478	0.5074	0.4783	0.4562	0.4244	0.3862	0.3376	0.2685
26	0.7205	0.6073	0.5451	0.5045	0.4752	0.4529	0.4209	0.3823	0.3330	0.2625
27	0.7187	0.6051	0.5427	0.5017	0.4723	0.4499	0.4176	0.3786	0.3287	0.2569
28	0.7171	0.6030	0.5403	0.4992	0.4696	0.4471	0.4146	0.3752	0.3248	0.2516
29	0.7155	0.6011	0.5382	0.4969	0.4671	0.4444	0.4117	0.3720	0.3211	0.2466
30	0.7141	0.5994	0.5362	0.4947	0.4648	0.4420	0.4090	0.3691	0.3176	0.2419
60	0.6933	0.5738	0.5073	0.4632	0.4311	0.4064	0.3702	0.3255	0.2654	0.1644
∞	0.6729	0.5486	0.4787	0.4319	0.3974	0.3706	0.3309	0.2804	0.2085	0

Values of ν_2

Appendix Table 7   5 per cent points of the variance ratio F
(values at which the d.f. = 0.95)

(Reproduced from Sir Ronald Fisher and Dr F. Yates: *Statistical Tables for Biological, Medical and Agricultural Research*, Oliver and Boyd Ltd., Edinburgh, by kind permission of the authors and publishers)

ν_2 \ ν_1	1	2	3	4	5	6	8	12	24	∞
1	161.40	199.50	215.70	224.60	230.20	234.00	238.90	243.90	249.00	254.30
2	18.51	19.00	19.16	19.25	19.30	19.33	19.37	19.41	19.45	19.50
3	10.13	9.55	9.28	9.12	9.01	8.94	8.84	8.74	8.64	8.53
4	7.71	6.94	6.59	6.39	6.26	6.16	6.04	5.91	5.77	5.63
5	6.61	5.79	5.41	5.19	5.05	4.95	4.82	4.68	4.53	4.36
6	5.99	5.14	4.76	4.53	4.39	4.28	4.15	4.00	3.84	3.67
7	5.59	4.74	4.35	4.12	3.97	3.87	3.73	3.57	3.41	3.23
8	5.32	4.46	4.07	3.84	3.69	3.58	3.44	3.28	3.12	2.93
9	5.12	4.26	3.86	3.63	3.48	3.37	3.23	3.07	2.90	2.71
10	4.96	4.10	3.71	3.48	3.33	3.22	3.07	2.91	2.74	2.54
11	4.84	3.98	3.59	3.36	3.20	3.09	2.95	2.79	2.61	2.40
12	4.75	3.88	3.49	3.26	3.11	3.00	2.85	2.69	2.50	2.30
13	4.67	3.80	3.41	3.18	3.02	2.92	2.77	2.60	2.42	2.21
14	4.60	3.74	3.34	3.11	2.96	2.85	2.70	2.53	2.35	2.13
15	4.54	3.68	3.29	3.06	2.90	2.79	2.64	2.48	2.29	2.07
16	4.49	3.63	3.24	3.01	2.85	2.74	2.59	2.42	2.24	2.01
17	4.45	3.59	3.20	2.96	2.81	2.70	2.55	2.38	2.19	1.96
18	4.41	3.55	3.16	2.93	2.77	2.66	2.51	2.34	2.15	1.92
19	4.38	3.52	3.13	2.90	2.74	2.63	2.48	2.31	2.11	1.88
20	4.35	3.49	3.10	2.87	2.71	2.60	2.45	2.28	2.08	1.84
21	4.32	3.47	3.07	2.84	2.68	2.57	2.42	2.25	2.05	1.81
22	4.30	3.44	3.05	2.82	2.66	2.55	2.40	2.23	2.03	1.78
23	4.28	3.42	3.03	2.80	2.64	2.53	2.38	2.20	2.00	1.76
24	4.26	3.40	3.01	2.78	2.62	2.51	2.36	2.18	1.98	1.73
25	4.24	3.38	2.99	2.76	2.60	2.49	2.34	2.16	1.96	1.71
26	4.22	3.37	2.98	2.74	2.59	2.47	2.32	2.15	1.95	1.69
27	4.21	3.35	2.96	2.73	2.57	2.46	2.30	2.13	1.93	1.67
28	4.20	3.34	2.95	2.71	2.56	2.44	2.29	2.12	1.91	1.65
29	4.18	3.33	2.93	2.70	2.54	2.43	2.28	2.10	1.90	1.64
30	4.17	3.32	2.92	2.69	2.53	2.42	2.27	2.09	1.89	1.62
40	4.08	3.23	2.84	2.61	2.45	2.34	2.18	2.00	1.79	1.51
60	4.00	3.15	2.76	2.52	2.37	2.25	2.10	1.92	1.70	1.39
120	3.92	3.07	2.68	2.45	2.29	2.17	2.02	1.83	1.61	1.25
∞	3.84	2.99	2.60	2.37	2.21	2.09	1.94	1.75	1.52	1.00

Lower 5 per cent points are found by interchange of ν_1 and ν_2, i.e. ν_1 must always correspond to the greater mean square.

Appendix Table 8   1 per cent points of the distribution of z
(values at which the d.v. = 0.99)

(Reprinted from Table VI of Sir Ronald Fisher's *Statistical Methods for Research Workers*, Oliver and Boyd Ltd., Edinburgh, by kind permission of the author and publishers)

		Values of ν_1									
		1	2	3	4	5	6	8	12	24	∞
Values of ν_2	1	4.1535	4.2585	4.2974	4.3175	4.3297	4.3379	4.3482	4.3585	4.3689	4.3794
	2	2.2950	2.2976	2.2984	2.2988	2.2991	2.2992	2.2994	2.2997	2.2999	2.3001
	3	1.7649	1.7140	1.6915	1.6786	1.6703	1.6645	1.6569	1.6489	1.6404	1.6314
	4	1.5270	1.4452	1.4075	1.3856	1.3711	1.3609	1.3473	1.3327	1.3170	1.3000
	5	1.3943	1.2929	1.2449	1.2164	1.1974	1.1838	1.1656	1.1457	1.1239	1.0997
	6	1.3103	1.1955	1.1401	1.1068	1.0843	1.0680	1.0460	1.0218	0.9948	0.9643
	7	1.2526	1.1281	1.0672	1.0300	1.0048	0.9864	0.9614	0.9335	0.9020	0.8658
	8	1.2106	1.0787	1.0135	0.9734	0.9459	0.9259	0.8983	0.8673	0.8319	0.7904
	9	1.1786	1.0411	0.9724	0.9299	0.9006	0.8791	0.8494	0.8157	0.7769	0.7305
	10	1 1535	1.0114	0.9399	0.8954	0.8646	0.8419	0.8104	0.7744	0.7324	0.6816
	11	1.1333	0.9874	0.9136	0.8674	0.8354	0.8116	0.7785	0.7405	0.6958	0.6408
	12	1.1166	0.9677	0.8919	0.8443	0.8111	0.7864	0.7520	0.7122	0.6649	0.6061
	13	1.1027	0.9511	0.8737	0.8248	0.7907	0.7652	0.7295	0.6882	0.6386	0.5761
	14	1.0909	0.9370	0.8581	0.8082	0.7732	0.7471	0.7103	0.6675	0.6159	0.5500
	15	1.0807	0.9249	0.8448	0.7939	0.7582	0.7314	0.6937	0.6496	0.5961	0.5269
	16	1.0719	0.9144	0.8331	0.7814	0.7450	0.7177	0.6791	0.6339	0.5786	0.5064
	17	1.0641	0.9051	0.8229	0.7705	0.7335	0.7057	0.6663	0.6199	0.5630	0.4879
	18	1.0572	0.8970	0.8138	0.7607	0.7232	0.6950	0.6549	0.6075	0.5491	0.4712
	19	1.0511	0.8897	0.8057	0.7521	0.7140	0.6854	0.6447	0.5964	0.5366	0.4560
	20	1.0457	0.8831	0.7985	0.7443	0.7058	0.6768	0.6355	0.5864	9.5253	0.4421
	21	1.0408	0.8772	0.7920	0.7372	0.6984	0.6690	0.6272	0.5773	0.5150	0.4294
	22	1.0363	0.8719	0.7860	0.7309	0.6916	0.6620	0.6196	0.5691	0.5056	0.4176
	23	1.0322	0.8670	0.7806	0.7251	0.6855	0.6555	0.6127	0.5615	0.4969	0.4068
	24	1.0285	0.8626	0.7757	0.7197	0.6799	0.6496	0.6064	0.5545	0.4890	0.3967
	25	1.0251	0.8585	0.7712	0.7148	0.6747	0.6442	0.6006	0.5481	0.4816	0.3872
	26	1.0220	0.8548	0.7670	0.7103	0.6699	0.6392	0.5952	0.5422	0.4748	0.3784
	27	1.0191	0.8513	0.7631	0.7062	0.6655	0.6346	0.5902	0.5367	0.4685	0.3701
	28	1.0164	0.8481	0.7595	0.7023	0.6614	0.6303	0.5856	0.5316	0.4626	0.3624
	29	1.0139	0.8451	0.7562	0.6987	0.6576	0.6263	0.5813	0.5269	0.4570	0.3550
	30	1.0116	0.8423	0.7531	0.6954	0.6540	0.6226	0.5773	0.5224	0.4519	0.3481
	60	0.9784	0.8025	0.7086	0.6472	0.6028	0.5687	0.5189	0.4574	0.3746	0.2352
	∞	0.9462	0.7636	0.6651	0.5999	0.5522	0.5152	0.4604	0.3908	0.2913	0

Appendix Table 9　　1 per cent points of the variance ratio F
(values at which the d.f. $= 0.99$)

(Reproduced from Sir Ronald Fisher and Dr F. Yates: *Statistical Tables for Biological, Medical and Agricultural Research*, Oliver and Boyd Ltd., Edinburgh, by kind permission of the authors and publishers)

ν_2 \ ν_1	1	2	3	4	5	6	8	12	24	∞
1	4052	4999	5403	5625	5764	5859	5981	6106	6234	6366
2	98.49	99.00	99.17	99.25	99.30	99.33	99.36	99.42	99.46	99.50
3	34.12	30.81	29.46	28.71	28.24	27.91	27.49	27.05	26.60	26.12
4	21.20	18.00	16.69	15.98	15.52	15.21	14.80	14.37	13.93	13.46
5	16.26	13.27	12.06	11.39	10.97	10.67	10.27	9.89	9.47	9.02
6	13.74	10.92	9.78	9.15	8.75	8.47	8.10	7.72	7.31	6.88
7	12.25	9.55	8.45	7.85	7.46	7.19	6.84	6.47	6.07	5.65
8	11.26	8.65	7.59	7.01	6.63	6.37	6.03	5.67	5.28	4.86
9	10.56	8.02	6.99	6.42	6.06	5.80	5.47	5.11	4.73	4.31
10	10.04	7.56	6.55	5.99	5.64	5.39	5.06	4.71	4.33	3.91
11	9.65	7.20	6.22	5.67	5.32	5.07	4.74	4.40	4.02	3.60
12	9.33	6.93	5.95	5.41	5.06	4.82	4.50	4.16	3.78	3.36
13	9.07	6.70	5.74	5.20	4.86	4.62	4.30	3.96	3.59	3.16
14	8.86	6.51	5.56	5.03	4.69	4.46	4.14	3.80	3.43	3.00
15	8.68	6.36	5.42	4.89	4.56	4.32	4.00	3.67	3.29	2.87
16	8.53	6.23	5.29	4.77	4.44	4.20	3.89	3.55	3.18	2.75
17	8.40	6.11	5.18	4.67	4.34	4.10	3.79	3.45	3.08	2.65
18	8.28	6.01	5.09	4.58	4.25	4.01	3.71	3.37	3.00	2.57
19	8.18	5.93	5.01	4.50	4.17	3.94	3.63	3.30	2.92	2.49
20	8.10	5.85	4.94	4.43	4.10	3.87	3.56	3.23	2.86	2.42
21	8.02	5.78	4.87	4.37	4.04	3.81	3.51	3.17	2.80	2.36
22	7.94	5.72	4.82	4.31	3.99	3.76	3.45	3.12	2.75	2.31
23	7.88	5.66	4.76	4.26	3.94	3.71	3.41	3.07	2.70	2.26
24	7.82	5.61	4.72	4.22	3.90	3.67	3.36	3.03	2.66	2.21
25	7.77	5.57	4.68	4.18	3.86	3.63	3.32	2.99	2.62	2.17
26	7.72	5.53	4.64	4.14	3.82	3.59	3.29	2.96	2.58	2.13
27	7.68	5.49	4.60	4.11	3.78	3.56	3.26	2.93	2.55	2.10
28	7.64	5.45	4.57	4.07	3.75	3.53	3.23	2.90	2.52	2.06
29	7.60	5.42	4.54	4.04	3.73	3.50	3.20	2.87	2.49	2.03
30	7.56	5.39	4.51	4.02	3.70	3.47	3.17	2.84	2.47	2.01
40	7.31	5.18	4.31	3.83	3.51	3.29	2.99	2.66	2.29	1.80
60	7.08	4.98	4.13	3.65	3.34	3.12	2.82	2.50	2.12	1.60
120	6.85	4.79	3.95	3.48	3.17	2.96	2.66	2.34	1.95	1.38
∞	6.64	4.60	3.78	3.32	3.02	2.80	2.51	2.18	1.79	1.00

Lower 1 per cent points are found by interchange of ν_1 and ν_2, i.e. ν_1 must always correspond to the greater mean square.

Appendix Table 10 5 per cent points for the Durbin–Watson d-test

n	$k' = 1$		$k' = 2$		$k' = 3$		$k' = 4$		$k' = 5$	
	d_L	d_U	d_L	d_U	d_L	d_U	d_L	d_U	d_L	d_U
15	1.08	1.36	0.95	1.54	0.82	1.75	0.69	1.97	0.56	2.21
16	1.10	1.37	0.98	1.54	0.86	1.73	0.74	1.93	0.62	2.15
17	1.13	1.38	1.02	1.54	0.90	1.71	0.78	1.90	0.67	2.10
18	1.16	1.39	1.05	1.53	0.93	1.69	0.82	1.87	0.71	2.06
19	1.18	1.40	1.08	1.53	0.97	1.68	0.86	1.85	0.75	2.02
20	1.20	1.41	1.10	1.54	1.00	1.68	0.90	1.83	0.79	1.99
21	1.22	1.42	1.13	1.54	1.03	1.67	0.93	1.81	0.83	1.96
22	1.24	1.43	1.15	1.54	1.05	1.66	0.96	1.80	0.86	1.94
23	1.26	1.44	1.17	1.54	1.08	1.66	0.99	1.79	0.90	1.92
24	1.27	1.45	1.19	1.55	1.10	1.66	1.01	1.78	0.93	1.90
25	1.29	1.45	1.21	1.55	1.12	1.66	1.04	1.77	0.95	1.89
26	1.30	1.46	1.22	1.55	1.14	1.65	1.06	1.76	0.98	1.88
27	1.32	1.47	1.24	1.56	1.16	1.65	1.08	1.76	1.01	1.86
28	1.33	1.48	1.26	1.56	1.18	1.65	1.10	1.75	1.03	1.85
29	1.34	1.48	1.27	1.56	1.20	1.65	1.12	1.74	1.05	1.84
30	1.35	1.49	1.28	1.57	1.21	1.65	1.14	1.74	1.07	1.83
31	1.36	1.50	1.30	1.57	1.23	1.65	1.16	1.74	1.09	1.83
32	1.37	1.50	1.31	1.57	1.24	1.65	1.18	1.73	1.11	1.82
33	1.38	1.51	1.32	1.58	1.26	1.65	1.19	1.73	1.13	1.81
34	1.39	1.51	1.33	1.58	1.27	1.65	1.21	1.73	1.15	1.81
35	1.40	1.52	1.34	1.58	1.28	1.65	1.22	1.73	1.16	1.80
36	1.41	1.52	1.35	1.59	1.29	1.65	1.24	1.73	1.18	1.80
37	1.42	1.53	1.36	1.59	1.31	1.66	1.25	1.72	1.19	1.80
38	1.43	1.54	1.37	1.59	1.32	1.66	1.26	1.72	1.21	1.79
39	1.43	1.54	1.38	1.60	1.33	1.66	1.27	1.72	1.22	1.79
40	1.44	1.54	1.39	1.60	1.34	1.66	1.29	1.72	1.23	1.79
45	1.48	1.57	1.43	1.62	1.38	1.67	1.34	1.72	1.29	1.78
50	1.50	1.59	1.46	1.63	1.42	1.67	1.38	1.72	1.34	1.77
55	1.53	1.60	1.49	1.64	1.45	1.68	1.41	1.72	1.38	1.77
60	1.55	1.62	1.51	1.65	1.48	1.69	1.44	1.73	1.41	1.77
65	1.57	1.63	1.54	1.66	1.50	1.70	1.47	1.73	1.44	1.77
70	1.58	1.64	1.55	1.67	1.52	1.70	1.49	1.74	1.46	1.77
75	1.60	1.65	1.57	1.68	1.54	1.71	1.51	1.74	1.49	1.77
80	1.61	1.66	1.59	1.69	1.56	1.72	1.53	1.74	1.51	1.77
85	1.62	1.67	1.60	1.70	1.57	1.72	1.55	1.75	1.52	1.77
90	1.63	1.68	1.61	1.70	1.59	1.73	1.57	1.75	1.54	1.78
95	1.64	1.69	1.62	1.71	1.60	1.73	1.58	1.75	1.56	1.78
100	1.65	1.69	1.63	1.72	1.61	1.74	1.59	1.76	1.57	1.78

k' denotes the number of regressor variables, excluding the constant.

By permission of authors and Biometrika Trust from J. Durbin and G. S. Watson (1951), "Testing for Serial Correlation in Least Squares Regression, II", *Biometrika*, **38**, 159.

Appendix Table 11   1 per cent points for the Durbin–Watson d-test

n	$k' = 1$		$k' = 2$		$k' = 3$		$k' = 4$		$k' = 5$	
	d_L	d_U	d_L	d_U	d_L	d_U	d_L	d_U	d_L	d_U
15	0.81	1.07	0.70	1.25	0.59	1.46	0.49	1.70	0.39	1.96
16	0.84	1.09	0.74	1.25	0.63	1.44	0.53	1.66	0.44	1.90
17	0.87	1.10	0.77	1.25	0.67	1.43	0.57	1.63	0.48	1.85
18	0.90	1.12	0.80	1.26	0.71	1.42	0.61	1.60	0.52	1.80
19	0.93	1.13	0.83	1.26	0.74	1.41	0.65	1.58	0.56	1.77
20	0.95	1.15	0.86	1.27	0.77	1.41	0.68	1.57	0.60	1.74
21	0.97	1.16	0.89	1.27	0.80	1.41	0.72	1.55	0.63	1.71
22	1.00	1.17	0.91	1.28	0.83	1.40	0.75	1.54	0.66	1.69
23	1.02	1.19	0.94	1.29	0.86	1.40	0.77	1.53	0.70	1.67
24	1.04	1.20	0.96	1.30	0.88	1.41	0.80	1.53	0.72	1.66
25	1.05	1.21	0.98	1.30	0.90	1.41	0.83	1.52	0.75	1.65
26	1.07	1.22	1.00	1.31	0.93	1.41	0.85	1.52	0.78	1.64
27	1.09	1.23	1.02	1.32	0.95	1.41	0.88	1.51	0.81	1.63
28	1.10	1.24	1.04	1.32	0.97	1.41	0.90	1.51	0.83	1.62
29	1.12	1.25	1.05	1.33	0.99	1.42	0.92	1.51	0.85	1.61
30	1.13	1.26	1.07	1.34	1.01	1.42	0.94	1.51	0.88	1.61
31	1.15	1.27	1.08	1.34	1.02	1.42	0.96	1.51	0.90	1.60
32	1.16	1.28	1.10	1.35	1.04	1.43	0.98	1.51	0.92	1.60
33	1.17	1.29	1.11	1.36	1.05	1.43	1.00	1.51	0.94	1.59
34	1.18	1.30	1.13	1.36	1.07	1.43	1.01	1.51	0.95	1.59
35	1.19	1.31	1.14	1.37	1.08	1.44	1.03	1.51	0.97	1.59
36	1.21	1.32	1.15	1.38	1.10	1.44	1.04	1.51	0.99	1.59
37	1.22	1.32	1.16	1.38	1.11	1.45	1.06	1.51	1.00	1.59
38	1.23	1.33	1.18	1.39	1.12	1.45	1.07	1.52	1.02	1.58
39	1.24	1.34	1.19	1.39	1.14	1.45	1.09	1.52	1.03	1.58
40	1.25	1.34	1.20	1.40	1.15	1.46	1.10	1.52	1.05	1.58
45	1.29	1.38	1.24	1.42	1.20	1.48	1.16	1.53	1.11	1.58
50	1.32	1.40	1.28	1.45	1.24	1.49	1.20	1.54	1.16	1.59
55	1.36	1.43	1.32	1.47	1.28	1.51	1.25	1.55	1.21	1.59
60	1.38	1.45	1.35	1.48	1.32	1.52	1.28	1.56	1.25	1.60
65	1.41	1.47	1.38	1.50	1.35	1.53	1.31	1.57	1.28	1.61
70	1.43	1.49	1.40	1.52	1.37	1.55	1.34	1.58	1.31	1.61
75	1.45	1.50	1.42	1.53	1.39	1.56	1.37	1.59	1.34	1.62
80	1.47	1.52	1.44	1.54	1.42	1.57	1.39	1.60	1.36	1.62
85	1.48	1.53	1.46	1.55	1.43	1.58	1.41	1.60	1.39	1.63
90	1.50	1.54	1.47	1.56	1.45	1.59	1.43	1.61	1.41	1.64
95	1.51	1.55	1.49	1.57	1.47	1.60	1.45	1.62	1.42	1.64
100	1.52	1.56	1.50	1.58	1.48	1.60	1.46	1.63	1.44	1.65

k' denotes the number of regressor variables, excluding the constant.

By permission of authors and Biometrika Trust from J. Durbin and G. S. Watson (1951), "Testing for Serial Correlation in Least Squares, II", *Biometrika*, **38**, 159.

REFERENCES

Note. References to W. G. Cochran, R. A. Fisher, J. Neyman, E. S. Pearson, K. Pearson, "Student", A. Wald, S. S. Wilks and G. U. Yule that are marked with an asterisk are reproduced in the following collections:

W. G. Cochran, *Contributions to Statistics.* Wiley, New York 1982.
R. A. Fisher, *Contributions to Mathematical Statistics.* Wiley, New York, 1950.
A Selection of Early Statistical Papers of J. Neyman. Cambridge Univ. Press 1967.
Joint Statistical Papers of J. Neyman and E. S. Pearson. Cambridge Univ. Press, 1967.
The Selected Papers of E. S. Pearson. Cambridge Univ. Press, 1966.
Karl Pearson's Early Statistical Papers. Cambridge Univ. Press, London, 1948.
"Student's" Collected Papers. Biometrika Office, University College London, 1942.
Selected Papers in Statistics and Probability by Abraham Wald. McGraw-Hill, New York, 1955; reprinted by Stanford Univ. Press, 1957.
S. S. Wilks: Collected Papers. Contributions to Mathematical Statistics. Wiley, New York, 1967.
Statistical Papers of George Udny Yule. Griffin, London, 1971.

AFIFI, A. A. and ELASHOFF, R. M. (1966). Missing observations in multivariate statistics. I. Review of literature. *J. Amer. Statist. Ass.*, **61**, 595.
AITCHISON, J. and SILVEY, S. D. (1958). Maximum-likelihood estimation of parameters subject to restraints. *Ann. Math. Statist.*, **29**, 813.
AITKEN, A. C. (1933). On the graduation of data by the orthogonal polynomials of least squares. *Proc. Roy. Soc. Edin.*, **A, 53**, 54. On fitting polynomials to weighted data by least squares. *Ibid.*, **54**, 1. On fitting polynomials to data with weighted and correlated errors. *Ibid.*, **54**, 12.
AITKEN, A. C. (1935). On least squares and linear combination of observations. *Proc. Roy. Soc. Edin.*, **A, 55**, 42.
AITKEN, A. C. (1948). On the estimation of many statistical parameters. *Proc. Roy. Soc. Edin.*, **A, 62**, 369.
AITKEN, A. C. and SILVERSTONE, H. (1942). On the estimation of statistical parameters. *Proc. Roy. Soc. Edin.*, **A, 61**, 186.
AITKIN, M. (1978). The analysis of unbalanced cross-classification. *J. R. Statist. Soc.*, **A, 141**, 195.
AKAIKE, H. (1969). Fitting autoregressive models for prediction. *Ann. Inst. Statist. Math.*, **21**, 243.
AKAIKE, H. (1982). On the fallacy of the likelihood principle. *Statist. Prob. Letters*, **1**, 75.
ALALOUF, I. S. and STYLAN, G. P. H. (1979). Characterizations of estimability in the general linear model. *Ann. Statist.*, **7**, 194.
ALI, M. M. and SILVER, J. L. (1985). Tests for equality between sets of coefficients in two linear regressions under heteroscedasticity. *J. Amer. Statist. Ass.*, **80**, 730.
ALLAN, F. E. (1930). The general form of the orthogonal polynomials for simple series with proofs of their simple properties. *"Proc. Roy. Soc. Edin.*, **A, 50**, 310.
ALLDREDGE, J. R. and GILB, N. S. (1976). Ridge regression: an annotated bibliography. *Inst. Statist. Rev.*, **44**, 355.
ALLING, D. W. (1966). Closed sequential tests for binomial probabilities. *Biometrika*, **53**, 73.
ALMON, S. (1965). The distributed lag between capital appropriations and expenditures. *Econometrica*, **30**, 178.
AMARI, S. (1982a). Geometrical theory of asymptotic ancillarity and conditional inference. *Biometrika*, **69**, 1.

AMARI, S. (1982b). Differential geometry of curved exponential families—curvatures and information loss. *Ann. Statist.*, **10**, 357.

AMEMIYA, T. (1980). The n'th order mean squared errors of the maximum likelihood and the minimum logit chi-squared estimators. *Ann. Statist.*, **8**, 488; *Correction*, **12**, 783.

AMOS, D. E. (1964). Representations of the central and non-central t distributions. *Biometrika*, **51**, 451.

ANDERSEN, E. B. (1970). Sufficiency and exponential families for discrete sample spaces. *J. Amer. Statist. Ass.*, **65**, 1248.

ANDERSON, R. L. and ANDERSON, T. W. (1956). Distribution of the circular serial correlation coefficient for residuals from a fitted Fourier series. *Ann. Math. Statist.*, **21**, 59.

ANDERSON, R. L. and BANCROFT, T. A. (1952). *Statistical Theory in Research.* McGraw-Hill, New York.

ANDERSON, T. W. (1960). A modification of the sequential probability ratio test to reduce sample size. *Ann. Math. Statist.*, **31**, 165.

ANDERSON, T. W. (1962a). The choice of the degree of a polynomial regression as a multiple decision problem. *Ann. Math. Statist.*, **33**, 255.

ANDERSON, T. W. (1962b). Least squares and best unbiased estimates. *Ann. Math. Statist.*, **33**, 266.

ANDREWS, D. F. (1971). A note on the selection of data transformations. *Biometrika*, **58**, 249.

ANSCOMBE, F. J. (1948). The transformation of Poisson, binomial and negative-binomial data. *Biometrika*, **35**, 246.

ANSCOMBE, F. J. (1949a). Tables of sequential inspection schemes to control fraction defective. *J. R. Statist. Soc.*, **A, 112**, 180.

ANSCOMBE, F. J. (1949b). Large-sample theory of sequential estimation. *Biometrika*, **36**, 455.

ANSCOMBE, F. J. (1950). Sampling theory of the negative bimomial and logarithmic series distributions. *Biometrika*, **37**, 358.

ANSCOMBE, F. J. (1952). Large-sample theory of sequential estimation. *Proc. Camb. Phil. Soc.*, **48**, 600.

ANSCOMBE, F. J. (1953). Sequential estimation. *J. R. Statist. Soc.*, **B, 15**, 1.

ANSCOMBE, F. J. and PAGE, E. S. (1954). Sequential tests for binomial and exponential populations. *Biometrika*, **41**, 252.

ANTLE, C., KLIMKO, L. and HARKNESS, W. (1970). Confidence intervals for the parameters of the logistic distribution. *Biometrika*, **57**, 397.

ARMITAGE, P. (1947). Some sequential tests of Student's hypothesis. *J. R. Statist. Soc.*, **B, 9**, 250.

ARMITAGE, P. (1957). Restricted sequential procedures. *Biometrika*, **44**, 9.

ARNOLD, S. F. (1979). Linear models with exchangeably distributed errors. *J. Amer. Statist. Ass.*, **74**, 194.

ARVESEN, J. N. and LAYARD, M. W. J. (1975). Asymptotically robust tests in unbalanced variance component models. *Ann. Statist.*, **3**, 1122.

ASKOVITZ, S. I. (1975). A short-cut graphic method for fitting the best straight line to a series of points according to the criterion of least squares. *J. Amer. Statist. Ass.*, **52**, 13.

ASPIN, A. A. (1948). An examination and further development of a formula arising in the problem of comparing two mean values. *Biometrika*, **35**, 88.

ASPIN, A. A. (1949). Tables for use in comparisons whose accuracy involves two variances, separately estimated. *Biometrika*, **36**, 290.

ATKINSON, A. C. (1970). A method of discriminating between models. *J. R. Statist. Soc.*, **B, 32**, 323.

ATKINSON, A. C. (1973). Testing transformations to normality. *J. R. Statist. Soc.*, **B, 35**, 473.

ATKINSON, A. C. (1981). Two graphical displays for outlying and influential observations in regression. *Biometrika*, **68**, 13.

ATKINSON, A. C. (1983). Diagnostic regression analysis and shifted power transformations. *Technometrics*, **25**, 23.

ATKINSON, A. C. (1985). *Plots, Transformations and Regression.* Oxford Univ. Press, Oxford.

ATKINSON, A. C. (1986a). Diagnostic tests for transformations. *Technometrics*, **28**, 29.

ATKINSON, A. C. (1986b). Masking unmasked. *Biometrika*, **73**, 533.

BAHADUR, R. R. (1964). On Fisher's bound for asymptotic variances. *Ann. Math. Statist.*, **35**, 1545.

BAHADUR, R. R. (1967). Rates of convergence of estimates and test statistics. *Ann. Math. Statist.*, **38**, 303.

BAIN, L. J. (1967). Reducing a random sample to a smaller set, with applications. *J. Amer. Statist. Ass.*, **62**, 510.

BAIN, L. J. (1969). The moments of a noncentral *t* and noncentral *F*-distribution. *Amer. Statistician*, **23(4)**, 33.

BAKSALARY, J. K. and KALA, R. (1983). Estimation via linearly combining two given statistics. *Ann. Statist.*, **11**, 691.

BALMER, D. W., BOULTON, M. and SACK, R. A. (1974). Optimal solutions in parameter estimation problems for the Cauchy distribution. *J. Amer. Statist. Ass.*, **69**, 238.

BANERJEE, K. S. (1964). A note on idempotent matrices. *Ann. Math. Statist.*, **35**, 880.

BANERJEE, S. (1967). Confidence interval of preassigned length for the Behrens–Fisher problem. *Ann. Math. Statist.*, **38**, 1175.

BARANKIN, E. W. (1949). Locally best unbiased estimates. *Ann. Math. Statist.*, **20**, 477.

BARANKIN, E. W. (1960a). Application to exponential families of the solution of the minimal dimensionality problem for sufficient statistics. *Bull Int. Statist. Inst.*, **38**, **(4)**, 141.

BARANKIN, E. W. (1960b). Sufficient parameters: solution of the minimal dimensionality problem. *Ann. Inst. Statist. Math.*, **12**, 91.

BARANKIN, E. W. (1961). A note on functional minimality of sufficient statistics. *Sankhyā*, **A**, **23**, 401.

BARANKIN, E. W. and KATZ, M., Jr. (1959). Sufficient statistics of minimal dimension. *Sankhyā*, **21**, 217.

BARANKIN, E. W. and MAITRA, A. P. (1963). Generalization of the Fisher–Darmois–Koopman–Pitman theorem on sufficient statistics. *Sankhyā*, **A**, **25**, 217.

BARLOW, R. E., BARTHOLOMEW, D. J., BREMNER, J. M. and BRUNK, H. D. (1972). *Statistical Inference Under Order Restrictions*, Wiley, New York.

BARNARD, G. A. (1946). Sequential tests in industrial statistics. *Suppl. J. R. Statist. Soc.*, **8**, 1.

BARNARD, G. A. (1947a). Significance tests for 2×2 tables. *Biometrika*, **34**, 123.

BARNARD, G. A. (1947b). 2×2 tables. A note on E. S. Pearson's paper. *Biometrika*, **34**, 168.

BARNARD, G. A. (1950). On the Fisher–Behrens test. *Biometrika*, **34**, 168.

BARNARD, G. A. (1963). The logic of least squares. *J. R. Statist. Soc.*, **B**, **25**, 124.

BARNARD, G. A. and GODAMBE, V. P. (1982). Allan Birnbaum, A memorial article. *Ann. Statist.*, **10**, 1033.

BARNARD, G. A. and SPROTT, D. A. (1971). A note on Basu's examples of anomalous ancillary statistics (with discussion). In *Foundations of Statistical Inference*, V. P. Godambe and D. A. Sprott (eds.), Holt Rinehart and Winston, Toronto, 163.

BARNARD, G. A. and SPROTT, D. A. (1983). The generalised problem of the Nile: robust confidence sets for parametric functions. *Ann. Statist.*, **11**, 104.

BARNDORFF-NIELSEN, O. (1978). *Information and Exponential Families in Statistical Theory*. John Wiley, New York and Chichester.

BARNDORFF-NIELSEN, O. (1980). Conditionality resolutions. *Biometrika*, **67**, 293.

BARNDORFF-NIELSEN, O. (1983). On a formula for the distribution of the maximum likelihood estimator. *Biometrika*, **70**, 343.

BARNDORFF-NIELSEN, O. E. (1986). Inference on full or partial parameters based on the standardized signed log likelihood ratio. *Biometrika*, **73**, 307.

BARNDORFF-NIELSEN, O. E. and COX, D. R. (1984). Bartlett adjustment to the likelihood ratio statistic and the distribution of the maximum likelihood estimator. *J. R. Statist. Soc.*, **B**, **46**, 483.

BARNDORFF-NIELSEN, O. E. and HALL, P. (1988). On the level-error after Bartlett adjustment of likelihood ratio statistic. *Biometrika*, **75**, 374.

BARNETT, V. D. (1966a). Evaluation of the maximum-likelihood estimator where the likelihood equation has multiple roots. *Biometrika*, **53**, 151.

BARNETT, V. D. (1966b). Order statistics estimators of the location of the Cauchy distribution. *J. Amer. Statist. Ass.*, **61**, 1205.

BARNETT, V. (1982). *Comparative Statistical Inference*. Wiley, London. (Second edition).

BARR, D. R. (1966). On testing the equality of uniform and related distributions. *J. Amer. Statist. Ass.*, **61**, 856.

BARTHOLOMEW, D. J. (1961). Ordered tests in the analysis of variance. *Biometrika*, **48**, 325.

BARTHOLOMEW, D. J. (1967). Hypothesis testing when the sample size is treated as a random variable. *J. R. Statist. Soc.*, **B, 29**, 53.

BARTKY, W. (1943). Multiple sampling with constant probability. *Ann. Math. Statist.*, **14**, 363.

BARTLETT, M. S. (1935). The effect of non-normality on the *t*-distribution. *Proc. Camb. Phil. Soc.*, **31**, 223.

BARTLETT, M. S. (1936). The square root transformation in analysis of variance. *Suppl. J. R. Statist. Soc.*, **3**, 68.

BARTLETT, M. S. (1936). The information available in small samples. *Proc. Camb. Phil. Soc.*, **32**, 560.

BARTLETT, M. S. (1937). Properties of sufficiency and statistical tests. *Proc. Roy. Soc.*, **A, 160**, 268.

BARTLETT, M. S. (1938). The characteristic function of a conditional statistic. *J. Lond. Math. Soc.*, **13**, 62.

BARTLETT, M. S. (1939). A note on the interpretation of quasi-sufficiency. *Biometrika*, **31**, 391.

BARTLETT, M. S. (1946). The large-sample theory of sequential tests. *Proc. Camb. Phil. Soc.*, **42**, 239.

BARTLETT, M. S. (1947). The use of transformations. *Biometrics*, **3**, 39.

BARTLETT, M. S. (1949). Fitting a straight line when both variables are subject to error. *Biometrics*, **5**, 207.

BARTLETT, M. S. (1951). An inverse matrix adjustment arising in discriminant analysis. *Ann. Math. Statist.*, **22**, 107.

BARTLETT, M. S. (1953, 1955). Approximate confidence intervals. *Biometrika*, **40**, 12, 306 and **42**, 201.

BARTLETT, M. S. and KENDALL, D. G. (1946). The statistical analysis of variance-heterogeneity and the logarithmic transformation. *Suppl. J. R. Statist. Soc.*, **8**, 128.

BARTMANN, F. C. and BLOOMFIELD, P. (1981). Inefficiency and correlation. *Biometrika*, **68**, 67.

BASU, A. P. and GHOSH, J. K. (1980). Asymptotic properties of a solution to the likelihood equation with life-testing applications. *J. Amer. Statist. Ass.*, **75**, 410.

BASU, D. (1955, 1958). On statistics independent of a complete sufficient statistic. *Sankhyā*, **15**, 377; *correction* **20**, 223.

BASU, D. (1964). Recovery of ancillary information. *Sankhyā*, **A, 26**, 3.

BASU, D. (1975). Statistical information and likelihood (with discussion). *Sankhyā*, **A, 37**, 1.

BASU, D. (1977). On the elimination of nuisance parameters. *J. Amer. Statist. Ass.*, **72**, 355.

BATEMAN, G. I. (1949). The characteristic function of a weighted sum of non-central squares of normal variables subject to *s* linear restraints. *Biometrika*, **36**, 460.

BATES, D. M. and WATTS, D. G. (1980). Relative curvature measures of nonlinearity. *J. R. Statist. Soc.*, **B, 42**, 1.

BATES, D. M. and WATTS, D. G. (1981). Parameter transformations for improved approximate confidence regions in non-linear least squares. *Ann. Statist.*, **9**, 1152.

BAYES, T. (1764). An essay towards solving a problem in the doctrine of chances. *Phil. Trans.*, **53**, 370. (Reprinted in *Biometrika*, **45**, 293 (1958), edited and introduced by G. A. Barnard.)

BEALE, E. M. L. (1960). Confidence regions in non-linear estimation. *J. R. Statist. Soc.*, **B, 22**, 41.

BEALE, E. M. L. and LITTLE, R. J. A. (1975). Missing values in multivariate analysis. *J. R. Statist. Soc.*, **B, 37**, 129.

BEALE, E. M. L., KENDALL, M. G. and MANN, D. W. (1967). The discarding of variables in multivariate analysis. *Biometrika*, **54**, 357.

BECHHOFER, R. (1960). A note on the limiting relative efficiency of the Wald sequential probability ratio test. *J. Amer. Statist. Ass.*, **55**, 660.

BECHHOFER, R. E., KIEFER, J. and SOBEL, M. (1968). *Sequential Identification and Ranking*

Procedures. Univ. of Chicago Press.

BECKER, N. and GORDON, I. (1983). On Cox's criterion for discriminating between alternative ancillary statistics. *Int. Statist. Rev.*, **51**, 89.

BECKMAN, R. J. and COOK, R. D. (1983). Outliers (with discussion). *Technometrics*, **25**, 119.

BELSLEY, D. A., KUH, E. and WELSH, R. E. (1980). *Regression Diagnostics.* Wiley, New York.

BEMENT, T. R. and WILLIAMS, J. S. (1969). Variance of weighted regression estimators when sampling errors are independent and heteroscedastic. *J. Amer. Statist. Ass.*, **64**, 1369.

BEMIS, K. G. and BHAPKAR, V. P. (1983). On BAN estimating for Chi squared test criteria. *Ann. Statist.*, **11**, 183.

BERGER, A. (1961). On comparing intensities of association between two binary characteristics in two different populations. *J. Amer. Statist. Ass.*, **56**, 889.

BERGER, J. (1985). The frequentist viewpoint and conditioning. *Proceedings, Berkeley Conference in Honor of J. Kiefer and J. Neyman.* L. LeCam and R. Olshen (eds.) Wadsworth, Belmont, California, 15.

BERGER, J. O. and WOLPERT, R. L. (1985). *The Likelihood Principle.* Inst. of Math. Statistics, Hayward, CA. (Second edition).

BERK, R. H. (1975). Comparing sequential and non-sequential tests. *Ann. Statist.*, **3**, 991.

BERK, R. H. (1976). Asymptotic efficiencies of sequential tests. *Ann. Statist.*, **4**, 891.

BERK, R. H. (1978). Asymptotic efficiences of sequential tests II. *Ann. Statist.*, **6**, 813.

BERK, R. H. and BROWN, L. D. (1978). Sequential Bahadar efficiency. *Ann. Statist.*, **6**, 567.

BERKSON, J. (1938). Some difficulties of interpretation encountered in the application of the chi-square test. *J. Amer. Statist. Ass.*, **33**, 526.

BERKSON, J. (1955). Maximum likelihood and minimum χ^2 estimates of the logistic function. *J. Amer. Statist. Ass.*, **50**, 130.

BERKSON, J. (1956). Estimation by least squares and by maximum likelihood. *Proc. 3rd Berkeley Symp. Math. Statist. and Prob.*, **1**, 1.

BERNARDO, J. M. (1979). Reference posterior distributions for Bayesian inference (with discussion). *J. R. Statist. Soc.*, **B, 41**, 113.

BERNARDO, J. M. (1980). A Bayesian analysis of classical hypothesis testing. In *Bayesian Statistics: Proceedings of the First International Meeting.* Valencia Univ. Press, Valencia, Spain.

BERNSTEIN, S. (1928). Fondements géométriques de la théorie des corrélations. *Metron*, **7, (2)**, 3.

BEST, D. J. (1974). The variance of the inverse binomial estimator. *Biometrika*, **61**, 385.

BHATTACHARJEE, G. P. (1965). Effect of non-mormality on Stein's two sample test. *Ann. Math. Statist.*, **36**, 651.

BHATTACHARYYA, A. (1943). On some sets of sufficient conditions leading to the normal bivariate distribution. *Sankhyā*, **6**, 399.

BHATTACHARYYA, A. (1946-7-8). On some analogues of the amount of information and their use in statistical estimation. *Sankhyā*, **8**, 1, 201, 315.

BICKEL, P. J. and DOKSUM, K. A. (1981). An analysis of transformations revisited. *J. Amer. Statist. Ass.*, **76**, 296.

BICKEL, P. J. and FREEDMAN, D. A. (1981). Some asymptotic theory for the bootstrap. *Ann. Statist.*, **9**, 1196.

BIGGERS, J. D. (1959). The estimation of missing and mixed-up observations in several experimental designs. *Biometrika*, **46**, 91.

BILLARD, L. (1972). Properties of some two-sided sequential tests for the normal distribution. *J. R. Statist. Soc.*, **B, 34**, 417.

BILLINGSLEY, P. (1961). *Statistical Inference for Markov Processes.* Univ. of Chicago Press, Chicago, Illinois.

BINKLEY, J. K. and NELSON, C. H. (1988). A note the efficiency of seemingly unrelated regression. *Amer. Statist.*, **42**, 137.

BINNS, M. (1975). Sequential estimation of the mean of a negative binomial distribution. *Biometrika*, **62**, 433.

BIRCH, M. W. (1964). A new proof of the Pearson–Fisher theorem. *Ann. Math. Statist.*, **35**, 817.

BIRNBAUM, A. (1962). On the foundations of statistical inference. *J. Amer. Statist. Ass.*, **57**, 269.

BIRNBAUM, A. (1970). On Durbin's modified principle of conditionality. *J. Amer. Statist. Ass*, **65**, 402.

BIRNBAUM, A. (1972). More on concepts of statistical evidence. *J. Amer. Statist. Ass.*, **67**, 858.

BIRNBAUM, A. (1977). The Neyman-Pearson theory as decision theory, and as inference theory; with a criticism of the Lindley-Savage argument for Bayesian theory. *Synthèse*, **36**, 19.

BIRNBAUM, A. and HEALY, W. C. Jr. (1960). Estimates with prescribed variance based on two-stage sampling. *Ann. Math. Statist.*, **31**, 662.

BIRNBAUM, Z. W. (1952). Numerical tabulation of the distribution of Kolmogorov's statistic for finite sample size. *J. Amer. Statist. Ass.*, **47**, 425.

BIRNBAUM, Z. W. and TINGEY, F. H. (1951). One-sided confidence contours for probability distribution functions. *Ann. Math. Statist.*, **22**, 592.

BJÖRCK, A. (1978). Comment on the iterative refinement of least-squares solutions. *J. Amer. Statist. Ass.*, **73**, 161.

BLACKWELL, D. (1946). On an equation of Wald. *Ann. Math. Statist.*, **17**, 84.

BLACKWELL, D. (1947). Conditional expectation and unbiased sequential estimation. *Ann. Math. Statist.*, **18**, 105.

BLACKWELL, D. and GIRSHICK, M. A. (1954). *Theory of Games and Statistical Decisions.* Wiley, New York.

BLISCHKE, W. R., TRUELOVE, A. J. and MUNDLE, P. B. (1969). On non-regular estimation. I. Variance bounds for estimators of location parameters. *J. Amer. Statist. Ass.*, **64**, 1056.

BLOCH, D. A. (1966). A note on the estimation of the location parameter of the Cauchy distribution. *J. Amer. Statist. Ass.*, **61**, 852.

BLOM, G. (1954). Transformations of the binomial, negative binomial, Poisson and χ^2 distributions. *Biometrika*, **41**, 302.

BLOM, G. (1978). A property of minimum variance estimates. *Biometrika*, **65**, 642.

BLOMQVIST, N. (1950). On a measure of dependence between two random variables. *Ann. Math. Statist.*, **21**, 593.

BLOOMFIELD, P. and WATSON, G. W. (1975). The inefficiency of least squares. *Biometrika*, **62**, 121.

BLYTH, C. R. (1986). Approximate binomial confidence limits. *J. Amer. Statist. Ass.*, **81**, 843.

BLYTH, C. R. and HUTCHINSON, D. W. (1960). Table of Neyman-shortest unbiased confidence intervals for the binomial parameter. *Biometrika*, **47**, 381.

BLYTH, C. R. and HUTCHINSON, D. W. (1961). Table of Neyman-shortest unbiased confidence intervals for the Poisson parameter. *Biometrika*, **48**, 191.

BLYTH, C. R. and STILL, H. A. (1983). Binomial confidence intervals. *J. Amer. Statist. Ass.*, **78**, 108.

BOHRER, R. (1967). On sharpening Scheffé bounds. *J. R. Statist. Soc.*, **B, 29**, 110.

BOHRER, R. (1973). An optimality property of Scheffé bounds. *Ann. Statist.*, **1**, 766.

BOHRER, R. and FRANCIS, G. K. (1972). Sharp one-sided confidence bounds for linear regression over intervals. *Biometrika*, **59**, 99.

BONDESSON, L. (1975). Uniformly minimum variance estimation in location parameter families. *Ann. Statist.*, **3**, 637.

BORGEN, O. (1979). Comparison of two sequential tests for two-sided alternatives. *J. R. Statist. Soc.*, **B, 41**, 101.

BOWDEN, D. C. (1970). Simultaneous confidence bands for linear regression models. *J. Amer. Statist. Ass.*, **65**, 413.

BOWKER, A. H. (1946). Computation of factors for tolerance limits on a normal distribution when the sample is large. *Ann. Math. Statist.*, **17**, 238.

BOWKER, A. H. (1947). Tolerance limits for normal distributions. *Selected Techniques of Statistical Analysis.* McGraw-Hill, New York.

BOWKER, A. H. (1948). A test for symmetry in contingency tables. *J. Amer. Statist. Ass.*, **43**, 572.

BOWMAN, K. O. (1972). Tables of the sample size requirement. *Biometrika*, **59**, 234.

BOWMAN, K. O. and SHENTON, L. R. (1968). Properties of estimators for the Gamma distribution. *Union Carbide Corpn Report CTC*-1, Oak Ridge, Tennessee.

BOWMAN, K. O. and SHENTON, L. R. (1969). Remarks on maximum likelihood estimators for the Gamma distribution. *1st Internat. Conf. Quality Control, Tokyo, 1969*, 519.

BOWMAN, K. O. and SHENTON, L. R. (1970). Small sample properties of estimators for the Gamma distribution. *Union Carbide Corpn Report CTC-28*, Oak Ridge, Tennessee.

BOWMAN, K. O. and SHENTON, L. R. (1988). *Properties of Estimators for the Gamma Distribution*. Decker, New York.

BOX, G. E. P. (1949). A general distribution theory for a class of likelihood criteria. *Biometrika*, **36**, 317.

BOX, G. E. P. (1980). Sampling and Bayes inference in scientific modelling and robustness (with discussion). *J. R. Statist. Soc.*, A, **143**, 383.

BOX, G. E. P. and COX, D. R. (1964). An analysis of transformations. *J. R. Statist. Soc.*, B, **26**, 211.

BOX, G. E. P. and TIAO, G. C. (1973). *Bayesian Inference in Statistical Analysis*. Addison-Wesley, Reading.

BOX, G. E. P. and TIDWELL, P. W. (1962). Transformation of the independent variables. *Technometrics*, **4**, 531.

BOX, M. J. (1971). Bias in non-linear estimation. *J. R. Statist. Soc.*, B, **33**, 171.

BOYLES, R. A. (1983). On the convergence of the EM algorithm. *J. R. Statist. Soc.*, B, **45**, 47.

BRADU, D. (1965). Main effect analysis of the general non-orthogonal layout with any number of factors. *Ann. Math. Statist.*, **36**, 88.

BRADU, D. and GABRIEL, K. R.. (1974). Simultaneous statistical inference on interactions in two-way analysis of variance. *J. Amer. Statist. Ass.*, **69**, 428.

BRADU, D. and MUNDLAK, Y. (1970). Estimation in lognormal linear models. *J. Amer. Statist. Ass.*, **65**, 198.

BRANDNER, F. A. (1933). A test of the significance of the difference of the correlation coefficients in normal bivariate samples. *Biometrika*, **25**, 102.

BREIMAN, L. and FREEDMAN, D. (1983). How many variables should be entered in a regression? *J. Amer. Statist. Ass.*, **78**, 131.

BRESLOW, N. (1970). Sequential modification of the UMP test for binomial probabilities. *J. Amer. Statist. Ass.*, **65**, 639.

BRETH, M. (1982). Nonparametric estimation for a symmetric distribution. *Biometrika*, **69**, 625.

BRILLINGER, D. R. (1962). Examples bearing on the definition of fiducial probability with a bibliography. *Ann. Math. Statist.*, **33**, 1349.

BRILLINGER, D. R. (1964). The symptotic behaviour of Tukey's general method of setting approximate confidence limits (the jackknife) when applied to maximum likelihood estimates. *Rev. Int. Statist. Inst.*, **32**, 202.

BROERSON, P. M. T. (1986). Subset regression with stepwise directed search. *Appl. Statist.*, **35**, 168.

BROFITT, J. D. and RANDLES, R. H. (1977). A power approximation for the chi-square goodness-of-fit test: simple hypothesis case. *J. Amer. Statist. Ass.*, **72**, 604.

BROWN, L. D. (1964). Sufficient statistics in the case of independent random variables. *Ann. Math. Statist.*, **35**, 1456.

BROWN, L. D. (1971). Non-local asymptotic optimality of appropriate likelihood ratio tests. *Ann. Math. Statist.*, **42**, 1206.

BROWN, M. B. (1977). Algorithm AS116: The tetrachoric correlation and its asymptotic standard error. *Appl. Statist.*, **26**, 343.

BROWN, P. J. (1982). Multivariate calibration. *J. R. Statist. Soc.*, B, **44**, 287.

BROWN, R. L., DURBIN, J. and EVANS, J. M. (1975). Techniques for testing the constancy of regression relationships over time. *J. R. Statist. Soc.*, B, **37**, 149.

BULGREN, W. G. (1971). On representations of the doubly non-central *F* distribution. *J. Amer. Statist. Ass.*, **66**, 184.

BULGREN, W. G. and AMOS, D. E. (1968). A note on representations of the doubly non-central *t* distribution. *J. Amer. Statist. Ass.*, **63**, 1013.

BURMAN, J. P. (1946). Sequential sampling formulae for binomial population. *J. R. Statist. Soc.*, **B, 8**, 98.

BUTLER, R. W. (1984). The significance attained by the best-fitting regressor variable. *J. Amer. Statist. Ass.*, **79**, 341.

CANE, G. J. (1974). Linear estimation of parameters of the Cauchy distribution based on sample quantiles. *J. Amer. Statist. Ass.*, **69**, 243.

CARNAP, R. (1962). *Logical Foundations of Probability*. Univ. of Chicago Press. Chicago. (2nd edition).

CARROLL, R. J. (1982). Prediction and power transformations when the choice of power is restricted to a finite set. *J. Amer. Statist. Ass.*, **77**, 908.

CARROLL, R. J. and RUPPERT, D. (1982). A comparison between maximum likelihood and generalized least squares in a heteroscedastic linear model. *J. Amer. Statist. Ass.*, **77**, 878.

CARROLL, R. J. and RUPPERT, D. (1985). Transformations: a robust analysis. *Technometrics*, **27**, 1.

CARROLL, R. J. and RUPPERT, D. (1987). Diagnostics and robust estimation when transforming the regression model and the response. *Technometrics*, **29**, 287.

CARROLL, R. J. and RUPPERT, D. (1988). *Transformation and Weighting in Regression*. Chapman and Hall, London and New York.

CAUSEY, B. D. (1980). A frequentist analysis of a class of ridge regression estimators. *J. Amer. Statist. Ass.*, **75**, 736.

CHAMBERS, J. R. (1973). Fitting nonlinear models: numerical techniques. *Biometrika*, **60**, 1.

CHAN, L. K. (1970). Linear estimation of the location and scale parameters of the Cauchy distribution based on sample quantiles. *J. Amer. Statist. Ass.*, **65**, 851.

CHAND, N. (1974). Sequential tests of composite hypotheses. *J. Amer. Statist. Ass.*, **69**, 394.

CHANDA, K. C. (1971). Asymptotic distribution of the sample size for a sequential probability ratio test. *J. Amer. Statist. Ass.*, **66**, 178.

CHANDLER, K. N. (1950). On a theorem concerning the secondary subscripts of deviations in multivariate correlation using Yule's notation. *Biometrika*, **37**, 451.

CHANDRA, M., SINGPURWALLA, N. D. and STEPHENS, M. A. (1981). Kolmogorov statistics for tests of fit for the extreme-value and Wiebull distributions. *J. Amer. Statist. Ass.*, **76**, 729.

CHANT, D. (1974). On asymptotic tests of composite hypotheses in non-standard conditions. *Biometrika*, **61**, 291.

CHAO, M. T. and GLASER, R. E. (1978). The exact distribution of Bartlett's test statistic for homogeneity of variances with unequal sample sizes. *J. Amer. Statist. Ass.*, **73**, 422.

CHAPMAN, D. G. (1950). Some two sample tests. *Ann. Math. Statist.*, **21**, 601.

CHAPMAN, D. G. and ROBBINS, H. (1951). Minimum variance estimation without regularity assumptions. *Ann. Math. Statist.*, **22**, 581.

CHAPMAN, J.-A. W. (1976). A comparison of the X^2, $-2 \log R$ and multinomial probability criteria for significance tests when expected frequencies are small. *J. Amer. Statist. Ass.*, **71**, 854.

CHARNES, A., FROME, E. L. and YU, P. L. (1976). The equivalence of generalized least squares and maximum likelihood estimates in the exponential family. *J. Amer. Statist. Ass.*, **71**, 169.

CHASE, G. R. (1972). On the chi-square test when the parameters are estimated independently of the sample. *J. Amer. Statist. Ass.*, **67**, 609.

CHÂTILLON, G. (1984). The balloon rules for a rough estimate of the correlation coefficient. *Amer. Statistician*, **38**, 58.

CHENG, R. C. H. and AMIN, N. A. K. (1983). Estimating parameters in continuous univariate distributions with a shifted origin. *J. R. Statist. Soc.*, **B, 45**, 394.

CHERNOFF, H. (1949). Asymptotic studentisation in testing of hypotheses. *Ann. Math. Statist.*, **20**, 268.

CHERNOFF, H. (1951). A property of some Type A regions. *Ann. Math. Statist.*, **22**, 472.

CHERNOFF, H. (1952). A measure of asymptotic efficiency for tests of a hypothesis based on the sum of observations. *Ann. Math. Statist.*, **23**, 493.

CHERNOFF, H. (1954). On the distribution of the likelihood ratio. *Ann. Math. Statist.*, **25**, 573.

CHERNOFF, H. and LEHMANN, E. L. (1954). The use of maximum likelihood estimates in χ^2 tests for goodness of fit. *Ann. Math. Statist.*, **25**, 579.

CHEW, V. (1970). Covariance matrix estimation in linear models. *J. Amer. Statist. Ass.*, **65**, 173.

CHHIKARA, R. S. and FOLKS, J. L. (1976). Optimum test procedures for the mean of first passage time distribution in Brownian motion with positive drift (Inverse Gaussian distribution). *Technometrics*, **18**, 189.

CHIBISOV, D. M. (1971). Certain chi-square type tests for continuous distributions. *Theory Prob. Applic.*, **16**, 1.

CHIPMAN, J. S. (1964). On least squares with insufficient observations. *J. Amer. Statist. Ass.*, **59**, 1078.

CHOI, S. C. (1977). Tests of equality of dependent correlation coefficients. *Biometrika*, **64**, 645.

CHOI, S. C. and WETTE, R. (1969). Maximum likelihood estimation of the parameters of the gamma distribution and their bias. *Technometrics*, **11**, 683.

CHOW, G. C. (1960). Test of equality between sets of coefficients in two linear regressions. *Econometrica*, **28**, 591.

CHOW, W. K. and HODGES, J. L., Jr. (1975). An approximation for the distribution of the Wilcoxon one-sample statistic. *J. Amer. Statist. Ass.*, **70**, 648.

CHOW, Y. S., ROBBINS, H. and TEICHER, H. (1965). Moments of randomly stopped sums. *Ann. Math. Statist.*, **36**, 789.

CLARK, R. E. (1953). Percentage points of the incomplete beta function. *J. Amer. Statist. Ass.*, **48**, 831.

CLARKE, G. P. Y. (1980). Moments of the least squares estimators in a non-linear regression model. *J. R. Statist. Soc.*, **B, 42**, 227.

CLARKE, G. P. Y. (1987a). Approximate confidence limits for a parameter function in non-linear regression. *J. Amer. Statist. Ass.*, **82**, 221.

CLARKE, G. P. Y. (1987b). Marginal curvatures and their usefulness in the analysis of nonlinear regression models. *J. Amer. Statist. Ass.*, **82**, 844.

CLEVELAND, W. S. and DEVLIN, S. J. (1988). Locally weighted regression: An approach to regression analysis by local fitting. *J. Amer. Statist. Ass.*, **83**, 596.

CLIFF, A. D. and ORD, J. K. (1981). *Spatial Processes: Models and Applications*. Pion, London.

CLOPPER, C. J. and PEARSON, E. S. (1934)*. The use of confidence or fiducial limits illustrated in the case of the binomial. *Biometrika*, **26**, 404.

COCHRAN, W. G. (1937)*. The efficiencies of the binomial series tests of significance of a mean and of a correlation coefficient. *J. R. Statist. Soc.*, **100**, 69.

COCHRAN, W. G. (1952)*. The χ^2 test of goodness of fit. *Ann. Math. Statist.*, **23**, 315.

COCHRAN, W. G. (1954)*. Some methods for strengthening the common χ^2 tests. *Biometrics*, **10**, 417.

COCHRANE, D. and ORCUTT, G. H. (1949). Application of least-squares regression to relationships containing auto-correlated error terms. *J. Amer. Statist. Ass.*, **44**, 32.

COHEN, A. (1972). Improved confidence intervals for the variance of a normal distribution. *J. Amer. Statist. Ass.*, **67**, 382.

COHEN, A. and STRAWDERMAN, W. E. (1971). Unbiasedness of tests for homogeneity of variances. *Ann. Math. Statist.*, **42**, 355.

COHEN, J. D. (1988). Noncentral chi-square: some observations on recurrence. *Amer. Statistician*, **42**, 120.

CONOVER, W. J. (1972). A Kolmogorov goodness-of-fit test for discontinuous distributions. *J. Amer. Statist. Ass.*, **67**, 591.

COOK, R. D. (1977). Detection of influential observations in linear regression. *Technometrics*, **19**, 15.

COOK, R. D. (1986). Assessment of local influence. *J. R. Statist. Soc.*, **B, 48**, 133.

COOK, R. D. and WANG, P. C. (1983). Transformations and influential cases in regression. *Technometrics*, **25**, 337.

COOK, R. D. and WEISBERG, S. (1982). *Residuals and Influence in Regression*. Chapman and Hall, London and New York.

COOK, R. D. and WEISBERG, S. (1983). Diagnostics for heteroscedsticity in regression. *Biometrika*, **70**, 1.

COOK, R. D. and WITMER, J. A. (1985). A note on parameter-effects curvature. *J. Amer. Statist. Ass.*, **80**, 872.

COOK, R. D., TSAI, C. L. and WEI, B. C. (1986). Bias in non-linear regression. *Biometrika*, **73**, 615.

COPAS, J. B. (1975). On the unimodality of the likelihood for the Cauchy distribution. *Biometrika*, **62**, 701.

COPAS, J. B. (1983). Regression, prediction and shrinkage. *J. R. Statist. Soc.*, **B, 45**, 311.

CORDEIRO, G. M. (1983). Improved likelihood ratio statistics for generalized linear models. *J. R. Statist. Soc.*, **B, 45**, 404.

CORDEIRO, G. M. (1987). On the corrections to the likelihood ratio statistic. *Biometrika*, **74**, 265.

COX, C. P. (1958). A concise derivation of general orthogonal polynomials. *J. R. Statist. Soc.*, **B, 20**, 406.

COX, D. R. (1952a). Sequential tests for composite hypotheses. *Proc. Camb. Phil. Soc.*, **48**, 290.

COX, D. R. (1952b). A note on the sequential estimation of means. *Proc. Camb. Phil. Soc.*, **48**, 447.

COX, D. R. (1952c). Estimation by double sampling. *Biometrika*, **39**, 217.

COX, D. R. (1956). A note on the theory of quick tests. *Biometrika*, **43**, 478.

COX, D. R. (1958). Some problems connected with statistical inference. *Ann. Math. Statist.*, **29**, 357.

COX, D. R. (1961). Tests of separate families of hypotheses. *Proc. 4th Berkeley Symp. Math. Statist. and Prob.*, **1, 105**.

COX, D. R. (1962). Further results on tests of separate families of hypotheses. *J. R. Statist. Soc.*, **B, 24**, 406.

COX, D. R. (1963). Large sample sequential tests for composite hypotheses. *Sankhyā*, **A, 25**, 5.

COX, D. R. (1967). Fieller's theorem and a generalization. *Biometrika*, **54**, 567.

COX, D. R. (1971). The choice between alternative ancillary statistics. *J. R. Statist. Soc.*, **B, 33**, 251.

COX, D. R. and HINKLEY, D. V. (1968). A note on the efficiency of least-squares estimates. *J. R. Statist. Soc.* **B, 30**, 284.

COX, D. R. and HINKLEY, D. V. (1974). *Theoretical Statistics.* Chapman and Hall, London.

COX, D. R. and MCCULLAGH, P. (1982). Some aspects of analysis of covariance. *Biometrics*, **38**, 541.

COX, D. R. and REID, N. (1987). Parameter orthogonality and approximate conditional inference. *J. R. Statist. Soc.*, **B, 49**, 1.

COX, D. R. and STUART, A. (1955). Some quick sign tests for trend in location and dispersion. *Biometrika*, **42**, 80.

CRAMÉR, H. (1946). *Mathematical Methods of Statistics.* Princeton Univ. Press.

CRESSIE, N. (1981). Transformations and the jackknite. *J. R. Statist. Soc.*, **B, 43**, 177.

CRESSIE, N. and READ, T. R. C. (1984). Multinomial goodness-of-fit tests. *J. R. Statist. Soc.*, **46**, 440.

CROW, E. L. (1956). Confidence intervals for a proportion. *Biometrika*, **43**, 423.

CROW, E. L. and GARDNER, R. S. (1959). Confidence intervals for the expectation of a Poisson variable. *Biometrika*, **46**, 441.

CURTISS, J. H. (1943). On transformations used in the analysis of variance. *Ann. Math. Statist.*, **14**, 107.

DAHIYA, R. C. (1981). An improved method of estimating an integer-parameter by maximum likelihood. *Amer. Statistician*, **35**, 34.

DAHIYA, R. C. and GURLAND, J. (1972). Pearson chi-squared test of fit with random intervals. *Biometrika*, **59**, 147.

DAHIYA, R. C. and GURLAND, J. (1973). How many classes in the Pearson chi-square test? *J. Amer. Statist. Ass.*, **68**, 707.

DALLAL, G. E. and WILKINSON, L. (1986). An analytical approximation to the distribution of Lilliefors' test statistic for normality. *Amer. Statistician*, **40**, 294.

DANIELS, H. E. (1948). A property of rank correlations. *Biometrika*, **35**, 416.

DANIELS, H. E. (1951-2). The theory of position finding. *J. R. Statist. Soc.*, **B, 13**, 186 and **14**, 246.

DANIELS, H. E. (1961). The asymptotic efficiency of a maximum likelihood estimator. *Proc. 4th Berkeley Symp. Math. Statist. and Prob.*, **1**, 151.

DANIELS, H. E. (1983). Saddlepoint approximations for estimation equations. *Biometrika*, **70**, 89.

DANIELS, H. E. and KENDALL, M. G. (1958). Short proof of Miss Harley's theorem on the correlation coefficient. *Biometrika*, **45**, 571.

DANTZIG, G. B. (1940). On the non-existence of tests of "Student's" hypothesis having power functions independent of σ. *Ann. Math. Statist.*, **11**, 186.

DAR, S. N. (1962). On the comparison of the sensitivities of experiments. *J. R. Statist. Soc.*, **B, 24**, 447.

DARLING, D. A. (1952). On a test for homogenetity and extreme values. *Ann. Math. Statist.*, **23**, 450.

DARMOIS, G. (1935). Sur les lois de probabilité à estimation exhaustive. *C.R. Acad, Sci., Paris*, **200**, 1265.

DASGUPTA, P. (1968). Tables of the non-centrality parameter of F-test as a function of power. *Sankhy⁻a*, **B, 30**, 73.

DASGUPTA, S. and PERLMAN, M. D. (1974). Power of the noncentral F-test: effect of additional variates on Hotelling's T-test. *J. Amer. Statist. Ass.*, **69**, 174.

DAVID, F. N. (1937). A note on unbiased limits for the correlation coefficient. *Biometrika*, **29**, 157.

DAVID, F. N. (1938). *Tables of the Correlation Coefficient.* Cambridge Univ. Press.

DAVID, F. N. (1950). An alternative form of χ^2. *Biometrika*, **37**, 448.

DAVID, F. N. and JOHNSON, N. L. (1948). The probability integral transformation when parameters are estimated from the sample. *Biometrika*, **35**, 182.

DAVID, H. A. and GROENEVELD, R. A. (1982). Measures of local variation in a distribution: Expected length of spacings and variances of order statistics. *Biometrika*, **69**, 227.

DAVIDSON, A. C. and HINKLEY, D. V. (1988). Saddlepoint approximations in resampling methods. *Biometrika*, **75**, 417.

DAVIS, A. W. and SCOTT, A. J. (1971). On the k-sample Behrens-Fisher distribution. (Div. of Math. Statist. Tech. Paper No. 33, C.S.I.R.O., Australia.)

DAVIS, L. (1984). Comments on a paper by T. Amemiya on estimation in a dichotomous logit regression model. *Ann. Statist.*, **12**, 778.

DAVIS, R. C. (1951). On minimum variance in nonregular estimation. *Ann. Math. Statist.*, **22**, 43.

DAWID, A. P. (1975). On the concepts of sufficiency and ancillarity in the presence of nuisance parameters. *J. R. Statist. Soc.*, **B, 37**, 248.

DAWID, A. P. (1977). Conformity of inference patterns. In *Recent Developments in Statistics*, J. R. Varva et al. (eds.). North-Holland, Amsterdam.

DAWID, A. P. (1984). Present position and potential developments: some personal views. Statistical theory, the prequential approach (with discussion). *J. R. Statist. Soc.*, **A, 147**, 278.

DAWID, A. P. and STONE, M. (1982). The functional-model basis of fiducial inference (with discussion). *Ann. Statist.*, **10**, 1054.

DE GROOT, M. H. (1959). Unbiased sequential estimation for binomial populations. *Ann. Math. Statist.*, **30**, 80.

DE GROOT, M. H. (1970). *Optimal Statistical Decisions.* McGraw-Hill, New York.

DEMPSTER, A. P. (1963). Further examples of inconsistencies in the fiducial argument. *Ann. Math. Statist.*, **34**, 884.

DEMPSTER, A. P. (1964). On the difficulties inherent in Fisher's fiducial argument. *J. Amer. Statist. Ass.*, **59**, 56.

DEMPSTER, A. P., LAIRD, N. M. and RUBIN, D. B. (1977). Maximum likelihood from incomplete data via the EM algorithm. *J. R. Statist. Soc.*, **B, 39**, 1.

DEMPSTER, A. P. and RUBIN, D. R. (1983). Rounding error in regression: the appropriateness of Sheppard's corrections. *J. R. Statist. Soc.*, **B, 45**, 51.

DEMPSTER, A. P. and SCHATZOFF, M. (1965). Expected significance level as a sensitivity index for test statistics. *J. Amer. Statist. Ass.*, **60**, 420.

DEMPSTER, A. P., SCHATZOFF, M. and WERMUTH, N. (1977). A simulation study of alternatives to ordinary least squares (with comments following). *J. Amer. Statist. Ass.*, **72**, 77.

DENNY, J. L. (1967). Sufficient conditions for a family of probabilities to be exponential. *Proc. Nat. Acad. Sci., U.S.A.*, **57**, 1184.

DENNY, J. L. (1972). Sufficient statistics and discrete exponential families. *Ann. Math. Statist.*, **43**, 1320.

DENT, W. T. and CASSING, S. (1978). On modified maximum likelihood estimators of the autocorrelation parameter in linear models. *Biometrika*, **65**, 211.

DIACONIS, P. and FREEDMAN, D. A. (1986a). On the consistency of Bayes estimates (with discussion). *Ann. Statist.*, **14**, 1.

DIACONIS, P. and FREEDMAN, D. A. (1986b). On inconsistent Bayes estimates of location. *Ann. Statist.*, **14**, 68.

DIACONIS, P. and ZABELL, S. L. (1982). Updating subjective probability. *J. Amer. Statist. Ass.*, **77**, 822.

DICICCIO, T. J. and ROMANO, J. P. (1988). A review of bootstrap confidence intervals. *J. R. Statist. Soc.*, **B, 50**, 338.

DICICCIO, T. and TIBSHIRANI, R. (1987). Bootstrap confidence intervals and bootstrap approxima-tions. *J. Amer. Statist. Ass.*, **82**, 163.

DIGBY, P. G. N. (1983). Approximating the tetrachoric correlation coefficient. *Biometrics*, **39**, 753.

DIXON, W. J. (1953). Power functions of the Sign Test and power efficiency for normal alternatives. *Ann. Math. Statist.*, **24**, 467.

DODGE, H. F. and ROMIG, H. G. (1944). *Sampling Inspection Tables.* Wiley, New York.

DOKSUM, K. A., FENSTAD, G. and AABERGE, R. (1977). Plots and tests for symmetry. *Biometrika*, **64**, 473.

DOLBY, J. L. (1963). A quick method for choosing a transformation. *Technometrics*, **5**, 317.

DONNER, A. (1986). A review of inference procedures for the intraclass correlation coefficient in the one way random effects model. *Int. Statist. Rev.*, **54**, 67.

DONNER, A. and WELLS, G. (1986). A comparison of confidence interval methods for the intraclass correlation coefficient. *Biometrics*, **42**, 401. (Corr. **43**, 1035).

DOWNTON, F. (1953). A note on ordered least-squares estimation. *Biometrika*, **40**, 457.

DOWNTON, F. (1966a). Linear estimates with polynomial coefficients. *Biometrika*, **53**, 129.

DOWNTON, F. (1966b). Linear estimates of parameters in the extreme value distribution. *Technometrics*, **8**, 3.

DRAPER, N. R. and COX, D. R. (1969). On distributions and their transformation to normality. *J. R. Statist. Soc.*, **B, 31**, 472.

DRAPER, N. R., GUTTMAN, I. and KANEMASU, H. (1971). The distribution of certain regression statistics. *Biometrika*, **58**, 295.

DRAPER, N. R. and HUNTER, W. G. (1969). Transformations: some examples revisited. *Technometrics*, **11**, 23.

DROST, F. C., KALLENBERG, W. C. M., MOORE, D. S. and OOSTERHOFF, J. (1989). Power approximations to multinominal tests of fit. *J. Amer. Statist. Ass.*, **84**, 130.

DUMOUCHEL, W. H. (1983). Estimating the stable index in order to measure tail thickness: critique. *Ann. Statist.*, **11**, 1019.

DUNCAN, A. J. (1957). Charts of the 10% points and 50% points of the operating characteristic curves for fixed effects analysis of variance F-tests, $\alpha = 0.10$ and 0.05. *J. Amer. Statist. Ass.*, **52**, 345.

DUNN, O. J. (1961). Multiple comparisons among means. *J. Amer. Statist. Ass.*, **56**, 52.

DUNN, O. J. (1968). A note on confidence bands for a regression line over a finite range. *J. Amer. Statist. Ass.*, **63**, 1028.

DUNN, O. J. and CLARK, V. (1969, 1971). Correlation coefficients measured on the same individuals; *and* Comparisons of tests of the equality of dependent correlation coefficients. *J. Amer. Statist. Ass.*, **64**, 366 and **66**, 904.

DUNNETT, C. W. (1980a, b). Pairwise multiple comparisons (a) in the homogeneous variance, unequal sample size case; (b) in the unequal variance case. *J. Amer. Statist. Ass.*, **75**, 789 and 796.

DURBIN, J. (1953). A note on regression when there is extraneous information about one of the coefficients. *J. Amer. Statist. Ass.*, **48**, 799.

DURBIN, J. (1960). Estimation of parameters in time-series regression models. *J. R. Statist. Soc.*, **B, 22,** 139.

DRUBIN, J. (1961). Some methods of constructing exact tests. *Biometrika*, **48,** 41.

DURBIN, J. (1979). On Birnbaum's theorem on the relation between sufficiency, conditionality and likelihood. *J. Amer. Statist. Ass.*, **65,** 395.

DURBIN, J. (1972). *Distribution theory for tests based on the sample distribution function.* Society for Industrial and Applied Mathematics.

DURBIN, J. (1975). Kolmogorov-Smirnov tests when parameters are estimated with applications to tests of exponentiality and tests on spacings. *Biometrika*, **62,** 5.

DURBIN, J. (1980). Approximations for densities of sufficient estimators. *Biometrika*, **67,** 311.

DURBIN, J. (1988). Is a philosophical consensus for statistics attainable? *J. Econometrics*, **37,** 51.

DURBIN, J. and KENDALL, M. G. (1951). The geometry of estimation. *Biometrika*, **38,** 150.

DURBIN, J. and WATSON, G. S. (1950, 1951, 1971). Testing for serial correlation in least squares regression, I, II, III. *Biometrika*, **37,** 409; **38,** 159; **58,** 1.

DYER, A. R. (1973). Discrimination procedures for separate families of hypotheses. *J. Amer. Statist. Ass.*, **68,** 970.

DYER, A. R. (1974). Hypothesis testing procedures for separate families of hypotheses. *J. Amer. Statist. Ass.*, **69,** 140.

DYER, D. D. and KEATING, J. P. (1980). On the determination of critical values for Bartlett's test. *J. Amer. Statist. Ass.*, **75,** 313.

DYNKIN, E. B. (1951). Necessary and sufficient statistics for a family of probability distributions. (Russian). *Usp. Mat. Nauk (N.S.)*, **6,** No. 1 (41), 68.

EDWARDS, A. W. F. (1972). *Likelihood.* Cambridge Univ. Press, Cambridge.

EDWARDS, A. W. F. (1974). The history of likelihood. *Int. Statist. Rev.*, **42,** 9.

EFRON, B. (1967). The power of the likelihood ratio test. *Ann. Math. Statist.*, **38,** 802.

EFRON, B. (1975). Defining the curvature of a statistical problem (with applications to second order efficiency). *Ann. Statist.*, **3,** 1189.

EFRON, B. (1978a). The geometry of exponential families. *Ann. Statist.*, **6,** 362.

EFRON, B. (1979). Bootstrap methods: another look at the jackknife. *Ann. Statist.*, **7,** 1.

EFRON, B. (1981). Nonparametric estimates of standard error: the jackknife, the bootstrap and other methods. *Biometrika*, **68,** 589.

EFRON, B. (1982). Maximum likelihood and decision theory. *Ann. Statist.*, **10,** 341.

EFRON, B. (1984). Comparing non-nested linear models. *J. Amer. Statist. Ass.*, **79,** 791.

EFRON, B. (1985). Bootstrap confidence intervals for a class of parametric problems. *Biometrika*, **72,** 45.

EFRON, B. (1987). Beeter bootstrap confidence intervals. *J. Amer. Statist. Ass.*, **82,** 171.

EFRON, B. and HINKLEY, D. V. (1978). Assessing the accuracy of the maximum likelihood estimator; observed versus expected Fisher information (with following comments). *Biometrika*, **65,** 457.

EFRON, B. and STEIN, C. (1981). The jackknife estimate of variance. *Ann. Statist.*, **9,** 586.

EFROYMSON, M. A. (1960). Multiple regression analysis. In A. Ralston and H. S. Wilf (ed.). *Mathematical Models for Digital Computers.* Wiley, New York.

EISENHART, C. (1938). The power function of the χ^2 test. *Bull. Amer. Math. Soc.*, **44,** 32.

EISENHART, C. (1947). The assumptions underlying the analysis of variance. *Biometrics*, **3,** 1.

ELLISON, B. E. (1964). On two-sided tolerance intervals for a normal distribution. *Ann. Math. Statist.*, **35,** 762.

ELSTON, R. C. (1975). On the correlation between correlations. *Biometrika*, **62,** 133.

EPSTEIN, B. and SOBEL, M. (1954). Some theorems relevant to life testing from an exponential distribution. *Ann. Math. Statist.*, **25,** 373.

EVANS, I. G. and SHABAN, S. A. (1974). A note on estimation in lognormal models. *J. Amer. Statist. Ass.*, **69,** 779.

EVANS, M. J., FRASER, D. A. S. and MONETTE, G. (1985). On the role of principles in statistical inference. In *Statistical Theory and Data Analysis*, K. Matsuita (ed.). North-Holland, Amsterdam.

EVANS, M. J., FRASER, D. A. S. and MONETTE, G. (1986). On principles and arguments to likelihood. *Canad. J. Statist.*, **14**, 181.

EZEKIEL, M. J. B. and FOX, K. A. (1959). *Methods of Correlation and Regression Analysis: Linear and Curvilinear.* Wiley, New York.

FAIRFIELD SMITH, H. (1957). Interpretation of adjusted treatment means and regressions in analysis of covariance. *Biometrics*, **13**, 282.

FAREBROTHER, R. W. (1976). Further results on the mean square error of ridge regression. *J. R. Statist. Soc.*, **B**, **38**, 248.

FEDER, P. I. (1968). On the distribution of the log likelihood ratio test statistic when the true parameter is "near" the boundaries of the hypothesis regions. *Ann. Math. Statist.*, **39**, 2044.

FEDER, P. I. (1975). The log likelihood ratio in segmented regression. *Ann. Statist.*, **3**, 84.

FELLER, W. (1938). Note on regions similar to the sample space. *Statist. Res. Mem.*, **2**, 117.

FELZENBAUM, A., HART, S. and HOCHBERG, Y. (1983). Improving some multiple comparison procedures. *Ann. Statist.*, **11**, 121.

FEND, A. V. (1959). On the attainment of Cramer-Rao and Bhattacharyya bounds for the variance of an estimate. *Ann. Math. Statist.*, **30**, 381.

FERGUSON, T. S. (1967). *Mathematical Statistics: A Decision Theoretic Approach.* Academic Press, New York.

FERGUSON, T. S. (1978). Maximum likelihood estimation of the parameters of the Cauchy distribution for samples of size 3 and 4. *J. Amer. Statist. Ass.*, **73**, 211.

FERGUSON, T. S. (1982). An inconsistent maximum likelihood estimate. *J. Amer. Statist. Ass.*, **77**, 831.

FÉRON, R. and FOURGEAUD, C. (1952). Quelques proprietés caractéristiques de la loi de Laplace-Gauss. *Publ. Inst. Statist. Paris*, **1**, 44.

FIELLER, E. C. (1940). The biological standardisation of insulin. *Suppl. J. R. Statist. Soc.*, **7**, 1.

FIELLER, E. C. (1954). Some problems in interval estimation. *J. R. Statist. Soc.*, **B**, **16**, 175.

FINNEY, D. J. (1941). On the distribution of a variate whose logarithm is normally distributed. *Suppl. J. R. Statist. Soc.*, **7**, 155.

FINNEY, D. J. (1949). On a method of estimating frequencies. *Biometrika*, **36**, 233.

FINSTER, M. (1985). Estimation in the general linear model when the accuracy is specified before data collection. *Ann. Statist.*, **13**, 663.

FISHBURN, P. C. (1986). The axioms of subjective probability. *Statist. Sci.*, **1**, 335.

FISHBURN, P. C. (1989). Foundations of decision analysis: Along the way. *Management Science*, **35**, 387.

FISHER, N. I. and LEE, A. J. (1986). Correlation coefficients for random variables on a unit sphere or hypersphere. *Biometrika*, **73**, 159.

FISHER, R. A. (1921a)*. On the mathematical foundations of theoretical statistics. *Phil. Trans.*, **A**, **222**, 309.

FISHER, R. A. (1921b)*. Studies in crop variation. I. An examination of the yield of dressed grain from Broadbalk. *J. Agric. Sci.*, **11**, 107.

FISHER, R. A. (1921c). On the "probable error" of a coefficient of correlation deduced from a small sample. *Metron*, **1**, (4), 3.

FISHER, R. A. (1922a)*. On the interpretation of chi-square from contingency tables, and the calculation of P. *J. R. Statist. Soc.*, **85**, 87.

FISHER, R. A. (1922b)*. The goodness of fit of regression formulae and the distribution of regression coefficients. *J. R. Statist. Soc.*, **85**, 597.

FISHER, R. A. (1924a). The distribution of the partial correlation coefficient. *Metron*, **3**, 329.

FISHER, R. A. (1924b). The influence of rainfall on the yield of wheat at Rothamsted. *Phil. Trans.*, **B, 213,** 89.

FISHER, R. A. (1924c).* The conditions under which χ^2 measures the discrepancy between observation and hypothesis. *J. R. Statist. Soc.,* **87,** 442.

FISHER, R. A. (1925-). *Statistical Methods for Research Workers.* Oliver and Boyd, Edinburgh.

FISHER, R. A. (1925).* Theory of statistical estimation. *Proc. Camb. Phil. Soc.,* **22,** 700.

FISHER, R. A. (1928a).* The general sampling distribution of the multiple correlation coefficient. *Proc. Roy. Soc.,* **A, 121,** 654.

FISHER, R. A. (1928b).* On a property connecting the χ^2 measure of discrepancy with the method of maximum likelihood. *Atti Congr. Int. Mat., Bologna,* **6,** 94.

FISHER, R. A. (1935a). *The Design of Experiments.* Oliver and Boyd, Edinburgh.

FISHER, R. A. (1935b).* The fiducial argument in statistical inference. *Ann. Eugen.,* **6,** 391.

FISHER, R. A. (1939).* The comparison of samples with possibly unequal variances. *Ann. Eugen.,* **9,** 174.

FISHER, R. A. (1941).* The negative binomial distribution. *Ann. Eugen.,* **11,** 182.

FISHER, R. A. (1956). *Statistical methods and scientific inference.* Oliver and Boyd, Edinburgh.

FISHER, R. A. (1966). *The Design of Experiments.* Oliver and Boyd, Edinburgh (8th edn).

FIX, E. (1949). Tables of noncentral χ^2. *Univ. Calif. Publ. Statist.,* **1,** 15.

FLETCHER, R. H. (1975). On the iterative refinement of least squares solutions. *J. Amer. Statist. Ass.,* **70,** 109.

FOLKS, J. L. and ANTLE, C. E. (1967). Straight line confidence regions for linear models. *J. Amer. Statist. Ass.,* **62,** 1365.

FOLKS, J. L. and CHHIKARA, R. S. (1978). The Inverse Ganssian distribution and its statistical application—a review. *J. R. Statist. Soc.,* **B, 40,** 263.

FOUTZ, R. V. (1977). On the unique consistent root to the likelihood equations. *J. Amer. Statist. Ass.,* **72,** 147.

FOUTZ, R. V. and SRIVASTAVA, R. C. (1977). The performance of the likelihood ratio test when the model is incorrect. *Ann. Statist.,* **5,** 1183.

FOX, M. (1956). Charts of the power of the *F*-test. *Ann. Math. Statist.,* **27,** 484.

FRANGOS, C. C. (1980). Variance estimation for the second-order jackknife. *Biometrika,* **67,** 715.

FRASER, D. A. S. (1950). Note on the χ^2 smooth test. *Biometrika,* **37,** 447.

FRASER, D. A. S. (1963). On sufficiency and the exponential family. *J. R. Statist. Soc.,* **B, 25,** 115.

FRASER, D. A. S. (1964). Fiducial inference for location and scale parameters. *Biometrika,* **51,** 17.

FRASER, D. A. S. (1968). *The Structure of Inference.* Wiley, New York.

FRASER, D. A. S. (1976). Necessary analysis and adaptive inference. *J. Amer. Statist. Ass.,* **71,** 99.

FRASER, D. A. S. and GUTTMAN, I. (1956). Tolerance regions. *Ann. Math. Statist.,* **27,** 162.

FRASER, D. A. S. and REID, N. (1989). Adjustments to profile likelihood. *Biometrika,* **76,** 477.

FRASER, D. A. S. and WORMLEIGHTON, R. (1951). Non-parametric estimation IV. *Ann. Math. Statist.,* **22,** 294.

FREEDMAN, D. A. (1981). Bootstrapping regression models. *Ann. Statist.,* **9,** 1218.

FREEDMAN, D. A. and PETERS, S. C. (1984). Bootstrapping a regression equation: some empirical results. *J. Amer. Statist. Ass.,* **79,** 97.

FREEMAN, D. and WEISS, L. (1964). Sampling plans which approximately minimize the maximum expected sample size. *J. Amer. Statist. Ass.,* **59,** 67.

FREEMAN, G. H. (1972). Experimental designs with many classifications. *J. R. Statist. Soc.,* **B, 34,** 84.

FREEMAN, G. H. (1975). Row-and-column designs with two groups of treatments having different replications. *J. R. Statist. Soc.,* **B, 37,** 114.

FREEMAN, G. H. and JEFFERS, J. N. R. (1962). Estimation of means and standard errors in the analysis of non-orthogonal experiments by electronic computer. *J. R. Statist. Soc.,* **B, 24,** 435.

FREEMAN, M. F. and TUKEY, J. W. (1950). Transformations related to the angular and the square root. *Ann. Math. Statist.,* **21,** 607.

FREUND, R. J., VAIL, R. W. and CLUNIES-ROSS, C. W. (1961). Residual analysis. *J. Amer. Statist. Ass.*, **56**, 98 (corrigenda: 1005).

FRYDENBERG, M. and JENSEN, J. L. (1989). Is the improved likelihood ratio statistic really improved in the discrete case? *Biometrika*, **76**, 655.

FU, J. C. (1973). On a theorem of Bahadur on the rate of convergence of point estimators. *Ann. Statist.*, **1**, 745.

FUJINO, Y. (1980). Approximate binomial confidence limits. *Biometrika*, **67**, 677.

FULLER, W. A. (1976). *Introduction to Statistical Time Series*. Wiley, New York.

FULLER, W. A. and RAO, J. N. K. (1978). Estimation for a linear regression model with unknown diagonal covariance matrix. *Ann. Statist.*, **6**, 1149.

GABRIEL, K. R. (1963). Analysis of variance of proportions with unequal frequencies. *J. Amer. Statist. Ass.*, **58**, 1133.

GABRIEL, K. R. (1964). A procedure for testing the homogeneity of all sets of means in analysis of variance. *Biometrics*, **20**, 459.

GABRIEL, K. R. (1969). Simultaneous test procedures—some theory of multiple comparisons. *Ann. Math. Statist.*, **40**, 224.

GABRIEL, K. R. (1978). A simple method of multiple comparisons of means. *J. Amer. Statist. Ass.*, **73**, 724.

GABRIEL, K. R. and PERITZ, E. (1973). Least squares and maximal contrasts. *Int. Statist. Rev.*, **41**, 155.

GAFARIAN, A. V. (1964). Confidence bands in straight line regression. *J. Amer. Statist. Ass.*, **59**, 182.

GALLANT, A. R. (1987). *Nonlinear Statistical Models*. Wiley, New York.

GARDNER, M. J. (1972). On using an estimated regression line in a second sample. *Biometrika*, **59**, 263.

GART, J. J. and PETTIGREW, H. M. (1970). On the conditional moments of the k-statistics for the poisson distribution. *Biometrika*, **57**, 661.

GARWOOD, F. (1936). Fiducial limits for the Poisson distribution. *Biometrika*, **28**, 437.

GEARY, R. C. (1942). The estimation of many parameters. *J. R. Statist. Soc.*, **105**, 213.

GEARY, R. C. (1944). Comparison of the concepts of efficiency and closeness for consistent estimates of a parameter. *Biometrika*, **33**, 123.

GEISSER, S. and CORNFIELD, J. (1963). Posterior distributions for multivariate normal parameters. *J. R. Statist. Soc.*, **B, 25**, 368.

GEMES, K. (1989). A refutation of Popperian inductive scepticism. *Brit. J. Phil. Sci.*, **40**, 183.

GENEST, C. and SCHERVISH, M. J. (1985). Resolution of Godambe's paradox. *Canad. J. Statist.*, **4**, 293.

GENIZI, A. and HOCHBERG, Y. (1978). On improved extensions of the T-method of multiple comparisons for unbalanced designs. *J. Amer. Statist. Ass.*, **73**, 879. (Corrigendum **74**, 744).

GHOSH, B. K. (1964). Simultaneous test by sequential methods in hierarchical classifications. *Biometrika*, **51**, 439.

GHOSH, B. K. (1969). Moments of the distribution of sample size in SPRT. *J. Amer. Statist. Ass.*, **64**, 1560.

GHOSH, B. K. (1975). A two-stage procedure for the Behrens-Fisher problem. *J. Amer. Statist. Ass.*, **70**, 457.

GHOSH, B. K. (1979). A comparison of some approximate confidence intervals for the binomial parameter. *J. Amer. Statist. Ass.*, **74**, 894.

GHOSH, J. K. and SINGH, R. (1966). Unbiased estimation of location and scale parameters. *Ann. Math. Statist.*, **37**, 1671.

GHOSH, J. K. and SINHA, B. K. (1981). A necessary and sufficient condition for second order admissibility with applications to Berkson's bioassay problem. *Ann. Statist.*, **9**, 1334.

GHOSH, J. K. and SUBRAMANYAM, K. (1974). Second order efficiency of maximum likelihood estimators. *Sankhyā*, **A, 36**, 325.

GHOSH, M. (1975). Admissibility and minimaxity of some maximum likelihood estimators when the parameter space is restricted to integers. *J. R. Statist. Soc.*, **B**, **37**, 264.

GHOSH, M. N. and SHARMA, D. (1963). Power of Tukey's test for non-additivity. *J. R. Statist. Soc.*, **B**, **25**, 213.

GIBBONS, D. G. (1981). A simulation study of some ridge estimators. *J. Amer. Statist. Ass.*, **76**, 131.

GIBBONS, J. D., OLKIN, I. and SOBEL, M. (1977). *Selecting and Ordering Populations: A New Statistical Methodology.* Wiley, New York.

GIERE, R. N. (1977). Allan Birnbaum's conception of statistical evidence. *Synthèse*, **36**, 5.

GIESBRECHT, F. and KEMPTHORNE, O. (1976). Maximum likelihood estimation in the three-parameter lognormal distribution. *J. R. Statist. Soc.*, **B**, **38**, 257.

GIRSHICK, M. A. (1946). Contributions to the theory of sequential analysis. *Ann. Math. Statist.*, **17**, 123 and 282.

GJEDDEBAEK, N. F. (1949-61). Contribution to the study of grouped observations. I-VI. *Skan. Aktuartidskr.*, **32**, 135; **39**, 154; **40**, 20; *Biometrics*, **15**, 433; *Skand. Aktuartidskr.*, **42**, 194; **44**, 55.

GLASER, R. R. (1976). Exact critical values for Bartlett's test for homogeneity of variance. *J. Amer. Statist. Ass.*, **71**, 488.

GLASER, R. E. (1980). A characterization of Bartlett's statistic involving incomplete beta functions. *Biometrika*, **67**, 53.

GLESER, L. J. (1985). Exact power of goodness-of-fit tests of Kolmogorov type for discontinuous distributions. *J. Amer. Statist. Ass.*, **80**, 954.

GLESER, L. J. and MOORE, D. S. (1983). The effect of dependence on chi-squared and empiric distribution tests of fit. *Ann. Statist.*, **11**, 1100.

GODAMBE, V. P. (1960). An optimum property of regular maximum likelihood estimation. *Ann. Math. Statist.*, **31**, 1208.

GODAMBE, V. P. (1976). Conditional likelihood and unconditional estimating equations. *Biometrika*, **63**, 277.

GODAMBE, V. P. (1982). Ancillarity principle and a statistical paradox. *J. Amer. Statist. Ass.*, **77**, 931.

GODAMBE, V. P. and SPROTT, D. A., eds. (1971). *Foundations of Statistical Inference.* Holt, Rinehart and Winston, Toronto, Canada.

GOLDBERGER, A. S. (1961). Stepwise least squares: residual analysis and specification error. *J. Amer. Statist. Ass.*, **56**, 998.

GOLDBERGER, A. S. and JOCHEMS, D. B. (1961). Note on stepwise least squares. *J. Amer. Statist. Ass.*, **56**, 105.

GOLDMAN, A. S. and ZEIGLER, R. K. (1966). Comparisons of some two stage sampling methods. *Ann. Math. Statist.*, **37**, 891.

GOLUB, G. H. and NASH, S. G. (1982). Non-orthogonal analysis of variance using a generalized conjugate-gradient algorithm. *J. Amer. Statist. Ass.*, **77**, 109.

GOOD, I. J. (1950). *Probability and the Weighing of Evidence.* Griffin, London.

GOOD, I. J. (1976). The Bayesian influence, or how to sweep subjectivism under the carpet. In *Foundations of Probability Theory, Statistical Inference and Statistical Theories of Science*, Vol. 2. C. A. Hooker and W. Harper (eds.). Reidel, Dordrecht, Holland, 125.

GOOD, I. J. (1983). *Good Thinking: The Foundations of Probability and its Applications.* U. Minnesota Press, Minneapolis.

GOOD, I. J. (1988). The interface between statistics and philosophy of science. *Statist. Sci.*, **3**, 386.

GOOD, I. J., GOVER, T. N. and MITCHELL, G. J. (1970). Exact distributions for X^2 and for the likelihood-ratio statistic for the equiprobable multinomial distribution. *J. Amer. Statist. Ass.*, **65**, 267.

GOODMAN, L. A. and MADANSKY, A. (1962). Parameter-free and nonparametric tolerance limits: the exponential case. *Technometrics*, **4**, 75.

GRANGER, C. W. J. (1969). Investigating causal relations by econometric models and cross-spectral methods. *Econometrica*, **37**, 424.

GRAY, H. L. and SCHUCANY, W. R. (1972). *The Generalized Jackknife Statistic.* Dekker, New York.

GRAYBILL, F. A. and BOWDEN, D. C. (1967). Linear segment confidence bands for simple linear models. *J. Amer. Statist. Ass.,* **62,** 403.

GRAYBILL, F. A. and CONNELL, T. L. (1964). Sample size required for estimating the variance within d units of the true value. *Ann. Math. Statist.,* **35,** 438.

GRAYBILL, F. A. and MARSAGLIA, G. (1957). Idempotent matrices and quadratic forms in the general linear hypothesis. *Ann. Math. Statist.,* **28,** 678.

GRAYBILL, F. A. and MILLIKEN, G. (1969). Quadratic forms and idempotent matrices with random elements. *Ann. Math. Statist.,* **40,** 1430.

GREEN, P. J. (1984). Iteratively reweighted least squares for maximum likelihood estimation, and some robust and resistant alternatives (with discussion). *J. R. Statist. Soc.,* B, **46,** 149.

GREENBERG, E. (1975). Minimum variance properties of principal component regression. *J. Amer. Statist. Ass.,* **70,** 194.

GRICE, J. V. and BAIN, L. J. (1980). Inference concerning the mean of the Gamma distribution. *J. Amer. Statist. Ass.,* **75,** 929.

GRILICHES, Z. (1967). Distributed lags: a survey. *Econometrica,* **35,** 16.

GROENEBOOM, P. and OOSTERHOFF, J. (1981). Bahadur efficiency and small-sample efficiency. *Int. Statist. Rev.,* **49,** 127.

GUENTHER, W. C. (1964). Another derivation of the non-central chi-square distribution. *J. Amer. Statist. Ass.,* **59,** 957.

GUENTHER, W. C. and WHITCOMB, M. G. (1966). Critical regions for tests of interval hypotheses about the variance. *J. Amer. Statist. Ass.,* **61,** 204.

GUEST, P. G. (1954). Grouping methods in the fitting of polynomials to equally spaced observations. *Biometrika,* **41,** 62.

GUEST, P. G. (1956). Grouping methods in the fitting of polynomials to unequally spaced observations. *Biometrika,* **43,** 149.

GUILKEY, D. K. and MURPHY, J. L. (1975). Directed ridge regression techniques in cases of multicollinearity. *J. Amer. Statist. Ass.,* **70,** 769.

GUMBEL, E. J. (1943). On the reliability of the classical chi-square test. *Ann. Math. Statist.,* **14,** 253.

GUNST, R. F. (1983). Regression analysis with multocollinear predictor variables: definition, detection and effects. *Commun. Statist.,* A, **12,** 2217.

GUNST, R. F. and MASON, R. L. (1977). Biased estimation in regression: an evaluation using mean squared error. *J. Amer. Statist. Ass.,* **72,** 616.

GURLAND, J. (1968). A relatively simple form of the multiple correlation coefficient. *J. R. Statist. Soc.,* B, **30,** 276.

GURLAND, J. and MILTON, R. (1970). Further consideration of the distribution of the multiple correlation coefficient. *J. R. Statist. Soc.,* B, **32,** 381.

GUTTMAN, I. (1957). On the power of optimum tolerance regions when sampling from normal distributions. *Ann. Math. Statist.,* **28,** 773.

GUTTMAN, I. (1970). *Statistical Tolerance Regions: Classical and Bayesian.* Griffin, London.

HAAS, G., BAIN, L. and ANTLE, C. (1970). Inferences for the Cauchy distribution based on maximum likelihood estimators. *Biometrika,* **57,** 403.

HABERMAN, S. J. (1975). How much do Gauss–Markov and least squares estimators differ? A coordinate-free approach. *Ann. Statist.,* **3,** 982.

HABERMAN, S. J. (1988). A warning on the use of chi-squared statistics with frequency tables with small expected cell counts. *J. Amer. Statist. Ass.,* **83,** 555.

HAJNAL, J. (1961). A teo-sample sequential t-test. *Biometrika,* **48,** 65.

HALDANE, J. B. S. (1938). The approximate normalization of a class of frequency distributions. *Biometrika,* **29,** 392.

HALDANE, J. B. S. (1955). Substitutes for χ^2. *Biometrika,* **42,** 265.

HALDANE, J. B. S. and SMITH, S. M. (1956). The sampling distribution of a maximum-likelihood estimate. *Biometrika,* **43,** 96.

HALL, P. (1981). Asymptotic theory of triple sampling for sequential estimation of the mean. *Ann. Statist.*, **9**, 1229.

HALL, P. (1982). Improving the normal approximation when constructing one-sided confidence intervals for binomial or Poisson parameters. *Biometrika*, **69**, 647.

HALL, P. (1986a). On the bootstrap and confidence intervals. *Ann. Statist.*, **14**, 143.

HALL, P. (1986b). On the number of bootstrap simulations required to construct a confidence interval. *Ann. Statist.*, **14**, 1453.

HALL, P. (1987). On the bootstrap and continuity correction. *J. R. Statist. Soc.*, B, **49**, 82.

HALL, P. (1988a). On symmetric bootstrap confidence intervals. *J. R. Statist. Soc.*, B, **50**, 35.

HALL, P. (1988b). Theoretical comparison of bootstrap confidence intervals. *Ann. Statist.*, **16**, 927.

HALL, P. and MARTIN, M. A. (1988). On bootstrapping resampling and iteration. *Biometrika*, **75**, 661.

HALL, W. J., WIJSMAN, R. A. and GHOSH, J. K. (1965). The relationship between sufficiency and invariance with applications in sequential analysis. *Ann. Math. Statist.*, **36**, 575.

HALMOS, P. R. and SAVAGE, L. J. (1949). Application of the Radon-Nikodym theorem to the theory of sufficient statistics. *Ann. Math. Statist.*, **20**, 225.

HALPERIN, M. (1963). Confidence interval estimation in non-linear regression. *J. R. Statist. Soc.*, B, **25**, 330.

HALPERIN, M. (1964). Interval estimation of non-linear parametric functions, II. *J. Amer. Statist. Ass.*, **59**, 168.

HALPERIN, M. and GURIAN, J. (1968). Confidence bands in linear regression with constraints on the independent variables. *J. Amer. Statist. Ass.*, **63**, 1020.

HALPERIN, M. and MANTEL, N. (1963). Interval estimation of non-linear parametric functions. *J. Amer. Statist. Ass.*, **58**, 611.

HALPERIN, M., RASTOGI, S. C., HO, I. and YANG, Y. Y. (1967). Shorter confidence bands in linear regression. *J. Amer. Statist. Ass.*, **62**, 1050.

HAMDAN, M. A. (1963). The number and width of classes in the chi-square test. *J. Amer. Statist. Ass.*, **58**, 678.

HAMDAN, M. A. (1968). Optimum choice of classes for contingency tables. *J. Amer. Statist. Ass.*, **63**, 291.

HAMDAN, M. A. (1970). The equivalence of tetrachoric and maximum likelihood estimates of ρ in 2×2 tables. *Biometrika*, **57**, 212.

HAMILTON, D. and WIENS, D. (1987). Correction factors for F-ratios in nonlinear regression. *Biometrika*, **74**, 423.

HAMILTON, D. C., WATTS. D. G. and BATES, D. M. (1982). Accounting for intrinsic nonlinearity in nonlinear regression parameter inference regions. *Ann. Statist.*, **10**, 386.

HAMMERSLEY, J. M. (1950). On estimating restricted parameters. *J. R. Statist. Soc.*, B, **12**, 192.

HAN, C. P. (1975). Some relationships between noncentral chi-squared and normal distributions. *Biometrika*, **62**, 213.

HANNAN, E. J. (1956). The asymptotic power of certain tests based on multiple correlations. *J. R. Statist. Soc.*, B, **18**, 227.

HANNAN, E. J. and QUINN, B. G. (1979). The determination of the order of an autoregression. *J. Roy. Statist. Soc.*, B, **41**, 190.

HANNAN, J. F. and TATE, R. F. (1965). Estimation of the parameters for a multivariate normal distribution when one variable is dichotomized. *Biometrika*, **52**, 664.

HARLEY, B. I. (1956). Some properties of an angular transformation for the correlation coefficient. *Biometrika*, **43**, 219.

HARLEY, B. I. (1957). Further properties of an angular transformation of the correlation coefficient. *Biometrika*, **44**, 273.

HARRIS, P. (1985). An asymptotic expansion for the null distribution of the efficient score static. *Biometrika*, **72**, 653.

HARRIS, P. (1986). A note on Bartlett adjustments to likelihood ratio tests. *Biometrika*, **73**, 735.

HARRIS, P. and PEERS, H. W. (1980). The local power of the efficient score test statistic. *Biometrika*, **67**, 525.

HARSAAE, E. (1969). On the computation and use of a table of percentage points of Bartlett's *M*. *Biometrika*, **56**, 273.

HARTER, H. L. (1963). Percentage points of the ratio of two ranges and power of the associated test. *Biometrika*, **50**, 187.

HARTER, H. L. (1964). Criteria for best substitute interval estimators, with an application to the normal distribution. *J. Amer. Statist. Ass.*, **59**, 1133.

HARTER, H. L. and OWEN, D. B. (Editors) (1970, 1974, 1975). *Selected Tables in Mathematical Statistics*, Vols I, II and III. American Mathematical Society, Providence, R.I., U.S.A.

HARTLEY, H. O. (1964). Exact confidence regions for the parameters in non-linear regression laws. *Biometrika*, **51**, 347.

HARTLEY, H. O. and BOOKER, A. (1965). Nonlinear least squares estimation. *Ann. Math. Statist.*, **36**, 638.

HARVEY, A. C. and PHILLIPS, G. D. A. (1979). Maximum likelihood estimation of regression models with autoregressive-moving average disturbances. *Biometrika*, **66**, 49.

HAWKINS, D. M. (1975). From the noncentral *t* to the normal integral. *Amer. Statistician*, **29**, 42.

HAYAKAWA, T. (1975). The likelihood ratio criterion for a composite hypothesis under a local alternative. *Biometrika*, **62**, 451.

HAYAKAWA, T. (1977). The likelihood ratio criterion and the asymptotic expansion of its distribution. *Ann. Inst. Statist. Math. Tokyo*, **29, A**, 359.

HAYTER, A. J. (1984). A proof of the conjecture that the Tukey–Kramer multiple comparisons procedure is conservative. *Ann. Statist.*, **12**, 61.

HEALY, M. J. R. (1955). A significance test for the difference in efficiency between two predictors. *J. R. Statist. Soc.*, **B, 17**, 266.

HEALY, M. J. R. and TAYLOR, L. R. (1962). Tables for power-law transformations. *Biometrika*, **49**, 557.

HEDAYAT, A. and ROBSON, D. S. (1970). Independent stepwise residuals for testing homoscedasticity. *J. Amer. Statist. Ass.*, **65**, 1573.

HEGEMANN, V. and JOHNSON, D. E. (1976). The power of two tests for nonadditivity. *J. Amer. Statist. Ass.*, **71**, 945.

HEITMANN, G. J. and ORD, J. K. (1985). An interpretation of the least squares regression surface. *Amer. Statistician*, **39**, 120.

HEMMERLE, W. J. (1974). Nonorthogonal analysis of covariance using iterative improvement and balanced residuals. *J. Amer. Statist. Ass.*, **69**, 772.

HEMMERLE, W. J. (1976). Iterative nonorthogonal analysis of variance. *J. Amer. Statist. Ass.*, **71**, 195.

HEMMERLE, W. J. (1982). Analysis of covariance algorithms. *Biometrics*, **38**, 725.

HERNANDEZ, F. and JOHNSON, R. A. (1980). The large-sample behaviour of transformations to normality. *J. Amer. Statist. Ass.*, **75**, 855.

HILL, B. M. (1963a). The three-parameter lognormal distribution and Bayesian analysis of a point-source epidemic. *J. Amer. Statist. Ass.*, **58**, 72.

HILL, B. M. (1963b). Information for estimating the proportions in mixtures of exponential and normal distributions. *J. Amer. Statist. Ass.*, **58**, 918.

HINKLEY, D. V. (1975). On power transformations to symmetry. *Biometrika*, **62**, 101.

HINKLEY, D. V. (1980). Likelihood as approximate, pivotal distribution. *Biometrika*, **67**, 287.

HINKLEY, D. V. (1988). Bootstrap methods. *J. R. Statist. Soc.*, **B, 50**, 321.

HINKLEY, D. V. and WANG, S. (1988). More about transformations and influential cases in regression. *Technometrics*, **30**, 435.

HOAGLIN, D. C. (1975). The small-sample variance of the Pitman location estimators. *J. Amer. Statist. Ass.*, **70**, 880.

HOCHBERG, Y. (1975). An extension of the T-method to general unbalanced models of fixed effects. *J. R. Statist. Soc.*, **B, 37**, 426.

HOCHBERG, Y. (1976). A modification of the T-method of multiple comparisons for a one-way layout with unequal variances. *J. Amer. Statist. Ass.*, **71**, 200.

HOCHBERG, Y. and QUADE, D. (1975). One-sided simultaneous confidence bounds on regression surfaces with intercepts. *J. Amer. Statist. Ass.*, **70**, 889.

HOCHBERG, Y. and TAMHANE, A. C. (1987). *Multiple Comparison Procedures.* Wiley, New York.

HOCHBERG, Y., WEISS, G. and HART, S. (1982). On graphical procedures for multiple comparisons. *J. Amer. Statist. Assoc.*, **77**, 767.

HODGES, J. L., Jr. and LEHMANN, E. L. (1956). The efficiency of some nonparametric competitors of the t-test. *Ann. Math. Statist.*, **27**, 324.

HODGES, J. L., Jr. and LEHMANN, E. L. (1967). Moments of chi and power of t. *Proc. 5th Berkeley Symp. Math. Statist. and Prob.*, **1**, 187.

HODGES, J. L., Jr. and LEHMANN, E. L. (1968). A compact table for power of the t-test. *Ann. Math. Statist.*, **39**, 1629.

HODGES, J. L., Jr. and LEHMANN, E. L. (1970). Deficiency. *Ann. Math. Statist.*, **41**, 783.

HOEFFDING, W. (1940). Maszstabinvariante Korrelationstheorie. *Schr. Math.*, *Univ. Berlin*, **5**, (3), 181.

HOEFFDING, W. (1952). The large sample power of tests based on permutations of observations. *Ann. Math. Statist.*, **23**, 169.

HOEL, D. G. and MAZUMDAR, M. (1969). A class of sequential tests for an exponential parameter. *J. Amer. Statist. Ass.*, **64**, 1549.

HOEL, P. G. (1951). Confidence regions for linear regression. *Proc. 2nd Berkeley Symp. Math. Statist. and Prob.*, 75.

HOERL, A. E. and KENNARD, R. W. (1970). Ridge regression: biased estimation for non-orthogonal problems *and Applications*. *Technometrics*, **12**, 55, 69.

HOERL, R. W., SCHUENMEYER, J. H. and HOERL, A. E. (1986). A simulation of biased estimation and subset selection regression techniques. *Technometrics*, **28**, 369.

HOGBEN, D., PINKHAM, R. S. and WILK, M. B. (1961). The moments of the non-central t-distribution. *Biometrika*, **48**, 465.

HOGG, R. V. (1956). On the distribution of the likelihood ratio. *Ann. Math. Statist.*, **27**, 529.

HOGG, R. V. (1961). On the resolution of statistical hypotheses. *J. Amer. Statist. Ass.*, **56**, 978.

HOGG, R. V. (1962). Iterated tests of the equality of several distributions. *J. Amer. Statist. Ass.*, **57**, 579.

HOGG, R. V. and CRAIG, A. T. (1956). Sufficient statistics in elementary distribution theory. *Sankhyā*, **17**, 209.

HOGG, R. V. and TANIS, E. A. (1963). An iterated procedure for testing the equality of several exponential distributions. *J. Amer. Statist. Ass.*, **58**, 435.

HOLLAND, P. W. (1986). Statistics and causal inference (with discussion). *J. Amer. Statist. Ass.*, **81**, 945.

HOLM, S. (1977). On a conjecture about the limiting minimal efficiency of sequential tests. *Ann. Statist.*, **5**, 375.

HORA, R. B. and BUEHLER, R. J. (1966). Fiducial theory and invariant estimation. *Ann. Math. Statist.*, **37**, 643.

HORA, R. B. and BUEHLER, R. J. (1967). Fiducial theory and invariant prediction. *Ann. Math. Statist.*, **38**, 795.

HORN, S. D., HORN, R. A. and DUNCAN, D. B. (1975). Estimating heteroscedastic variances in linear models. *J. Amer. Statist. Ass.*, **70**, 380.

HOTELLING, H. (1940). The selection of variates for use in prediction, with some comments on the general problem of nuisance parameters. *Ann. Math. Statist.*, **11**, 271.

HOTELLING, H. (1953). New light on the correlation coefficient and its transforms. *J. R. Statist. Soc.*, **B, 15**, 193.

HOUGAARD, P. (1985). The appropriateness of the asymptotic distribution in a nonlinear regression model in relation to curvature. *J. Roy. Statist. Soc.*, B, **47**, 103.

HOUGAARD, P. (1988). The asymptotic distribution of nonlinear regression parameter estimates: Improving the approximation. *Int. Statist. Rev.*, **56**, 221.

HOWE, W. G. (1969). Two-sided tolerance limits for normal populations—some improvements. *J. Amer. Statist. Ass.*, **64**, 610.

HOWSON, C. and URBACH, P. (1989). *Scientific Reasoning: The Bayesian Approach*. Open Court, La Salle, Illinois.

HOYLE, M. H. (1968). The estimation of variances after using a Gaussianating transformation. *Ann. Math. Statist.*, **39**, 1125.

HOYLE, M. H. (1971). Spoilt data—an introduction and bibliography. *J. R. Statist. Soc.*, A, **134**, 429.

HOYLE, M. H. (1973). Transformations—an introduction and a bibliography. *Int. Statist. Rev.*, **41**, 203.

HSU, P. L. (1941). Analysis of variance from the power function standpoint. *Biometrika*, **32**, 62.

HUBER, P. (1964). Robust estimation of a loction parameter. *Ann. Math. Statist.*, **35**, 73.

HUBER, P. J. (1967). The behaviour of maximum likelihood estimates under nonstandard conditions. *Proc. 5th Berkeley Symp. Math. Statist. and Prob.*, **1**, 221.

HUBER, P. J. (1972). Robust statistics: a review. *Ann. Math. Statist.*, **43**, 1041.

HUZURBAZAR, V. S. (1948). The likelihood equation, consistency and the maxima of the likelihood function. *Ann. Eugen.*, **14**, 185.

HUZURBAZAR, V. S. (1949). On a property of distributions admitting sufficient statistics. *Biometrika*, **36**, 71.

HUZURBAZAR, V. S. (1955). Confidence intervals for the parameter of a distribution admitting a sufficient statistic when the range depends on the parameter. *J. R. Statist. Soc.*, B, **17**, 86.

IMAN, R. L. (1982). Graphs for use with the Lilliefors test for normal and exponential distributions. *Amer. Statistician*, **36**, 109.

IWASE, K. and SETO, N. (1983). Uniformly minimum variance unbiased estimation for the inverse Gaussian distribution. *J. Amer. Statist. Ass.*, **78**, 660.

JACKSON, J. E. (1960). Bibliography on sequential analysis. *J. Amer. Statist. Ass.*, **55**, 516.

JAMES, W. and STEIN, C. (1961). Estimation with quadratic loss. *Fourth Berkeley Symposium Math. Statist. and Prob.*, **1**, 361, U. of California Press, Berkeley.

JAMSHIDIAN, M. and JENNRICH, R. I. (1988). Nonorthogonal analysis of variance using gradient methods. *J. Amer. Statist. Ass.*, **83**, 483.

JARRETT, R. G. (1984). Bounds and expansions for Fisher information when the moments are known. *Biometrika*, **71**, 101.

JAYNES, E. T. (1968). Prior probabilities. *IEEE Transactions on Systems and Cybernetics*, SSC-4, 227.

JEFFREYS, H. (1961). *Theory of Probability*. 3rd ed, Oxford Univ. Press.

JENSEN, D. R. and SOLOMON, H. (1972). A Gaussian approximation to the distribution of a definite quadratic form. *J. Amer. Statist. Sss.*, **67**, 898.

JENSEN, J. L. (1986). Similar tests and the standardized log likelihood ratio statistic. *Biometrika*, **73**, 567.

JOANES, D. N. (1972). Sequential tests of composite hypotheses. *Biometrika*, **59**, 633.

JOBSON, J. D. and FULLER, W. A. (1980). Least squares estimation when the covariance matrix and parameter vector are functionally related. *J. Amer. Statist. Ass.*, **75**, 176.

JOHN, S. (1975). Tables for comparing two normal variances or two Gamma means. *J. Amer. Statist. Ass.*, **70**, 344.

JOHNS, M. V., Jr. (1974). Nonparametric estimation of location. *J. Amer. Statist. Ass.*, **69**, 453.

JOHNSON, D. E. and GRAYBILL, F. A. (1972a, b). The estimation of σ^2 in a two-way classification model with interaction *and* An analysis of a two-way model with interaction and no replication. *J. Amer. Statist. Ass.*, **67**, 388 and 862.

JOHNSON, N. L. (1950). On the comparison of estimators. *Biometrika*, **37**, 281.

JOHNSON, N. L. (1959a). On an extension of the connexion between Poisson and χ^2 distributions. *Biometrika*, **46**, 352.

JOHNSON, N. L. (1959b). A proof of Wald's theorem on cumulative sums. *Ann. Math. Statist.*, **30**, 1245.

JOHNSON, N. L. (1961). Sequential analysis: a survey. *J. R. Statist. Soc.*, A, **124**, 372.

JOHNSON, N. L. and PEARSON, E. S. (1969). Tables of percentage points of non-central χ. *Biometrika*, **56**, 255.

JOHNSON, N. L. and WELCH, B. L. (1939). Applications of the non-central t distribution. *Biometrika*, **31**, 362.

JOHNSTONE, D. J. (1989). On the necessity for random sampling. *Brit. J. Phil. Sci.*, **40**, 443.

JOINER, B. L. (1969). The median significance level and other small sample measures of test efficacy. *J. Amer. Statist. Ass.*, **64**, 971.

JORESKÖG, K. G. (1981). Analysis of covariance structures (with discussion). *Scand. J. Statist.*, **8**, 65.

JOSHI, V. M. (1976). On the attainment of the Cramér–Rao lower bound. *Ann. Statist.*, **4**, 998.

JOSHI, V. M. (1989). A counter-example against the likelihood principle. *J. Roy. Statist. Soc.*, B, **51**, 215.

KAC, M., KEIFER, J. and WOLFOWITZ, J. (1955). On tests of normality and other tests of goodness of fit based on distance methods. *Ann. Math. Statist.*, **26**, 189.

KADIYALA, K. R. and OBERHELMAN, H. D. (1984). Alternative tests for heteroscadascticity of disturbances: A comparative study. *Commun. Statist.*, A, **13**, 987.

KAGAN, A. M., LINNIK, Yu. V. and RAO, C. R. (1973). *Characterization Problems in Mathematical Statistics*. Wiley, New York.

KAILATH, T. (1974). A view of three decades of linear filtering theory. *IEEE Trans. Info. Theory*, IT-20, 145.

KALBFLEISCH, J. D. (1975). Sufficiency and conditionality. (With discussion by G. A. Barnard, O. Barndorff-Nielsen, A. Birnbaum and A. D. Maclaren, and author's reply.) *Biometrika*, **62**, 251.

KALE, B. K. (1961). On the solution of the likelihood equation by iteration processes. *Biometrika*, **48**, 452.

KALE, B. K. (1962). On the solution of likelihood equations by iteration processes. The multiparametric case. *Biometrika*, **49**, 479.

KALLENBERG, W. C. M. (1983). Intermediate efficienty theory and examples. *Ann. Statist.*, **11**, 170.

KALLENBERG, W. C. M. (1985). On moderate and large deviations in multinomial distributions. *Ann. Statist.*, **13**, 1554.

KALLENBERG, W. C. M. and LEDWINA, T. (1987). On local and nonlocal measures of efficiency. *Ann. Statist.*, **15**, 1401.

KALLENBERG, W. C. M., OOSTERHOFF, J. and SCHRIEVER, B. F. (1985). The number of classes in chi-squared goodness-of-fit tests. *J. Amer. Statist. Ass.*, **80**, 969.

KAMAT, A. R. and SATKE, Y. S. (1962). Asymptotic power of certain tests criteria (based on first and second differences) for serial correlation between successive differences. *Ann. Math. Statist.*, **33**, 186.

KANOH, S. (1988). The reduction of the width of confidence bands in linear regression. *J. Amer. Statist. Ass.*, **83**, 116.

KARAKOSTAS, K. X. (1985). On minimum variance unbiased estimates. *Amer. Statistician*, **39**, 303.

KARIYA, T. (1980). Note on a condition for equality of sample variances in a linear model. *J. Amer. Statist. Ass.*, **75**, 701.

KASTENBAUM, M. A., HOEL, D. G. and BOWMAN, K. O. (1970a). Sample size requirements: one-way analysis of variance. *Biometrika*, **57**, 421.

KASTENBAUM, M. A., HOEL, D. G. and BOWMAN, K. O. (1970b). Sample size requirements: randomized block designs. *Biometrika*, **57**, 573.

KATTI, S. K. and GURLAND, J. (1962). Efficiency of certain methods of estimation for the negative binomial and the Neyman type A distributions. *Biometrika,* **49,** 215.

KEMPERMAN, J. H. B. (1956). Generalized tolerance limits. *Ann. Math. Statist.,* **27,** 180.

KEMPTHORNE, O. (1952). *The Design and Analysis of Experiments.* Wiley, New York.

KEMPTHORNE, O. (1967). The classical problem of inference—goodness of fit. *Proc. 5th Berkeley Symp. Math. Statist. and Prob.,* **1,** 235.

KENDALL, M. G. (1949). On the reconciliation of theories of probability. *Biometrika,* **36,** 101.

KENDALL, M. G. and GIBBONS, J. D. (1990). *Rank Correlation Methods* (fifth edition). Edward Arnold, London and Oxford U.P., New York.

KENDALL, M. G. and ORD, J. K. (1990). *Time Series* (3rd edition). Edward Arnold, London.

KENT, J. T. (1982). Robust properties of likelihood tests. *Biometrika,* **69,** 19.

KENT, J. T. (1983). Information gain and a general measure of correlation. *Biometrika,* **70,** 163.

KENT, J. T. (1986). The underlying structure of non-nested hypothesis tests. *Biometrika,* **73,** 333.

KESELMAN, H. J. and ROGAN, J. C. (1978). A comparison of the modified-Tukey and Scheffé methods of multiple comparisons for pairwise contrasts. *J. Amer. Statist. Ass.,* **73,** 47.

KESELMAN, H. J., GAMES, P. A. and ROGAN, J. C. (1979). An addendum to Keselman and Rogan (1978). *J. Amer. Statist. Ass.,* **74,** 626.

KETTENRING, J. R. (1971). Canonical analysis of several sets of variables. *Biometrika,* **58,** 433.

KEYNES, J. M. (1911). The principal averages and the laws of error which lead to them. *J. R. Statist. Soc.,* **74,** 322.

KEYNES, J. M. (1921). *A Treatise on Probability.* Macmillan, London.

KHAN, R. A. (1973). On some properties of Hammersley's estimator of an integer mean. *Ann. Statist.,* **1,** 756.

KHATRI, C. G. (1978). A remark on the necessary and sufficient conditions for a quadratic form to be distributed as chi-squared. *Biometrika,* **65,** 239.

KIEFER, J. (1952). On minimum variance estimators. *Ann. Math. Statist.,* **23,** 627.

KIEFER, J. (1977). Conditional confidence statements and confidence estimators. *J. Amer. Statist. Ass.,* **72,** 789.

KIEFER, J. and WEISS, L. (1957). Some properties of generalized sequential probability ratio tests. *Ann. Math. Statist.,* **28,** 57.

KIEFER, J. and WOLFOWITZ, J. (1956). Consistency of the maximum likelihood estimator in the presence of infinitely many incidental parameters. *Ann. Math. Statist.,* **27,** 887.

KIMBALL, B. F. (1946). Sufficient statistical estimation functions for the parameters of the distribution of maximum values. *Ann. Math. Statist.,* **17,** 299.

KNAFL, G., SACKS, J. and YLVISAKER, D. (1985). Confidence bands for regression functions. *J. Amer. Statist. Ass.,* **80,** 683.

KNIGHT, W. (1965). A method of sequential estimation applicable to the hypergeometric, binomial, Poisson, and exponential distributions. *Ann. Math. Statist.,* **36,** 1494.

KNOTT, M. (1975). On the minimum efficiency of least squares. *Biometrika,* **62,** 129.

KOEHLER, K. J. and LARNTZ, K. (1980). An empirical investigation of goodness-of-fit statistics for sparse multinomials. *J. Amer. Statist. Ass.,* **75,** 336.

KOENKER, R. and BASSETT, G. (1978). Robust tests for heteroscedasticity based on regression quantiles. *Econometrica,* **50,** 43.

KÖLLERSTRÖM, J. and WETHERILL, G. B. (1979). SPRT's for the normal correlation coefficient. *J. Amer. Statist. Ass.,* **74,** 815.

KOLMOGOROV, A. (1933). Sulla determinazione empirica di una legge di sitribuzione. *G. Ist. Ital. Attuari,* **4,** 83.

KOLODZIEJCZYK, S. (1935). On an important class of statistical hypotheses. *Biometrika,* **27,** 161.

KONIJN, H. S. (1956, 1958). On the power of certain tests for independence in bivariate populations. *Ann. Math. Statist.,* **27,** 300 and **29,** 935.

KOOPMAN, B. O. (1936). On distributions admitting a sufficient statistic. *Trans. Amer. Math. Soc.,* **39,** 399.

KOOPMAN, R. F. (1983). On the standard error of the modified biserial correlation. *Psychometrika*, **48**, 639.

KORN, E. L. (1984). The range of limiting values of some partial correlations under conditional independence. *Amer. Statistician*, **38**, 61.

KOSCHAT, M. A. (1987). A characterization of the Fieller solution. *Ann. Statist.*, **15**, 462.

KOUROUKLIS, S. and PAIGE, C. C. (1981). A constrained least squares approach to the general Gauss–Markov linear model. *J. Amer. Statist. Ass.*, **76**, 620.

KOZIOL, J. A. (1978). Exact slopes of certain multivariate tests of hypotheses. *Ann. Statist.*, **6**, 546.

KOZIOL, J. A. (1980). On a Cramer–Von Miseses-type statistic for testing symmetry. *J. Amer. Statist. Ass.*, **75**, 161.

KRAEMER, H. C. (1981). Modified biserial correlation coefficients. *Psychometrika*, **46**, 275.

KRAFT, C. H. and LECAM, L. M. (1956). A remark on the roots of the maximum likelihood equation. *Ann. Math. Statist.*, **27**, 1174.

KRAMER, K. C. (1963). Tables for constructing confidence limits on the multiple correlation coefficient. *J. Amer. Statist. Ass.*, **58**, 1082.

KRAMER, W. (1980). Finite sample efficiency of ordinary least squares in the linear regression model with autocorrelated errors. *J. Amer. Statist. Ass.*, **75**, 1005.

KREUGER, R. G. and NEUDECKER, H. (1977). Exact linear restrictions on parameters in the general linear model with a singular covariance matrix. *J. Amer. Statist. Ass.*, **72**, 430.

KRISHNAN, M. (1966). Locally unbiased type M test. *J. R. Statist. Soc.*, B, **28**, 298.

KRISHNAN, M. (1967). The moments of a doubly non-central *t* distribution. *J. Amer. Statistl. Ass.*, **62**, 278.

KRISHNAN, M. (1968). Series representations of the doubly noncentral *t*-distribution. *J. Amer. Statist. Ass.*, **63**, 1004.

KRUSKAL, J. B. (1965). Analysis of factorial experiments by estimating monotone transformations of the data. *J. R. Statist. Soc.*, B, **27**, 251.

KRUSKAL, W. H. (1958). Ordinal measures of association. *J. Amer. Statist. Ass.*, **53**, 814.

KRUSKAL, W. (1961). The coordinate-free approach to Gauss–Markov estimation, and its application to missing and extra observations. *Proc. 4th Berkeley Symp. Math. Statist. and Prob.*, **1**, 435.

KRUTCHKOFF, R. G. (1967). Classical and inverse regression methods of calibration. *Technometrics*, **9**, 425.

KUHN, T. S. (1970). *The Structure of Scientific Revolutions*. Univ. of Chicago Press, Chicago (second edition).

KULLDORF, G. (1958). Maximum likelihood estimation of the mean/standard deviation of a normal random variable when the sample is grouped. *Skand. Aktuartidskr.*, **41**, 1 and 18.

KULLDORF, G. (1961). *Contributions to the Theory of Estimation from Grouped and Partially Grouped Samples*. Almqvist and Wiksell, Stockholm.

LAI, T. L. (1975). Termination moments and exponential boundedness of the stopping rule for certain invariant sequential probability ratio tests. *Ann. Statist.*, **3**, 581.

LAI, T. L. (1978). Pitman efficiencies of sequential tests and uniform limit theorems in non-parametric statistics. *Ann. Statist.*, **6**, 1027.

LAIRD, N., LANGE, N. and STRAM, D. (1987). Maximum likelihood computations with repeated measures: Applications of the EM algorithm. *J. Amer. Statist. Ass.*, **82**, 97.

LAKATOS, I. (1974). Falsification and the methodology of scientific reasearch programs. In *Criticism and the Growth of Knowledge*, I. Lakatos and A. E. Musgrave (eds.). Cambridge Univ. Press, Cambridge, 91.

LAMBERT, D. and HALL, W. J. (1982). Asymptotic lognormality of P-values. *Ann. Statist.*, **10**, 44.

LAND, C. E. (1971). Confidence intervals for linear functions of the normal mean and variance. *Ann. Math. Statist.*, **42**, 1187.

LAND, C. E. (1974). Confidence interval estimation for means after data transformations to normality. *J. Amer. Statist. Ass.*, **69**, 795.

LARNTZ, K. (1978). Small-sample comparisons of exact levels for chi-squared goodness-of-fit statistics. *J. Amer. Statist. Ass.*, **73**, 253.

LAUBSCHER, N. F. (1961). On stabilizing the binomial and negative binomial variances. *J. Amer. Statist. Ass.*, **56**, 143.

LAWAL, H. B. and UPTON, G. J. G. (1980). An approximation to the distribution of the X^2 goodness-of-fit statistic for use with small expectations. *Biometrika*, **67**, 447.

LAWLESS, J. F. (1981). Mean squared error properties of generalized ridge estimators. *J. Amer. Statist. Ass.*, **76**, 462.

LAWLEY, D. N. (1956). A general method for approximating to the distribution of likelihood ratio criteria. *Biometrika*, **43**, 295.

LAWRANCE, A. J. (1987). The score statistic for regression transformation. *Biometrika*, **74**, 275.

LAWRANCE, A. J. (1988). Regression transformation diagnostics using local influence. *J. Amer. Statist. Ass.*, **83**, 1067.

LAWTON, W. H. (1965). Some inequalities for central and non-central distributions. *Ann. Math. Statist.*, **36**, 1521.

LAYARD, M. W. J. (1973). Robust large-sample tests for homogeneity of variances. *J. Amer. Statist. Ass.*, **68**, 195.

LEAMER, E. E. (1981). Coordinate-free ridge regression bounds. *J. Amer. Statist. Ass.*, **76**, 842.

LECAM, L. (1953). On some asymptotic properties of maximum likelihood estimates and related Bayes' estimates. *Univ. Calif. Publ. Statist.*, **1**, 277.

LECAM, L. (1970). On the assumptions used to prove asymptotic normality of maximum likelihood estimates. *Ann. Math. Statist.*, **41**, 802.

LEE, A. F. S. and GURLAND, J. (1975). Size and power of tests for equality of means of two normal populations with unequal variances. *J. Amer. Statist. Ass.*, **70**, 933.

LEE, Y. S. (1971). Some results on the sampling distribution of the multiple correlation coefficient. *J. R. Statist. Soc.*, **B, 33**, 117.

LEE, Y. S. (1972). Tables of upper percentage points of the multiple correlation coefficient. *Biometrika*, **59**, 175.

LEECH, D. and COWLING, K. (1982). Generalized regression estimation from grouped observations: A generalization and an application to the relationship between diet and mortality. *J. Roy. Statist. Soc.*, **A, 145**, 208.

LEHMANN, E. L. (1947). On optimum tests of composite hypotheses with one constraint. *Ann. Math. Statist.*, **18**, 473.

LEHMANN, E. L. (1949). Some comments on large sample tests. *Proc. 1st Berkeley Symp. Math. Statist. and Prob.*, 451.

LEHMANN, E. L. (1950). Some principles of the theory of testing hypotheses. *Ann. Math. Statist.*, **21**, 1.

LEHMANN, E. L. (1983a). Estimation with inadequate information. *J. Amer. Statist. Ass.*, **78**, 624.

LEHMANN, E. L. (1983b). *Theory of Point Estimation.* Wiley, New York.

LEHMANN, E. L. (1986). *Testing Statistical Hypotheses.* 2nd edn. Wiley, New York.

LEHMANN, E. L. and SCHEFFÉ, H. (1950, 1955). Completeness, similar regions and unbiased estimation. *Sankhyā*, **10**, 305 and **15**, 219.

LEHMANN, E. L. and STEIN, C. (1948). Most powerful tests of composite hypotheses. I. Normal distributions. *Ann. Math. Statist.*, **19**, 495.

LEHMANN, E. L. and STEIN, C. (1950). Completeness in the sequential case. *Ann. Math. Statist.*, **21**, 376.

LEHMER, E. (1944). Inverse tables of probabilities of errors of the second kind. *Ann. Math. Statist.*, **15**, 388.

LESLIE, J. R., STEPHENS, M. A. and FOTOPOULOS, S. (1986). Asymptotic distribution of the Shapiro-Wick W for testing normality. *Ann. Statist.*, **14**, 1497.

LEWIS, T. O. and ODELL, P. L. (1966). A generalization of the Gauss–Markov theorem. *J. Amer. Statist. Ass.*, **61**, 1063.

LEWONTIN, R. C. and PROUT, T. (1956). Estimation of the number of different classes in a population. *Biometrics*, **12**, 211.

LIDDELL, I. G. and ORD, J. K. (1978). Linear-circular correlation coefficients: some further results. *Biometrika*, **65**, 448.

LILLIEFORS, H. W. (1967). On the Kolmogorov-Smirnov test for normality with mean and variance unknown. *J. Amer. Statist. Ass.*, **62**, 399.

LIN, C. C. and MUDHOLKAR, G. S. (1980). A simple test for normality against asymmetric alternatives. *Biometrika*, **67**, 455.

LINDLEY, D. V. (1950). Grouping corrections and Maximum Likelihood equaitons. *Proc. Camb. Phil. Soc.*, **46**, 106.

LINDLEY, D. V. (1953). Statistical inference. *J. R. Statist. Soc.*, B, **15**, 30.

LINDLEY, D. V. (1958). Fiducial distributions and Bayes' theorem. *J. R. Statist. Soc.*, B, **20**, 102.

LINDLEY, D. V. (1971). *Bayesian Statistics Review.* S.I.A.M., Philadelphia.

LINDLEY, D. V. and SMITH, A. F. M. (1972). Bayesian estimates for the linear model. *J. R. Statist. Soc.*, B, **34**, 1.

LINDLEY, D. V., EAST, D. A. and HAMILTON, P. A. (1960). Tables for making inferences about the variance of a normal distribution. *Biometrika*, **47**, 433.

LINDSAY, B. (1982). Conditional score functions: some optimality results. *Biometrika*, **69**, 503.

LINDSAY, B. G. and ROEDER, K. (1987). A unified treatment of integer parameter models. *J. Amer. Statist. Ass.*, **82**, 758.

LINNIK, YU. V. (1964). On the Behrens-Fisher problem. *Bull. Int. Statist. Inst.*, **40**(2), 833.

LINNIK, YU, V. (1967). On the elimination of nuisance parameters in statistical problems. *Proc. 5th Berkeley Symp. Math. Statist. and Prob.*, **1**, 267.

LITTLE, R. J. and RUBIN, D. B. (1987). *Statistical Analysis with Missing Data.* Wiley, New York and London.

LLOYD, E. H. (1952). Least-squares estimation of location and scale parameters using order stastics. *Biometrika*, **39**, 88.

LOCKS, M. O., ALEXANDER, M. J. and BYARS, B. J. (1963). New tables of the noncentral *t*-distribution. *Aeronaut. Res. Lab. (Ohio)*, no. ARL 63–19.

LOH, W-Y. (1985). A new method for testing separate families of hypotheses. *J. Amer. Statist. Ass.*, **80**, 362.

LOH, W-Y. (1989). Bounds on the size of the χ^2 text of independence in a contingency table. *Ann. Statist.*, **17**, 1709.

LOUIS, T. A. (1982). Finding the observed information matrix when using the EM algorithm. *J. R. Statist. Soc.*, B, **44**, 226.

LOWERRE, J. (1974). Some relationships between BLUEs, WLSEs and SLSEs. *J. Amer. Statist. Ass.*, **69**, 223.

LWIN, T. (1975). Exponential family distribution with a truncation parameter. *Biometrika*, **62**, 218.

McCULLAGH, P. (1984). Local sufficiency. *Biometrika*, **71**, 233.

McCULLAGH, P. and COX, D. R. (1986). Invariants and likelihood ratio statistics. *Ann. Statist.*, **14**, 1419.

McCULLAGH, P. and NELDER, J. A. (1983). *Generalized Linear Models.* Chapman and Hall, London and New York.

McDONALD, G. C. and GALARNEAU, D. I. (1975). A Monte Carlo evaluation of some ridge-type estimators. *J. Amer. Statist. Ass.*, **70**, 407.

McELROY, F. W. (1967). A necessary and sufficient condition that ordinary least-squares estimators can be best linear unbiased. *J. Amer. Statist. Ass.*, **62**, 1302.

McNOLTY, F. (1962). A contour-integral derivation of the non-central chi-square distribution. *Ann. Math. Statist.*, **33**, 796.

MADANSKY, A. (1962). More on length of confidence intervals. *J. Amer. Statist. Ass.*, **57**, 586.

MADSEN, R. (1974). A procedure for tuncating SPRT's. *J. Amer. Statist. Ass.*, **69**, 403.

MÄKELÄINEN, T., SCHMIDT, K. and STYAN, G. P. H. (1981). On the existence and uniqueness of the maximum lkelihood estimate of a vector-valued parameter in fixed-size samples. *Ann. Statist.*, **9**, 758.

MALINVAUD, E. (1966). *Statistical Methods of Econometrics.* North-Holland, Amsterdam.

MALLOWS, C. L. (1973). Some comments on C_p. *Technometrics*, **15**, 661.

MALLOWS, C. L. (1975). *On Some Topics in Robustness.* Bell Telephone Labs., Murray Hill, New Jersey, USA (unpublished).

MANLY, B. F. J. (1970a). The choice of a Wald test on the mean of a normal population. *Biometrika*, **57**, 91.

MANLY, B. F. J. (1970b). On the distribution of the decisive sample number of certain sequential tests. *Biometrika*, **57**, 367.

MANN, H. B. (1949). *Analysis and Design of Experiments: Analysis of Variance and Analysis of Variance Designs.* Dover, New York.

MANN, H. B. and WALD, A. (1942). On the choice of the number of intervals in the application of the chi-square test. *Ann. Math. Statist.*, **13**, 306.

MARASINGHE, M. G. and JOHNSON, D. E. (1981). Testing subhypotheses in the multiplicative interaction model. *Technometrics*, **23**, 385.

MARASINGHE, M. G. and JOHNSON, D. E. (1982). A test of incomplete additivity in the multiplicative interaction model. *J. Amer. Statist. Ass.*, **77**, 869.

MARDIA, K. V. and SUTTON, T. W. (1978). A model with cylindrical variables with applications. *J. R. Statist. Soc.*, **B**, **40**, 229.

MARGOLIN, B. H. and MAURER, W. (1976). Tests of the Kolmogorov–Smirnov type for exponential data with unknown scale, and related problems. *Biometrika*, **63**, 149.

MARITZ, J. S. (1953). Estimation of the correlation coefficient in the case of a bivariate normal population when one of the variables is dichotomised. *Psychometrika*, **18**, 97.

MARITZ, J. S. and LWIN, T. (1989). *Empirical Bayes Methods.* Chapman and Hall, London. (second edition).

MARSAGLIA, G. (1964). Conditional means and covariances of normal variables with singular covariance matrix. *J. Amer. Statist. Ass.*, **59**, 1203.

MASSEY, F. J., Jr. (1950). A note on the estimation of a distribution function by confidence limits. *Ann. Math. Statist.*, **21**, 116.

MASSEY, F. J., Jr. (1951). The Kolmogorov–Smirnov test of goodness of fit. *J. Amer. Statist. Ass.*, **46**, 68.

MATHEW, T. (1983). Linear estimation with an incorrect diagonal matrix in linear models with a common linear part. *J. Amer. Statist. Ass.*, **78**, 468.

MATLOFF, N., ROSE, R. and TAI, R. (1984). A comparison of two methods for estimating optimal weights in regression analysis. *J. Statist. Comp. Simul.*, **19**, 265.

MAULDON, J. G. (1955). Pivotal quantities for Wishart's and related distributions, and a paradox in fiducial theory. *J. R. Statist., Soc.*, **B**, **17**, 79.

MEAD, R., BANCROFT, T. A. and HAN, C-P. (1975). Power of analysis of variance test procedures for incompletely specified fixed models. *Ann. Statist.*, **3**, 797.

MEHTA, J. S. and SRINIVASAN, R. (1970). On the Behrens-Fisher problem. *Biometrika*, **57**, 649.

MICKEY, M. R. and BROWN, M. B. (1966). Bounds on the distribution functions of the Behrens-Fisher statistic. *Ann. Math. Statist.*, **37**, 639.

MILLER, A. J. (1984). Selection of subsets of regression variables. *J. R. Statist. Soc.*, **A**, **147**, 389.

MILLER, D. (1990). A restoration of Popperian inductive scepticism. *Brit. J. Phil. Sci.*, **41**, 137.

MILLER, L. H. (1956). Table of percentage points of Kolmogorov statistics. *J. Amer. Statist. Ass.*, **51**, 111.

MILLER, R. G., Jr. (1964). A trustworthy jackknife. *Ann. Math. Statist.*, **35**, 1549.

MILLER, R. G. (1966). *Simultaneous Statistical Inference.* McGraw-Hill, New York.

MILLER, R. G. (1974a). The jackknife—a review. *Biometrika*, **61**, 1.

MILLER, R. G., Jr. (1974b). An unbalanced jackknife. *Ann. Statist.*, **2**, 880.

MILLER, R. G., Jr. (1977). Developments in multiple comparisons 1966-1976. *J. Amer. Statist. Ass.*, **72**, 779.

MITCHELL, T. J. and BEAUCHAMP, J. J. (1988). Bayesian variable selection in linear regression (with discussion). *J. Amer. Statist. Ass.*, **83**, 1023.

MITRA, S. K. and RAO, C. R. (1969). Conditions for optimality and validity of simple least squares theory. *Ann. Math. Statist.*, **40**, 1617.

MIYASHITA, H. and NEWBOLD, P. (1983). On the sensitivity to non-normality of a test for outliers in linear models. *Commun. Statist.*, **A, 12**, 1413.

MOLINARI, L. (1977). Distribution of the chi-squared test in nonstandard situations. *Biometrika*, **64**, 115.

MOORE, D. S. (1971). A chi-square statistic with random cell boundaries. *Ann. Math. Statist.*, **42**, 147.

MOORE, D. S. (1977). Generalized inverses, Wald's method, and the construction of chi-squared tests of fit. *J. Amer. Statist. Ass.*, **72**, 131.

MOORE, D. S. (1982). The effect of dependence on chi-squared tests of fit. *Ann. Statist.*, **10**, 1163.

MOORE, D. S. and SPRULL, M. D. (1975). Unified large-sample theory of general chi-squared statistics for tests of fit. *Ann. Statist.*, **3**, 599.

MORAN, P. A. P. (1950). The distribution of the multiple correlation coefficient. *Proc. Camb. Phil. Soc.*, **46**, 521.

MORAN, P. A. P. (1970). On asymptotically optimal tests of composite hypotheses. *Biometrika*, **57**, 47-55.

MOSHMAN, J. (1958). A method for selecting the size of the initial sample in Stein's two sample procedure. *Ann. Math. Statist.*, **29**, 1271.

MOSTELLER, F. and YOUTZ, C. (1961). Tables of the Freeman–Tukey transformations for the binomial and Poisson distributions. *Biometrika*, **48**, 433.

MURPHY, R. B. (1948). Non-parametric tolerance limits. *Ann. Math. Statist.*, **19**, 581.

MUDHOLKAR, G. S. and TRIVEDI, M. C. (1981). A Gaussian approximation to the distribution of the sample variance for nonnormal populations. *J. Amer. Statist. Ass.*, **76**, 479.

MUIRHEAD, R. J. (1985). Estimating a particular function of the multiple correlation coefficient. *J. Amer. Statist. Ass.*, **80**, 923.

MULLER, K. E. and BARTON, C. N. (1989). Approximate power for repeated-measures ANOVA lacking sphericity. *J. Amer. Statist. Ass.*, **84**, 549.

MYERS, M. H., SCHNEIDERMAN, M. A. and ARMITAGE, P. (1966). Boundaries for closed (wedge) sequential *t* test plans. *Biometrika*, **53**, 431.

NADDEO, A. (1968). Confidence intervals for the frequency function and the cumulative frequency function of a sample drawn from a discrete random variable. *Rev. Int. Statist. Inst.*, **36**, 313.

NAGARSENKER, P. B. (1980). On a test of equality of several exponential distributions. *Biometrika*, **67**, 475.

NAGARSENKER, P. B. (1984). On Bartlett's test for homogeneity of variances. *Biometrika*, **71**, 405.

NAIMAN, D. Q. (1983). Comparing Scheffé-type to constant-width confidence bounds in regression. *J. Amer. Statist. Ass.*, **78**, 906.

NAIMAN, D. Q. (1986). Conservative confidence bands in curvilinear regression. *Ann. Statist.*, **14**, 896.

NAIMAN, D. Q. (1987). Simultaneous confidence bounds in multiple regression using prediction variable constraints. *J. Amer. Statist. Ass.*, **82**, 214.

NARULA, S. C. (1979). Orthogonal polynomial regression. *Int. Statist. Rev.*, **47**, 31.

NARULA, S. C. and WELLINGTON, J. F. (1979). Selection of variables in linear regression using the minimum of weighted absolute errors criterion. *Technometrics*, **21**, 299.

NARULA, S. C. and WELLINGTON, J. F. (1982). The minimum sum of absolute errors regression: A state of the art survey. *Int. Statist. Rev.*, **50**, 317.

NAYLOR, J. C. and SMITH, A. F. M. (1988). Econometric illustrations of novel numerical integration strategies for Bayesian inference. *J. Econometrics*, **38**, 103.

NELDER, J. A. and WEDDERBIRN, R. W. M. (1972). Generalized linear models. *J. R. Statist. Soc.*, **A**, **135**, 370.

NEYMAN, J. (1935).* Sur la vérification des hypothèses statistiques composées. *Bull. Soc. Math. France*, **63**, 1.

NEYMAN, J. (1937).* Outline of a theory of statistical estimation based on the classical theory of probability. *Phil. Trans.*, **A**, **236**, 333.

NEYMAN, J. (1938a).* On statistics the distribution of which is independent of the parameters involved in the original probability law of the observed variables. *Statist. Res. Mem.*, **2**, 58.

NEYMAN, J. (1938b). Tests of statistical hypotheses which are unbiased in the limit. *Ann. Math. Statist.*, **9**, 69.

NEYMAN, J. (1949).* Contribution to the theory of the χ^2 test. *Proc. 1st Berkeley Symp. Math. Statist. and Prob.*, 239.

NEYMAN, J. (1959). Optimal asymptotic tests of composite statistical hypotheses. In *Probability and Statistics*, U. Grenander (ed). Almqvist and Wiksell, Stockholm, Sweden.

NEYMAN, J. (1962). Two breakthroughs in the theory of statistical decision making. *Rev. Int. Statist. Inst.*, **30**, 11.

NEYMAN, J. and PEARSON, E. S. (1928).* On the use and interpretation of certain test criteria for the purposes of statistical inference. *Biometrika*, **20A**, 175 and 263.

NEYMAN, J. and PEARSON, E. S. (1931).* On the problem of k samples. *Bull. Acad. Polon. Sci.*, **3**, 460.

NEYMAN, J. and PEARSON, E. S. (1933a).* On the testing of statistical hypotheses in relation to probabilities *a priori*. *Proc. Camb. Phil. Soc.*, **29**, 492.

NEYMAN, J. and PEARSON, E. S. (1933b).* On the problem of the most efficient tests of statistical hypotheses. *Phil. Trans.*, **A**, **231**, 289.

NEYMAN, J. and PEARSON, E. S. (1936a).* Sufficient statistics and uniformly most powerful tests of statistical hypotheses. *Staist. Res. Mem.*, **1**, 113.

NEYMAN, J. and PEARSON, E. S. (1936b).* Unbiassed critical regions of Type A and Tye A_1. *Statist. Res. Mem.*, 1, 1.

NEYMAN, J. and PEARSON, E. S. (1938).* Certain theorems on unbiassed critical regions of Type A, *and* Unbiassed tests of simple statistical hypotheses specifying the values of more than one unknown parameter. *Statist. Res. Mem.*, **2**, 25.

NEYMAN, J. and SCOTT, E. L. (1948). Consistent estimates based on partially consistent observations. *Econometrica*, **16**, 1.

NEYMAN, J. and SCOTT, E. L. (1960). Correction for bias introduced by transformation of variables. *Ann. Math. Statist.*, **31**, 643.

NEYMAN, J., IWASEKIEWICZ, K. and KOLODZIEJCZYK, S. (1935).* Statistical problems in agricultural experimentation. *Suppl. J. R. Statist. Soc.*, **2**, 107.

NOETHER, G. E. (1955). On a theorem of Pitman. *Ann. Math. Statist.*, **26**, 64.

NOETHER, G. E. (1957). Two confidence intervals for the ratio of two probabilities, and some measures of effectiveness. *J. Amer. Statist. Ass.*, **52**, 36.

OBENCHAIN, R. L. (1978). Good and optimal ridge estimators. *Ann. Statist.*, 6, 1111.

ODEH, R. E. (1982). Critical values of the sample product-moment correlation coefficient in the bivariate normal distribution. *Commun. Statist.*, **B**, **11**, 1.

OGBURN, W. G. (1935). Factors in the variation of crime among cities. *J. Amer. Statist. Ass.*, **30**, 12.

O'HAGAN, A. (to be published). *Kendall's Advanced Theory of Statistics: Bayesian Inference and Relationship*. Arnold, London and Oxford, Univ. Press, New York.

OLIVER, E. H. (1972). A maximum likelihood oddity. *Amer. Statistician*, **26**, 43.

OLKIN, I. and PRATT, J. W. (1958). Unbiased estimation of certain correlation coefficients. *Ann. Math. Statist.*, **29**, 201.

OLSHEN, R. A. (1973). The conditional level of the *F*-test. *J. Amer. Statist. Ass.*, **68**, 692.

OMAN, S. D. (1981). A confidence bound approach to choosing the biasing parameter in ridge regression. *J. Amer. Statist. Ass.*, **76**, 542.

O'NEILL, R. and WETHERILL, G. B. (1971). The present state of multiple comparison methods. *J. R. Statist. Soc.*, **B, 33**, 218.

ORCHARD, T. and WOODBURY, M. A. (1972). A missing formation principle: theory and applications. *Proc. 6th Berkeley Symp. on Math. Statist. and Prob.*, **1**, 697.

OTTEN, A. (1973). Note on the Spearman rank correlation coefficient. *J. Amer. Statist. Ass.*, **68**, 585.

OWEN, D. B. (1962). *Handbook of Statistical Tables*. Addison-Wesley, Reading, Mass., U.S.A.

OWEN, D. B. (1963). Factors for one-sided tolerance limits and for variables sampling plans. *Sandia Corp. Monogr.* (Off. Techn. Serv., Dept. Commerce, Washington), SCR-607.

OWEN, D. B. (1965). The power of Student's *t*-test. *J. Amer. Statist. Ass.*, **60**, 320.

OWEN, D. B. and ODEH, R. E. (eds.) (1977). *Selected Tables in Mathematical Statistics*, Vol. V, Amer. Math. Soc., Prondence, R.I., USA.

PACHARES, J. (1960). Tables of confidence limits for the binomial distribution. *J. Amer. Statist. Ass.*, **55**, 521.

PACHARES, J. (1961). Tables for unbiased tests on the variance of a normal population. *Ann. Math. Statist.*, **32**, 84.

PAIK, U. B. and FEDERER, W. T. (1974). Analysis of non-orthogonal *n*-way classifications. *Ann. Statist.*, **2**, 1000.

PARR, W. C. (1983). A note on the jackknife, the bootstrap and the delta method estimators of bias and variance. *Biometrika*, **70**, 719.

PARR, W. C. and SCHUCANY, W. R. (1980). The jackknife: a biliography. *Int. Statist. Rev.*, **48**, 73.

PATIL, G. P. and SHORROCK, R. (1965). On certain properties of the exponential-type families. *J. R. Statist. Soc.*, **B, 27**, 94.

PATNAIK, P. B. (1949). The non-central χ^2- and *F*-distributions and their applications. *Biometrika*, **36**, 202.

PAULSON, E. (1941). On certain likelihood-ratio tests associated with the exponential distribution. *Ann. Math. Statist.*, **12**, 301.

PAULSON, E. (1952). An optimum solution to the *k*-sample slippage problem for the normal distribution. *Ann. Math. Statist.*, **23**, 610.

PEARCE, S. C. (1963). The use and classification of non-orthogonal designs. *J. R. Statist. Soc.*, **A, 126**, 353.

PEARCE, S. C. and JEFFERS, J. N. R. (1971). Block designs and missing data. *J. R. Statist. Soc.*, **B, 33**, 131.

PEARCE, S. C., CALIŃSKI, T. and MARSHALL, T. F. de C. (1974). The basic contrasts of an experimental design with special reference to the analysis of data. *Biometrika*, **61**, 449.

PEARSON, E. S. (1959). Note on an approximation to the distribution of non-central χ^2. *Biometrika*, **46**, 364.

PEARSON, E. S. and HARTLEY, H. O. (1951). Charts of the power function for analysis of variance tests derived from the non-central *F*-distribution. *Biometrika*, **38**, 112.

PEARSON, E. S. and PLEASE, N. W. (1975). Relation between the shape of population distribution and the robustness of four simple test statistics. *Biometrika*, **62**, 223.

PEARSON, E. S. and TIKU, M. L. (1970). Some notes on the relationship between the distributions of central and non-central *F*. *Biometrika*, **57**, 175.

PEARSON, E. S., D'AGOSTINO, R. B. and BOWMAN, K. O. (1977). Tests for departure for normality: comparison of powers. *Biometrika*, **64**, 231.

PEARSON, K. (1897). On a form of spurious correlation which may arise when indices are used in the measurement of organs. *Proc. Roy. Soc.*, **60**, 489.

PEARSON, K. (1900).* On a criterion that a given system of deviations from the probable in the case of a correlated system of variables is such that it can be reasonably supposed to have arisen in random sampling. *Phil. Mag.*, **(5), 50,** 157.

PEARSON, K. (1904).* On the theory of contingency and its relation to association and normal correlation. *Drapers' Co. Memoirs, Biometric Series,* No. 1, London.

PEARSON, K. (1909). On a new method for tetermining correlation between a measured character *A* and a character *B*, of which only the percentage of cases wherein *B* exceeds (or falls short of) a given intensity is recorded for each grade of *A*. *Biometrka,* **7,** 96.

PEARSON, K. (1913). On the probable error of a correlation coefficient as found from a fourfold table. *Biometrika,* **9,** 22.

PEARSON, K. (1915). On the probable error of a coefficient of mean square contingency. *Biometrika,* **10,** 570.

PEDERSON, J. G. (1978). Fiducial inference. *Int. Statist. Rev.,* **46,** 147.

PEERS, H. W. (1965). On confidence points and Bayesian probability points in the case of several parameters. *J. R. Statist. Soc.,* **B, 27,** 9.

PEERS, H. W. (1971). Likelihood ratio and associated test criteria. *Biometrika,* **58,** 577.

PEERS, H. W. (1978). Second-order sufficiency and statistical invariants. *Biometrika,* **65,** 489.

PEREIRA, B. de B. (1977). Discriminating among separate models: a bibliography. *Int. Statist. Rev.* **45,** 163.

PERLMAN, M. D. (1972). On the strong consistency of approximate maximum likelihood estimates. *Proc. 6th Berkeley Symp. Math. Statist. and Prob.,* **1,** 263.

PERNG, S-S. (1978). Exponentially bounded stopping times of invariant SPRT's in general linear models: finite m.g.f. case. *Ann. Statist.,* **6,** 85.

PETRONDAS, D. A. and GABRIEL, K. R. (1983). Multiple comparison by rerandomization tests. *J. Amer. Statist. Ass.,* **78,** 949.

PFANZAGL, J. (1973). Asymptotic expansions related to minimum contrast estimators. *Ann. Statist.,* **1,** 993.

PFANZAGL, J. (1974). On the Behrens-Fisher problem. *Biometrika,* **61,** 39.

PFEFFERMAN, D. (1984). On extensions of the Gauss–Markov theories to the case of stochastic regression coefficients. *J. R. Statist. Soc.,* **B, 46,** 139.

PHILLIPS, P. C. B. (1982). The true characteristic function of the *F* distribution. *Biometrika,* **69,** 261.

PICARD, R. R. and COOK, R. D. (1984). Cross-validation of regression residuals. *J. Amer. Statist. Ass.,* **79,** 575.

PIEGORSCH, W. W. (1985a). Average-width optimality for confidence bands in simple linear regression. *J. Amer. Statist. Ass.,* **80,** 692.

PIEGORSCH, W. W. (1985b). Admissible and optimal confidence bands in simple lindar regression. *Ann. Statist.,* **13,** 801.

PIEGORSCH, W. W. (1987). Model robustness for simultaneous confidence bands. *J. Amer. Statist. Ass.,* **82,** 879.

PIERCE, D. A. and KOPECKY, K. J. (1979). Testing goodness of fit for the distribution of errors in regression models. *Biometrika,* **66,** 1.

PITMAN, E. J. G. (1936). Sufficient statistics and intrinsic accuracy. *Proc. Camb. Phil. Soc.,* **32,** 567.

PITMAN, E. J. G. (1937). The "closest" estimates of statistical parameters. *Proc. Camb. Phil. Soc.,* 33, 212.

PITMAN, E. J. G. (1938). The estimation of the location and scale parameters of a continuous population of any given form. *Biometrika,* **30,** 391.

PITMAN, E. J. G. (1939a). A note on normal correlation. *Biometrika,* **31,** 9.

PITMAN, E. J. G. (1939b). Tests of hypotheses concerning location and scale parameters. *Biometrika,* **31,** 200.

PITMAN, E. J. G. (1948). *Non-Parametric Statistical Inference.* University of North Carolina Institute of Statistics (mimeographed lecture notes).

PITMAN, E. J. G. (1957). Statistics and science. *J. Amer. Statist. Ass.,* **52,** 322.

PLACKETT, R. L. (1949). A historical note on the method of least squares. *Biometrika*, **36**, 458.

PLACKETT, R. L. (1950). Some theorems in least squares. *Biometrika*, **37**, 149.

PLACKETT, R. L. (1972). The discovery of the method of least squares. *Biometrika*, **59**, 239.

POPPER, K. R. (1968). *The Logic of Scientific Discovery*. Hutchinson, London.

POPPER, K. R. (1969). *Conjectures and Refutations*. Routledge and Kegan Paul, London.

POPPER, K. R. and MILLER, D. (1983). A proof of the impossibility of inductive probability.. *Nature*, **302**, 687.

PORTNOY, S. (1977). Asymptotic efficiency of minimum variance unbiased estimators. *Ann. Statist.*, **5**, 522.

PRASAD, G. and SAHAI, A. (1982). Sharper variance upper bound for unbiased estimation in inverse sampling. *Biometrika*, **69**, 286.

PRATT, J. W. (1961). Length of confidence intervals. *J. Amer. Statist. Ass.*, **56**, 549.

PRATT, J. W. (1962). Discussion on paper by Birnbaum. *J. Amer. Statist. Ass.*, **57**, 314.

PRATT, J. W. (1963). Shorter confidence intervals for the mean of a normal distribution with known variance. *Ann. Math. Statist.*, **34**, 574.

PRESS, S. J. (1966). A confidence interval comparison of two test procedures proposed for the Behrens-Fisher problem. *J. Amer. Statist. Ass.*, **61**, 454.

PRICE, R. (1964). Some non-central F-distributions expressed in closed form. *Biometrika*, **51**, 107.

PRZYBOROWSKI, J. and WILÉNSKI, M. (1935). Statistical principles of routine work in testing clover seed for dodder. *Biometrika*, **27**, 273.

QUENOUILLE, M. H. (1956). Notes on bias in estimation. *Biometrika*, **43**, 353.

QUENOUILLE, M. H. (1958). *Fundamentals of Statistical Reasoning*. Griffin, London.

QUENOUILLE, M. H. (1959). Tables of random observations from standard distributions. *Biometrika*, **46**, 178.

QUINE, M. P. and ROBINSON, J. (1985). Efficiencies of chi-square and likelihood ratio goodness-of-fit tests. *Ann. Statist.*, **13**, 727.

RAHMAN, M. and SALEH, A. K. M. E. (1974). Explicit form of the distribution of the Behrens-Fisher d-statistic. *J. R. Statist. Soc.*, **B, 36**, 54; *corrections*, 466.

RAMACHANDRAN, K. V. (1958). A test of variances. *J. Amer. Statist. Ass.*, **53**, 741.

RAMSEY, F. P. (1931). Truth and probability. In *The Foundations of Mathematics and Other Essays*. Kegan, Paul Trench, Tubner. Reprinted in H. E. Kyburg, Jr. and H. E. Smokler (eds. 1964). *Studies in Subjective Probability*. Wiley, New York, 61.

RANDLES, R. H., FLIGNER, M. A., POLICELLO, G. E. II and WILFE, D. A. (1980). An asymptotically distribution-free test for symmetry versus asymmetry. *J. Amer. Statist. Ass.*, **75**, 168.

RANKIN, N. O. (1974). The harmonic mean method for one-way and two-way analysis of variance. *Biometrika*, **61**, 117.

RAO, B. R. (1958). On an analogue of Cramér–Rao's inequality. *Skand. Aktuartidskr.*, **41**, 57.

RAO, C. R. (1945). Information and accuracy attainable in the estimation of statistical parameters. *Bull. Calcutta Math. Soc.*, **37**, 81.

RAO, C. R. (1947). Minimum variance and the estimation of several parameters. *Proc. Camb. Phil. Soc.*, **43**, 280.

RAO, C. R. (1952). *Advanced Statistical Methods in Biometric Research*. Wiley, New York.

RAO, C. R. (1957). Theory of the method of estimation by minimum chi-square. *Bull. Int. Statist. Inst.*, **35(2)**, 25.

RAO, C. R. (1961). Asymptotic efficiency and limiting information. *Proc. 4th Berkeley Symp. Math. Statist. and Prob.*, **1**, 531.

RAO, C. R. (1962a). Efficient estimates and optimum inference procedures in large samples. *J. R. Statist. Soc.*, **B, 24**, 46.

RAO, C. R. (1962b). Apparent anomalies and irrigularities in maximum likelihood estimation. *Sankhyā*, **A, 24**, 72.

RAO, C. R. (1967). Least squares theory using an estimated dispersion matrix and its application to measurement of signals. *Proc. 5th Berkeley Symp. Math. Statist. and Prob.*, **1**, 355.

RAO, C. R. (1970). Estimation of heteroscadastic variances in linear models. *J. Amer. Statist. Ass.*, **65**, 161.

RAO, C. R. (1974). Projectors, generalized inverses and the BLUE's. *J. R. Statist. Soc.*, **B, 36**, 442.

RAO, J. N. K. (1980). Estimating the common mean of possibly different normal populations: a simulation study. *J. Amer. Statist. Ass.*, **75**, 447.

RAO, P. V., SCHUSTER, E. F. and LITTELL, R. C. (1975). Estimation of shift and center of symmetry based on Kolmogorov–Smirnov statistics. *Ann. Statist.*, **3**, 862.

READ, T. R. C. (1984). Small-sample comparisons for the power divergence goodness-of-fit statistics. *J. Amer. Statist. Ass.*, **79**, 929.

REEDS, J. A. (1978). Jackknifing maximum likelihood estimates. *Ann. Statist.*, **6**, 727.

REEDS, J. A. (1985). Asymptotic number of roots of Cauchy location likelihood equations. *Ann. Statist.*, **13**, 775.

REES, D. H. (1966). The analysis of variance of designs with many non-orthogonal classifications. *J. R. Statist. Soc.*, **B, 28**, 110.

REISS, R. D. and RÜSCHENDORF, L. (1976). On Wolks' distribution-free confidence intervals for quantile intervals. *J. Amer. Statist. Ass.*, **71**, 940.

RESNIKOFF, G. J. (1962). Tables to facilitate the computation of percentage points of the non-central t-distribution. *Ann. Math. Statist.*, **33**, 580.

RESNIKOFF, G. J. and LIEBERMAN, G. J. (1957). *Tables of the non-central t-distribution.* Stanford Univ. Press.

RIDOUT, M. S. (1988). An improved branch and bound algorithm for feature subset selection. *Appl. Statist.*, **37**, 139.

RIVEST, L-P. (1986). Bartlett's, Cochran's and Hartley's tests on variances are liberal when the underlying distribution is long-tailed. *J. Amer. Statist. Ass.*, **81**, 124.

ROBBINS, H. (1944). On distribution-free tolerance limits in random sampling. *Ann. Math. Statist.*, **15**, 214.

ROBBINS, H. (1964). The empirical Bayes approach to statistical decision problems. *Ann. Math. Statist.*, **35**, 1.

ROBBINS, H. (1956). An empirical Bayes approach to statistics. *Proceedings Third Brekeley Symposium on Math. Statist. and Prob.*, **1**, 157. U. California Press, Berkeley.

ROBINSON, G. K. (1976). Properties of Student's t and of the Behrens-Fisher solution to the two means problem. *Ann. Statist.*, **4**, 963.

ROBISON, D. E. (1964). Estimates for the points of intersection of two polynomial regression. *J. Amer. Statist. Ass.*, **59**, 214.

ROBSON, D. S. (1959). A simple method for constructing orthogonal polynomials when the independent variable is unequally spaced. *Biometrics*, **15**, 187.

ROBSON, D. S. and WHITLOCK, J. H. (1964). Estimation of a truncation point. *Biometrika*, **51**, 33.

ROOTZÉN, H. and SIMMONS, G. (1977). On the exponential boundedness of stopping times of invariant SPRT's. *Ann. Statist.*, **5**, 571.

ROSCOE, J. T. and BYARS, J. A. (1971). An investigation of the restraints with respect to sample size commonly imposed on the use of the chi-square statistic. *J. Amer. Statist. Ass.*, **66**, 755.

ROSS, W. H. (1987). The expectation of the likelihood ratio criterion. *Int. Statist. Rev.*, **55**, 315.

ROTHE, G. (1981). Some properties of the asymptotic relative Pitman efficiency. *Ann. Statist.*, **9**, 663.

ROTHENBERG, T. J., FISHER, F. M. and TILANUS, C. B. (1964). A note on estimation from a Cauchy sample. *J. Amer. Statist. Ass.*, **59**, 460.

ROUSSEEUW, P. J. (1984). Least median of squares regression. *J. Amer. Statist. Ass.*, **79**, 871.

ROY, A. R. (1956). On χ^2-statistics with variable intervals. Technical Report, Stanford University, Statistics Department.

ROY, K. P. (1957). A note on the asymptotic distribution of likelihood ratio. *Bull. Calcutta Statist. Ass.*, **7**, 73.

ROY, S. N. (1954). Some further results in simultaneous confidence interval estimation. *Ann. Math. Statist.*, **25**, 752.

ROY, S. N. and BOSE, R. C. (1953). Simultaneous confidence interval estimation. *Ann. Math. Statist.*, **24**, 513.

RUBIN, D. B. (1976). Noniterative least squares estimates, standard errors and *F*-tests for analyses of variance with missing data. *J. R. Statist. Soc.*, **B, 38**, 270.

RUBIN, D. B. and WEISBERG, S. (1975). The variance of a linear combination of independent estimators using estimated weights. *Biometrika*, **62**, 708.

RUKHIN, A. L. (1986). Improved estimation in lognormal models. *J. Amer. Statist. Ass.*, **81**, 1046.

RUSHTON, S. (1951). On least squares fitting of orthonormal polynomials using the Choleski method. *J. R. Statist. Soc.*, **B, 13**, 92.

RUYMGAART, F. H. (1975). A note on chi-square statistics with random cell boundaries. *Ann. Statist.*, **3**, 965.

SAMPSON, A. R. (1974). A tale of two regressions. *J. Amer. Statist. Ass.*, **69**, 682.

SANIGA, E. M. and MILES, J. A. (1979). Power of some standard goodness-of-fit tests of normality against asymmetric stable alternatives. *J. Amer. Statist. Ass.*, **74**, 861.

SANKARAN, M. (1964). On an analogue of Bhattacharya bound. *Biometrika*, **51**, 268.

SARHAN, A. E. (1954). Estimation of the mean and standard deviation by order statistics. *Ann. Math. Statist.*, **25**, 317.

SARKADI, K. (1975). The consistency of the Shapiro–Francia test. *Biometrika*, **62**, 445.

SATHE, Y. S. and LINGRAS, S. R. (1981). Bounds for the confidence coefficients of outer and inner confidence intervals for quantile intervals. *J. Amer. Statist. Ass.*, **76**, 473.

SAVAGE, L. J. (1954). *The Foundations of Statistics.* Methuen, London.

SAVAGE, L. J. (1970). Comments on a weakened principle of conditionality. *J. Amer. Statist. Ass.*, **65**, 399.

SCHEFFÉ, H. (1942a). On the theory of testing composite hypotheses with one constraint. *Ann. Math. Statist.*, **13**, 280.

SCHEFFÉ, H. (1942b). On the ratio of the variances of two normal populations. *Ann. Math. Statist.*, **13**, 371.

SCHEFFÉ, H. (1943). On solutions of the Behrens-Fisher problem based on the *t*-distribution. *Ann. Meth. Statist.*, **14**, 35.

SCHEFFÉ, H. (1944). A note on the Behrens-Fisher problem. *Ann. Math. Statist.*, **15**, 430.

SCHEFFÉ, H. (1953). A method for judging all contrasts in the analysis of variance. *Biometrika*, **40**, 87.

SCHEFFÉ, H. (1959). *The Analysis of Variance.* Wiley, New York.

SCHEFFÉ, H. (1970a). Multiple testing versus multiple estimation. Improper confidence sets. Estimation of directions and ratios. *Ann. Math. Statist.*, **41**, 1.

SCHEFFÉ, H. (1970b). Practical solutions of the Behrens-Fisher problem. *J. Amer. Statist. Ass.*, **65**, 1501.

SCHEFFÉ, H. and TUKEY, J. W. (1945). Non-parametric estimation: I. Validation of order statistics. *Ann. Math. Statist.*, **16**, 187.

SCHENKER, N. (1985). Qualms about bootstrap confidence intervals. *J. Amer. Statist. Ass.*, **80**, 360.

SCHEUER, E. M. and SPURGEON, R. A. (1963). Some percentage points of the noncentral *t*-distribution. *J. Amer. Statist. Ass.*, **58**, 176.

SCHLESSELMAN, J. (1971). Power families: a note on the Box and Cox transformation. *J. R. Statist. Soc.*, **B, 33**, 307.

SCHMETTERER, L. (1960). On a problem of J. Neyman and E. Scott. *Ann. Math. Statist.*, **31**, 656.

SCHUSTER, E. F. (1973). On the goodness-of-fit problem for continuous symmetric distributions. *J. Amer. Statist. Ass.*, **68**, 713; *corrigenda*, **69**, 288.

SCHUSTER, E. F. (1975). Estimating the distribution function of a symmetric distribution. *Biometrika*, **62**, 631.

SCHWARZ, G. (1978). Estimating the dimension of a model. *Ann. Statist.*, **6**, 461.

SCOTT, D. W. (1985). Frequency polygons theory and application. *J. Amer. Statist. Ass.*, **80**, 348.

SEAL, H. L. (1948). A note on the χ^2 smooth test. *Biometrika*, 35, 202.

SEAL, H. L. (1967). The historical development of the Gauss linear model. *Biometrika*, **54**, 1.

SEARLE, S. R., SPEED, F. M. and HENDERSON, H. V. (1981). Some computational and model equivalences in analysis of variance of unequal-subclass-numbers data. *Amer. Statistician*, **35**, 16.

SEBER, G. A. F. (1964). Orthogonality in analysis of variance. *Ann. Math. Statist.*, **35**, 705.

SEELBINDER, B. M. (1953). On Stein's two-stage sampling scheme. *Ann. Math. Statist.*, **24**, 640.

SEIDENFELD, T. (1979). *Philosophical Problems of Statistical Inference.* Reidel, Boston.

SELF, S. G. and LIANG, K-Y. (1987). Asymptotic properties of maximum likelihood estimators and likelihood ratio tests under nonstandard conditions. *J. Amer. Statist. Ass.*, **82**, 605.

SEN, P. K. (1969). A generalization of the T-method of multiple comparisons for interactions. *J. Amer. Statist. Ass.*, **64**, 290.

SEN, P. K. (1977). Some invariance principles relating to jackknifiing and their role in sequential analysis. *Ann. Statist.*, **5**, 316.

SEN, P. K. and GHOSH, B. K. (1976). Comparison of some bounds in estimation theory. *Ann. Statist.*, **4**, 755.

SHAH, B. K. and ODEH, R. E. (1986). *Selected Tables in Mathematical Statistics*, Vol. X. Amer. Math. Soc., Providence, R.I., USA.

SHAPIRO, S. S. and FRANCIA, R. S. (1972). An approximate analysis of variance test for normality. *J. Amer. Statist. Ass.*, **67**, 215.

SHAPIRO, S. S. and WILK, M. B. (1965). An analysis of variance test for normality (complete samples). *Biometrika*, **52**, 591.

SHAPIRO, S. S., WILK, M. B. and CHEN, H. J. (1968). A comparative study of various tests for normality. *Amer. Statist. Ass.*, **63**, 1343.

SHAROT, T. (1976). Sharpening the jackknife. *Biometrika*, **63**, 71.

SHENTON, L. R. (1949). On the efficiency of the method of moments and Neyman's Type A contagious distribution. *Biometrika*, **36**, 450.

SHENTON, L. R. (1950). Maximum likelihood and the efficiency of the method of moments. *Biometrika*, **37**, 111.

SHENTON, L. R. (1951). Efficiency of the method of moments and the Gram-Charlier Type A distribution. *Biometrika*, **38**, 58.

SHENTON, L. R. and BOWMAN, K. (1963). Higher moments of a maximum-likelihood estimate. *J. R. Statist. Soc.*, **B, 25**, 305.

SHENTON, L. R. and BOWMAN, K. O. (1977). *Maximum Likelihood Estimation in Small Samples.* Griffin, London & High Wycombe.

SHIBATA, R. (1981). An optimal selection of regression variables. *Biometrika*, **68**, 45.

SHIBATA, R. (1984). Approximate efficiency of a selection procedure for the number of regression variables. *Biometrika*, **71**, 43.

SHIRAHATA, S. (1981). Intraclass rank tests for independence. *Biometrika*, **68**, 451.

SHOUKRI, M. M. and WARD, R. H. (1984). On the estimation of the intraclass correlation coefficient. *Commun. Statist.*, **A, 13**, 1239.

SICHEL, H. S. (1951-2). New methods in the statistical evaluation of mine sampling data. *Trans. Inst. Mining Metallurgy*, **61**, 261.

SIEGEL, A. F. (1979). The noncentral chi-squared distribution with zero degrees of freedom and testing for uniformity. *Biometrika*, **66**, 381.

SIEGMUND, D. (1975). Error probabilities and average sample number of the sequential probability ratio test. *J. R. Statist. Soc.*, **B, 37**, 394.

SIEGMUND, D. (1978). Estimation following sequential tests. *Biometrika*, **65**, 341.

SIEVERS, G. L. (1969). On the probability of large deviations and exact slopes. *Ann. Math. Statist.*, **40**, 1908.

SILVERSTONE, H. (1957). Estimating the logistic curve. *J. Amer. Statist. Ass.*, **52**, 567.

SILVEY, S. D. (1959). The Lagrange multiplier test. *Ann. Math. Statist.*, **30**, 389.

SILVEY, S. D. (1969). Multicollinearity and imprecise estimation. *J. R. Statist. Soc.*, **B, 31**, 539.

SIMON, G. A. and SIMONOFF, J. S. (1986). Diagnostic plots for missing data in least squares regression. *J. Amer. Statist. Ass.*, **81**, 501.

SIMS, C. A. (1975). A note on exast tests for serial correlation. *J. Amer. Statist. Ass.*, **70**, 162.

SINGH, K. (1981). On the asymptotic accuracy of Efron's bootstrap. *Ann. Statist.*, **9**, 1187.

SIOTANI, M. (1964). Interval estimation for linear combinations of means. *J. Amer. Statist. Ass.*, **59**, 1141.

SKOVGAARD, I. M. (1985). A second-order investigation of asymptotic ancillarity. *Amer. Statist.*, **13**, 534.

SLAKTER, M. J. (1966). Comparative validity of the chi-square and two modified chi-square goodness-of-fit tests for small but equal expected frequencies. *Biometrika*, **53**, 619.

SLAKTER, M. J. (1968). Accuracy of an approximation to the power of the chi-square goodness of fit test with small but equal expected frequencies. *J. Amer. Statist. Ass.*, **63**, 912.

SMIRNOV, N. V. (1948). Table for estimating the goodness of fit of empirical distributions. *Ann. Math. Statist.*, **19**, 279.

SMITH, G. and CAMPBELL, F. (1980). A critique of some ridge regression methods. *J. Amer. Statist. Ass.*, **75**, 74.

SMITH, K. (1916). In the "best" values of the constants in frequency distributions. *Biometrika*, **11**, 262.

SMITH, J., RAE, D. S., MANDERSCHEID, R. W. and SILBERGELD, S. (1981). Approximating the moments and distribution of the likelihood ratio statistic for multinomial goodness of fit. *J. Amer. Statist. Ass.*, **76**, 737.

SMITH, R. L. (1985). Maximum likelihood estimation in a class of nonregular cases. *Biometrika*, **72**, 67.

SNEDECOR, G. W. (1946). *Statistical Methods: Applied to Experiments in Agriculture and Biology.* Iowa State College Colegiate Press, Ames, Iowa (4th edn).

SNEE, R. D. (1982). Nonadditivity in a two-way classification: it is interaction or non-homoeneous variance? *J. Amer. Statist. Ass.*, **77**, 515.

SOPER, H. E. (1914). On the probable error of the biserial expression for the correlation coefficient. *Biometrika*, **10**, 384.

SOUVAINE, D. L. and STEELE, J. M. (1987). Time and space-efficient algorithms for least median of squares regression. *J. Amer. Statist. Ass.*, **82**, 794.

SPEED, F. M. and MONZELUN, C. J. (1979). Exact F-tests for the method of unweighted means in a 2^k experiment. *Amer. Statistician*, **33**, 15.

SPEED, F. M., HOCKING, R. R. and HACKNEY, O. P. (1978). Methods of analysis of linear models with unbalanced data. *J. Amer. Statist. Ass.*, **73**, 105.

SPEED, T. P. (1987). What is an analysis of variance? *Ann. Statist.*, **15**, 885.

SPITZER, J. J. (1978). A Monte Carlo investigation of the Box–Cox transformation in small samples. *J. Amer. Statist. Ass.*, **73**, 488.

SPJØTVOLL, E. (1968). Most powerful tests for some non-exponential families. *Ann. Math. Statist.*, **39**, 772.

SPJØTVOLL, E. (1972a). Multiple comparison of regression functions. *Ann. Math. Statist.*, **43**, 1076.

SPJØTVOLL, E. (1972b). Unbiasedness of likelihood ratio confidence sets in cases without nuisance parameters. *J. R. Statist. Soc.*, **B, 34**, 268.

SPJØTVOLL, E. (1972c). Joint confidence intervals for all linear functions of means in the one-way layout with unknown group variances. *Biometrika*, **59**, 683.

SPJØTVOLL, E. and STOLINE, M. R. (1973). An extension of the T-method of multiple comparison to include the cases with unequal sample sizes. *J. Amer. Statist. Ass.*, **68**, 975.

SPROTT, D. A. (1960). Necessary restrictions for distributions *a posteriori*. *J. R. Statist. Soc.*, **B, 22**, 312.

SPROTT, D. A. (1961). An example of an ancillary statistic and the combination of two samples by Bayes' theorem. *Ann. Math. Statist.*, **32**, 616.

STADJE, W. (1985). Estimation problems for samples with measurement errors. *Ann. Statist.*, **13**, 1592.

STARK, A. E. (1975). Some estimators of the integer-valued parameter of a Poisson variate. *J. Amer. Statist. Ass.*, **70**, 685.

STARR, N. and WOODROOFE, M. (1972). Further remarks on sequential estimation: the exponential case. *Ann. Math. Statist.*, **43**, 1147.

STEIN, C. (1945). A two-sample test for a linear hypothesis whose power is independent of the variance. *Ann. Math. Statist.*, **16**, 243.

STEIN, C. (1946). A note on cumulative sums. *Ann. Math. Statist.*, **17**, 498.

STEIN, C. (1956). Inadmissibility of the usual estimator for the mean of a multivariate normal distribution. *Proc. Third Berkeley Symp. Math. Statist. and Prob.*, **1**, 197. U. California Press, Berkeley.

STEPHENS, M. A. (1970). Use of the Kolmogorov–Smirnov, Cramér–von Mises and related statistics without extensive tables. *J. R. Statist. Soc.*, B, **32**, 115.

STEPHENS, M. A. (1974). EDF statistics for goodness of fit and some comparisons. *J. Amer. Statist. Ass.*, **69**, 730.

STEPHENS, M. A. (1975). Asymptotic properties for covariance matrices of order statistics. *Biometrika*, **62**, 23.

STEPHENS, M. A. (1976). Asymptotic results for goodness-of-fit statistics with unknown parameters. *Ann. Statist.*, **4**, 357.

STEPHENS, M. A. (1977). Goodness of fit for the exteme value distribution. *Biometrika*, **64**, 583.

STEPHENS, M. A. (1979a). Tests of fit for the logistic distribution based on the empineal distribution function. *Biometrika*, **66**, 591.

STEPHENS, M. A. (1979b). Vector autocorrelation. *Biometrika*, **66**, 41.

STERNE, T. E. (1954). Some remarks on confidence or fiducial limits. *Biometrika*, **41**, 275.

STEVENS, W. L. (1939). Distribution of groups in a sequence of alternatives. *Ann. Eugen.*, **9**, 10.

STEVENS, W. L. (1948). Statistical analysis of a non-orthogonal tri-factorial experiment. *Biometrika*, **35**, 346.

STEVENS, W. L. (1950). Fiducial limits of the parameter of a discontinuous distribution. *Biometrika*, **37**, 117.

STEWART, G. W. (1987). Collinearity and least squares regression. *Statist. Science*, **2**, 68.

STIGLER, S. M. (1986). *The History of Statistics: The Measurement of Uncertainty Before 1900.* Harvard Univ. Press, Cambridge, Mass., and London.

STIGLER, S. M. (1990). The 1990 Neyman memorial lecture: A Galtonian perspective on shrinkage estimators. *Statist. Sci.*, **5**, 147.

STOLINE, M. R. (1978). Tables of the studentized augmented range and applications to problems of multiple comparison. *J. Amer. Statist. Ass.*, **73**, 656.

STONE, M. (1976). Strong inconsistency from uniform priors. *J. Amer. Statist. Ass.*, **71**, 114.

STRAND, O. N. (1974). Coefficient errors caused by using the wrong covariance matrix in the general linear model. *Ann. Statist.*, **2**, 935.

STROUD, T. W. F. (1972). Fixed alternatives and Wald's formulation of the noncentral asymptotic behaviour of the likelihood ratio statistic. *Ann. Math. Statist.*, **43**, 447.

STROUD, T. W. F. (1973). Noncentral convergence of Wald's large-sample test statistic in exponential families. *Ann. Statist.*, **1**, 161.

STUART, A. (1954). Too good to be true? *Appl. Statist.*, **3**, 29.

STUART, A. (1955). A paradox in statistical estimation. *Biometrika*, **42**, 527.

STUART, A. (1958). Equally correlated variates and the multinormal integral. *J. R. Statist. Soc.*, B, **20**, 273.

STUART, A. (1967). The average critical value method and the asymptotic relative efficiency of tests. *Biometrika*, **54**, 308.

STUDDEN, W. J. (1980). D-optimal designs for polynomial regression using continued fractions. *Ann. Statist.*, **8**, 1132.

STUDDEN, W. J. (1982). Some robust-type D-optimal designs in polynomial regression. *J. Amer. Statist. Ass.*, **77**, 916.

"STUDENT" (1908).* The probable error of a mean. *Biometrika*, **6**, 1.

STYAN, G. P. H. (1970). Note on the distribution of quadratic forms in singular normal variables. *Biometrika*, **57**, 567.

SUBRAHMANIAN, K. and SUBRAHMANIAN, K. (1983). Some extensions to Miss F. N. David's tables of the sample correlation coefficient. *Sankhyā*, **B, 45**, 75.

SUBRAHMANIAM, K., GAJJAR, A. V. and SUBRAHMANIAM, K. (1981). Polynomial representations for the distribution of the sample correlation and its transformations. *Sankhyā*, **B, 43**, 319.

SUKHATME, P. V. (1936). On the analysis of k samples from exponential populations with special reference to the problem of random intervals. *Statist. Res. Mem.*, **1**, 94.

SUNDRUM, R. M. (1954). On the relation between estimating efficiency and the power of tests. *Biometrika*, **41**, 542.

SWED, F. S. and EISENHART, C. (1943). Tables for testing randomness of grouping in a sequence of alternatives. *Ann. Math. Statist.*, **14**, 66.

SWINDEL, B. F. (1968). On the bias of some least-squares estimators of variance in a general linear model. *Biometrika*, **55**, 313.

TAGUTI, G. (1958). Tables of tolerance coefficients for normal populations. *Rep. Statist. Appl. Res. (JUSE)*, **5**, 73.

TAKEUCHI, K. (1969). A note on the test for the location parameter of an exponential distribution. *Ann. Math. Statist.*, **40**, 1838.

TAMHANE, A. C. (1979). A comparison of procedures for multiple comparisons of means with unequal variances. *J. Amer. Statist. Ass.*, **74**, 471.

TANG, P. C. (1938). The power function of the analysis of variance tests with tables and illustrations of their use. *Statist. Res. Mem.*, **2**, 126.

TATE, R. F. (1953). On a double inequality of the normal distribution. *Ann. Math. Statist.*, **24**, 132.

TATE, R. F. (1954). Correlation between a discrete and a continuous variable. Point-biserial correlation. *Ann. Math. Statist.*, **25**, 603.

TATE, R. F. (1955). The theory of correlation between two continuous variables when one is dichotomised. *Biometrika*, **48**, 205.

TATE, R. F. (1959). Unbiased estimation: functions of location and scale parameters. *Ann. Math. Statist.*, **30**, 341.

TATE, R. F. and KLETT, G. W. (1959). Optimal confidence intervals for the variance of a normal distribution. *J. Amer. Statist. Ass.*, **54**, 674.

TEICHER, J. (1961). Maximum likelihood characterization of distributions. *Ann. Math. Statist.*, **32**, 1214.

TERRELL, C. D. (1983). Significance tables for the biserial and the point biserial. *Educ. Psych. Meas.*, **42**, 475.

THATCHER, A. R. (1964). Relationships between Bayesian and confidence limits for predictions. *J. R. Statist. Soc.*, **B, 26**, 176.

THEOBALD, C. M. (1974). Generalizations of mean square error applied to ridge regression. *J. R. Statist. Soc.*, **B, 36**, 103.

THOMPSON, J. R. (1968). Some shrinkage techniques for estimating the mean. *J. Amer. Statist. Ass.*, **63**, 113.

THOMPSON, M. L. (1978a, b). Selection of variables in multiple regression. *Int. Statist. Rev.*, **46**, 1 and 129.

THOMPSON, R. O. R. Y. (1979). Bias and monotonicity for goodness-of-fit tests. *J. Amer. Statist. Ass.*, **74**, 875.

THOMPSON, W. R. (1936). On confidence ranges for the median and other expectation distributions for populations of unknown distribution form. *Ann. Math. Statist.*, **7**, 122.

THÖNI, H. (1969). A table for estimating the mean of a lognormal distribution. *J. Amer. Statist. Ass.*, **64**, 632.

THORBURN, D. (1976). Some asymptotic properties of jackknife statistics. *Biometrika*, **63**, 305.

TIBSHIRANI, R. (1988). Variance stabilization and the bootstrap. *Biometrika*, **75**, 433.

TIKU, M. L. (1965). Laguerre series forms of non-central χ^2 and F distributions. *Biometrika*, **52**, 415.

TIKU, M. L. (1966). A note on approximating the non-central F distribution. *Biometrika*, **53**, 606.

TIKU, M. L. (1967, 1972). Tables of the power of the F-test. *J. Amer. Statist. Ass.*, **62**, 525 and **67**, 709.

TITTERINGTON, D. M. (1984). Recursive parameter estimation using incomplete data. *J. R. Statist. Soc.*, **B, 46**, 257.

TOCHER, K. D. (1952). The design and analysis of block experiments. *J. R. Statist. Soc.*, **B, 14**, 45.

TOOTHAKER, L. E., HICKS, J. L. and PRICE, J. M. (1978). Optimum subsample sizes for the Bartlett-Kendall homogeneity of variance test. *J. Amer. Statist. Ass.*, **73**, 53.

TRICKETT, W. H., WELCH, B. L. and JAMES, G. S. (1956). Further critical values for the two-means problem. *Biometrika*, **43**, 203.

TUKEY, J. W. (1947). Non-parametric estimation, II. Statistically equivalent blocks and tolerance regions—the continuous case. *Ann. Math. Statist.*, **18**, 529.

TUKEY, J. W. (1948). Non-parametric estimation, III. Statistically equivalent blocks and multivariate tolerance regions—the discontinuous case. *Ann. Math. Statist.*, **19**, 30.

TUKEY, J. W. (1949). One degree of freedom for non-additivity. *Biometrics*, **5**, 232.

TUKEY, J. W. (1951). Quick and dirty methods in statistics. II: Simple analyses for standard designs. *Proc. 5th Annu. Conf. Amer. Soc. Qual. Contr.*, 189.

TUKEY, J. W. (1952). Allowances for various types of error rate. Unpublished, Princeton University.

TUKEY, J. W. (1953). The problem of multiple comparisons. Unpublished, Princeton University.

TUKEY, J. W. (1957a). On the comparative anatomy of transformations. *Ann. Math. Statist.*, **28**, 602.

TUKEY, J. W. (1957b). Some examples with fiducial relevance. *Ann. Math. Statist.*, **28**, 687.

TUKEY, J. W. (1962). The future of data analysis. *Ann. Math. Statist.*, **33**, 1 (corr. **33**, 812).

UUSIPAIKKA, E. (1985). Exact simultaneous confidence intervals for multiple comparisons among three or four mean values. *J. Amer. Statist. Ass.*, **80**, 196.

VAN, DER PARREN, J. L. (1970). Tables for distribution-free confidence limits for the median. *Biometrika*, **57**, 613.

VAN EEDEN, C. (1963). The relation between Pitman's asymptotic relative efficiency of two tests and the correlation coefficient between their test statistics. *Ann. Math. Statist.*, **34**, 1442.

VENABLES, W. (1975). Calculation of confidence intervals for noncentrality parameters. *J. R. Statist. Soc.*, **B, 37**, 406.

VERRILL, S. and JOHNSON, R. A. (1987). The asymptotic equivalence of some modified Shapiro-Wilk statistics—complete and censored sample cases. *Ann. Statist.*, **15**, 413.

VERRILL, S. and JOHNSON, R. A. (1988). Tables and large-sample distribution theory for censored-data correlation statistics for testing normality. *J. Amer. Statist. Ass.*, **83**, 1192.

VILLEGAS, C. (1969). On the least squares estimation of non-linear relations. *Ann. Math. Statist.*, **40**, 462.

VINOD, H. D. (1976). Application of new ridge regression methods to a study of Bell System scale economies. *J. Amer. Statist. Ass.*, **71**, 835.

VOORN, W. J. (1981). A class of variate transformations causing unbounded likelihood. *J. Amer. Statist. Ass.*, **76**, 709.

WALD, A. (1940).* The fitting of straight lines if both variables are subject to error. *Ann. Math. Statist.*, **11**, 284.

WALD, A. (1941).* Asymptotically most powerful tests of statistical hypotheses. *Ann. Math. Statist.*, **12**, 1.

WALD, A. (1942).* On the power function of the analysis of variance test. *Ann. Math. Statist.*, **13**, 434.

WALD, A. (1943a).* Tests of statistical hypotheses concerning several parameters when the number of observations is large. *Trans. Amer. Math. Soc.*, **54**, 426.

WALD, A. (1943b). An extension of Wolks' method for setting tolerance limits. *Ann. Math. Statist.*, **14**, 45.

WALD, A. (1947). *Sequential Analysis.* Wiley, New York.

WALD, A. (1949).* Note on the consistency of the maximum likelihood estimate. *Ann. Math. Statist.*, **20**, 595.

WALD, A. (1950). *Statistical Decision Functions.* Wiley, New York.

WALD, A. and WOLFOWITZ, J. (1939).* Confidence limits for continuous distribution functions. *Ann. Math. Statist.*, **10**, 105.

WALD, A. and WOLFOWITZ, J. (1940).* On a test whether two samples are from the same population. *Ann. Math. Statist.*, **11**, 147.

WALD, A. and WOLFOWITZ, J. (1946).* Tolerance limits for a normal distribution. *Ann. Math. Statist.*, **17**, 208.

WALKER, A. M. (1963). A note on the asymptotic efficiency of an asymptotically normal estimator sequence. *J. R. Statist. Soc.*, **B, 25**, 195.

WALLACE, D. L. (1958). Asymptotic approximations to distributions. *Ann. Math. Statist.*, **29**, 635.

WALLACE, T. D. (1964). Efficiencies for stepwise regressions. *J. Amer. Statist. Ass.*, **59**, 1179.

WALLACE, T. D. and TORO-VIZCARRONDO, C. E. (1969). Tables for the mean square error test for exact linear restrictions in regression. *J. Amer. Statist. Ass.*, **64**, 1649.

WALLIS, K. F. (1972). Testing for fourth-order autocorrelation in quarterly regression equations. *Econometrica*, **40**, 617.

WALLIS, W. A. (1951). Tolerance intervals for linear regression,. *Proc. 2nd Berkeley Symp. Math. Statist. and Prob.*, 43.

WALSH, J. E. (1947). Concerning the effect of intraclass correlation on certain significance tests. *Ann. Math. Statist.*, **18**, 88.

WALSH, J. E. (1962). Distribution-free tolerance intervals for continuous symmetrical populations. *Ann. Math. Statist.*, **33**, 1167.

WALTON, G. S. (1970). A note on nonrandomized Neyman-shortest unbiased confidence intervals for the binomial and Poisson parameters. *Biometrika*, **57**, 223.

WATSON, G. S. (1957a). Sufficient statistics, similar regions and distribution-free tests. *J. R. Statist. Soc.*, **B, 19**, 262.

WATSON, G. S. (1957b). The χ^2 goodness-of-fit test for normal distributions. *Biometrika*, **44**, 336.

WATSON, G. S. (1958). On chi-square goodness-of-fit tests for continuous distributions. *J. R. Statist. Soc.*, **B, 20**, 44.

WATSON, G. S. (1959). Some recent results in chi-square goodness-of-fit tests. *Biometrics*, **15**, 440.

WATSON, G. S. (1967). Linear least squares regression. *Ann. Math. Statist.*, **38**, 1679.

WEILER, H. (1972). Inverse sampling of a Poisson distribution. *Biometrics*, **28**, 959.

WEISBERG, S. (1974). An empirical comparison of the percentage points of W and W'. *Biometrika*, **61**, 644.

WEISS, L. (1962). On sequential tests which minimize the maximum expected sample size. *J. Amer. Statist. Ass.*, **57**, 551.

WEISS, L. and WOLFOWITZ, J. (1972). An asymptotically efficient sequential equivalent of the *t*-test. *J. R. Statist. Soc.*, **B, 34**, 456.

WEISS, L. and WOLFOWITZ, J. (1973). Maximum likelihood estimation of a translation parameter of a truncated distribution. *Ann. Statist.*, **1**, 944.

WEISSBERG, A. and BEATTY, G. H. (1960). Tables of tolerance-limit factors for normal distributions. *Technometrics*, **2**, 483.

WELCH, B. L. (1938). The significance of the difference between two means when the population variances are unequal. *Biometrika*, **29**, 350.

WELCH, B. L. (1939). On confidence limits and sufficiency, with particular reference to parameters of location. *Ann. Math. Statist.*, **10**, 58.

WELCH, B. L. (1947). The generalisation of "Student's" problem when several different population variances are involved. *Biometrika*, **34**, 28.

WELCH, B. L. (1965). On comparisons between confidence point procedures in the case of a single parameter. *J. R. Statist. Soc.*, **B**, **27**, 1.

WELCH, B. L. and PEERS, H. W. (1963). On formulae for confidence points based on integrals of weighted likelihood. *J. R. Statist. Soc.*, **B**, **25**, 318.

WERMUTH, N. (1980). Linear recursive equations, covariance selection and path analysis. *J. Amer. Statist. Ass.*, **75**, 963.

WHITAKER, L. (1914). On Poisson's law of small numbers. *Biometrika*, **10**, 36.

WHITE, H. (1981). Consequences and detection of misspecified nonlinear regression models. *J. Amer. Statist. Ass.*, **76**, 419.

WHITE, H. and MACDONALD, G. M. (1980). Some large sample tests for non-normality in the linear regression model. *J. Amer. Statist. Ass.*, **75**, 16.

WHITEHEAD, Y. (1986). On the bias of maximum likelihood estimation following a sequential test. *Biometrika*, **73**, 573.

WHITTAKER, J. (1973). The Bhattacharyya matrix for the mixture of two distributions. *Biometrika*, **60**, 201.

WICKSELL, S. D. (1917). The correlation function of Type A. *Medd. Lunds Astr. Obs.*, Series 2, No. 17.

WICKSELL, S. D. (1934). Analytical theory of regression. *Medd. Lunds. Astr. Obs.*, Series 2, No. 69.

WIEAND, H. S. (1976). A condition under which the Pitman and Bahadur approaches to efficiency coincide. *Ann. Statist.*, **4**, 1003.

WIJSMAN, R. A. (1977). A general theorem with applications on exponentially bounded stopping time, without moment conditions. *Ann. Statist.*, **5**, 292.

WILKINSON, G. N. (1957). The analysis of covariance with incomplete data. *Biometrics*, **13**, 363.

WILKINSON, G. N. (1958a). Estimation of missing values for the analysis of incomplete data. *Biometrics*, **14**, 257.

WILKINSON, G. N. (1958b). The analysis of variance and derivation of standard errors for incomplete data. *Biometrics*, **14**, 360.

WILKINSON, G. N. (1960). Comparison of missing value procedures. *Aust. J. Statist.*, **2**, 53.

WILKINSON, G. N. (1970). A general recursive procedure for analysis of variance. *Biometrika*, **57**, 19.

WILKINSON, G. N. (1977). On resolving the controversy in statistical inference (with discussion). *J. R. Statist. Soc.*, **B**, **39**, 119.

WILKINSON, L. and DALLAL, G. E. (1981). Tests of significance in forward selection regression with an F-to-enter stopping rule. *Technometrics*, **23**, 377.

WILKS, S. S. (1938a).* The large-sample distribution of the likelihood ratio for testing composite hypotheses. *Ann. Math. Statist.*, **9**, 60.

WILKS, S. S. (1938b).* Shortest average confidence intervals from large samples. *Ann. Math. Statist.*, **9**, 166.

WILKS, S. S. (1941).* Determination of sample sizes for setting tolerance limits. *Ann. Math. Statist.*, **12**, 91.

WILKS, S. S. (1942).* Statistical prediction with special reference to the problem of tolerance limits. *Ann. Math. Statist.*, **13**, 400.

WILKS, S. S. and DALY, J. F. (1939).* An optimum property of confidence regions associated with the likelihood function. *Ann. Math. Statist.*, **10**, 225.

WILLIAMS, C. A., Jr. (1950). On the choice of the number and width of classes for the chi-square test of goodness of fit. *J. Amer. Statist. Ass.*, **45**, 77.

WILLIAMS, E. J. (1959). The comparison of regression variables. *J. R. Statist. Soc.*, **B, 21**, 396.

WILLIAMS, E. J. (1969). A note on regression methods in calibration. *Technometrics*, **11**, 189.

WILLIAMSON, J. A. (1984). A note on the proof by H. E. Daniels of the asymptotic efficiency of a maximum likelihood estimator. *Biometrika*, **71**, 651.

WINTERBOTTOM, A. (1979). Cornish-Fisher expansions for confidence limits. *J. R. Statist. Soc.*, **B, 41**, 69.

WISE, M. E. (1963). Multinomial probabilities and the χ^2 and X^2 distributions. *Biometrika*, **50**, 145.

WISE, M. E. (1964). A complete multinomial distribution compared with the X^2 approximation and an improvement to it. *Biometrika*, **51**, 277.

WISHART, J. (1931). The mean and second moment coefficient of the multiple correlation coefficient, in samples from a normal population. *Biometrika*, **22**, 353.

WISHART, J. (1932). A note on the distribution of the correlation ratio. *Biometrika*, **24**, 441.

WOLD, H. O. A. (1960). A generalization of causal chain models. *Econometria*, **28**, 443.

WOLFOWITZ, J. (1947). The efficiency of sequential estimates and Wald's equation for sequential processes. *Ann. Math. Statist.*, **18**, 215.

WOLFOWITZ, J. (1949). The power of the classical tests associated with the normal distribution. *Ann. Math. Statist.*, **20**, 540.

WOOD, C. L. and ALTAVELA, M. M. (1978). Large-sample results for Kolmogorov–Smirnov statistics for discrete distributions. *Biometrika*, **65**, 235.

WOODCOCK, E. R. and EAMES, A. R. (1970). *Confidence limits for numbers from 0 to 1200 based on the Poisson distribution*. Authority Health and Safety Branch, AHSB(S) R.179, H.M.S.O., London.

WOODROOFE, M. (1978). Large deviations of likelihood ratio statistics with applications to sequential testing. *Ann. Statist.*, **6**, 72.

WORKING, H. and HOTELLING, H. (1929). The application of the theory of error to the interpretation of trends. *J. Amer. Statist. Ass.*, **24** (Suppl), 73.

WRIGHT, S. (1923). The theory of path coefficients: A reply to Niles' criticism. *Genetics*, **8**, 239.

WRIGHT, S. (1934). The method of path coefficients. *Ann. Math. Statist.*, **5**, 161.

WU, C. F. J. (1983). On the convergence properties of the EM algorithm. *Ann. Statist.*, **11**, 95.

WU, C. F. J. (1986). Jackknife, bootstrap and other resampling plans in regression analysis (with discussion). *Ann. Statist.*, **14**, 1261.

WYNN, H. P. (1984). An exact confidence band for one-dimensional polynomial regression. *Biometrika*, **71**, 375.

WYNN, H. P. and BLOOMFIELD, P. (1971). Simultaneous confidence bands in regression analysis. *J. R. Statist. Soc.*, **B, 33**, 202.

YARNOLD, J. K. (1970). The minimum expectation in X^2 goodness of fit tests and the accuracy of approximations for the null distribution. *J. Amer. Statist. Ass.*, **65**, 864,

YATES, F. (1933).* The analysis of replicated experiments when the field results are incomplete. *Empire J. Exper. Agric.*, **1**, 129.

YATES, F. (1934a). The analysis of multiple classifications with unequal numbers in the different classes. *J. Amer. Statist. Ass.*, **29**, 1.

YATES, F. (1934b). Contingency tables involving small numbers and the χ^2 test. *Suppl. J. R. Statist. Soc.*, **1**, 217.

YATES, F. (1939a). An apparent inconsistency arising from tests of significance based on fiducial distributions of unknown parameters. *Proc. Camb. Phil. Soc.*, **35**, 579.

YATES, F. (1939b). Tests of significance of the differences between regression coefficients derived from two sets of correlated variates. *Proc. Roy. Soc. Edin.*, **A, 59**, 184.

YATES, F. (1984). Tests of significance for 2×2 contingency tables. *J. R. Statist. Soc.*, **A, 147**, 426.

YULE, G. U. (1907).* On the theory of correlation for any number of variables treated by a new system of notation. *Proc. Roy. Soc.*, **A, 79**, 182.

YULE, G. U. (1926).* Why do we sometimes get nonsense-correlations between time-series?—A study in sampling and the nature of time-series. *J. R. Statist. Soc.*, **89**, 1.

ZACKS, S. (1970). Uniformly most accurate upper tolerance limits for monotone likelihood ratio families of discrete distributions. *J. Amer. Statist., Ass.*, **65**, 307.

ZAHN, D. A. and ROBERTS, G. C. (1971). Exact χ^2 criterion tables with cell expectations one: an application to Coleman's measure of consensus. *J. Amer. Statist. Ass.*, **66**, 145.

ZAR, J. H. (1972). Significance testing of the Spearman rank correlation coefficient. *J. Amer. Statist. Ass.*, **67**, 578.

ZELLNER, A. (1962). An efficient method of estimating seemingly unrelated regressions and tests for aggregation bias. *J. Amer. Statist. Ass.*, **57**, 348.

ZELLNER, A. (1977). Maximal data information prior distributions. In *New Developments in the Application of Bayesian Methods*, A. Aykae and C. Brumat (eds.). North-Holland, Amsterdam.

ZYSKIND, G. (1963). A note on residual analysis. *J. Amer. Statist. Ass.*, **58**, 1125.

ZYSKIND, G. (1969). Parametric augmentations and error structures under which certain simple least squares and analysis of variance procedures are also best. *J. Amer. Statist. Ass.*, **64**, 1353.

INDEX OF EXAMPLES IN TEXT

Chapter	17	18	19	20	21	22	23	24	25	26	27	28	29	30	31
Examples															
.1	.7	.8	.4	.5	.7	.7	.2	.3	.4	.6	.18	.8	.11	.7	.5
.2	.7	.8	.4	.6	.11	.10	.2	.5	.7	.6	.19	.11	.24	.22	.7
.3	.9	.9	.4	.9	.12	.14	.8	.9	.10	.7	.20	.12	.25	.27	.9
.4	.10	.13	.5	.12	.12	.15	.9	.11	.11	.9		.27	.29	.29	.9
.5	.12	.15	.5	.17	.13	.18	.19	.13	.13	.9		.39	.36	.32	.9
.6	.17	.16	.5	.18	.14	.18	.22	.14	.21	.12		.41	.60	.32	.13
.7	.17	.16	.9	.40	.15	.20	.24	.15		.12		.49		.41	.13
.8	.17	.17	.9	.40	.19	.20	.28	.21		.32		.55			.16
.9	.17	.21	.15	.43	.19	.20		.21		.34		.60			.19
.10	.18	.21	.21	.44	.22	.21		.22		.40		.61			.20
.11	.22	.24	.33	.44	.23	.23		.32		.48		.62			.21
.12	.24	.25		.45	.24	.24		.35				.68			.32
.13	.29	.28		.45	.25	.25		.35							.38
.14	.30	.29			.31	.33									.38
.15	.33	.30				.33									.44
.16	.33	.31				.33									.45
.17	.38	.32													.54
.18	.41	.35													.55
.19	.41	.37													
.20	.41	.38													
.21	.41	.41													
.22	.41														
.23	.41														

Each entry in the table gives the section number that contains the Example numbered in the left margin, in the chapter at the head of the column. Thus the entry **.31** with coordinates (.14,21), means that Example 21.14 appears in section **21.31**.

INDEX

(References are to chapter-sections, displayed at the tops of pages. Examples in the text are indexed by the chapter-section in or immediately after which they appear. Exercises appear at the ends of chapters.)